Communications
in Computer and Information Science 86

Luo Qi (Ed.)

Information and Automation

International Symposium, ISIA 2010
Guangzhou, China, November 10-11, 2010
Revised Selected Papers

Springer

Volume Editor

Luo Qi
Wuhan Institute of Technology
Wuhan, Hubei, China
E-mail: witluo@ieee.org

ISSN 1865-0929
e-ISSN 1865-0937
ISBN 978-3-642-19852-6
ISBN 978-3-642-19853-3 (eBook)
DOI 10.1007/978-3-642-19853-3
Springer Heidelberg Dordrecht London New York

Library of Congress Control Number: 2011924882

CR Subject Classification (1998): C.2.4, C.3, H.4, I.2, I.2.6, K.4.3, K.6.5

Typesetting: Camera-ready by author, data conversion by Scientific Publishing Services, Chennai, India

Printed on acid-free paper

Springer is part of Springer Science+Business Media (www.springer.com)

Preface

We are delighted to present the proceedings of the 2010 International Symposium on Information and Automation (ISIA 2010) that was held in Guangzhou, China, during November 10–11, 2010. The objective of ISIA 2010 was to provide a forum for researchers, educators, engineers, and government officials involved in the general areas of information and automation to disseminate their latest research results and exchange views on the future research directions of these fields.

This year, ISIA 2010 invited high-quality recent research results in the areas of information and automation.

The main goal of the conference is to bring together scientists and engineers who work on information and automation aspects. ISIA 2010 provided an opportunity for academic and industry professionals to discuss the latest issues and progress in the area of information and automation. Furthermore, we expect that the conference and its publications will be a trigger for further related research and technology improvements in this important subject.

ISIA 2010 also included presentations of contributed papers and state-of-the-art lectures by invited keynote speakers. We would like to thank the Program Chairs, organization staff, and the members of the Program Committees for their hard work. Special thanks go to Springer.

We look forward to seeing all of you next year at the ISIA 2011.

Luo Qi

Organization

ISIA 2010 Organizing Committee

Honorary Conference Chairs

Chin-Chen Chang	IEEE Fellow, Feng Chia University, Taiwan
Jun Wang	The Chinese University of Hong Kong, Hong Kong
Chris Price	Aberystwyth University, UK

Organizing Chairs

Honghua Tan	Wuhan Instititue of Technology, China
Qihai Zhou	Southwestern University of Finance and Economics, China
Junwu Zhu	Yangzhou University, China
Peide Liu	ShangDong Economic University, China

Program Chairs

Xueming Zhang	Beijing Normal University, China
Mark Zhou	International Industrial Electronics Center, Hong Kong
Yi-chuan Zhang	Henan Institute of Science and Technology, China

Publication Chair

Luo Qi	Wuhan Instititue of Technology, China

International Committees

Ying Zhang	Wuhan University, China
Xueming Zhang	Beijing Normal University, China
Peide Liu	Shangdong Economic University, China
Dariusz Krol	Wroclaw University of Technology, Poland
Jason J. Jung	Yeungnam University, Republic of Korea
Paul Davidsson	Blekinge Institute of Technology, Sweden
Cao Longbing	University of Technology Sydney, Australia
Huaifeng Zhang	University of Technology Sydney, Australia
Qian Yin	Beijing Normal University, China

ISIA 2010 Reviewers

Qian Yin	Beijing Normal University, China
Zhigang Chen	Central South University, China
Hoi-Jun Yoo	Korea Advanced Institute of Science and Technology, Republic of Korea
Chin-Chen Chang	Feng Chia University, Taiwan
Jun Wang	The Chinese University of Hong Kong, Hong Kong

Table of Contents

Spectrum Allocation Based on Game Theory in Cognitive Radio Networks

Anna Auguste Anghuwo, Yutao Liu, Xuezhi Tan, and Shuai Liu

Communication Research Center, Harbin Institute of Technology,
Harbin 150080, P.R. China

Abstract. This paper proposes a new spectrum allocation scheme based on detection probability of cognitive users. This new scheme takes a look at a cognitive radio network system composed of one primary system with several cognitive users which is combined with game theory to compensate the detection cost through detection probability. Nash Equilibrium (NE) was used to determine the cost based on the detection probability. Outcomes data revealed that in the utility function of the game, NE was stable through price adjustment. The findings shows that NE is related to the detection probability and the higher the detection probability, the more spectrum resources are in dynamic allocation and thus the higher the quality communication services for the user will gained, thus the detection cost influences the quality of the system. Finally a comparison was made between NE and Pareto Optimality to look at the necessity and conditions of possible conversion from NE to Pareto Optimality.

Keywords: detection probability; utility function; Nash Equilibrium; detection probability; Pareto Optimality.

1 Introduction

In recent years, spectrum resource scarcity has become more serious as the rapid development of wireless communication technology increases [1]. In addition, a measure report from national radio network research testbed showed that the average utilization of spectrum below 3GHz is only 5.2% [2]. Many experts have analyzed and dealt with the spectrum allocation for cognitive radio using game theory. A research group of Manitoba University proposed a dynamic game theory model in which the cognitive users could adjust their strategy repeatedly according to spectrum requirement, and finally get a stable equilibrium [5]. Huang applied the auction mechanism of game theory to the distributed spectrum sharing, and get Nash Equilibrium [6]. Wang designed a channel/power assignment scheme introducing price mechanism into the spectrum allocation which overcome the ineffectiveness of iterative water-filling and could be applied to distributed network [7]. Although these mechanisms yielded some results in cognitive radio spectrum allocation, they always neglected the detection spending in cognitive users' communication and thus were under improvement and enhancement.

L. Qi (Ed.): ISIA 2010, CCIS 86, pp. 1–9, 2011.

In wireless network, the licensed users often does not fully use the spectrums that they owned, so as a result to maximize the spectrum efficiency they could lease some or all of their idle spectrums to cognitive users who could use the leased spectrum for communication task without interfering with the licensed users [3,4]. Therefore, spectrum resource scarcity is not due to lack of physical spectrums, but because of low spectrum utilization caused by the unreasonable existing spectrum management and allocation. Cognitive radio can implement dynamic spectrum access and allocation, and also utilizes the wireless spectrum resource more reasonably and efficiently. This paper also studied the process in which the numbers of cognitive users allocated spectrum dynamically, and according to the pricing of licensed users in Gauss fading environment, in which the detection ability of cognitive users has been taken into account to compensate for their communication overhead.

1.1 Cognitive Users' Detection Probability

In cognitive radio, spectrum sensing is an essential part and cognitive users' detection overhead is the major part of communication cost. The function of cognitive users' spectrum detection is to detect and analyze spectrums of licensed users in specific region, in order to find the spectrum holes for communication, and to make the cognitive users work without causing interference to the existing communication system. The spectrum detection methods mainly include energy detection method, static cyclical test method, high-order spectral detection method and cooperative detection method [8,9]. The important yardsticks for measuring the detection performance are the magnitude of detection probability and the length of detection time. The detection performance is not only related with the physical location of cognitive users, but also related with the number of cognitive symbols for spectrum detection and the energy of detection signal. The higher the detection probability of cognitive users', results in more signal detection, better detection performance and less interference to licensed users. Therefore, the cognitive users' utility function presented in this paper dynamically allocates spectrum according to the magnitude of detection probability, and makes the cognitive users who have higher or better detection performance to get more spectrums, which not only embodies the principle of fairness of the cognitive system, but also simulate the cognitive users input more energy to spectrum detection, and improve the overall performance of the system.

1.2 Cognitive Radio Spectrum Allocation Model

Wireless Transmission Model

Assume that cognitive users use QAM modulation mode and ideal phase detection, and transmission rate can be adjusted adaptively according to the channel state, so we can consider spectrum utility as a function of *SNR* and *BER* [10,11].

$$k = \log_2 \left(1 + K \cdot \gamma\right) \tag{1}$$

Where γ denotes the *SNR* at receiver, and K is a constant denoting the users' *BER*, which can be considered as the difference of *SNR* between M-QAM and Shannon capacity. For example, K can be expressed as below in Gauss fading Channels.

$$K = \frac{1.5}{-\ln(5BER)} \tag{2}$$

Assume that the licensed users' idle spectrum can be divided into M orthogonal channels whose bandwidths are all b_0, so the cognitive users' communication rate is $v_i = k_i b_0$. Assume that the bandwidth demand of user i is b_i, and the bit error rates of all receivers are BER, so we can obtain that a cognitive user's benefit at each channel is $r_i = v_i$.

1.3 System Model

Assuming that there is a distributed hybrid network consisting of one licensed user and N cognitive users, where each cognitive user cannot acquire the spectrum demands of other cognitive users. Cognitive users adjust their spectrum demand strategies in real time according to licensed user's pricing and their detection probability, and then reach a stable equilibrium through the game among users. Also let's assume that the bandwidth demands of every cognitive users system are $B = \{b_i \mid i = 1, 2, \cdots, N; b_i > 0\}$, and the licensed user determines the spectrum price according to the amount of bandwidths that cognitive users need. Considering price demand function, we can obtain

$$\sum_{i=1}^{N} b_i = \rho \ln P + \eta \tag{3}$$

Where $\sum_{i=1}^{N} b_i$ is the amount of bandwidths demand which the cognitive users system asks the licensed user for, and P is the price per unit bandwidth that the licensed user system asks the cognitive users for. For cognitive users, we can assume that there is no difference between the idle spectrums that the licensed user leases, so the prices per unit bandwidth for different cognitive users are the same. In expression (3), ρ and η are the adjustment factors of the demand price function respectively. From expression (3), we can obtain the price per unit bandwidth

$$P = e^{(\sum_{i=1}^{N} b_i - \eta)\frac{1}{\rho}} \tag{4}$$

In cognitive users system, according to the differences between detection probabilities of cognitive users, the normalized detection probabilities can be expressed as $d_i, i = 1, 2, \cdots, N$

$$d_i = c_i / \sum_{i=1}^{N} c_i \tag{5}$$

Where c_i is the detection probability and

$$\sum_{i=1}^{N} d_i = 1, d_i > 0 \cdot$$

As the cognitive users system has different detection probabilities and different detection costs, when allocating spectrum, the system could set an inner spectrum pricing which is irrespective to the pricing of licensed user but related to the corresponding detection probabilities. This pricing can be expressed as $P_i = P/(\alpha d_i + \beta)$, where $\alpha d_i + \beta$ is a cognitive user's normalized weighted detection probability impact factor, and α and β are the adaptive adjustment coefficient of weighted detection probability, neither of which is negative. When the detection probability of cognitive user has a great influence to the whole system, α is greater than 1, and when the detection probability of cognitive user has an inconspicuous influence to the whole system, α is less than 1. The utility function of cognitive user can be expressed as follow,

$$\pi_i(b_i) = r_i - p_i \tag{6}$$

Where $\pi_i(b_i)$ is the utility of cognitive user i, r_i is the benefit of user i who occupies a certain bandwidth of spectrum, and p_i is the cost of user i for this spectrum. Therefore the utility function of the cognitive users system is

$$\pi_i(b_i) = v_i b_i - b_i e^{(\sum_{i=1}^{N} b_i - \eta)\frac{1}{\rho}} \frac{1}{\alpha d_i + \beta} \tag{7}$$

In distributed network, assume that every cognitive user is selfish and rational, and they only adjust their strategies in real time according to the pricing information of licensed user in order to reach the system Nash Equilibrium. The marginal utility function of cognitive user i can be expressed as in formula (8).

$$\frac{\partial \pi_i(b_i)}{\partial b_i} = v_i - (e^{(\sum_{i=1}^{N} b_i - \eta)\frac{1}{\rho}} + b_i e^{(\sum_{i=1}^{N} b_i - \eta)\frac{1}{\rho}} \frac{\partial \sum_{i=1}^{N} b_i}{\partial b_i} \frac{1}{\rho}) \frac{1}{\alpha d_i + \beta} \tag{8}$$

The cognitive system finally adjusts its state according to the change in marginal utility in order to converge to a stable Nash Equilibrium. Assume that $b_i(t)$ is the bandwidth that user i occupies at time t, and λ_i is its cognitive coefficient, so we can obtain the bandwidth at time $t+1$ as in formula (9).

$$b_i(t+1) = F(b_i(t)) = b_i(t) + \lambda_i(v_i - (e^{(\sum_{i=1}^{N} b_i - \eta)\frac{1}{\rho}} + b_i e^{(\sum_{i=1}^{N} b_i - \eta)\frac{1}{\rho}} \frac{\partial \sum_{i=1}^{N} b_i}{\partial b_i} \frac{1}{\rho}) \frac{1}{\alpha d_i + \beta}) \tag{9}$$

If the utility function is nonlinear, when a number of cognitive users are in equilibrium we can know that $b_i(t+1) = F(b_i(t)) = b_i(t) = b_i$, and therefore we can obtain the bandwidth of each cognitive user in Nash Equilibrium.

1.4 Simulation Performance Analysis Nash Equilibrium State

Consider a distributed network consisting of one licensed user and five cognitive users. Assume that the channel bandwidth is $b_0 = 25\,\text{Hz}$, and the SNR and BER at the receivers of cognitive users are respectively $\gamma = 12\text{dB}$ and $BER = 10^{-4}$, so from expression (1) and (2) we can obtain the cognitive user's information transmission rate $v_i = 50\,\text{bps}$. Assume that the adjustment coefficients of demand pricing function are $\rho = 1$ and $\eta = 0$, and the normalized detection probability of the five cognitive users is $\mathbf{d} = [0.22\ 0.21\ 0.20\ 0.19\ 0.18]$. At the same time assume that the adjustment coefficients of normalized detection function are $\alpha = 1$ and $\beta = 0$. When the cognitive ability of a cognitive user is $\lambda_i = 0.017$ (which is variable), the bandwidths allocated to the five cognitive users in equilibrium are shown in figure 1. Cognitive users constantly adjust their strategies and reach the state of Nash Equilibrium. Meanwhile, the greater the normalized detection probability is, the more bandwidths the cognitive user gets, and as a result the higher utility the cognitive user gains, just as shown in figure 2.

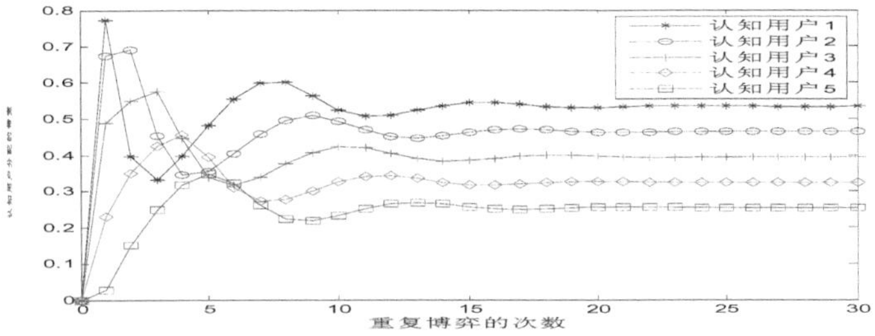

Fig. 1. Nash Equilibrium bandwidth allocation map

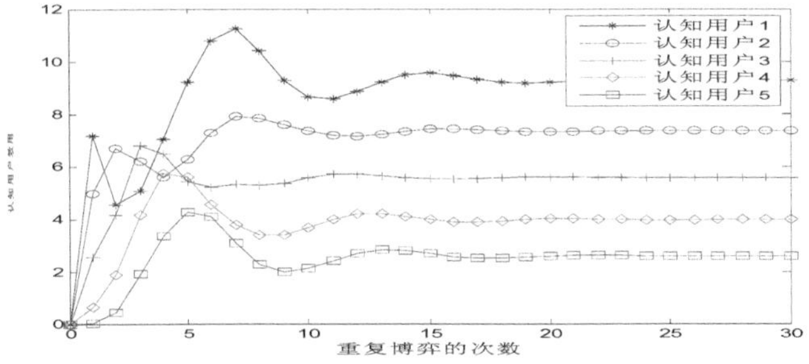

Fig. 2. Nash Equilibrium cognitive user utility

1.5 Relationship between Nash Equilibrium and Cognitive Ability

For a cognitive user, the convergence rate is related to its own cognitive ability (the equilibrium bandwidth unit is kHz). Taking cognitive user 1 for example, the number of repeatedly gaming reaches convergence along with its cognitive ability, as shown in figure 3. Cognitive ability reflects the ability that a cognitive user depends on the current information of licensed user. When cognitive ability is high, the cognitive user has a great dependence on the information applied by the licensed user, such as pricing information, and the reverse is also true. At the same time, the number of repeatedly gaming could be reduced by reasonably adjusting the cognitive ability, which makes the cognitive users make decision more rapidly and improve real-time of the whole system.

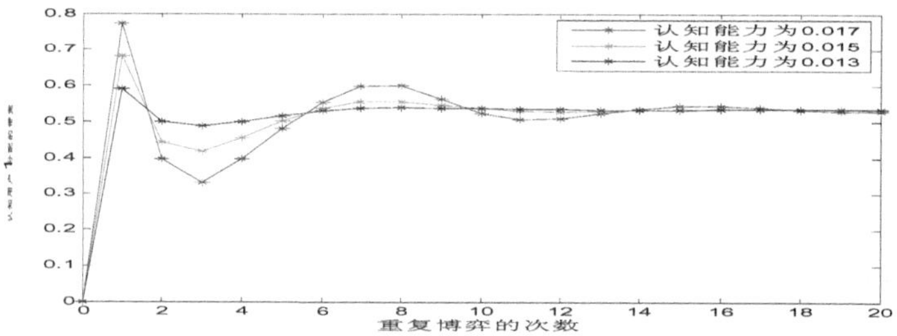

Fig. 3. Relationship between Nash Equilibrium and cognitive ability

1.6 Impact of Normalized Detection Probability

In order to better illustrate the impact of detection probability to cognitive users, consider a distributed network including only two cognitive users. Assume that the channel bandwidth is $b_0 = 10$ Hz, BER is the same as above, and thus the information transmission rate after accessing the idle spectrum of licensed user is $v_i = 20$ bps, i=1, 2. When the detection probabilities of cognitive user 1 and user 2 are respectively 0.51, 0.49 and 0.53, 0.47, we can obtain the bandwidths and utilities as shown in figure 4 and figure 5.

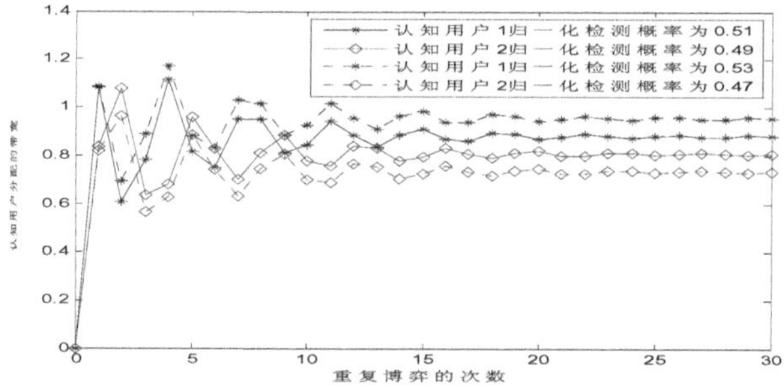

Fig. 4. Detection probability impact on broadband users

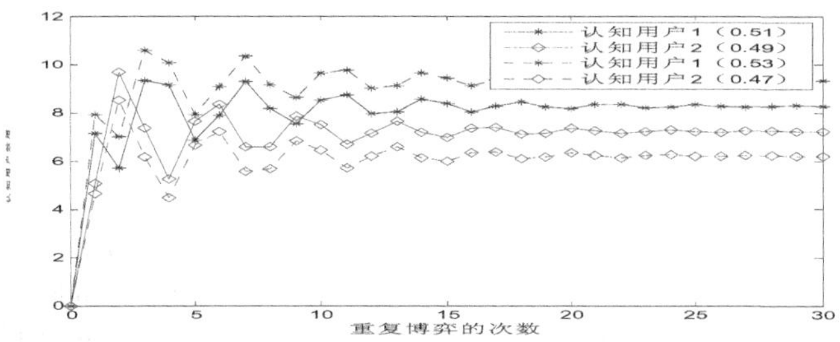

Fig. 5. Probability of detection of user utility

From the figures we will see that the bandwidths which cognitive users get and the utilities which they produce change with the normalized detection probability. The cognitive user who has a greater normalized detection probability can get more bandwidths and higher utility, which can compensate for its detection cost. At the same time, when the difference of normalized detection probabilities gets larger progressively, the bandwidths that different cognitive users get increase further. Even if the normalized detection probability changes a little, the utility of cognitive users changes greatly.

1.7 Impact of Differences in Detection Probabilities

The change of detection probability of cognitive user will directly impact its utility, and when probability adjustment coefficient $\alpha = \sum_{i=1}^{N} b_i$, the normalized detection probability becomes the detection probability.

When cognitive user 1 and user 2 have the same detection probability which respectively equals 0.8, 0.9, 1.0, the utilities of cognitive users, licensed user and the whole system are shown in figure 6; when the detection probabilities of cognitive user 1 and user 2 are both 1.0, which is the case that the utility does not consider the detection probability of cognitive user, the system utility should be the highest. From the figures we can see that when detection probability becomes larger, the utilities of cognitive users system, licensed user system and the whole system all increase in varying degrees. Therefore, in order to improve the system utility, the licensed user system carries out incentive by allocating more bandwidths to cognitive users, and meanwhile if detection probability is higher, the cognitive user need more detection symbols and energy, and of course need be compensated for. Note that the normalized detection probabilities of cognitive users are all 0.5 and remain constant.

1.8 Nash Equilibrium and Pareto Optimality

The analysis above considered the system utility when the two cognitive users reached Nash Equilibrium, but actually Nash Equilibrium is the case that cognitive users reach a stable state by gaming but not the case of optimal system utility. Figure 7 shows the whole utility of Nash Equilibrium and Pareto Optimality when the detection probability of cognitive user 1 unchanged and that of user 2 rises from 0.9 to 1.0. From the figure we can see that the utility of Nash Equilibrium by cognitive users' non-cooperative game is lower than that of Pareto Optimality by cognitive users' cooperative game. The arrowhead is the room to improve from Nash Equilibrium to cooperative optimal state.

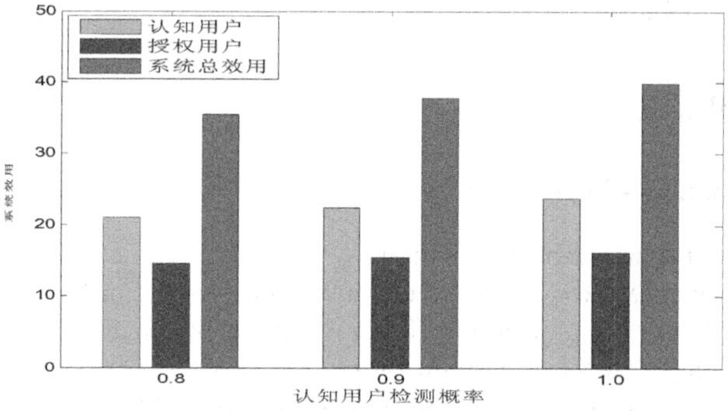

Fig. 6. Detection probability impact on the system utility

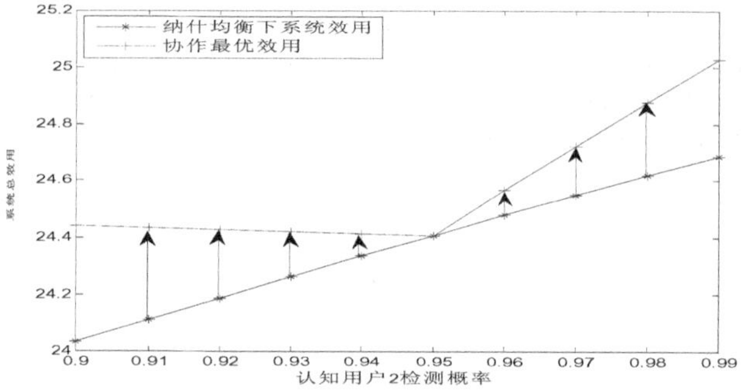

Fig. 7. Nash Equilibrium and Pareto optimality

1.9 Conclusion

Dynamic spectrum allocation is an important part of cognitive radio. This paper proposed a cognitive user utility function combined with game theory according to a wireless network including a wireless network in AWGN including one primary user system and several cognitive users, so as to solve the problem of spectrum allocation in cognitive radio. The cognitive users who have different detection abilities will be allocated different bandwidths of spectrum, and the users who have stronger detection abilities will pay extra detection costs but gain more spectrum, so they can achieve a higher communication quality, which embodied the fairness of the system. Moreover, the analysis showed that it could reduce the convergence times and improves the system's real-time to choose a reasonable cognitive coefficient, and that the change of detection probability had an obvious impact to system utility. Meanwhile we can see that the system utility in Nash Equilibrium was not optimal, and that the system could reach Pareto Optimality through cooperation among users.

References

[1] Tan, X.Z., Liu, Y.T., Xu, G.S.: Dynamic Spectrum Allocation in Cognitive Radio: Auction and Equilibrium In: 2009 International Forum on Information Technology and Applications, Chengdu, China, pp. 15–17 (2009)

[2] Mchenry, M.: Spectrum Occupancy Measurements at National Radio Astronomy Observatory [EB/OL], `http://www.sharedspectrum.com/inc/content/measurements/nsf/5_NSF_NRAO_Report.doc.2005.8`

[3] Ji, Z., Ray, K.J.: Belief-assisted pricing for dynamic spectrum allocation in wireless networks with selfish users. In: IEEE SECON 2006, pp. 119–127. IEEE Press, Linthicum (2006)

[4] Wang, W., Liu, X.: List-coloring based channel allocation for open-spectrum wireless networks. In: IEEE Fall VTC 2005, pp. 690–694. IEEE Press, Dallas (2005)

[5] Dusit, N., Ekram, H.: Competitive Spectrum Sharing in Cognitive Radio Networks: A Dynamic Game Approach. IEEE Transactions on Wireless Communications 7(7), 2651–2660 (2008)

[6] Huang, J.W., Berry, R.A., Honig, M.L.: Auction Mechanisms for Distributed Spectrum Sharing. Mobile Networks and Applications 11(3), 405–418 (2006)

[7] Wang, F., Krunz, M., Cui, S.G.: Price-Based Spectrum Management in Cognitive Radio Networks. IEEE Journal on Selected Topics in Signal Processing 2(1), 74–87 (2008)

[8] Han, N., Shon, S.H., Chung, J.H., et al.: Spectral Correlation Based Signal Detection Method for Spectrum Sensing in IEEE 802.22 WRAN Systems. In: 8th International Conference Advanced Communication Technology, vol. 3, pp. 1765–1770 (2006)

[9] Jiang, J., Tan, X.Z., Sun, H.J.: Cooperative Algorithm for Cognitive Radio Networks Which is Based on Adaptive Election. In: IEEE Region 10 Annual International Confer. (2006)

[10] Nie, N., Comanniciu, C.: Adaptive channel allocation spectrum etiquette for cognitive radio networks. In: 2005 1st IEEE International Symposium on New Frontiers in Dynamic Spectrum Access Networks, pp. 269–278 (2005)

[11] Niyato, D., Hossain, E.: Competitive pricing for spectrum sharing in cognitive radio networks: Dynamic game, inefficiency of Nash equilibrium, and collusion. IEEE Journal on Selected Areas in Communications 26(1), 192–202 (2008)

Dynamic Modeling and Simulation of a Manipulator with Joint Inertia

Shi-xiang Tian and Sheng-ze Wang

College of Mechanical Engineering, Donghua University
2999, North People Road, Songjiang District, 201620, Shanghai, China
tianshixiang@126.com,
wasz@dhu.edu.cn

Abstract. This work deals with the dynamic modelling, analysis and simulation of a three degree of freedom spatial manipulator using symbolic and numerical method. A specially designed concise and novel algorithm based on Newton-Euler equation is proposed to establish the dynamic equation in form of state space automatically. Through the method not only the kinematic and dynamic parameters of the manipulator are obtained automatically, but also the simulation equation is produced concurrently. The algorithm is implemented in the well known algebraic system Maple and simulated in the Matlab/simulik. if extended, the program can be adopted for more complex manipulator.

Keywords: Algebraic system, Simulation, Modelling, Robot manipulator, Dynamics, Newton-Euler equation.

1 Introduction

Understanding the dynamic structure of the plant is particularly important and dynamical modeling of robot has been attracting the researchers all around the world [2, 4, 10, 16, 15, 17, 3, 6]. The inability to specify joint torques in the manipulators would deem most advanced control strategies unsuitable, since almost all of them are invariably based on the capability to control joint torques [5, 15, 16, 12]. This article established the dynamic model of a three degree of freedom robot, and getting the closed form solution to dynamics of the robot by means of the algebraic system Maple to lay a foundation for the subsequent controller design.

There are a number of procedures for generating the dynamic equations of motion for a manipulator, i.e., the equations, which relate joint forces and torque set $\tau(t)$ to positions $\theta(t)$, velocities $\dot{\theta}(t)$ and accelerations $\ddot{\theta}(t)$, in terms of the specified kinematic and inertial parameters of the links. At present, a number of ways have been proposed for this purpose, such as Lagrange-Euler (L-E) method [5], Newton-Euler (N-E) method [5], Recursive Lagrangian method [9], Kane's method [11], Appel's method [11] and Generalized D'Alambert principle method [7]. If only they describe the dynamic behavior of the same physical robot manipulator, these methods

L. Qi (Ed.): ISIA 2010, CCIS 86, pp. 10–16, 2011.

are "equivalent" to each other. However, the structure of these equations and, particularly, the computational efficiency of the equations may differ, as they are obtained for various reasons and purposes, such as suitability for simulation, real-time control, parameter significance, controller design, etc.

Among these methods, the L-E and the N-E formulation have been generally used. These methods, based on Lagrangian and Newtonian methods, respectively, have their own characteristics [14]. In this article, The focus of attention will be on deriving a specific N-E formulation and the state-space form of a 3R serial manipulator with the joint mass inertia included.

2 Newton Euler Formulation with Joint Inertia

The derivation of the dynamic equations of an n DoF manipulator (including the masses of the electronic actuating devices) is based on the understanding of the N-E force equation, described as

$$\begin{aligned} \mathbf{F} &= m\ddot{\mathbf{c}} \\ \mathbf{n}_c &= \mathbf{I}_c\dot{\boldsymbol{\omega}} + \boldsymbol{\omega} \times \mathbf{I}_c \end{aligned} \tag{1}$$

Where $\ddot{\mathbf{c}}$ is the acceleration of the rigid body center of mass, $\mathbf{F}$ is the superposition of all forces exerted on the rigid body, m is the total mass of the body. $\mathbf{n}_c$ is the total moment exerted on the rigid body, $\dot{\boldsymbol{\omega}}$ is the angular acceleration, $\boldsymbol{\omega}$ is the angular velocity, and $\mathbf{I}_c$ is the inertia tensor of the body written in the frame originated the center of the mass of the body.

The prerequisite to apply N-E equation to dynamics is that relevant kinematic and inertial parameters. There are mainly three ways in which the kinematic parameters and the inertial parameters can be obtained, which are direct computation, experimentally kinematic calibration [1][2] and system Identification [13, 8]. Although, theoretically, there is no method that can get the absolutely precision model, the mathematical computation and simulation method is cheaper, more efficient, more testable and more generalizable than any other methods. Therefore, the mathematical computation model should be established and used firstly as could as possible. In this paper, we assume that the material is isotropic and uniform, the joint actuators are viewed as point of mass. the inertia moment of links including the joint inertia is computed by the equation

$$I_c^i = \frac{1}{12}m_i l_i^2 + m_i d_i^2 (0.5l_i - d_i)^2 \tag{2}$$

Where I_c^i is the inertial moment of ith link with respect to its center of mass; m_i is the mass of i th link; l_i is the length of i th link, m_{ji} is the lump mass of i th joint; d_i is the offset distance from geometric center of i th link.

Having obtained the relevant parameters, the N-E formulation can then be applied, which yields the necessary torque to drive the manipulator. As mentioned above, the obtained torque is not taken the state-space form, so in this work we propose a specially simple algorithm to make the solution have the state space form to facilitate designing controller.

The solution to dynamics can be in a matrix-vector equation form [5] as:

$$\mathbf{MD}(\theta)\mathbf{D}(\ddot{\theta}) + \mathbf{MC}(\theta)\mathbf{C}(\dot{\theta}) + \mathbf{MG}(\theta)\mathrm{g} = \boldsymbol{\tau} \tag{3}$$

Where $\mathbf{MD}(\theta) \in \mathbf{R}^n \times \mathbf{R}^n$ denotes the inertia matrix associated with the distribution of mass, $\mathbf{D}(\ddot{\theta}) \in \mathbf{R}^n \times \mathbf{R}^1$ is angular acceleration vector, $\mathbf{MC}(\theta) \in \mathbf{R}^n \times \mathbf{R}^{n(1+n)/2}$ denotes centripetal and coriolis forces matrix associated with the coupling items of angular velocity, $\mathbf{C}(\dot{\theta}) \in \mathbf{R}^{n(1+n)/2} \times \mathbf{R}^1$ is angular velocity product vector and where $\mathbf{MG}(\theta) \in \mathbf{R}^n \times \mathbf{R}^1$ represents the gravity force terms. Joint torques are included in vector $\boldsymbol{\tau} \in \mathbf{R}^n \times \mathbf{R}^1$.

The equation can be further expand as

$$\begin{bmatrix} \tau_1 \\ \tau_2 \\ \vdots \\ \tau_n \end{bmatrix} = \begin{bmatrix} MD_{11} & MD_{12} & \cdots & MD_{1n} \\ MD_{21} & MD_{22} & \cdots & MD_{2n} \\ \vdots & \vdots & \cdots & \vdots \\ MD_{n1} & MD_{n2} & \cdots & MD_{nn} \end{bmatrix} \begin{bmatrix} \ddot{\theta}_1 \\ \ddot{\theta}_2 \\ \vdots \\ \ddot{\theta}_n \end{bmatrix} + \begin{bmatrix} G_1 \\ G_2 \\ \vdots \\ G_n \end{bmatrix} g + \underbrace{\begin{bmatrix} MC_{11} & MC_{12} & \cdots & MC_{1,n\bullet(1+n)/2} \\ MC_{21} & MC_{22} & \cdots & MC_{2,n\bullet(1+n)/2} \\ \vdots & \vdots & \cdots & \vdots \\ MC_{n1} & MC_{n2} & \cdots & MC_{n,n\bullet(1+n)/2} \end{bmatrix}}_{\frac{n\times(1+n)}{2}} \begin{bmatrix} \dot{\theta}_1^2 \\ \dot{\theta}_1 \bullet \dot{\theta}_2 \\ \vdots \\ \dot{\theta}_1 \bullet \dot{\theta}_n \\ \dot{\theta}_2^2 \\ \dot{\theta}_2 \bullet \dot{\theta}_3 \\ \vdots \\ \dot{\theta}_2 \bullet \dot{\theta}_n \\ \vdots \\ \dot{\theta}_{n-1}^2 \\ \dot{\theta}_{n-1} \bullet \dot{\theta}_n \\ \dot{\theta}_n^2 \end{bmatrix} \tag{4}$$

3 Implementing Algorithm of Computing Dynamics Based on Maple

The algorithm of computing dynamics is implemented by the well known algebraic system Maple, because of the limits to the length of the article, herein will give the main command flow to compute the joint torque. Following is the piece of program for computing joint torque.

```
#mi: link mass; mji: joint mass such as motor, reducer. etc; aPci set mass central #acceleration; aPi: set joint mass acceleration; Fmji: gravity from joint weight; EON: #Newton equation; FI: the force exert on i th joint; EEU: Euler equation; Zd: joint #axis direction cosine
for ni from sn by -1 to 0 do
    mi := m[ni]; mji := mj[ni+1];aPci := aPc[ni];aPi := aP[ni];
    FG := Matrix(3, 1, [0, 0, -mi*g]); FE1 := Matrix(3, 1, [fe1x, fe1y, fe1z]);
    FE2 := -F[ni+1]; Fmji := Matrix(3, 1, [0, 0, -mji*g]);
    #begin to solve end forces
    EON := FE1+FE2+FG+Fmji-mi*aPci-mji*aPi;
    sEON := solve(EON[1, 1], EON[2, 1], EON[3, 1], [fe1x, fe1y, fe1z]);
    FI := Matrix(3, 1, [rhs(sEON[1, 1]), rhs(sEON[1, 2]), rhs(sEON[1, 3])]);
    F[ni] := FI; F[ni] := combine(F[ni], trig);
    #begin to solve joint torque Ti
    FE1 := FI; TE1 := Matrix(3, 1, [te1x, te1y, te1z]); TE2 := -T[ni+1];
    Rfe1 := P[ni]-Pc[ni]; Rfe2 := P[ni+1]-Pc[ni];
    CRfe1 := Matrix(3, 3, [0, -Rfe1[3, 1], Rfe1[2, 1], Rfe1[3, 1], 0, -Rfe1[1, 1], -Rfe1[2, 1], Rfe1[1, 1], 0]);
    CRfe2 := Matrix(3, 3, [0, -Rfe2[3, 1], Rfe2[2, 1], Rfe2[3, 1], 0, -Rfe2[1, 1], -Rfe2[2, 1], Rfe2[1, 1], 0]);
    Tfe1 := MatrixMatrixMultiply(CRfe1, FE1);
    Tfe2 := MatrixMatrixMultiply(CRfe2, FE2);
    Tmji := MatrixMatrixMultiply(CRfe2, Fmji); Ii := IM[ni];
    vPHi := Matrix(3, 1, [vPH[ni][1], vPH[ni][2], vPH[ni][3]]);
    aPHi := Matrix(3, 1, [aPH[ni][1], aPH[ni][2], aPH[ni][3]]);
    IVW := MatrixMatrixMultiply(Ii, aPHi); CvPHi := CvPH[ni];
    WIW := MatrixMatrixMultiply(MatrixMatrixMultiply(CvPHi, Ii), vPHi);
    EEU := Tfe1+Tfe2+TE1+TE2+Tmji-IVW-WIW;
    sEEU := solve(EEU[1, 1], EEU[2, 1], EEU[3, 1], [te1x, te1y, te1z]);
    TI := Matrix(3, 1, [rhs(sEEU[1, 1]), rhs(sEEU[1, 2]), rhs(sEEU[1, 3])]);
    T[ni] := TI; Tz[ni] := MatrixMatrixMultiply(Transpose(T[ni]), Zd[ni]);
end do;
```

4 Dynamic Simulation and Results

To simulate the motion of a manipulator we must make use of a model of the dynamics, which we have just developed. simulation requires solving the dynamic equation for acceleration:

$$\ddot{\mathbf{\Theta}} = \mathbf{MD}^{-1}(\boldsymbol{\tau} - \mathbf{MC}(\theta)\mathbf{C}(\dot{\theta}) - \mathbf{MG}(\theta)g \tag{5}$$

Where

$$\ddot{\mathbf{\Theta}} = \begin{bmatrix} \ddot{\theta}_1 \\ \ddot{\theta}_2 \\ \ddot{\theta}_3 \end{bmatrix} \tag{6}$$

We then apply Matlab/simulink to integrate the acceleration to compute future positions and velocities.

Manipulator system is a highly nonlinear system. in order to understand the properties of the manipulator to lay a foundation for the controller design, we have arranged a series driven torques to investigate its dynamical response to the load. The relevant parameters and simplified notations for simulation are given in table 1. The simulation results are shown from Fig. 1 until Fig. 3.

Table 1. Parameters for simulation

Item	Value	Unit	Item		Unit
l_0	0.45	m	l_1	0.26	m
l_2	0.26	m	m_0	9.928	kg
m_1	0.658	kg	m_2	0.494	kg
I_{0xx}	0.164	kg • m^2	I_{0yy}	0.174	kg • m^2
I_{0zz}	0.699	kg • m^2	I_{1xx}	0.374	kg • m^2
I_{1yy}	0.393×10^{-2}	kg • m^2	I_{1zz}	0.258	kg • m^2
I_{2xx}	0.162×10^{-3}	kg • m^2	I_{2yy}	0.291×10^{-2}	kg • m^2
I_{2zz}	0.172	kg • m^2			
m_{j0}	10	kg	m_{j1}	15	kg
m_{j2}	10	kg	m_{j3}	10	kg

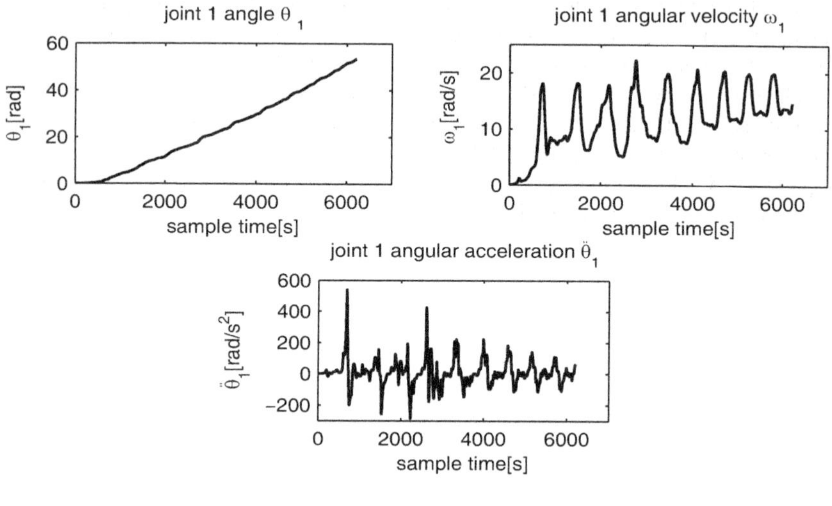

Fig. 1. Joint 1 response to constant torque

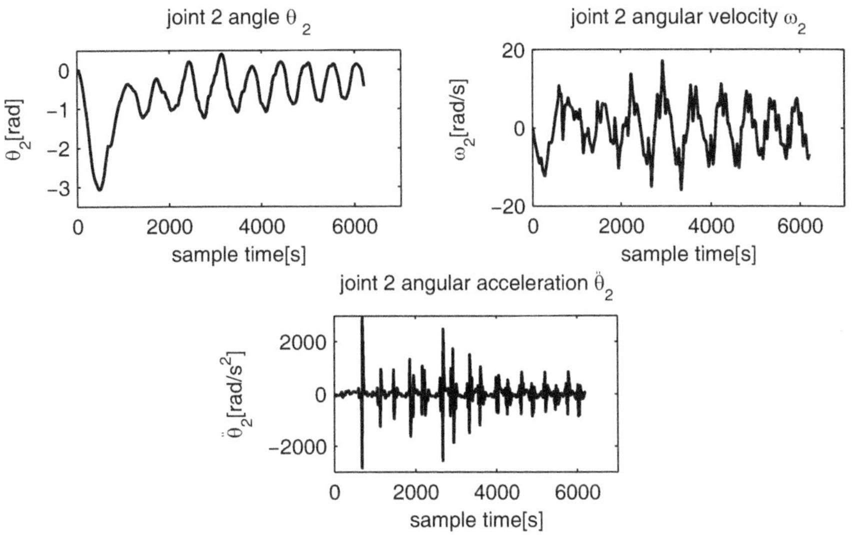

Fig. 2. Joint 2 response to constant torque

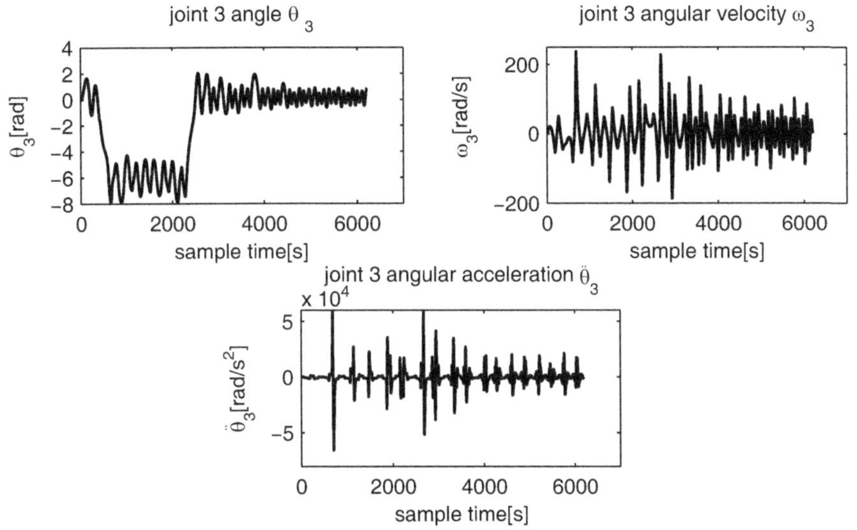

Fig. 3. Joint 3 response to constant torque

5 Conclusion

A specific state-space model of a three degree of freedom manipulator is derived by algebraic system Maple and simulation of the model is carried in Matlb/simulink. By applying symbolic computation the dynamic model and simulation model is obtained automatically, and the relevant programme can be extended to more complex manipulator systems. The work done in the article reveals that the reasonable application of the algebraic system to the specific dynamic problem can lead to a rapid establishment of the dynamic model, so as to save the time and cost of design.

References

1. An, C., Atkeson, C., Hollerbach, J.: Model Based Control of a Robot Manipulator. MIT Press, Cambridge (1988)
2. Armstrong, B., Khatib, O., Burdick, J.: The explicit dynamic model and inertial parameters of the puma 560 arm. In: IEEE International Conference on Robotics and Automation, pp. 510–518 (1986)
3. Carrera, E., Serna, M.A.: Inverse dynamics of flexible robots. Mathematics and Computers in Simulation 41, 485–508 (1996)
4. Corke, P.: An automated symbolic and numeric procedure for manipulator rigid body dynamic signicance analysis and simplication. In: IEEE International Conference on Robotics and Automation, pp. 1018–1023 (1986)
5. Craig, J.: Introduction to Robotics: Mechanics and Control, 3rd edn. Prentice Hall, Englewood Cliffs (2005)
6. Dwivedy, S.K., Eberhard, P.: Dynamic analysis of flexible manipulators, a literature review. Mechanism and Machine Theory 41, 749–777 (2006)
7. Fu, K., Gonzalez, R., Lee, C.: Robotics: Control, Sensing, Vision and Intelligence. McGraw-Hill, New York (1987)
8. José Antonio Martin, H., JavierdeLope, MatildeSantos: MatildeSantos: A method to learn the inverse kinematics of multilink robots by evolving neuro-controllers. Neurocomputing 72, 2806–2814 (2009)
9. Hollerbach, J.: A recursive formulation of lagrangian manipulator dynamics. IEEE Trans. Systems, Man Cybern. SMC-10(11), 730–736 (1980)
10. Korayem, M.H., Basu, A.: Automated fast symbolic modeling of robotic manipulators with compliant links. Mathl. Comput. Modelling 22(9), 41–55 (1995)
11. Li, C.: A new lagrangian formulation of dynamics for robotic manipulators. J. Dynamic Systems, Measurement and Control, Trans. ASME 111, 559–567 (1989)
12. Li, Y., Xu, Q.: Dynamic modeling and robust control of a 3-prc translational parallel kinematic machine. Robotics and Computer-Integrated Manufacturing 25, 630–640 (2009)
13. Ljung, L.: From data to model: a guided tour. In: IEE CONTROL 1994, pp. 422–430 (1994)
14. Miro, J., White, A.: Modelling an industrial manipulator a case study. Simulation Practice and Theory (9), 293–319 (2002)
15. Moosavian, S.A.A., Papadopoulos, E.: Cooperative object manipulation with contact impact using multiple impedance control. International Journal of Control, Automation, and Systems 8(2), 314–327 (2010)
16. Wang, H., Xie, Y.: Adaptive inverse dynamics control of robots with uncertain kinematics and dynamics. Automatica (45), 2114–2119 (2009)
17. Zhao, Y., Gao, F.: Dynamic formulation and performance evaluation of the redundant parallel manipulator. Robotics and Computer-Integrated Manufacturing 25, 770–781 (2009)

Shear Failure Criteria of Soft Soil under Complex Stress Condition

Weimin Jin, Wei Wang*, Baolin Wang, and Zeng Pan

Department of Civil Engineering, Shaoxing University, Shaoxing 312000, China
wellsking.wang@gmail.com

Abstract. Shear behavior is an interesting subject of soft soil engineering, and it is important to determine shear critical state under complex stress conditions. In this paper, eight failure criteria are established according to Mohr stress cycle and Coulomb strength theory. Three criteria are based on conventional soil mechanics, and the other three criteria are based on geometry-algebra theory. Last two criteria are on the basis of interdiscipline involving soil mechanics and geometry-algebra theory. Finally, example is introduced to verify correctness of these criteria and good agreements have been found.

Keywords: Failure criteria; shear behavior; soil mechanics; geometry theory.

1 Introduction

With the development of economy, high-rise buildings and highway on soft soil foundation have gradually increased, which makes a more strict demands for the foundation stability. Soft soil foundation has some distinct characteristics, such as low strength, large and long time deformation, high water content and poor water permeability [1, 2, 3]. These characteristics lead to a certain particularity of soft soil foundation. If not handled properly, it will produce a series engineering problems, such as excessive ground settlement, ground cracking, serious embankment deformation and even collapse. It is necessary to establish failure criterion for soft soil.

2 Basis Theory of Shear Strength

The basis theory of shear strength mainly consists of Mohr theory and Coulomb criterion [4, 5]. Mohr theory describes the relationship between shear stress τ and normal stress σ in the shear surface of a certain point, and shows the basic characteristics of friction behavior of soil, as granular materials. According to the theory of material mechanics, it usually can be expressed as:

$$\sigma = \frac{\sigma_1 + \sigma_3}{2} + \frac{\sigma_1 - \sigma_3}{2}\cos 2\theta ; \tau = \frac{\sigma_1 - \sigma_3}{2}\sin 2\theta \qquad (1)$$

* Corresponding author.

L. Qi (Ed.): ISIA 2010, CCIS 86, pp. 17–23, 2011.

where σ_1 denotes the major principal stress, σ_3 denotes the minor principal stress, and θ denotes the included angle between shear plane and major principal stress.

According to the trigonometric functions, Eq. 1 can be transformed as:

$$\left(\sigma-\frac{\sigma_1+\sigma_3}{2}\right)^2+\tau^2=\left(\frac{\sigma_1-\sigma_3}{2}\right)^2 \tag{2}$$

Coulomb criterion, describing the relationship between shear strength τ_f and normal stress σ, can be expressed as follows

$$\tau_f = c + \sigma \tan\phi \tag{3}$$

Among them, φ is friction angle, c is cohesive force.

Combining Mohr theory Eq. 2 and Coulomb criterion Eq. 3, the state of limit equilibrium can be obtained, by which the failure of soil can be determined. When the soil is under the state of limit equilibrium, the Mohr's circle touches the Coulomb strength line, as show in Fig.1. Relationship between the principal stresses and the shear strength parameters is as follows [6, 7]:

$$\begin{cases}\sigma_3=\sigma_1\tan^2(45^0-\phi/2)-2c\tan(45^0-\phi/2)\\ \sigma_1=\sigma_3\tan^2(45^0+\phi/2)+2c\tan(45^0+\phi/2)\end{cases} \tag{4}$$

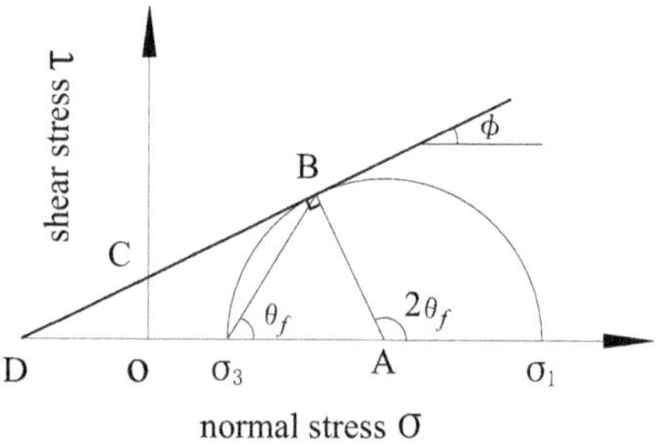

Fig. 1. Stress sketch for soil state of limit equilibrium

Through the understanding of the basic law, and based on the condition of limit equilibrium, whether the soil shear failure will occurred can be easily determined.

3 Shear Failure Criteria

Given one point of soil, its practical principal stress σ_1 and σ_3 can be calculated, and parameter c and φ can be obtained by lab test. Following discussion assumes that: (1) above four parameters are known; (2) limit state is considered failure state, too.

3.1 Major Principal Stress Criterion

This criterion assumes that the value of major principal stress with limit state, σ_{1f}, is equal to that of known major principal stress, that is, $\sigma_{1f}=\sigma_1$. According to Eq. 4, we can obtain the corresponding σ_{3f} with limit state. If the $\sigma_{3f} < \sigma 3$, then the point will not be damaged, or else it will be destroyed.

3.2 Minor Principal Stress Criterion

This criterion assumes that the value of minor principal stress with limit state, $\sigma_{3f,}$ is equal to that of known minor principal stress, that is, $\sigma_{3f} = \sigma_3$. According to Eq. 4, we can obtain the corresponding $\sigma 1f$ with limit state. If the $\sigma_{1f} > \sigma_1$, then the point will not be damaged, or else it will be destroyed.

When using the above two criteria, following conclusions can be drawn. When the major principal stress is constant, the smaller the minor principal stress is, the stress circle is nearer strength envelope, the more it tends to be destroyed. On the other hand, when the minor principal stress is constant, the bigger the major principal stress is, the stress intensity circle is nearer strength envelope, the more it tends to be destroyed.

3.3 Theoretical Shear Failure Angle Criterion

For a given soil, the friction angle and cohesion force are constant. When the straight line DB is fixed, regardless of how the stress circle moves, once with the straight line DB tangent, $\angle BAF = \angle B'A'F'$ is the fixed value, shown in Fig.2, then the theoretical shear failure angle is also a fixed value, namely:

$$\theta_f = 45^0 + \phi / 2 \tag{5}$$

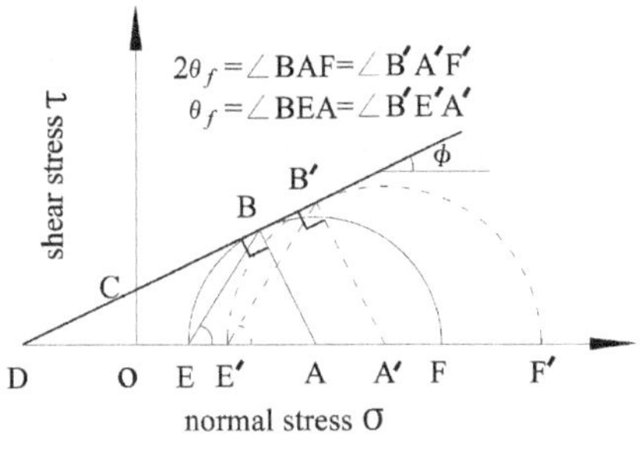

Fig. 2. Sketch for theory shear failure angle criterion

When the soil shear failure happens, its failure plane firstly occurs in the theoretical shear failure surface corresponding to the theoretical shear angle. Comparing with theoretical shear stress and shear strength of the shear break surface can be needed. According to the known major and minor principal stress, putting the Eq. 5 into Eq. 1 gives rise to the normal stress σ and shear stress τ of shear break surface, and then its shear strength, τ_f, can be determined by Eq. 3 and be contrasted to the previously calculated shear stress τ. If $\tau \leq \tau_f$, then the shear failure occurs, otherwise failure dose not occur.

3.4 Graphical Chart Criterion

According to the relationship between the stress circle and strength envelope, graphical chart criterion can be used. Based on the known conditions, Mohr circle A and line BD in proportion can be mapped together, respectively, show as in Fig.3, which can easily determine whether the failure will occur. If the straight line leaves circle, it will not be failed. If the straight line touches the circle, which is a limit state, it will be failed; if the straight line cuts the circle, it has been failed. It's receptive to transform theoretical mechanics problem into a simple plane geometry problems.

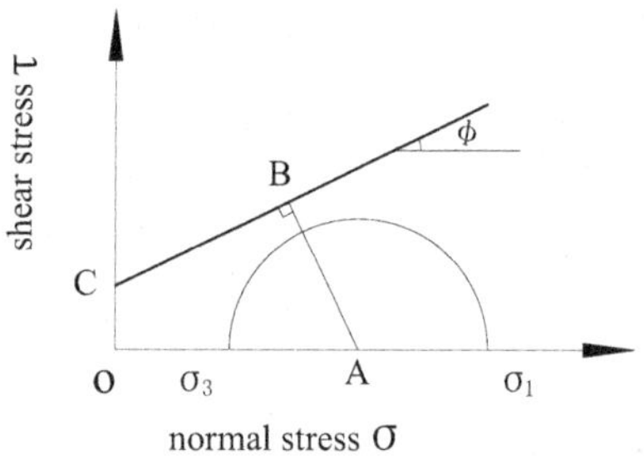

Fig. 3. Sketch for graphical chart criterion

3.5 Point-Line Distance Criterion

Graphical chart criterion is intuitive, but there are still some shortcomings of proportion controlled and the angle determined with many difficulties. At this time, we can judge the relationship between Mohr circle and strength envelope from another point of view, which is calculating the point-line distance from circle center to the strength line. Based on the mathematical relationship, the point-line distance, d, and the circle radius r, can be respectively expressed as:

$$d = \frac{\left|2c + (\sigma_1 + \sigma_3)\tan\phi\right|}{2\sqrt{1+\tan^2\phi}}; \quad r = \frac{\sigma_1 - \sigma_3}{2} \tag{6}$$

It is easy to determine soil stress condition. $d>r$, $d=r$ and $d<r$ denote safety state, limit state and failure state, respectively.

3.6 Discriminant of Quadratic Form Criterion

Although point-line distance criterion overcomes the constraints of graphical chart criterion, its formula is complex and difficult to remember. In fact, according to the characteristics of Eq. 2 and Eq. 3, the limit equilibrium condition can be determined from the perspective of a quadratic equation. Ordering $\tau_f = \tau$, putting Eq. 3 into Eq. 2 produces following expression:

$$(1+\tan^2\phi)\sigma^2 - (\sigma_1 + \sigma_3 - 2c\tan\phi)\sigma + \sigma_1\sigma_3 + c^2 = 0 \tag{7}$$

Assuming that soil is failure, Eq. 7 should have real solutions. Discriminant of the quadratic root can be obtained as follows:

$$\Delta = (\sigma_1 + \sigma_3 - 2c\tan\phi)^2 - 4(1+\tan^2\phi)(\sigma_1\sigma_3 + c^2) \tag{8}$$

According to the known major principal stress, minor principal stress, friction angle and cohesive force, it is easy to get the discriminant value to judge whether destruction condition is reached. (1) if $\Delta < 0$, there is no solution for Eq. 8, namely safety state; (2) if $\Delta = 0$, there is only one solution for Eq. 8, namely limit state; (3) if $\Delta > 0$, there are two solutions Eq. 8, namely failure state.

This method transforms the complex and difficult mechanics theory into simple elementary mathematics knowledge, which is simple and convenient.

3.7 Friction Angle Criterion

According to Eq.4, friction angle with limit state can be written as:

$$\phi_f = \arcsin\frac{\sigma_1 - \sigma_3}{\sigma_1 + \sigma_3 - 2c\cot\phi} \tag{9}$$

Stress state can be judged by Eq. 9 associated with Fig. 4. φ>φ, φ=φ and φ<φ denote safety state, limit state and failure state, respectively.

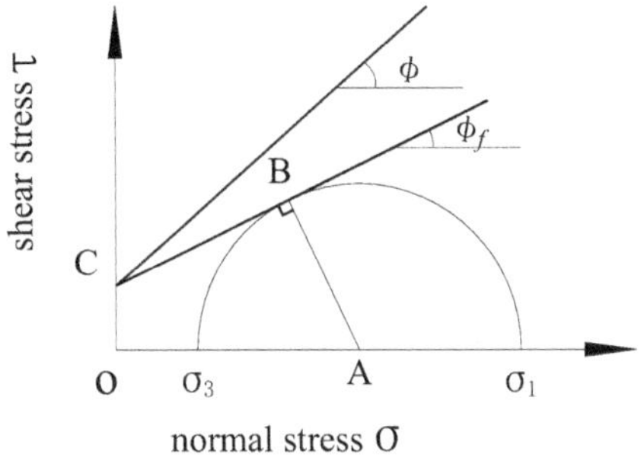

Fig. 4. Sketch for friction angle criterion

3.8 Cohesive Force Criterion

Using the relationship Eq. 4 between the principal stress and shear strength parameters, we can obtain the expression of cohesive force, c, with limit state:

$$c_f = \frac{\sigma_1\tan(45^0 - \phi/2) - \sigma_3\cot(45^0 - \phi/2)}{2} \tag{10}$$

Similar to friction angle criterion, stress state can be judged by Eq. 11 associated with Fig. 5. $c>c_f$, $c=c_f$ and $c<c_f$ denote safety state, limit state and failure state, respectively.

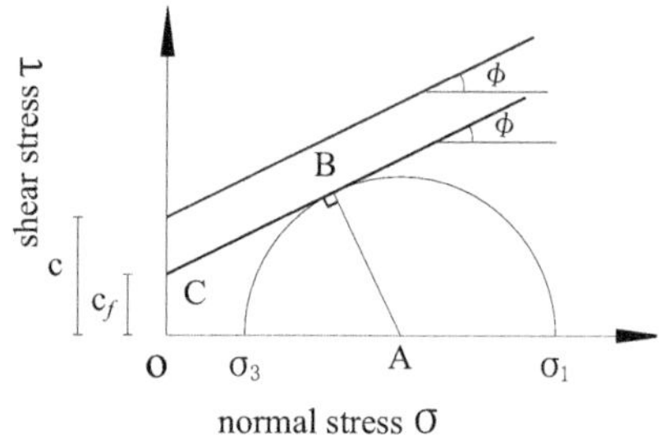

Fig. 5. Sketch for Cohesive force criterion

4 Examples

Example is employed to verify correctness of proposed eight criteria. Major principal stress and minor principal stress are σ_1=120.0 kPa and σ_3=30.0 kPa, respectively. Cohesive force and friction angle are 8.0 kPa and φ =30^0, respectively. Judgments of eight criteria are listed in Table 1 with same results, namely failure state.

Table 1. Judgments of eight criteria

Criteria	Failure standard	Calculation result
major principal stress criterion	$\sigma_{3,f} > \sigma_3$	σ_3=30.0 kPa, $\sigma_{3,f}$=30.8 kPa
minor principal stress criterion	$\sigma_{1,f} < \sigma_1$	σ_1=120.0 kPa, $\sigma_{1,f}$=117.7 kPa
theory shear failure angle criterion	$\tau_f < \tau$	τ_f =38.3 kPa, τ =39.0 kPa
graphical chart criterion	point-circle	point-circle intercross
point-line distance criterion	d<r	d=43.5, r=45.0
discriminant of quadratic form criterion	$\Delta > 0$	Δ =272.3
friction angle criterion	$\varphi < \varphi_f$	φ =30^0, φ_f =30.4^0
cohesive force criterion	$c < c_f$	c =8.0 kPa, c_f =8.7 kPa

5 Summaries

In this paper, eight criteria are proposed for shear failure of soil based on Mohr stress cycle and Coulomb strength theory. Main conclusions are as follows.

(1) Major principal stress criterion, minor principal stress criterion and theory shear failure angle criterion are based on conventional soil mechanics. With constant major

principal stress, the smaller the minor principal stress is, the more it tends to be failure. With constant minor principal stress, the bigger the major principal stress is, the more it tends to be failure.

(2) Graphical chart criterion, point-line distance criterion and discriminant of quadratic form criterion are based on geometry-algebra theory. They convert the complex and difficult mechanics theory into simple elementary mathematics knowledge, which are simple and convenient.

(3) Friction angle criterion and cohesive force criterion are on the basis of interdiscipline involving soil mechanics and geometry-algebra theory. Stress state can be easily determined by comparing values of corresponding parameters.

(4) Correctness and consistence of eight criteria are verified by on example.

Acknowledgement

The authors thank the reviewers who gave a through and careful reading to the original manuscript. Their comments are greatly appreciated and have help to improve the quality of this paper. This work is supported in part by a grant from Nature Science Foundation of Zhejiang Province with NO. Y1080839.

References

1. Powrie, W.: Soil Mechanics, E&FNSpon, London, UK (1997)
2. Lu, T.: Soil Mechanics. Houhai University Press, Nanjing (2002)
3. Wang, W., Lu, T., Zhou, G.: Improved tangent modulus of nonlinear soil model. Chinese Journal of Geotechnical Engineering 29, 458–462 (2007)
4. Gao, H., Zheng, Y., Feng, X.: Deduction of failure criterion for geomaterials based on maximum principal shear strain. Chinese Journal of Rock Mechanics and Engineering 26, 518–524 (2007)
5. Liu, G., Shi, C., Huang, L.: Analyses of the failure criteria of masonry in compression-shear. Journal of Hunan University Natural Sciences 34, 19–23 (2007)
6. Yang, X., Ling, P., Xiang, S.: Comments on slope stability based on a series of Drucker-Prager failure criteria. Rock and Soil Mechanics 30, 865–870 (2009)
7. Li, Z., Tang, X.: Deduction and verification of a new strength criterion for soils. Rock and Soil Mechanics 28, 1247–1249 (2007)

Conflict Detection Based on Improved Unscented Particle Filter

Lianzhi Yu, Shilei Zhang, and Xiaofei Zhu

University of Shanghai for Science and Technology, No. 516 of Jungong Road, Shanghai, China
zhangshilei007@163.com

Abstract. With increasing air traffic flow, the increasingly complex air traffic situation has raised possibility of conflicts, which requires higher timeliness and accuracy for conflict detection. Based on the rapidly growing air traffic control technologies, such as radars with excellent accuracy, an improved unscented particle filter (MUPF) algorithm is proposed to perform real-time aircraft status estimation. Compared to traditional particle filter algorithms (UPF), our MUPF uses fewer particles and generates higher accuracy, resulting in considerable reduction in amount of computation. Therefore, it is then implemented to perform conflict detection. The simulation results show that our algorithm has higher accuracy than UPF, and in conflict detection, it also surpass UPF with lower rate of false alert, higher rate of success alert and smaller error.

Keywords: Air traffic control; conflict detection; improved unscented particle filter; the state estimates.

1 Introduction

With the rapid develop of economy, the air traffic is becoming heavier and heavier and making the possibility of air conflict greatly increases. So It is crucial to take the acts of the efficient prevention from conflict, the predict of air conflict become the key of improving the safety of flight. It contains two aspects: conflict detection and resolution. In the stage of conflict detection, based on the current flight plan and aircraft position to predict the location of aircraft in a future time. If the distance of any two plans less than the safety, then it will conflict. The next step is the stage of conflict resolution, based on the current situation of the region, gives a safe route.

At present, most documents did not consider radar error, but the radar error must be considered in the actual conflict detection. We need to estimate the flight conditions by using the radar observations, so we put into use the conflict detection algorithm, and get a more realistic value, which purpose is reduce the rate of error and increase accuracy. Although there were some literatures application UPF algorithms which considered the radar noise, but it require of more particulates and a lot of calculation that maybe receive a higher precision. Therefore we need to improve the algorithm.

In this paper, first, I introduce the improved unscented particle filter (MUPF). Second, I give the conflict probability model of the terminal area. And then, I use the MUPF algorithm and UPF algorithm respectively into aircraft flight state estimation

L. Qi (Ed.): ISIA 2010, CCIS 86, pp. 24–31, 2011.

and detection of conflict, and compare with the estimation accuracy, the difference of successful prediction rate and false alarm rate. Final, I give the conclusion.

2 The Improved Unscented Particle Filter(MUPF)

The unscented particle filter:

$$\begin{cases} x_t = f(x_{t-1}, v_{t-1}) \\ y_t = h(x_t, u_t) \end{cases} \tag{1}$$

x_i denotes the states of the model, f(.) represent the deterministic process, y_t denotes the observations, h(.) represent the measurement models, v_{t-1} denotes the process noise, u_t denotes the measurement noise.

PF common problem is the degradation, that after several iterations, almost all the particles have a negative weight. It has proved impossible to eliminate the degradation because the variance of the important weight which increased with increasing time. Generally, there are three ways to reduce degradation. One is to increase the number of particles, but in many cases this is unrealistic. One is re-sampling method, now there are lots of re-sampling algorithms, such as residual sampling, minimum variance sampling, and polynomial sampling and so on, the above re-sampling results have little difference. What's more, one is the importance of optimal density function. The PF does not use the latest state of the system measurement information, makes the particles to depend on the model, and hence the actual posterior distribution of samples produces larger error. In the framework of the PF , using of unscented Kalman filter (UKF), get the importance density function generated, and then make UPF algorithm.

To summarize, a complete algorithm for a generic UPF is given as follows:

Step 1: Initialization

$$\begin{cases} \bar{x}0 = E[x_0], \boldsymbol{P}_0 = Cov(x_0), \\ \boldsymbol{x}_0^a = [x_t^T\ v_t^T\ u_t^T]^T = [\bar{x}0\ 0\ 0], \\ \bar{\boldsymbol{x}}_0^a = E[x_0^a], \boldsymbol{P}_0^a = diag(P_0, Q_0, R_0) \end{cases} \tag{2}$$

$$\begin{cases} \lambda = \alpha^2(n_a + \kappa) - n_a, W_0^m = \lambda/(n_a + \lambda) \\ W_0^c = \lambda/(n_a + \lambda) + (1 - \alpha^2 + \beta) \\ W_j^m = W_j^c = 1/2(n_a + \lambda), j = 1, \ldots, 2n_a \end{cases} \tag{3}$$

Step 2: For t = 1, 2…do

1) Calculate sigma points

$$\boldsymbol{X}_{j,t-1}^a = [\bar{\boldsymbol{x}}_{t-1}^a, \bar{\boldsymbol{x}}_{t-1}^a \pm \sqrt{(n_a + \lambda)\boldsymbol{P}_{t-1}^a}_j] \tag{4}$$

2) Time update

$$\begin{cases} \boldsymbol{X}^x_{t|t-1} = f(\boldsymbol{X}^x_{t-1}, \boldsymbol{X}^v_{t-1}),\ \bar{x}_{t|t-1} = \sum_{j=0}^{2n_a} W^m_j \boldsymbol{X}^x_{j,t|t-1} \\ \boldsymbol{P}_{t|t-1} = \sum_{j=0}^{2n_a} W^c_j [\boldsymbol{X}^x_{j,t|t-1} - \bar{x}_{t|t-1}][\boldsymbol{X}^x_{j,t|t-1} - \bar{x}_{t|t-1}]^T \\ \boldsymbol{Y}_{t|t-1} = h(\boldsymbol{X}^x_{t|t-1}, \boldsymbol{X}^u_{t-1}),\ \bar{y}_{t|t-1} = \sum_{j=0}^{2n_a} W^m_j \boldsymbol{Y}_{j,t|t-1} \end{cases} \tag{5}$$

3) Measurement update

$$\begin{cases} x^i_t \sim q(x_{0:t} \mid y_{1:t}) = N(\bar{x}^i_{t|t-1}, \boldsymbol{P}^i_t) \\ w^i_t = w^i_{t-1} \cdot \dfrac{p(y_t \mid x^i_t) p(x^i_t \mid x^i_{t-1})}{q(x^i_t \mid x^i_{t-1}, y_t)},\ w^i_t = \dfrac{w^i_t}{\sum_{i=1}^{N} w^i_t} \end{cases} \tag{6}$$

Step 3: Get the particle recommendations drawn from distribution function, and weight to update and normalized values:

$$\begin{cases} x^i_t \sim q(x_{0:t} \mid y_{1:t}) = N(\bar{x}^i_{t|t-1}, \boldsymbol{P}^i_t) \\ w^i_t = w^i_{t-1} \cdot \dfrac{p(y_t \mid x^i_t) p(x^i_t \mid x^i_{t-1})}{q(x^i_t \mid x^i_{t-1}, y_t)},\ w^i_t = \dfrac{w^i_t}{\sum_{i=1}^{N} w^i_t} \end{cases} \tag{7}$$

Step 4: Re-sampling and outputs the result:

$$\bar{\boldsymbol{x}}_t = \sum_{i=1}^{N} w^i_t \boldsymbol{x}^i_t,\ \boldsymbol{P}_t = \sum_{i=1}^{N} w^i_t (\boldsymbol{x}^i_t - \bar{\boldsymbol{x}}_t)(\boldsymbol{x}^i_t - \bar{\boldsymbol{x}}_t)^T \tag{8}$$

The advantage of UKF can effectively solve the problem of filtering diverge due to the intensification of nonlinear system, but UKF only applicable to the situation of system state posterior probability density function by the Gaussian distribution to approximate. UKF also has shortcomings, first, in the time and surveying update, remains the only to the linear gauss Kalman filter the best of the structural upgrading; second, only using the second-order matrix for effective Gaussian distribution; third, sigma is too small, may not fully represent the complex distribution structure. When the observing noise is relatively small, UKF algorithms and PF algorithm to effectively integrate, we can approach the proposal distribution function by using the Gaussian distribution.

So, we have the formula (1) as the main model, the introduction of auxiliary model:

$$\begin{cases} r_t = r_{t-1} + m_t, \\ z_t = h(r_t) + n_t \end{cases} \tag{9}$$

Where m_t denotes the process noise which is a zero-mean and small variance, observing noise n_t the same as the main model. In the MPF arithmetic, the moment of t-1,UKF combines with the main model, likes UKF arithmetic, generates the proposal distribution from x^1_{t-1} particle and extract the next particle x^i_t in the t moment.

3 Probabilistic Model of Conflict

Assume aircraft have the same navigational height, the aircraft's flight plan obtained from the ATC, and each aircraft try to fly in accordance with its flight plan. The number i aircraft flight plan make up of route node sequence $\{P_j^i\}_{j=1,\ldots n,} P_j \in R^2$ and speed serial $\{V_j^i\}_{j=1,\ldots,n,} V_j \in R^+$. We can get the theory of time-series $\{T_j^i\}_{j=1\ldots n}$,if Aircraft straight flight with the speed of V_j^i from P_{j-1}^i to P_j^i. If the distance of any two plans less than the safety, then it will conflict.

Consider the nonlinear model:

$$\begin{cases} X_t^i = X_{t-1}^i + V_t^i \Delta t . \dfrac{\mathbf{X}_{t-1}^i \mathbf{P}_j^i}{\left\| \mathbf{X}_{t-1}^i \mathbf{P}_j^i \right\|} + v_{t-1}^i \\ Y_t^i = X_t^i + \omega_t^i \end{cases} \tag{10}$$

Where X_t^i denotes the location of the i flight and the t moment, v_t^i denotes the process noise, Y_t^i denotes radar observations, ω_t^i denotes observation noise of radar, V_t^i denotes the speed of plan at this moment, Δt Radar scan interval about four seconds. Flight campaign tracking error is subject to Gaussian random announced. The variance can be decomposed into two parts independent of each other: Flight direction increase the variance $\sigma_a^2(t) \sim r_a^2 t^2$ with the time; the variance of the vertical flight direction with the increase of flight distance until a fixed value $\sigma_c^2(t) \sim \min\{r_c^2 s^2(t), \overline{\sigma}_c^2\}$, $r_a, r_c, \overline{\sigma}_c$ is empirical constants. Convert to the inertial coordinate system covariance matrix:

$$R_t = \begin{bmatrix} \cos\theta_t & -\sin\theta_t \\ \sin\theta_t & \cos\theta_t \end{bmatrix} \cdot \begin{bmatrix} \sigma_a^2(\Delta t) & 0 \\ 0 & \sigma_c^2(\Delta t) \end{bmatrix} \cdot \begin{bmatrix} \cos\theta_t & -\sin\theta_t \\ \sin\theta_t & \cos\theta_t \end{bmatrix}^T \tag{11}$$

Where θ_t is the aircraft heading angle of the present moment, Δt is radar scan cycle. Suppose the aircraft independent of each other, as the aircraft's position in the next distribution is normally distributed, so the distance between the aircraft Z_t has a normal distribution, the probability of conflict at the t moment can always be expressed as:

$$P_t = \int_{z_t \in C} p(z_t) dz_t \tag{12}$$

Where $C = \{d \in R^2 : \|d\| \le \rho\}$, ρ is 3 sea mile, $P(Z_t)$ denotes the interval probability density function at this moment, in the time of [0,T], by the conflict detection algorithm, obtained the greatest probability of conflict, that this time the probability of conflict, conflict if it beyond the threshold of conflict. Hu and Prandini et al proposed a random-based algorithm, but they did not consider radar noise, only to calculate the actual probability of conflict by aircraft position, so it just a value of theory. In this paper, conflict model is more realistic because consider the radar noise.

4 Simulation Experiments and Results

This section is divided into two parts: the first part compared the state estimation accuracy difference of UPF algorithm and the MUPF algorithm, the second part of the algorithm, respectively MUPF and UPF algorithm applied to the detection probability of conflict, and compares the two differences of the successful prediction rate and rate of false alert.

4.1 Difference of State Accuracy

To compare the difference of MUPF and UPF, highlighting MUPF algorithm to get higher precision by taking small particles, suppose in the same process noise variance, simulated in different noise variance for observations, based on aircraft flying along the horizontal direction. The same observation noise variance in the flight and vertical direction, separate choose 0.01^2 nmi^2, 0.001^2 nmi^2, 0.0001^2 nmi^2, the number is 50, run 50 seconds later, and reached the absolute error of simulation diagram as shown in Figure 1-6.

From a moment of the particle distribution of the two algorithms, we can see MUPF algorithm convergence of fast particles, and more close to the true value of the particle.

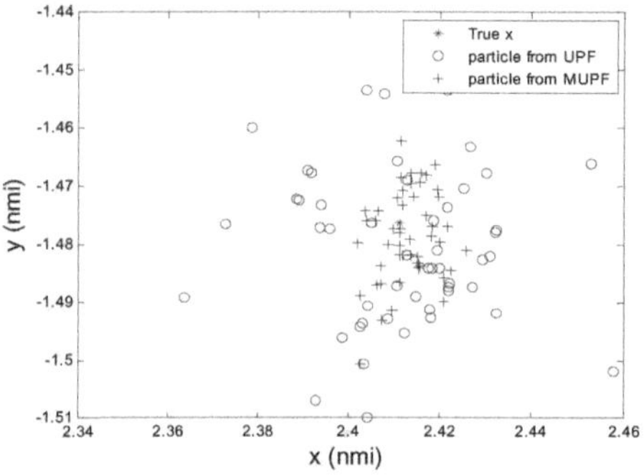

Fig. 1. (0.01^2 nmi^2)

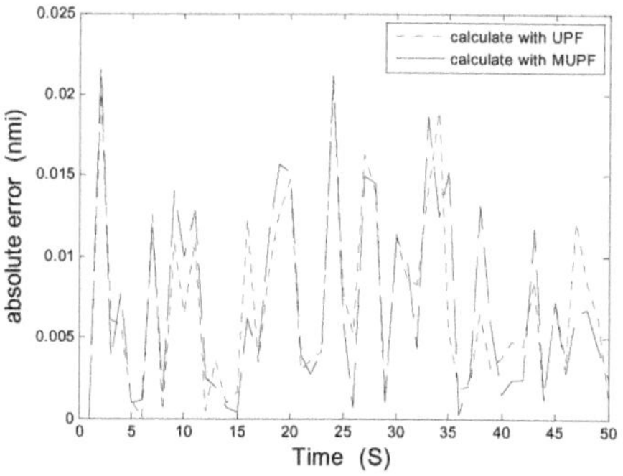

Fig. 2. (0.01^2 nmi^2)

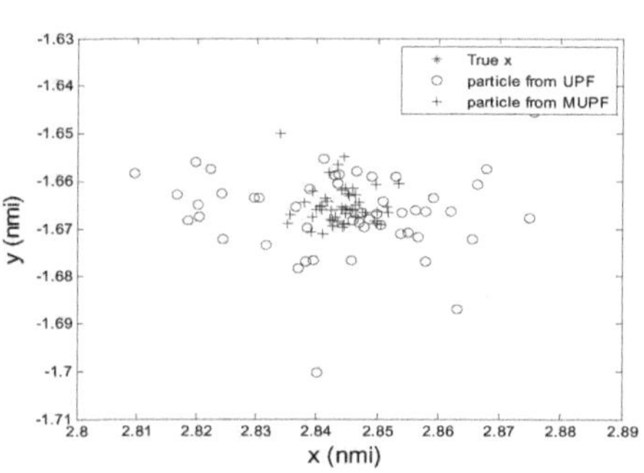

Fig. 3. (0.001^2 nmi^2)

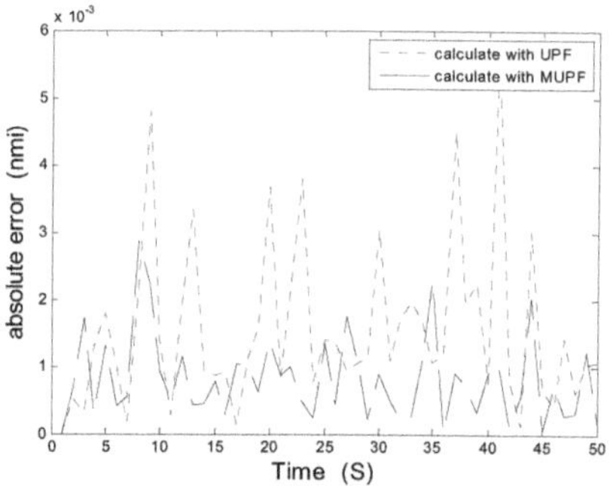

Fig. 4. (0.001^2 nmi^2)

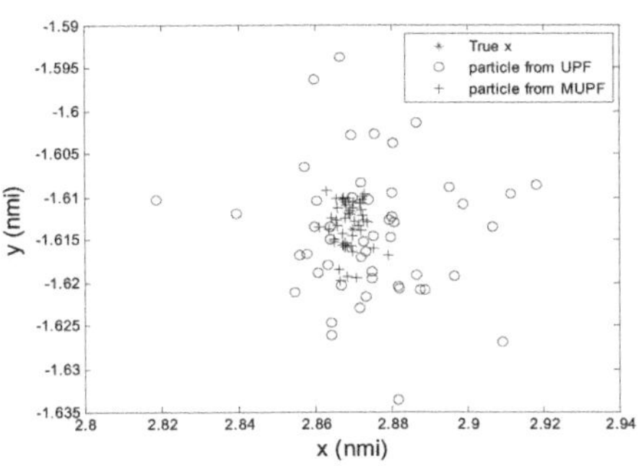

Fig. 5. (0.0001^2 nmi^2) a moment of the particle distribution of the two algorithms

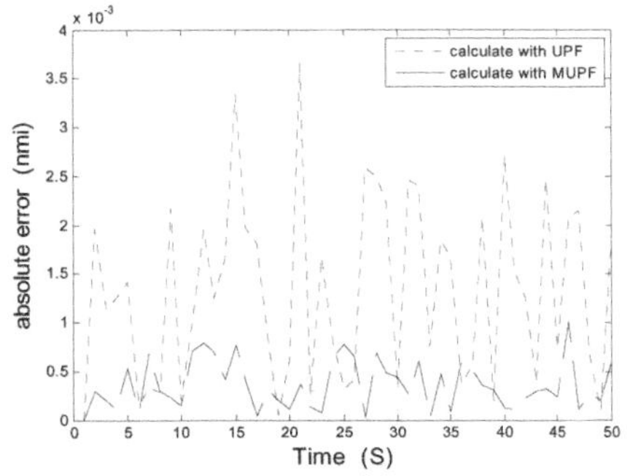

Fig. 6. (0.0001^2 nmi^2) comparison of two algorithms absolute error

When observation noise variance is 0.01^2 nmi^2 time, the absolute error of MUPF and UPF algorithms are close, due to measurement noise and process noise variance approaches and the observation values greater influenced by the noise. so the results have no significant difference; but when observation noise variance is 0.001^2 nmi^2, MUPF and UPF algorithm significantly different, MUPF algorithm has smaller absolute error. We can draw the conclusion, the smaller of the measurement noise error, the estimates by MUPF algorithm is closer to true value and a smaller error.

When the particle numbers (N) equal 10,50,200 respectively, get the data table of algorithm (Alg) MUPF with UPF in the X, Y axis absolute error (RX, RY) and standard deviation (R) data, shown as table 1-3.

Table 1. MUPF compared with the estimated accuracy of UPF (R0 =0.01^2 nmi^2)

Alg	R-X	R-Y	R	N
	0.0084	0.0069	1e-5*diag(3.239,1.281)	10
UPF	0.0080	0.064	1e-6*diag(6.479,2.563)	50
	0.0062	0.0058	1e-6*diag(1.62,0.641)	200
	0.0077	0.0066	1e-6*diag(4.571,4.075)	10
MUPF	0.0077	0.0065	1e-7*diag(7.192,6.976)	50
	0.0072	0.0062	1e-7*diag(1.662,1.661)	200

Table 2. MUPF compared with the estimated accuracy of UPF (R0 =0.001^2 nmi^2)

Alg	R-X	R-Y	R	N
	0.0036	0.0030	1e-5*diag (2.783,0.88)	10
UPF	0.0015	0.0017	1e-6*diag (5.566,1.761)	50
	8.7868e-4	0.0011	1e-6*diag (1.391,0.44)	200
	0.0011	0.0013	1e-6*diag (1.077,1.073)	10
MUPF	7.8961e-4	8.6313e-4	1e-7*diag (2.098,2.097)	50
	7.4006e-4	6.9529e-4	1e-8*diag (5.244,5.243)	200

Table 3. MUPF compared with the estimated accuracy of UPF (R0 =0.0001^2 nmi^2)

Alg	R-X	R-Y	R	N
	0.026	0.0025	1e-5* diag (2.778,0.876)	10
UPF	0.0015	0.0015	1e-6* diag (5.556,1.751)	50
	6.7248e-4	7.9311e-4	1e-6*diag (1.389,0.438)	200
	9.1885e-4	8.3048e-4	1e-6* diag (1.001,1.001)	10
MUPF	4.6607e-4	3.4805e-4	1e-7* diag (2.001,2.001)	50
	1.6972e-4	1.8078e-4	1e-8* diag (5.002,5.0)	200

From the table, in the MUPF algorithm, taking 10 particles are close to the absolute error in the UPF algorithm of using 200 particles. Take equal particles, whether in the direction X or Y direction of the absolute error, MUPF algorithm have smaller absolute error and the estimated value closer to the true value.

4.2 Difference between Terminal Area Forecast Conflicts

Suppose the initial minimum distance between aircraft is 3 nmi and minimum distances of time to reach is 4 min, the radar noise variance is 0.0012 nmi2, heading angle is 0 degrees and 90 degrees, the speed is 480 nmi / h, 420 nmi / h, change the heading angle to 45 degrees after intersection, combined into one route. We use MUPF and UPF algorithm to conflict detection, simulation 400 times, get the absolute error map in the false alert rate (P (FA)) and successful alarm rate (P (SA)) of shown in Figure 7-8:

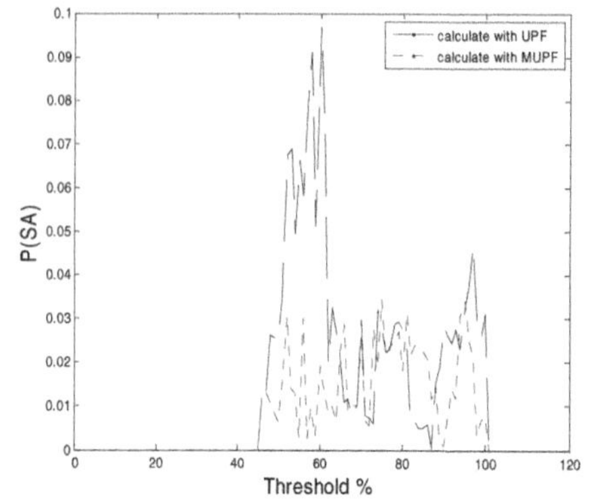

Fig. 7. Absolute error about rate of success alert

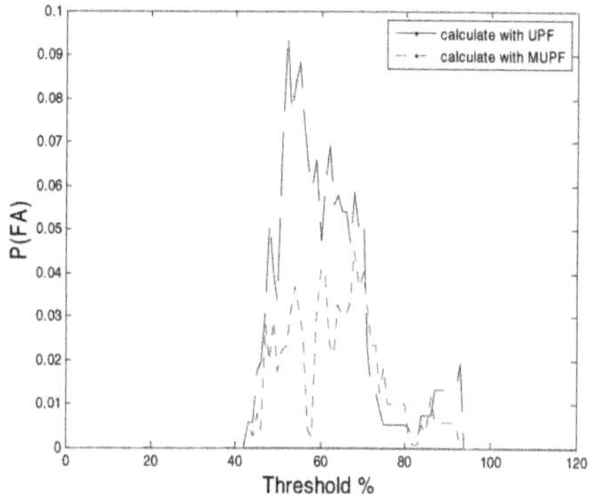

Fig. 8. Absolute error about rate of false alert

As the threshold increases, MUPF algorithm error rate of change is little, while the UPF significantly changed. Compared with the UPF algorithm, MUPF algorithm obtained a higher rate of successful prediction, false alarm rate lower, closer to the theoretical value of the probability of conflict.

5 Conclusions

This article introduced an improved UPF algorithm, and applied for the terminal conflict detection. When the observation noise is small, we can get a more accurate estimates, and MUPF need particles more less than UPF and reduce computation effectively. In the conflict detection, because the MUPF has a high accuracy and the smaller the variance, so it has a smaller error in the false rate and rate of success alert.

Acknowledgments. This work was supported by Education Commission of Shanghai (No.: 10YZ103).

References

1. Bashllari, A., Kaciroti, N., Nace, D., et al.: Conflict probability estimations based on geometrical and Bayesian approaches. In: 10th International IEEE Conference on Intelligent Transportation Systems, Seattle, WA, USA, pp. 479–484 (2007)
2. Li, D., Cui, D.G.: Air traffic control conflict detection algorithm based on Brownian motion. J. Tsinghua Univ. (Sci. &Tech.) 48(4), 477–481 (2008)
3. Hou, D.W., Yin, F.L.: A Dual Particle Filter for State and Parameter Estimation in Nonlinear System. Journal of Electronics and Information Technology 30(9), 2128–2133 (2008)
4. Hong, S.H., Shi, Z.G., Chen, K.S.: Simplified Algorithm and Hardware Implementation for Particle Filter Applied to Bearings-only Tracking. Journal of Electronics and Information Technology 31(1), 96–100 (2009)
5. Han, C.Z., Zhu, H.Y., Duan, Z.S.: Multi-source Information Fusion, pp. 29–68. Tsinghua University Press, Beijing (2006)
6. Feng, C., Wang, M., Ji, Q.B.: Analysis and Comparison of Resampling Algorithms in Particle Filter. Journal of System Simulation 21(4), 1101–1105 (2009)

A Multi-dimensional Coordinate Factorization Algorithm for Network Distance Prediction

Shanxiong Chen[1,*], Ya Li[1], Maoling Pen[2], and Rui Zhang[2]

[1] College of Computer and Information Science
Southwest University,
Chongqing, China
[2] Chongqing City Management College
Chongqing, China

Abstract. Large-scale Internet applications can benefit from an ability to predict round-trip times to other hosts without having to contact them first. In various predictable model, network coordinates system is an efficient mechanism for internet distance prediction with limited measurements. In this paper, we identify a coordinate matrix which consists of measured value between benchmark node and other common node, so convert computation between nodes into question of factorizing coordinate matrix. We present an algorithm of matrix factorization which factorize coordinate a matrix into non-negative matrix U,V. Through the factorization of the matrix greatly reduces the dimension from the calculation, make for fast convergence of the prediction algorithm.

Keywords: non-negative matrix, coordinate, distance predication.

1 Introduction

Currently, network distance prediction technology can be categorized into two classes in computing method, namely virtual coordinate prediction and network topology prediction. Network distance space is embedded into geometric space and every node is assigned a coordinate in coordinate-based prediction. The coordinate is used to calculate the distances between nodes by the formula of the distance in geometric space. As a example of coordinate based approach called "Global Network Positioning (GNP)" has been proposed by Ng [1]. This approach models the Internet as a geometric space and computes geometric coordinates to characterize the position of nodes on the Internet. In this way, network distance is converted into geographic distance. Frank Dabek simulates Space embedding as minimization of potential energy of spring and proposed Vivaldi [2], a distributed distance prediction mechanism. Internet topology map is constructed via AS linking relation or measured data in Network topology based prediction, and the map can be utilized to predict the path among nodes, then in accordance with the predicted result to compute the distance between nodes. IDMaps, a

---

* Supported by the Fundamental Research Funds for the Central Universities, supported by Natural Science Foundation Project of CQ CSTC (CSTC, 2010BB2006).

L. Qi (Ed.): ISIA 2010, CCIS 86, pp. 32–39, 2011.

kind of network topology prediction mechanism, proposed by Francis P etc, involves address prefix, AP and Tracer. AP measures the distance of a adjacent Tracer in a cycle. So the distance between two nodes is the sum of the distance that each node's AP reaches their respective Tracer and distance between two related Tracer nodes.[3],[4],[5]

Network Distance Prediction Based on Non-negative Matrix Factorization proposed in this paper is a coordinate-based prediction mechanism. We construct a characteristic matrix (non-negative matrix) for the RTT between common node and landmark node, and decompose the matrix into a product of a base vector and a weight vector. The calculation of distance between any two nodes means a change in the original feature matrix, combining the base vector and weight vector with the distance computational formula reconstructed distance matrix for solving the distance between the nodes.

2 Distance Prediction Based on Virtual Coordinate

We consider that there are N nodes $H = (H_1, H_2, H_3, \ldots H_N)$, in a network, and the distances of any node H_i to all the nodes in the network form a N-dimensional distance vectors $(D_{i1}, D_{i2}, D_{i3} \ldots D_{iN})$,so that all nodes constitutes an $N \times N$ dimensional distance matrix D, where D_{ij} shows the measured distance of H_i node to node H_j.

We construct maps p: $H^N \rightarrow R^M$ of a network node to the M dimensional vector space, and map the N nodes $H = (H_1, H_2, H_3, \ldots H_N)$ mentioned above to N-dimensional coordinate points in the M dimensional geometric space. The key to making the distance prediction lies in minimum the distance error between the distances calculated according to coordinate values and that of the actual measured values. The distance of the network nodes can be approximated by the nodes distance in the coordinate space as follows:

$$D_{ij} \approx \hat{D}_{ij} = \| p(H_i) - p(H_j) \|, \forall i, j = 1,2,\ldots,N \tag{1}$$

In Euro-style space, we have forecasted distance value:

$$\hat{D}_{ij} = \| p(H_i) - p(H_j) \| = \left(\sum_{k=1}^{M} (H_{ik} - H_{jk})^2 \right)^{1/2} \tag{2}$$

$\hat{D}_{ij}$ expresses the predicted distance between node H_i and H_j, and p (H_i) is a M dimensional vector $(H_{i1}$, H_{i2}, $H_{iM})$, which represents the virtual coordinate of H_i in embedded space.

3 Coordinate Factorization Based on Non-negative Matrix

In this paper we introduce Non-negative Matrix Factorization (NMF) proposed by Lee etc to deal with distance matrix[6-7], which is a method of feature extraction, mainly used in reduction of dimensionality, data compression and local feature extraction etc.

The basic principle we can be described as: for any Non-negative Matrix Factorization $A=[x_{i,j}]_{n\times m}$, NMF can find Non-negative Matrix $U=[u_{i,j}]_{n\times d}$ and $V=[v_{i,j}]_{d\times m}$ to accord with

$$A\approx UV, \tag{3}$$

NMF can be applied to the statistical analysis of multivariate data in the following manner [13]. Given a set of multivariate n-dimensional data vectors, the vectors are placed in the columns of an n×m matrix A where m is the number of examples in the data set. This matrix is then approximately factorized into an n×d matrix U and an d×m matrix V. Usually d is chosen to be smaller than n or m, so that U and V are smaller than the original matrix A . This results in a compressed version of the original data matrix

If the matrix U, V, rewritten as follows:

$$U=[u_{i,j}]_{n\times d}=[U_1,U_2,\ldots U_d] \tag{4}$$

$$V=[v_{i,j}]_{d\times m}=[V_1,V_2\ldots V_m] \tag{5}$$

After non-negative matrix factorization, the distance vector X_j is expressed as

$$X_j\approx UV,\quad \text{where } V_j=[v_{ij},v_{2j},\ldots v_{dj}]^T \tag{6}$$

$$X_j\approx v_{1,j}U_1+v_{2,j}U_2+\ldots+v_{d,j}U_d \tag{7}$$

Distance vector X_j is approximated by a linear combination of column vector U_1, U_2, ... U_d. To find an approximate factorization A≈UV, we first need to define cost functions that quantifies the quality of the approximation. Such a cost function can be constructed using some measure of distance between two non-negative matrices A and UV. One useful method is the square of the Euclidean distance between A and UV .

$$\|A-UV\|^2=\sum_{ij}(x_{i,j}-\sum_k u_{i,k}v_{k,j}) \tag{8}$$

another useful measure is

$$D(A\|UV)=\sum_{i,j}\left[\log\frac{x_{i,j}}{\sum_k u_{i,k}v_{k,j}}-x_{i,j}+\sum_k u_{i,k}v_{k,j}\right] \tag{9}$$

Where U, $V\geqq 0$ $\sum_i u_{i,j}=1$

The process finding the optimal U and V is just the process that to minimize the value of $\|A\text{-}UV\|^2$and *D (A // UV)* [8,[9], so we need to find a way which can quickly factorize A to U and V, two non-negative matrix, by the same time, ensure that the value of$\|$A-UV$\|^2$and *D (A // UV)*is the minimum.

We now consider two alternative formulations of NMF as optimization problems:

Problem 1 : Minimize$\|$A-UV$\|^2$ with respect to U and V, subject to the constraints $U,V\geqq 0$.

Problem 2 : Minimize *D (A || UV)*with respect to U and V, subject to the constraints U,V≧0

The functions $\|A-UV\|^2$ and *D (A//UV)* are convex in U only or V only, but they are not convex in both variables together. Therefore it is difficult to present an algorithm to solve Problems 1 and 2 in the sense of finding global minima.

We have introduce the following "multiplicative update rules", it is a good equilibrium between speed and ease of implementation for solving Problems 1 and 2.

Rule 1: The Euclidean distance $\|A-UV\|^2$ is nonincreasing under the update rules

$$V_{i,j} = V_{i,j} \frac{(U^T A)_{i,j}}{(U^T UV)_{i,j}} \tag{10}$$

$$U_{k,i} = U_{k,i} \frac{(AV^T)_{k,i}}{(UVV^T)_{k,i}} \tag{11}$$

Rule 2: The divergence D(A||UV) is nonincreasing under the update rules.

$$V_{ij} = V_{ij} \frac{\sum_k U_{ki} A_{kj} /(UV)_{kj}}{\sum_m U_{mj}} \tag{12}$$

$$U_{kj} = U_{kj} \frac{\sum_j V_{ij} A_{kj} /(UV)_{kj}}{\sum_n V_{in}} \tag{13}$$

According to rules 1 and 2, we can ensure $\|A-UV\|^2$ and D(A||UV) converge to a local minimum. Proofs of these theorems are given in [10 .For now, we note that each update consists of multiplication by a factor. In particular, it is straightforward to see that this multiplicative factor is unity when A = UV, so that perfect reconstruction is necessarily a fixed point of the update rules.

In the process of distance prediction, u, v can select any non-negative matrix as the initial value. The distance matrix, which must be a non-negative matrix (negative is insignificant for distance matrix), consist of the measured value between benchmark node and other common nodes. For n×m distance matrix A, we factorize it into U'and V' according to rule 1, then carry through factorization by rule 2 again. These processes of factorizations execute until $\|A-UV\|^2$ and D(A||UV)converge to a local minimum at same time .Finally introduce result of factorization into formula (2) to calculate distance between any nodes, as follows:

$$\hat{D} = \left(\sum_{k=1}^{d} \left(\left(\sum_{l=1}^{d} \left(u_{i,l} v_{l,k} - u_{j,l} v_{l,k} \right) \right)^2 \right) \right)^{1/2} \tag{14}$$

After the distance matrix has been factorized according to non-negative matrix factorization theory, M dimensional coordinate space has been broken up as the product of D dimensional space coordinate (D<M) and its weight coefficient, reducing the dimensionality of prediction distance. Simultaneously accuracy of $\hat{D}$ is dependent on the accuracy that the Non-negative matrix breaks up, therefore minimizing $\|A-UV\|^2$ and D(A||UV) is the key to accurate prediction.

4 Complexity of Computing

In the M-dimensional vector space, the distance between the nodes using Euclidean space calculation formula, and the computational complexity is $O(M^2)$. In this section, we focus on computational complexity about using multiplicative update rules factorize coordinate matrix.

When considering $A_{n\times m} \approx U_{n\times d} V_{d\times m}$, the iterative computation for V is (10) and (12), multiply operation is much slower than the add operation, therefore we just discuss the times of multiply operation. The iterative multiply operation for formula (10) is nd+2n, formula (12) is ndn, hereby the multiplicative times for V iterative once is nd(n+2/d+1). the iterative computation for U is (11) and (13), where multiply operation times for formula (11) is md+2m and formula (13) is mdm,so the multiplicative times for V iterative once is md(m+2/d+1),therefore the multiplicative operation times for NMF iterative once is T= nd(n+2/d+1)+ md(m+2/d+1).

According to analysis, the computational complexity for NMF iterative can be $O(M^4)$, and computational complexity in [8] is $O(M^2)$, while computational complexity of distance prediction for formula(14)is just $O(d^2)$. In the network calculation the calculation of U, V is invariant, so it can be done in a high-performance server, then send the result to the nodes, which make use of the coordinate of the sum between U, V matrix to calculate the distance between them. This has realized the distributed distance prediction in fact, while computational complexity of distance calculation on every node may reach $O(M^2)$ in pure Euclidean space's distance prediction.

5 Prediction Accuracy Analysis

We use the actual node overhead in the Internet as a dataset to testify the prediction accuracy. In this paper, three datasets are involved. The first dataset comes from King Dataset in P2PSim project, which measure the two-way RTT between 1143 connected Internet DNS server. The DNS server is constructed in Gnutella structure, with an average RTT of about 144ms, a maximum of 971ms. The second dataset involves 169 RTT between nodes on the PlanetLab network test bed, with an average RTT of about 1005ms. The third dataset comes from NLANR Laboratory AMP (Active Measurement Project) project. In this data set, 110 nodes are connected via a high-performance network, with an average RTT of about 55ms, a maximum of 373ms. Here we use the relative prediction error (RPF) to describe the accuracy of prediction.

$$RPF = \left| \frac{prediction \quad distance - measure \quad distance}{measured \quad distance} \right| \tag{15}$$

Fig.1 shows the cumulative distribution function (CDF) curve of RPF in the three data sets. The prediction accuracy of the three data sets is regional consensus after the using non-negative matrix factorization.

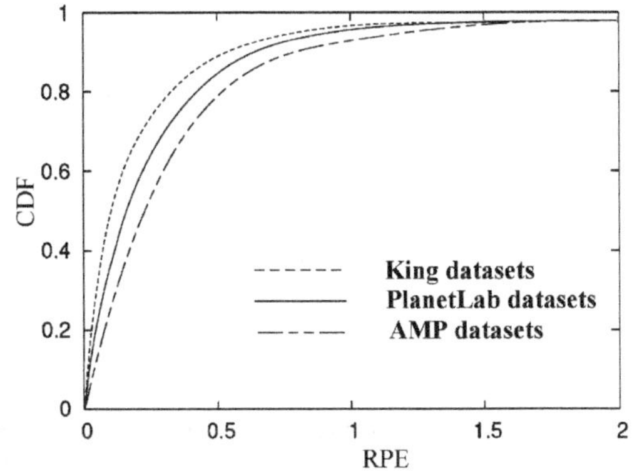

Fig. 1. CDF curve for the three datasets

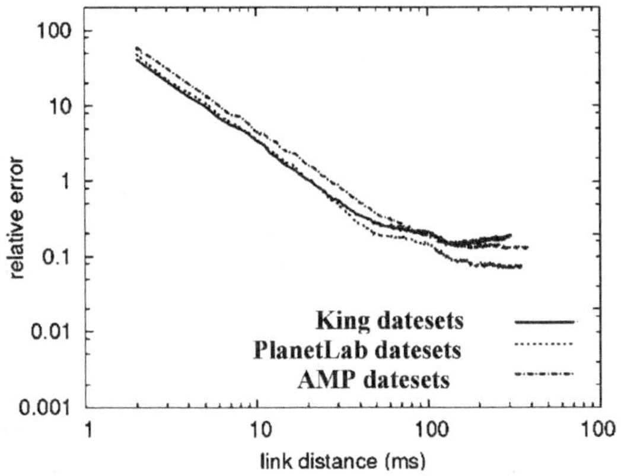

Fig. 2. Average relative error for the three data sets

Fig. 2 shows the average prediction error under links of different latency, and it also demonstrates that a short distance has a bigger error compared to a long distance.

The relative prediction error can effectively express the size of prediction error. In order to get the further details about the relative deviation between predicted RTT and measured value, we define the trend of the Directional Relative Prediction Error (DRPE).

$$DRPF = \frac{prediction \quad dis\tan ce - measure \quad dis\tan ce}{measured \quad dis\tan ce} \tag{16}$$

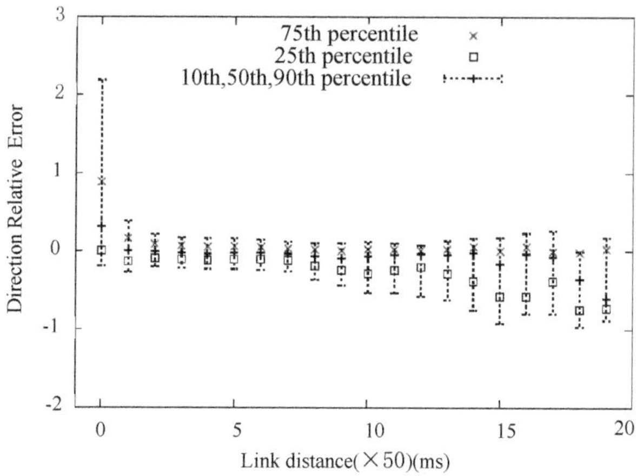

Fig. 3. DRPE distribution for king dataset

Fig .3 plots the trend of the Directional Relative Prediction Error of King dataset. The link can be grouped according to the link distance between nodes. Group i represents node sets with a distance ranging of [50i, 50 (i +1)] ms. In this way, all links are covered into groups. The trend of the Directional Relative Prediction Error range for the number of nodes ranging from 10% , 25% , 50% , 75% to 90% has been plotted in the figure respectively. In accordance with our analysis we can conclude that the predicted value of a shorter link distance is overestimated, whereas longer link distance is underestimated.

From the theoretical point of view, prediction algorithm could indirectly calculate the cost of distance between one node and another node by directly measuring distance the node and benchmark node. Distance from benchmark node to two other node accord with triangle inequality, consequently, in the case of short link, RTT value is relatively small and predicted value is overestimated relatively.

6 Conclusion

This paper analyzes the virtual coordinate based distance prediction, and proposes Non-negative Matrix Factorization to reduce the dimension of distance calculation and to improve the Euclidean coordinate space method. Simultaneity, the computation complexity of the decomposed coordinate matrix reduced to O (d2). The prediction accuracy analysis clearly shows that the accuracy of the proposed method is consistent when apply in different data sets. So it is in line with the prediction error trends distribution model.

References

[1] Ng, T.S., Zhang, H.: Predicting Internet network distance with coordinates-based approaches. In: Proc. of the IEEE INFOCOM. IEEE Press, Piscataway (2002)

[2] Dabek, F., Cox, R., Kaashoek, F., Morris, R.: Vivaldi: A decentralized network coordinate system. In: Proc. of the ACM SIGCOMM, pp. 15–26. ACM Press, New York (2004)

[3] Francis, P., Jamin, S., Jin, C., Jin, Y., Raz, D., Shavitt, Y., Zhang, L.: IDMaps: A global internet host distance estimation service. IEEE/ACM Trans. on Networking 9(5), 525–540 (2001)

[4] Eades, P., Lin, X.: Spring algorithm and symmetry. Theoretical Computer Science, 379–405 (2000)

[5] Guyton, J.D., Schwartz, M.F.: Locating nearby copies of replicated Internet servers. In: Proc. of the ACM SIGCOMM, pp. 288–298. ACM Press, New York (1995)

[6] Leed, D., Seungh, S.: Learning the parts of objects by non-negative matrix factorization. Nature 401(6755), 788–791 (1999)

[7] Leed, D., Seungh, S.: Algorithms for non2negative matrix factorization. In: Proc. of Neural Information Processing Systems, pp. 556–562 (2000)

[8] Inderjit, S., Dharmendra, S.: Concept decompositions for large sparse text using clustering. Machine Learning 42(1), 143–175 (2001)

[9] Wild, S., Curry, J., Dougherty, A.: Improving non-negative matrix factorizations through structured initialization. Pattern Recognition 37(11), 2217–2232 (2004)

[10] Wang, Y., Jia, Y., Hu, C., et al.: Fisher non2negative matrix factorization for learning local features. In: Proc. of Asian Conf. on Comp. Vision, Jeju Island, Korea (2004)

Numerical Simulation Research of the Laminated RC Shear Walls with Different Concrete Ages

Hongmei Zhang* and Xilin Lu

State Key Laboratory of Disaster Reduction in Civil Engineering,
Tongji University Shanghai 200092, China
Tel.: 086-021-65984945; Fax: 086-021-65984945
zhanghongmei@tongji.edu.cn

Abstract. The structural member quality can be effectively insured in factory, but the joints of the fabricated members are weak to actual load. A new partially prefabricated laminated wall is introduced here. To investigate the seismic performance of this kind of laminated wall, the specimens with and without opening were tested under inversed cyclic loading. Simulation program was performed by the general finite element program—ANSYS. The property of the interface of the prefabricated part and the cast-in-place part was investigated through the simulation. Research results indicate that: 1) the simulate results can agree with the test results using three-dimension springs to simulate the interface property; 2) the half prefabricated and half cast-in-place concrete laminated wall can coordinately work together fairly well.

Keywords: Laminated wall, prefabricated wall, numerical simulation, FE analysis.

1 Introduction

A kind of partial prefabricate concrete shear wall start to be applied in Japan and Germany in 1990s [1]. One part of the laminated concrete wall is fabricated in factory, and the other is cast in place (see Fig. 1). In order to enhance the bonding capability of the two concrete parts, horizontal and vertical truss is set between the prefabricated part and the cast in place part. Besides, the fabricated part can be used as mould when cast the other concrete part in site which saves construction time and material. Since the fabricated part can be manufactured in factory, the quality can be controlled and managed more efficiently, and the construction pollution can be lessened greatly. Moreover, the windows, doors and facing decoration could be installed on the wall in factory also. The factory prefabricated structure can demonstrate more economical efficiency large scale constructions like reconstruction after seismic disaster.

But up to day, seldom evidence has been made by relative experiment or numerical analysis. Most fabricated walls are used as heat insulation or heat prevention members [2]. While reinforced concrete shear wall can offer great potential for both lateral load

* Corresponding author.

L. Qi (Ed.): ISIA 2010, CCIS 86, pp. 40–47, 2011.

resistance and drift control. If the bond and joint strength can be controlled efficiently, the prefabricated or half prefabricated shear wall structures could have identical seismic resistance capability also. There are laminated concrete shear wall similar to the half prefabricated laminated wall have already put into actual projects so far. Since there is no standard or regulation of the structural designing method of the laminated RC shear wall to abide by, this laminated wall faces many uncertainties or waste. How to apply the partially prefabricated concrete shear wall into practice is a challenge now. Aiming at the new type of partially prefabricated laminated concrete shear wall, research is conducted on the bearing capacity in real project. This paper laid importance on: 1) give out experimental evidence to the new half prefabricated half cast in site laminated shear walls; 2) give out precise analysis of this kind of laminated shear walls.

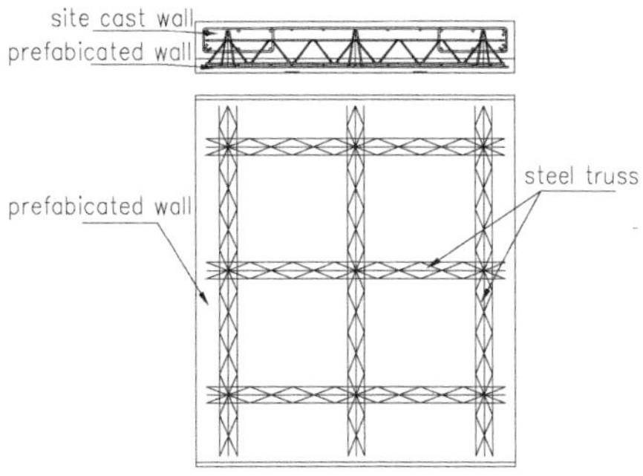

Fig. 1. Construction schematic diagram of the composite wall

2 Experimental Program

The two typical tested specimens mentioned below involve the rectangular specimen and the specimens with an opening. The testing procedure, results and relative details are described also. Attention focused on the lateral resistance capacity, the failure modes and other seismic behavior of the two specimens.

2.1 Specimen Design

Details of the two specimens are listed in Fig.2. Both the two type of specimens were tested under reversed cyclic load. Fig.2 shows the nominal dimensions of the specimen WKSWA-I, WKSWB-I and the vertical and horizontal reinforcement bars. Horizontal and vertical distributed reinforcement of is Φ^R8 @150, and the main longitudinal bar at the boundary area is 6Φ^R14 [3-5]. The specimen with an opening is constructed to simulate the wall with a window or door.

2.2 Testing Procedure and Loading Sequence

Material properties listed in table 3 and 4 were tested in the State Key Laboratory of Disaster Reduction in Civil Engineering in Tongji University.

Each specimen was subjected to a combination of axial and horizontal load. The vertical load was applied by 3 hydraulic jacks and was adjusted during the test to balance the vertical load to a constant value. Horizontal load was applied by a hydraulic actuator fixed on the concrete reaction wall. Specimens were instrumented with Linear Variable Differential Transducers (LVDTs) for displacement measuring.

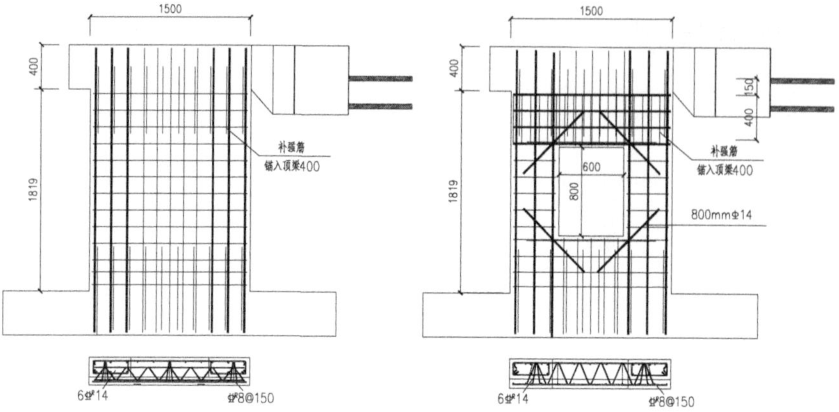

Fig. 2. Reinforcement details of WKSWA-I, WKSWB-I

Electric resistance strain gauges were placed on the reinforcing bars, the surface of the concrete wall to measure the strains. The load cells and the stroke of actuators were connected to data acquisition systems for data processing.

Specimens were tested under reversed cyclic quasi-static loading. A mixed loading mode was employed. Each test was first conducted under load control until specimen yield, and then continued with a displacement-controlled scheme. The vertical axial load on the solid specimen and specimen with an opening was designed to the value of 1200 kN and 600 kN individually. This load level can represent that of the wall of a low or medium-height building. After the axial load was applied, the specimen was incrementally loaded in the horizontal direction. The load increment was controlled by 10 kN at first, when the specimen become yield, the load increment was controlled by displacement and each step is increased by 2mm. Every load stage maintain for 3 cycles until the specimen failed. Pressure transducers in the hydraulic supply line of the rams provided the means of accurate measurement of the applied load. Deformation response was monitored by LVDTs calibrated before each test. Three LVDTs were positioned at selected wall elevations to monitor the displacements of the foundation, the middle height and the top beam. Strain gages were employed to measure the strain of the longitudinal reinforcing bars at the 4 corner of the wall and the wall center. Fig. 3 and 4 demonstrate the failure picture of the two specimens.

3 Finite Element Analysis Program

The pushover analysis of the two specimens was emphasized in light of the load conditions of the experiment. The FE method and the ANSYS software have been employed to obtain the numerical results in the cases. Analysis model is set up according to the tested model considering the bond property of the interface of the prefabricated part and the cast in place part.

Fig. 3. Failure picture of WKSWA-I

Fig. 4. Failure picture of WKSWB-I

3.1 Definition of Real Parameters and Materials

The real parameter applied in the simulation is defined according to the material and the member property. In general, the laminated shear wall can be divided into three pars: the prefabricated part, the cast in place boundary part and the cast in place inner part. Element Solid 65 is applied to simulate for the concrete material. Link 8 is selected to simulate the main steel bars, and Combin 39 is used to simulate the bond property of the interface of the prefabricated concrete part and the cast in place concrete part, the bond property is simulated by the three-direction springs. The spring property is defined according to interface shear and splitting tensile strength test [6, 7]. Details of the element type and the real parameters are shown in table 1.

Table 1. Real type of the models

Real set No.	Element type	Location of the real set	Description of the real set
Real 1	Solid 65		Inner part of the cast in place part
Real 2	Solid 65		Boundary element of the cast in place part
Real 3	Solid 65		The prefabricated part of the composite wall
Real 5	Link 8		Vertical steel bars in the boundary area
Real 6	Combin 39	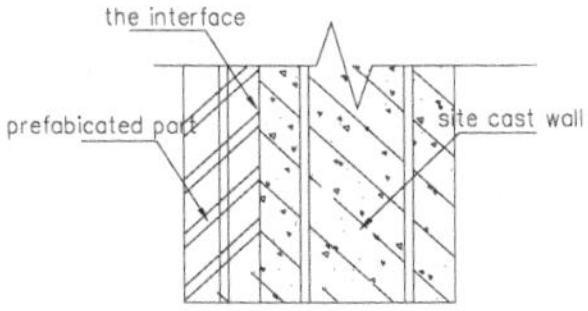	Interface spring in X direction
Real 7	Combin 39		Interface spring in Y direction
Real 8	Combin 39		Interface spring in Z direction

3.2 Material Property

Material properties are shown in table 2 and table 3. The material property applied in the analysis model is referred to the test results. The concrete cube strength is 27.7 MPa, and the ultimate strength of ϕ14 is 539 MPa.

3.3 FE Model

The FE model was set up by ANSYS. Each specimen was divided into 3 parts in accordance with the material property or element type. A mixed type reinforcement concrete model was applied in the modeling, to the detail the uniformly distributed steel bars was set up considering uniformly scattering in the concrete, the real type was defined according the horizontal and vertical distribution steel bars, and the main vertical steel bars in the boundary area was build up separately to consider the contribution of the bars at the boundary area to the lateral bearing capacity. In this paper, the bond relationship was simulated by three dimension springs which can consider the shear property and the tension property. Combin 39 was applied as bond spring which has no dimension, each node pair was connected by three Combin 39 springs which was in X, Y, Z directions separately. The contact element location diagram applied in the model is shown in Fig.5, but essentially, the contact element occupied no space. The bond property was defined by the relative experiment including the shear test and the splitting tensile strength test of the surface between new and old concrete [8, 9]. Besides, the action of the steel truss cross the interface was also considered in the shear and splitting tensile strength defined in the contact element of the model. Loading schedule was drawn according to the test condition. The bottom of the specimen was totally constrained. The surface pressure was loaded on the top area of the specimen and lateral load was applied on the top and controlled by displacement.

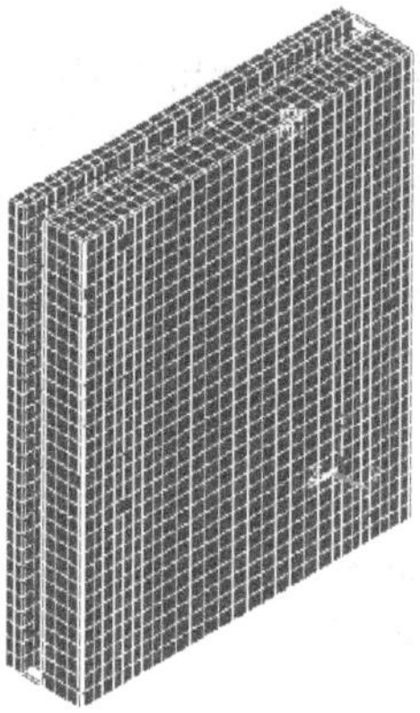

Fig. 5. Diagram of meshed element

4 Analysis Results

According to the experiment case, pushover analysis was performed using the FE model. In general, the tested hysteretic skeleton agrees with the pushover test result in the rectangular shear wall static load test, and there is still difficulty in the concrete shear wall hysteretic analysis, so the pushover analysis is selected to find the discipline of the composite shear wall under vertical and lateral load. The efficiency of the contact element using in the new-old interface can be verified through the comparison between the test result and the FE analysis results.

4.1 Pushover Analysis

The calculated pushover curve and the tested skeleton curve of the top drift-bottom shear force are drawn in Fig. 6. From Fig. 6, it can be found that the calculated curve agree with the test result, the lateral bearing peak point of the calculated result of the two model is within 20% to the tested result either. Both the two pushover curves indicate that the calculated result is higher than the test result, that may attribute to the idealized model, the actual specimen may exist steel bar offset, concrete is not density and so on which will deduce the strength of the test result. It could still indicate that the model and the load condition agree to the actual case since the difference could be controlled within 20%.

From Fig. 6, it can be found that the calculate result of WKSWA-I extend to obviously degrading stage, and the value is close to the test value, which indicates that the calculated result can be considered close to the actual case. To the case of WKSWB-I which there is an opening in the wall center, since there is no distinct degrading stage in both calculate and test results.

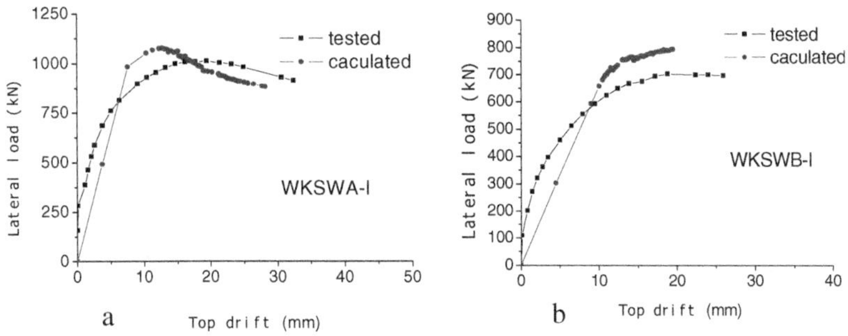

Fig. 6. (a) Top displacement-load curve, tested and calculated, WKSWA-I; (b) Top displacement-load curve, tested and calculated, WKSWB-I

4.2 Compatibility of the Laminated RC Walls

The node strain distribution of the calculation model can be given out by the ANSYS software. And Fig. 7 demonstrates the major principal strain of the two models WKSWA-I and WKSWB-I. The major principal strain of the inner surface between the prefabricated part and the cast in place part are also given out.

From the strain distribution contour of WKSWA-I, the corner at the bottom was firstly damaged, which agree with the actual failure mode in the experiment (Fig. 7 (a)). The strain distribution of WKSWB-I (the model with hole, Fig. 7 (a)) shows that crack will firstly emerge at the wall edge at the bottom level of the opening and the corners of the opening. Since the axial compression value is not very high, the location where the section area varies abruptly will have stress concentration, which will induce local crush.

The contact element between the prefabricated concrete surface and the cast in place concrete surface is used to simulate the shear and tensile property of the interface. The strain distribution at the two sides of the interface is also shown in Fig. 7. Both the two specimen indicate that there are little difference in the strain distribution on the two sides even at the ultimate loading stage. The difference of the strain on the two sides of the interface has not induced severely failure according to the calculation results.

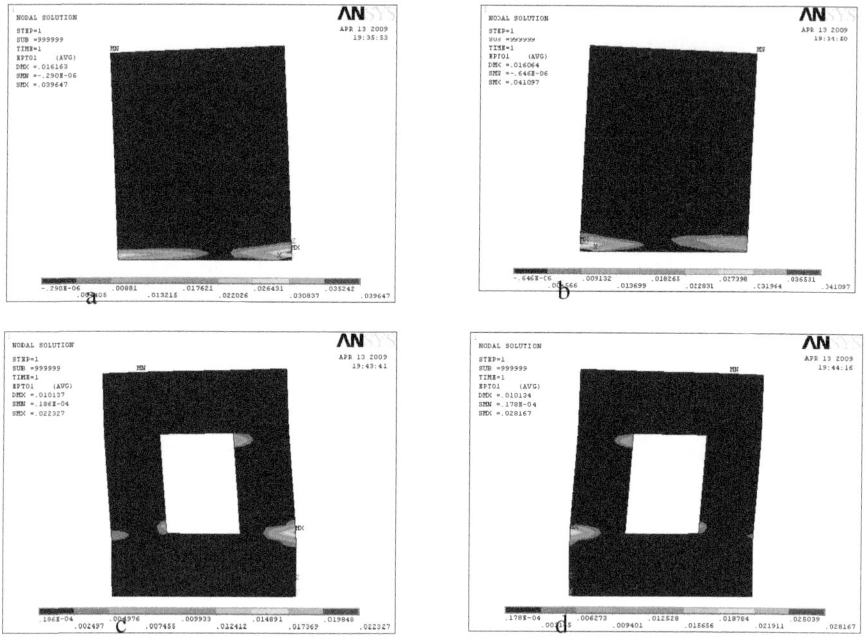

Fig. 7. First principle strain contour of the specimens (a) The prefabricated part of the Solid specimen – WKSWA-I; (b) The cast-in-place part of the Solid specimen – WKSWA-I; (c) The prefabricated part of the specimen with an opening – WKSWB-I; (d) The cast-in-place part of the specimen with an opening – WKSWB-I

5 Conclusion

In this paper, the Combin 39 spring element was used in the property simulation of the contact surface of the prefabricated concrete and the in site cast concrete. Two models subject to the axial and lateral load using FE method and the ANSYS software. It can be observed that the simulated results agree to the test results and the discordance between the two sides of the contact interface is so severely as to induce obvious failure. In addition, the experiment of the new and old concrete interface property test result is referred to the relative data, the result may exist some difference to the actual case. Nonetheless, the experiment results and the analysis results both indicate that the new and old concrete part can work together coordinately, at lest under the test condition. The contact element applied in the model calculation can give out an acceptable simulation results. The strain of the concrete inner surface is not severely different even in the ultimate stage. Comparing with experimental, the strain distribution results obtained from the calculation are approximately in accordance with the practical case.

Acknowledgement

The authors would like to thank the referees for their valuable comments, and also thanks for the financial support of Kwang-Hua Fund for College of Civil Engineering, Tongji University.

References

1. Jiang, Q.J., Liu, R.N.: Precast concrete energy efficient wall technology and engineering applications. Wall Materials Innovation & Energy Saving in Buildings 20(5), 44–47 (2010)
2. Holden, T., Restrepo, J., Mander, J.B.: Seismic performance of precast reinforced and prestressed concrete walls. Journal of Structural Engineering (29), 286–296 (2003)
3. Schmitz, R.P.: Fabric-formed concrete panel design. In: 17th Analysis and Computation Specialty Conference, St. Louis, Mo, United States, May 18-21 (2006)
4. West, M.: Fabric-formed concrete structures. In: Proceedings First International Conference on Concrete and Development, Tehran, Iran, April 30-May 2, pp. 133–142 (2001)
5. Saito, M., Yoshimatsu, T., Homma, K., Nakajima, R., et al.: Trusses and precast concrete slabs reinforced thereby. U.S. 5448866, U.S., September 12 (1995)
6. Wang, Z.L., Lin, Y.J., Qian, Y.J.: Experimental Research on Shear Properties of New-to-Old Concrete Interface. Journal of Southwest Jiaotong University 40(5), 600–604 (2005)
7. Zhao, Z.F., Zhao, G.F., Liu, J., et al.: Experimental Study on Adhesive Tensile Performance of Young on Old Concrete. Journal of Building Structures 22(2), 51–55 (2001)
8. Zhang, H.M., Lu, X.L., Wu, X.H.: Cyclic Loading Experiment and Numerical Simulation of RC Walls. In: 2009 WRI World Congress on Computer Science and Information Engineering, Los Angeles, CA, USA, pp. 642–647 (2009)
9. Zhang, H.M., Lu, X.L., Wu, X.H.: Experimental Study and Numerical Simulation of the Reinforced Concrete Walls with Different Stirrup in the Boundary Element. Journal of Asian Architecture and Building Engineering 17(2), 447–454 (2010)

Penalty-Optimal Brain Surgeon Process and Its Optimize Algorithm Based on Conjugate Gradient

Cuijuan Wu[1,2], Dong Li[1,2], and Tian Song[3]

[1] JiangSu Province Support Software Engineering R&D Center for Modern Information Technology Application in Enterprise, Suzhou, China, 215104
[2] Department of Mechanics & Electronics, Suzhou Institute of Trade and Commerce, Xuefu Road, No. 287, Suzhou 215009, China
[3] College of Computer, Beijing Institute of Technology, Beijing 100081, China
{cjuan_w,tsonglee}@126.com, songtian@bit.edu.cn

Abstract. In view of the high complexity of pruning algorithm for OBS (optimal brain surgery) process and the deficiency of its match usage with training algorithm, this paper presents a penalty OBS computational model, in which the pruning condition is considered as a penalty term integrated in the objective function of NN (neural network). Based on its theoretical convergence, this model is realized by adopting the conjugate gradient method. Moreover, the effectiveness of this model is validated by a simulation test. The parallelization of network training process and OBS process ensures the accuracy and the efficiency of regularization so as to improve the generalization capacity of NN.

Keywords: neural network; optimal brain surgery process; model; conjugate gradient; convergence; network pruning.

1 Introduction

Network pruning is a typical method of decreasing the network structure and improving network generalization performance. It is generally employed top-down design method, construct a complex network sufficient to meet the problem first, and after the convergence of neural network training, according to the various connection weights or node on the network error function contribution to the extent of the right , remove the contribution of the smallest connected or node. The extent of its contribution, also known as sensitivity [1], so the sensitivity of such methods is also known as sensitivity algorithms. Reed [2] had summarized early sensitivity algorithms in the 1990s. Mozer and Smolensky [3] had proposed a method to remove the hidden nodes. Kartin [4] pointed out that the weight is removed should have a low sensitivity, and defined the weights of sensitivity. Le Cun [5] also used the low sensitivity of the right of the ideological values of the first to be removed, using the second derivative of error function to calculate the sensitivity of the weights (or significant). In the optimal brain damage (OBD) process, the formula for calculating significance was defined. As the OBD procedure based on the premise Hessian matrix is a diagonal matrix, Hassibi [6,7] and so on raised a optimal brain surgery (OBS) process based on the general form

L. Qi (Ed.): ISIA 2010, CCIS 86, pp. 48–57, 2011.

Hessian matrix, to be as the OBD's improved method. Optimal brain surgery [7] is a better performance of the network pruning techniques [8], which uses the second derivative of error function of information, analytical weight predicted effects of disturbance on the function of the extent to which top-down way to reduce or eliminate part of the connection weights to achieve the network structure optimization. But the OBS pruning algorithm is a time-consuming process, the second training series of problems directly affect the usefulness of the algorithm. For this, by introducing the OBS pruning conditions in the form of punishment into the neural network training function, employing weight decay strategies and conjugate gradient(CG) method to obtain inverse Hessian information, deriving a new class of structural optimization algorithm to achieve the parallel study of right value and structure. By simulation experiments on classic function the effectiveness and feasibility of the algorithm was verified.

2 Penalty OBS Model

OBS pruning process is fully trained neural network post training algorithm including extremum approximation and the quadratic approximation assumptions [4]. The objective function of neural networks $\xi\ (\boldsymbol{w})$ at the operating point $\boldsymbol{w}_k$ of the Taylor can be approximately as follows:

$$\begin{aligned}\xi(\boldsymbol{w}_k+\Delta\boldsymbol{w}) &= \xi(\boldsymbol{w}_k)+\boldsymbol{g}^{\mathrm{T}}(\boldsymbol{w}_k)\Delta\boldsymbol{w}+\frac{1}{2}\Delta\boldsymbol{w}^{\mathrm{T}}\boldsymbol{H}(\boldsymbol{w}_k)\Delta\boldsymbol{w}+O(\|\Delta\boldsymbol{w}\|^3)\\ &\approx \xi(\boldsymbol{w}_k)+\frac{1}{2}\Delta\boldsymbol{w}^{\mathrm{T}}\boldsymbol{H}(\boldsymbol{w}_k)\Delta\boldsymbol{w}\end{aligned} \tag{1}$$

Where, $\Delta\boldsymbol{w}$ is $\boldsymbol{w}_k$ increments, $\boldsymbol{g}(\boldsymbol{w}_k)$ as a gradient vector at the $\boldsymbol{w}_k$, denoted by $\boldsymbol{g}_k$; $\boldsymbol{H}(\boldsymbol{w}_k)$ as a function of the Hessian matrix, denoted by $\boldsymbol{H}_k$. The basic idea of OBS is to set a weight value to zero, making the incremental function of $\Delta\xi=\xi(\boldsymbol{w}_k+\Delta\boldsymbol{w})-\xi(\boldsymbol{w}_k)$ is the smallest. Marking $\boldsymbol{w}_q$ as special weights, the weight of the deletion can be equivalent of

$$\Delta w_q + w_q = \boldsymbol{u}_q^{\mathrm{T}}\Delta\boldsymbol{w} + w_q = 0 \tag{2}$$

Where, $\boldsymbol{u}_q$ is the $\boldsymbol{q}_{th}$ vector of unit length whose value is 1. Therefore, OBS process can be described as incremental changes on the weight vector to minimize $\Delta\boldsymbol{w}$ quadratic $\Delta\boldsymbol{w}^{\mathrm{T}}\boldsymbol{H}_k\Delta\boldsymbol{w}/2$, satisfy the equation (2) constraints, and then on the next subscript q minimized. Its mathematical description as follows:

$$\min\left(\frac{1}{2}\Delta\boldsymbol{w}^{\mathrm{T}}\boldsymbol{H}_k\Delta\boldsymbol{w}\right)\ \text{s.t.}\ \boldsymbol{u}_q^{\mathrm{T}}\Delta\boldsymbol{w}+w_q=0 \tag{3}$$

For the above model equation (3) constrained optimization problems, build Lagrange operator

$$L(\Delta\boldsymbol{w},\lambda)=\frac{1}{2}\Delta\boldsymbol{w}^{\mathrm{T}}\boldsymbol{H}_k\Delta\boldsymbol{w}-\lambda(\boldsymbol{u}_q^{\mathrm{T}}\Delta\boldsymbol{w}+w_q) \tag{4}$$

λ is Lagrange factor. L operator of the derivative $\Delta\boldsymbol{w}$ is zero, the best weight changes obtained

$$\Delta\boldsymbol{w}=-\frac{w_q}{[\boldsymbol{H}_k^{-1}]_{q,q}}\boldsymbol{H}_k^{-1}\boldsymbol{u}_q \tag{5}$$

W_q corresponds to the Lagrange operator *L's* optimal value S_q is the weight of significance, that is,

$$S_q = \frac{w_q^2}{2[\boldsymbol{H}_k^{-1}]_{q,q}} \tag{6}$$

Where, H_k^{-1} *for* the inverse Hessian matrix; $\left[H_k^{-1}\right]_{q,q}$ is H_k^{-1}'s $(q,q)_{th}$ elements. For the general situation of the inverse matrix calculation, when the network error function ξ $(\boldsymbol{w})$ take the form of MSE, the

$$\xi(\boldsymbol{w}) = \frac{1}{2}\sum_{i=1}^{N}\left(F(x_i,\boldsymbol{w}) - d_i\right)^2 \tag{7}$$

Employing the approximation relation of Hessian outer product

$$\boldsymbol{H}(N) = \frac{\partial^2 \xi(\boldsymbol{w})}{\partial \boldsymbol{w}^2} \approx \frac{1}{N}\sum_{i=1}^{N}\left(\frac{\partial F(x_i,\boldsymbol{w})}{\partial \boldsymbol{w}}\right)\left(\frac{\partial F(x_i,\boldsymbol{w})}{\partial \boldsymbol{w}}\right)^{\mathrm{T}} \tag{8}$$

Noting $\boldsymbol{\varphi}(i) = \frac{1}{\sqrt{N}}\frac{\partial F(x_i,\boldsymbol{w})}{\partial \boldsymbol{w}}$, the equation (8) transform to Woodbury recursive equations

$$\boldsymbol{H}(n) = \sum_{i=1}^{n}\boldsymbol{\varphi}(i)\boldsymbol{\varphi}^{\mathrm{T}}(i) = \boldsymbol{H}(n-1) + \boldsymbol{\varphi}(n)\boldsymbol{\varphi}^{\mathrm{T}}(n) \qquad (n = 1,2,\cdots,N) \tag{9}$$

Employing the matrix inverse theorem again, obtain the inverse Hessian matrix

$$\boldsymbol{H}^{-1}(n) = \boldsymbol{H}^{-1}(n-1) - \frac{\boldsymbol{H}^{-1}(n-1)\boldsymbol{\varphi}(n)\boldsymbol{\varphi}^{\mathrm{T}}(n)\boldsymbol{H}^{-1}(n-1)}{1+\boldsymbol{\varphi}^{\mathrm{T}}(n)\boldsymbol{H}^{-1}(n-1)\boldsymbol{\varphi}(n)} \tag{10}$$

Inverse iteration initial $\boldsymbol{H}^{-1}(0) = \alpha^{-1}\boldsymbol{I}$, where α is a small positive number, in order to meet the continuous decreasing positive definite iteration requirements.
Trust region [6], a class of unconstrained optimization methods was introduced. At the operating point $\boldsymbol{x}_k$ in the neighborhood Ω_k ($\Omega_k = \{\ \boldsymbol{x} \mid \| \boldsymbol{x} - \boldsymbol{x}_k \| \le v_k\}$), the second expansion of $q_k\ (s)$ is $f(\boldsymbol{x}_k + \boldsymbol{s})$'s reasonable approximation, mathematical model as

$$\min\ q^k(\boldsymbol{s}) = f(\boldsymbol{x}_k) + \boldsymbol{g}_k^{\mathrm{T}}\boldsymbol{s} + \frac{1}{2}\boldsymbol{s}^{\mathrm{T}}\boldsymbol{H}_k\boldsymbol{s} \qquad \text{s.t.}\ \ \|\boldsymbol{s}\| \le v_k \tag{11}$$

Where, $\boldsymbol{g}_k$ is $\boldsymbol{x}_k$'s gradient vector, $\boldsymbol{H}_k$ is $\boldsymbol{x}_k$'s quadratic Hessian matrix, v_k is trust region radius, select appropriate values to ensure $f(\boldsymbol{x}_k + \boldsymbol{s})$ and $q_k(\boldsymbol{s})$ consistency. Defined r_k as the ratio of the actual decline volume and forecast decline volume. It can be the degree of approximation of quadratic model $q_k\ (s_k)$ for the objective function $f(\boldsymbol{x}_k+s_k)$. The mathematical model is

$$r_k = \frac{\Delta f_k}{\Delta q^{(k)}} = \frac{f(\boldsymbol{x}_k) - f(\boldsymbol{x}_k + \boldsymbol{s}_k)}{f(\boldsymbol{x}_k) - q^k(\boldsymbol{s}_k)} \tag{12}$$

Here, the more r_k closer to 1, the more approximate level better. The objective function in the neural network weights on the deletion of the conditions imposed constraints, constitute the basic model of penalty OBS

$$\min\ \xi(\boldsymbol{w}) \qquad \text{s.t.}\ w_q = 0 \tag{13}$$

To facilitate the solution and take full advantage of second derivative information, trust region method can be applied to transform above-mentioned model

$$\begin{aligned} &\min\ q^k(\Delta w) = \xi(w_k) + g_k^T \Delta w + \frac{1}{2}\Delta w^T H_k \Delta w \\ &\text{s.t.}\ \|\Delta w\| \le v_k, \quad u_q^T \Delta w + \eta w_q = 0 \end{aligned} \tag{14}$$

As the norm equivalence, we take $\|\cdot\|$ for the 2-norm, and assuming that the conditions for trust region was effective, conversing the inequality constraints into equality constraints, and construct the Lagrange function L

$$L(\Delta w, \mu, \lambda) = g_k^T \Delta w + \frac{1}{2}\Delta w^T H_k \Delta w + \frac{\mu}{2}(\Delta w^T \Delta w - v_k^2) - \lambda(u_q^T \Delta w + \eta w_q) \tag{15}$$

Where, μ and λ are the Lagrange factor, η call attenuation factor; For constructor function is only the function of $\boldsymbol{\Delta w}$ with the Lagrange factor, and $\xi(\boldsymbol{w}_k)$ has nothing to do with $\boldsymbol{\Delta w}$, it will not be considered. L function $\boldsymbol{\Delta w}$ derivative to zero, there

$$\frac{\partial L}{\partial \Delta w} = g_k + H_k \Delta w + \mu \Delta w - \lambda u_q = 0 \tag{16}$$

After simplification, the equivalent equation based on penalty OBS model is obtained:

$$(H_k + \mu I)\Delta w = -g_k + \lambda u_q \tag{17}$$

3 The Convergence of Penalty OBS Model

There are following lemma [7] for the unconstrained trust region problem model.

Lemma 1. Let s_k=0 is solution to the equation (14), if and only if x_k is minimum point to satisfy $f(x)$ of the second-order necessary conditions.

Lemma 2. Let x_k is not minimum point to satisfy $f(x)$ of the second-order necessary conditions, then there exists v_k>0 makes the problem solution s_k to meet the $f(x_k + s_k) < f(x_k)$.

Based on above lemma, convergence theorem based on trust region methods can be obtained.

Theorem 1. Let $f(x)$ be second-order continuously differentiable, given the initial point $\boldsymbol{x}_1$. If the level set $L(x_1) = \{x \in R^n \mid f(x) \le f(x_1)\}$ bounded, and there is a constant M> 0, makes any $\boldsymbol{x}_k \in L(\boldsymbol{x}_1)$ there is $\|\boldsymbol{H}_k\| \le M$, then the trust region algorithm produces a sequence $\{\boldsymbol{x}_k\}$ has at least one cluster point to satisfy $f(\boldsymbol{x})$ of a first-order and second-order necessary conditions.

In addition, trust region method has second-order convergence rate [6] under stronger trust assumptions. The unconstraint model extends to the linear constraint trust region model, its convergence guaranteed by the following theorem.

Theorem 2. Let $f(\boldsymbol{x})$ be continuously differentiable in feasible region; matrix sequence $\{\boldsymbol{H}_k\}$ uniformly bounded, that is, there is a normal number of M, making

$\|H_k\| \leq M$ for all k betenable; The sequence $\{w_k\}$ generated by the trust region method has at least one accumulation point; then the accumulation point must be the original question's K-T point (to meet the Kuhn-Tucker theorem points).

The verifying process of theorem 1 and 2 see reference [6] and [7]. According to theorem 1 and 2, there has following inference.

Inference. For penalty OBS model of equation (14), there is at least one accumulation point of the sequence produced by trust region methods, and it must be the K-T point of the model.

In the penalty OBS model, $\xi(w)$ parameters for the neural network weights constitutes the objective function, thus satisfying theorem 2, continuously differentiable on the feasible region conditions, as well as the Hessian matrix sequence H_k of uniformly bounded conditions. Trust region method can guarantee the sequence at least one accumulation point, so the accumulation point of penalty OBS model must be K-T point. Namely, the model has convergence.

4 CG Method Algorithms

4.1 CG Method

Set $A \in R^{n\times n}$ symmetric positive definite, $d_1, d_2, \ldots, d_n$ is group of non-zero vector of R^n. If $d_j^T H d_k = 0 (j \neq k)$, $d_1, d_2, \ldots, d_n$ are H-conjugate mutually. For the extreme problem of quadratic function $f(x) = (1/2)x^T H x - g^T x$, where H is $n\times n$ symmetric positive definite Hessian matrix. The direction vector of CG method can be deduced by linear combination of the gradient g_k at current point x_k and the previous direction vector d_{k-1}. The form of the standard algorithms is shown as following

$$
\begin{aligned}
x_{k+1} &= x_k + \alpha_k d_k \\
d_k &= \begin{cases} -g_k & k = 1 \\ -g_k + \beta_k d_{k-1} & k \geq 2 \end{cases}
\end{aligned} \tag{18}
$$

Where, α_k is the one-dimensional step length; β_k is the direction of the structure factors conjugate. β_k can be determined in accordance with the basic conditions $d_k^T H d_{k-1} = 0$ of conjugate direction.

According to β_k's different structural forms, there are some standard CGs,such as HS(Hestenes-Stiefel), PR (Polak-Ribière), FR (Fletcher-Reeves), DY (Dai-Yuan) and DY&HS etc types. Four structure type is exactly to be the combination of a pair of numerators $g_k^T g_k$ and $g_k^T y_{k-1}$ and the combination of a pair of denominator $g_{k-1}^T g_{k-1}$ and $d_{k-1}^T y_{k-1}$. Under exact line search, there exists the following relation

$$
\begin{aligned}
g_k^T y_{k-1} &= g_k^T (g_k - g_{k-1}) = g_k^T g_k, \quad \because g_k^T g_{k-1} = 0 \\
d_{k-1}^T y_{k-1} &= (-g_{k-1}^T + \beta_{k-1} d_{k-2}^T)(g_k - g_{k-1}) = g_{k-1}^T g_{k-1}, \quad \because g_k^T d_{k-2} = 0
\end{aligned} \tag{19}
$$

Thus, HS, PR, FR and DY in the positive definite quadratic function has the equivalence of optimization problems. HS may be taken as for further study, its conjugate directions can be written as

$$d_k = -g_k + \frac{g_k^{\mathrm{T}} y_{k-1}}{d_{k-1}^{\mathrm{T}} y_{k-1}} d_{k-1} = -(I - \frac{d_{k-1} y_{k-1}^{\mathrm{T}}}{d_{k-1}^{\mathrm{T}} y_{k-1}}) g_k = -P_k g_k \tag{20}$$

Where

$$P_k = I - \frac{d_{k-1} y_{k-1}^{\mathrm{T}}}{d_{k-1}^{\mathrm{T}} y_{k-1}} \tag{21}$$

Since $P_k^{\mathrm{T}} y_{k-1} = 0$, P_k can be regarded as null-space affine transformation from R^n to y_{k-1}. Consider a more general transformation matrix $\hat{P}_k$ meets

$$d_k = -\hat{P}_k g_k \tag{22}$$

Substitute expression (22) into the conjugate condition expression d_k

$$d_k y_{k-1} = -g_k^{\mathrm{T}} \hat{P}_k y_{k-1} = 0 \tag{23}$$

In the exact line search, if the relationship-type (24) , equation (23) must be betenable.

$$\hat{P}_k^{\mathrm{T}} y_{k-1} = \alpha_{k-1} d_{k-1} = s_{k-1} \tag{24}$$

Expression (24) is called quasi-Newton condition. With mature quasi-Newton formula can easily construct $\hat{P}_k$. Here select Quasi-Newton formula to calculate the best-performing BFGS (Broyden-Fletcher-Goldfarb-Shanno) inverse Hessian matrix structure expression Z_k

$$Z_k = Z_{k-1} - \frac{Z_{k-1} s_{k-1} y_{k-1}^{\mathrm{T}} + y_{k-1} s_{k-1}^{\mathrm{T}} Z_{k-1}}{s_{k-1}^{\mathrm{T}} y_{k-1}} + (1 + \frac{y_{k-1}^{\mathrm{T}} Z_{k-1} y_{k-1}}{s_{k-1}^{\mathrm{T}} y_{k-1}}) \frac{s_{k-1} s_{k-1}^{\mathrm{T}}}{s_{k-1}^{\mathrm{T}} y_{k-1}} \tag{25}$$

Assuming $Z_{k-1} =$ I of the Shanno no memory quasi-Newton form, $\hat{P}_k$ to be formulated

$$\hat{P}_k = I - \frac{d_{k-1} y_{k-1}^{\mathrm{T}} + y_{k-1} d_{k-1}^{\mathrm{T}}}{d_{k-1}^{\mathrm{T}} y_{k-1}} + (\alpha_{k-1} + \frac{\| y_{k-1} \|^2}{d_{k-1}^{\mathrm{T}} y_{k-1}}) \frac{d_{k-1} d_{k-1}^{\mathrm{T}}}{d_{k-1}^{\mathrm{T}} y_{k-1}} \tag{26}$$

As the equation (26) of the constructed $\hat{P}_k$ is symmetric positive definite matrix, the CG direction construct expression be available by substituting it back to equation (22).

$$d_k = -g_k + \frac{d_{k-1}^{\mathrm{T}} g_k}{d_{k-1}^{\mathrm{T}} y_{k-1}} y_{k-1} - [(\alpha_{k-1} + \frac{\| y_{k-1} \|^2}{d_{k-1}^{\mathrm{T}} y_{k-1}}) \frac{d_{k-1}^{\mathrm{T}} g_k}{d_{k-1}^{\mathrm{T}} y_{k-1}} - \frac{y_{k-1}^{\mathrm{T}} g_k}{d_{k-1}^{\mathrm{T}} y_{k-1}}] d_{k-1} \tag{27}$$

4.2 Penalty OBS CG Algorithm

Substituting Levenberg-Marquardt weight update calculation formula and $\tilde{H}_k$ in significance equator formula for $\alpha_k \hat{P}_k$, namely

$$\Delta w = -\alpha_k \hat{P}_k g_k + \frac{[\alpha_k \hat{P}_k g_k]_q - \eta w_q}{[\alpha_k \hat{P}_k]_{q,q}} \alpha_k \hat{P}_k u_q = s_k - \frac{s_{kq} + \eta w_q}{e_{qq}} e_q \tag{28}$$

$$S_q = \frac{\eta^2 w_q^2 - 2\eta w_q [\alpha_k \hat{P}_k g_k]_q}{2[\alpha_k \hat{P}_k]_{q,q}} = \frac{\eta^2 w_q^2 + 2\eta w_q \alpha_k d_{kq}}{2\alpha_k e_{qq}} = \frac{\eta^2 w_q^2 + 2\eta w_q s_{kq}}{2\alpha_k e_{qq}} \tag{29}$$

Where, α_k for the one-dimensional step length, s_k for the CG search step length. $s_k = \alpha_k d_k$, where $s_{k,q}$ for s_k of the first q elements, e_q is $\hat{P}_k$'s first q row column vector. Namely

$$e_q = \hat{P}_k u_q = u_q - \frac{s_{k-1,q}}{s_{k-1}^{\mathrm{T}} y_{k-1}} y_{k-1} + \left[(1 + \frac{\| y_{k-1} \|^2}{s_{k-1}^{\mathrm{T}} y_{k-1}}) \frac{s_{k-1,q}}{s_{k-1}^{\mathrm{T}} y_{k-1}} - \frac{y_{k-1,q}}{s_{k-1}^{\mathrm{T}} y_{k-1}}\right] s_{k-1} \tag{30}$$

Here, u_q for the first q elements of a unit vector, e_{qq} as e_q the first q elements, $y_{k-1,\,q}$ for y_{k-1} of the first q elements. The penalty OBS CG algorithm based on symmetric CG method and re-launch strategy is shown in table 1.

Table 1. Penalty OBS model CG algorithm

function penalty-OBS.CG ()

(1) initialize the network weights w_1. dimension lw=length (w_1), deleted factor η, control factor α and accuracy ε, the number of iterations *epochs*, the conjugate number of iterations *times*= 0, and re-start *flag*=true;

(2) for k = 1, 2, ..., *epochs* *times* = *times* + 1;

call BP method back-propagation(), calculation neural network gradient g_k and error ξ_k under the weights w_k;

if $\|g_k\|$ or $|\xi_k|$ allow the precision ε, then stop iteration, output $w^* = w_k$;

(3)if flag==true, then $d_k = \tilde{g}_k$; flag=false; else calculate the CG direction d_k in accordance with formula (27)

(4) utilize one-dimensional line-search technology to get the best step length α_k, and then $s_k = A_k d_k$;

(5) update the CG direction, $w_{k+1} = w_k + s_k$;

(6) if *times* == *lw* set *flag* = true

for j=1:lw using equation (30) to calculate j column row vectors of $\hat{P}_k$, and significance S_j by expression (29)

end

obtain the least significance index q, which makes S_q=Min S_j;

(7) based on index of q, substituting into equation (28) get the OBS excision update weight Δw;

set status $w^{new} = w_k + \Delta w$, and calculate the error of indicator ξ^{new};

(8) if $\xi^{new} < \alpha \cdot \xi_k$, shows that OBS attenuation feasible, $w_{k+1} \leftarrow w^{new}$;

end

4.3 Once Regulation Penalty OBS CG Algorithm

Start from the unit matrix I iteration, the search direction for the function is negative gradient direction $d_0 = -g_0$. After a one-dimensional step-size α_0 search and be updated step length $s_0 = \alpha_0 d_0$, and then before next iteration make the following transformation $H_0^{-1} = \alpha_0 H_0^{-1} = \alpha_0 I$, that is, approximate inverse Hessian matrix is the $\alpha_0 I$ made on the basis of iteration, and this ratio is called once regulation operation. According to this idea, the symmetry CG method construction matrix $\hat{P}_k$ amended as follows

$$\hat{P}_k = \alpha_0 (I - \frac{d_{k-1} y_{k-1}^T + y_{k-1} d_{k-1}^T}{d_{k-1}^T y_{k-1}}) + (\alpha_{k-1} + \frac{\alpha_0 \| y_{k-1} \|^2}{d_{k-1}^T y_{k-1}}) \frac{d_{k-1} d_{k-1}^T}{d_{k-1}^T y_{k-1}} \tag{31}$$

The corresponding CG vector is also amended to read:

$$d_k = -\alpha_0 g_k + \frac{\alpha_0 d_{k-1}^T g_k}{d_{k-1}^T y_{k-1}} y_{k-1} - [(\alpha_{k-1} + \frac{\alpha_0 \| y_{k-1} \|^2}{d_{k-1}^T y_{k-1}}) \frac{d_{k-1}^T g_k}{d_{k-1}^T y_{k-1}} - \frac{\alpha_0 y_{k-1}^T g_k}{d_{k-1}^T y_{k-1}}] d_{k-1} \tag{32}$$

Equation (30) also correspondingly made once regulation and amended to read:

$$e_q = \alpha_0 u_q - \frac{\alpha_0 d_{k-1,q}}{d_{k-1}^T y_{k-1}} y_{k-1} + [(\alpha_{k-1} + \frac{\alpha_0 \| y_{k-1} \|^2}{d_{k-1}^T y_{k-1}}) \frac{d_{k-1,q}}{d_{k-1}^T y_{k-1}} - \frac{\alpha_0 y_{k-1,q}}{d_{k-1}^T y_{k-1}}] d_{k-1} \tag{33}$$

Thus, the penalty OBS model once regulation symmetry CG method acquired after once regulation ratio correction on OBS.CG based on once regulation symmetry CG method, the basic processes such as table 2 shows

Table 2. Penalty OBS model once regulation symmetry CG method

function penalty-OBS.CG1 ()

(1)initialize the weight $\boldsymbol{w}_1$, dimension lw=length($\boldsymbol{w}_1$), excised factor η, control factor α and accuracy of ε, once regulation step length $\alpha_0=1$, iteration *epochs*, the conjugate number of iterations *times*=0, and re-start *flag=true*;

(2) for k=1, 2, ...,*epochs*   *times*=*times*+1;

call BP method back-propagation (), calculate neural network gradient $\boldsymbol{g}_k$ and error ξ_k under weights $\boldsymbol{w}_k$;

if $\|\boldsymbol{g}_k\|$ or $|\xi_k|$ reach to the allowable precision ε, stop iteration, then output $\boldsymbol{w}^* = \boldsymbol{w}_k$;

(3) if *flag* == true, then $\boldsymbol{d}_k=-\boldsymbol{g}_k$; else in accordance with formula (32) calculate the conjugate gradient direction $\boldsymbol{d}_k$;

(4) apply one-dimensional line-search technology to get the best step length α_k, and there $\boldsymbol{s}_k=A_k\boldsymbol{d}_k$;

if *flag* == true, then $\alpha_0=\alpha_k$; *flag*=false;

(5) update in the direction of conjugate gradient, $\boldsymbol{w}_{k+1}=\boldsymbol{w}_k+\boldsymbol{s}_k$;

(6) if *times* ==*lw* set *flag*=true

for *j*=1: *lw*

using equation (33) calculate $\hat{P}_k$, column vector $\boldsymbol{e}_j$ of *j*th row;

calculate S_j according to (29)

end

Obtain the least significance index q, so that S_q=min S_j;

(7) based on indicators of q, substituting into equation (28) to compute OBS excision update weight $\Delta\boldsymbol{w}$; set the state of $\boldsymbol{w}^{new}=\boldsymbol{w}_k+\Delta\boldsymbol{w}$, and calculate the error indicator ξ^{new};

(8) if $\xi^{new}<\alpha\cdot\xi_k$, that shows OBS attenuation feasible, $\boldsymbol{w}_{k+1}\leftarrow\boldsymbol{w}^{new}$;

end

4.4 Simulation of Penalty OBS Model CG Algorithm

By Freidman #1 function regression question simulation to verify algorithm generalization capacities.

$$y(x_1, x_2, x_3, x_4, x_5) = y(x_i) = 10\sin(\pi x_1 x_2) + 20(x_3 - 0.5)^2 + 10x_4 + 5x_5, \quad x_i \in [0, 1]$$

Here select 500 groups of random numbers x_i (i=1,2,...,5) distributed in [0.2, 0.8] to calculate function $d=y(x_i)+N(0,1)$. Training input and output sample set T = $\{x_i; d\}_{i=1}^{500}$, where N (0, 1) is the normal distribution of analog noise term. And then, select randomly 1000 group of random numbers x_i distributed in [0, 1] to calculate the noise-free function $y(x_i)$ as test set $T = \{x_i; y\}_{i=1}^{1000}$. The design structural parameters are as follows:

Network structure: 5-15-1; hidden layer activation function: tansig (hyperbolic sine); output layer activation function: purelin (linear function).

In order to compare with the effect of CG control, we select the number of iterations *epoch* = 3000 times, use PR method and FR method in MATLAB neural network toolbox, conjugate gradient algorithm and once regulation symmetry CG algorithm to implement penalty OBS.CG and OBS.CG1, training and testing error values as shown in table 3.

From Table 3 we can see, training error from CGS to penalty OBS.CG increased 0.3696×10^{-5}, measure errors decrease 0.8363×10^{-4}; manner, CGS1 to penalty training error increases 0.3796×10^{-5}, measure errors decrease 0.6772×10^{-4} correspondingly. By sacrificing part of the training error, generalization error of penalty OBS algorithm CG and CG1 methods are less than the other four kinds of conjugate gradient training algorithm. Penalty OBS model keeps higher generalization ability in nature from quantitative view. Then, compare OBS.CG with CGS, OBS.CG1 with CGS1, two pairs of algorithms using the same conjugate direction optimization, relatively speaking, the two penalty OBS algorithms maintained the balanced development of training error and testing error. The subtle changes reflects penalty OBS' implications and penalty-OBS.CGx series of algorithms in maintaining the neural network training and pruning the validity and usefulness of parallel behavior.

Table 3. Simulation error of different conjugation method

Algorithm Type	Training error	Test error
PR	1.43390×10^{-4}	1.73530×10^{-3}
FR	0.89417×10^{-4}	0.37105×10^{-3}
CGS	0.88713×10^{-4}	0.41669×10^{-3}
CGS1	1.00480×10^{-4}	0.42982×10^{-3}
penalty-OBS.CG	0.92409×10^{-4}	0.33306×10^{-3}
penalty-OBS.CG1	0.96684×10^{-4}	0.36210×10^{-3}

5 Conclusions

OBS process is improvement on optimal brain damage process. It achieved structural optimization by using the second derivative of error function information, resolving predicted weight disturbance effects on the function to the extent, reducing or eliminating some certain link weight by top-down way. It is a typical representative of the network pruning, and has very good pruning performance. However, the computational complexity of the process is very high, and need to coordinate with network training algorithm, undergo training-pruning- re-training tedious process; it is difficult to meet the practical needs. Integrated the advantages OBS process and regularization, the penalty OBS model is proposed by using the idea of trust region method and regarding OBS pruning conditions as penalty entry into the neural network training function. It implemented the parallel of training and pruning, and raised the generalization performance of the network model, overcame the time-consuming nature of the OBS pruning and inefficiency. The model's correctness is verified by analysis and contrast on different algorithms simulation data, as well as achieving the model's effectiveness and feasibility.

Acknowledgement

This research was supported by the National Science Foundation of China (Grant No. 60803002) and the opening project of jiangsu province support software engineering R&D center for modern information technology application in enterprise (Grant No. SX201004).

References

1. Nong, S., Yantao, X.: Feature Selection Based on Neuro-Fuzzy Networks. Pattern Recognition and Artificial Intelligence 06, 739–745 (2006)
2. Hassibi, B., Stork, D.G., Wolff, G.: Optimal brain surgeon and general network pruning. In: Proceeding of IEEE International Conference on Neural Networks, pp. 293–299 (1993)
3. Stahlberger, A., Riedmiller, M.: Fast network pruning and feature extraction by removing complete units. In: Advances in Neural Information Processing Systems, vol. 9, pp. 655–661. MIT Press, Cambridge (1997)
4. Harkin, S.: Neural network (the original book version 2). Mechanical Industry Press, Beijing (2004)
5. Zuoyong, L., Penglai, H.: BP network learning ability and generalization ability to meet the uncertainty relation. Science in China (E Series) 33(10), 887–895 (2003)
6. Yaxiang, Y., Yu, W.: Optimization theory and methods, pp. 154–162. Science Press, Beijing (1997)
7. Zhongping, W., Pusheng, F.: Optimization theory and methods, pp. 186–200. Science Press, Beijing (2002)
8. Wanfu, Z., Shijun, Z.: Rough set-based neural network structure optimization design. Computer Engineering and Design 28(17), 4210–4212 (2007)

Research Based on Edge Feature Detection in Computer Image Processing

Peng Wu, Huijuan Lv, and Shilei Shen

Computing Center, Henan University,
Kaifeng, China
hehe9840@126.com, 33237509@qq.com

Abstract. Edge being one of the most basic features of image, the features of it have important uses in computer image processing, and are widely employed in feature extraction, image recognition and other fields. This paper first introduces several major traditional and modern feature detections of edges, analyzes their respective advantages and disadvantages; and then conducts an analysis and research of the development trends of edge detection.

Keywords: Features of the Edge, Computer Image Processing, Operator.

1 Introduction

In terms of digital image processing, the edge of an image refers to the discontinuity of an image's features. Since the main features of an image include gray scale, texture and so on, the edge of the image can be interpreted as the mutation of gray scale and changes in texture. The variation of gray scale can be divided into two types: one is the step change in which the gray scale changes dramatically, the other is the roof change that the gray level changes occur excessively slowly.

The information of an image's edge is very useful. In the application of computer vision, mostly an object can be identified through a rough edge, therefor edge feature detection provides an important basis for the identification and description of a target object and plays an significant role. Since the edge feature detection has been introduced in the field of computer image processing, domestic and international researchers have been carrying out the improvement of its methods and efficiency, trying to find a more general edge detection. However, in addition to the edge of a target object, texture features of an image as well as noises and so forth will be wrongly detected as the edge of the image, so it s difficult to find a suitable algorithm of image edge detection for various requirements .

Traditional edge feature detection still has some shortcomings and needs further improvement. This article next will introduce several kinds of traditional and modern edge feature detection, present the experimental results of each edge detection, and then describe the development trends of edge detection.

2 Traditional Edge Feature Detection

Methods of traditional edge detection mainly are differential method and fitting method. In terms of the frequency domain, the edge of a digital image is a high-frequency

L. Qi (Ed.): ISIA 2010, CCIS 86, pp. 58–62, 2011.

component, and the nature of edge detection can be understood as high-frequency enhancement. Since differentiation is a method of high frequency enhancement, it can be applied to edge detection of the image.

2.1 Image Edge Feature Detection of the Differential Method

Edge detection based on differential method is mainly achieved through the operator, and its main idea is to calculate the value of its gradient for each image pixel, find the maximum gradient direction, then people can easily get the direction of the edge. As the calculation of the gradient is relatively complex, we commonly use a template in a small area to obtain the approximate value. Operators under the differential image edge detection mainly include edge detection operator of Roberts, edge detection operator of Laplace, etc.

(1) Edge detection operator of Roberts. Differential operation usually can not be directly used in digital images, instead the difference method being adopted. The operator of Roberts's edge detection is a very common difference operator using the cross-difference algorithm, and the specific operator is as follows:

$$\Delta_x f=f(i,j)-f(i+1,j+1)$$

$$\Delta_y f=f(i,j+1)-f(i+1,j)$$

The visual form of the convolution operator is:

$$\Delta_x f: \begin{bmatrix} 1 & 0 \\ 0 & -1 \end{bmatrix}$$

$$\Delta_y f: \begin{bmatrix} 0 & 1 \\ -1 & 0 \end{bmatrix}$$

As Roberts's edge detection operator needs at least two lines of image pixels, there is no way to carry out the convolution algorithm in the last row and the last column during the course of the template's convolution operation, and at this time we can replace them with the values of the previous row or column. Because this method adopts the difference values of the pixels in the diagonal direction in order to detect the edge of an object in a digital image, so the effects of vertical and horizontal directions are better, and the accuracy of edge detection is relatively high. The result that we make use of Roberts' edge detection operator to detect the edge of a motor tyre is as follows:

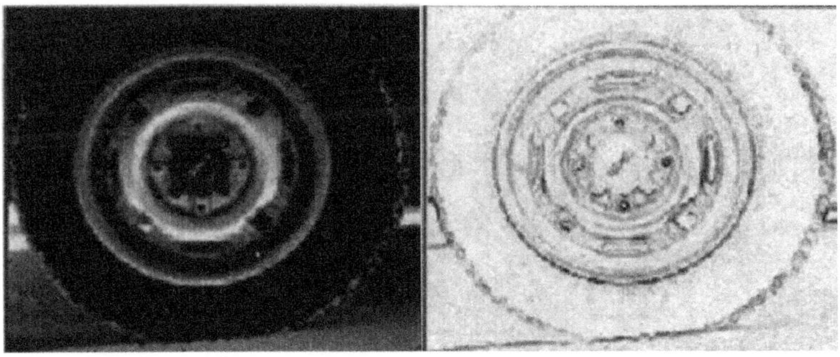

Fig. 1. The original motor tyre and the effect of the edge detection of Roberts' s operator

(2) Edge detection operator of Laplace. Being the same as Roberts's edge detection operator, Laplace's edge detection operator is also a method of edge detection using the differential method, but the difference of them is that Laplace's edge detection operator is a second order differential operator, an isotropic operation, so the operator is very sensitive to the mutation of gray scale. Laplace's edge detection operator is as follows:

$$\nabla^2 f(i,j)=\Delta^2_x f(i,j)+\Delta^2_y f(i,j)$$
$$=f(i+1,j)+f(i-1,j)+f(i,j-1)+f(i,j+1)-4*f(i,j)$$

Laplace's edge detection operator can accurately detect the edge of the step-type image. The second derivative on both sides of the point of the step-type edge is the contrary sign, so Laplace's edge detection operator is making use of this nature to detect the second order difference of each pixel dot in the x and y directions in order to determine whether the pixel point is the edge dot or not.

Laplace's edge detection operator is the operator after the calculus of differences of the second-order, which counts on the principle that the second derivatives of a pixel dot have zero-crossing when detecting the edge of an image, and because this method has no directivity, thus its accuracy of positioning the edge is relatively high, few false edges existing.

Although Laplace's edge detection operator has certain advantages in the process of image edge detection over Roberts's edge detection operator, but it also has some shortcomings: one is that it loses some information of the edge, thus resulting that the edge is not very continuous; the other shortcoming lies in its nature of second order, although this nature can eliminate directivity, but it also can make the noises of the image double extra strong, increase the image's noise information, which is not conducive to the follow-up image processing.

Based on the above shortcomings of Laplace's edge detection operator, Ji Hu, Sun Xiang and some other people made some improvements to this operator. As the original Laplace's edge detection operator took the detected pixel dot as the central point and only had 8 directions, they added another eight directions, gave different weights to these directions, then employed the new template for edge detection, thereby effectively improving the accuracy of detection and reducing the number of false edges.

2.2 Image Edge Feature Detection of the Fitting Method

Different from the image edge detection of the differential method, the fitting method requires the image's surface fitting with a mathematical method, then conducts the image's edge detection using the parameters of surface-fitting.

Using the method of surface-fitting to detect the edge of an image was first put forward by Prewitt. According to the principle of the least square method, Prewitt used the n-order polynomial to fit the image and obtained the parameters of the n-order polynomial, then determined the gray level based on the fitted image, thus detecting the edge of the image. The computation process of fitting the image with n-order polynomial was relatively complex, so Haralick turned to employ the orthogonal polynomial for fitting the image, detecting the image's edge in light of the zero-crossing nature of the fitted image's second derivative. Compared to the differential method for image edge detection, the fitting method has the anti-noise function which is at the cost of many mathematic operations, so the speed of edge detection may be slower than the differential method.

Making use of the fitting method for image edge detection essentially is using the statistical properties of the image. Statistical properties are generally associated with the image as a whole, so the outcome will be better; and some intermediate results of the detection also can be used in image classification and other follow-up treatment, therefore the fitting method is applied more widely than the differential method.

3 Modern Edge Feature Detection

With the practical application demanding detection accuracy, traditional edge detection methods have been unable to meet the requirements of application, domestic and foreign researchers therefore are trying to improve traditional edge detection methods, and has obtained several new methods of edge feature detection, which are as follows:

(1) Edge feature detection of relaxation. Relaxation method was first applied to image processing in the last century, which carries out the initial detection of the image's edge by a simple operator, then makes use of the space distribution of the edge to enhance and optimize the result of the detection so as to obtain a global optimal edge. Relaxation method effectively improves the precision of edge detection, and it is the advancement of traditional edge detection methods.

There are mainly two kinds of relaxation methods, the discrete relaxation which detects the non-continuous data and the continuous relaxation (probability relaxation) that detects the continuous data. When using the discrete relaxation method to detect the image's edge, the affiliation is absolute. A pixel belonging to this edge or the other edge, there is no intermediate fuzzy state. However, when using the continuous relaxation method to detect the image's edge, the pixel dot's affiliation is relative, maybe belonging to multiple edges. As the iterative processes continue and the fuzzy factors becoming lower and lower, the accuracy of edge detection will be increasingly higher.

Schatcher put forward a continuous relaxation method for edge detection: firstly, calculate the gradient of each pixel dot in the image, and judge the initial edge probability and non-edge probability depending on the gradient; then on the basis of these two probabilities determine the degree of compatibility of adjacent pixel dots. If they are in a certain compatibility range, link the two pixel dots and take it as the candidate for the edge; continue the above process until fully eliminate the fuzzy information, and at length accurately detect the edge despite the trouble of the noise.

(2) Multi-scale edge detection method. The traditional edge detection can really handle the problem of edge location well, but yet overlooks some inherent disturbing edges of the actual images, so its accuracy of edge detection is not very high. The method of multi-scale edge detection detects the edge respectively in different scales under the requirements of different details, thereby well eliminating the disturbing edges. Combining various sizes of operators, Rosenfeld first proposed the thinking of multi-scale edge detection, and achieved good results.

Wavelet method can be effectively applied to multi-scale edge detection. Since the wavelet transform can make a multi-scale analysis of images by means of translation as well as lengthening and shortening that have experienced Fourier transform, the wavelet transform has many advantages that are not available to other methods in edge detection : by selecting operators of the appropriate sizes, irrelevance among the detected edges can be reduced to the greatest extent; detection of high frequency, low frequency and so on can be conducted.

Multi-scale image edge detection can achieve good effects. It can not only quickly identify the image's features like the edge, but can describe these features with different scales, thus removing the disruptive edge, interference noise being prevented, the edge being detected effectively.

4 Development of Edge Feature Detection

Edge is one of the basic features of the image, which plays an important role in accurately detecting the edge of the image. This paper describes several traditional and modern edge detection methods, and presents their respective advantages and disadvantages. Edge detection in essence is to find the mutation of the gray scale so as to solve the problem of the classification of the edge points or non-edge points. However, due to the influence of the image's inherent distortion and high-frequency noise, results of the original image detection method are very fuzzy and inaccurate, so the introduction of artificial intelligence, machine learning and other new methods are in need of. In general, edge detection needs to meet certain standards for good detection results. First of all, to be able to remove the noise, and separate the interference information like noise from the real edge; then the accuracy of edge detection should be high and the positioning result should be good; eventually the edge detection also have to be capable of parallel processing in order to raise the speed of edge detection. In these standards of detection, the relationship of noise-immune and the accuracy of the detection is mutual restraint, and it's hard to get good results of both of them at the same time, so people should more or less split the difference during the practical application.

References

1. Lv, T., Liu, B., Bi, X.: Simple algorithm for image edge extraction and its application. Computer Simulation 20(4) (2003)
2. Ji, H., Sun, X., Shao, X., et al.: Image edge detection and its prospects. Computer Engineering and Application 14 (2004)
3. Zhou, D., Pan, Q.: Improved Algorithms of Fuzzy Image Edge. China Journal of Image and Graphics 6(4), 353–358 (2001)
4. Rosenfeld, A.: Computer Vision: A Source of Models for Biological Visual Process. IEEE Trans. on Biomed. Eng. 36(1), 83–94 (2006)
5. Wang, Q., Ruan, H.: An Fast fuzzy edge detection algorithm. China Journal of Image Graphics 6(1), 92–95 (2001)

Adaptive Fuzzy Path Following Control for Mobile Robots with Model Uncertainty

Yang Zhou, Wu-xi Shi, Mu Zhang, Li-jin Guo, and Wen-cheng Guo

School of Electrical Engineering and Automation,
Tianjin Polytechnic University, Tianjin, 300160

Abstract. This paper addresses an adaptive fuzzy path following control scheme for mobile robots with model uncertainty. In this scheme, the fuzzy logic system is used to approximate the uncertainty function of the controller, the parameters in fuzzy logic system are adjusted by time-varying dead-zone which its size is adjusted adaptively. The proposed design scheme guarantees that the tracking error converges to a small neighborhood of the origin. A simulation example is given to demonstrate the effectiveness of the proposed scheme.

Keywords: path following control; adaptive fuzzy control; time-varying dead-zone.

1 Introduction

In recent years, the tracking control problems of nonholonomic mobile robot have attracted considerable attention. Conceptually, there are two distinct approaches that have been formulated in the tracking control problems: trajectory tracking and path following. In the trajectory tracking problem, based on the kinematic model, the backstepping method, neural networks method, fuzzy neural networks method and input-output linearization method are proposed to meet the required control objective [1-4]. With the path following problem, In Ref. [5], the authors have developed their path following scheme when the center of mass of robot is exactly located at the geometric center of axle. When the center of mass of robot is on the axis of the two driving wheels instead of located at the geometric center of axle, the geometric path following scheme have been presented in [6]. It is well known that the designed robot systems will be eventually loaded, and the position of loads will directly affect the center of mass position of the whole robot system. Usually, the center of mass position doesn't lie on the axis of the two driving wheels, moreover, it's difficult to determine its exact position, so it's not suitable to assume that the centroid is center of mass at the geometric center of axle or on the axis of the two driving wheels for loaded robot [5-6]. Therefore, the loaded mobile robot system is a typical uncertain nonlinear system owing to the difficulty confirmation of the center of mass.

Ref. [7] proved that the fuzzy systems are universal approximators. Based on this property, many researchers have presented an adaptive fuzzy control scheme for nonlinear systems [8-11], and the stability analysis in such schemes is performed by using the Lyapunov synthesis method. Moreover, adaptive fuzzy system has been used to design the tracking controller for mobile robot with uncertainty [12-13].

L. Qi (Ed.): ISIA 2010, CCIS 86, pp. 63–70, 2011.

In this paper, the geometric path following problem for mobile robot system with uncertainty is developed. Firstly, the fuzzy logic system is used to approximate the uncertainty function of the controller. Then, the unknown parameters in fuzzy logic system are adjusted adaptively based on time-varying dead-zone which its size is adjusted adaptively. Finally, it is proved that the proposed design scheme guarantees the tracking error converges to a small neighborhood of the origin.

2 Problem Description

The configuration of three-point mobile robot which is studied in this paper is given in Fig. 1

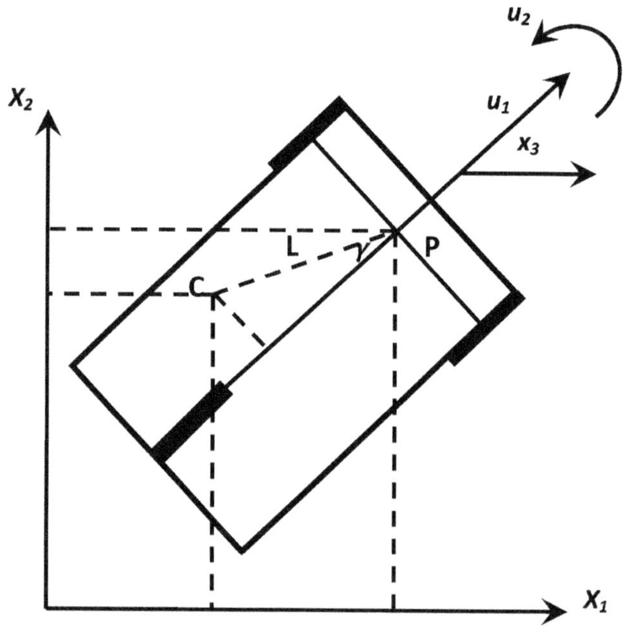

Fig. 1. Three-point mobile robot

where the front two wheel are coaxial independent driving wheel, they drive two motors to achieve the motion and orientation. The rear wheel is a no-power steering wheel, it can support the robot body.

In Fig. 1, point P is the midpoint of two driving wheel axle, the center of mass of robot isn't usually on the axis of the two driving wheels due to the load effect, assuming that point C is the center of mass of robot. The distance between point C and point P is denoted by L (L is positive when the centroid is located at the first half of the axis, otherwise L is negative), γ represents the angle which point C deviates from the axis.

This paper assumes that $\gamma \in \left(-\frac{\pi}{2}, \frac{\pi}{2}\right)$, so L is negative. u_1 is the forward speed of robot, x_3 is the orientation angle of robot with respect to x_1 axis, $u_2 = \dot{x}_3$ is the angular velocity of robot. Then we discuss the kinematic equation of the center of mass C, setting point C coordinates $C = (x_1, x_2, x_3)$, we can obtain the three speed components of the center of mass C, i.e. kinematic model describing the mobile robot which C is treated as a reference point

$$\begin{cases} \dot{x}_1 = u_1 \cos x_3 + L u_2 \sin(x_3 - \gamma) \\ \dot{x}_2 = u_1 \sin x_3 - L u_2 \cos(x_3 - \gamma) \\ \dot{x}_3 = u_2 \end{cases} \tag{1}$$

The objective of this paper is to make the kinematic model of the robot shown in Eq. (1) track the given smooth geometric path $f(x_1, x_2) = 0$ under the control of u_1 and u_2 which parameters L and γ are unknown, namely, that there exists a $t_1 > 0$ such that the tracking error $z = f(x_1, x_2) < \delta$ when $t > t_1$, here δ is a arbitrarily small positive constant.

3 Design of the Adaptive Fuzzy Controller

In this paper, we assume that the robot move forward at a constant speed u_1, since $z = f(x_1, x_2)$, then we have

$$\dot{z} = h(x_1, x_2, x_3) + g(x_1, x_2, x_3)u \tag{2}$$

where $u = u_2$, $h(x_1, x_2, x_3) = (f_{x_1} \cos x_3 + f_{x_2} \sin x_3) u_1$,

$$g(x_1, x_2, x_3) = (f_{x_1} \sin(x_3 - \gamma) - f_{x_2} \cos(x_3 - \gamma)) L$$

When the exact position of the center of mass C of mobile robot system shown in Fig. 1 is known, i.e. $h(x_1, x_2, x_3)$ and $g(x_1, x_2, x_3)$ are both known, then it is well known that there exists an controller

$$u = \frac{-h(x_1, x_2, x_3) - kz}{g(x_1, x_2, x_3)} \tag{3}$$

where $k > 0$, that drives the tracking error z converges to the origin , this means that the objective is achieved.

However, the center of mass C cannot be exactly located on the axis under the effect of loads in the robot practical operation, so the exact position of C is unknown. Consequently, the distance L and departure angle γ are unknown, i.e. $g(x_1, x_2, x_3)$

in Eq. (3) is unknown, the above controller (3) can no longer be implemented. In this paper, we assume that the unknown function $g(x_1, x_2, x_3)$ would be approximated by the fuzzy logic system [7] as $\hat{g}(\bar{x} \mid \theta_g) = \bar{\theta}_g^T \bar{\xi}(\bar{x})$. In order to take care the case where the estimated value $\bar{\theta}_g^T \bar{\xi}(\bar{x})$ becomes zero at some points of time, we design the control as

$$u = \frac{-h(x_1, x_2, x_3) - kz}{\hat{g} + \varepsilon\, sgn(\hat{g})} \tag{4}$$

where ε is a given small positive, $sgn(a) = \begin{cases} 1 & a \geq 0 \\ -1 & a < 0 \end{cases}$.

Substituting Eq. (4) into Eq. (2) yields

$$\dot{z} = -kz - (\hat{g} - g)u - \varepsilon\, sgn(\hat{g})u \tag{5}$$

Let us define the optimal approximation parameter θ_g as follows

$$\bar{\theta}_g^* = arg\ min_{\theta \in \Omega_g} \left\{ sup_{\bar{x} \in U} \left| \hat{g}(\bar{x} | \bar{\theta}_g) - g(\bar{x}) \right| \right\}$$

where Ω_g is the compact set of allowable controller parameter. Define the parameter error $\Phi = \theta_g - \bar{\theta}_g^*$, and $\omega = \hat{g}(\bar{x} \mid \bar{\theta}_g^*) - g(\bar{x})$ as the minimum approximation error, so the error equation (5) can be written as

$$\dot{z} = -kz - \Phi^T \bar{\xi}(\bar{x})u - \omega u - \varepsilon\, sgn(\hat{g})u \tag{6}$$

In this paper, we assume that the used fuzzy system do not violate the universal approximation theorem on the compact set U, which is assumed large enough so that state variables remain within U under closed-loop control. So it is reasonable to assume that the minimum approximation error is bounded for all $\bar{x} \in U$, accordingly, we can make the following assumption

Assumption: There is a unknown constant $\rho^* > 0$ such that $|\omega| \leq \rho^*$, and $\rho(t)$ is its estimated value.

In order to compensate for the lumped error terms in Eq. (6), in this paper, a dead-zone with time-varying size is used to design the adaptation laws. The used time-varying dead-zone is defined as

$$z_\Delta = \begin{cases} z - \varphi(t) sign(z) & |z| > \varphi(t) \\ 0 & |z| \leq \varphi(t) \end{cases} \tag{7}$$

where $\varphi(t)$ is the size of time-varying dead-zone, its adaptive law will be designed later, $sign(*)$ is a sign function.

Using the following adaptation law to adjust the parameter $\bar{\theta}_g$

$$\dot{\bar{\theta}}_g = \lambda z_\Delta \bar{\xi}(\bar{x}) u \tag{8}$$

where $\lambda > 0$.The time-varying dead-zone size $\varphi(t)$ is adjusted by the following adaptation law

$$\dot{\varphi}(t) = -k_\varphi \varphi(t) + (\varepsilon + \rho(t))|u| \tag{9}$$

where $k_\varphi > 0$. $\rho(t)$ is adjusted by the adaptive law as

$$\dot{\rho}(t) = -\beta |z_\Delta| |u| \tag{10}$$

where $\beta > 0$.

Remark: From Eq. (10) ,we obtain $\varphi = exp(-k_\varphi t)\varphi(0) + \int_0^t (\varepsilon|u| + \rho(t)|u|)d\tau$, so if $\varphi(0) \geq 0$, $\rho(0) \geq 0$, then φ is positive, and obviously φ is bounded due to the δ-modify adaptation law of Eq. (10).

Theorem: Given the kinematic model of the robot defined by Eq. (1) satisfying Assumption, when $k > k_\varphi$, the control law Eq. (4) with adaptation laws Eq. (8)-Eq. (10) will ensure that the tracking error converge to a small neighborhood of origin.

Proof

Setting $\rho^* - \rho = \bar{\rho}$. Consider the following Lyapunov function

$$V = \frac{1}{2} z_\Delta^2 + \frac{1}{2\lambda} \Phi^T \Phi + \frac{1}{2\beta} \bar{\rho}^2$$

differentiating the above equation

$$\dot{V} = z_\Delta \dot{z}_\Delta + \frac{1}{\lambda} \Phi^T \dot{\Phi} + \frac{1}{\beta} \bar{\rho}\dot{\bar{\rho}} \tag{11}$$

If $|z| \leq \varphi(t)$, then $z_\Delta = 0$, thus $\dot{V} = 0$, therefore only the region $|z| > \varphi(t)$ is considered in the subsequent proof. If $|z| > \varphi(t)$, from Eq. (6) and Eq. (7),we obtain

$$\dot{z}_\Delta = -kz - \Phi^T \bar{\xi}(\bar{x}) u - \omega u - \varepsilon\, sgn(\hat{g}) u - \dot{\varphi}(t) sign(z) \tag{12}$$

substituting Eq. (12) into Eq. (11), and using Eq. (8), the Eq. (12) can be expressed as

$$\dot{V} = -kz_{\Delta}z - z_{\Delta}\omega u - z_{\Delta}\varepsilon\, sgn(\hat{g})u - z_{\Delta}\dot{\varphi}(t)sign(z) + \frac{1}{\beta}\bar{\rho}\dot{\bar{\rho}} \tag{13}$$

from Eq. (7), if $|z| > \varphi(t)$, $z_{\Delta}sign(z) = |z_{\Delta}|$, then Eq. (13) can be rewritten as

$$\dot{V} \le -kz_{\Delta}z + |z_{\Delta}|\rho^{*}|u| + |z_{\Delta}|\,\varepsilon\,|u| - |z_{\Delta}|\dot{\varphi}(t) + \frac{1}{\beta}\bar{\rho}\dot{\bar{\rho}} \tag{14}$$

using Eq. (9) and Eq. (10), Eq. (14) can be written as

$$\dot{V} \le -kz_{\Delta}z + k_{\varphi}|z_{\Delta}|\varphi(t) \tag{15}$$

from Eq. (7), it is noted that when $|z| > \varphi(t)$, $z = z_{\Delta} + \varphi(t)sign(z)$ and $z_{\Delta}sign(z) = |z_{\Delta}|$, thus Eq. (15) can be rewritten as

$$\dot{V} \le -kz_{\Delta}^{2} - |z_{\Delta}|\,(k - k_{\varphi})\varphi(t) \tag{16}$$

Since $k > k_{\varphi}$, when $|z| > \varphi(t)$, $\dot{V} \le 0$, therefor z_{Δ}, Φ and $\bar{\rho}$ are all bounded.

Using the fact that $|\hat{g} + \varepsilon\, sgn(\hat{g})| > \varepsilon$, from Eq. (4) we have u is bounded. From Eq.(9) we get $\dot{\varphi}(t)$ is bounded, thus $\dot{z}_{\Delta}$ is bounded from Eq. (12). Then Eq. (16) can be given as

$$\dot{V} \le -kz_{\Delta}^{2} \tag{17}$$

integrating both sides of Eq. (17), we get

$$\int_0^t z_{\Delta}^{2}(\tau)d\tau \le \frac{1}{k}V(0)$$

This implies that $z_{\Delta} \in L_2$, therefore, from Barbalat's lemma, we conclude that $\lim_{t\to\infty} z_{\Delta} = 0$. From Eq. (7), we have $|z| \le \varphi(t)$ as $t \to \infty$.

4 Simulation

In this section, we test the adaptive fuzzy path following control schemes of the mobile robot, the kinematic model of such system are given by Eq. (1).The control objective is to design a adaptive fuzzy tracking controller such that the robot tracking the desired path $f(x_1, x_2) = x_1^2 + x_2^2 - 1$, so the tracking error $z = f(x_1, x_2) = x_1^2 + x_2^2 - 1$, from Eq. (2),we get

$$h(x_1, x_2, x_3) = 2u_1(x_1 \cos x_3 + x_2 \sin x_3)$$

$$g(x_1, x_2, x_3) = 2L[x_1 \sin(x_3 - \gamma) - x_2 \cos(x_3 - \gamma)]$$

Since the exact position of the center of mass C is unknown, so that $g(x_1, x_2, x_3)$ is unknown as well. This paper uses the fuzzy system $\hat{g}(\bar{x} | \bar{\theta}_g) = \bar{\theta}_g^T \bar{\xi}(\bar{x})$ to approximate $g(x_1, x_2, x_3)$, the membership function of its each input are all taken as Gaussian type

$$\mu_{F_i^l}(x_i) = \exp(-(\frac{x_i + 1.5}{2})^2),\ \mu_{F_i^l}(x_i) = \exp(-(\frac{x_i}{2})^2),\ \mu_{F_i^l}(x_i) = \exp(-(\frac{x_i - 1.5}{2})^2)$$

Thus, the fuzzy logic system we designed has totally 27 rules.

Setting the robot speed $u_1 = 1$, we assume that $L = 0.3$, $\gamma = \pi / 6$ in model Eq. (1). The each component of the parameter initial value $\bar{\theta}_g$ is randomly chosen in [-1, 1]. Let the initial conditions $(x_1, x_2, x_3) = (0.4, 0.2, \pi / 8)$, $\rho(0) = 0.1$, $\varphi(0) = 0.1$. Other parameters are chosen as $k = 4.5$, $k_\varphi = 4$, $\varepsilon = 0.1$, $\lambda = 3.8$, $\beta = 0.6$.

The simulation result is presented in Fig.2 and Fig.3, where Fig.2 shows the path which the robot tracks the unit circle, Fig.3 shows the tracking error. This means that the tracking error converges to a small neighborhood of the origin.

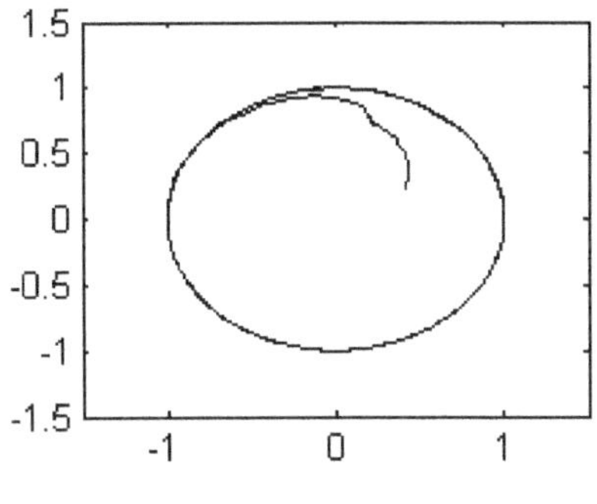

Fig. 2. The tracking path of robot

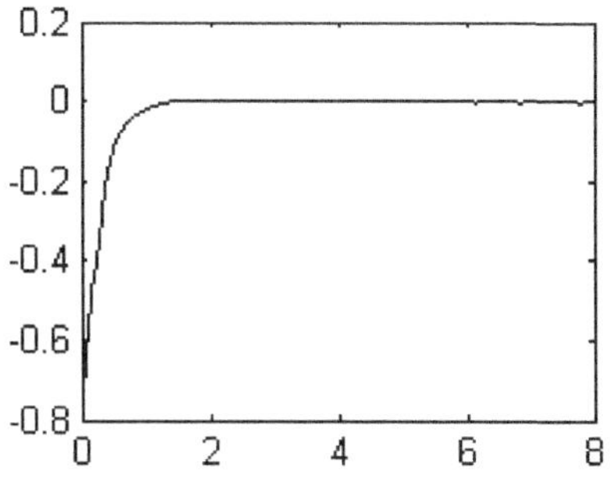

Fig. 3. The tracking error

5 Conclusion

This paper addresses the geometric path following problem for mobile robot system with uncertainty of the centroid position. In the paper, the fuzzy logic system is used to approximate the uncertainty function of the controller, and the unknown parameters are adjusted adaptively by time-varying dead-zone in order to compensate the error. The proposed design scheme guarantees that the tracking error converges to a small neighborhood of the origin.

Acknowledgment

This work was supported by Natural Science Foundation of Tianjin under Grant 10JCYBJC07400.

References

[1] Kanayama, Y., Kimura, Y., Miyazaki, F., Noguchi, T.: A stable tracking control method for an autonomous mobile robot. In: Proc. IEEE Int. Conf. Robot. Autom. Cincinnati, OH, vol. 1, pp. 384–389 (1990)
[2] Yuan, G., Yang, S.X., Mittal, G.S.: Tracking control of a mobile robot using a neural dynamics based approach. In: Proc. IEEE Int. Conf. Robot. Autom. Port Island, Kobe, Japan, vol. 1, pp. 163–168 (2001)
[3] Hu, Y., Yang, S.X.: A fuzzy neural dynamics based tracking controller for a nonholonomic mobile robot. In: Proc. IEEE Int. Conf. Adv. Intell. Mechatron., pp. 205–210 (2003)
[4] Kim, D.H., Oh, J.H.: Tracking control of a two-wheeled mobile robot using input-output linearization. Control Eng. Practice 7(3), 369–373 (1999)
[5] Ma, B.L., Huo, W.: Path tracking control and stabilization of mobile car. Robot 17(6), 359–362 (1995)
[6] Sun, D.Q., Huo, W., Yang, X.: Path following control of mobile robots with model uncertainty based on hierarchical fuzzy system. Control Theory and Applications 21(4), 489–500 (2004)
[7] Wang, L.X.: Adaptive fuzzy systems and control-design and stability analysis. Prentice Hall, New Jersey (1994)
[8] Koo, K.M.: Stable adaptive fuzzy controller with time-varying dead-zone. Fuzzy Sets and Systems 121, 161–168 (2001)
[9] Nounou, H.N., Passin, K.M.: Stable auto-tuning of adaptive fuzzy/neural controllers for nonlinear discrete-time systems. IEEE Trans. Fuzzy Systems 12(1), 70–83 (2004)
[10] Tong, S.H., Li, H.X.: Direct adaptive fuzzy output tracking control of nonlinear systems. Fuzzy Sets and Systems 128(1), 107–115 (2002)
[11] Parka, J.H., Seob, S.J., Parka, G.T.: Robust adaptive fuzzy controller for nonlinear system using estimation of bounds for approximation errors. Fuzzy Sets and Systems 133, 19–36 (2003)
[12] Das, T., Kar, I.N.: Design and Implementation of an Adaptive Fuzzy Logic-Based Controller for Wheeled Mobile Robots. IEEE Trans. Control Systems Technology 14(3), 501–510 (2006)
[13] Hou, Z.G., Zou, A.M., Cheng, L., Tan, M.: Adaptive Control of an Electrically Driven Nonholonomic Mobile Robot via Backstepping and Fuzzy Approach. IEEE Trans. Control Systems Technology 17(4), 803–815 (2009)

A Partition-Based Model Checking Method for Verifying Communication Protocols with SPIN

Xinchang Zhang[1,2], Meihong Yang[1,2], Xingfeng Li[3], and Huiling Shi[1,2]

[1] Shandong Provincial Key Laboratory of computer networks, Jinan 250014, China
[2] Shandong Computer Science Centre, Jinan 250014, China
[3] Computer Network Information Center, Chinese Academy of Sciences, Beijing 100190, China
zhangxc@keylab.net

Abstract. The state explosion is a well-known problem in the field of model checking, which confines the application of model checking to some extent. This paper proposes a partition-based model checking method, which can be employed to address the state explosion problem in some procedures of verifying complex communication protocols with SPIN. The proposed method partitions the design model of a communication protocol into different sub-models by a message-based way, and verifies the design model through validating the sub-models with relatively low resource consumption.

Keywords: model checking, state explosion, partition, SPIN.

1 Introduction

As the rapid development of network, more and more new kinds of protocols or variations of existing protocols have been developed or designed. However, ensuring the correctness of communication protocols can be challenging, due to their complexity and inherent distribution and concurrency. In addition to traditional techniques such as testing, model checking has been viewed as a promising technique for validating the correctness of complex communication protocols.

Model checking is a method for formally verifying finite-state concurrent systems. In model checking, properties about the system under verification are usually expressed as temporal logic formulas, and efficient algorithms are used to traverse the system model to check whether the properties hold or not. Model checking is particularly attractive for communication protocols in which problems of concurrency and distribution make traditional testing challenging. In recent years, there have been many papers (e.g. [1], [2], [3], [4] and [5]) which report the successful instances of using model checking to validate communication protocols.

SPIN is a general model checking tool for verifying the correctness of distributed system design in a rigorous and mostly automated fashion [6][7]. SPIN accepts design specifications written in promela language, and accepts correctness claims written in Linear Temporal Logic (LTL). Like other model checking tools, SPIN might face the state explosion problem when it is used to verify the large-scale systems. There are several approaches to combat this problem, which can be classified into two

L. Qi (Ed.): ISIA 2010, CCIS 86, pp. 71–77, 2011.

categories, i.e. reducing the resource consumption in the process of model checking (e.g. [8] and [9]) and simplifying the system model by higher abstraction. The latter is often a methodology, and many users may find no good ideas to simplify the model when they face the state explosion problem. This paper will propose a partition-based model checking method for verifying communication protocols with SPIN. When the process of validating the model of a communication protocol fails because of the limited resource (e.g. RAM), the proposed method partitions the model into different modules (called sub-models) by a message-based way, and verifies the model through validating the sub-models with relatively low resource consumption.

The rest of the paper is organized as follows. Section 2 presents the partition-based model checking method for verifying communication protocols with SPIN. In Section 3, we evaluate the performance of our proposed method by analyzing the experiment results. Finally, we summarize the paper in Section 4.

2 A Partition-Based Model Checking Method

Since model checking attempts to search each reachable state, the state explosion might inevitably happen in the process of validating some whole models of complex communication protocols. Some optimization methods may alleviate the above problem, but these methods cannot completely solve the problem, i.e. they sometimes fail to finish the verification because of the intrinsic high complexity of communication protocols. Abstraction at higher level is the key approach to solve the state explosion problem. However, the simplified system usually does not satisfy exactly the same properties as the original one so that a process of refinement may be necessary. Additionally, refining a desirable abstraction of a complex communication protocol often wastes much time and labor. This section will present a partition-based model checking method (called PBMC) for reducing the complexity of verifying communication protocols with SPIN. When the state explosion happens in the verification process, the proposed method can be used to divide the model under verification into many modules (called sub-models) with little labor. Each divided sub-model is independently validated, whereby the original model is validated.

2.1 Message Relativity

In this paper, we summarize three kinds of relativity between different kinds of messages of the communication protocol. Let M1 and M2 be two different kinds of messages in a communication protocol, then the three kinds of relativities can be explained as follows:

- Independency: M_1 is independent of M_2 if (1) the sending and receiving actions of M_1 is independent of the sending or receiving actions of M_2, and (2) M_1 does not make use of any information provided by M_2 or produced by M_2.
- Direct influence: Assume that the event of sending and/or receiving M_2 has a direct influence on the content of M_1, or the operation of sending or receiving M_1, then we say that M_2 has a direct influence on M_1.

- Indirect influence: M_2 has an indirect influence on M_1 if there are m ($m \geq 1$) different kinds of messages, donated by A_1, A_2, ..., A_m, such that (1) A_m has a direct influence on M_1, (2) M_2 has a direct influence on A_1, and (3) A_i has a direct influence on A_j ($1 \leq i < j \leq m$).

We further explain the above three kinds of relativities through the example shown in Fig.1. In the figure, we can see that the off-hook message is independent of other messages. The off-hook message has a direct influence on the dial tone message, because the switch sends the dial tone message only after it receives the off-hook message. We also can notice that the off-hook message has an indirect influence on the number, ring tone, busy tone and on-hook messages.

For dividing the model based on the messages, this paper introduces a term called dependency set. We use denotation $R(x)$ to mean the dependency set of message x, then $R(x)$ is recursively defined as follows:

(1) $x \in R(x)$;
(2) Assume that message y has a direct influence on x, then $y \in R(x)$;
(3) For any element z in $R(x)$, $R(z) \subseteq R(x)$.

For the example shown in Fig.1, R(ring tone)={off-hook, dial tone, number, ring tone}. Specifically, R(off-hook)= {off-hook}.

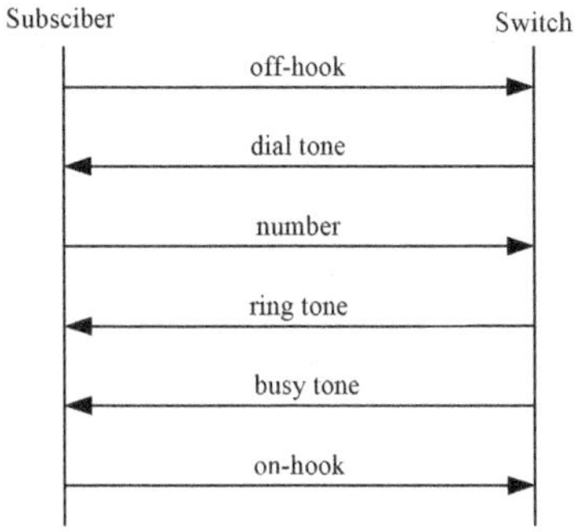

Fig. 1. Typical scenario for a POTS call (source from [10])

2.2 PBMC Design

The SPIN tool uses the promela language to describe the design model, and the description is called promela model. The promela model is constructed from three basic types of objects, i.e. processes, data objects and message channel. Message channels are used to model the exchange of data (i.e. messages) between processes. For communication protocols, the message means a data packet in most cases. Since the packets are the main data objects in a communication protocol, and the most protocol behaviors are based on the packets, it is suitable to divide the whole promela model, which describes the design of a communication protocol, into different partitions in terms of the packets such that each partition only contains a part of types of packets. As mentioned previously, each divided partition is called sub-model. Comparing with the whole promela model, each sub-model has low complexity.

Based on the dependency set, we introduce a new term called included dependency set. Given any message x, the included dependency set of x is denoted by $I(x)$, which is defined as

$$I(x)=\{k|(k\in R(x))\wedge(R(k)\subseteq R(x))\}. \quad (1)$$

Specially, $x\in I(x)$. According to the above definition, if $y\in I(x)$ holds, then each message, that has direct or indirect influence on y, is included in $R(x)$. A promela model can be equivalently divided into two sub-models in terms of $R(x)$ and $I(x)$, denoted by $\Gamma(x)$ and $\Gamma'(x)$, respectively. $\Gamma(x)$ is a part of the promela model, and it is only corrective to the messages in $R(x)$, while $\Gamma'(x)$ indicates the modified promela model that some messages in $I(x)$ are refined in original promela model. In PBMC, a message x is refined in a promela model P as follows:

(1) For any message y in $I(x)$ which satisfies $y\notin R'(M-I(x))$, where $R'(M-I(x))=\bigcup_{Z\in M-I(x)} R(z)$, the information of y is deleted in model P.

(2) For any message y in $I(x)$ which satisfies $y\in R'(M-I(x))$ and $R(y)\bigcap(M-I(x))=\Phi$, then the sending and receiving operations are deleted in model P, and the influences of these operations are simulated by some equivalent promela statements.

According to the above descriptions, we can have the following theorem.

Theorem 1. Assume that PBMC divides a promela model P into two sub-models—$\Gamma(x)$ and $\Gamma'(x)$, then P passes the verification of model checking if and only if all the sub-models pass the verification.

Proof 1. Since the all messages corrective to $I(x)$ is included in $\Gamma(x)$, the part of P which is relative to $I(x)$ passes the verification of model checking if and only if $\Gamma(x)$ passes the verification. According to the above steps of the refinement of messages, we can see that all the information of the messages, which are not included in $I(x)$, is completely included in $\Gamma'(x)$. Thus it is safety to say that the design on the messages involving the messages in $I(x)$ passes the verification of model checking if and only if $\Gamma'(x)$ passes the verification. Additionally, PBMC divides P based on messages, and all the information which is not relative to messages is remained. Thus the theorem has been proven.

We explain how to simulate the influence of a sending operation in a promela model through the following instance. The left part of Fig.2 illustrates the agent process of a client-server model that will be further introduced in next section. In the process, sending the grant message has a direct influence on sending the hold message because the former will result in an empty agent which is a prerequisite of sending the hold message. In the right part of Fig.2, the sending operation of the grant message is deleted, and the influence of the operation is simulated by bolded part. The receiving operation also can be simulated by a similar way.

Let M mean the set of all messages in a promela model P, then PBMC can be described by the following steps:

Step 1: $T \leftarrow M;\ W \leftarrow P$.

Step 2: For any element x in T, compute dependency set $R(x)$ according to W, and compute included dependency set $I(x)$ in terms to W.

Step 3: Find an element d in T if T is not null, such that: $\forall y(y \in T)$, $|R(d)| \geq |R(y)|$. Then delete d from T and obtain two sub-models, i.e. $\Gamma(d)$ and $\Gamma'(d)$. If T is null, then end the validation procedure.

Step 4: Validate $\Gamma(d)$ with SPIN. If the state explosion happens, then: $W \leftarrow \Gamma(d)$; $T \leftarrow$ all the messages in $\Gamma(d)$; goto Step 2.

Step 5: Validate $\Gamma'(d)$ with SPIN. If the state explosion happens, then: $W \leftarrow \Gamma'(d)$; $T \leftarrow$ all the messages in $\Gamma'(d)$; goto Step 2.

Note that T and W in Step 1 mean set of messages and promela model (or sub-model), respectively. According to the above steps, we can see that PBMC will partition the promela model, which is verifying, into two or more sub-models. PBMC provides an effective method for reducing the complexity of a verification process, which is powerful enough to solve the state explosion in most cases.

```
proctype Agent(chan listen, talk)
{
        ....
        :: talk!grant(listen) ->
wait:      listen?return; break
        ....
}
```

```
proctype Agent(chan listen, talk)
{
        ....
        :: nfull(talk) ->
wait:      listen?return; break
        ...
}
```

Fig. 2. An example of simulating sending operation

3 Experiments

We evaluated proposed PBMC method by validating the client-server model described in Chapter 15 of [10]. The related experiments were executed on Intel Core 2 Duo 2.83GHz, 1.5GB of RAM, running linux. The client-server model includes five kinds of messages, i.e. request, deny, hold, grant and return. More detail can be seen in [10].

According to PBMC method, we divided the client-server model described in [10] into two sub-models. One sub-model (denoted by sub-model 1) includes the request, deny, grant and return messages, and another sub-model (denoted by sub-model 2) contains the request and hold messages. Note that we simulated the grant message in sub-model 2 because it has some influence on the hold message. Next, the client-server model described in [10] is called the whole model. We validated the two sub-models and the whole model with SPIN in seven different scenarios. In these scenarios, the numbers of clients are from 3 to 9. Additionally, the server only provides one agent in all the scenarios.

In this paper, we introduce some performance measurement metrics for the reduction of complexity, i.e. state reduction ratio (srr), matched state reduction ratio (msrr), transition reduction ratio (tr) and total memory usage ratio (tmur). These metrics are defined as

$$srr = \frac{\text{number of reached states in sub - model}}{\text{number of reached states in whole model}}, \tag{2}$$

$$msrr = \frac{\text{number of matched states in sub - model}}{\text{number of mached states in whole model}}, \tag{3}$$

$$tr = \frac{\text{number of transitions in sub - model}}{\text{number of transitions in whole model}}, \tag{4}$$

$$tumr = \frac{\text{total memory usage in sub - model}}{\text{total memory usage in whole model}}. \tag{5}$$

According to the above definitions, we can see that these metrics are not larger than 1. Additionally, the degree of reducing the complexity in corresponding respect is higher with lower value of a given measurement metric.

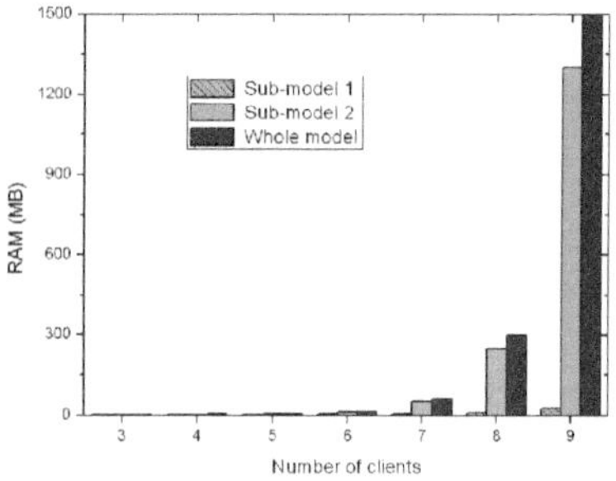

Fig. 3. PBMC's capacity of solving the state explosion problem

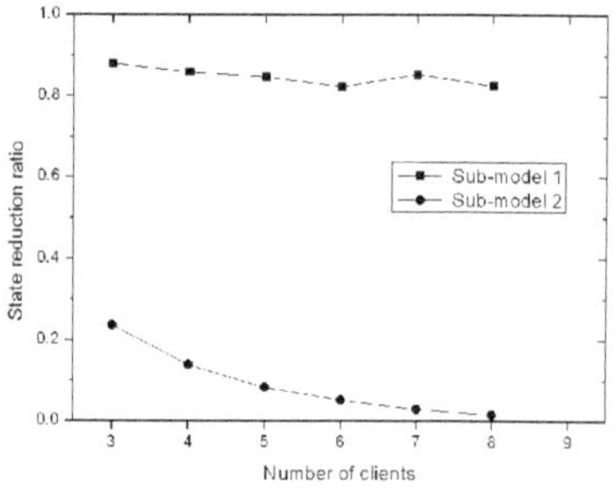

Fig. 4. Reduction of reached states in PBMC

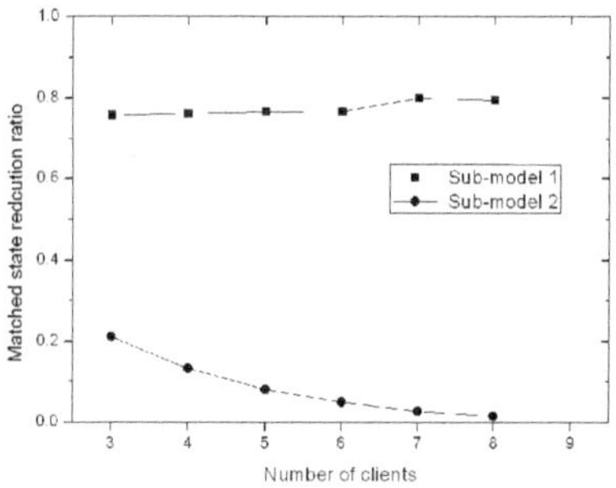

Fig. 5. Reduction of matched states in PBMC

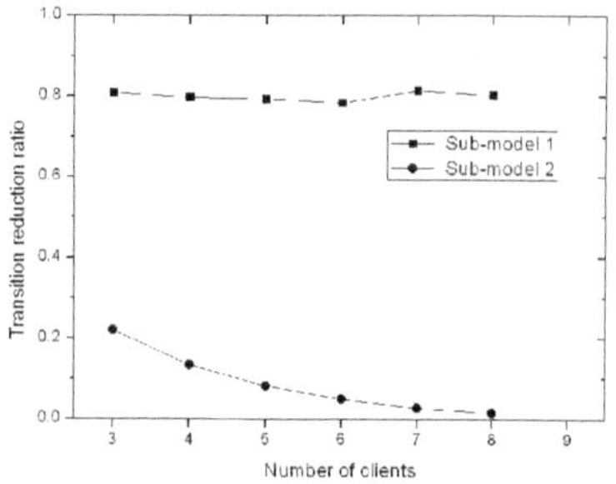

Fig. 6. Reduction of searched transiti-ons in PBMC

Fig.3 illustrates the consumptions of RAM in the seven scenarios. From the figure, we can notice that the validation, without PBMC, run out of the memory in the scenario with 9 clients. However, the validation can be completed when we used PBMC to divide the model into two sub-models (i.e. sub-model 1 and sub-model 2).

Fig.4-Fig.6 show some details on the performance of PBMC in the experiments. From Fig.4 to Fig.6, we can notice that the values of srr, msrr and tr in sub-model 2 are less than 0.3, and the value of each measurement metric in sub-model 1 is about 0.83, which is obviously higher than that of corresponding metric in sub-model 1. We attribute the above feature to the high complexity of the design relative to the grant message. In general, the complexity of validating each sub-model is distinctly lower than that of validating the whole model.

4 Conclusions

SPIN can be used to effectively verify the correctness of distributed system design. However, like other model checkers, SPIN also faces the state explosion problem when it is used to verify some complex systems. This paper presented a partition-based model checking method called PBMC, to address the state explosion problem in the process of verifying complex communication protocols with SPIN. The proposed method partitions the design model of a communication protocol into different sub-models by a message-based way. PBMC verifies the model of communication protocols through validating the divided sub-models with relatively low resource consumption, which can effectively overcome the state explosion problem without refining the whole abstraction of the design model.

Acknowledgments

This work was supported by the Fund for the Doctoral Research of Shandong Academy of Sciences under Grant No. 2010-12 and the National Natural Science Foundation of China under Grant No. 61070039.

References

1. Shanbhag, V.K., Gopinath, K.: A SPIN-Based Model Checker for Telecommunication Protocols. In: Proceedings of the 8th International SPIN Workshop on Model Checking of Software, pp. 252–271 (2001)
2. Islam, S.M.S., Sqalli, M.H., Khan, S.: Modeling and Formal Verification of DHCP Using SPIN. International Journal of Computer Science & Application 2(6), 145–159 (2006)
3. de Renesse, R., Aghvami, A.H.: Formal verification of Ad-hoc routing protocols using spin model checker. In: Proceedings of IEEE MELECON 2004 (2004)
4. Simei, L., Jianlin, Z., Liming, L.: The Automatic Verification and Improvement of SET Protocol Model with SMV. In: Proceedings of Information Engineering and Electronic Commerce (2009)
5. McInnes, A.I.: Model-checking the Flooding Time Synchronization Protocol Control and Automation. In: Proceedings of ICCA 2009, pp. 422–429 (2009)
6. Holzmann, G.J.: The model checker spin. IEEE Transactions on Software Engineering 23(5), 279–295 (1997)
7. The spin tool, `http://spinroot.com/spin/whatispin.html`
8. Biere, A., Cimatti, A., Clarke, E.M., Strichman, O., Zhu, Y.: Bounded Model Checking. Advances in Computer 58, 117–148 (2003)
9. Flanagan, C., Godefroid, P.: Dynamic partial-order reduction for model checking software. ACM SIGPLAN Notices 40(1), 110–121 (2005)
10. Spin Model Checker, The: Primer and Reference Manual

Fitting of Fuzzy Fractal Interpolation for Uncertain Data

Xiaoping Xiao[1], Zisheng Li[2], and Shiliang Yan[1]

[1] Engineering and Technology Center, Southwest University of Science and Technology, Mianyang, 621010 Sichuan, China
xxp8010@126.com
[2] School of Manufacturing Science and Engineering, Southwest University of Science and Technology, Mianyang, 621010 Sichuan, China

Abstract. Tackling of uncertain data is a major problem in data analysis and processing. The fuzzy theory with fuzzy numbers and fractal interpolation is employed to solve the issue of uncertainty. Sample data is used as the kernel of Gaussian fuzzy membership function and its fuzzy numbers are obtained by specifying λ-cut. These fuzzy numbers are used as uncertain data and defined as a new kind of fuzzy interpolation points. With these interpolation points fractal interpolation method is applied to fit curve of sample data. By these definitions, the flow of interpolation approach is given, and example is illustrated to show that a novel interpolation scheme is proposed for manipulating uncertain data.

Keywords: Uncertain data processing, Fuzzy set and fuzzy number, Fuzzy interpolation point, Iterated function system.

1 Introduction

Collected data are affected by noise, data approximation or part missing in experiments. We can't assure whether these data are true in turn. They are uncertain data and their tackling is a major problem in data analysis and processing. Byung [1] applied union operation of fuzzy set and six-neighbor weighted average method to recover A/D converted data. Abbasbandy [2] used global approximation algorithm and center-average defuzzification scheme to interpolate data in solving fuzzy differential equation, and there are other algorithms for interpolation and approximation [3,4]. All of these schemes are either approaches to reduction the errors between precision data and sample data or ways for smoothing fitting curves [5,6,7], but they can't simulate abnormity shape nor stochastic data. Although Chand [8] and Bouboulis [9] etc. applied fractal principle to generate fractal like interpolation surfaces and simulate irregular data, interpolation itself requires curves pass through interpolation points, i.e., interpolation points must be specified unambiguously.

In engineering application, characteristic data of object always not only fluctuates in a certain range, but also are time-dependent. So sample data are reasonable in an interval and vary within it randomly. However a small group of data only describe the properties of object at acquiring time, they doesn't illustrate the overall features of object. In this paper, each one of sample data is applied to the kernel of Gaussian fuzzy membership function, and its fuzzy numbers can be computed by specifying λ. Then the left fuzzy number and right fuzzy number is defined as the upper bound and

L. Qi (Ed.): ISIA 2010, CCIS 86, pp. 78–84, 2011.

lower bound of an interval respectively. Therefore, fuzzy numbers are converted into crisp data, and fuzzy interpolation points are obtained in turn. Then the fractal interpolation is imposed on these points to plot a group of fitting curves to represent the overall features of object.

2 Fuzzy Interpolation Point

2.1 Fuzzy Number

The introduction of fuzzy set theory by Zadeh in 1965's has brought in multi-value logic, which changes values of "yes" or "no" of normal set into more values. Fuzzy set theory is used in pattern recognition, fuzzy inference and fuzzy control etc. natural science domain widely, and its concept had been generalized. Fuzzy number is a particular subset of fuzzy set. A fuzzy set A in real number field R which is called a fuzzy number must satisfy the following conditions

(1) There exist $x \in R$ and $\mu_A(x) = 1$

(2) $\forall \lambda \in (0,1], \{x \mid x \in R, \mu_A(x) \geq \lambda\}$ is a closed interval marked with $[A_\lambda^-, A_\lambda^+]$, i.e. $A_\lambda = [A_\lambda^-, A_\lambda^+]$.

Generally, suppose that the collectivity of the fuzzy numbers is denoted by $\overset{\square}{R}$. If $A \in \overset{\square}{R}$, $\forall \lambda \in (0,1]$ and A_λ is bounded, A is called a bounded fuzzy number. The left fuzzy number and right fuzzy number of A are denoted by L_A and R_A respectively. L_A and R_A are defined as the lower bound and upper bound of the data P in question, there is $P \in [L_A, R_A]$.

Fuzzy number has several forms depending on fuzzy membership function, for example, triangular fuzzy number, trapezoid fuzzy number and Gaussian fuzzy number involved in this paper. Fig.1 is the structure of Gaussian fuzzy number using function $\exp(-(x-x_i)^2 / 2\sigma_i^2)$.

2.2 The Extension Principle

The extension principle proposed by Zadeh in 1975 builds mapping, which is called fuzzy function, between two fuzzy sets. Especially if a mapping acts on two fuzzy sets, each one of which is the Cartesian product of two normal fuzzy sets and its membership grade is determined by two joint variables, it is called fuzzy relation. The extension principle provides standard operations and arithmetic operations for fuzzy sets operation. In this paper, crisp data are mapped into fuzzy system with their own lambda cut, and fuzzy relation is used to build mapping from L_A to lower bound, from R_A to upper bound of the closed interval respectively. Its mapping relation is illustrated by Fig.2.

2.3 The Fuzzy Logic

Fuzzy logic which is close to human logic contains an infinite number of membership values. We use fuzzy logic to reduce, infer and transform fuzzy set in order to

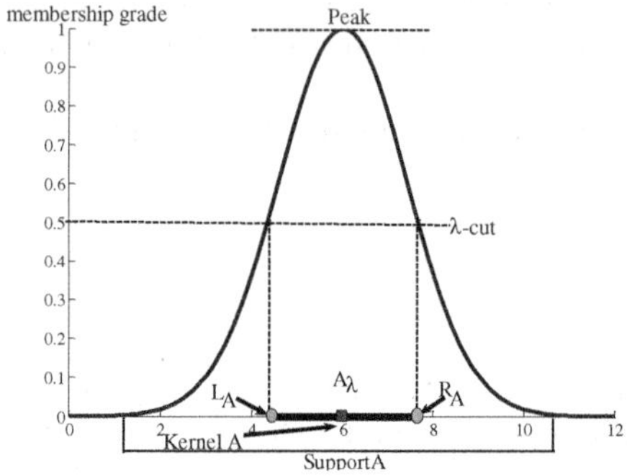

Fig. 1. Structure of Gaussian fuzzy number about "6"

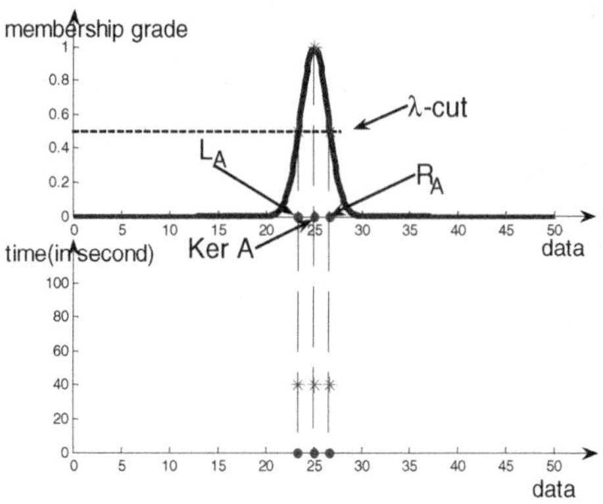

Fig. 2. Mapping of Gaussian fuzzy numbers to data

manipulate data conveniently. By using fuzzy logic, we can manipulate data set which involves the uncertainty condition precisely. In this paper, if there is data P, whose left fuzzy number is L_A and right fuzzy number is R_A, for any data $\tilde{P} \in [L_A, R_A]$, $\tilde{P}$ is a reasonable data, then plots it, otherwise deletes $\tilde{P}$. For example, if we apply fuzzy logic expression "IF…THEN…"to data $\tilde{P}$, we can get the following expression "IF $\tilde{P} \in [L_A, R_A]$ THEN Plot $\tilde{P}$ ELSE Remove $\tilde{P}$".

2.4 The Fuzzy Interpolation Points

In engineering practice, the properties of object are always time-correlated, so the values of data are transient. They can't be used to represent the overall properties of objects. It is inaccurate to master the features of objects only by a small group of data points. Fuzzy interpolation points extend these data to a closed interval by λ-cut, hence there are more data values to illustrate the properties of object. These data can represent object's properties completely in turn. Fig.3 depicts that how the original interpolation data evolves into fuzzy interpolation points.

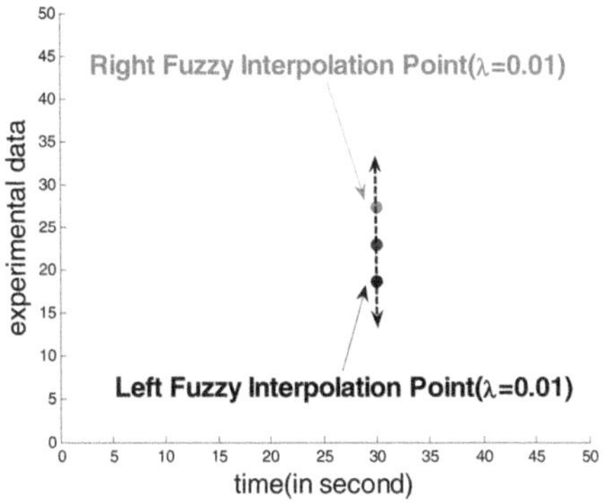

Fig. 3. Evolution of fuzzy interpolation points

3 Fractal Interpolation

3.1 Common Fractal Interpolation

Fractal interpolation is based on the concepts of affine transform and iterated function systems (IFS) [10], which is firstly introduced by Hutchinson in 1981 and popularized by Barnsley and Demko in 1985. It is an easy way to realize the fractal interpolation.

Generally iterated function system $\{W_n:R^2 \to R2, n=1,2\cdots N\}$is a linear fractal interpolation over partition$\Pi = \{x_i\}_{i=0}^N : a = x_0 < x_1 < \cdots < x_{n-1} < x_N = b$ and ordered data serial$\{y_i\}_{i=0}^N$, if

$$W_n \begin{pmatrix} x \\ y \\ 1 \end{pmatrix} = \begin{pmatrix} a_n & c_n & 0 \\ 0 & d_n & 0 \\ e_n & f_n & 1 \end{pmatrix} \begin{pmatrix} x \\ y \\ 1 \end{pmatrix} \quad n = 1, \cdots, N \ . \tag{1}$$

Where W_n is the homogenous affine transform matrix, $|d_n|<1$ is free parameter called vertical scaling factor, which plays a critical role in the implementation of fractal interpolation and affects the shape of curve directly. The coefficients a_n, c_n, e_n and f_n are determined by conditions $W_n(x_0,y_0,1)=(x_{n-1},y_{n-1},1)$ and $W_n(x_N,y_N,1)=(x_n,y_n,1)$. Their equations are defined as follows.

$$a_n =(x_n-x_{n-1})/(x_n-x_0) \ . \tag{2}$$

$$e_n =(x_N\, x_{n-1}- x_0x_n)/(x_n-x_0) \ . \tag{3}$$

$$c_n=[(y_n-y_{n-1})-d_n(y_N-y_0)]/(x_n-x_0) \ . \tag{4}$$

$$f_n=[(x_Ny_{n-1}- x_0y_n)-d_n(x_Ny_0- x_0y_N)]/(x_n-x_0) \ . \tag{5}$$

3.2 Fractal Interpolation Based on Fuzzy Interpolation Points

After the fuzzy interpolation points have been computed, fractal interpolation is imposed on these points and it is easy to implement interpolation in turn. According to the scheme of computing fuzzy interpolation points and IFS, the flow of interpolation algorithm is expressed as follows.

Step 1. Input sample data (t_i, p_i), λ_i and d_i, i=0,1,2…N, starting point (t, p, 1) (t ∈ $[t_0, t_0]$) randomly, matrix D to store the result of affine transform and D = [];

Step 2. Take p_i as kernel, compute L_A^i and R_A^i;

Step 3. Compute a_n, c_n, e_n and f_n on the basis of equations (2)-(5), construct W_n;

Step 4. Specify iterative times L and iterative variable m =1, subscript of W_n denoted by n, set n=1;

Step 5.According to expression (1) compute interpolation point P_n, D = [D P_n], n=n+1; if $n \leq N$ goto step 5; else (t, p, 1) = D;

Step 6. m=m+1, if $m \leq L$ goto Step 5;

Step 7. Plot all the points in D. End of algorithm.

4 Numerical Example

In this paper, one group of sample data with five points are provided to illustrate the constructive process of fuzzy fractal interpolation in detail. The sample data and their Gaussian fuzzy numbers show in table 1, and fractal interpolation of sample data is depicted by Fig.4, while their relation are depicted by Fig.5. Fig.6 is the result of fuzzy fractal interpolation.

Table 1. Sample data and its Gaussian fuzzy numbers (λ_i =0.1)

number	1	2	3	4	5
time(s)	10	20	30	40	50
data	8.5	19.8	34.6	25.4	14.7
L_A	5.4651	16.7651	31.5651	22.3651	11.6651
R_A	11.5349	22.8349	37.6349	28.4349	17.7349

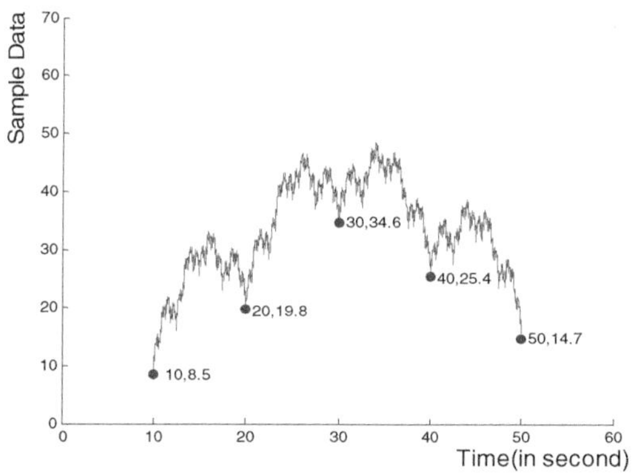

Fig. 4. Fractal interpolation curve of sample data

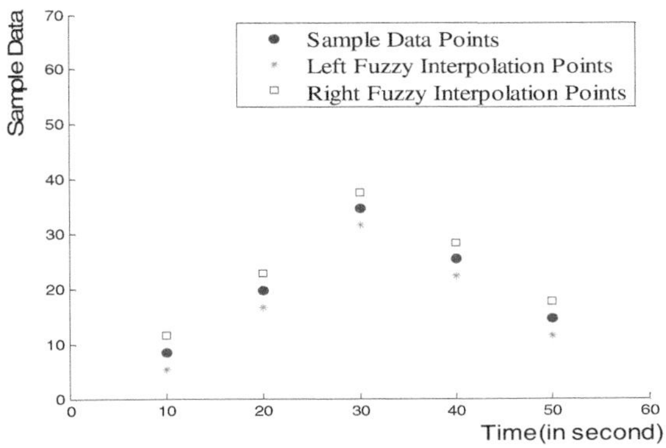

Fig. 5. Fuzzy interpolation of sample data

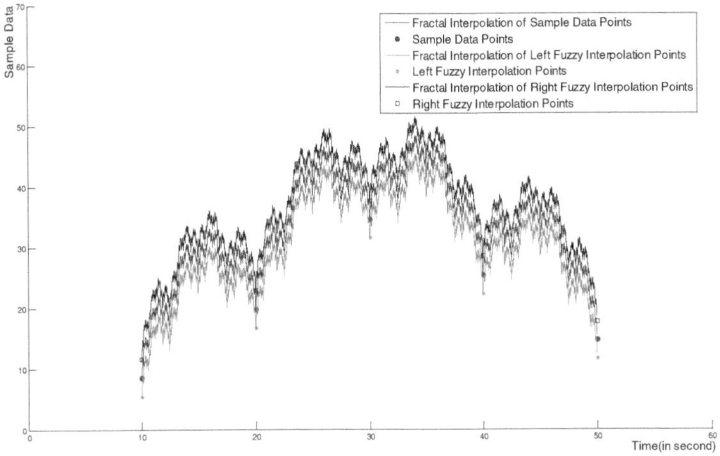

Fig. 6. Fractal interpolation with left and right fuzzy interpolation points

5 Conclusion

There are lots of uncertain data and errors in data acquisition and processing. It is necessary to distinct uncertainty from errors. For the errors assume that true values exist, uncertainty means all the knowledge for their existing can not be mastered, perhaps parts of them are missing or they are in unstable state. Whether those sample data can or not represent the true properties of object is unknown. Fuzzy set and its corresponding principle provide a solution for manipulating these uncertain data. By solving the fuzzy numbers of sample data, fuzzy interpolation points is obtained, then fractal interpolation method is applied to these points, and more than one fitting curves representing overall features of object are achieved. The scheme proposed in this paper can be used to simulate irregular data such as electroencephalogram signal, time-dependent flame temperature etc.

References

1. Byung, S.M.: A Curve Smoothing Method by Using Fuzzy Sets. J. Fss. 96, 353–358 (1998)
2. Abbasbandy, S., Adabitabar Firozja, M.: Fuzzy Linguistic Model for Interpolation. Chaos. Soliton. Fract. 34, 551–556 (2007)
3. Kaleva, O.: Interpolation of Fuzzy Data. Fss. 61, 63–70 (1994)
4. Lowen, R.: A Fuzzy Lagrange Interpolation Theorem. Fss. 34, 33–38 (1990)
5. Fatah, W.A., Jamaludin, M.A., Ahmad, A.M., Osman, A.M.T.: Fuzzy Set in Geometric Modeling. In: International Conference on Computer Graphics, Imaging and Visualization, pp. 227–232. IEEE Computer Society, Washington (2004)
6. Jacas, J., Monreal, A., Recasens, J.: A Model for CAGD Using Fuzzy Logic. Int. J. Approx. Reason. 16, 289–308 (1997)
7. Lodwick, W.A., Santos, J.: Constructing Consistent Fuzzy Surfaces From Fuzzy Data. Fss. 135, 259–277 (2003)
8. Chand, A.K.B., Navascués, M.A.: Natural Bicubic Spline Fractal Interpolation. Nonlinear. Anal.-Theor. 69(11), 3679–3691 (2008)
9. Bouboulis, P., Dalla, L.: Fractal Interpolation Surfaces Derived From Fractal Interpolation Functions. J. Math. Aanl. Appl. 336(2), 919–936 (2007)
10. Barnsley, M.F.: Fractal Functions and Interpolation. Constr. Approx. 2, 303–329 (1986)

Research and Application of Query Rewriting Based on Materialized Views

Yaying Hu[1,*], Weifang Zhai[2], Yulong Tian[2], and Tao Gao[3]

[1] Hebei University Computer Center, Baoding, 071000, China
[2] China University of Geosciences Great wall College, Baoding, 071000, China
[3] Electronic Information Products Inspection Institute of Hebei Province (河北省电子信息产品监督检验院), Shijiazhuang 050071, China
hehyy8080@139.com

Abstract. It is very important to answer queries quickly in database environment. Queries are transparently rewritten using materialized views and the time consumed by queries is reduced by avoiding accessing huge raw records and performing time-consuming operations in order to improve query speed. This paper discussed the extended query rewriting algorithm based on join relation with foreign key, studied Query System Based on Materialized Views for small databases, and proved its validity.

Keywords: Query Rewriting, Materialized View, Foreign Key.

1 Introduction

Nowadays, there are a lot of useful information that are not mined and understood in data warehouse and OLTP system. User's queries often involve a lot of raw data, and may include time-consuming operations such as join and aggregate calculation. So It becomes a serious problem to speed up query response and avoid user's long waiting [1]. Now a good way to improve query speed is rewriting the query using materialized views, but small databases don't have the mechanism of materialized view and query rewriting. Therefore, this paper researches query system based on materialized views so as to use the query rewriting technology in the small database to improve its query efficiency.

2 Query Rewriting Based on Foreign Key

The purpose of rewriting queries using materialized views is to re-organize the user's query. This avoids direct access to the large amount of raw data and time-consuming calculation.

2.1 Concept of Query Rewriting

Shown in Fig 1, the plan is a complete rewriting, meaning the rewriting result contains only materialized view; else if the result also contains the basic tables, then it is called part rewriting.

* Corresponding author.

L. Qi (Ed.): ISIA 2010, CCIS 86, pp. 85–91, 2011.

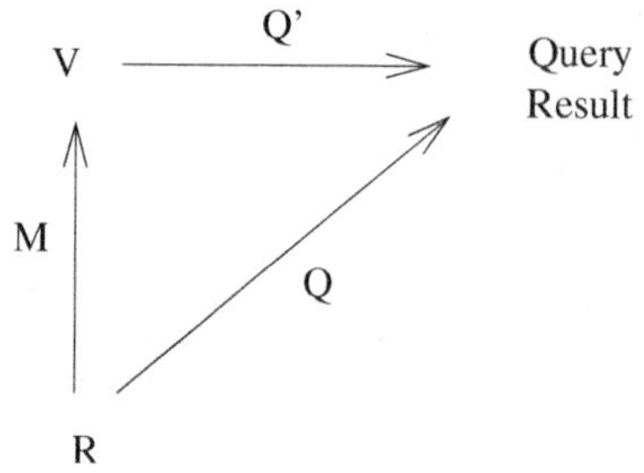

Fig. 1. Query Rewriting. *R* represents the basic table in database, *V* represents materialized view, *Q* represents the user's query for basic table, *M* represents the definition of the materialized view, *Q'* represents rewriting result of *Q* obtained under *M*.

If the query Q' is the rewriting of query Q using materialized views, then satisfy [2]:

1. Q and Q' is on the same in multi-set;
2. Q' contains at least one materialized view.

Query rewriting is conditional. When the query can be derived from materialized views, the rewriting is possible. If the materialized view V is useful to rewrite query Q, then V must be able to replace part of tables or conditions of Q, and the rest of tables and conditions of Q still remain in the rewriting result Q'. Therefore, query rewriting using materialized views must satisfy the following conditions [3]:

a. Tables in materialized view V is a subset of tables in query Q;
b. The output columns and grouping columns of the table replaced and the columns in the compensatory conditions of rewriting should be in V as output columns;
c. Conditions of V can not be stricter than conditions of replaced table;
d. Aggregate output columns must satisfy derivative relations between aggregate materialized views and aggregate columns of query.

If two queries are equal, they should return the same results. Suppose a database contains the following two tables:

```
DEPT(dept_id, dept_name, dept_manager) ;
EMP(dept_id, emp_id, emp_name, salary) ;
```

Materialized view V is as follows:

```
SELECT dept_id, emp_id, emp_name, salary
FROM  EMP
WHERE salary>1500
```

Query Q is as follows:

```
SELECT emp_id, emp_name, salary
FROM  EMP
WHERE salary>3000
```

The rewriting results Q' obtained from Q using materialized view V is as follows:

```
SELECT emp_id, emp_name, salary
FROM  V
WHERE salary>3000
```

2.2 Query Rewriting Algorithm Based on Foreign Key

The primary key of other tables is called foreign key of a table. The table that has a foreign key is a slave table. The table referenced by foreign key is a master table. Foreign key join relation is many-to-one relationship between tables. Master table and slave table are joined with foreign key in query. Query result and slave table have the same number of tuples. In addition, column values of master table can be obtained. Using the characteristic of foreign key several concepts can be introduced to improve the conditions that query rewriting satisfies.

Lossless Join Table: In materialized view or query, if a few tables and conditions that contain them are removed and the number of output tuples is unchanged, then the removed tables are called lossless join table. If a table is in a materialized view but not in a query, and the table is lossless join table of the materialized view, then the materialized view can also be used for rewriting the query.

Extensible Table: In materialized view, master table and slave table are joined with foreign key, and the foreign key is in the output of the materialized view, then if the materialized view and its master table are joined, it can be considered that all the columns of the master table are in the output of the materialized view. This master table in materialized views is called extensible table. If the columns of materialized view, that query requires, are not in the output of the materialized view, but as long as the extensible table contains these columns, they can be available when the materialized view and its extensible table are joined.

Equivalence Classes: In materialized view or query, equal columns set built up according to the equal conditions of columns are called equivalence classes.

Here are examples of lossless join table and extensible table. Still use the database previously mentioned, in which dept_id, a field of table EMP, is a foreign key.

Materialized view V is as follows:

```
SELECT EMP.dept_id, DEPT.dept_name,
EMP.emp_name, SUM (EMP.salary) as MNY
FROM  EMP, DEPT
WHERE EMP.dept_id=DEPT.dept_id AND
DEPT.dept_id>1000
GROUP BY EMP.dept_id, DEPT.dept_name,
EMP.emp_name
```

As EMP.dept_id and DEPT.dept_id belong to the same equivalence class, when removing table DEPT and its corresponding conditions, DEPT.dept_id>1000 can be turned into EMP.dept_id>1000. Therefore, the minimum core table is {EMP}. Table DEPT is a lossless join table.

V can be used to rewrite the query Q as follows:

```
SELECT EMP.emp_name, SUM (EMP.salary)
FROM  EMP
WHERE EMP.dept_id>1000
GROUP BY EMP.emp_name
```

The rewriting result Q' is as follows:

```
SELECT V.emp_name, SUM (V.MNY)
FROM  V
```

On the other hand, EMP.dept_id is in the output of V, so table DEPT is an extensible table. If you want to provide dept_manager information from V, then join V with DEPT:

```
SELECT V.dept_id, V.dept_name, V.emp_name, V.MNY,
DEPT.dept_manager
FROM  DEPT, V
WHERE DEPT.dept_id=V.dept_id
```

In addition to column dept_manager, the query results and V are the same.

Applying the join relation of foreign key to extend the general query rewriting, we can draw the following extended query rewriting steps based on foreign key:

1. To establish table mapping and column mapping for Materialized views and queries;
2. To compute equivalence classes of materialized views, find corresponding output columns of equivalence classes to simplify these equivalence classes of materialized views, and remove columns not involved in mapping;
3. To compute equivalence classes of queries, and find the corresponding output columns of equivalence classes in queries;
4. To test conditions deriving. The compensatory equivalence classes referenced by compensatory conditions are replaced with their corresponding output columns or extensible output columns, as compensatory conditions of the result rewritten with materialized views. If extensible output columns are used, note the corresponding extensible tables;
5. To determine whether output columns and group columns involved in mapping in queries can be derived from the output column of materialized views, and note the extensible tables used;
6. On the SELECT-PROJECT-JOIN materialized views, judge whether the columns involved in aggregation can be derived from the output columns of materialized views. If columns of extensible tables are used, write down these extensible tables;
7. On the aggregate materialized views, see whether the columns involved in aggregation can be derived from the aggregate output columns of materialized views, but columns of extensible tables can not be considered in this case;
8. Tables in mapping are replaced with materialized views, grouping columns and output columns in mapping are replaced with corresponding output columns or extensible output columns of equivalence classes in materialized views, new aggregate output columns are generated, the compensatory conditions are added into rewritten conditions, extensible tables used are also added, other tables and columns of queries remain the same.

Compare to general query rewriting, the extended query rewriting algorithm based on foreign key has improved significantly in availability and ease of realization of materialized views. The introduction of lossless join table and extensible table make some materialized views available that was abandoned by other rewriting methods. The use of equivalence classes and its mapping throughout the entire query rewriting process greatly simplifies the complexity of query rewriting. Deriving conditions test method and compensatory conditions deriving method are also given in the extended query rewriting algorithm.

3 Query System Based on Materialized Views

In view of reducing the cost of development and use, many small and medium applications often use small databases for free. Small database usually could not provide the mechanisms of materialized views and query rewriting; nevertheless the data scale of applications may not be small. When queries contain join and aggregate operations with a large amount of data, query speed problem becomes a bottleneck of the applications.

3.1 Design and Implementation

The aims of query rewriting system based on materialized views are as follows:

Users connect their own database to query system, and enter the desired query with SQL. Query system searches automatically materialized views stored in database in advance. If there are not materialized views in database, users can establish materialized views in database via query system. The relationship of materialized views and query are analyzed automatically with query rewriting mechanism to determine whether the user's query can be derived from the materialized views. If materialized views are usable, the query for basic table will be rewritten to the form for the materialized views and query result will be obtained through materialized views.

The core of query system based on materialized views is the extended foreign key-based query rewriting algorithm. The system model is shown in Fig 2.

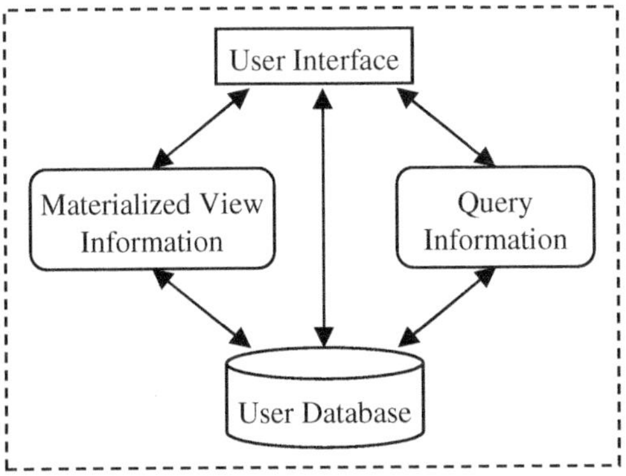

Fig. 2. Query System Model

First of all, user connects the database to query system, system judges whether the materialized view already exists in database. If there is not any materialized view in the database, then system allows the user to input SQL statement to create the materialized view. The system automatically creates the materialized view and analyzes all parts of it and saves the analysis result to the database. Else if there is materialized view in the database, then system extracts the SQL statement of the materialized view for user to modify. System automatically analyzes the revised SQL statement and saves it to the database and updates the materialized view in database. Secondly, Users input SQL statement of query. The system automatically analyzes all parts of the query information, and saves them to the database. Finally, query system extracts the information both of materialized view and query, through analysis and comparison determines whether the materialized view can derive the query. If the materialized view includes all or part of the contents required by the query, then the query will be rewritten with materialized view applying the query rewriting algorithm and the rewriting result will be output. Query system executes the rewritten query on materialized view and basic tables and returns query result to the user.

Query system uses B/S model structure and MySQL database in language Java, in which materialized views can be created, maintained and used. Query system creates temporary table as materialized view in user database. Materialized view updates with thread monitoring and incremental approach.

3.2 Efficiency Analysis

The following experiment is to analyze the query efficiency of query system based on materialized views:

There are three experimental programs: to query in the basic tables, to create a materialized view and query on the materialized view, to query on the materialized view which already exists. The three programs perform separately on databases of different number of records and execute the same query work flow which are 100 queries related to materialized view and containing join and aggregate operations. The average executive times of each query are shown in Table 1.

Table 1. Query Efficiency Comparison Result

Records	Basic Table Query Time	Create Materialized View and Query Time	Materialized View Query Time
20990	51.25 ms	34.25 ms	19.35 ms
41980	55.71 ms	37.34 ms	21.63 ms
83960	61.01 ms	41.26 ms	24.75 ms
167920	68.28 ms	46.61 ms	29.78 ms

In table 1, the experimental results show that the use of materialized views can greatly improve the query efficiency. Compare to query in the basic tables, time to create a materialized view and query is reduced by 32.58% on average, and time to query on the exist materialized view is reduced by 60.8% on average. The average time to create a materialized view is 15.99ms. Along with the increase of records, the time advantage of materialized view is more and more obvious, which proves that the materialized view strategy is effective.

4 Conclusion

The problem of query response speed is an important issue of database application systems. This paper, aiming at small databases, discusses query system based on materialized views, in which foreign key-based query rewriting algorithm as the core query is rewritten using materialized views to avoid direct access to a lot of raw data and time-consuming calculation so as to improve query efficiency. Experimental results demonstrate the correctness of the foreign key-based query rewriting algorithm and the effectiveness of the query rewriting system based on materialized views.

References

1. Grumbach, S., Rafanelli, M., Tininini, L.: Querying Aggregate Data. In: Proceedings of the Eighteenth ACM SIGACT-SIGMOD-SIGART Symposium on Principles of Database Systems, pp. 174–184. ACM Press, New York (1999)
2. Srivastava, D., Dar, S., Jagadish, H.V.: Answering Queries with Aggregation Using Views. In: Proc. of the 22nd VLDB Conf., Mumbai, India, pp. 318–329 (1996)
3. Goldstein, J., Larson, P.-A.: Optimizing Queries Using Materialized Views A Practical, Scalable Solution. In: ACM SIGMOD, Santa Barbara, California, pp. 21–24 (2001)
4. Dobra, A., Garofalakis, M., Gehrke, J.: Complex Aggregate Queries over Data Streams. In: Proceedings of the 2002 ACM SIGMOD International Conference on Management of Data, pp. 61–72. ACM Press, New York (2002)
5. Grumbach, S., Tininini, L.: On the Content of Materialized Aggregate Views. In: Proceedings of the Nineteenth ACM SIGMOD-SIGACT-SIGART Symposium on Principles of Database Systems, pp. 47–57. ACM Press, New York (2000)
6. Cohen, S., Nutt, W., Serebrenik, A.: Algorithms for Rewriting Aggregate Queries Using Views. In: Proceedings of the East-European Conference on Advances in Databases and Information Systems, London, pp. 65–78. Springer, UK (2000)

Research and Implementation of License Plate Character Segmentation Based on Tilt Correction

Weifang Zhai[1,*], Tao Gao[2], Yaying Hu[3], and Yulong Tian[1]

[1] China University of Geosciences Great wall College,
Baoding, 071000, Hebei, China
9136027@qq.com
[2] Electronic Information Products Inspection Institute of Hebei Province
(河北省电子信息产品监督检验院), Shijiazhuang, 050071, Hebei, China
[3] Hebei University Computer Center, Baoding 071000, Hebei, China

Abstract. License plate location and character segmentation are the basis for license plate recognition systems, the accuracy of license plate location and character segmentation directly affect the accuracy of character recognition rate. The article used the polar coordinates transformation method for image tilt correction, and segmented characters accurately through line-by-column searching black points. Experimental results showed that the method is simple and practical, and the accuracy of character segmentation is higher.

Keywords: license plate location; character segmentation; license plate recognition; polar coordinates transformation.

1 Introduction

Along with the enhancement of computer performance and the development of computer vision technology, the technology of Automatic License Plate Recognition has been already mature day by day, and the system has been consummated day by day. But there are still some problems when it is applied in practice. For example, because of the particularity of shooting the car license, which usually causes inclines to the car license. But the inclined license is very difficultly to be divided and recognized, therefore the solution to the inclined license is extremely essential. In addition, when characters are identified only according to the characteristics of each character in the vehicle license plate recognition module, so character segmentation is essential. The function of character segmentation is mainly to split each character from binary license image which has been established. Due to long-term exposure of license plate, license plate portion is likely to be spot on, wear and other interferences. Therefore, this paper has done a pretreatment before partitioning, such as gradient sharpening and noise removal, etc. then through line by line-by-column searching black points, spited the characters from the license plate exactly.

[*] Corresponding author.

L. Qi (Ed.): ISIA 2010, CCIS 86, pp. 92–97, 2011.

2 License Plate Location

License plate recognition system consists of several parts, such as image acquisition, license plate location, license plate character segmentation and database management, etc [1]. License plate image collection is the premise of subsequent processing. After the vehicles image and the background image are obtained by image acquisition system, the license plate image that we are interested in need to be obtained immediately. First of all, the collected vehicles image which include license plate images should be preprocessed before location, including graying, binaryzation, edge detection and median filtering, etc. then we will get a black and white binary image which include prominent information of license plate part, and the rules of plate region showing a texture, as shown in Figure 1,2.

Fig. 1. The image before pretreatment

Fig. 2. The image after pretreatment

There are many kinds of methods of car license image localization. Such as projection, the template matching and the neural network method, etc [2]. From the perspective of simple and positioning accuracy without reducing, In the black and white binaryzation image that has been obtained, according to textural characteristic presented in the car license region and a horizontal direction jump characteristic, we can determined the top and bottom boundary of the car license, based on this, we can obtain the left and right boundary of car license according to the vertical projection. More than 80 automobile image's experiment proved that the algorithm used for accurate positioning plate has a good adaptability.

3 License Plate Tilt Correction

Regarding the inclined car license, its characters are also inevitably inclined, so there is need to make the tilting adjustment to the character, causes the character to be in the identical horizontal position, which is advantageous to character division, and also may improve the accuracy of character recognition. The adjustment algorithm is based on the average height of the black image pixels of the left and right sides , generally speaking, the image that is composed of numerous character, its pixel height of the character of the left and right sides should be at a near horizontal position, if there is a relatively large fluctuations,it shows that the image existence incline, and corrective treatment is needed. Specifically, we must first calculate average height of the pixels of left side and right side of image, and then seek the slope, and the specific process is as follows:

1. Progressive-by-column scan the left side of the image, if we encounter black spots, then counts accumulates its height;
2. Calculate the average height of the left part of image;
3. Progressive-by-column scan right side of the image, if we encounter black spots, then counts accumulates its height;
4. Calculate the average height of the right part of image;
5. According to the average height of the left side and right side of average height, calculate the slope of the image generally.

According to the image inclined slope, we can reorganized the image ,this article used polar coordinate transformation method to realize image revolving, which includes the pixel map of a new image to the old image [3]. If the new image pixel is mapped to the scope of the old image, sets the pixels at new image white. Set image pixel coordinates (x_0, y_0), θ degrees clockwise rotation, the coordinate becomes (x_1, y_1).we can obtain following relationship by Figure 3. Polar coordinate of pixels before rotation is expressed as:

$$\begin{cases} x_0 = r\cos(\alpha) \\ y_0 = r\sin(\alpha) \end{cases} \tag{1}$$

After the rotation θ:

$$\begin{cases} x_1 = r\cos(\alpha-\theta) = r\cos(\alpha)\cos(\theta) + r\sin(\alpha)\sin(\theta) = x_0\cos(\theta) + y_0\sin(\theta) \\ y_1 = r\sin(\alpha-\theta) = r\sin(\alpha)\cos(\theta) - r\cos(\alpha)\sin(\theta) = -x_0\sin(\theta) + y_0\cos(\theta) \end{cases} \tag{2}$$

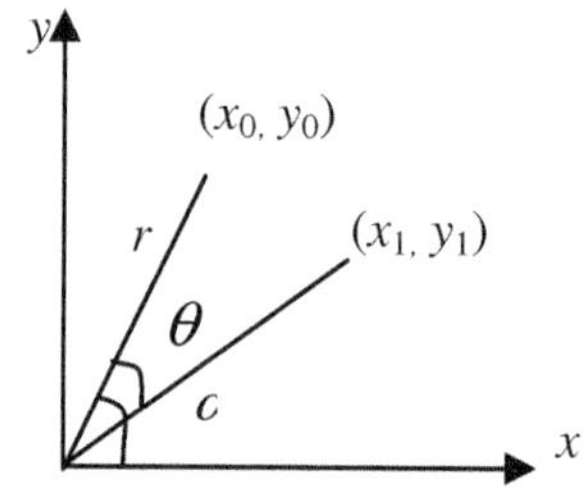

Fig. 3. Schematic diagram of pixel rotation

Matrix expression is:

$$\begin{bmatrix} x_1 \\ y_1 \\ 1 \end{bmatrix} = \begin{bmatrix} \cos\theta & \sin\theta & 0 \\ -\sin\theta & \cos\theta & 0 \\ 0 & 0 & 1 \end{bmatrix} \begin{bmatrix} x_0 \\ y_0 \\ 1 \end{bmatrix} \tag{3}$$

According to the matrix formula above we can introduce following matrix expression:

$$\begin{bmatrix} x_0 \\ y_0 \\ 1 \end{bmatrix} = \begin{bmatrix} \cos\theta & -\sin\theta & 0 \\ \sin\theta & \cos\theta & 0 \\ 0 & 0 & 1 \end{bmatrix} \begin{bmatrix} x_1 \\ y_1 \\ 1 \end{bmatrix} \tag{4}$$

After rotating the image, for each point, according to the formula (4), the corresponding points in the original image would be obtained, then get the corresponding gray value. If the coordinate has surpassed the original image scope, fills by the white. And the coordinates take integer after calculating.

After revolving adjustment, part of experimental result as shown in Figure 4:

Fig. 4. The result of comparation between incline correctiong and after

4 License Plate Character Segmentation

4.1 The Character Image Gradient Sharpening

After localization, the car license after binaryzation and inclined correction processing, regarding black on white, and for no other interference with the license plate image, the vertical projection of all intervals should be zero. However, in practice, the vertical projection of interval place of the characters and character is often greater than zero because of the existence of license plate frame, so break point can not be determined simply according to the projection. Moreover, some of licenses may be left shadow because of light interference in the shooting; and it is possible that there is rivet disturbance. Information on the character of these disturbances will have a negative impact on segmentation. Therefore, it is necessary to carry on the essential pretreatment after the binaryzation, in order to facilitate accurate of character segmentation.

Based on the above, sometimes it is necessary to sharpen the character images to cause the fuzzy image becomes clear; simultaneously it may also play certain elimination role to removing the noise. There are many ways of the image sharpening,

We used the Robert gradient operator to carry on to the image sharpening [4].

Defined as follows: Set point on the original image $f(x, y)$, define $f(x, y)$ in (x, y), the gradient vector for the Department is as follow:

$$G[f(x,y)] = |f(i,j) - f(i+1,j)| + |f(i,j) - f(i,j+1)| \tag{5}$$

Supposes a determination threshold value for Δ, the definition of the image $g(x, y)$ after the change is:

$$g(x,y) = \begin{cases} G[f(x,y)] & (G[f(x,y)] \geq \Delta) \\ f(x,y) & (G[f(x,y)] < \Delta) \end{cases} \tag{6}$$

Through the gradient sharpening, blurred edges can become clear, simultaneously choosing the appropriate threshold value also to be possible to weaken and to eliminate some tiny noises. The fact proved that the gradient sharpening has certain denoising ability, but simultaneously will cause damage to the character edge. So when relatively small characters are in the picture, the gradient sharpening should not be used to the image.

Part of the gradient sharpening results before and after as shown in Figure 5, 6:

Fig. 5. Gradient sharpening before

Fig. 6. After the gradient sharpening

4.2 Miscellaneous Points of Discrete Noise Removal

Although most of the noise the in the license plate image has been removed through the gradient sharpening, but there still be a small part of Miscellaneous points noise existence to affect character segmentation. Therefore, it is necessary to remove the residual noise. We commonly use filter method to denoise sound, such as value filter, average value filter and so on, but this algorithm is not suitable to use in processing goal long and narrow image such as character image, because the filtering process is likely to remove the character itself pixels [5].

In this paper, we used the approach of removal of miscellaneous point to treatment to noise, the specific denoising algorithm is to scan the entire image, when we discover a black spot ,we will make investigation to the number of black points connected with the black points, directly or indirectly connected. If the number is more than a certain value, it shows that the point is not a discrete point; otherwise it is a discrete points, and get rid of it. In addition, we used recursive method to investigate connected black point.

4.3 Character Segmentation Implementation

For the convenience of following work, When we identify characters in the car license recognition module, each character can only to be judged according to its characteristic, therefore the work for character segmentation is needed [6]. This step is to divide characters in the image independently.

The specific algorithm is described as follows:

1. Carry on bottom-up line-by-line scanning to the image until to meet the first black element spot, and record it, then carry on the line-by-line scanning downward until to meet the first black element spot, and we can obtain approximate altitude scope of the image.

2. In this altitude scope, carry on the scanning from left to right again, when meets the first black picture pixel, we can consider that it is the starting position of character segmentation, and then continue to scan until encounters a column that have no black pixel, this column can be considered the end position of character segmentation, then continue to scan to the most right margin of the image according to the above method. And then we obtained quite precise width scope of each character.

3. Within more precise width of each character, according to the first step of the way, respectively downward and bottom-up progressive-scan to get the exact height of each character area. Some experimental results are shown in Figure 7:

Fig. 7. Character segmentation results

Through the 95 license plate characters to detect, the test results are as follows:

Table 1. Character segmentation results

Test records	Correct segmentation	Error segmentation	Accuracy
Plate image(95)	92	3	96.84%

5 Conclusion

Character segmentation is an important part of License plate recognition system, the quality of character segmentation has an immediate influence to the character recognition, this article fully mainly use the apriori knowledge under the characteristic of image itself, selecting appropriate threshold value, vertical projection and so on to separate the character. Through the experiment example analysis, we found that in good condition of car license localization and when the character is in a quite clear situation, the division condition was good, the rate of accuracy has reached 96.84%; but it is not easy for precise segmentation if there is serious pollution, low light conditions, a more serious character adhesion, so it is need to be further studied.

References

1. ter Brugge, M.H., Stevens, J.h., Nijhuits, J.A.G.: License Plate Recognition Using DTCNNs. In: Proceeding of the IEEE International Workshop on Cellular Neural NetWorks and their Application, pp. 212–217 (1998)
2. Chen, Y., Ding, X.: The method of car license localization and character division in complex vehicles image. Infrared and Laser Engineering 33(1), 29–33 (2004)
3. Paolo, C., Paolo, F., Notturno, G.M., et al.: Optical recognition of motor vehicle license plates. IEEE Transactions on Vehicular Technology 44, 790–799 (1995)
4. Zhou, J., Chen, F., Chen, W., Wang, J.: The research and implementation of License plate character segmentation. Computer Engineering 32(5), 238–240 (2006)
5. Chen, L., Huang, X., Wang, M., Li, W.: A car license character division method Based on cluster analysis. Computer Project and Application 38(6), 221–256 (2002)
6. Li, W., Liang, D., Wang, X., Yu, D.: The license plate character segmentation of Quality degradation. Computer Aided Design & Computer Graphics 16(5), 697–700 (2004)

Simulation of Tumor Detection Based on Bioelectrical Impedance Measurement

Yulong Tian[1,*], Weifang Zhai[1], Xinfeng Li[1], Yaying Hu[2], and Tao Gao[3]

[1] China University of Geosciences Great wall College, Baoding, 071000, Hebei, China
[2] Hebei University Computer Center, Baoding, 071000, Hebei, China
[3] Electronic Information Products Inspection Institute of Hebei Province (河北省电子信息产品监督检测院), Shijiazhuang, 050071, Hebei, China
tianyl_001@sina.com

Abstract. Bioelectrical impedance measurement can be used for clinic in early detection and diagnosis of many diseases or organ functional evaluation, has the merits such as non-traumatic, low-cost, continuous monitoring, simple operating and informative etc. By the use of injected alternating current signal and measured voltage data, according to the different conductive characteristics of tissues, we can calculate tissue's bioelectrical impedance value in organism. In this article, the authors developed a simplified tumor model, measured the electrical impedance value changes in the model's different levels, located the tumor's position and size, simulated tumor detection based on bioelectrical impedance measurement, provided theoretical basis for the tumor impedance detection system.

Keywords: bioelectrical impedance, tumor detection, simulation, functional imaging.

1 Introduction

Electrical Impedance Tomography (EIT) is a new functional imaging technology of more efficient and non-traumatic, developed from computer tomography (CT) in recent years. By the use of bioelectrical impedance characteristics in tissues or organs, extracted the related physiological and pathological impedance information, reconstructed the internal structure image of organism. It is essentially injected the current field to be measured in the organism through electrodes, measured the voltage distribution of internal organs, reconstructed the internal resistivity distribution according to a image reconstructing algorithm.

Electrical Impedance Tomography was able to rebuild internal electrical parameters distribution, because biogenic tissue has electromagnetic characteristics. Studies show that electrical characteristics has great difference not only in the normal tissues but also between tissues and lesions. Gabriel measured 30 different tissues, found that permittivity and conductivity in high water tissues(such as muscle or malignant tumors etc.) are high about one magnitude than low water tissues (such as

* Corresponding author.

L. Qi (Ed.): ISIA 2010, CCIS 86, pp. 98–104, 2011.

fat or normal breast tissue). In the microwave band, electrical characteristics of breast cancer and healthy organizations have a very clear difference [1], the impedance of breast tumors is smaller 20 to 40 times than normal tissue. Joines and Surowier experimentally found that contrast of dielectric constant in the breast between malignant tissue and normal tissue is the largest, which help to determine the location of tumor [2]. Thus, the dielectric constant and conductivity varies greatly in normal tissues because of their different moisture content. The permittivity and conductivity difference are very large between malignant tumors and normal tissues. Electrical impedance tomography provides an advantageous basis in the use of electromagnetic technology to detect different tissues, as well as to detect the presence of lesions [3].

2 Tumor Model

2.1 Problem Description and Assumptions

The electrical characteristics in human blood and tissues is a dynamic function of three dimensional space and time, but also involves physical parameter changes, the calculating complexity is self-evident. Biological tissues performance with conductive and dielectric properties in the electric field. The motivating frequency and tissue temperature or moisture can affect dielectric constant and conductivity. In general, the dielectric constant decreased with increasing motivated frequency, conductivity increased with increasing frequency. Biological tissue conductivity and dielectric constant related to the external electric field frequency, biological tissue structure and composition. A variety of tissues have different compositions, conductivity and dielectric constant at the same frequencies also has considerable differences. Water in tissues can affect the dielectric constant, in the high moisture tissues, the dielectric constant is larger than the low moisture tissues. For example, fat and bone are low moisture tissues, skin and muscle are high moisture tissues, heart and brain are the most high moisture organs, so their dielectric constant is also the largest.

In bioelectric research, we should consider how to easablish an equivalent model representing the irregular shape, uneven organisms, suitable for solving the electrical characteristics. Initially researchers used regular shape media to simulate human or animal body. For example, using lossy dielectric sphere or ellipsoid and multi-layers media to simulate human or animal body. The outer layer is on behalf of skin, the inner layer is on behalf of fat, then to the muscles and bones. We can also use simple geometric shape medium combination to simulate complex organisms, using sphere to simulate the head, with different dimensions cylinders to simulate the arm or leg and torso. In order to improve the calculating accuracy, we will try to make the model's shape, dimensions and electromagnetic characteristics better approximation for the various components of organism. An unit accumulated simulation model spaces the human body into many units, each unit's electrical parameters are assigned equivalent value to the real organs. Therefore, this model looks from the outside dimension and the electromagnetic characteristics are similar to the human body. Its equivalent accuracy is related to the unit size, the smaller the size, the higher the equivalent accuracy. With the development of science and technology, now we can use magnetic resonance scanning method to obtain a precise human body model. But this approach requires special equipment to scan a particular person, does not have general [4]. Some simplified model does not affect our deliberations, we analyze and solve problems in the pursuit of a goal is to get correct results, the easy solution, the high computing efficiency. It noted that not only the establishment of simulation models, but also analysis of specific problems, in order to obtain good results and efficiency.

We use simplified model in the article [5]. The electrical parameters of human body are equivalent to 2/3 muscles, regard it as linear organization, tumors (breast cancer) resistivity is smaller than other tissues in modeling and analysis, established of a single organizational model, i.e. human muscle model.

2.2 Material Characteristics

To simplify the analysis, this model has a single material properties of resistivity, three material types are electrode, the whole model and tumor, their resistivity respectively take 2E-8Ω.m, 20Ω.m and 2Ω.m. The area of the entire measuring electrodes is 67*67mm^2, around the electrodes is a 7cm width protecting rings, the area of each electrode is 3*3mm^2, distance between electrode centers is 4mm, the interval of electrodes is 1mm, the protecting ring and electrodes height are 2mm. The entire model is a 10*10*5cm^3 cuboid, internal tumor is a 0.8*0.8*0.8cm^3 cube.

The tumor model is shown by Figure 1.

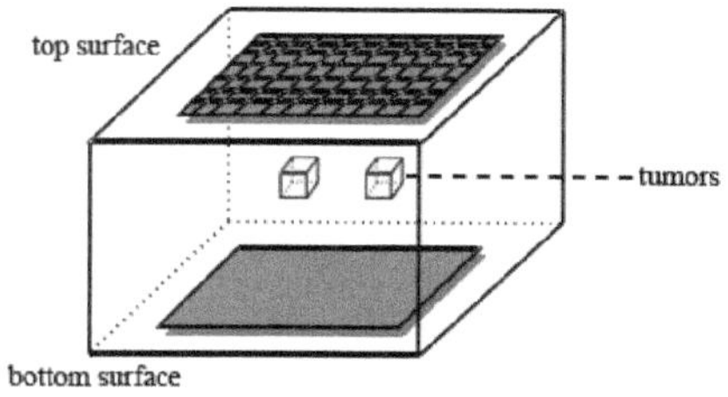

Fig. 1. The sketch of tumor model

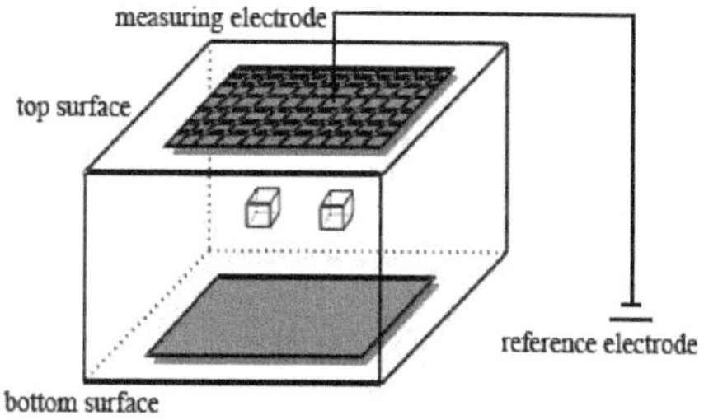

Fig. 2. Model loading current diagram

2.3 Model Loading Current

Loading 5v DC voltage into the measuring electrodes, 0v into reference electrodes, as shown in Figure 2.

2.4 Incentive Method

In overseas, T-SCAN (this system is mainly used for breast cancer detection) using contralateral incentive mode has been applied in clinic [6-9]. That is , when detecting human right breast, the stimulating electrode should be wrung by left hand, using mapped modes to detect breast cancer, the simulation is based on the above conclusions are also using contralateral incentive mode.

3 Simulating and Analyzing

To simplify the simulation, the simulating experiment uses the measuring electrodes values to reflect the location of tumors. In order to obtain sufficient data, we select 16*16 electrodes measurement values. The model is $10*10*5cm^3$ cuboid, The internal tumor is $0.8*0.8*0.8cm^3$. Tumors were placed beneath the measuring electrodes, as shown in Figure 2. Measure the current density values when the distance from the tumors to the model top surface is 1.0cm, 1.5cm, 2.0cm respectively.

3.1 Simulated Data

Selected the current density measured values of the upper surface center 11*11 electrodes as the analysis values. Suppose $I_k n = J_k n - J_0 n$, in the formula, J represents the current density of electrode, k represents the distance form the tumor to the model top surface, respectively takes 1.0cm, 1.5cm, 2.0cm, n respectively takes 1, 2, 3, 4, 5, 6, 7, 8, 9, 10, 11.

3.2 Results Processing and Analyzing

Drawing graphics of experimental data by the use of MATLAB. The measuring electrodes current density difference distribution, k respectively takes 1.0cm, 1.5cm and 2.0cm, as shown by Figure 3,Figure 4, Figure 5.

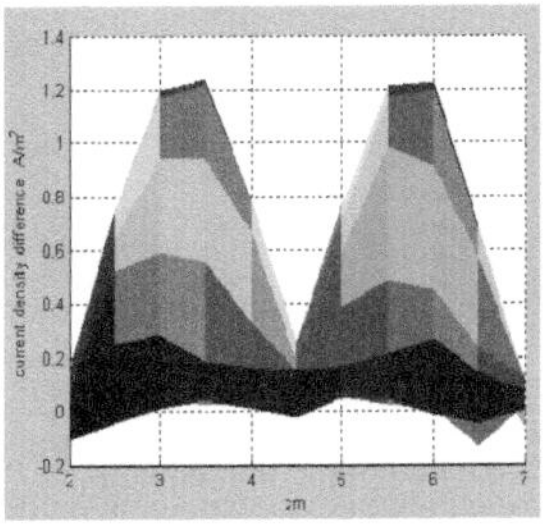

Fig. 3. k takes 1.0cm

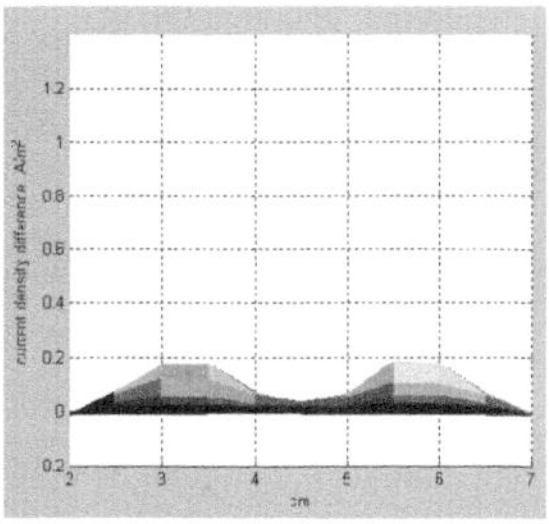

Fig. 4. k takes 1.5cm

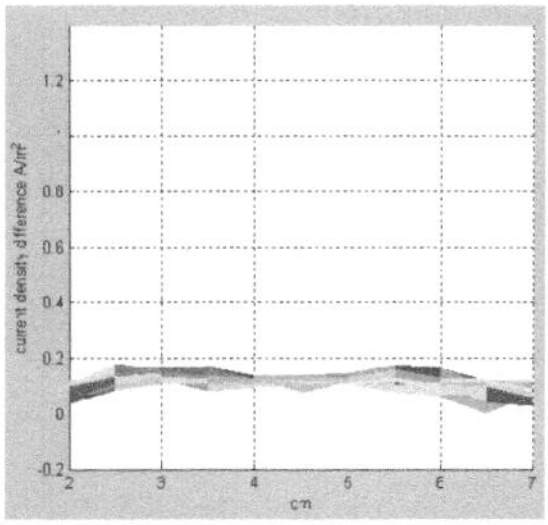

Fig. 5. k takes 2.0cm

As Figure 3,Figure 4, Figure 5 is shown, the distance of the tumor to the model top surface is shorter, the electrodes current density difference is greater, formed the more sharp graphics. The tumor is closer to model top surface, image reflects better, on the other hand, image reflects worse. When the distance is 2.0cm, the effect is already quite evident.

4 Tumor Location Analyzing

The above analysis shows that the current density differences of the tumor locations are greater than others. Therefore, the experimental results can be mapped to a three-dimensional X-Y-Z, established a functional relationship, be evaluated through the function extreme value to determine the tumor coordinates. This experiment using

MATLAB *interp2* function bicubic interpolated. In the interpolated fields, we changed this issue into calculating *interp2* function maximum values to determine the tumor location. Interpolated results are shown by Figure 6, Figure 7 and Figure 8. In order to facilitate comparing and analyzing, amplified the interpolated results when the tumor distance is 2.0cm, as shown in Figure 9.

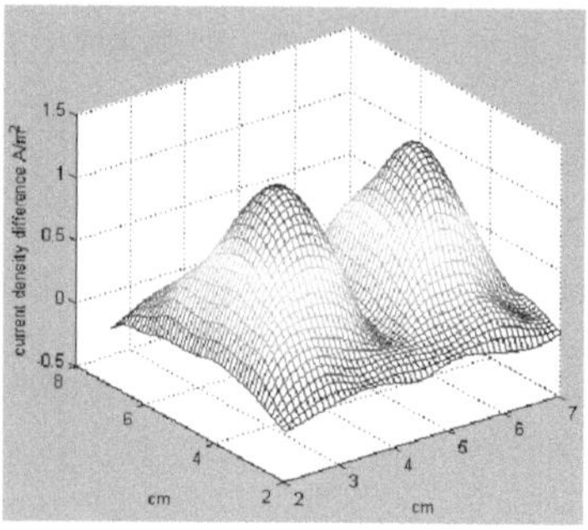

Fig. 6. *Interp2* results when k takes 1.0cm

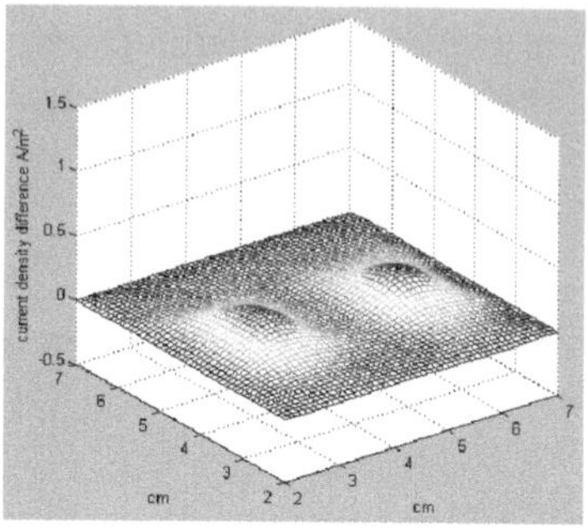

Fig. 7. *Interp2* results when k takes 1.5cm

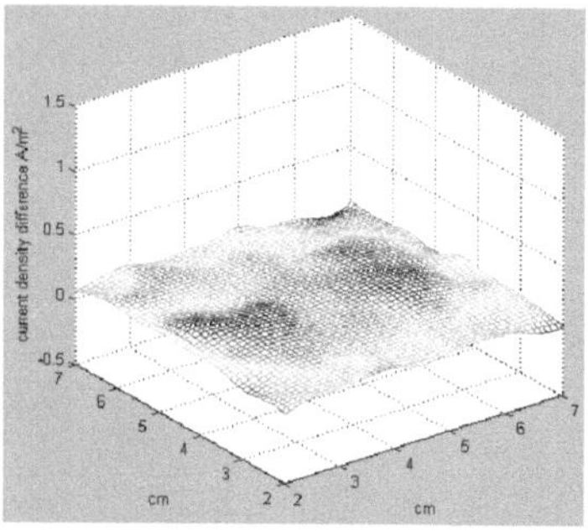

Fig. 8. *Interp2* results when k takes 2.0cm

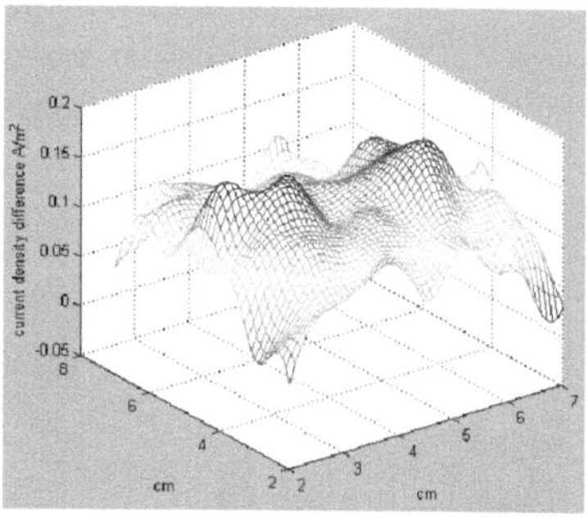

Fig. 9. Enlarged results when k takes 2.0cm

Thus, we changed this issue into calculating multimodal function local maximum values. we can use local search algorithm or genetic algorithm to find the optimal solution, that is the tumor location. Due to genetic algorithm's parallel random search mechanism, single running is not necessarily guaranteed to find the optimum solution for the function. Genetic algorithm can only find one optimum solution at a time, so we should repeatedly execute the genetic algorithm to find more than one point of great values. This experiment's current density difference distribution between the different levels is shown by Figure 9, Figure 11 and Figure 12.

Figure 9, Figure 11 and Figure 12 show that, when the distance takes 1.0cm and 1.5cm, the current density difference distribution is well-distributed, so it's easy to determine the tumor location by genetic algorithm. When the distance takes 2.0cm, the current density difference distribution becomes heterogeneous, applying genetic algorithm to determine the tumor location appeared larger error. This also explains when the distance takes greater than 1.5cm, interference and other reasons will have great effect on the measuring results.

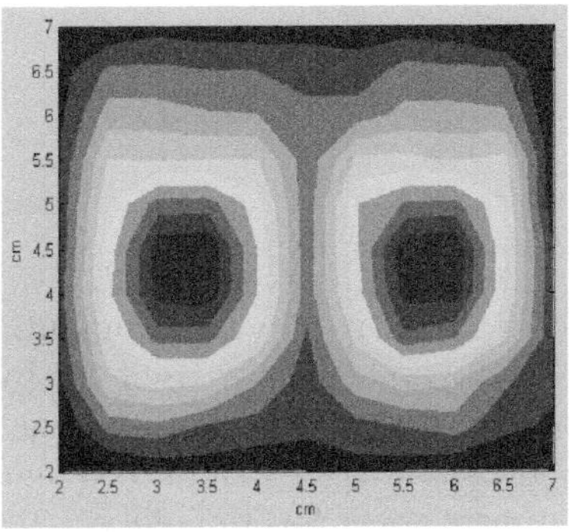

Fig. 10. *k* takes 1.0cm

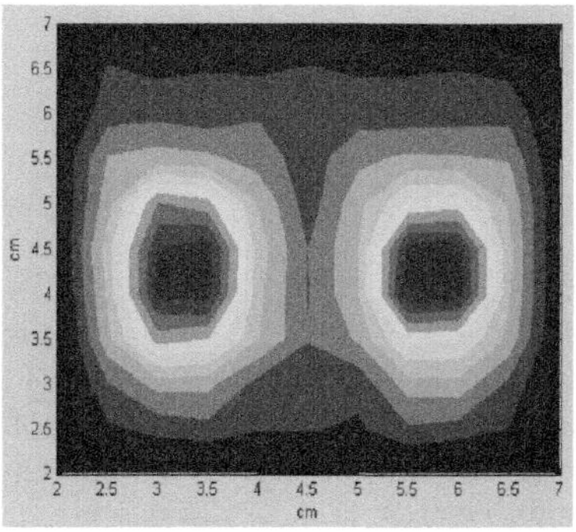

Fig. 11. *k* takes 1.5cm

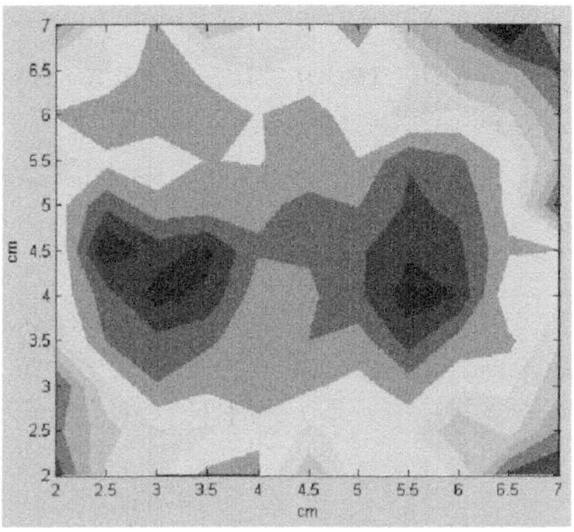

Fig. 12. *k* takes 2.0cm

5 Conclusion

In this article, We established a simplified model, measured the tumor impedance changes between different levels, reflected the tumor position and size by current density difference distribution characteristics, provides theoretical basis to electrical impedance tomography. Compared with CT and ultrasound etc. the EIT imaging resolution is relatively low, but it has the advantage of functional imaging, that is CT or ultrasound can not be compared. EIT provides predictive or forward-looking information, have important clinical implications. So despite the bioelectrical impedance technologies are not mature enough, but it has incomparable advantages to other medical means, as a complementary means of medical imaging, it can be applied to the low resolution requirements.

References

1. Fear, E.C., Hagness, S.C., Meaney, P.M., Okoniewski, M., Stuchly, M.A.: Enhancing breast tumor detection with near-field imaging. IEEE Microwave Magazine, 48–56 (March 2002)
2. Barter, S.J., Hicks, I.P.: Electrical impedance imaging of the breast (TranScan TS 2000): initial UK experience Bdfordshire and Hertfordshire Breast Screening Unit, Loton, UK. In: Symposium Mammographicum 2000, York, UK, October 4-6 (2000)

3. Zhu, S., Lu, S.: Tumor basic theory, vol. 12, pp. 16–20. Shanghai Medical college Publishing house, Shanghai (2005)
4. Qu, L.: The Research and Realize of Electromagnetic Field Finite Element Analysis Technology. Zhejiang University master's degree paper (2002)
5. Zhang, X.: Electromagnetic Analysis of Bioelectric Impedance Measurement and Simulation of Breast-detection. Tianjin University master's degree paper, Tianjin (2006)
6. Seo, J.K., Kwon, O., Ammari, H.: A Mathematical Model for Breast Cancer Lesion Estimation: Electrical Impedance Technique Using TS 2000 Commercial System. IEEE Transactions On Biomedical Engineering 51(11), 1898–1906 (2004)
7. Tang, M., Wang, W., Wheeler, J.: Effects of Incompatible Boundary Information in EIT on the Convergence Behavior of an Iterative Algorithm. IEEE Transactions On Biomedical Engineering 21(6), 6–47 (2002)
8. Krol, A., Coman, I.L.: Inter-Modality Non-Rigid Breast Image Registration Using Finite-Element Method, pp. 1958–1961. IEEE, Los Alamitos (2004)
9. Kao, T.-J., Newel1, J.C., Saulnier, G.J.: Distinguish ability of inhomogene-ities using planar, electrode arrays and different patterns of applied excitation. Physiol. Meas. 24, 403–411 (2003)

A Framework for Embedded Software Testability Measurement

Jianping Fu, Bin Liu, and Minyan Lu

School of Reliability and Systems Engineering, Beihang University, No. 37, Xueyuan Road, Haidian District, Beijing, China
jianping.fu.cn@gmail.com, liubin@china.com, luminyan@vip.sina.com

Abstract. Testability is the ability of software to facilitate testing and expose defects. Nowadays numerous methods for software testability measurement are based on the framework. As the limitation of the framework no existed method can satisfy the measurement requirements of embedded software testability completely. To solve this problem a measurement framework for embedded software testability is proposed. Four classes of universal elements: testability, testability characteristics, affecting factors, and relations are used to construct a steady framework structure. Elements in the framework can be updated according to the development of the software technology. The framework is all-purpose to embedded software testability measurement and an application indicates its feasibility.

Keywords: embedded software, testability, measurement, framework.

1 Introduction

Software testability is a quality factor that attempts to predict how much effort will be required for software testing and to estimate the difficulty of causing a fault to result in a failure. A quantitative value, gotten from the testability measurement, is usually used to evaluate and allocate costs before test execution, help designers decide whether to modify the software design to increase testability, and help project managers determine when to stop testing and release a product.

Many approaches to measure the testability value have been proposed. Pu-Lin Yeh measures the program testability with the data flow graph [1]. Traon uses the information transfer graph to analyze the software testability [2]. Voas and Jin-Cherng Lin calculate the program testability based on the process model of software fault detection [3, 4]. All the methods measure the testability with a framework, such as the data flow graph, the information transfer graph and the fault detection model.

Because of the limitation of the framework there is no one method can satisfy all kinds of requirements for embedded software testability measurement. For example, the method based on the data flow graph [1] or information transfer graph [2] can predict the testing effort, but it cannot evaluate the probability of fault revealing of the embedded software. To solve this problem this paper proposes a general framework to meet the requirements of embedded software testability measurement.

L. Qi (Ed.): ISIA 2010, CCIS 86, pp. 105–111, 2011.

2 Framework Elements

Final destinations of embedded software testability measurement are to predict and allocate test resources, assist the testability design, and help the project decision-making. Therefore three kinds of measures: factors, characteristics, and testability should be gotten from the framework [5]. To calculate one measure from other measures the relationships of these measures should also be included in the framework. Therefore four groups of elements should be contained in the final framework: testability, testability characteristics, affecting factors, and relations.

2.1 Testability

Several definitions for software testability have been published. According to the 1990 IEEE standard glossary [6], testability is the degree to which a component facilitates the establishment of test criteria and the performance of tests to determine whether those criteria have been met. Freedman [7] defined testability as the ease of modifying a program so that it was observable and controllable (observability and controllability). Observability referred to the ease of determining if specified inputs affect the outputs and controllability referred to the ease of producing a specified output from a specified input. Voas and Miller's [8] definition of software testability focuses on the probability that a pieces of software will fail on its next execution during testing (with a particular assumed input distribution) if the software includes a fault. These definitions express the software testability from different views. Synthesizing these definitions we think software testability is the ability of the software to facilitate testing and expose defects.

2.2 Testability Characteristics

Software testability is a composite quality character which consists of several testability characteristics. A testability characteristic is the software feature that has some relationships with software testability and represents one facet of software testability respectively. Observability and controllability [7] are two familiar testability characteristics. Gao divided component testability into five testability characteristics: understandability, observability, controllability, traceability and testing support capability [9]. More testability characteristics are introduced in other papers [10, 11, 12].

2.3 Affecting Factors

Many factors can affect software testability. Binder believes software testability is a result of six high-level factors: characteristics of the representation, characteristics of the implementation, built-in test capabilities, the test suite, the test support environment, and the software development process [13]. Bruce groups the factors which can affect the testability of object-oriented software into three classes: structure factors, communication factors and inheritance factors [14]. Software attributes which can impact software testability are called affecting factor here. The number of methods in a class, the cohesion among methods and the depth in the inheritance tree are all affecting factors.

2.4 Relations

The relations that exist in the testability, testability characteristics and affecting factors are very complex. As seen in Fig. 1: testability is divided into testability characteristics, some affecting factors impact software testability directly, some affecting factors impact testability characteristics, and some affecting factors have relations with other affecting factors.

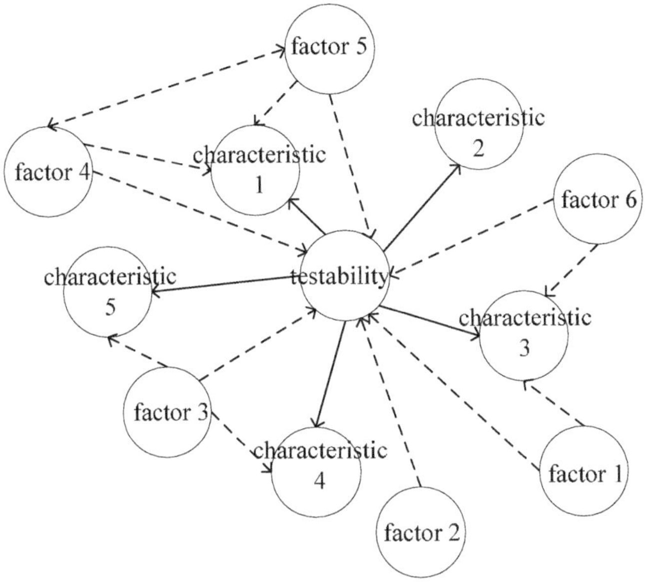

Fig. 1. Relations of testability, testability characteristics and affecting factors

In these relations the relation between testability and testability characteristics and the relation between testability characteristics and affecting factors are more important. The relation between testability and affecting factors can be taken as the composition of the two kinds of relations. At the same time the relation between affecting factors can be weakened by some special methods. Therefore only these two kinds of relations are referred in the framework.

3 Framework Structure

By collecting testability characteristics, affecting factors and relations from lots of embedded software, a measurement framework for embedded software testability is constructed. Fig. 2 expresses the structure of the framework. The top element is the testability. It is divided into nine testability characteristics: understandability, controllability, observability, testing support capability, simplicity, decomposability, applicability, traceability, and sensitivity. Fifty-six affecting factors, listed in Table 1, are in the bottom of the framework. A testability characteristic is impacted by one or more affecting factors.

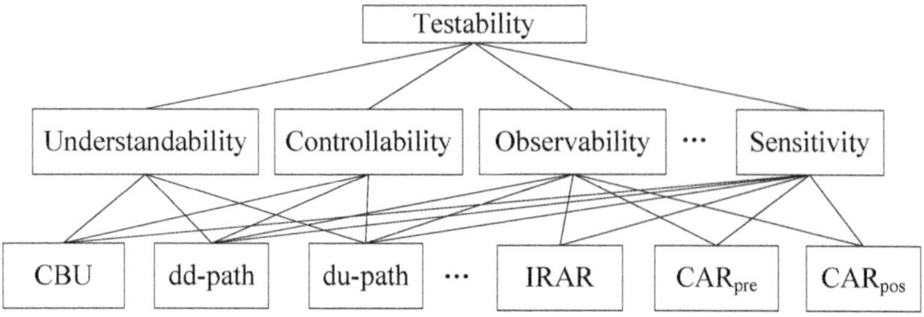

Fig. 2. Testability measurement framework for embedded software

Table 1. Affecting factors of embedded software testability

No.	Factor	No.	Factor	No.	Factor	No.	Factor
1	FPI	15	CODAI	29	NEO	43	NIT
2	NNFR	16	COR	30	AEIDL	44	IP_{min}
3	NSU	17	COD	31	LOEIDS	45	RC
4	CBU	18	FRR	32	AN	46	MCC
5	KLOC	19	ERR	33	LOSVDS	47	ISCC
6	dd-path	20	QRR	34	DPC	48	OP_{min}
7	du-path	21	EUR	35	DIR	49	ISOC
8	CC	22	QUR	36	DN	50	EOI
9	CN	23	ARD	37	DRR	51	FOI
10	SN	24	ARI	38	NFs	52	OFS
11	CoN	25	AUD	39	NFO	53	IRAR
12	CoC	26	AUI	40	NCS	54	CAR_{Pre}
13	Sen	27	URT	41	SOR	55	CAR_{Pos}
14	CORAD	28	NEI	42	SIR	56	Red

4 Framework Features

The framework has two significant features.

4.1 Stability

In the framework, testability and testability characteristics are all general attributes. Affecting factors exist in any embedded software. The relations that testability includes testability characteristics and affecting factor impact testability indirectly by impacting some testability characteristics directly are common, too. With these four kinds of common elements the framework would never change its basic structure in the testability measurement of any embedded software. The stability can guarantee the measurement process and calculation method based on the framework is applicable to any embedded software.

4.2 Expansibility

As the development of the software technology, no framework can be used to measure testability of embedded software without any modification. Although the structure of the framework proposed in this paper is fixed, testability characteristics, affecting factors and relations in the framework are changeable. New elements can be added into the framework and outdated elements can be removed from the framework, too. Therefore the framework is expansible. The expansibility makes the framework is able to meet any requirements of embedded software testability measurement. When the information is enough in the framework the software testability can be measured with the framework directly. Otherwise it is necessary to analyze the software under measurement to expand the framework first and then measure the software testability.

5 Framework Application

This section states how to measure the testability of embedded software with the framework proposed in this paper.

Mission computer application (MC) is an embedded program in an airplane. System testing was planned to execute in the testing phase. In order to arrange the testing staffs and time the testability of MC was needed before executing the test to predict the testing efforts.

According to the features of MC software and its testing conditions, twenty-seven affecting factors were derived from the framework and formed an affecting factor set, AF_{MC}={FPI, NNFR, KLOC, CORAD, CODAI, COR, COD, FRR, ERR, QRR, ARD, ARI, NEI, NEO, AEIDL, LOEIDS, NCS, SOR, SIR, NIT, IP_{min}, RC, MCC, ISCC, OP_{min}, ISOC, EOI}.

Furthermore the set of relations between testability characteristics and affecting factors $R_{affect\text{-}MC}$, the set of testability characteristic TC_{MC} and the set of relations between the testability and testability characteristics $R_{include\text{-}MC}$ were also obtained from the framework. $R_{affect\text{-}MC}$ can be expressed by Table 2, where '+' represents a positive relation between the affecting factor and testability characteristic and '-' represents a negative relation. $TC_{MC}=\{Und, Con, Obv, TSC, Sim, Dec, App, Tra, Sen\}$, where *Und* denotes understandability, *Con* denotes controllability, *Obv* denotes Observability, *TSC* denotes Testing Support Capacity, *Sim* denotes Simpleness, *Dec* denotes Decomposability, *App* denotes Applicability, *Tra* denotes Traceability and *Sen* denotes Sensitivity. $R_{include\text{-}MC}=\{<ST, Und>, <ST, Con>, <ST, Obv>, <ST, TSC>, <ST, Sim>, <ST, Dec>, <ST, App>, <ST, Tra>, <ST, Sen>\}$, where *ST* denotes the software testability. With these elements the testability measurement framework for MC can be established, $Framework_{ST}(MC)=<ST, TC_{MC}, AF_{MC}, R_{include\text{-}MC}, R_{affect\text{-}MC}>$.

All testability characteristics in $Framework_{ST}(MC)$ were calculated by the fuzzy comprehensive evaluation method [15]. Results were listed as follows: $|Und| = 0.8942$, $|Con| = 0.6838$, $|Obv| = 0.7867$, $|TSC| = 0.9875$, $|Sim| = 0.7157$, $|Dec| = 0.5486$, $|App| = 0.9966$, $|Tra| = 0.632$, $|Sen| = 0.6199$. Again using the fuzzy comprehensive evaluation model the software testability of MC was calculated, $|ST| = 0.7694$.

Table 2. Relations between affecting factors and testability characteristics of MC

No.	Affecting factor	Testability characteristic								
		Und	*Con*	*Obv*	*TSC*	*Sim*	*Dec*	*App*	*Tra*	*Sen*
1	FPI					-				
2	NNFR					-				
3	KLOC					-				
4	CORAD	+								
5	CODAI	+								
6	COR	+								
7	COD	+								
8	FRR	+								
9	ERR	+								
10	QRR	+								
11	ARD	+								
12	ARI	+								
13	NEI		-			-				
14	NEO			-		-			-	+
15	AEIDL	+								
16	LOEIDS	-	-			-	-			-
17	NCS		-	-						
18	SOR			+	+			+		
19	SIR		+		+			+		
20	NIT		-	-	-			-		
21	IP_{min}		+		+					
22	RC		+							
23	MCC		+	+	+				+	
24	ISCC		+							
25	OP_{min}			+	+				+	
26	ISOC			+					+	+
27	EOI			+						

6 Summary

A measurement framework for embedded software testability is proposed in this paper. Four classes of elements: testability, testability characteristics, affecting factors, and relations are included in the framework. The structure and features of the framework are analyzed. Stability and expansibility both ensure the framework applicable to any embedded software. In the end an example is given to indicate how to measure the embedded software testability with the proposed framework.

References

1. Yeh, P.L., Lin, J.C.: Software Testability Measurements Derived from Data Flow Analysis. In: 2th Euromicro Conference on Software Maintenance and Reengineering, pp. 96–102. IEEE Press, New York (1998)
2. Traon, Y.L., Robach, C.: From Hardware to Software Testability. In: International Test Conference, pp. 710–719. IEEE Press, New York (1995)

3. Voas, J.M.: PIE: A Dynamic Failure-Based Technique. IEEE Transactions on Software Engineering, 717–727 (1992)
4. Lin, J.C., Lin, S.W.: An Analytic Software Testability Model. In: 11th Asian Test Symposium, pp. 278–283. IEEE Press, New York (2002)
5. Fu, J.P.: Method Research of Software Testability Measurement (in Chinese). Ph.D. dissertation, Beihang University, Beijing, China (2008)
6. IEEE Standard Glossary of Software Engineering Terminology. IEEE Press, New York (1990)
7. Freedman, R.S.: Testability of Software Components. IEEE Transactions on Software Engineering 17(6), 553–564 (1991)
8. Voas, J.M., Miller, K.W.: Software Testability: The New Verification. IEEE Software 12(3), 17–28 (1995)
9. Gao, J.Z., Tsao, J., Wu, Y.: Testing and Quality Assurance for Component-based Software. Artech House, Massachusett (2003)
10. Dssouli, R., Karoui, K., Saleh, K., Cherkaoui, O.: Communication Software Design of Testability: Specification Transformations and Testability Measures. Information and Software Technology 41, 729–743 (1999)
11. Yu, J., Yang, H.Y., Gao, Z.Y., Li, C.Y.: Design for Software Testability (in Chinese). Computer Engineering and Applications, 124–126 (2003)
12. Liu, F.F., Shan, J.H., Jiang, Y.: A Review of Approaches for Software Testability Analysis (in Chinese). Computer Science 32(40), 212–215 (2005)
13. Binder, R.V.: Design for Testability in Object-Oriented Systems. Communication of the ACM, 87–101 (1994)
14. Bruce, W.N.L., Shi, H.F.: A Preliminary Testability Model for Object-Oriented Software. In: International Conference on Software Engineering: Education & Practice, pp. 330–337. IEEE Press, New York (1998)
15. Fu, J.P., Lu, M.Y.: Software Testability Measurement Based on Fuzzy Comprehensive Evaluation (in Chinese). Computer Engineering and Applications 45(27), 69–71, 122 (2009)

Research on the Text Length's Effect of the Text Similarity Measurement

Yan Niu and Yongchao Chen

Computer School, Hubei University of Technology, Wuhan,
Hubei Province, P.R. China
ny@mail.hbut.edu.cn, cyc_hbut@tom.com

Abstract. Similarity measurement plays the fundamental role in the classification of information resources and transmission of network information. According to the research of text-based similarity algorithm on three-layer structure, add the word difference factors to the measurement method of the original text similarity factor, thereby reducing the similarity measurement error resulted by semantics and words difference. The results demonstrate that compare with the improved algorithm and the similarity measurement method base on the original three-layer structure, the measurement accuracy can be improved.

Keywords: semantic; text similarity; text 3-layer structure; words different factors.

1 Introduction

With the development of information technology, and the widely use of a new internet generation, so that make the amount of current information growing explosively. How to remove interference in the network space, and search for objective information effectively, it is the basic factor to make full use of network information resources. Text Similarity measurement applied widely in many areas of information retrieval, data excavation, machine translations and document clustering.

Currently, there are many scholars study at home and abroad are doing research of the text similarity measurement, and a lot of document similarity models have been proposed and widely used: as Nirenburg [1]etc, they proposed the words string matching method, Levenshtein distance [2] and LikeIt method [3]; Lambros [4] etc, they proposed the similarity measurement method according to both of the surface structure and content of the sentences; Gerard Salton and McGill [5] [6] proposed the vector space model in early stage (Vector Space Model, VSM); Carbon ell J, who's proposed the Maximal Marginal Relevance method; Chris HQ Ding [7] adopted the Latent Semantic Indexing Model (Latent Semantic Indexing, LSI), Belkin and Croft [8] proposed the probabilistic model. There are some types of the similarity measurement method are used as followed: vector space model, generalized vector space model, latent semantic indexing model, based attributes method, based Hamming distance measurement method, the reconstruction method of digital based,

L. Qi (Ed.): ISIA 2010, CCIS 86, pp. 112–117, 2011.

all of them based on statistics measured method. That needed the support of large-scale corpus and the long training process, but that has some limitations. Compare with the similarity measurement method based on statistical, the similarity measurement method based on semantic understanding do not need the support of large-scale corpus and long-term training, with the characteristics of high accuracy, these are the main related research: Kim Bo [9] proposed a text-based similarity algorithm based on semantic understanding; Yugang [10] proposed the text-based similarity research based on the words and semantic measurement; the similarity measurement method according to use of synonym forest. And Chewanxiang proposed the sentence similarity measurement method, etc. [11]; use CNK knowledge structure to do the research of similarity measurement, such as Liu Qun, Su-Jian proposed the words similarity and sentence similarity research [12]. At present, the similarity measurement based on the semantic understanding are mostly limited to the range of words or sentences.

In this paper, it based on the text-based semantic similarity algorithm proposed by Kim Bo [9], and research on the effect of text words in the text similarity measurement. We try to solve the problem of different degree text of words coverage rate and semantic similarity. So as to improve the accuracy of similarity measurement based on the semantic understanding.

2 Text Similarity Analysis of Three-Layer Structure Based

The researched texts in the paper are the words in the different degree of paragraph. according to literature [9] proposed the text similarity of three -layer structure, paragraphs are make of sentences, sentences are make up of words. Paragraphs can be broken down into sentences with punctuation marks. Sentences can be broken down into words with segmentalized terms. Set paragraphs as t, set the sentence of the paragraph as s, word which be segmentalized as w, the number of sentences as m, there are two paragraphs: t_1 and t_2.

$$t_1 = (s_{11}, s_{12}, ..., s_{1m}) \tag{1}$$

$$t_2 = (s_{21}, s_{22}, ..., s_{2m}) \tag{2}$$

The sentence of the paragraph can be expressed as:

$$s_1 = (w_{11}, w_{12}, ..., w_{1n}) \tag{3}$$

$$s_2 = (w_{21}, w_{22}, ..., w_{2n}) \tag{4}$$

n is the number of words after the processing of sentence segmentation.

After the paragraph cut into sentences. At first, it should measure the words similarity according to the semantic analysis, then analyses the sentences structure and calculate the sentence similarity, at last, it can get the paragraph similarity according to the relationship between sentences and paragraphs.

In the expression of algorithm, set the two words w_1, w_2, if $w1$ contain n-items s_{11}, s_{12}, ..., s_{1n}, w_2 contains m-items, s_{21}, s_{22}, ..., s_{2m}, ruled the similarity between w_1 and w_2 is the similarity maximum of each semantic item. The similarity of the two words is which said that the two meanings of the similarity.

$$simW(w_1, w_2) = \max_{i=1,\dots,n,\, j=1,\dots,m} = simWS(s_{1i}, s_{2i}) \tag{5}$$

$simW(w_1, w_2)$ refer to the similarity between two words. $simWS(s_{1i}, s_{2j})$ refer to the similarity of two semantic items. The semantic item mean as primitive. According to literature [12], it can get the similarity between two semantic items.

$$simWP(p_1, p_2) = \alpha(d + \alpha) \tag{6}$$

p_1 and p_2 represent two primitive, d is the path length of primitive degree system of p_1 and p_2. It is a positive integer, where the primitive distance between two primitives; α is an adjustable parameter, which means the similarity is the path length of 0.5, accordance to the depth of the primitive, here α can take $\alpha = 1.6$. The similarity between two substantive words is calculated as:

$$simWS(p_1, p_2) = \beta_1 simWP(p_1, p_2) + \sum_{i=2}^{4} \beta_1 simWP(p_1, p_2) \times \beta_i simWP_i(p_1, p_2) \tag{7}$$

The two sentences s_1, s_2, $s_1 = (w_{11}, w_{12}, \dots, w_{1m})$, $s_2 = (w_{21}, w_{22}, \dots, w_2 n)$. Set N_{12} as the sentence s_1, the characteristic of s_2 similarity is followed:

$$N_{12} = N_1 \times N_2^T = \begin{pmatrix} w_{11}w_{21} & \cdots & w_{1m}w_{21} \\ \vdots & \ddots & \vdots \\ w_{11}w_{2n} & \cdots & w_{1m}w_{2n} \end{pmatrix} \tag{8}$$

$$W_{1i}W_{2j} = simW(W_{1i}, W_{2j}) \tag{9}$$

The similarity of sentences s_1 and s_2:

$$simS_1(s_1, s_2) = \frac{1}{k}\sum_{i=1}^{k} simW_{\max i} = \sum_{i=1}^{6} \beta_i simS_i(s_1, s_2) \tag{10}$$

Here β_i is the weight coefficient; its value is selected according to linguistic knowledge and experiments.

The paragraph similarity is composed by the maximum similarity sentences that get the formula:

$$\max L\_S = \{simS_{\max 1}, simS_{\max 2}, \dots, simS_{\max k}\} \tag{11}$$

$$sim(t_1, t_2) = \frac{1}{k}\sum_{i=1}^{k} simW_{\max i} \tag{12}$$

3 Text Coverage and Semantic Similarity

Some conditions often happen in the paragraph similarity measurement: suppose there are three paragraphs *A*, *B*, *C* , and they are *A 50* word, *B 50* word, *C 500* word respectably, according to the measurement method of three-tier structure, because the words number of paragraph A words are fewer, after segmentalized the word in the sentence, it is inclined to find the words in paragraph *A* is included in document *C*, the measure results is the high degree of similarity, that is the long text corpus coverage on the short text corpus. But from the text content, *A* and *B* with *50* words are closed in semantics degree.

To solve this problem, it needs to add the words difference of text to the text similarity measurement. According to the characteristics of Chinese language itself, only substantive words have real meaning, in the process of semantic similarity measurement, it should only consider the role of substantive words, but the function words without specific semantics are also the part of the text, when check the effect of words on similarity measurement, it should take properly consideration into it.

We define a word different factor η, set two paragraphs s_1, s_2, total words number of the paragraph were O_1, O_2, after segmentalized the sentences and words, it can get the content words respectively m, n, according to the characteristic matrix of formula 8, there are:

$$\eta = \frac{O_1}{O_2} \sum_{i=1}^{m+n} simW_{\max i} \tag{13}$$

After added the word different factor η to the text similarity measurement, there are:

$$sim(t_1, t_2) = \frac{1}{k} \sum_{i=1}^{k} \eta simW_{\max i} \tag{14}$$

After add the words factor, the paragraphs similarity measurement shown in Figure 1.

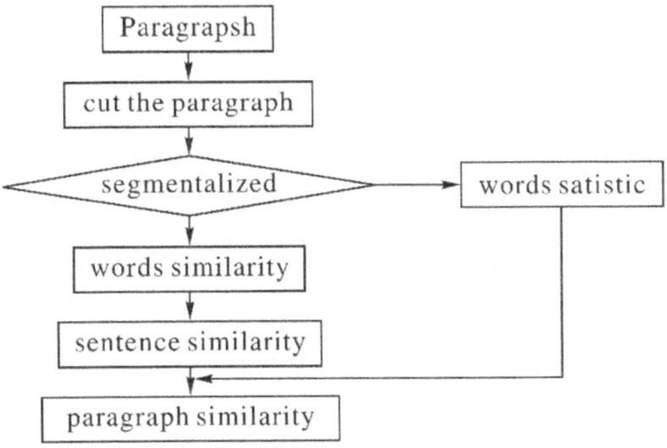

Fig. 1. This the process of paragraph similarity calculation after add the words factor

4 Test and Results Analysis

Using this method, this paper has realized the text semantic similarity measurement module based on the CNKI, with the base of the three -layer structure of original words, sentences and paragraphs, add the function of words in paragraph, it can analysis the distinguish function through experiments to words.

The experiment text corpus adopt the self-built small collection as the test subjects, it include six areas' abstract: economic, legal, news, chemical, computer, physics, astronomy and so on, there are *300* articles in all, they were taken from the published public materials in the journals of related fields , the words of each paragraph generally ranged between *50-500* words. It extracted the material from any two paragraphs when do experiments, used the method *1, 2, 3* to do similarity measurement. Method *1* adopted the word semantic similarity measurement method in literature [10]. Method *2* used word semantic similarity measurement method in literature [12]. Method *3* used the modified text similarity measurement method base on 3-layer structure. The parameters in the formula adopted the parameter values in reference [9]. Here $\alpha = 1.6, \beta_1 = 0.5, \beta_2 = 0.2, \beta_3 = 0.17, \beta_4 = 0.13, \gamma = 0.2, \delta = 0.2$.

Methods of results evaluation reference the evaluation methods of information retrieval [13], the main evaluation indicators include the recall rate (R), precision (P) and so on. Recall rate is the percentage for the correct results of actual identification and the total correct results in the database; precision result is the percentage for the returned results and correct results. The experimental results shown in Table 1, we can see, compare with the traditional vector space model and a simple measurement model of semantic understanding, this article prove that it can increase the recall rate and precision rate after using the improved text structure similarity measurement methods based on three-layer structure.

Table 1. Calculations results of paragraphs similarity

Document	R/%			P/%		
Type	Method 1	Method 2	Method 3	Method 1	Method 2	Method 3
economic	27	33	48	31	61	70
legal	22	35	47	28	60	66
news	32	47	52	42	55	61
chemical	26	37	46	31	41	53
computer	29	38	51	35	44	57
physics	26	39	48	40	48	52
astronomy	19	24	39	39	36	41

5 Conclusion

This paper try to use the CNKI knowledge mode, it base on the text similarity measurement method in the literature [9], and improve the original three-layer structure similarity measurement, then add words different factors to the text similarity measurement, so that to make the similarity measurement of different degree paragraph more accurate .

There are still some issues of the algorithm and need resolved, in different weight classes, especially for the solved aspect of less than *50* words and more than *500* words, it also needs further verification.

References

1. Nirenburg, S., Domashnev, C., Grannes, D.J.: Two approaches to matching in example-based machine translation. In: Proceedings of TMI 1993, Kyoto, Japan, vol. 7, pp. 47–57 (1993)
2. Levenshtein, V.I.: Binary codes capable of correcting spurious insertions and deletions of ones (orginal in Russian). Russian Problemy Peredachi informatsii 1, 12–25 (1965)
3. Peter, N.Y., Kirk, G.K.: The like it intelligent string comparison facility. NEC Institute Tech. Report, 093 (1997)
4. Lambros, C., Harris, P., Stelios, P.: A Matching Technique in Example-based Machine Translation. In: Proceeding of COLING 1994 (1994)
5. Salton, G., Mcgill, M.: Introduction to Modern Information Retrival. McGraw-Hill, New York (1983)
6. Salton, G., Chris, B.: Term Weighting Approaches in Automatic Text. Retrieval Information Processing and Management 24(5), 513–523 (1988)
7. Ding, C.H.Q., He, X., Zha, H.y., Gu, M., Simon, H.D.: A Min-max Cut Algorithm for Graph partitioning and Data Clustering. IEEE, Los Alamitos (2001)
8. Belkin, N., Croft, W.B.: Information filtering and information retrieval, two sides of the same coin. Communications of the ACM 33(12), 29–38 (1992)
9. Yu, J.-l., Zhou, C., Liu, R.-j.: Experimental research on premixed gases explosion in overpressure. Journal of Dalian University of Technology 45(2), 291–297 (2005)
10. Yu, G., Pei, Y.-j., Zhu, Z.-y., Chen, H.-y.: Research of text similarity based on word similarity computing. Computer Engineering and Design 27(2), 241–244 (2006)
11. Che, W., Liu, T., Qin, B., Li, S.: Chinese Sentences Similarity Computation Oriented the Searching in Bilingual Sentence Pair [A]. JSCL, 81–88 (2003)
12. Liu, Q., Li, S.-J.: The words similarity and sentence similarity research. In: Third Chinese Lexical Semantics Workshop TECHNOLOGY, pp. 59–76 (2002)
13. Yu, S., Duan, H., Tian, J.: Machinery Digest principle and implementation of automatic evaluation. In: LeQuan, W. (ed.) Intelligent Computer Interface and Application——Third China Computer Intelligent Interface and Intelligent Applications Conference Technology, pp. 230–233. Electronic Industry Press, BeiJing (1998)

Vertex Distinguishing Total Coloring of Ladder Graphs*

Shitang Bao**, Zhiwen Wang***, and Fei Wen

[1] School of Information Science and Engineering, Lanzhou City University, Lanzhou 730070, P.R. China
[2] School of Mathematics and Computer Sciences, Ningxia University, Yinchuan 750021, P.R. China
[3] Institute of Applied Mathematic Lanzhou Jiaotong University Lanzhou 730070, P.R. China
bst642@sohu.com,
w.zhiwen@163.com,
wenfei1998@126.com

Abstract. *Let G be a simple and connected graph, and $|V(G)| \geq 2$. A proper k-total-coloring of a graph G is a mapping f from $V(G) \bigcup E(G)$ into $\{1, 2, \cdots, k\}$ such that every two adjacent or incident elements of $V(G) \bigcup E(G)$ are assigned different colors. Let $C(u) = f(u) \bigcup \{f(uv)|uv \in E(G)\}$ be the neighbor color-set of u, if $C(u) \neq C(v)$ for any two vertices u and v of $V(G)$, we say that f is a vertex-distinguishing proper k-total-coloring of G, or a k-VDT-coloring of G for short. The minimal number of all over k-VDT-colorings of G is denoted by $\chi_{vt}(G)$, and it is called the $VDTC$ chromatic number of G.* In this paper, we obtain a new sequence of all combinations of 4 elements selected from the set $\{1, 2, \cdots, n\}$ by changing some combination positions appropriately on the lexicographical sequence, we call it the new triangle sequence. Using this technique, we obtain vertex distinguishing total chromatic number of ladder graphs.$L_m \cong P_m \times P_2$ as follows: For ladder graphs L_m and for any integer $n = 9 + 8k(k = 1, 2, \cdots)$. If $\frac{\binom{n-1}{4}}{2} + 2 < m \leq \frac{\binom{n}{4}}{2} + 2$, then $\chi_{vt}(L_m) = n$.

Keywords: Vertex-distinguishing total coloring, Vertex-distinguishing total chromatic number, New triangle, Ladder graphs.

1 Introduction

The coloring problem is one of the most important problems of the graph theory.Because of its theoretical and practical significance, the coloring problem of

* Supported by the NSFC of China (No.10771091) and Science Research Found of Ningxia University (No.(E)ndzr09-15).
** Shitang Bao(1970-), Male, Associate professor.
*** Corresponding author.

L. Qi (Ed.): ISIA 2010, CCIS 86, pp. 118–124, 2011.

graph is one of the primary fields studied by many scholars all over the world. According to computer science, information science, travel net, light transmission and so on, a series of difficult problems are presented, such as vertex distinguishing edge coloring[1-3], adjacent vertex distinguishing edge coloring[4-6], $D(\beta)$ vertex distinguishing edge coloring[7-8], total coloring[9-10], adjacent vertex distinguishing total coloring[11-16], vertex distinguishing total coloring[17], $D(\beta)$ vertex distinguishing total coloring[18], adjacent vertex strong distinguishing total coloring[19]. In this paper, we put forward a method of combinational sequence-New triangle sequence, this can make sure the necessary condition of vertex distinguishing total coloring of ladder graphs.

For a 4−combination from set $\{1, 2, \cdots, n\}$, we can sequence as follows:

$$\begin{array}{c}
123n \quad 124n \quad 125n \quad \cdots \quad 12(n-2)n \quad 12(n-1)n \\
123(n-1) \quad 124(n-1) \quad 125(n-1) \quad \cdots \quad 12(n-2)(n-1) \\
1345 \quad 1346 \\
1356 \\
\cdots \\
13(n-1)n \quad 13(n-2)n \quad 13(n-3)n \quad \cdots \quad 134n \\
145n \quad 146n \quad 147n \quad \cdots \quad 14(n-2)n \quad 14(n-1)n \\
145(n-1) \quad 146(n-1) \quad 147(n-1) \quad \cdots \quad 14(n-2)(n-1) \\
\cdots \\
1456 \\
1567 \\
\cdots \\
1(n-3)(n-1)n \quad 1(n-3)(n-2)n \\
1(n-2)(n-1)n \\
2(n-2)(n-1)n \\
2(n-3)(n-1)n \quad 2(n-3)(n-2)n \\
\cdots \\
2345 \\
3456 \\
3467 \quad 3457 \\
\cdots \\
(n-4)(n-3)(n-2)(n-1) \\
(n-3)(n-2)(n-1)n
\end{array}$$

this sequence is called the New Triangle Sequence of all combinations.

Obviously, the new triangle sequence should include all 4−combinations from set $\{1, 2, \cdots, n\}$, every 4− combination has 4 number, and every number sequence from small to large array. S_i denotes a 4−combination, and be called as *Coloring Meta*, S_{ij} denotes the $jth(j = 1, 2, 3, 4)$ number in S_i. Namely, $S_{i1} < S_{i2} < S_{i3} < S_{i4}$. According to above coloring sequence, we queue by S_{i1}: For integer $n = 9 + 8k(k = 1, 2, 3, \cdots)$, and $S_{i1} < n - 9$. If $S_{i1} \equiv 1(mod 8)$, those combinations be queued in first line; If $S_{i1} \equiv 2(mod 8)$, those combinations be queued in second line; $\cdots$; If $S_{i1} \equiv 0(mod 8)$, those combinations be queued in 8th line. When $S_{i1} \geq n-9$, we queue as 9th line, 10th line, $\cdots$, 15th line. Similarly, we grouped the same coloring sequence by S_{i2}. When $S_{i2} < n-8$, we give a number for those 4-combinations such as first group, 2th group, $\cdots$, 7th group and 8th group circularly. When $S_{i2} \geq n-8$, we make a number for those 4-combinations such as 9th group, 10th group, $\cdots$, 14th group and 15th group in turn. We make every two coloring meta matching from front to back in coloring sequence. In every coloring meta matching, the first coloring meta color vertices u_i, the second color vertices v_i. Furthermore, a pair combinations in inverse 4th coloring meta, and 3 pairs after it be called *Connected Vertex Area.* Other continuous coloring sequence be called *Proper Area.* If there is no proper area after connected vertex area, then we grouped all the coloring meta to connected vertex area. For two coloring meta S_i and S_j in the same proper area, then:

$$S_{i1} = S_{j1}(i \neq j), S_{i2} = S_{j2}(i \neq j);$$
$$\{\{S_{i3}\} \bigcup \{S_{i4}\}\} \bigcap \{\{S_{i+1,3}\} \bigcup \{S_{i+1,4}\} = \{m_i\}.$$

Since $\{m_i\} \neq \emptyset$ of new triangle sequence, so it is easy to know the $\{m_i\}$ unique. Denote $K_i = \{S_{i3}\} \bigcup \{S_{i4}\} - \{m_i\}$, then $K_i \neq K_{i+1}$.

For the sake of convenience, the coloring of vertex u_i, v_i and incident edges be denoted by T_k. The coloring of edges $u_{i-1}u_i$ and $v_{i-1}v_i$ be denoted by T_{k1}; The coloring of vertices u_i and v_i be denoted by T_{k2}; The coloring of edges u_iu_{i+1} and v_iv_{i+1} be denoted by T_{k3}; The coloring of comment edges u_iu_i be denoted by T_{k4}; Where k denotes the inferior index of S_k which corresponding to vertex u_i or v_i. Usually, when coloring for ladder graphs L_m, u_2 be colored by 1th coloring mate, v_2 be colored by 2th coloring mate, the rest may be deduced by analogy.

Definition 1.1. For two paths $P_m = u_1u_2 \cdots u_m$ of order m, and $P'_m = v_1v_2 \cdots v_m$. The ladder graphs L_m are defined as:

$$V(L_m) = \{u_i | i = 1, 2, \cdots, m\} \bigcup \{v_j | j = 1, 2, \cdots, m\}; E(L_m) =$$
$$E(P_m) \bigcup E(P'_m)\{u_iv_i | i = 1, 2, \cdots, m\}.$$

Definition 1.2.[9, 10] Let $G(V, E)$ be a simple graph and k a positive integer. The k-proper total coloring f of the graph $G(V, E)$ is defined as a mapping from $E(G) \bigcup V(G)$ to a set with cardinality k such that:$\forall uv, uw \in E(G), v \neq w$, $f(uv) \neq f(uw)$; and $\forall uv \in E(G), f(u) \neq f(v), f(u) \neq f(uv), f(v) \neq f(uv)$. The minimum integer k that there exists a k-proper total coloring of the graph $G(V, E)$ is called the total chromatic number of graph $G(V, E)$ and denoted by

$$\chi_t(G) = min\{k | there\ is\ a\ k - proper\ total\ coloring\ of\ G\}.$$

Definition 1.3.[15] Let G be a simple and connected graph, and $|V(G)| \geq 2$. A proper k-total-coloring of a graph G is a mapping f from $V(G) \bigcup E(G)$ into

$\{1, 2, \cdots, k\}$ such that every two adjacent or incident elements of $V(G) \bigcup E(G)$ are assigned different colors. Let $C(u) = f(u) \bigcup \{f(uv)|uv \in E(G)\}$ be the neighbor color-set of u, if $C(u) \neq C(v)$ for any two vertices u and v of $V(G)$, we say that f is a vertex-distinguishing proper k-total-coloring of G, or a k-$VDTC$ of G for short, and

$$\chi_{vt}(G) = min\{k|there\ exists\ a\ k - VDTC\ of\ G\}$$

is called the vertex-distinguishing edge chromatic number.

Definition 1.4. Let $G(V, E)$ be a simple graph and n_i the number of vertices with degree i. The combinatorial total degree of graph G is defined as

$$\mu_t(G) = min\{\lambda \Big| \tbinom{\lambda}{i+1} \geq n_i, \delta \leq i \leq \Delta\}.$$

where δ and Δ denote the minimum and maximum degree of graph $G(V, E)$, respectively.

Conjecture 1.1.[9] For a connected graph $G(V, E)$ with $|V| \geq 3$, we have

$$\mu_t(G) \leq \chi_{vt}(G) \leq \mu_t(G) + 1.$$

It is clear that the left hand inequality of the Conjecture is true.

Definition 1.5. If two coloring mate S_{1_i} and S_{2_i}, as well as corresponding to vertex coloring S_{1_i} and S_{2_i} are satisfy:

$(1) S_{1_{i1}} = S_{2_{i1}}, S_{1_{i2}} = S_{2_{i2}}, S_{1_{i3}} = S_{2_{i3}}, S_{1_{i4}} = S_{2_{i4}};$

$(2) T_{2_{ij}} = \begin{cases} T_{1_{i,(4-i)\equiv 4}}, & if\ j \neq 1; \\ S_{2_{i1}}, & if\ j = 1. \end{cases}$ The two coloring method be called Symmetrical Coloring.

According to the new triangle sequence, the even groups and odd groups can be colored by Symmetrical Coloring.

The undefined terminologies and notations in this paper are refereed to references [20]-[23].

2 Conclusion

Since there are $2(2m-2)$ degree vertices of ladder graphs L_m, it is easy to have:

Lemma 2.1. $\chi_{vt}(L_m) \geq \min\{n|\binom{n}{4} \geq 2(m-2)\}$.

Theorem 2.1. For ladder graphs $L_m(38 \leq n < 66\ )$, then $\chi_{vt}(L_m) = 9$.

Proof. When $n = 9$, According to the New Triangle Sequence and correspondence coloring to obtain:

	1239		1259		1279		1238		1258		1278		1247		1267	
3	2	1	5	2	7	1	3	2	5	1	8	2	4	1	7	2
	9		9		9		8		8		7		7		6	
1	4	2	6	1	8	2	4	1	6	2	3	1	5	2	3	1
	1249		1269		1289		1248		1268		1237		1257		1236	
	1246		1235		1234		1356		1367		1347		1368		1348	
2	4	1	3	2	1	3	5	1	6	3	4	1	6	3	4	1
	6		5		4		6		7		7		8		8	
1	5	2	4	1	5	3	4	1	5	3	8	1	5	3	9	1
	1256		1245		1345		1346		1357		1378		1358		1389	
	1379		1359		1459		1479		1458		1478		1467		1567	
1	7	3	1	5	4	1	7	4	5	1	8	4	7	1	5	6
	9		9		9		9		8		7		6		7	
1	6	3	4	1	6	4	8	1	6	4	5	1	4	5	8	1
	1369		1349		1469		1489		1468		1457		1456		1578	
	1568		1579		1679		1678		2789		2689		2569		2589	
6	1	5	7	1	9	7	6	8	7	9	8	2	6	5	9	2
	8		9		6		1		2		6		9		8	
1	9	5	6	1	8	9	7	8	6	7	9	2	7	5	2	6
	1589		1569		1689		1789		2678		2679		2579		2568	
	2578		2456		2457		2468		2489		2469		2349		2369	
2	8	5	4	2	5	4	6	2	8	4	6	2	4	3	6	2
	7		6		7		8		9		9		9		9	
6	5	2	7	4	8	2	5	4	7	2	4	5	2	3	7	2
	2567		2467		2478		2458		2479		2459		2359		2379	
	2389		2358		2378		2357		2346		2345		3467		3478	
2	9	3	5	2	8	3	5	2	3	4	2	3	6	4	7	3
	8		8		7		7		6		5		7		8	
2	4	3	6	2	4	3	6	2	5	3	6	4	5	3	6	4
	2348		2368		2347		2367		2356		3456		3457		3468	
	3458		3479		3459		3579		3568		3567		3689		3789	
3	5	4	7	3	4	5	7	3	6	5	3	6	9	8	3	9
	8		9		9		9		8		7		3		7	
4	9	3	4	6	5	3	8	5	7	3	8	6	7	9	4	8
	3489		3469		3569		3589		3578		3678		3679		4789	
	4679		4678		4578		4589		4569		5689		5789			
9	7	6	8	4	7	5	8	4	5	6	9	5	7	8		
	4		7		8		9		9		8		9			
8	9	6	4	5	6	4	7	5	6	7	5	6	8	7		
	4689		4567		4568		4579		5679		5678		6789			

Obviously, the conclusion is true.

Theorem 2.2. For any integer $n = 9 + 8k(k = 1, 2, 3 \cdots)$ and $\frac{\binom{n-1}{4}}{2} + 2 < m \leq \frac{\binom{n}{4}}{2} + 2$, then

$\chi_{vt}(L_m) = n.$

Proof. When $n = 9$, According to the New Triangle Sequence and correspondence coloring, we use Microsoft Visual C++6.0 to obtain the results. When $n = 25$, partly vertex distinguishing total coloring of ladder graphs as follows:

	1181925		1182125		1182325		1181924		1182124		1182324		1182023		1182223	
19	18	1	21	18	23	1	19	18	21	1	24	18	20	1	23	18
	25		25		25		24		24		23		23		22	
1	20	18	22	1	24	18	20	1	22	18	19	1	21	18	19	1
	1182025		1182225		1182425		1182024		1182224		1181923		1182123		1181922	
	1212224		1212325		1222325		1222324		2232425		2222425		2212225		2212425	
22	1	21	23	1	25	23	22	24	23	25	24	2	22	21	25	2
	24		25		22		1		2		22		25		24	
1	25	21	22	1	24	25	23	24	22	23	25	2	23	21	2	22
	1212425		1212225		1222425		1232425		2222324		2222325		2212325		2212224	
	3212224		3212223		3222425		3232425		4222325		4222324		4212324		4212425	
3	22	21	3	22	25	24	3	25	23	22	24	4	23	21	24	4
	24		23		3		23		4		23		24		25	
21	23	3	24	22	23	25	4	24	25	22	4	21	22	4	23	21
	3212324		3222324		3222325		4232425		4222425		4212223		4212224		4212325	
	451425		451225		451025		45825		45625		56825		561025		561225	
4	14	5	12	4	10	5	8	4	5	6	8	5	10	6	12	5
	25		25		25		25		25		25		25		25	
5	13	4	11	5	9	4	5	7	6	5	9	6	11	5	13	6
	451325		451125		45925		45725		56725		56925		561125		561325	
	7212324		7222324		7222325		8232425		8222425		8212223		8212224		8212325	
21	24	7	23	22	7	23	24	8	25	22	21	8	22	21	23	8
	23		24		25		25		24		23		24		25	
7	21	22	25	7	24	23	8	22	23	8	24	21	25	8	21	22
	7212223		7222425		7232425		8222325		8222324		8212324		8212425		8212225	
	892325		892125		891925		891725		891525		891325		891125		9101125	
8	23	9	21	8	19	9	17	8	15	9	13	8	9	11	10	9
	25		25		25		25		25		25		25		25	
8	22	9	20	8	18	9	16	8	14	9	12	8	10	9	12	10
	892225		892025		891825		891625		891425		891225		891025		9101225	

Obviously, the conclusion is true.

Acknowledgment

The authors would like to thank the supported by the NSFC of China (No. 10771091) and Science Research Found of Ningxia University (No.(E)ndzr09-15).

References

1. Burris, A.C., Schelp, R.H.: Vertex-distinguishing Proper Edge-colorings. J. of Graph Theory 26(2), 3–82 (1997)
2. Bazgan, C., Harkat-Benhamdine, A., Li, H., Woźniak, M.: On the vertex-distinguishing proper edge-coloring of graph. J. of Combine. Theory, Ser. B 75, 288–301 (1999)
3. Balister, P.N., Riordan, O.M., Schelp, R.H.: Vertex-distinguishing coloring of graphs. J. of Graph Theory 42, 95–109 (2003)
4. Zhang, Z.F., Liu, L.Z., Wang, J.F.: Adjacent strong edge coloring of graphs. J. Applied Mathematics Letters 15, 623–626 (2002)
5. Hatami, H.: Δ+300 is a bound on the adjacent vertex distinguishing edge chromatic number. J. of Combinatorial Theory 95, 246–256 (2005)
6. Balister, P.N., Györi, E., Lehel, J., Schelp, P.H.: Adjacent vertex distinguishing edgecolorings. SIAM Journal On Discrete Mathematics 21, 237–250 (2006)
7. Zhang, Z.F., Li, J.W., Chen, X.E., et al.: D(β) Vertex-distinguishing proper edge-coloring of graphs. Acta Mathematica Sinica, Chinese Series 49(3), 703–708 (2006)
8. Akbari, S., Bidkhori, H., Nosrati, N.: r-Strong Edge Colorings of Graphs. Discrete Mathematics 306(23), 3005–3010 (2006)
9. Zhang, Z., Zhang, J., Wang, J.: The total chromatic number of some graphs. Sci. China Ser. A 31, 1434–1441 (1988)
10. Zhang, Z., Wang, J.: A summary of the progress on total colorings of graphs. Adv. in Math (China) 21, 90–397 (1992)
11. Zhang, Z., Chen, X., Li, J., et al.: On adjacent-vertex-distinguishing total coloring of graphs. Sci. China Ser. A 48(3), 289–299 (2005)
12. Wang, H.: On the adjacent vertex-distinguishing total chromatic numbers of the graphs with $\Delta(G) = 3$. Journal of Combinatorial Optimization 14(1), 87–109 (2007)
13. Chen, X.: On the adjacent vertex distinguishing total coloring numbers of graphs with $\Delta = 3$. Discrete Mathematics 308, 4003–4007 (2008)
14. Hulgan, J.: Concise proofs for adjacent vertex-distinguishing total colorings. Discrete Mathematics 309(8), 2548–2550 (2009)
15. Wang, W., Wang, Y.: Adjacent vertex distinguishing total coloring of graphs with lower average degree. Taiwanese J. Math. 12, 979–990 (2008)
16. Wang, Y., Wang, W.: Adjacent vertex distinguishing total colorings of outerplanar graphs. Journal of Combinatorial Optimization (2008)
17. Zhang, Z., Qiu, P., Xu, B., et al.: Vertex-distinguishing total coloring of graphs. Ars Combinatoria. 87, 33–45 (2008)
18. Zhang, Z., Li, J., Chen, X., et al.: D(β)-vertex-distinguishing total colorings of graphs. Science in China Series A: Mathematics 48(10), 1430–1440 (2006) (in Chinese)
19. Zhang, Z., Cheng, H., Yao, B., et al.: On The adjacent-vertex-strongly-distinguishing total coloring of graphs. Science in China Series A: Mathematics 51(3), 427–436 (2008)
20. Zhang, Z., Qiu, P., Xu, B., et al.: Vertex distinguishing total coloring of graphs. Ars Combinatoria 87, 33–45 (2008)
21. Chartrand, G., Linda, L.F.: Graphs and Diagraphs, 2nd edn. Wadswirth Brooks/Cole, Monterey, CA (1986)
22. Hansen, P., Marcotte, O.: Graph coloring and application. AMS providence, Rhode Island USA (1999)
23. West, D.B.: Introduction to Graph Theory, 2nd edn. Person Education, Inc., Prentice Hall (2006)

A New Cluster Based Real Negative Selection Algorithm

Wen Chen, Tao Li, Jian Qin, and Hui Zhao

Department of Computer Science, Sichuan University,
610065 Chengdu, China
{cwcwcwcw2006,qingjianswu}@163.com,
{litao,zhaohui}@scu.edu.cn

Abstract. In this article, based on our previous work CB-RNSA, we proposed an improved algorithm ICB-RNSA: using Principal Component Analysis (PCA) method to reduce data dimension in the data pre-treatment process; the distances between antigens are calculated by fractional norm distance to increase the detection discriminations. The experiment result shows that the efficiency and detection ability of ICB-RNSA are superior to CB-RNSA and other traditional negative selection algorithms.

Keywords: artificial immune; negative selection; detector; cluster.

1 Introduction

Negative selection algorithm (NSA) is an important paradigm for detectors creation in artificial immunology. NSA was proposed by Forrest [1], which simulated the T cell clipping process in the marrow of biological body to create mature detectors which avoid self immune reactive, and these detectors are used in the classification, pattern recognition and intrusion detection, etc [2].

In traditional negative selection algorithms NSA[1], RNSA[2], V-Detector[3] each randomly new created candidate detector has to calculate distance with all of the self data, which leads to very low efficiency [4][5].In our previous work [6] we proposed a negative selection algorithm CB-RNSA that based on the hierarchical cluster of self set, it employed cluster centers to calculate nearest neighbor distances to initial detectors and through random value range to focus the new detectors on the low coverage regions. However, the efficiency of CB-RNSA and the antigen discriminations are much affected by the data dimension. In this article we proposed an improved algorithm ICB-RNSA: first using principle component analysis method to reduce data dimension to improve the efficiency; second applied fractional norm distance to increase the antigen discriminations. This article is organized as: section 2 analyzed the shortage of CB-RNSA; section 3 described the new algorithm; section 4 showed the simulated experiment result and section 5 made conclusion.

L. Qi (Ed.): ISIA 2010, CCIS 86, pp. 125–131, 2011.

2 The Cost of Detectors Creation Process

The total number of initial detectors is used to evaluate the price of detectors creation process in the NSA algorithms [7] [8]. Let p be the probability of a random initial detector is self reactive, reference [1] argued that more initial detectors are needed with bigger p.

For RNSA and V-detector, p is equal to the proportion of total volume of self hyperspheres to the unit hypercube in formula (1):

$$P_1 = \frac{n \cdot V_{self}}{V_{cube}} = \frac{n \cdot r_s^d \cdot \Pi^{d/2}}{\Gamma(d/2+1)}. \tag{1}$$

Where n is the number of self data, V_{self} is self volume, $V_{cube}=1$ is the volume of unit hypercube.As for CB-RNSA, the random vector is confined in the hypercube with edge length equal to 4r, where r is the cluster radius, p is the proportion of total volume of cluster hypershperes to the random value volume in formula (2):

$$P_2 = \frac{m \cdot V_{clu}}{V_{random\_cube}} = \frac{m \cdot r^d \cdot \Pi^{d/2}}{(4 \cdot r)^d \cdot \Gamma(d/2+1)}. \tag{2}$$

To compare the cost of detectors creation process of RNSA [2], V-detector [3] and CB-RNSA [6], by formula (1), (2) we get the scale coefficient ρ in formula (3):

$$\rho = \frac{P_1}{P_2} = \frac{n \cdot r_s^d \cdot \Pi^{d/2}}{\Gamma(d/2+1)} \Bigg/ \frac{m \cdot r^d \cdot \Pi^{d/2}}{(4 \cdot r)^d \cdot \Gamma(d/2+1)} = \frac{n}{m}(4r_s)^d. \tag{3}$$

From formula (3) get $\rho = (\sqrt[d]{\frac{n}{m}} \cdot 4r_s)^d$, when ρ is bigger than 1, the price of detectors creation of CB-RNSA is smaller than traditional methods. As fig.1 shows, just in the low dimensional data space (d<15) and rs>0.01 the efficiency of CB-RNSA is better.

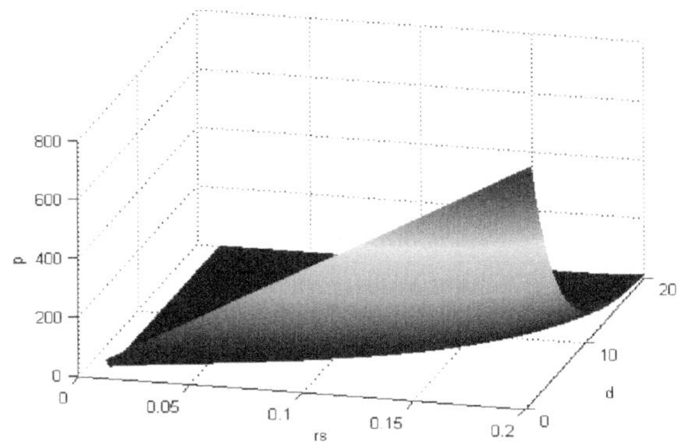

Fig. 1. The variation curve of scale coefficient ρ and data dimension d, n/m=1000

3 The Description of ICB-RNSA

As the analyzed result of section 2, when the data dimension is higher than 15, the cost of CB-RNSA will be bigger than other NSA algorithms, so data pre-treatment process is needed to reduce data dimension when deals with high dimensional data. We employed Principle Component Analysis (PCA) to reduce dimension:
$x_1, x_2 \ldots x_p$ is the native attributes of vectors, $z_1, z_2 \ldots z_m$ $(m \le p)$ is new attributes

Step 1: select n antigen data samples $x_1, x_2 \ldots x_n$ • $x_i = (x_{i1}, x_{i2} \ldots x_{ip})$, $1 \le i \le n$

Step 2: calculate correlation coefficient matrix R by formula (4)

$$R_{ij} = \frac{\sum_{k=1}^{n} (x_{ki} - \bar{x}_i)(x_{kj} - \bar{x}_j)}{\sqrt{\sum_{k=1}^{n} (x_{ki} - \bar{x}_i)^2 \sum_{k=1}^{n} (x_{kj} - \bar{x}_j)^2}} . \tag{4}$$

Step 3: calculate the eigenvalue of R: $\lambda_1 \ldots \lambda_d$, *and* $\lambda_1 \ge \lambda_2 \ldots \ge \lambda_d$

Step 4: select the first m eigenvector as principle components whose accumulative contribution rate is more than 85%

As the reference [7] pointed out that in the high dimensional data space, all of the data points have the nearly same distance:

$$\lim_{d \to \infty} E\left[\frac{D\max x - D\min x}{d^{(1/k)-(1/2)}} \right] = C_k . \tag{5}$$

From formula (5) we can see Dmax x-Dmin x is increasing with $d^{(1/k)-(1/2)}$, which inspires us to use fractional norm distance in formula (6) to increase the data discrimination

$$\mathrm{dis}(x,c)=\sqrt[l]{\sum_{i=1}^{d}|x.fi-c.fi|^{l}},0<l<1. \tag{6}$$

As in reference [3], if the detector coverage is bigger than expected coverage while $z > z\alpha$ in formula (7), where $Z\alpha$ is the Z score for a confidence level of $1-\alpha$ for a standard normal distribution, then exit the detector creation process.

$$z=\frac{m}{\sqrt{np(1-p)}}-\sqrt{\frac{np}{1-p}}. \tag{7}$$

We call the improved CB-RNSA algorithm as ICB-RNSA which employed PCA method to treat high dimensional data and fractional norm distance:

```
range: the value range of detectors
C(1): the cluster centers in level 1
p: the expected nonself coverage
r_1: the cluster radius of level 1
Step 1: for each cluster level 1
Step 2: x=random (range)
Step 3: for each c in C(1)
dis(x,c) is calculated by formula(3) where c, x⊆R^n
            ⎧ true, go to step2, if dis(x, c) ≤ r_1
drop(x) =   ⎨
            ⎩ false, put x in Td, otherwise
Step 4: for each k in detector set D, n=n+1
```

$$dis(x,\ k.c)=\sqrt[l]{\sum_{i=1}^{d}|x.fi-k.c.fi|^{l}}$$

```
if dis(x,k.c) ≤ r_k then m=m+1 end
Step 5: if n! = N then go to step2 end
      Z is calculated by formula (7)
      If z > zα then 1=1+1 goto step1
      Else copy x from Td into D, goto step2
     end
   end
```

In step2 the new created candidate detectors is restrained in the random value range, which reduced a lot of redundancy detectors compared with creating detectors in the value range $[0, 1]^d$ [2] [3]. In step3, the cluster centers replaced self data to calculate distance with candidate detector x, the time complexity of this step is O(*m_numofclu*), compared with O(|S|) of the traditional methods. As the number of cluster centers *m_numofclu* is far less than the size of self data set |S|, so the efficiency of negative selection could be improved. In step4, checks weather x is covered by detectors in D, and the time complexity is O(|D|).In Step5 when the sampling times equals to N and $z > z\alpha$, then exit the detector creation process in level l. Because the coverage changes as new detectors are added into D, so we keep the detector set unchanged during the sampling process. As mentioned above, the time complexity of ICB-RNSA is O(k(avgclu +|D|)), avegclu is the average number of clusters in each level, k is the number of recursion and k= $\left\lfloor \log_2 \frac{1}{rs} \right\rfloor$, rs represents the self radius.

4 Experiment

In the comparative experiment, we use four types of negative selection algorithms: CB-RNSA, ICB-RNSA, V-Detector and RNSA to train detectors based on three groups of UCI data: Abalone Data Set (ADS), Breast Cancer Wisconsin (BCW) and Spambase Data Set (SDS).

The data characters are described in tab.1. All of the data is normalized into $[0, 1]^d$ space and data with class label 1 is self antigens and others is non-self antigens. The four algorithms are used to create detectors with the expect coverage 90% and self radius $r_s = 0.03$, and the experiment is run 20 times. The recorded average number of initial detectors is shown in tab.2.

Table 1. UCI data attributes

data set	attribute number	type	dimension	class
ADS	4177	real, integer	8	1 for male, 0 for others
BCW	699	integer	9	1 for malignant, 0 for others
SDS	4601	real, integer	56	1 for spam, 0 for not spam

From tab.2 we can see, on the data sets ADS and BCW, when the data dimension is low (d<10), the required number of initial detectors of CB-RNSA and ICB-RNSA are less than RNSA and V-Detector. While the dimension increased to 56, the efficiency of CB-RNSA is poor and needs more initial detectors than traditional algorithms but ICB-RNSA always needed the least number of initial detectors as the PCA method has reduced the data dimension, and the fractional norm distance makes the antigen discrimination more clear to reduce the number of invalidate detectors. The result shows that ICB-RNSA is suitable for the high dimensional applications.

Table 2. The number of initial detector

dataset	NSA	Initial detector number		ratio of self reactive detectors	
		avg (10 thousand)	sd (10 thousands)	avg (%)	sd(%)
ADS	RNSA	17.09	1.306	95.6	17.5
	V-Detector	6.351	0.579	87.4	9.32
	CB-RNSA	2.165	0.372	70.3	5.9
	ICB-RNSA	1.79	0.11	67.2	5.5
BCW	RNSA	3.462	0.423	91.62	19.35
	V-Detector	2.309	0.272	84.31	11.7
	CB-RNSA	1.995	0.137	68.3	5.2
	ICB-RNSA	1.87	0.12	66.5	3.9
SDS	RNSA	110.975	12.369	99.993	0.03
	V-Detector	83.693	6.931	99.92	0.06
	CB-RNSA	396.242	49.683	99.9996	0.01
	ICB-RNSA	43.6	1.2	90.3	0.01

5 Conclusion

This article presents an improved algorithm of CB-RNSA to improve the poor algorithm efficiency in high dimensional data space. In ICB-RNSA, the PCA method is used to choose less number of new data attributes to form new antigens, while most of the classification characters are preserved. On the other hand, the fractional norm distance was used to increase the discrimination between antigens to reduce the number of self reactive detectors. The experiment result showed that ICB-RNSA is suitable for high dimensional applications and superior to other methods.

Acknowledgments. This work is sponsored by National Natural Science Foundation of China (No.60873246) and the National Research Foundation for Doctoral Program of Higher Education of China (No. 20070610032).

References

1. Forrest, S., Perelson, A.S., Allen, L., Cherukuri, R.: Self-nonself discrimination in a computer. In: Proceedings of the IEEE Symposium on Research in Security and Privacy, pp. 202–212. IEEE Press, Washington (1994)
2. Gonzalez, F., Dasgupta, D.: Anomaly detection using real-valued negative selection. Genetic Programming and Evolvable Machine 4, 383–403 (2003)
3. Ji, Z., Dasgupta, D.: Real-valued negative selection algorithm with variable-sized detectors. In: Proceedings Genetic and Evolutionary Computation Conference (GECCO), pp. 287–298. IEEE Press, Berlin (2004)
4. Stibor, T., Mohr, P., Timmis, J.: Is negative selection appropriate for anomaly detection? In: Proceedings Genetic and Evolutionary Computation Conference (GECCO), pp. 321–328. ACM Press, New York (2005)

5. Li, T.: Computer immunology. Publishing House of Electronics Industry, Beijing (2004)
6. Chen, W., Li, T.: A negative selection algorithm based on hierarchical clustering of self set. In: Proceedings of CNCE, pp. 50–53. IEEE Press, Qingdao (2010)
7. Stibor, T., Timmis, J., Eckert, C.: On the use of hyperspheres in artificial immune systems as antibody recognition regions. LNCS, vol. 41, pp. 215–228 (2006)
8. Stibor, T., Timmis, J., Eckert, C.: A comparative study of real-valued negative selection to statistical anomaly detection techniques. In: Jacob, C., Pilat, M.L., Bentley, P.J., Timmis, J.I. (eds.) ICARIS 2005. LNCS, vol. 3627, pp. 262–275. Springer, Heidelberg (2005)

Quantitive Evaluation on the Preservation of Polarimetric Information for PolSAR Filters

Niu Chaoyang, Sheng Guangming, Ma Debao, and Zhang Junhua

Zhengzhou Information Science and Technology Institute, Zhengzhou, China
nuecy@126.com

Abstract. Several rules are presented in the paper to evaluate quantitively the preservation performance of polarimetric information for PolSAR filters. These rules comprise the matching coefficient, the difference plot, and the mean square errors between the polarization signatures of the original PolSAR data and those of the filtered data. They are tested in terms of experimental PolSAR data to show how they can be exploited in evaluating the preservation ability of polarimetric information. Results make evident that the rules more efficient to describe the maintainability in polarimetric information for speckle filters compared to polarization signatures.

Keywords: quantitive evaluation; polarimetric information; PolSAR filter; polarization synthesis; polarization signature.

1 Introduction

Speckle must be carefully addressed and filtered from PolSAR (Polarimetric Synthetic Aperture Radar) data to grant access to the information of interest. Speckle filtering techniques for PolSAR data are normally designed to achieve optimality in three following criteria. The first is to maintain the spatial resolution and the spatial details for textured scenarios of PolSAR images. The second is to reduce the speckle noise in the homogeneous areas. The third is to avoid the information mixing from areas with different polarimetric content. A deficient reduction of speckle noise may produce errors on the estimated information [1, 2], which appear on the retrieved physical information [3].

For the third criterion, a speckle filter is desired to well preserve the polarimetric information for PolSAR data. The tool used to evaluate the preservation of polarimetric information is the polarization signature. The polarization signature has, as a basis, the important concept of polarization synthesis; by means of which, it is possible to synthesize the received power for any pair of transmitting and receiving polarizations [4, 5]. In order to evaluate the preservation performance of polarimetric information for a PolSAR filter, we need to observe visually and find how the shape of polarization signature for the filtered data is different from that for the original data. This subjective way is so arbitrary that it would not distinguish in detail the difference between two similar polarization signatures.

L. Qi (Ed.): ISIA 2010, CCIS 86, pp. 132–138, 2011.

The objective of this paper is to evaluate quantitively the preservation ability of polarimetric information for PolSAR filters. Section 1introduces the basic concepts of polarization synthesis and polarization signature. Section 3 presents several rules to evaluate quantitively the preservation ability of polarimetric information for PolSAR filters. These rules comprise the matching coefficient C, the difference plot D, and the mean square errors M between the polarization signature of the original PolSAR data and that of the filtered data. They are tested in terms of experimental PolSAR data in Section 4 to show how they can be exploited in evaluating the preservation ability of polarimetric information for PolSAR filters. A final discussion and conclusions close this paper in Section 5.

2 Polarization Synthesis

For PolSAR, the individual power measurements for each radar resolution cell are related only statistically; therefore, polarization synthesis is usually expressed in terms of the Kennaugh matrix [***K***] [4, 6, 7]:

$$p(\chi_r,\psi_r,\chi_t,\psi_t)=\kappa \boldsymbol{J}_r^{\mathrm{T}}(\chi_r,\psi_r)[\boldsymbol{K}]\boldsymbol{J}_t(\chi_t,\psi_t), \tag{1}$$

where κ is a factor relating with the antenna gain function, the permittivity and permeability of free space, $\boldsymbol{J}_r$ and $\boldsymbol{J}_t$ are the Stokes vectors describing the polarizations of the receiving and transmitting antennas, $\chi_{r(t)}$ and $\psi_{r(t)}$ are ellipticity and orientation angles which describe the polarizaton ellipse of the receiving (transmitting) antennas [8], and the definitions of the Kennaugh matrix [***K***] in the linear basis is [7]

$$[\boldsymbol{K}]=[\boldsymbol{U}][\boldsymbol{R}][\boldsymbol{W}][\boldsymbol{R}]^{-1}. \tag{2}$$

Assuming that a normalized electric field is radiated, $\boldsymbol{J}_r$ and $\boldsymbol{J}_t$ are given by

$$\begin{cases}\boldsymbol{J}_r(\chi_r,\psi_r)=(1,\cos 2\chi_r\cos 2\psi_r,\cos 2\chi_r\sin 2\psi_r,\sin 2\chi_r)^{\mathrm{T}}\\ \boldsymbol{J}_t(\chi_t,\psi_t)=(1,\cos 2\chi_t\cos 2\psi_t,\cos 2\chi_t\sin 2\psi_t,\sin 2\chi_t)^{\mathrm{T}}\end{cases}. \tag{3}$$

In the frame of (1), two different power measurements are defined. One is the co-polarized power p_{CO}, assuming that the transmitting and receiving antennas are characterized by the same polarization state, i.e. $\psi_r=\psi_t=\psi$ and $\chi_r=\chi_t=\chi$. The other is the cross-polarized power p_X in the case that the receiver antenna receives with the orthogonal polarization of the transmitting system, i.e. $\psi_r=\pi-\psi_t=\pi-\psi$ and $\chi_r=-\chi_t=-\chi$. The two power measurements are evaluated respectively by:

$$p_{CO}(\chi,\psi)=\kappa \boldsymbol{J}_r^{\mathrm{T}}(\chi,\psi)[\boldsymbol{K}]\boldsymbol{J}_t(\chi,\psi), \tag{4}$$

$$p_X(\chi,\psi)=\kappa \boldsymbol{J}_r^{\mathrm{T}}(-\chi,\pi-\psi)[\boldsymbol{K}]\boldsymbol{J}_t(\chi,\psi). \tag{5}$$

The information contained in (4) and (5), which depends on the elliptiticy and the orientation angle of the transmitting and receiving antennas, is generally displayed, as tri-dimensional figure called polarization signature [6, 10, 11]. Although many targets can produce similar plots, the plot shapes provide information about the scattering mechanism dominant from the target [12]. Polarization signatures are very useful to describing polarization properties of terrains for PolSAR images, and have been successfully employed to evaluate the preservation of polarimetirc information for PolSAR filters [4, 13-16].

3 Measuring the Preservation of Polarimetric Information

This chapter presents several rules to evaluate quantitively the preservation performance of polarimetric information of a speckle filter for PolSAR data. These rules comprise the matching coefficient *C*, the difference plot *D*, and the mean square errors *M* between the polarization signature of the original PolSAR data and the one of the filtered data.

3.1 Matching Coefficient

Setting p_1 and p_2 are respectively the polarization signature of the original PolSAR data and the one of the filtered data, the matching coefficient is introduced as

$$C = \frac{\iint\left[p_1(\chi,\psi)-\bar{p}_1\right]\left[p_2(\chi,\psi)-\bar{p}_2\right]d\chi d\psi}{\sqrt{\iint\left[p_1(\chi,\psi)-\bar{p}_1\right]^2 d\chi d\psi}\cdot\sqrt{\iint\left[p_2(\chi,\psi)-\bar{p}_2\right]^2 d\chi d\psi}} \tag{6}$$

where, $\bar{p}_1$ and $\bar{p}_2$ are respectively the average power of two data sets, which are calculated by

$$\begin{cases} \bar{p}_1 = \dfrac{2\kappa}{\pi^2}\iint p_1(\chi,\psi)d\chi d\psi = \dfrac{2\kappa}{\pi^2}\iint \boldsymbol{J}_r^{\mathrm{T}}[\boldsymbol{K}_1]\boldsymbol{J}_t, d\chi d\psi \\ \bar{p}_2 = \dfrac{2\kappa}{\pi^2}\iint p_2(\chi,\psi)d\chi d\psi = \dfrac{2\kappa}{\pi^2}\iint \boldsymbol{J}_r^{\mathrm{T}}[\boldsymbol{K}_2]\boldsymbol{J}_t, d\chi d\psi \end{cases}. \tag{7}$$

The magnitude range of *C* is obviously [0, 1], and the unity holds if the shape of p_1 perfectly matches that of p_2. In fact, with the error and disturbance, matching coefficient is always between 0 and 1. For PolSAR filters, the preservation performance of polarimetric information is evaluated by the matching coefficient of polarization signatures of the filtered data and the original data. The bigger value of the matching coefficient means a better performance in maintaining the polarimetric information for a PolSAR filter.

3.2 Difference Plot

The second measure to evaluate the preservation performance of polarimetric information for PolSAR filters is the plot of the difference between two polarization signatures p_1 and p_2. It is defined as:

$$D(\chi,\psi) = p_1(\chi,\psi) - p_2(\chi,\psi) = \kappa \boldsymbol{J}_r^{\mathrm{T}}([\boldsymbol{K}_1]-[\boldsymbol{K}_2])\boldsymbol{J}_t. \tag{8}$$

The difference plot can show more details of the difference between two polarization signatures. We can obviously see the diversity between them. So it indicates in all around how a filter maintains the polarimetric information of PolSAR data.

3.3 Mean Squared Error

The difference between the polarization signatures p_1 and p_2 can also be found by using the mean-squared error of the difference above, which is calculated by

$$\begin{aligned} M &= 10\lg\left[\frac{2}{\pi^2}\iint\left(p_1(\chi,\psi)-p_2(\chi,\psi)\right)^2 d\chi d\psi\right] \\ &= 10\lg\left[\frac{2\kappa^2}{\pi^2}\iint\left(\boldsymbol{J}_r^{\mathrm{T}}([\boldsymbol{K}_1]-[\boldsymbol{K}_2])\boldsymbol{J}_t\right)^2 d\chi d\psi\right]. \end{aligned} \tag{9}$$

The *MSE* can reflect integrally the difference between polarization signatures, while the difference plot shows in detail the diversity between them.

3.4 Summary

These three rules above can describe the relationship between two polarization signatures from different points of view. The matching coefficient C presents the similarity between them as a whole, the difference plot D shows in a plot the details of the difference related to the ellipticity angle χ and the orientation angle ψ, and the mean squared error M addresses the errors caused by PolSAR filters in polarization signatures. Further more, in one case that p_1 of the original PolSAR data and p_2 of the filtered data are both the co-polarized power p_{CO}, C_{CO}, D_{CO}, and M_{CO} can be derived; in the other case that p_1 and p_2 are both the cross-polarized power p_X, C_X, D_X, and M_X are obtained.

4 Simulation and Analysis

Simulation experiment in terms of experimental SAR data is carried out in the chapter to present how the rules above are exploited to evaluate the preservation ability of polarimetric information of PolSAR filters. The Lee's adaptive and polarimetric filter [13] is tested, which is proposed by Jong-Sen Lee in 1999. The experimental SAR data is acquired by the Danish airborne EMISAR [17] on April 17, 1998. For this study, only the L-band data has been applied. The measured area is located near Research Centre Foulum in the central part of Denmark, and comprises the main types of terrains. The forest field and the crop field, respectively labeled as Z1 and Z2, have been selected from the area, as shown in Fig. 1(a). The data has been processed with Lee's filter by a 7×7 sliding window, and the result image is given in Fig. 1 (b).

The polarization signatures calculated and plotted from the filtered data are compared with those of the original data. Fig. 2 and Fig. 3 show these polarization signatures respectively for two fields in 3-D. Each field exhibits nearly identical 3-D plots, and minor differences are not observed. So the preservation of polarization signatures in general is good.

(a) (b)

Fig. 1. Filtering of EMISAR image by Lee's filter. (a) The Pauli-decomposed image of the original data. (b) The Pauli-decomposed image of the filtered data.

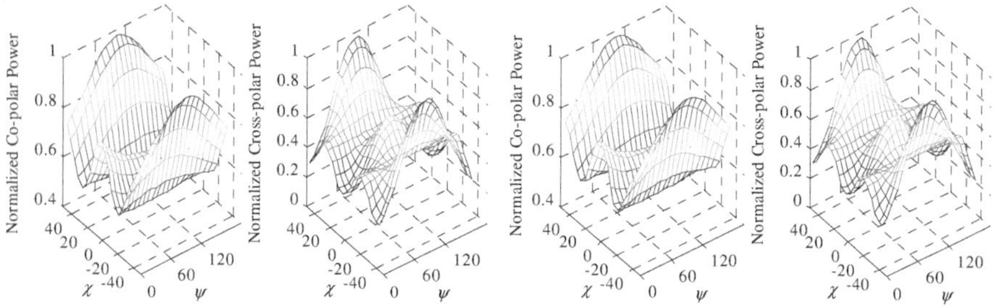

(a) (b)

Fig. 2. Polarization signatures of Z1. (a) Original polarization signatures. (b) Filtered polarization signatures.

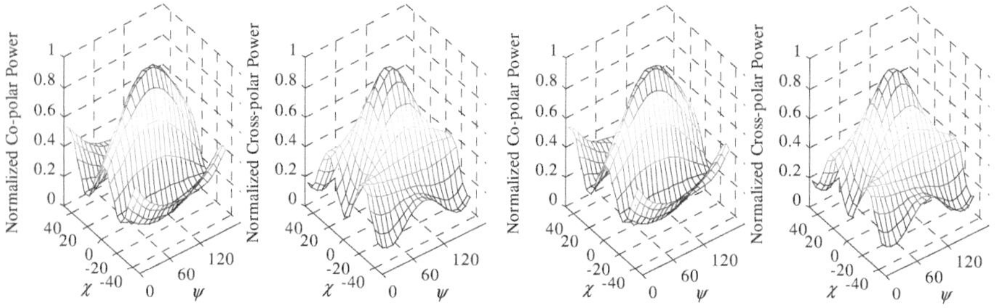

(a) (b)

Fig. 3. Polarization signatures of Z2. (a) Original polarization signatures. (b) Filtered polarization signatures.

Table 1. gives matching coefficients and mean squared errors of polarization signatures for two fields. C_{CO} and C_X in Table 1. are both greater than 99.9% for both fields. So Lee's filter is good at the preservation of polarimetric information for them.

Also, it seems that the polarimetric information of Z1 is better preserved than Z2 since the former has larger matching coefficients than the later. Mean squared errors M_{CO} and M_X show a reverse result that both of errors for Z1 are larger than Z2. In fact, mean squared errors for Z1 and Z2 are both neglectable since they are below -27dB.

Table 1. Statistics of Polarization Signature for Two Fields

Fields	Matching Coefficients		Mean Squared Errors (dB)	
	C_{CO}	C_X	M_{CO}	M_X
Z1	0.9999	0.9999	-27.6955	-29.2082
Z2	0.9994	0.9997	-61.8849	-60.5547

Fig. 4 gives difference plots of the polarization signature for two fields. These difference plots in Fig. 4 obviously show where the difference is large and where it is small, and the shapes are very similar to polarization signatures of the original data and the filtered data. Compared to them of Z2, difference plots of Z1 both have higher values everywhere.

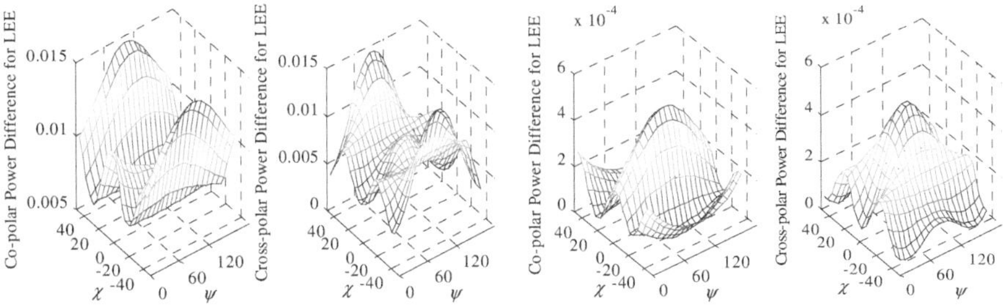

(a) (b)

Fig. 4. Difference Plots of the Selected Fields. (a) Difference Plots of Z1, (b) Difference Plots of Z2.

5 Conclusion

The matching coefficient *C*, the difference plot *D*, and the mean square errors *M* between the Polarization signatures of filtered sets and those of the original, are derived and compared with the corresponding Polarization signatures. The minor differences are not observed from the polarization signatures, while the corresponding *C*, *D* and *M* show obviously the differences. The results make evident that the rules more efficient to describe the maintainability in polarization signatures for speckle filters.

References

1. Touzi, R., Lopes, A., Bruniquel, J., Vachon, P.W.: Coherence estimation for SAR imagery. IEEE Trans. Geosci. Remote Sens. 37(1), 135–149 (1999)
2. López-Martínez, C., Pottier, E., Cloude, S.R.: Statistical assessment of eigenvector-based target decomposition theorems in radar polarimetry. IEEE Trans. Geosci. Remote Sens. 43(9), 2058–2074 (2005)
3. López-Martínez, C., Hajnsek, I., Lee, J.S., Pottier, E., Fàbregas, X.: Polarimetric speckle noise effects in quantitative physical parameters retrieval. Proc. Inst. Electr. Eng.—Radar Sonar Navig. 153(3), 250–259 (2006)
4. Van Zyl, J.J., Zebker, H.A., Elachi, C.: Imaging radar polarization signatures: Theory and applications. Radio Sci. 22(4), 529–543 (1987)
5. Touzi, R., Boerner, W.M., Lee, J.S., Lueneburg, E.: A review of polarimetry in the context of synthetic aperture radar: Concepts and information extraction. Can. J. Remote Sens. 30(3), 380–407 (2004)
6. Zebker, H.A., van Zyl, J.J.: Imaging Radar Polarimetry: A Review. Proc. of the IEEE 79(11), 1583–1606 (1991)
7. Guissard: Mueller and Kennaugh matrices in radar polarimetry. IEEE Trans. Geosci. Remote Sens. 32, 590–597 (1994)
8. Ulaby, F.T., Sarabandi, K., Nashashibi, A.: Statistical properties of the Mueller matrix of distributed targets. IEE Proc. F 139, 136–146 (1992)
9. Guissard: Phase calibration of polarimetric radars from slightly rough surfaces. IEEE Trans. Geosci. Remote Sens. 32, 712–715 (1994)
10. van Zyl, J.J.: Imaging radar polarization signatures: Theory and observation. Radio Sci. 22(4), 529–543 (1987)
11. Evans, L.D., Farr, T.G., van Zyl, J.J., Zebker, H.A.: Radar polarimetry: analysis tools and applications. IEEE Trans. Geosci. Remote Sens. 26(6) (1988)
12. McNairn, H., Duguay, C., Brisco, B., Pultz, T.J.: The effect of soil and crop residue characteristics on polarimetric radar response. Remote Sens. Environ. 80, 308–320 (2001)
13. Lee, J.S., Grunes, M.R., De Grandi, G.: Polarimetric SAR speckle filtering and its impact on terrain classification. IEEE Trans. Geosci. Remote Sens. 37(5), 2363–2373 (1999)
14. Lee, J.S., De Grandi, G., Grunes, M.R., et al.: Polarimetric Signature Preservation in SAR Speckle Filtering. In: IGARSS 1996, pp. 1574–1576 (1996)
15. Gu, J., Yang, J., Zhang, H., Peng, Y., Wang, C., Zhang, H.: Speckle Filtering in Polarimetric SAR Data Based on the Subspace Decomposition. IEEE Trans. Geosci. Remote Sens. 42(8), 1635–1641 (2004)
16. López-Martínez, C., Fàbregas, X.: Model-based polarimetric SAR speckle filter. IEEE Trans. Geosci. Remote Sens. 46(11), 3894–3907 (2008)
17. Christensen, E., Skou, N., Dall, J., Woelders, et al.: EMISAR: An absolutely calibrated polarimetric L- and C-band SAR. IEEE Trans. Geosci. Remote Sens. 36(6), 1852–1865 (1998)

A New Method for Knowledge Acquisition from Incomplete Information System Based on Rough Set

E. Xu[1,2], Wang Quantie[2], Sun Fuming[1], and Ren Yongchang[3]

[1] Electronic & Information Engineering School, Liaoning University of Technology, Jinzhou 121001, Liaoning Province, China
[2] Liaoning Engineering and Professional Technology Institute, Tieling 112000, Liaoning Province, China
[3] Institute of Information Science and Engineering, Bohai University, Jinzhou 121001, Liaoning Province, China
exu21@163.com

Abstract. Knowledge acquisition is an important research area of knowledge discovery database and machine leaning, which includs knowledge reduction and knowledge extraction from large number of original data. Researchers in these fields are very interested in this new research topic since it offers opportunities to discover useful knowledge in information systems.Many algorithms demand information system must be complete. To deal with the problem in an incomplete information system, this paper proposed a method based on rough set theory. Based on tolerance relationship, the concept of tolerance relationship similar matrix via using an extension of equivalence relationship of rough set theory are defined in incomplete information systems. It calculates the core attributes of incomplete information systems via the tolerance relationship similar matrix. To overcome its drawback of NP-hard time complexity,It applies attribute significance, which this paper puts forward based on attribute frequency in the tolerance relationship similar matrix, as the heuristic knowledge, makes use of binsearch heuristic algorithm to calculate the candidate attribute expansion so that it can reduce the expansion times to speed up reduction. Experiment results show that the algorithm is simple and effective.

Keywords: Knowledge Acquisition; incomplete information system; tolerance relationship.

1 Introduction

Rough set theory[1,2] was put forward by Prof Pawlak who was a Polish mathematician in 1980s, which is a tool to deal with uncertainty and vagueness of data. It can effectively analyze inaccurate, inconsistent and incomplete information. Based on the perspective of knowledge classification, rough set theory processes in the approximation space (knowledge base) and under the premise of maintaining the ability of classification. Rough set theory searches the implicit knowledge and reveals the potential rules through knowledge reduction, which does not require

L. Qi (Ed.): ISIA 2010, CCIS 86, pp. 139–145, 2011.

priori rules, avoiding the impact of personal preferences. Rough set theory has been widely used in despite of its late start.

In rough set theory, the two-dimension table called information system or information table is used as a special formal language to represent knowledge. The information system is called complete, in which all the attribute values of each object are known. Otherwise, it is called incomplete. Obviously, it is very difficult to ensure the completion of an information system due to the uncertainty of information. In many practical cases, data are often incomplete. There are two main approaches to deal with incomplete information systems using rough set theory [3-4]: One is the indirect approach, that is, incomplete information can be completed through some certain methods that are called data filling. The other is the direct method, that is, appropriate extensions are carried out in rough set theory to deal with the incomplete information systems. Indirect method is to deal with null values, where incomplete information system is firstly transformed into a complete information system via data filling method, and then is treated as complete information system. There are some indirect methods such as using statistical analysis to fill the null values, using other condition attribute values and decision attribute values or relationship attributes to estimate the null values, asking experts to give the estimated value of the null values in accordance with some certain conditions, using Bayesian model and evidence theory to filll the missing data. But there are many drawbacks in indirect methods. For examples, Bayesian model needs to know the probability density; evidence theory requires evidence functions, which are often difficult to get, subjectivity and arbitrariness are also big concerns in some of these methods. As the computation complexity is too high, the efficiency is extremely low. Some of these methods can not deal with incomplete information system when there are a lot of null values in information systems, and at the same time the knowledge we get may not be reliable. As the result in contrast with the indirect method, the direct method maintains the original structure of information systems, and avoids human subjectivity. Direct methods are also more effective and reliable in dealing with the situation with many missing data.

As we know, not all attributes are indispensable in an information system. The approach to remove the redundant attributes and preserve the classification or decision ability of the system is called attribute reduction or knowledge reduction. Though knowledge reduction has been an important topic in rough set theory, most of knowledge reductions focus on complete information systems. To deal with incomplete information, several extended models are proposed, in which equivalence relation is relaxed to tolerance relation, similarity relation and general binary relation, respectively. In rough computational methods based on tolerance matrix are studied. Several definitions of knowledge reduction based on tolerance relation are presented. However, the definition and approach to knowledge reduction based on similarity relation are not studied. Attribute reduction is one of the core contents in rough set theory. In information systems with massive data sets, due to huge number of attributes and examples, the attribute reduction algorithm efficiency is particularly important[7]. So far there is no accepted and efficient algorithm in reduction algorithms based on rough set theory[8]. In practical applications, it is required to obtain as a relative attribute reduction.

To deal with the above problems, this paper proposed a knowledge acquisition method based on rough set theory in incomplete information systems.

2 The Basic Concepts

In order to describe attributes reduction, we define some conception as follows.

Definition 2.1 Information System
In rough set, an information system can be represented as

$$S = (U, A, V, f) \tag{1}$$

Where U is the universe, a finite set of N objects $U = \{x_1, x_2, ..., x_n\}$, A is a finite set of attributes, which are divided into disjoint sets, i.e. $A = C \cup D$, where C is the set of condition attributes and D is the set of decision attribute. $V = \bigcup_{q \in A} V_q$ is the total decision function such that $f(x, q) \in V_q$ for every $q \in A$, $x \in U$.

Definition 2.2 Equivalence Relation
If an information system is $S = (U, A, V, f)$, then a subset of attributes $Q \subseteq A$ defines an equivalence relation (an indiscernible relation) on *U*,

$$IND(Q) = \{(x, y) \in U : \forall a \in Q, f(x, a) = f(y, a)\} \tag{2}$$

If an information system is $S = (U, A, V, f)$, a subset of attributes $Q \subseteq A$ determines the approximation space $U \mid IND(Q)$ in *S* as below.

Definition 2.3 Q-positive Region
In the relation *IND*(*D*) we can define the *Q*-positive region $POS_Q(D)$ as

$$POS_Q^U(D) = \bigcup\{\underline{Q}X : X \in IND(D)\} \tag{3}$$

Definition 2.4 Limited To Lerance Rrelationship
Suppose $P_B(x)=\{b \mid b \in B \wedge b(x) \neq *\}$, then definition,

$$\begin{aligned} &\forall_{x,y \in U \times U} (L_B(x, y) \Leftrightarrow \forall_{b_j \in B} (b(x) = b(y) = *) \vee ((P_B(x) \cap P_B(y) \neq \phi) \\ &\wedge \forall_{b \in B} (b(x) \neq *) \wedge (b(y) \neq *) \rightarrow (b(x) = b(y))))) \end{aligned} \tag{4}$$

is Limited Tolerance Relationship, which is reflexive and symmetric, but not necessarily transmitting.

Theorem 1. Give incomplete information system S = (U, $C \cup D$, f,V), C is condition attribute set and D is decision attribute set $A=C \cup D$, $C \cap D=\varnothing$. Under tolerance relationship in incomplete information system, if $B \subseteq A \subseteq C$, then $Pos_B \subseteq Pos_A$.

Proof: Suppose $U / ind(D) = \{Q_1, Q_2, ... Q_r\}$, $Q_j \in U / ind(D)$. According to the definition of tolerance relationship, we know that $T_A(x_i) \subseteq T_B(x_i)$, $x_i \in U$. According to the definition of lower approximation, we know

$D_T^B Q_j = \{x_i | x_i \in U \wedge T_B(x_i) \subseteq Q_j\}$, $D_T^A Q_j = \{x_i | x_i \in U \wedge T_A(x_i) \subseteq Q_j\}$, so we can deduce that $D_T^B Q_j \subseteq D_T^A Q_j$, so $\bigcup_{Q_j \in U/D} D_T^B Q_j \subseteq \bigcup_{Q_j \in U/D} D_T^A Q_j$. According to the definition of positive region, we get that $Pos_B \subseteq Pos_A$. Proof finishes.

Theorem 2. Let *S=(U,A,V,f)* where $U = U^0 \cup U'$, U^0 is the total sample set with complete attribute values, U' is the sample set that we only know the partial values. $A = C^0 \cup C' \cup D$, C^0 is the significant attribute set, C' is the redundant attribute set, D is the decision attribute set. *If* $\forall a \in U', \forall b \in U^0, \forall c \in C'$, $c(a) = c(b)$, then conclude that the information system's certainty is stable.

3 Description of the Algorithm Process

Input: An incomplete decision table S=(U,C∪D,V,F).
Output: Decision rules S—R

Step 1: Initialization min=1;max=card(C)-card(core$_D$ (C)). Incomplete Decision table S is converted into the corresponding tolerance relationship similarity matrix M_C^T and M_C^T is used to calculate core$_D$ (C) and the significance of the remaining attributes. And then arrange remaining attributes in accordance with descending sort of significance and put them into the array Z.

Step 2: Initialization: R=core$_D$ (C), if $\gamma_R = \gamma_c$,then execute Step 4, else execute Step 3.

Step 3 :According to attribute's importance, select and add attributes in array Z into R with binsearch method, then calculate γ_R. If ($\gamma_R < \gamma_c$), the put the important attributes array Z, go to Step 3; if($\gamma_R = \gamma_c$), then go to Step 4.

Step 4: The process end, R is the relative reduction which we required.

Step 5: Calculate the condition attribute importance in $U \setminus MOS$, sort them by the importance, and then place MOS at the end of the information table.

Step 6: Assumed that P^i is the breaking point set of anyone attribute A^i , $A^i \in U \setminus MOS$, the breaking point is P_j^i , (j=1,2,…,n-1) its nearest neighbours are x_j^i , x_{j+1}^i , furthermore, $x_j^i < P_j^i < x_{j+1}^i$,we can deal with A^i as: if $\alpha = x_j^i; x_j^i = x_{j+1}^i$,the information table has no conflicts, then $P^i = P^i \setminus \{P_j^i\}$; else $x_j^i = \alpha; x_{j+1}^i = x_{j+1}^i$.

Step 7: Calculate core values and reduce the initial rules, finally obtain the reduced knowledge.

4 Example Analysis of Algorithm

As shown in Table I, the incomplete decision table S= (U, A, V, F) is given, where U={a_1,a_2,a_3,...,a_{11},a_{12}}, A = { c_1,c_2,c_3,...,c_7,c_8,d} , condition attribute C={ c_1,c_2,c_3,...,c_7,c_8 }, decision attribute D={d}.

Table 1. Font sizes of headings. Table captions should always be positioned *above* the tables.

U	c_1	c_2	c_3	c_4	c_5	c_6	c_7	c_8	d
a_1	3	2	1	1	1	0	*	*	0
a_2	2	3	2	0	*	1	3	1	0
a_3	2	3	2	0	1	*	3	1	1
a_4	*	2	*	1	*	2	0	1	1
a_5	*	2	*	1	1	2	0	1	1
a_6	2	3	2	1	3	1	*	1	1
a_7	3	*	*	3	1	0	2	*	0
a_8	*	0	0	*	*	0	2	0	1
a_9	3	2	1	3	1	1	2	1	1
a_{10}	1	*	*	*	1	0	*	0	0
a_{11}	*	2	*	*	1	*	0	1	0
a_{12}	3	2	1	*	*	0	2	3	0

First of all, step1 is executed. m_C (i,j) and m_C^T (i,j) are calculated to obtain the tolerance relationship similar matrice. According to the elements of tolerance relationship similar matrice which has been built, we can find that m_C^T (4,1)=m_C^T (5,1)= $c_1\ c_2\ c_3\ c_4\ c_5\ c_7\ c_8$,then c_6 is a core attribute. m_C^T (2,6)= $c_1\ c_2\ c_3\ c_5\ c_6\ c_7\ c_8$,then c_4 is a core attribute. $Core_D$ (C)={ c_4, c_6 }. According to the definition of attribute significance, attribute significance of every attribute in C except core attribute are calculated. F(c_1)=-2.65, F(c_2)=-3.32, F(c_3)=-2.29,

F(c_5)=-1, F(c_7)=-3.01, F(c_8)=-2.92. Candidate attributes are arranged in accordance with attribute significance of descending sort and are added into the array Z. So we get Z={ $c_5, c_3, c_1, c_8, c_7, c_2$ };

Let R= Core $_D$ (C)={ c_4, c_6 }. After calculating, we get that $\gamma_R < \gamma_c$ .Access to step3;

Thirdly, step3 is executed. Access to the first cycle. min=1, max=6, mid=3, Tempt= { c_4, c_6 } . Attributes with number from min to mid in array Z are inserted into R, that is, c_5, c_3, c_1 are inserted into R. Then R={ c_4, c_6, c_5, c_3, c_1 }. After calculating, we get that $\gamma_R < \gamma_c$. As max-mid=3>1, min=mid+1=3+1=4, then access to the second cycle. mid=5, Tempt= { c_4, c_6, c_5, c_3, c_1 } . Attributes with number from min to mid in array Z are inserted into R, that is, c_8, c_7 are inserted into R. Then R={ $c_4, c_6, c_5, c_3, c_1, c_8, c_7$ }. After calculating, we get that $\gamma_R = \gamma_c$. As min=4, mid=5, min $\neq$ mid, so max=mid=5, R=Tempt= { c_4, c_6, c_5, c_3, c_1 } , then access to the third cycle. min=mid=4, Tempt=R={ c_4, c_6, c_5, c_3, c_1 }. Attributes with number from min to mid in array Z are inserted into R, that is, c_8 is inserted into R. Then R={ $c_4, c_6, c_5, c_3, c_1, c_8$ }. After calculating, we get that $\gamma_R = \gamma_c$. Since min = mid, quit the cycle. R={ $c_1, c_3, c_4, c_5, c_6, c_8$ } that is the relative reduction of C with respect to decision attribute D.

$a_1 = a_{12}$, $a_4 = a_5$, so delete a_{12} and a_5. And calculate the core values and delete redudant attributes values, then obtain the rules as below:

$c_1(1) \rightarrow d(0)$

$c_1(2) \wedge c_3(0) \rightarrow d(1)$

$c_5\ (1)\ \wedge c_6\ (0)\ \wedge c_8(3) \rightarrow d(1)$

$c_5\ (0)\ \wedge c_6\ (1)\ \wedge c_8(1) \rightarrow d(1)$

$c_5\ (1)\ \wedge c_6\ (2)\ \wedge c_8(1) \rightarrow d(1)$

$c_5\ (3)\ \wedge c_6\ (1)\ \wedge c_8(1) \rightarrow d(1)$

$c_5\ (1)\ \wedge c_6\ (0)\ \wedge c_8(1) \rightarrow d(0)$

$c_5\ (1)\ \wedge c_6\ (1)\ \wedge c_8(1) \rightarrow d(1)$

5 Conclusion

In this paper, for incomplete information system, this algorithm use tolerance relationship similar matrix to acequte knowledege, for example, calculate attribute reduction, core attributes of the incomplete information system and extract rules. It also can overcome the problem of inconsistency in incomplete decision system. In this paper, the frequency of attributes in the tolerance relationship similar matrix is defined as attribute significance which is used as the heuristic konwledge. It makes use of binsearch heuristic algorithm to calculate the candidate attribute expansion so that it can reduce the expansion times to speed up reduction. The more candidate attributes are, the more obvious the advantage of the algorithm is. Experiment results show that the algorithm is simple and effective.

Acknowledgement

This work is supported by the National Natural Science Foundation of China under Grant No. 60674056, 70771007, 70971059; Liaoning doctoral funds under Grant No. 20091034, Liaoning higher education funds under Grant No. 2008T090 and Chinese postdoctoral funds under Grant No. 20100471475.

References

1. Pawlak, Z.: Rough Sets and Intelligent Data Analysis. Information Sciences 147(124), 1212–1218 (2002)
2. Pawlak, Z.: Rough Set Theory and Its Application to Data Analysis. Cybernetics and Systems 29(9), 661–668 (1998)
3. Krysikiewicz, M.: Rough Set Approach to Incomplete Information System. Information Sciences 112, 39–49 (1998)
4. Wang, G.: Extention of Rough Set Under Incomplete Information System. Journal Of Computer Research And Development 39(10), 1240–1243 (2002)
5. Huang, H., Wang, G.: Direct Reduction Method For Incomplete Information System. Mini-Micro System 26(10), 1761–1765 (2005)
6. Xu, E., Shao, L., Ye, B., Li, S.: Algorithm for Rule Extraction Based on Rough Set. Journal of Harbin Institute of Technology 14, 34–37 (2007)
7. Wang, G.: Calculation Methods For Core Attributes of Decision Table. Chinese Journal Of Computers 26(6), 615–622 (2003)
8. Miao, D., Hu, G.: A Heuristic Algorithm For Reduction Of Knowledge. Journal Of Computer Research And Development 36(6), 681–684 (1999)

Minimum Risk Generalized Assignment Problem and Its Particle Swarm Optimization Algorithm

Xuejie Bai

College of Science, Agricultural University of Hebei, Baoding 071001, Hebei, China
baixuejie123@sina.com

Abstract. This paper addresses a new class of two-stage minimum risk generalized assignment problems, in which the resource amounts consumed are represented in the form of fuzzy variables with known possibility distributions. To calculate the credibility in the objective function, an approximation approach (AA) is employed to turn the fuzzy GAP model into an approximating one. Since traditional optimization methods cannot be used to solve the approximating GAP model, to overcome this difficulty, we design a hybrid algorithm integrating the approximation approach and particle swarm optimization (PSO). Finally, one numerical example with six tasks and three agents is given to illustrate the effectiveness of the designed intelligent algorithm.

Keywords: Generalized assignment problem; Minimum risk criteria; Two-stage fuzzy programming; Approximation approach; Particle swarm optimization.

1 Introduction

The generalized assignment problem (GAP) is concerned with optimally assigning n tasks to m agents such that each task is assigned to exactly one agent, while the total resource capacity of each agent is not exceeded. Both interesting and useful, GAP has received more and more researchers' attention in the literature [3,4,5,12]. As we know, the majority of its applications have a stochastic nature. For example, Albareda and Fernández [1] discussed the GAP where only a random subset of the given set of tasks distributed as a Bernoulli random vairable were required to be actually processed and proposed some model-based heuristics; Toktas et al. [13] focused on GAP with stochastic capacities.

With the presentation of fuzzy set theory and the concept of possibility measure introduced by Zadeh, some scholars employed these theories to reflect the vagueness and ambiguity of the resource amount in GAP. For instance, Chang et al. [2] generalized fuzzy rules of a fuzzy modeling method and evolved the fuzzy modeling method for due-date assignment problem in manufacturing by a genetic algorithm; Lin and Wen [8] concentrated on the assignment problem that the elements of the cost matrix were subnormal fuzzy intervals and simplified the problem into either a linear fractional programming problem or a bottleneck assignment problem by the max-min criterion suggested by Bellman and Zadeh.

L. Qi (Ed.): ISIA 2010, CCIS 86, pp. 146–152, 2011.

Recently, Liu and Liu [9] presented the credibility measure instead of possibility measure in the fuzzy decision system. Based on credibility theory, a two-stage fuzzy programming was studied in [11]. The purpose of this paper is to construct a class of two-stage minimum risk generalized assignment problems. In the proposed fuzzy GAP model, the resource consumed by agent i to perform task j is characterized by fuzzy variable with known possibility distribution, and the objective is to formulate an optimization problem via minimum risk criteria.

The rest of this paper is organized as follows. In Section 2 we propose two-stage fuzzy generalized assignment problem. In Section 3 a hybrid algorithm incorporating AA [10] and PSO is designed to solve the proposed model. One numerical example is given to illustrate the effectiveness of the designed algorithm in Section 4. Finally, Section 5 summarizes the main results.

2 Formulation of Fuzzy GAP Model

In this section, we apply two-stage optimization methods to model generalized assignment problem from a new point of view. For the sake of simplicity of presentation, we employ the notations listed in Table 1.

Table 1. List of notations

Notations	Definitions
i	index of agents, $i = 1, 2, \cdots, n$
j	index of tasks, $j = 1, 2, \cdots, m$
c_{ij}	qualification of processing task j by agent i
x_{ij}	binary variable indicating whether task j is assigned to agent i or not
r_i	capacity availability of agent i
ξ_{ij}	capacity consumption of task j processed by agent i
q_i^+	a penalty paid per unit of shortage resource
q_i^-	a penalty paid per unit of resource b_i in excess of $\sum_{j=1}^{m} \xi_{ij} x_{ij}$
y_i^+	the amount of unsatisfied require to agent i in state γ
y_i^-	the amount of remaining resource to agent i in state γ

GAP differs from the classical assignment problem in that each task j is assigned to a single agent, while each agent i can complete several tasks and the assignments have to be made taking into account the resource availability. When making decisions in the generalized assignment problems, the assignment of task j to agent i must be decided before the actual values of the demands for resource capacity are known. Thus we assume that the decisions are made in two stages. Some decision variables x_{ij} must be taken before knowing the realization values of fuzzy variable ξ_{ij}. So we call x_{ij} the first-stage decisions. As a result of the uncertainty of resource capacity, the total consumed amount for agent i may not equal to the capacity availability. Whether resource amount for agent i exceed the capacity availability or not, we all pay a penalty and introduce the second-stage decision variables y_i^+ and y_i^- indicating the insufficient amount and the under-utilization of capacity. The total penalty costs for extra or insufficient resources, i.e. the second-stage programming, can be described as follows

$$\begin{cases} Q(x,\xi(\gamma)) = \min \sum_{i=1}^{n} q_i^+ y_i^+ + \sum_{i=1}^{n} q_i^- y_i^- \\ \qquad \text{s.t. } r_i + y_i^+ - y_i^- = \sum_{j=1}^{m} \xi_{ij}(\gamma) x_{ij}, \quad i = 1,2,\cdots,n \\ \qquad y_i^+ \geq 0, y_i^- \geq 0, \quad i = 1,2,\cdots,n. \end{cases}$$

Observing the above expression, we find out that it can be converted into the following equation

$$Q(x,\xi(\gamma)) = \sum_{i=1}^{n} q_i^+ \Big(\sum_{j=1}^{m} \xi_{ij}(\gamma) x_{ij} - r_i\Big)^+ + \sum_{i=1}^{n} q_i^- \Big(\sum_{j=1}^{m} \xi_{ij}(\gamma) x_{ij} - r_i\Big)^- .$$

When addressing risk aversion, a preselected threshold value $\varphi_0 \in \Re_+$, which may be the level of bankruptcy or a budget limit, is useful in the formulation of the first-stage problem. Therefore, the two-stage minimum GAP model can be constructed as follows

$$\begin{cases} \min \text{Cr}\left\{ \sum_{i=1}^{n} \sum_{j=1}^{m} c_{ij} x_{ij} + Q(x,\xi) \leq \varphi_0 \right\} \\ \text{s.t. } \sum_{i=1}^{n} x_{ij} = 1, \quad j = 1,2,\cdots,m \\ \qquad x_{ij} \in \{0,1\}, \quad i = 1,2,\cdots,n; j = 1,2,\cdots,m, \end{cases} \tag{1}$$

where

$$Q(x,\xi(\gamma)) = \sum_{i=1}^{n} q_i^+ \Big(\sum_{j=1}^{m} \xi_{ij}(\gamma) x_{ij} - r_i\Big)^+ + \sum_{i=1}^{n} q_i^- \Big(\sum_{j=1}^{m} \xi_{ij}(\gamma) x_{ij} - r_i\Big)^- .$$

3 Hybrid PSO Algorithm

Since the minimum risk GAP model in Section 2 is not generally a convex programming, conventional optimization methods usually fail to find a global optimal solution. We suggest a hybrid PSO algorithm to solve the proposed fuzzy GAP model.

3.1 Approximation Approach

Suppose that $\xi = (\xi_{11}, \xi_{12}, \cdots, \xi_{nm})$ in the problem (1) is a continuous fuzzy vector with support $\Xi = \prod_{ij=11}^{nm} [a_{ij}, b_{ij}]$, where $[a_{ij}, b_{ij}]$ is the support of ξ_{ij}. In this case, we will try to use the AA to approximate the possibility distribution function of ξ by a sequence of possibility distribution functions of discrete fuzzy vectors $\{\zeta_s\}$. For the detailed approach, the interested reader may refer to [10].

The problem (1) is referred to as the original fuzzy GAP model. By generating a sequence $\{\zeta_s\}$ according to the distribution of ξ, one can obtain the approximating minimum risk GAP model

$$\begin{cases} \min \operatorname{Cr}\left\{ \sum\limits_{i=1}^{n} \sum\limits_{j=1}^{m} c_{ij}x_{ij} + Q(x, \zeta_s) \le \varphi_0 \right\} \\ \text{s.t. } \sum\limits_{i=1}^{n} x_{ij} = 1, \quad j = 1, 2, \cdots, m \\ \qquad x_{ij} \in \{0, 1\}, \quad i = 1, 2, \cdots, n; j = 1, 2, \cdots, m, \end{cases} \tag{2}$$

where

$$Q(x, \zeta_s(\gamma)) = \sum_{i=1}^{n} q_i^+ \Big(\sum_{j=1}^{m} \zeta_{s,ij}(\gamma)x_{ij} - r_i \Big)^+ + \sum_{i=1}^{n} q_i^- \Big(\sum_{j=1}^{m} \zeta_{s,ij}(\gamma)x_{ij} - r_i \Big)^- . \tag{3}$$

The objective value of the approximating fuzzy GAP model (2) provides an estimator for that of the original fuzzy GAP model (1).

Theorem 1. *Consider the fuzzy generalized assignment problem (1). Suppose ξ is a continuous fuzzy vector such that $Q(x, \xi)$ is not $-\infty$ for any feasible decision x. If ξ is a bounded fuzzy vector and $\{\zeta_s\}$ is the discretization of ξ, then for any given feasible decision x, we have*

$$\lim_{s\to\infty} \operatorname{Cr}\Big\{ \sum_{i=1}^{n} \sum_{j=1}^{m} c_{ij}x_{ij} + Q(x, \zeta_s) \le \varphi_0 \Big\} = \operatorname{Cr}\Big\{ \sum_{i=1}^{n} \sum_{j=1}^{m} c_{ij}x_{ij} + Q(x, \xi) \le \varphi_0 \Big\}.$$

provided $\varphi = \varphi_0$ is a continuity point of $\operatorname{Cr}\{\sum_{i=1}^{n} \sum_{j=1}^{m} c_{ij}x_{ij} + Q(x, \xi) \le \varphi\}$.

Proof. Since for any feasible decision x and every realization γ of fuzzy vector ξ, $Q(x, \gamma)$ is not $-\infty$, which together with the suppositions of the theorem satisfy the conditions of [10, Theorem 1]. The assertion of the theorem is valid.

3.2 PSO Algorithm

These last few years have witnessed the emergence of a new class of optimization algorithm based on the particle swarm optimization. Initially proposed by Kennedy and Eberhart [6], because of better intelligent background and theoretical framework, recently, the PSO algorithm has attracted much attention and been applied successfully in the fields of evolutionary computing, unconstrained continuous optimization problems and many others [7]. For the subsequent discussion, we will give more detailed explanations about the hybrid PSO algorithm for solving the approximating GAP model (2).

Representation Structure: Suppose there are *pop_size* particles to form a colony. In the two-stage GAP model, we use a vector $X = (x_{11}, x_{12}, \cdots, x_{nm})$ as a particle to represent a decision.

Initialization Process: Firstly, we generate *pop_size* initial feasible particles $X_1, X_2, \cdots, X_{pop\_size}$. For each particle X_k, $k = 1, 2, \cdots, pop\_size$, we can calculate the value of the recourse function $Q(x, \zeta_s)$ at X_k via the AA. Thus, the objective value $z(x)$ of the fuzzy GAP model (2) can be computed by

$$\begin{aligned} z(X_k) = \tfrac{1}{2}\big(1 &+ \max\{\mu_k \mid \textstyle\sum_{i=1}^n \sum_{j=1}^m c_{ij}x_{ij} + Q(x, \zeta_s) \leq \varphi_0\} \\ &- \max\{\mu_k \mid \textstyle\sum_{i=1}^n \sum_{j=1}^m c_{ij}x_{ij} + Q(x, \zeta_s) \geq \varphi_0\}\big) \end{aligned} \tag{4}$$

After that, we denote its current position of each particle by *pbest*, using P_k as abbreviation, $P_k = (p_{k,11}, p_{k,12}, \cdots, p_{k,n1}, \cdots, p_{k,nm})$ which represents the personal smallest objective value so far at time t. On the other hand, we set the global best particle of the colony as *gbest*, using P_g as abbreviation, $P_g = (p_{g,11}, p_{g,12}, \cdots, p_{g,n1}, \cdots, p_{g,nm})$ which represents the position of the best particle found so far at time t in the colony. Finally, we initialize the velocity V_k of the kth particle randomly, $V_k = (v_{k,11}, v_{k,12}, \cdots, v_{k,n1}, \cdots, v_{k,nm})$.

Updating Process: As mentioned above, the PSO algorithm is an evolutionary computation algorithm, and it searches for the optimal solution by renewing generations. Using the above notations, for the *pop_size* particles, the new velocity of the kth particle is updated by

$$V_k(t+1) = \omega_t V_k(t) + c_1 r_1 (P_k(t) - X_k(t)) + c_2 r_2 (P_g(t) - X_k(t)), \tag{5}$$

for $k = 1, 2, \cdots, pop\_size$, where ω_t is the inertia weight that decreases linearly from 0.9 to 0.4; c_1 and c_2 are nonnegative constants, called the cognitive and social parameter, respectively; and r_1 and r_2 are two independent random numbers generated from the unit interval $[0, 1]$. When the new velocity $V_k(t+1)$ is obtained, we can update the position of the kth particle by

$$X_k(t+1) = X_k(t) + V_k(t+1). \tag{6}$$

Summarizing the above process immediately yields the hybrid PSO algorithm as follows.

Step 1. Initialize *pop_size* particles and evaluate the objective values by the formula (4).
Step 2. Set *pbest* of each particle and its objective equal to its current position and objective value,and set *gbest* and its objective equal to the position and objective value of the best initial particle.
Step 3. Update the velocity and position of each particle.
Step 4. Calculate the objective values for for all particles.
Step 5. Renew *pbest* and its objective values with the current position and objective value.
Step 6. Renew *gbest* and its objective values with the position and objective value of the current best particle.
Step 7. Repeat the fifth to eighth steps for a given number of cycles.
Step 8. Return the *gbest* and its objective values.

4 One Numerical Example

To show the feasibility and effectiveness of the hybrid algorithm, consider the following generalized assignment problem with $n = 3, m = 6$. Capacity availability and penalty coefficient (r_i, q_i^+, q_i^-) are (50,40,80), (50,30,80) and (50,60,60), respectively. The cost c_{ij} and capacity consumption ξ_{ij} are displayed in Table 2. In addition, the fuzzy variables involved in this problem are supposed to be mutually independent.

Table 2. The parameters for the two-stage fuzzy GAP problem

Cost (c_{ij})						
	Task, j					
Agent, i	1.	2.	3.	4.	5.	6.
1.	130	30	510	30	340	20
2.	460	150	20	40	30	450
3.	40	370	120	390	40	30
Consumption (ξ_{ij})						
	Task, j					
Agent, i	1.	2.	3.	4.	5.	6.
1.	(28,30,32)	(48,50,51)	(8,10,13)	(10,11,13)	(10,13,14)	(7,9,15)
2.	(5,10,12)	(15,20,28)	(58,60,62)	(9,10,14)	(5,10,20)	(15,17,19)
3.	(68,70,72)	(8,10,12)	(6,10,12)	(12,15,18)	(6,8,10)	(8,12,16)

Based on the related data, the two-stage minimum risk generalized assignment problem model is built as follows

$$\begin{cases} \min \text{Cr}\{130x_{11} + 30x_{12} + 510x_{13} + 30x_{14} + 340x_{15} + 20x_{16} + 460x_{21} \\ \quad +150x_{22} + 20x_{23} + 40x_{24} + 30x_{25} + 450x_{26} + 40x_{31} + 370x_{32} \\ \quad +120x_{33} + 390x_{34} + 40x_{35} + 30x_{36} + Q(x,\xi) \le 470000\} \\ \text{s.t.} \ \sum_{i=1}^{3} x_{ij} = 1 \quad j = 1, 2, \cdots, 6 \\ \quad x_{ij} \in \{0, 1\} \quad i = 1, 2, 3; j = 1, 2, \cdots, 6, \end{cases} \tag{7}$$

where

$$\begin{aligned} Q(x,\xi(\gamma)) = {} & 40\Big(\sum_{j=1}^{6} \xi_{1j}(\gamma)x_{1j} - 50\Big)^+ + 80\Big(\sum_{j=1}^{6} \xi_{1j}(\gamma)x_{1j} - 50\Big)^- \\ & +30\Big(\sum_{j=1}^{6} \xi_{2j}(\gamma)x_{2j} - 50\Big)^+ + 80\Big(\sum_{j=1}^{6} \xi_{2j}(\gamma)x_{2j} - 50\Big)^- \\ & +60\Big(\sum_{j=1}^{6} \xi_{3j}(\gamma)x_{3j} - 50\Big)^+ + 60\Big(\sum_{j=1}^{6} \xi_{3j}(\gamma)x_{3j} - 50\Big)^-. \end{aligned}$$

In order to solve the minimum risk GAP model (7), for each fixed first-stage decision variable x, we generate 3000 sample points via the approximation approach to calculate the recourse function. For each sample point $\widehat{\zeta}^k$, we can use the formula (3) to obtain the optimal value as $Q(x, \widehat{\zeta}^k)$ for $k = 1, 2, \cdots, 3000$. After that, the value $z(x)$ of the objective function at x can be computed by Equation (4). If we set the population size in the implementation of the hybrid PSO algorithm is 30, then a run of the proposed algorithm with 600 generations gives the optimal solution. An optimal assigns tasks 1 and 6 to agent 1, tasks 2, 4 and 5 to agent 2, and task 3 to agent 3 with the credibility 0.9168.

5 Conclusions

In this paper, we took credibility theory as the theoretical foundation of fuzzy optimization and developed a two-stage minimum risk GAP model, where the resource amounts were uncertain and assumed to be fuzzy variables with known possibility distributions. Since it is inherently an infinite-dimensional optimization problem that can rarely be solved directly by conventional optimization algorithm, this paper designed a AA-based PSO algorithm to solve the approximating two-stage GAP model. Finally, we gave one numerical example to demonstrate the feasibility of the designed hybrid PSO algorithm.

References

1. Albareda-Sambola, M., Fernández, E.: The Stochastic Generalised Assignment Problem with Bernoulli Demands. TOP 8, 165–190 (2000)
2. Chang, P.C., Hieh, J.C., Liao, T.M.: Evolving Fuzzy Rules for due-date Assignment Problem in Semiconductor Manufacturing Factory. J. Intell. Manuf. 16, 549–557 (2005)
3. Chu, P.C., Beasley, J.E.: A Genetic Algorithm for the Generalized Assignment Problem. Comput. Oper. Res. 24, 17–23 (1997)
4. Diaz, J.A., Fernandez, E.: A Tabu Search Heuristic for the Generalized Assignment Problem. Eur. J. Oper. Res. 132, 22–38 (2001)
5. Haddadi, S.: Lagrangian Decomposition based Heuristic for the Generalised Assignment Problem. INFOR 37, 392–402 (1999)
6. Kennedy, J., Eberhat, R.C.: Particle Swarm Optimization. In: Proc. of the IEEE International Conference on Neural Networks, New York, pp. 1942–1948 (1995)
7. Kennedy, J., Eberhart, R.C., Shi, Y.: Swarm Intelligence. Morgan Kaufmann Publishers, San francisco (2001)
8. Lin, C.J., Wen, U.P.: A Labeling Algorithm for the Fuzzy Assignment Problem. Fuzzy Sets Syst. 142, 373–391 (2004)
9. Liu, B., Liu, Y.K.: Expected Value of Fuzzy Variable and Fuzzy Expected Value Models. IEEE Trans. Fuzzy Syst. 10, 445–450 (2002)
10. Liu, Y.K.: Convergent Results about the Use of Fuzzy Simulation in Fuzzy Optimization Problems. IEEE Trans. Fuzzy Syst. 14, 295–304 (2006)
11. Liu, Y.K., Zhu, X.: Capacitated Fuzzy Two-Stage Location-Allocation Problem. Int. J. Innov. Comput. Inf. Control 3, 987–999 (2007)
12. Ross, G.T., Soland, R.M.: A Branch-and-Bound Algorithm for the Generalized Assignment Problem. Math. Program. 8, 91–103 (1975)
13. Toktas, B., Yen, J.W., Zabinsky, Z.B.: Addressing Capacity Uncertainty in Resource-Constrained Assignment Problem. Comput. Oper. Res. 33, 724–745 (2006)

Visual Search Strategy and Information Processing Mode: An Eye-Tracking Study on Web Pages under Information Overload

Wanxuan Lu[1,*], Mi Li[1,2,*], Shengfu Lu[1,**], Yangyang Song[1], Jingjing Yin[1], and Ning Zhong[1,3]

[1] International WIC Institute, Beijing University of Technology, Beijing 100124, China
Tel.: +86-10-6739-6667
lusf@bjut.edu.cn
[2] The School of Comupter and Communication Engineering, Liaoning ShiHua University, Liaoning 113001, China
[3] Department of Life Science and Informatics, Maebashi Institute of Technology, Japan

Abstract. Most studies about visual search on Web page focused on factors such as information form or information layout. However, there are few about information quantity. This study investigates the search strategy and information processing mode of visual search on Web pages under information overload. Results show that 1) users' search strategies on Web pages with and without information overload are the same which is paying more attention on picture. Information overload make this picture-oriented strategy more significant. 2) Under information overload, users are more likely to use "deep search" information processing mode which is decreasing parallel processing and increasing serial processing. These results indicate that (1) information overload doesn't change but impact users' search strategy; (2) Information overload has a significant impact on users' information processing mode. This study provides some evidence both for cognitive psychology and human-computer interaction, especially Web page design.

Keywords: Information overload, Visual search, Web pages, Search strategy, Information processing mode.

1 Introduction

Internet has become one of the most important parts in people's daily life. As the carrier of on-line information, Web page is the main interface of human-computer interaction (HCI) on Internet. Researchers of HCI have described an adaptive and intelligent HCI that could predict and diagnose users' cognitive state, and then adapt itself to improve HCI [1]. To achieve this, it is necessary to investigate users' physical

* These authors contributed equally to this work.
** Corresponding author.

L. Qi (Ed.): ISIA 2010, CCIS 86, pp. 153–159, 2011.

and mental activities. Traditional methods of user behavior study on Web pages (e.g. feedback analyzing, data mining and user modeling) are "off-line" investigation. It's impossible to obtain users' cognitive activities through these methods. Eye-tracking technology allows researchers to record users' eye movement data when they interact with Web pages. These data including fixation, eye movement contrail and pupil size could directly and objectively reflect users' behavior on Web pages, which could reveal their cognitive activities [2].

There are two types of visual behavior on Web pages: visual search and browsing. Visual search refers to the behavior of searching on Web pages with a target, while browsing refers to the behavior of browsing on Web pages without target [3,4]. Previous studies have demonstrated that users' visual search on Web pages is impacted by many factors, such as colors [5,6], fonts [7], information forms [8,9] and information layout [10]-[16]. However, there are few studies focused on the impact of information quantity. Nowadays, users often face on Web pages with massive information. To deal with it, Web page designers have to know how users react when they search on Web pages under information overload.

In this study, an experiment with two groups was designed using eye-tracking to investigate users' search strategy and information processing mode of visual search on Web pages under information overload.

2 Experiments

2.1 Experimental Design

Ten different Chinese Web pages involving different topics (e.g. cell phone or clothing) were used in this study. We controlled information quantity by limiting the size of Web pages. In Group 1, Web pages were single-page without scrollbar which could be presented on one screen; in Group 2, Web pages were double-page with vertical scrollbar, half of which could be presented on one screen. As two groups used same information density (e.g. font, font size and line space), the information quantity in Group 2 was approximately twice as much as that in Group 1.

To eliminate the effects of information form and location, we used two target forms (text and picture) and arranged them on different locations. There were 10 different targets including 5 texts and 5 pictures. Texts were 3 to 6 Chinese characters which were companies' name (e.g. Dell) and pictures were companies' logo (e.g. the logo of Audi). Targets in Group 1 were arranged on 5 locations: upper-left, lower-left, central, upper-right and lower-right; Targets in Group 2 were arranged on 10 locations: the 5 locations on upper half of Web page and that on lower half. Thus, there were 10 Web pages in Group 1 (5 locations × 2 target forms) and 20 in Group 2.

2.2 Participants

There were 130 undergraduates or postgraduates from various majors in this study, half of which were female. All participants were native Chinese speakers, right-handed, skilled users of Internet and had normal or corrected-to-normal vision. 50 participants with age range of 21 to 25 (M = 23.0, SD = 1.3) were in Group 1 and 80 with age range of 20 to 27 (M = 23.2, SD = 1.3) were in Group 2.

2.3 Apparatus and Procedure

Eye movements were recorded at the rate of 120 HZ by Tobii T120 eye-tracker, which had a 17 LCD monitor with resolution set to 1024 × 768 pixels. Web pages were automatically presented on screen by software called Tobii Studio.

At the beginning of each task, the pre-page which had description of the target was presented (whether target was text or picture, it was described in Chinese so that participants didn't know the target form). Participants needed to click into the search page, found the target and clicked on it. After that, participants could have a rest and click into the next task.

3 Results

We firstly contrast the search time of two groups. The search time is defined as the time between Web page showing up and participant clicking on the search target. As shown in Fig. 1, the search time of Group 2 is significantly longer than Group 1, which means that the search efficiency of Group 2 is significantly lower than Group 1. These result indicates that the increase of information quantity reduces users' visual search efficiency, which is consistent with Hornof's finding [17,18] and Rau's [19]. Thus, we consider Group 1 as non information overload and Group 2 as information overload.

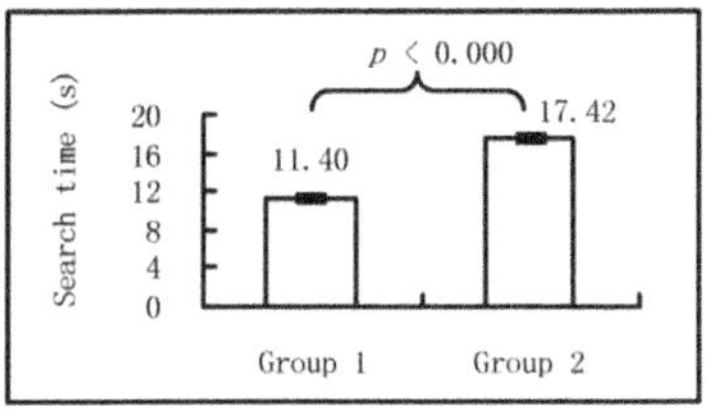

Fig. 1. The search time of two groups

3.1 Search Strategy

To investigate participants' search strategy under information overload, we look into their eye movement contrails. Fig. 2 shows some typical contrails (because some contrail has too many fixation points, it only shows no more than 20 fixation points) with different search target forms and locations under information overload.

It is likely that participants tend to pay more attention on picture than text under information. To find out what it likes under non information overload and how information overload impacts, we further count the distribution of the earliest 20 fixation points under non information overload and information overload, as shown in Fig. 3.

No matter information overload or not, the number of fixation point on picture is significantly larger than text. However, the number of fixation point on picture under information overload is significantly larger than non information overload. These results indicate that information overload doesn't change participants' search strategy which is paying more attention on picture but makes this picture-oriented search strategy more significant.

(a) Upper picture (b) Lower picture (c) Upper text (d) Lower text

Fig. 2. Some typical eye movement contrails with different search target forms and locations. (a) Target is picture on upper half of Web page. (b) Target is picture on lower half of Web page. (c) Target is text on upper half of Web page. (d) Target is text on lower half of Web page.

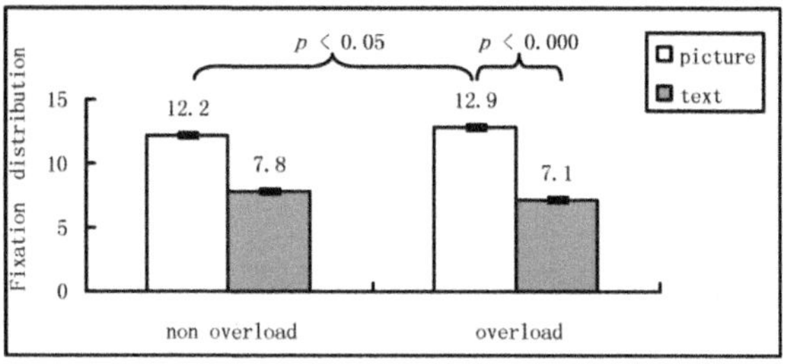

Fig. 3. The distribution of the earliest 20 fixation points

3.2 Information Processing Mode

There are two basic information processing type for visual search: parallel processing and serial processing [20]. Parallel processing refers to the behavior that comparing search target and all the distracters at the same time. Serial processing refers to the behavior that comparing search target and distracters one to another. When user searches target on a Web page, he uses parallel processing to find a target area in which he thinks the target is, and then uses serial processing to search in this area. If he doesn't find the target, he uses parallel processing again to find another target area and then uses serial processing again [21]. The average fixation duration could show how much participants use serial processing: the more the average fixation duration is, the more participants use serial processing. The fixation count could show how much participants use parallel processing: the more the fixation count is, the more participants use parallel processing.

To investigate the impact of information overload on information processing mode, we analyze participants' the average fixation duration and fixation count. The average fixation duration is defined as "fixation duration / fixation count", which refers the average duration of each fixation points. Fixation count is the number of fixation points.

10. Li, M., Song, Y.Y., Lu, S.F., Ning, Z.: The Layout of Web pages: a Study on the Relation Between Information Forms and Locations Using Eye-Tracking. In: Liu, J., Wu, J., Yao, Y., Nishida, T. (eds.) AMT 2009. LNCS, vol. 5820, pp. 207–216. Springer, Heidelberg (2009)
11. Maldonado, C.A., Resniek, M.L.: Do Common User Interface Design Patterns Improve Navigation? In: Proceedings of the Human Factors and Ergonomics Society 46th Annual Meeting, pp. 1315–1319. Human Factors and Ergonomics Society Press, Santa Monica (2002)
12. Buscher, G., Cutrell, E., Morris, M.R.: What do You See when You're Surfing? Using Eye Tracking to Predict Salient Regions of Web Pages. In: Proceedings of the 27th International Conference on Human Factors in Computing System, pp. 21–30. Association for Computing Machinery Press, New York (2009)
13. Byrne, M.D., Anderson, J.R., Douglass, S., Matessa, M.: Eye Tracking the Visual Search of Click-Down Menus. In: Proceedings of the SIGCHI Conference on Human Factors in Computing System: the CHI is the Limit, pp. 402–409. Association for Computing Machinery Press, New York (1999)
14. Halverson, T., Hornof, A.J.: Local Density Guides Visual Search: Sparse Goups are First and Faster. In: Proceedings of the Human Factors and Ergonomics Society 48th Annual Meeting, pp. 1860–1864. Human Factors and Ergonomics Society Press, Santa Monica (2004)
15. Schaik, P.V., Ling, J.: The Effects of Frame Layout and Differential Background Contrast on Visual Search Performance in Web Pages. Interacting with Computer 13, 513–525 (2001)
16. Ling, J., Schaik, P.V.: The Influence of Line Spacing and Text Alignment on Visual Search of Web Pages. Displays 28, 60–67 (2007)
17. Hornof, A.J.: Visual Search and Mouse-Pointing in Labeled Versus Unlabeled Two-Dimensional Visual Hierarchies. ACM Transactions on Computer-Human Interaction 8, 171–197 (2001)
18. Hornof, A.J., Halverson, T.: Cognitive Strategies and Eye Movements for Searching Hierarchical Computer Displays. In: Proceedings of the SIGCHI Conference on Human Factors in Computing Systems, pp. 249–256. Computing Machinery Press, New York (2003)
19. Rau, P.P., Gao, Q., Liu, J.: The Effect of Rich Web Portal Design and Floating Animation on Visual Search. International Journal of Human-Computer Interaction 22, 195–216 (2007)
20. Sternberg, S.: Memory Scanning: Mental Processes Revealed by Reaction-Time Experiment. American Scientist 57, 421–457 (1969)
21. Ren, Y.Z., Chen, J., Zhao, J.X.: The Pilot Study in the Relationship Between Visual Search and Web Design. Art and Design 10, 61–63 (2007) (in Chinese)
22. Carroll, P.J., Young, J.R., Guertin, M.S.: Visual Analysis of Cartoons: a View from the Far Side. In: Rayner, K. (ed.) Eye Movement and Visual Cognition: Scene Perception and Reading, pp. 444–461 (1992)
23. Rayner, K., Rotello, C.M., Steward, A.J., Keir, J., Duffy, S.A.: Integrating Text & Pictorial Information: Eye Movements when Looking at Print Advertisements. Journal of Experimental Psychology: Applied 7, 219–226 (2001)

The Application of Support Vector Machine in Surrounding Rock Classification

Dan Chen, Yongjie Li, and Zhiqiang Fu

Liaoning Technical University
Liaoning Fuxin 123000, China
dan218@sohu.com

Abstract. The surrounding rock stability classification of tunnel is an important basis for the engineering design, construction, risk assessment and to lay down appropriate engineering measures. The present paper gives a brief introduction to the common method of rock classification in today's and influence factors to the stability of surrounding rock. It lays emphasis on the nature of rock and rock structure impact on the surrounding rock stability. And then to take rock integrity coefficient, surface structure friction coefficient, coefficient of saturated rock firm and rock longitudinal wave velocity coefficient as Index of the surrounding rock stability classification, classify the surrounding rock stability with relevant data in historical documents by using support vector machine. The results prove that the classification of support vector machine in surrounding rock stability is feasible. It should be noted that support vector machine classification depend on the training sample data and optimal choice of sample data need further study.

Keywords: support vector machine; surrounding rock; stability classification; sample data.

1 Introduction

The surrounding rock stability classification of tunnel is an important basis for the engineering design, construction, risk assessment [1] and to lay down appropriate engineering measures. At present many kinds of surrounding rock stability Classification Methods had been proposed. These methods may be divided into two categories in general. One method of surrounding rock stability Classification is based on Index value of classification. These methods include Mainly two-step classification method proposed by 《Engineering rock mass classification》 GB 50218-94, CSIR method proposed by South Africa Scientific and Industrial Research [2], Barton rock mass classification method [3], and so on. Because Such methods involve many factors that affect the stability of surrounding rock and analysis comprehensive, they are widely used in engineering practice. However, the computing needs of different steps and more complex. At the same time because there are many influence factors to the stability of surrounding rock and these factors are with greater uncertainty and subjectivity, more and more people turn to study the surrounding rock stability classification of tunnel by using statistical theory.

L. Qi (Ed.): ISIA 2010, CCIS 86, pp. 160–165, 2011.

The other method of surrounding rock stability classification is based on statistical learning theory. Statistical theory is the only means in the face of a large number of observational data and the lack of suitable theoretical model. These methods include mainly fuzzy nearest neighbor principle [4], to consider multi-factor fuzzy rock classification method [5], rock classification method Fuzzy Pattern Recognition [6], artificial neural network surrounding rock classification [7], gray rock classification optimization theory method [8], and so on. Artificial neural networks and fuzzy rock classification theory in practical application need the number of data large enough to support the conclusion. In this context, it can be guaranteed only when the data set tends to infinity in theory. However, in real applications, the number of data is often limited. Especially in stability analysis of rock, monitoring costs will rapidly increase with the sample size increases. It even can not be achieved In some cases. So people began to look for a statistical learning theory of limited sample.

Support vector machine is targeted at a limited sample of cases, and its goal is to get the optimal solution under the existing information, not just the number of samples tends to infinity the optimal value. This feature of limited samples is just to be demand in surrounding rock stability classification and the advantages will be fully embodied.

2 Influence Factors to the Stability of Surrounding Rock

Based on current research, influence factors to the stability of surrounding rock includes the nature of the rock, rock structure and construction, groundwater, the natural stress state of rock, geological structure, excavation, support form and so on. In these factors, the nature of rock and rock structure play a leading role. So it is important to understand the nature of rock and rock structure impact on the surrounding rock stability and mechanism.

The strength of rock is the main factor affecting the stability of surrounding rock. Rock properties and rock mass structure determine the strength of rock directly. For the plastic rock that the main components are the clay rocks, clay and broken loose rock, usually they will quickly weathering and further broken. They also appear softening, disintegration, swelling and other objectionable when the water infiltrate the rock. So the plastic rock is low intensity and will produce a larger deformation when the secondary stress generate after the tunnel excavation. This is most unfavorable to the stability of surrounding rock tunnel.

For the brittleness rock that the main components are hard rock, usually their Intensity is much higher than the strength of rock discontinuities. So the stability of this kind of surrounding rock depends on the strength of rock structure. According to the perspective of rock structures, rock mass structure can be divided into the overall block structure, layered structure, fragmented structure and the granular structure. Overall block structure has the best stability of all. It will produce a very small deformation when the secondary stress generate after the tunnel excavation and supporting structure is relatively simple, even do not support in some cases. loose Structure of rock and broken structure's the stability is the worst. For thick layered and massive brittle rock, its strength is mainly affected by weak structure plane and the weak interlayer distribution. This affect is not only depends on the structure surface characteristics of themselves, but also structure surface composition.

3 Surrounding Rock Classification

According to reference[6], we take rock integrity coefficient, surface structure friction coefficient, coefficient of saturated rock firm and rock longitudinal wave velocity coefficient as Index of the surrounding rock stability classification. In the paper, stability of surrounding rock is divided into five grades. They are marked I, II, III, IV and V by the stability of the order from highest to lowest. And then the binary tree classification is used to the surrounding rock stability classification and RBF kernel function is used in support vector machines. In the first place, the value of C and g are determined by grid search method. The results shown in figure 1~ figure 4.

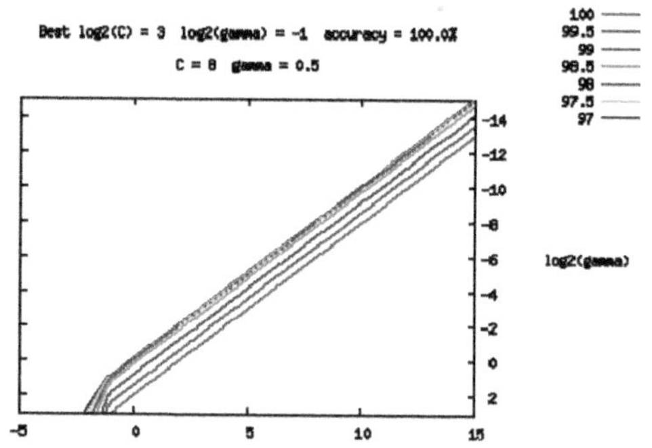

Fig. 1. Parameter optimization for I to II, III, IV, V

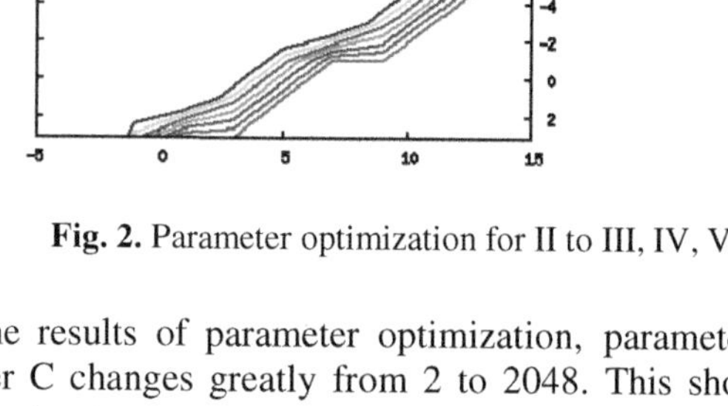

Fig. 2. Parameter optimization for II to III, IV, V

According to the results of parameter optimization, parameter g is 0.5 always. However parameter C changes greatly from 2 to 2048. This shows that number of samples affects penalty factor, but this effect is nonlinear. This phenomenon needs further study.

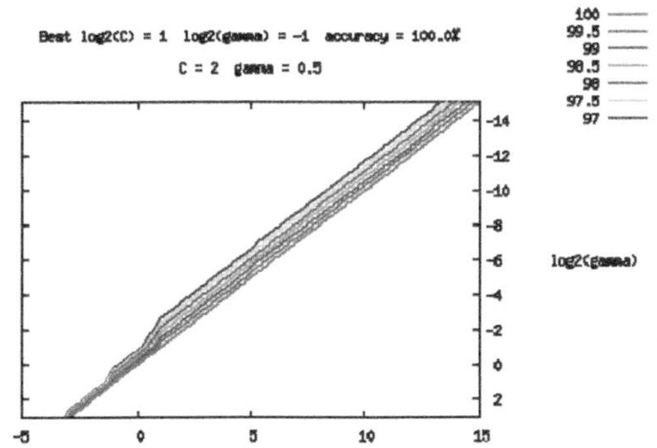

Fig. 3. Parameter optimization for III to IV, V

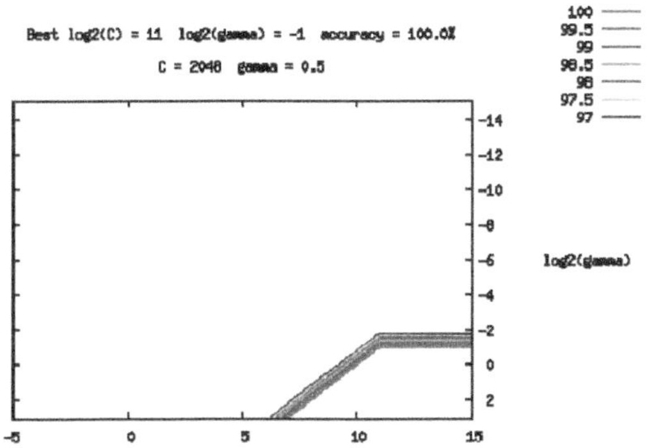

Fig. 4. Parameter optimization for IV to V

Using literature [6] data train vector machine model after parameter optimization and the results shown in figure 5~ figure 8.

Fig. 5. Stability classification for I to II, III IV, V

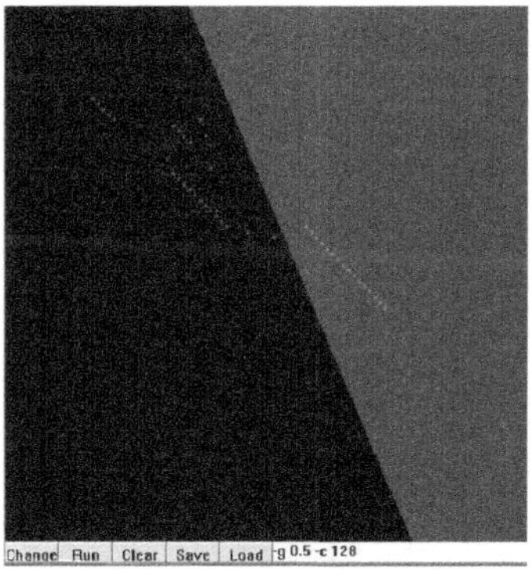

Fig. 6. Stability classification for II to III, IV, V

Fig. 7. Stability classification for III to IV, V

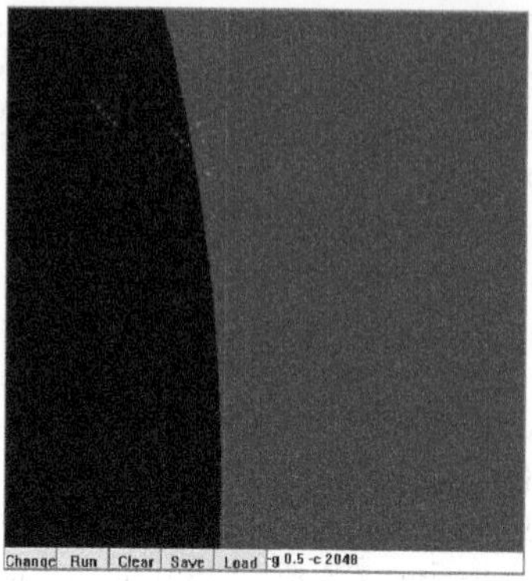

Fig. 8. Stability classification for IV to V

Then the recognition model received is tested by the 10 non-learning training samples and the results shown in table 1. The results of table 1 show that support vector machine is higher mapping accuracy than the traditional fuzzy recognition model in the case of small sample size. The model can correctly inference surrounding rock stability classification on the base of the nature of rock and rock structure.

Table 1. Results of the recognition by support vector machine

original classification	results of the recognition	rock integrity coefficient	surface structure friction coefficient	coefficient of saturated rock firm	rock longitudinal wave velocity coefficient
I	I	0.76	0.61	6.1	5.1
	I	0.87	0.82	7.2	8.2
II	II	0.59	0.44	7.1	3.7
	II	0.70	0.60	8.2	5.2
III	III	0.35	0.35	6.5	3.5
	III	0.55	0.46	5.5	3.6
IV	IV	0.36	0.21	3.1	1.5
	IV	0.37	0.24	3.2	1.9
V	V	0.20	0.21	2.5	1.4
	V	0.31	0.22	3.6	2.5

4 Conclusion

In the paper, it takes rock integrity coefficient, surface structure friction coefficient, coefficient of saturated rock firm and rock longitudinal wave velocity coefficient as Index of the surrounding rock stability classification, then tries to use the support vector machine multi-classification model in surrounding rock stability classification. During the course of parameter optimization of the grid search method, parameter g is 0.5 always but parameter C changes greatly from 2 to 2048. This phenomenon needs further study. The results of prediction show that support vector machine is higher mapping accuracy than the traditional fuzzy recognition model in the case of small sample size.

References

1. Qian, Q., Rong, X.: State Issues and Relevant Recommendations for Security Risk Management of China's Underground Engineering. Chinese Journal of Rock Mechanics and Engineering 27(04), 649–656 (2008)
2. Harrison, J.P.: Selection of the threshold value in RQD assessment. International Journal of Rock Mechanics & Mining Sciences 36, 673–685 (1999)
3. Choi, S.Y., Park, H.D.: Variation of rock quality designation (RQD) with scalene orientation and length: a case study in Korea. International Journal of Rock Mechanics & Mining Sciences 41, 207–221 (2004)
4. Han, G., Liu, B., Fan, H.: Application of Principle of Fuzzy Closeness Optimization to Classification of Surrounding Rock in Tunnel. Journal of Northeastern University (Natural Science) 27(07), 819–822 (2006)
5. Li, X., Zhai, C.: Fuzzy Theory Application in Surrounding Rock Classification and in Highway Tunnel. Journal of Nanjing Forestry University (Natural Sciences Edition) 30(03), 55–58 (2006)
6. Cai, B.: The APPlication of the FuzZy Pattem Recognitionin Classification of Surrounding Roek. Journal of Hebei University of Technology 29(03), 99–101 (2000)
7. Ye, B., Li, X., Zhang, J.: Study on artificial nerve network method of surrounding rock classification for hydraulic tunnel. Water Resources & Hydropower of Northeast 21(08), 1–3 (2003)
8. Feng, Y.: APPlication of Grey Optimal Theory Model in the Stability Classification Of Adioining Rock Of Underground Construction. Chinese Journal of Geotechnical Engineering 18(03), 62–66 (1996)

Detecting and Identification System about Water Environmental Pollutant Based on Fluorescence Theory

ShuTao Wang[1], YanYan Cui[1], Chuan Zhang[2], Liang Huang[1], Zhao Pan[1], and ZhongDong Wang[1]

[1] Key Laboratory of Measurement Technology and Instrument of Hebei, Yanshan University, Qinhuangdao, Hebei, 066004
[2] The Department of Mechanical & Electrical Integration, China Mining University in Beijing, Beijing, 100083

Abstract. This system mainly adapted a new fluorescent technique to monitor oil pollutant in water. And the problem was solved that fluorescent signal which obtain wealthy of information, but weak signal intensity, it used the method which combine fiber optical sensor technology with linear CCD in spectrum detecting application. In signal processing, the MATLAB program was apply to analyze three-dimensional fluorescent spectrum, including PCA algorithm and NNLS algorithm

Keywords: Fluorescence; CCD; PCA; NNLS.

1 Introduction

With the rapid development of our country economy, gradual expand of petroleum exploitation scale in the land and the sea, prosper of shipping enterprise, as well as a number of industrial enterprise, the massive oil hydrocarbon was divulged into the sea, discharged in the natural rivers. It not only destroyed seriously the natural environment and ecosystem, but also threatened the human being survival and ecological equilibrium. It is most important and valuable to detect oil hydrocarbon exactly and rapidly, then it was helpful and convenient to monitor the types and origins of pollutions and investigate accident reason. This will promote our social harmony.

2 Fluorescence Detecting Theory

Based on the spectral method, when the electronic, which emitted part of energy in molecular excited state though non-radiation form back to the lowest vibration level of the first excited state-S1, gone back to the basic state-S0, emitting abundant of energy by radiation, it could emit the fluorescence. Molecular luminescence includes photoluminescence and bioluminescence. In particularly, the photoluminescence could gain energy from outside. In this paper, mineral oil hydrocarbon could emit fluorescence by the methods of laser induced fluorescence and introduce spectrum in different wavelength range. In a general way, what molecular fluorescence spectra refer to is Ultraviolet-Visible emitting spectrum.

L. Qi (Ed.): ISIA 2010, CCIS 86, pp. 166–171, 2011.

No matter what fluorescence compound has two feature spectrums which are excitation spectrum and emission spectrum. Specially, excitation spectrum shows the relation between fluorescence intensity and excitation wavelength of measured substances, whereas, the imitation spectrum expresses the relation between emitting wavelength and emitting intensity.

For most of aromatic and conjugate compound which do not include heteroatom, transition from S0 energy level to S1 energy level means transition from π to π^*, that is to say substance absorb the more radiation energy, it emits the more fluorescence intensity [1].

The fluorescence intensity which the detected substance emits is relation to absorbing intensity and its fluorescence quantum yield. Based on the Lambert-beer principle, it is known as

$$I_f = AY_f I_0\left(1-10^{-\varepsilon cl}\right) \tag{1}$$

A is a lateral area of clear quartz slot which contains fluorescence material solution; Y_f is fluorescence quantum yield; I_0 is transmitted light intensity; εis molar absorption coefficient; c is fluorescence material concentration; l is optical distance of incident light. So formula (2) can be gain by formula (1),

$$I_f = Y_f AI_0\left[2.3\varepsilon cl - \frac{(2.3\varepsilon cl)^2}{2!} + \frac{(2.3\varepsilon cl)^3}{3!} - \cdots\cdots\right] \tag{2}$$

When the solution is dilute, the term of higher order can be ellipsis in formula (2). So that becomes formula (3)

$$I_{\mathrm{f}} = 2.3Y_f AI_0\varepsilon cl \tag{3}$$

Therefore, the result can be get from the above formulas. When the substance with certain fluorescent efficiency is excited by the light which had proper wavelength, proper intensity, its concentration can be detected by the photoelectric detection system of high sensitivity.

3 The System Structure and Key Technology Problems

The system is mainly composed of pulse xenon lamp illuminant, polychromator, the high- powered photoelectric detection device CCD and all kinds of optical components, signal acquisition and data transmission device etc. The figure 1 is system structure diagram.

Fig. 1. The radiation is emitted continuously by pulses xenon lamp, via excitation filter, we can get a certain wavelength exciting light which is coupled to optical fiber through the lens. Among them, one beam as reference direct into CCD via optical fiber, another beam stimulates sample solution to launch fluorescence. The polychromator split fluorescence collected by optical fiber probe, so as to make split light shoots linear CCD photosensitive face. Finally, the photoelectric conversion is achieved. Then after weak signal processing system, the detected signal is enlarger and sent into the

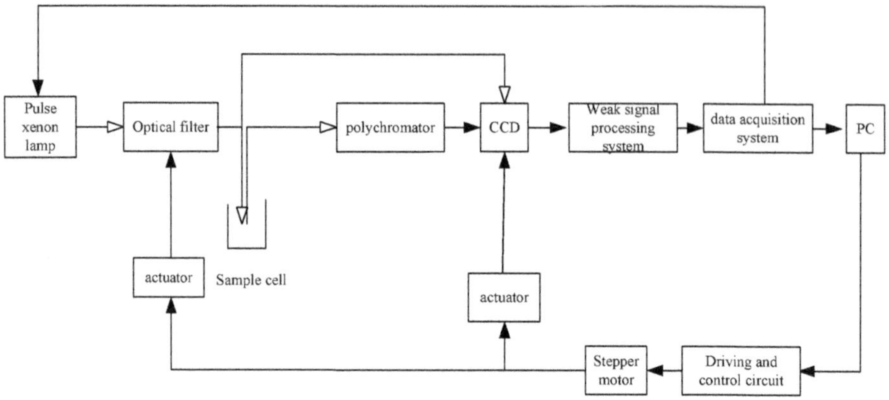

Fig. 1. Fluorescence detecting system structure diagram

: circuit :light path

computer via A/D transformation. Through analyze the fluorescence spectrum, the fluorescence intensity and sample concentration are gained.

3.1 Optical Fiber Probe

The measured substance which is stimulated by exciting light in solution can emit the fluorescence which emission direction is irrelevant to exciting direction. This fluorescence radiate in sphere. In order to effectively detect the fluorescence, designing a kind of high performance optical fiber probe is a key problem. The prober can not only collect fluorescence, but also had a high ability of resisting astigmatism interference and a high transmission efficiency in certain exciting and emitting spectrum scope. So, the transmission fiber should have transmission ability for the ultraviolet and visible light. In here, the figure (2) is the preliminary design of optical fiber probe, but some technology problem is searching.

3.2 The Design of Polychromator

Based on the traditional spectrograph principle, according to spatial dispersion and multichannel sensor technology, a microscopic polychromator is designed which adapts to on-site application. This instrument can efficiently indentified emitted fluorescence of oil hydrocarbon, finally, get spectrum signal which is collected and controlled by data acquisitions and control system. In this article, the polychromator adopt flat field holographic concave grating which had the function of dispersion and focusing, it has a smooth and level spectrum surface, and in addition, it can reduce the energy loss. This structure directly influent the measurement accuracy and functional reliability of the system, by which the emitting light is transmitted from optical fiber probe to CCD. The figure (3) shows the design of polychromator.

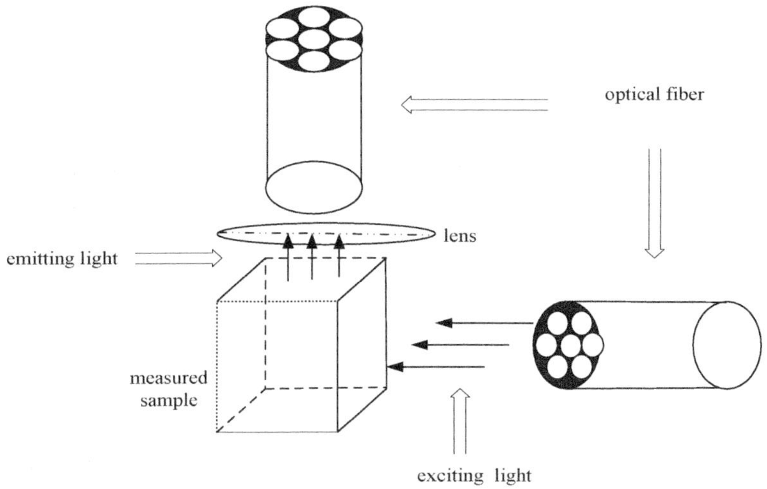

Fig. 2. Principle diagram of probe

3.3 The High Performance CCD

CCD [2] has some characters which are excellent photoelectric response quantum efficiency, wide frequency response, large dynamic range, low noise, small current, high sensitivity and good phase sensitive unit uniformity. Ordinary optical glasses absorb more light in ultraviolet band. In normal optical glass as window, this ends spectral response curve at about 350nm [3]. However, the device in quartz glass as window which absorbs little light in ultraviolet band, and its spectral response curve extend to outside ultraviolet area about 200nm. In this article, the detected aim of CCD is organic molecules, and its fluorescent wavelength range from ultraviolet to visible region. So we should collect the device with quartz glass window. The output charge in CCD is relation to phase sensitive unit exposure. Only at lower saturation exposure, do they have suitable linear relation. Therefore, we should control exposure time in application, ensuring it at range of linearity.

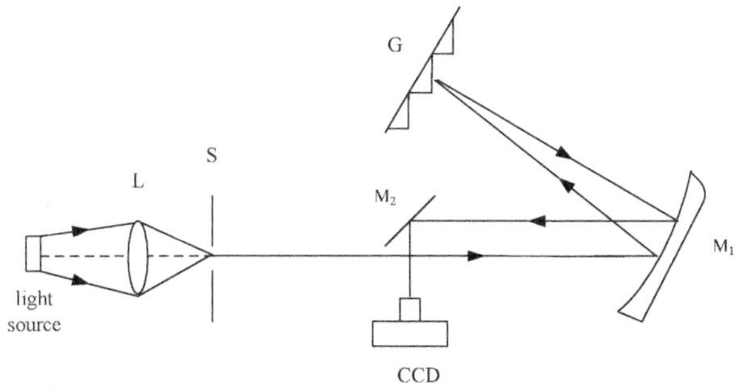

Fig. 3. The CCD structure of spectrum measurement

Figure 3 is the CCD structure of spectrum measurement. The light from light source image spectrum at the S position of incident slit through the condenser L, after G grating diffraction imaging on the CCD photos face. Own to cancel the exit slit, CCD can resave wider spectrum band, finally, the information delivered into computer by vision signal.

3.4 Fluorescent Signal Processing System

Fluorescent signal processing method based on principal component analysis method and traversal least-square method (NNLS) technology combining oils hydrocarbons type recognition technology. PCA [4] is a process of feature selection and feature extraction process. Its main target is to seek a suitable eigenvector in the largest input space and extract principle character in all characters. Feature selection is from data space to the feature space. In theory, the feature space and data space dimension is the same. However, when the transformed space obtains several vectors which include all main information, then can consider reducing the number of secondary feature to reduce the dimension of feature space.

By singular value decomposition the same hydrocarbon features spectrum are extracted, forming a standard oil hydrocarbon fluorescence spectrum. Accordingly, fluorescence spectrum bank is built. At a given temperature conditions, each special take out a spectrum from the character spectrum bank to compose a design matrix, which make up multivariate linear regression model with the spectrum of mixture, and adopt the nonnegative least squares to calculate regression coefficient and residual error. Then the ergodic characteristic spectral [5] of different species combined, choose the regression coefficient of minimum residual error as split outcome. Therefore, in the same conditions, only choose a pure specific spectrum, and choose m pure specific spectrum to fit by nonnegative least-square, finally, m coefficient are solved. Consequently, the coefficient in minimum residual error is considered as the different concentration. In this way, you can monitor oil hydrocarbon species in water environment.

4 The Experience Planning

According to the standard of chemical reagent preparation program, the sample solution is made up. First made up standard mother liquor (water and standard diesel), then gradually to dilute the solution of samples, gaining these proportion solution which are 10 ppt, 5ppt, 2.5ppt, 1ppt, 0.5ppt, 0.05ppt, 0.01ppt, 0.005ppt. In the sample preparation process, the use of ultrasonic oscillations device to ensure diesel dissolved uniformly, to avoid water and oil separated (oil droplets float on the surface) in high concentrations. In the experiment process, the optical fiber probe must be keep clean, and the laboratory temperature should be control at normal temperature.

Choose 200nm-400nm as excitation wavelength, in this band, the main saturated hydrocarbon in oil composition can emit intensity fluorescence. This system mainly analyzes the spectrum in software of MTLAB, including the algorithm of principal component analysis (PCA), non-negative least-square (NNLS), and traversal (zigzag).

5 Conclusions

This paper first introduces the fluorescence detection principle of water environmental pollutants. This system mainly lies in the innovation of forward more effective and reliable design of optical fiber probe, the main method is to use principal component analysis method and non-negative least-square method combined to identify the hydrocarbons. This system is more suitable for the online monitoring and is to carry conveniently.

Financing Projects

This work is supported by the National Nature Science foundation of China. (No.: 60974115) and by the Natural Science Foundation of Hebei Province. (F2010001313)

References

1. Jin, W., Ruan, S.: Optical fiber sensing technology progress (2005), http://www.sciencep.com
2. Zhang, H., Ma, X.: Spectrum measurement for linear CCD: physical experimentation, vol. 25(10) (2005)
3. Jeff, P.: Modern UV-VIS spectroscopy: Adecade of fiber-optica CCD array spectrophotometers. International Scientific Communication Inc. (2004)
4. Jeff, P.: Testing the water. Laurin Publishing Co. Inc. (2004)
5. Wang, S., Che, R., Liu, R., Wang, Y.: Study of fiber fluorescence spectrometer for monitoring carbamate imidacloprid insecticide, Board of Optronics Lasers (2004)

Design and Implementation on the New Method of Water-Containing Measure for Grain

Zhai Baofeng[1], E. Xu[2,3], Wang Quantie[2], and Zhang Yizhi[3]

[1] Software School, Liaoning University of Technology, Jinzhou 121001, Liaoning Province, China
[2] Liaoning Engineering and Professional Technology Institute, Tieling 112000, Liaoning Province, China
[3] Electronic & Information Engineering School, Liaoning University of Technology, Jinzhou 121001, Liaoning Province, China
exu21@163.com

Abstract. This paper introduces a newly-structured capacitance sensor which is used for water-containing measure for grain. The special circuit is designed and measure result is also given. It analyzes the factors that affect the measure and puts forward reasonable method.

Keywords: water-containing measure; capacitance sensor; C-V switch; BP nerve net.

The capacitive method is an usual water-containing method for grainy grain. Compared with the dielectric constant of dry grain's 2~5, the pure water's 81 [1] is quite bigger. As a result, water content in grain will influence its dielectric constant directly[1]. So, we can measure water content in grain according to this property reasonably. In addition, the sample's temperature and compactness affect the measurement as well.

Soybeans is chosen as the sample in the experiment [2,3]. The capacitance detection circuit adopts linear C-V switch circuit, then the output voltage varies directly with the capacity value of the sample; And the compactness is determined by fixed volume weight, pour the sample into a certain apparatus by free falling firstly, get some sample and weigh it secondly, then pour it into the sensor(cylindrical capacitance sensor) on the same way above mentioned; The sampling temperatures are 10℃, 15℃, 20℃, 25℃.

1 The Design of Capacitance Sensor and Detection Circuit

The experiment adopts coaxial cylindrical capacitance which includes two coaxial metallic(cupreous) cylinder as shown in figure 1.

L. Qi (Ed.): ISIA 2010, CCIS 86, pp. 172–177, 2011.

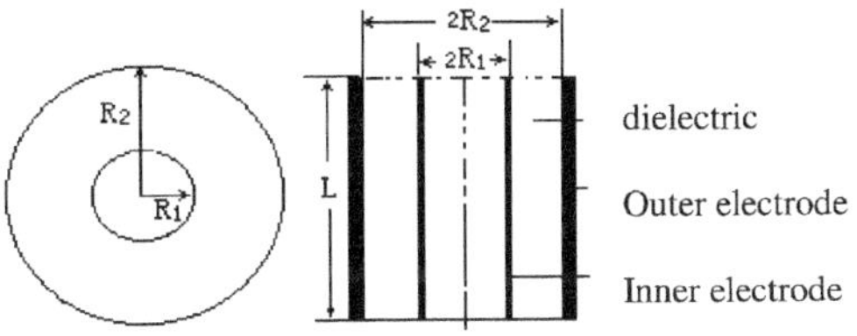

Fig. 1. The structure of coaxial cylindrical capacitance

It is vary easy to pour the grain into Sensor with this structure, and coaxial cylindrical sensor can minimize edge effect to capacitance. The formula of coaxial cylindrical capacitance sensor is as follows [2]:

$$C = \frac{2\pi\varepsilon L}{\ln\dfrac{R_2}{R_1}}$$

C--capacity ε--dielectric constant L--cylinder's height of capacitor R_2--outer electrode's diameter of capacitor R_1--inner electrode's diameter of capacitor

The variation of capacity with grain and not is(ε_0 is the dielectric constant without grain):

$$\Delta C = C - C_0 = \pi(\varepsilon - \varepsilon_0)\frac{L}{\ln\dfrac{R_2}{R_1}} = K\pi(\varepsilon - \varepsilon_0)\frac{R_2}{\ln\dfrac{R_2}{R_1}}$$

If $\dfrac{\partial \Delta C}{\partial R_2} = 0$, then $\left(\ln\dfrac{R_2}{R_1} - 1\right) = 0$, so when $R_2 = eR_1$,the capacity is maximum. Sensor used in experiment has a structure of R_2=40mm、R_1=15mm、L=80mm, thickness of copper-plate is 0.5mm.

C-V switch circuit, as follow figure 2

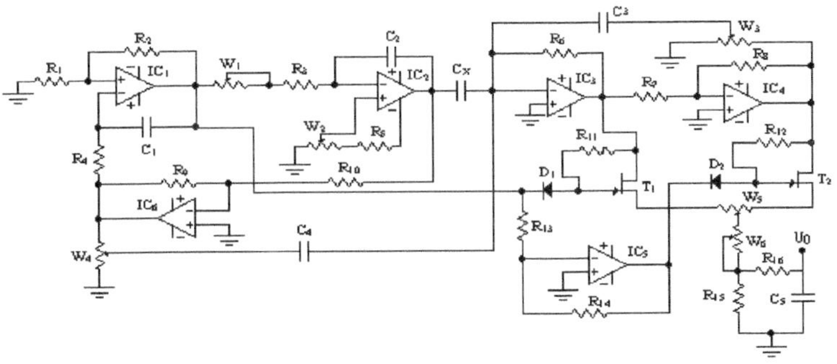

Fig. 2. The detection circuit of capacitance sensor

The ac method is used for capacity detecting, which is of a lot of advantages, good degree of linearity, strong ability of anti-interference and anti-parasitic. So the method is suitable for field test. In addition, the circuit has simple structure, good stability and small volume. It is a portable moisture meter. C_X is just the coaxial cylindrical sensor connected in figure 2.

2 Experimental Data and Analysis on It

In view of water content variation in grain affecting capacity value largely, the range of the detecting circuit is 10pF-350pF according to sensor's structure and the experimental result. C-V switch data show in table 1. According to the data in table 1, the C-V switch output as fairly ideal linear relation.

Table 1. Relation of C-V switch

capacity(mV)	10	15	20	25	30	35	40
voltage(mV)	14.7	23.5	32.3	41.3	49.9	58.9	67.6
capacity(pF)	50	75	100	125	150	175	200
voltage(mV)	85	129	173	217	261	305	350
capacity(pF)	225	250	275	300	325	350	375
voltage(mV)	394	438	482	526	571	615	659

With different water contained and temperature, the output voltage is in table 2.

Table 2. Relation between output voltage and water contained

10℃	water (%)	9.1	10.1	11.0	12.8	14.6	15.3	16.5	17.8
	voltage (mV)	45.3	62.4	77.4	97.9	104.4	112.5	140.3	177.6
15℃	water (%)	8.7	9.9	11.1	13.0	14.9	15.8	16.5	18.3
	voltage (mV)	52.6	76.5	96.8	109.5	120.8	137.6	160.5	212.0
20℃	water (%)	8.3	10.0	11.5	13.8	15.2	16.9	18.4	19.5
	voltage (mV)	49.1	79.7	107.6	121.6	132.4	181.5	232.2	266.4
25℃	water (%)	8.2	10.0	12.0	14.4	15.8	17.6	19.8	20.4
	voltage (mV)	58.0	96.7	113.3	128.2	143.6	182.3	250.4	296.1

Different relations between water contained and weight of certain volume shows in table 3.

Figure 3 is the fitted curve of output voltage and water contained at 15℃.Figure 4 is the fitted curve of water contained and weight of certain volume at 15℃. the fitted curves are similar to figure 3 and figure 4 at other temperatures.

Table 3. Relation between water contained and weight of certain volume

10°C	water (%)	1 9.	10.1	11.0	12.8	14.6	15.3	16.5	17.8
	weight (g)	231.8	230.0	230.1	228.6	226.9	223.5	221.5	217.3
15°C	water (%)	8.7	9.9	11.1	13.0	14.9	15.8	16.5	18.3
	weight (g)	234.7	232.2	232.6	230.5	228.6	223.8	220.6	215.4
20°C	water (%)	8.3	10.0	11.5	13.8	15.2	16.9	18.4	19.5
	weight (g)	235.3	232.4	231.8	230.7	227.9	223.4	218.9	214.8
25°C	water (%)	8.2	10.0	12.0	14.4	15.8	17.6	19.8	20.4
	weight (g)	235.4	232.5	230.6	229.8	226.5	222.9	214.5	212.2

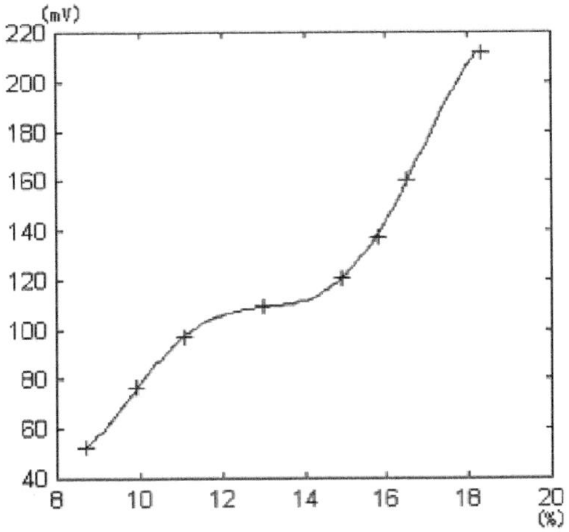

Fig. 3. Relation of output voltage and water contained at 15°C

At the same temperature, there are two inflexions in the curve according to figure 3. The voltage ascends steadily between two inflexions, whereas the curve changes notably out of two inflexions. It is truly that the sample's capacity become

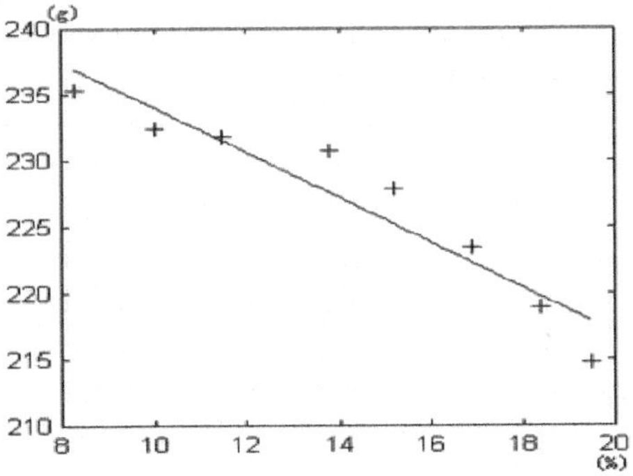

Fig. 4. Relation of water contained and weight of certain volume at 15°C

bigger with the increase of water contained in grain. From figure 4, we can see the water content and the compactness act as the monotonic decreasing relation. Namely, the compactness become looser with the increase of water, that is to say, the bigger the gap in sample is, the lighter the sample. Otherwise, the capacity of sample with same water amplifies when temperature rises. So the output voltage of detecting circuit rises.

3 Data Processing

During the detecting, there is the cross sensitivity(capacity and compactness are reflected by voltage and weight of the same volume) among three sensors of capacity, temperature and compactness. So it is wrong to measure water contained in grain only by the voltage at last.

BP is a fairly matured network model in neural net. It is a typical and a kind of widely used multi-layer network [3]. It contains input-layer, hidden-layer and output-layer. The layers are full meshed. The basic processing units(input-layer besides) are non-linear output relations. S function is usually chosen as the activity function. The input and output values of processing units can vary continuous.

There are 221 teams for test sample, each set of data corresponds to a set of calibrated data. With voltage, temperature and weight being input data, the corresponding value of water detected is expected value. It adopts two layers BP net, the number of hidden-layer's nerve cell is 15, the number of exercises is 50000, the study factor is 0.02, errors as follows' data. The experiment uses Matlab neural net toolbox.

Difference between net output result and expected value-δ's distribution shows in Table 4:

Table 4. Result and expected value-δ's distribution

error	>1.0	>0.5	>0.4	>0.3	>0.2	>0.1	<=0.1
0.0005	32	60	21	31	23	25	29
0.0004	24	53	24	28	32	31	29
0.0003	18	45	26	29	31	34	38
0.0002	9	38	20	29	37	47	41
0.0001	2	23	11	20	28	56	81

According to the experimental result, it is closer to expected result as the error becomes smaller.

If using the other 66 teams data in BP network with error of 0.0001, then the net output result and expected value-δ's distribution shows in Table 4:

Table 5. Testing error distribution ofδ's value at 0.0001

error	>1.0	>0.5	>0.4	>0.3	>0.2	>0.1	<=0.1
0.0001	1	5	5	7	9	8	31

It is obviously that BP net is of good generic function according to the data.

For the 9 teams data used to test on pointing pointer(bulky errors besides), all the difference of result and expected value below 0.5, after the output value of water contained by BP net can be calculate the average.

4 Data Processing

According to the experiment, there are many influence factors upon capacitive method for water-containing in grain .Several sensors are used to take sample of data with different factors on the hardware; because of the interaction of factors, there won't be a regular mathematical model. In the software's aspect, the data fusion of manual neural network can deal with sampling data synthetically is used. Then satisfactory and greater accuracy detecting results are achieved.

Acknowledgement

This work is supported by the National Natural Science Foundation of China under Grant No. 60674056, 70771007, 70971059; Liaoning doctoral funds under Grant No. 20091034, Liaoning higher education funds under Grant No. 2008T090 and Chinese postdoctoral funds under Grant No.20100471475.

References

1. Zong, W.-l.: The development of capacitance senor for buggy. Journal of China University of Mining and Technology 28(3), 24–26 (1999)
2. Chen, W.-r.: Electromagentics, p. 161. tsinghua university press, Beijing (1994)
3. Cong, S.: Nerve net theory and application in the direction of MATLAB tools, pp. 45–46. University of Science and Technology of China Press, He Fei (1998) ISBN7-312-01048-2/TP.218

An Image-Segmentation Method Based on Improved Spectral Clustering Algorithm

Chang-an Liu*, Zhen Guo, Chunyang Liu, and Hong Zhou

School of Control and Computer Engineering,
North China Electric Power University, Beijing, China
liuanhit@163.com

Abstract. An image-segmentation method based on the improved spectral clustering algorithm is proposed. Firstly, the improved hill-climbing is applied to searching for the optimal clustering center in order to avoid classical spectral clustering algorithm's heavy dependency on initial clustering center. The search directions are added to improve global search capability of hill-climbing method to avoid the local optimum. Secondly, to reduce cost of computation, the pixels that have the same gray scale values are merged in the image. Finally, the improved spectral clustering algorithm is applied to image-segmentation. The experiment results prove that the method proposed in the paper is more stable, fast and effective, and the effect of the segmentation is better and obvious.

Keywords: improved spectral clustering; hill-climbing; image-segmentation.

1 Introduction

Image-segmentation is a fundamental problem in image processing and computer vision area. It means separating the image into characteristic, mutually disjoint, similar connected sub-regions [1]. In recent years, scholars home and abroad have conducted much research on image-segmentation and proposed many methods, such as threshold segmentation [2], region growing [3] and watershed algorithm [4]. However, neither a common method nor an objective standard can judge the effect of segmentation now [5].

Spectral clustering based on image is proposed recently as a method of image segmentation. However, there are some limitations on accuracy. The image-segmentation method based on the improved spectral clustering algorithm is proposed in the paper, reducing the number of spectral clustering samples by merging the pixels that have the same gray scale values. In the space of L*a*b, spectral clustering is then used to cluster different samples, merging the pixels that belong to the same cluster. The improved hill-climbing is applied to searching for the optimal clustering center in order to overcome the disadvantage of heavy dependency of classical spectral clustering algorithm on initial clustering center and to enhance the accuracy of clustering.

* Liu Chang-an: Male, Professor, researching fields: Technology of intelligent robot, theory of artificial intelligence.

L. Qi (Ed.): ISIA 2010, CCIS 86, pp. 178–184, 2011.

2 Improved Spectral Clustering

2.1 Spectral Clustering

In recent years spectral clustering arises as a novel clustering method attracting widespread concerns in the field of pattern recognition [6]. Spectral clustering algorithm is derived from the spectra division theory, and its essence is to turn the clustering problem into the problem of graph's optimal partitioning. The algorithm can cluster in the sample space of any shapes, and it is easy to understand and implement, difficult to fall into local optimum. It has been successfully implemented in such fields as bioinformatics [7] and image segmentation [8].

Many different implementations of spectral clustering algorithm can be summarized in three main procedures:

Step 1: Calculation of data set of matrix Z;

Step 2: Calculation of Z by the first q eigenvalues and eigenvectors to construct the feature vector space;

Step 3: Using K-means or other clustering algorithms to cluster feature vectors in feature vector space.

In this paper, the improved hill-climbing is applied to searching for the optimal clustering center in order to overcome the disadvantage of the heavy dependency on initial clustering center of the classical spectral clustering algorithm and to improve the accuracy of classification.

2.2 Improved Hill-Climbing Method

Hill-climbing method is a local search algorithm. Before each step in the climbing, a climber first calculates the values after four steps to the east, south, west and north [9]. If the optimal value of these four is preferable to the current value, then the current value is replaced by the optimal value and the search continues. Otherwise, the current point is thought to be the peak and the search ends. Each step of mountain climbing is orients to the steepest gradient direction, which is not a blind search, so the path to the summit can be found quickly. However, if there are several peaks, the algorithm will be trapped into local optimum easily and can not find the real peak.

The global search ability of hill-climbing method is improved by increasing search directions in this paper. Assume current position of climber is X_i, n positions are generated random around, and the jth searching position is $Y_j (1 \le j \le n)$, then:

$$Y_j = X_i + rand(j) \bullet step \tag{1}$$

where $rand(j)$ is a random number uniformly distributed in the range [-1,1]; *step* is the length of searching step. Select the optimal one from n points around the climber as the direction of the next step.

Improved hill-climbing method avoided the disadvantage of hill-climbing falling into local optimum easily, and held good global search ability.

2.3 Improved Spectral Clustering

In order to improve the disadvantage of spectral clustering algorithm that holding high dependence on cluster centers, the K-means algorithm is applied to cluster the feature vectors in this paper. K-means algorithm [10] is a classical clustering algorithm to solve the problems and has the advantage of fast convergence. However, the K-means algorithm can easily fall into local optimal solution and is difficult to obtain global optimal solution. The improved hill-climbing has good global search capability, which can help the spectral clustering converge to the global optimum.

In order to eliminate the effect of data magnitude differences on classification, they should first be normalized. Generally, the normalization formula for *j*th datum in *i*th-dimensional space is as follows:

$$\frac{x_{ij} - \min(x_{i.})}{\max(x_{i.}) - \min(x_{i.})} \tag{2}$$

$$\frac{\max(x_{i.}) - x_{ij}}{\max(x_{i.}) - \min(x_{i.})} \tag{3}$$

where $x_{i.}$ is all the data sets in *ith*-dimensional space, $\max(x_{i.})$ and $\min(x_{i.})$ are maximum and minimum of $x_{i.}$.

Suppose a given data set $X = (x_1, x_2 ..., ..., x_n)$ of *n* data, and $x_i \in X\,(i = 1, 2, ..., n)$ is a *d*-dimensional vector, data sets are to be divided into *k* classes by K-means algorithm.

In the cluster analysis, clustering criterion function is a measure of cluster:

$$J = \sum_{j=1}^{k} \sum_{\forall x_l \in C_j} D(x_l, m_j) \tag{4}$$

where $D(x_l, m_j)$ is the distance between data x_l and cluster center m_j of the corresponding cluster C_j in the cluster results; J is sum of all distances. If the value J is smaller, the clustering effect is better.

Euclidean distance is adopted in K-means algorithm:

$$D(x_l, m_j) = \sum_{i=1}^{d} \sqrt{| x_{li}^2 - m_{ji}^2 |} \tag{5}$$

where x_{li} is ith value of x_l, m_{ji} is ith value of m_j.

Improved spectral clustering algorithm can be described as follows:

Step 1: Normalize data sets according to formula (2) ;

Step 2: Build the affinitive matrix A .If $i \neq j$, $A_{ij} = \exp(- \| x_i - x_j \|^2 / 2\sigma^2)$; else, $A_{ii} = 0$;

Step 3: Calculate laplacian matrix L, $L = D^{-1/2} A D^{-1/2}$, D is the diagonalizable matrix, $D_{ij} = \sum_j A_{ij}$;

Step 4: Calculate n-eigenvalues ($\lambda_1 \geq \lambda_2 \geq ... \lambda_n$) and corresponding eigenvectors ($\zeta_1, \zeta_2, ..., \zeta_n$) of L, then sort out the corresponding eigenvectors to constitute matrix Z based on the order of eigenvalues and select eigenvectors corresponding to the first q eigenvalues as classification samples;

Step 5: Choose arbitrary k data sets as the initial cluster centers;

Step 6: Centralize the data according to cluster centers and assign the objects to the most similar clusters;

Step 7: According to the results, update optimum clustering centers by improved climbing method with formula (4) as target function;

Step 8: Judge whether each clustering changes. If changed, go to Step 6; otherwise, end algorithm.

3 Image Segmentation Algorithm Based on Improved Spectral Clustering

The image with different colors can be separated effectively in L*a*b space. This RGB space of images is converted to L*a*b space and then spectral clustering algorithm is applied to segment images. The cost of eigenvector computations is high due to the large number of pixels in the image,. So the pixels of the same gray scale values are merged after RGB values of images are converted into gray scale values. The corresponding number of pixels with the same gray scale value is recorded. Calculate the centers $CC_i (1 \leq i \leq 255)$ of the pixels with the same gray scale values in L*a*b space and cluster these centers as samples. Then steps of the image segmentation algorithm based on improved spectral clustering can be summarized as follows:

Step 1: Calculate gray scale values of images, merge pixels of the same gray scale value together;

Step 2: Transform images to L*a*b space, calculate the centers of the pixels with the same gray scale value;

Step 3: Use improved spectral clustering algorithm to cluster center CC_i;

Step 4: Pixels corresponding to the center points of clusters are image segments of the same color characteristic. Thus the image segmentation is achieved.

4 The Experiment and Analysis

In order to verify the effect of improved spectral clustering algorithm, international public data sets Iris and Wine of UCI database were adopted in the experiment. The details of each data set are shown in table 1. In improved hill-climbing method, the number of searching direction n is set 10 and searching step length 0.005, the number of eigenvectors q is selected to be 3, σ in affinitive matrix is 2. K-means clustering algorithm, spectral clustering algorithm and improved spectral clustering algorithm cluster data sets 20 times each in MATLAB 7.0. The maximum, minimum, average values of clustering accuracy are compared in table 2.

Table 1. The Introduction of data sets

Data set	Quantity	Dimension	Cluster number
Iris	150	4	3
Wine	178	13	3

Table 2. The comparison of the three algorithms

Data	Type	K-means	Spectral clustering	Improved Spectral clustering
Iris	Maximum	88.67%	97.33%	98.00%
	Minimum	85.33%	90.00%	91.33%
	Average	88.50%	91.93%	92.47%
Wine	Maximum	95.51%	96.63%	97.19%
	Minimum	94.94%	96.07%	96.07%
	Average	95.22%	96.38%	96.57%

It can be inferred from Table 2 that maximum, minimum and average values of K-means algorithm are the least. It illustrates the accuracy of clustering is the lowest and has high error rate. Spectral clustering algorithm has been enhanced and its effect is obvious. Average values of improved spectral clustering algorithm in the two test data sets are the biggest, and the absolute difference between its accuracy maximum and minimum is the least, showing that its accuracy is the highest. Its clustering effect is obvious and its result is stable.

The paper applies spectral clustering algorithm to image-segmentation, and selects three organizational types of H&E images to separate. Figure 1 is original, and the images separated by K-means, spectral clustering and improved clustering algorithm are shown in figure 2, 3, and 4 respectively. As illustrated in image-segmentation effect, K-means has not separated the objective according to actual condition in original figure, as shown in figure 2. Image-segmentation of spectral clustering algorithm is shown in figure 3. The region is divided into 3 parts equally, in which there are only a few holes. As shown in figure 4, compared with image-segmentation of spectral clustering algorithm, the holes in image of improved spectral clustering algorithm are less. Its effect is more close to actual image and its objective is easy to separate. Therefore, improved spectral clustering algorithm is proper for separating images and its effect is obvious.

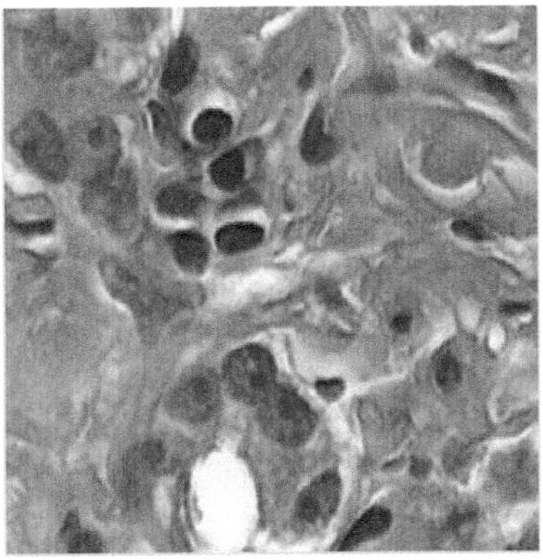

Fig. 1. Original image

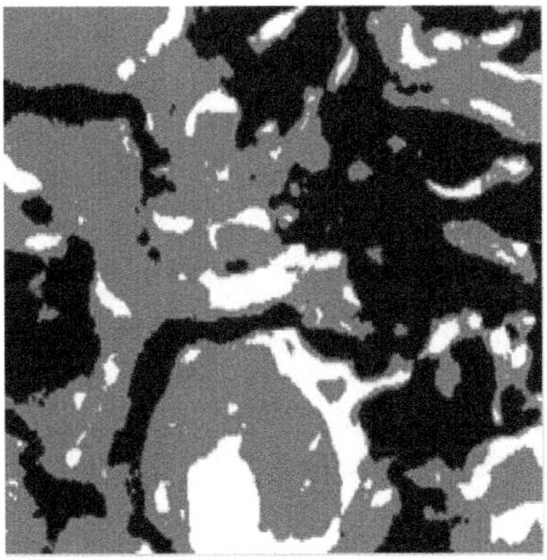

Fig. 2. Segmentation of K-means

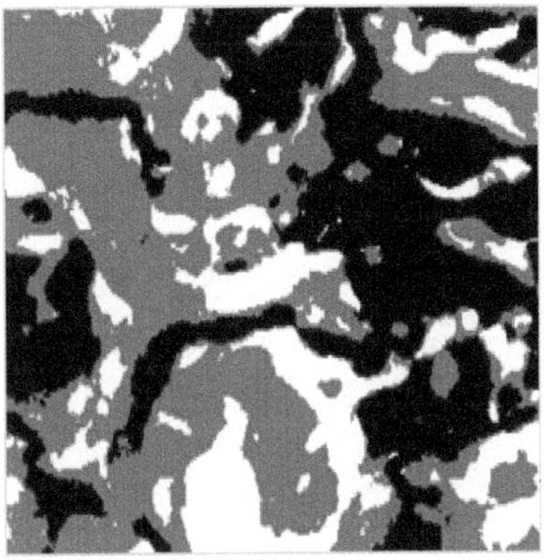

Fig. 3. Segmentation of spectral clustering

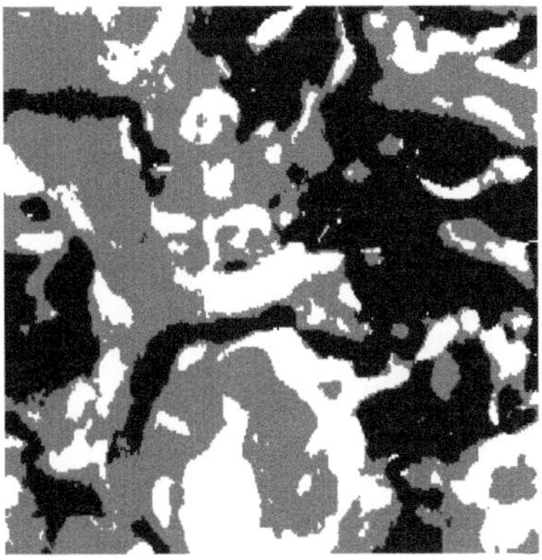

Fig. 4. Segmentation of improved spectral clustering

5 Conclusion

An image-segmentation method based on the improved spectral clustering algorithm is proposed. The improved hill-climbing method is introduced to enhance spectral clustering algorithm. It is designed to avoid heavy dependency of classical spectral clustering algorithm on initial clustering center. K-means clustering algorithm, spectral clustering algorithm and improved spectral clustering algorithm are compared by experimenting with international standard data sets and separating images respectively. It can be concluded as following: (1) n search directions are added to original hill-climbing method in order to improve its global search capability. (2) The improved hill-climbing is applied to searching for the optimal clustering center in order to enhance the accuracy of clustering. (3) Improved spectral clustering algorithm is more stable, fast and effective, and the effect of the segmentation is better and obvious.

Acknowledgments. This research was supported by the Fundamental Research Funds for the Central Universities.

References

1. Zhanga, H., Frittsb, J.E., Goldman, S.A.L.: Image segmentation evaluation: A survey of unsupervised methods. Computer Vision and Image Understanding 110, 260–280 (2008)
2. Taheri, S., Ong, S.H., Chong, V.F.H.: Level-set segmentation of brain tumors using a threshold-based speed function. Image and Vision Computing 28, 26–37 (2010)
3. Saeed, K., Albakoor, M.: Region growing based segmentation algorithm for typewritten and handwritten text recognition. Applied Soft Computing 9, 608–617 (2009)
4. Alaknanda, Anand, R.S., Kumar, P.: Flaw detection in radiographic weldment images using morphological watershed segmentation technique. NDT & E International 42, 2–8 (2009)
5. Kong, M., Sun, X.-P., Wang, Y.-J.: The algorithm of threshold image segmentation based on the variance between two classes. Journal of Huazhong University of Science and Technology 32, 46–47 (2004)
6. Wang, L., Bo, L.-F., Jiao, L.-C.: Density-Sensitive Semi-Supervised Spectral Clustering. Journal of Software 18, 2412–2422 (2007)
7. Higham, D.J., Kalna, G., Kibble, M.: Spectral clustering and its use in bioinformatics. Journal of Computational and Applied Mathematics 204, 25–37 (2007)
8. Wu, R., Huang, J.-H., Tang, X.-L., Liu, J.-F., et al.: Method of Text Image Binarization Processing Using Histogram and Spectral Clustering. Journal of Electronics & Information Technology 31, 2460–2464 (2009)
9. Storey, C.: Applications of a hill climbing method of optimization. Chemical Engineering Science 17, 45–52 (1962)
10. Tao, X.-m., Xu, J., Yang, L.-b., et al.: Improved Cluster Algorithm Based on K-Means and Particle Swarm Optimization. Journal of Electronics & Information Technology 32, 92–97 (2010)

An ACO and Energy Management Routing Algorithm for ZigBee Network

Peng You[1], Yang Huixian[1], and Man Sha[2]

[1] Faculty of Material and Photo electronic Physics, Xiangtan University, Xiangtan, Hunan 411105

[2] College of Information Engineering, Xiangtan University,

pengyou@live.cn

Abstract. Network performance and nodes energy problems are considered as one of the most important aspects of a wireless network based on ZigBee technology. How to prolong the life of ZigBee networks is an important purpose to design ZigBee routing protocol. Based on energy management, ACO-AODV (Ant Colony Optimization-Ad hoc On-Demand Distance Vector Routing) routing protocol can keep the good network performance and extend the life of ZigBee networks. Simulation results show that ACO-AODV algorithm is feasible and energy conservation. This approach can maintain a low the average end to end data packet delay while effectively reducing energy consumption, so the performance of ZigBee networks can be improved and achieve the low energy consumption, low-Delay design goals.

Keywords: ACO; ZigBee; AODV; Energy management.

1 Introduction

The main difference of ZigBee network and other networks is its mobility, ZigBee network performance largely depends on the efficiency of the battery. Therefore, to extend the ZigBee network nodes lifetime need to make full use of battery power. Compare to the computer and other communication technologies, battery technology have slow progress. Therefore, design a routing algorithm for ZigBee network from the key node features can large extent reduces the energy consumption. This article focuses on two aspects to improve the energy consumption of ZigBee networks. The first is that use the ant colony algorithm to optimize the AODV routing algorithm and reduces the routing overhead in order to maximize use of limited energy and prolong the life of ZigBee network; the second is that add energy management strategy to avoid the excessive use of low-energy nodes premature easy failure nodes, to balance the overall network energy consumption.

2 ZigBee Routing Algorithm and Ant Colony Algorithm

2.1 ZigBee Technology

ZigBee is an emerging low-cost, low-power short-range wireless communication protocol, mainly used for short-range wireless connection. It has lots of tiny sensors to

L. Qi (Ed.): ISIA 2010, CCIS 86, pp. 185–191, 2011.

achieve communication between the co-ordinations. The sensor requires very little energy to relay the data adopted by from a wireless sensor to another, the communication efficiency is very high. ZigBee protocol stack above the 802.15.4 standard, 802.15.4 standard defines MAC and PHY layer protocol standard, developed based on IEEE802.15.4, high reliability, high cost, and low power network application specifications. Fig. 1. is a ZigBee protocol stack diagram. Fig. 2. is a ZigBee network topology.

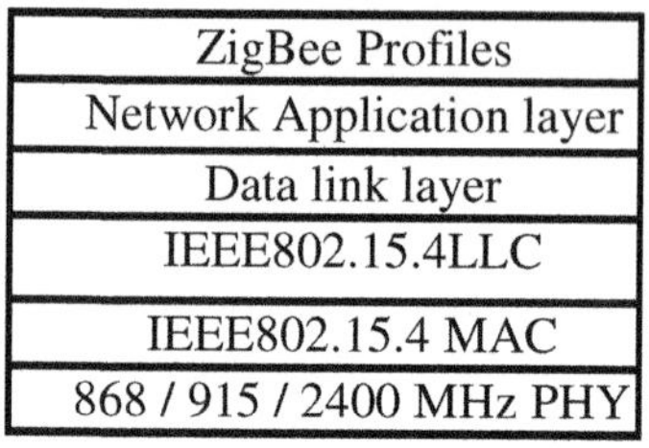

Fig. 1. ZigBee protocol stack diagram

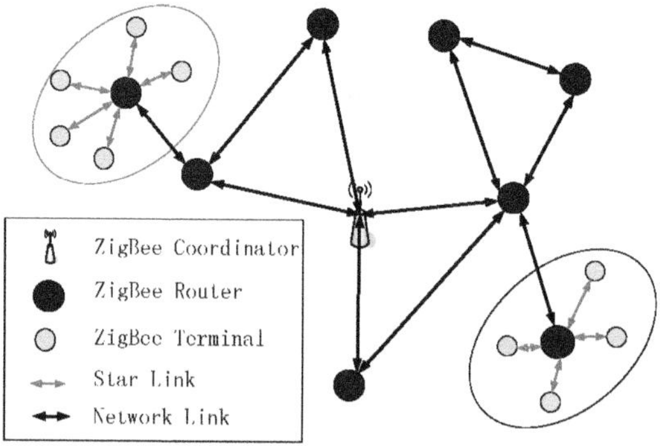

Fig. 2. ZigBee network topology

2.2 AODV Protocol and Its Advantages and Disadvantages

AODV protocol is a wireless ad hoc on-demand distance vector routing protocol, which is used in the wireless mesh network for routing. The routing protocol is generated on demand Ad Hoc network routing protocols and the typical reactive routing protocol. Only when sending data to the destination node, the source node before the network launched the process of routing lookup to find the appropriate route, and only in the communication process was the maintenance of routing, when the communication will be removed after the routing, nodes only need to save them as the source node or intermediate nodes required to reach the destination node communication routing, routing overhead low.

Demand routing protocols AODV has some shortcomings through simulating:

1. When a node sends data packets with no route exist, it can only wait until a suitable route can be set up to send. Therefore, there are longer AODV routing delays.

2. When establishment of routing, the neighbor nodes in turn broadcast the packet to the peripheral node until the packet is sent to a destination node routing information that the intermediate nodes. But by the intermediate nodes to find the route may be not the best route.

3. Does not support the radio nodes dynamically adjusted so energy saving strategy.

2.3 Ant Colony Algorithm and Its Advantages and Disadvantages

Ant Colony Optimization (ACO) is a method used to find optimal paths in the network the probability-based algorithm. Ant colony algorithm basic principle of ant colony algorithm is ants foraging time, between individuals through a known pheromone ("Pheromone") substances transfer information. Ants in the course of the campaign, left and after perception of this material to the path , and use it to guide their own direction of motion: the ants walking on a path to more information on the more left, which select the path more likely.

Basic ACO routing algorithm found shortcomings in the simulation:

1. Nodes are a separate agency to rely on ants to find the shortest route. When the network large changes and routing life-cycle is small, the performance is not good.

2. In the ACO routing algorithm does not consider energy management, and ZigBee node energy is limited, the key routing nodes vulnerable to node failure.

3. ACO routing algorithm implementation process, not as AODV algorithm similar to the maintenance of the local node to connect, so when a route break, the source node does not know the link has already been sent off to the data. This will result in the loss of large amounts of data.

3 ACO Routing Algorithm Design and Energy Management

3.1 ACO-AODV Algorithm Design Ideas

Taking into account the lower costs, reduce energy consumption and ease of use, ZigBee general use the AODV routing protocol. To further extend the network lifetime, save energy, to overcome the traditional shortcomings of AODV routing protocol, an improved ACO-AODV routing algorithm is designed. Algorithm is mainly designed in two ways:

1. Energy Management: Based on long-established use of energy management, low power routing pheromone reduced the probability of being selected, high-capacity pheromone increases the probability of priority, greatly reduce the failure rate of nodes.

2. Reduce the delay: According to the number of hops and delay, and residual energy function constructed pheromone increment, increase the short path and the chance to

select the optimal route. Before AODV algorithm multicast add the ant colony optimization, quickly identify qualified set of routing nodes, narrow multicast range. Access delay function T for the delay of each node:

$$T = \sum_{i=0}^{N} T_i \tag{1}$$

Hop N the routing transit number n, the remaining energy function is formed by the initial energy E_0, the time t and the parameter α. The parameter α for a specific factor, the role of initial energy E_0 is slowing the rate decreases

$$E = \frac{1}{t} \times \sqrt{E_0} \times \alpha \tag{2}$$

From equation (2) you can see the value of the node remaining energy is inversely proportional to the use of time. ACO-AODV algorithm makes the lower residual energy nodes that before avoid to connecting, when time t increases to a certain extent, some of the nodes can continue to use. When time t approaches infinity, that the residual energy E close to zero. Delay function, hops, and the composition of the remaining energy function increment function of pheromone that

$$\Delta\phi_{n,d} = \frac{1}{\omega_1 T + \omega_2 N + \omega_3 E^{-1}} \tag{3}$$

ω1, ω2, ω3 is the weight of each parameter. And through the pheromone iterative formula (4) real-time updates on the pheromone

$$\phi_{c,d}^{i+1} = (1 - \lambda)\phi_{c,d}^{i} + \Delta\phi_{c,d}^{i} \tag{4}$$

λ is the pheromone decay coefficient, $0 < \lambda < 1$, node c ants use pheromones to calculate the concentration of d choose the next node in the probability of $P_{c,d}$:

$$P_{c,d} = \frac{\left(\phi_{c,d}^{i}\right)^{\beta}}{\sum_{i \in M} \left(\phi_{c,d}^{i}\right)^{\beta}} \tag{5}$$

Where β is the adjustable weight of pheromone, M is the node set of all nodes one hop away from node C, all the probabilities and the next hop node 1:

$$\sum_{j \in M} p_{j,d} = 1 \tag{6}$$

3.2 Algorithm Design Process

Network initialization, the relevant node generates exploration ants, according to the formula (1), (2) calculate the time delay T, the remaining energy E and N hops and other related parameters. Based on formula (3) and (4) ants update pheromone increment and then in accordance with the formula (5) to calculate the relevant node selection probability of the next hop to find the optimal path node set, using AODV multicast, and then return later to the ant path, update the information through the optimal path hormone concentration. Routing flow chart of intermediate nodes shown in Fig. 3.

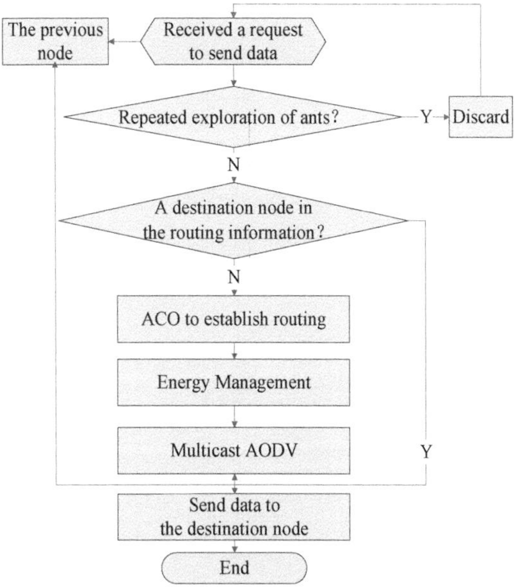

Fig. 3. Routing flow chart of intermediate nodes

4 Experimental Simulations and Analysis of Algorithms

4.1 Experimental Environment

Simulation tools with Linux + NS2, network coverage area 500 * 500m^2, the number of network nodes is set to 50, set the node's transmission distance of 50m. Transmission channel used in data transmission rate of 250kbps, the channel delay of 0.3s, packet length is 128bit. Network settings, all the nodes of the initial energy of 10000 units of energy, to receive a message consumes 1 energy unit, sending a message consumes two units of energy, receiving a packet consumes two units of energy, sends a packet consumption 4 units of energy. Each simulation runs 200s, using an average of 30 times the simulation data.

4.2 Simulation Results

Simulation results shown in Fig. 4, Fig. 5, Fig. 6.

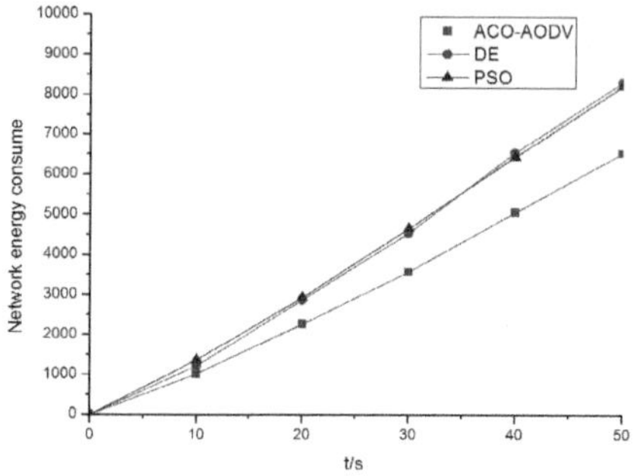

Fig. 4. Comparison of network energy consumption

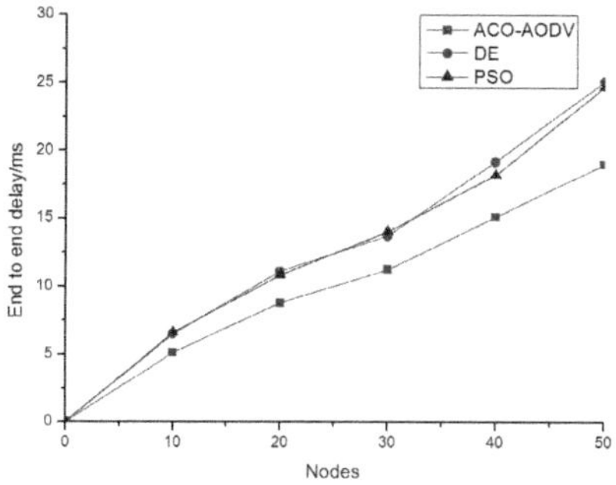

Fig. 5. Comparison of network delay

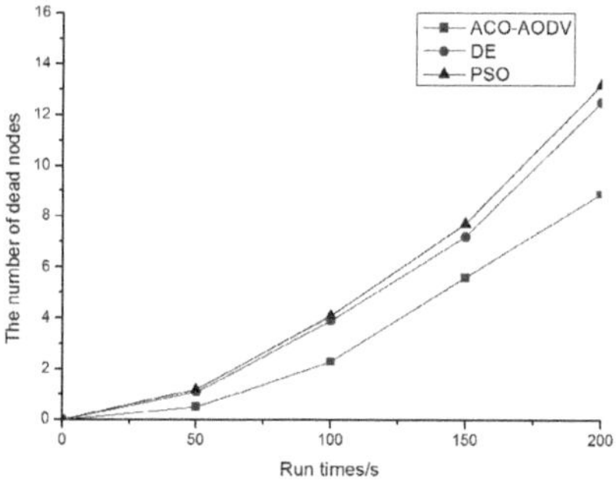

Fig. 6. Comparison of failure nodes

In Fig. 4. ACO-AODV routing algorithm explore the process of delay and hop on the control, and avoid low energy nodes connected, so a significant saving overall energy consumption of the network. Particle swarm optimization and differential evolution algorithm optimize the energy consumption of little effect.

In Fig. 5 when there are fewer nodes, ACO-AODV algorithm network delay is normal, the effect is not obvious. This is because the route when there are fewer nodes to optimize the computational overhead large; when the routing nodes increases, ACO-AODV algorithm delay prominent optimization. Fig. 6 shows the comparison of failure nodes with three algorithms. As can be seen from Figure 6 by adding the node energy management, so optimization occurs first failure node delay, and compared to non energy management of the differential evolution algorithm and particle swarm optimization algorithm failure occurred in operation of the number of nodes significantly reduce.

5 Conclusions

ACO-AODV algorithm based on the ZigBee traditional on-demand routing algorithm, adding ant colony optimization algorithm, and specifically for delay, hop count and residual energy optimization. That a certain extent reduces overall energy consumption while maintaining low latency network performance and reduce the failure node, and thus extend the life of ZigBee networks.

About the Author

Peng You (1983-), male, Hunan Ningxiang people, Xiangtan University, Institute of Materials and Optoelectronic Physics, Master, Research direction: digital signal processing. Tel.: +8615173256570. Yang Huixian, male, Professor, Xiangtan University, master instructor. Research direction: digital signal processing, digital image processing.

References

[1] Jiang, T., Zhao, C.: Purple bee technology and its applications. Beijing University of Posts and Telecommunications Press, Beijing (2006)
[2] Fang, et al.: Based on ZigBee network ZiCL improved algorithm. Computer Applications 02 (2009)
[3] Classes bright and so improve the ZigBee routing algorithm. Computer Engineering and Applications 05 (2009)
[4] ZigBee routing of. ZigBee protocol specification chapter - the network layer [EB / OL] [2007209223], http://www.huiyou98.com
[5] Wang, C., et al.: Based on tree structure of the ZigBee protocol energy balance. Computer Engineering and Design 30(15) (2009)
[6] Zhou, W., Luo, D., Zig, B.: Routing Protocol. Computer Engineering and Science 31(6), 12–14 (2009)
[7] Liang, H.: Based on ant colony optimization of wireless sensor network energy balanced routing algorithm. Pattern Recognition and Artificial Intelligence 04, 275–279 (2007)
[8] Wang, Y., et al.: Based on ant colony optimization routing algorithm. Computer Applications 01, 7–9 (2008)
[9] Friedman, R., Shotland, A., Simon, G.: Efficient route discovery in hybrid networks. Ad Hoc Networks 7, 1110–1124 (2009)
[10] Okdem, S., Karaboga, D.: Routing in Wireless Sensor Networks Using Ant Colony Optimization. In: Adaptive Hardware and Systems 2006, pp. 401–404 (2006)

Monitoring the Bridge's Health Status by GPS and Surveying Robot*

Bao-guo Qian[1], Chen-guang Jiang[2,1,**], and Jian-guo Peng[3]

[1] College of Civil Engineering, Shanghai University,
200444 Shanghai, P.R. China
[2] College of Environmental and Civil Engineering, Jiangnan University,
214122 Wuxi Jiangsu, P.R. China
[3] Communication Plan and Investigation and Design Institute of Hunan Province,
410008 Changsha, Hunan, P.R. China
jiangcg999@yahoo.cn

Abstract. The bridge's health status is affiliated with factors such as bridge's structure form, material adopted, rationality of design, quality of construction and action of load etc; as to the bridge in active service, its health status mainly depends on action extent and action process of various loads, as different bridges show diverse response features to load action, so it is feasible to take preliminary diagnosis on the bridge's health status according to the bridge's response features to load action. In terms of numerous monitoring data of bridge's original position, the diagnosis method of the bridge's health status based on load response effect is put forward, in addition to preliminary diagnosis elements and instances.

Keywords: GPS, surveying robot, bridge project, health examination, structure in active service, load response effect, diagnosis method, original position monitoring.

1 Introduction

There are innumerable and various bridges existing in all regions of the world. Along with increase of the bride's use age together with influence of various environmental factors (wind load, vehicle load, temperature and earthquake), about 30% of these bridges have been damaged with different extents. In order to ensure these bridges are secure and persistent, which further safeguard people's traffic safety, it has become more and more important to take health examination and health diagnosis on these bridges. The bridge scientists and researchers in each country pay more attention to the bridge's health examination and health diagnosis, and also put forward all sorts of theories and methods. The bridge's security monitoring has increasingly become an extremely active study trend in civil engineering subject [1]-[4].

* This work is supported by National Nature Science Foundation under Grant 79160173.
** Corresponding author.

L. Qi (Ed.): ISIA 2010, CCIS 86, pp. 192–198, 2011.

At present, there are many instruments used for monitoring the bridge's structure, mainly including GPS(Global Positioning System), surveying robot(Automatic electronic total station instrument), transit instrument, water level gauge and all kinds of sensors (displacement sensor, acceleration sensor, pressure sensor and strain sensor etc). The traditional examination methods only aim at monitoring the bridge's appearance and certain structure features, while examination results may merely reflect the structure's state of the time partly, difficult to discover damage of hidden components, and unable to make judgment to the bridge's whole health status, especially make systematic appraisal to the bridge's security reserve and degenerative way. To make diagnosis on the bridge's whole health status in virtue of examination data of various bridges, people have adopted all kinds of methods, including frequency region approach, time region approach, small wave transformation, NN technology, mode analysis method, mode modification method, fingerprint analysis method, limited element method, dark theory, and structure reliability theory and so on.

The bridge's damage mechanism belongs to typical nonlinearity issue. Frequency is the mode parameter with higher precision which is most easily obtained during the bridge's online monitoring, so it is a kind of simple and effective method to identify whether the structure is damaged or not through diversification of the bridge responding frequency. In addition, it is helpful as well for discovering damage of the structure to examine precision lower than amplitude of frequency, because the amplitude contains more damage information; in order to estimate occurring way of the structure's damage by right of amplitude, people offered some methods such as MAC method, COMAC method, CMS method, DI method and pliability matrix method.

The writer and seminar discovered that, the bridge's health status is affiliated with factors such as bridge's structure, material adopted, rationality of design, quality of construction and action of load etc; as to the bridge in active service, its health status mainly depends on action extent and process of various loads, as different bridges show diverse response features to load action, so it is feasible to take preliminary diagnosis on the bridge's health status according to the bridge's response features to load action. In terms of numerous monitoring data of bridge's original position, the diagnosis method of bridge's health status based on load response effect is put forward, in addition to preliminary diagnosis elements and instances.

2 Examination Method of the Bridge's Health Status Diagnosis

The examination method, put forward by the writer and seminar, which is a kind of diagnosis method of the bridge's health status based on load response effect, is RTK-GPS monitoring method.

GPS technology is satellite navigation system of the second generation developed by America. RTK-GPS technology is one of the important developments which makes use of receiving the navigational satellite's carrier phase to go on with real-time phase differentiation as well as take the real-time monitoring to all kinds of civil engineering structure's space displacement.

RTK-GPS differentiation system is made up of one GPS reference station, one or more GPS monitoring station and communication system. The reference station transfer the satellite's difference information to each monitoring station via wireless communications or wire communications. Every monitoring station receives satellite's signal or GPS reference station's information independently and after undertaking real-time differentiating they can measure the site's three-dimensional space coordinate. All the monitoring results transmit to bridge monitoring centre real-timely. The bridge monitoring centre undertakes the bridge deformation information's calculation and disposal according to the receiver's GPS differential signal outcome.

When the RTK-GPS technology monitors the bridge, the GPS reference station is mounted in the stable geological framework area and the GPS monitoring stations are mounted on every designed monitoring position on the bridge. As long as every GPS monitoring station can receive more than 6 GPS satellites and reference stations' GPS differential signal, it can undertake RTK-GPS differentiating localization. Every GPS monitoring station has no need to see each other, and the observed value is mutual independence and mutual disrelation. RTK-GPS technology is almost free from climatic conditions' impact and possesses supermatic and all weather monitoring ability. GPS monitoring station's instant three dimensional coordinate can automatic logging into the bridge monitoring centre server via the wireless communications or wire communication to undertake the bridge's safety analysis. The velocity of RTK-GPS's output localization resultant can reach to 10-20 Hz.

The load comes true via mounting the weight dynamic measurements instrument on the top of the bridge pier by which the dynamic load of the bridge floor is obtained. The monitoring data is transmitted to the bridge monitoring centre real-timely via the fiber optic cable or the wireless communications.

The bridge monitoring centre matches and analyses both the instant three dimensional coordinate of the GPS monitoring station and the instantaneous measurement outcome of the weight dynamic measurements instrument. According to the instant three dimensional coordinate and the instant weighing outcome, the load response curve of the bridge deformation can be drawn and in terms of the load response curve's change characteristic the bridge's health status could be diagnosed.

3 Diagnosis Method of Bridge's Health Status Based on Load Response Effect

According to the bridge's response feature to load action (the bridge's deformed load response curve), the preliminary diagnosis on the bridge's health status could be taken.

If the bridge works normally with good health status, the load response curve of bridge's vertical deformation is supposed to be acute and intense fluctuation, see A of Fig. 1.

If the bridge doesn't work normally being lack of health, the load response curve of bridge's vertical deformation ought to be smooth and slow fluctuation, see B of Fig. 1.

If the bridge is in bad working state and bad health status, the load response curve of bridge's vertical deformation should be slow and slight fluctuation, see C of Fig. 1.

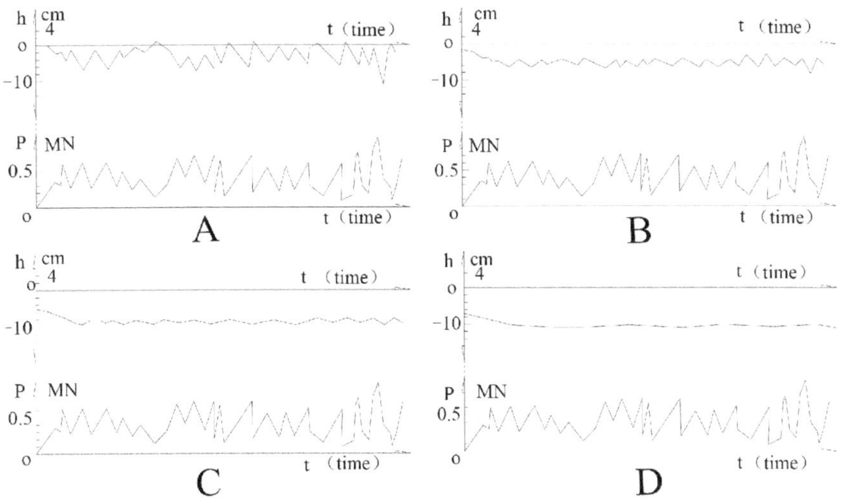

Fig. 1. Load response curve of vertical deformation of the bridge

The load response curve of vertical deformation of dangerous bridges should be invisible to-and-fro and fluctuation, see D of Fig. 1.

4 Instances for Diagnosis Method of Bridge's Health Status Based on Load Response Effect

The writer and seminar had taken devastating load response effect experiment on the old Nancha River Bridge ever.

The old Nancha River Bridge is founded in 1972, belongs to continuous simple-supported beam bridge with 4-span reinforced concrete, each span interval of 21m, and designed driving load of 20ton; due to route alteration of the highroad, there is a new bridge built 30m away from the highroad, so the old Nancha River Bridge is out of action, becoming an otiose bridge to be dismantled; through overall negotiation, the old Nancha River Bridge could be served as testing bridge for seminar to carry out devastating experiment of original position before dismantled.

The original position devastating experiment of old Nancha River Bridge was carried out in winter, when the riverbed had already been dry up; for the sake of guarantee security of experimental vehicle, equipment and personnel, under the bridge we set up full-cloth steel tube supporting protective system, the top of which is 26cm far away from bottom of the bridge, and the width of steel tube supporting protective system is 2m wider than the bridge's width, while there is preventive striking column with inner stow-wood which is 1m higher than level of the bridge at the edge of steel tube supporting protective system.

The devastating load response effect experiment of the old Nancha River Bridge made use of RTK-GPS technology, one GPS reference station is chosen at top of Waizi Mountain which is 300m away from the bridge, while seven GPS monitoring stations are respectively set at the sides of the spanning bridge boards of four bridge spans of old Nancha River as well as sides of bridge boards of three piers in the river, and seven GPS monitoring stations are connected with the bridge board together through connector made specially by the subject team. Seven GPS monitoring stations are connected into string passage of Data Taker System DT80 via fiber optic cable, and Data Taker System DT80 is connected with the notebook computer on the spot, through which GPS real-time monitoring data is disposed.

The bridge loading of load response effect experiment is carried out in virtue of a large freight-handling truck, and 20 concrete matching-weight blocks with size of 1m×1m×1m are specially made.

With total weight of 18ton, the large freight-handling truck got across the old Nancha River Bridge for four times respectively with speeds of 10km/h and 20km/h and 30km/h and 40km/h, accordingly obtained four overall experimental data of load response effect.

Successively, the large freight-handling truck got across the old Nancha River Bridge for six times respectively with total weight of 25ton, 32ton and 41ton, and speeds of 10km/h and 20km/h, obtaining six overall experimental data of load response effect. In 41-ton experiment, 2-span bridge boards of the bridge occur permanent approach perforation crack.

The Fig. 2 is diverse value (namely, subsidence value, or named as vertical displacement value) of GPS Geodetic height of seven GPS monitoring stations at each moment of the freight-handling truck passing the bridge.

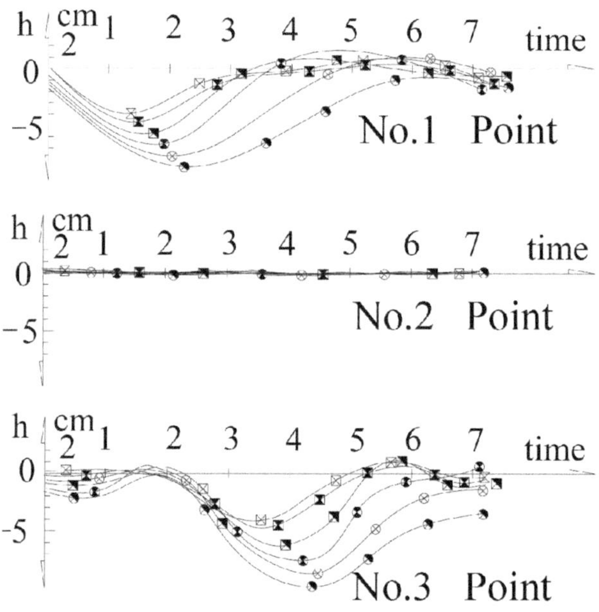

Fig. 2. Diverse value (vertical displacement value) of GPS Geodetic height of seven GPS monitoring stations at each moment of the freight-handling truck passing the bridge

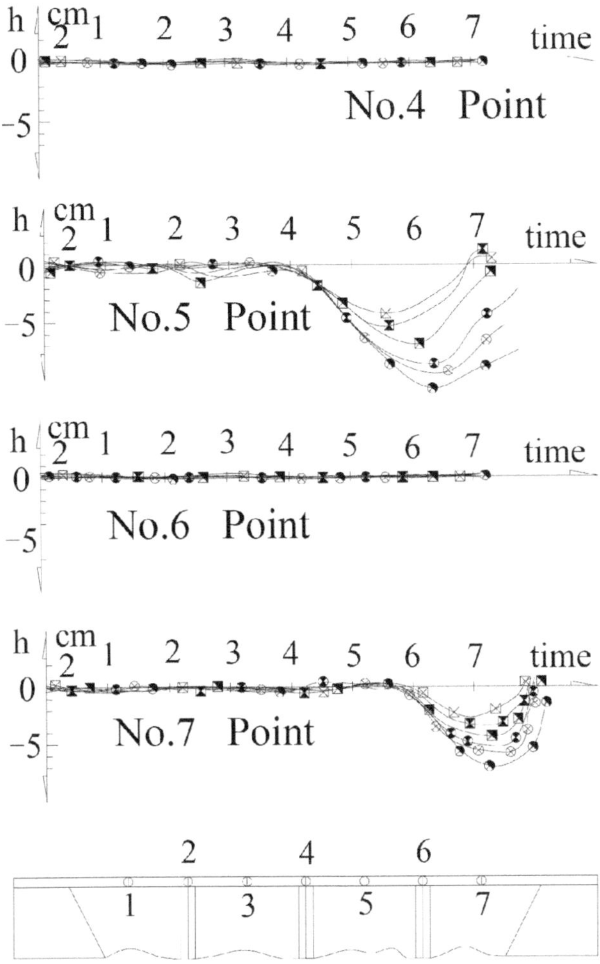

Fig. 2. (continued)

5 Conclusion

The original position devastation experiment of the old Nancha River Bridge shows that, it is effective and feasible to judge the bridge's health status using the bridge's response features to load action. The diagnosis method of bridge's health status based on load response effect, offered by the writer and subject team, has got basic satisfying effect in health status diagnosis trial on four bridges in active service, so we suggest that everyone could have a try in factual works. Certainly, being subject to level and knowledge, the method in this context may be very immature and even fallacious, we hope experts in this line could give more comments, and also wish this context could be helpful to diagnosis method of the bridge's health status.

Acknowledgment

The research project discussed in this paper has been funded by the National Natural Science Fund of China (No. 79160173). Sincere gratitude is presented to them.

References

1. Fu, G.K., Moosa, A.G.: Health monitoring of structures using optical instrumentation and probabilistic diagnosis. Condition Monitoring of Materials and Structures [s.l.]:[s.n.], 190–201 (2000)
2. Brown, C., Jkaruna, R., Ashkenazi, V., Roberts, G.W., Evans, R.A.: Monitoring of structures using the GPS. Proc. of the Institution Civil Engineers, Structural and Buildings 134(1), 97–105 (1999)
3. Dodson, A.H., Moore, T., Roberts, G.W.: Monitoring the movements of bridges by GPS. In: Proc. of ION GPS v2, 1997 1st of Navigation, September16-19, pp. 1165–1172 (1997)
4. Sohn, H., Czarnecki, J.A., Farrar, C.R.: Structural health monitoring using statistical process control. Journal of Struct. Engrg., ASCE, 126(11), 1356–1363 (2000)

Design on Integral Monitoring System for Subway Tunnel Construction Based on GPS and Surveying Robot*

Chen-guang Jiang[1,**], Jian-guo Peng[2], and Bao-guo Qian[3]

[1] College of Environmental and Civil Engineering, Jiangnan University, 214122 Wuxi Jiangsu, P.R. China
[2] Communication Plan and Investigation and Design Institute of Hunan Province, 410008 Changsha, Hunan, P.R. China
[3] College of Civil Engineering, Shanghai University, 200444 Shanghai, P.R. China
jiangcg999@yahoo.cn

Abstract. At present, the comparative deformation in the inner of the tunnel is attached more importance to during soft-soil tunnel construction, so when taking tunnel analysis and design, we think much of comparative deformation in the inner of the tunnel. As a matter of fact, inner and outer deformations of the tunnel are a uniform organism, none but considering and analyzing both inner and outer deformation of the tunnel conjunctly, it is much better to ensure the quality of tunnel construction. Therefore, the writer and research team put forward the thought concerning inner and outer deformation integrated monitoring for the tunnel construction as well as offer corresponding monitoring method and technical instrument. An integral monitoring system for external and internal deformation of tunnel based on GPS and Surveying Robot has been developed.

Keywords: Integral monitoring system, tunnel deformation, external and internal deformation, GPS technology, surveying robot.

1 Introduction

The tunnel construction belongs to underground construction, with complicated and hidden geological environment underground, we have to face various engineering geological disaster problem during our construction, notably for soft-soil tunnel. Therefore, the security of construction has always been paid more attention to in the process of tunnel construction. As for the security of soft-soil tunnel construction, it can be judged by inner and outer deformation of the tunnel. So inner and outer deformation monitoring of the tunnel have become one of most important works in the process of soft-soil tunnel construction.

At present, the comparative deformation in the inner of the tunnel is attached more importance to during soft-soil tunnel construction, so when taking tunnel analysis and

* This work is supported by National Nature Science Foundation under Grant 79160173.

** Corresponding author.

L. Qi (Ed.): ISIA 2010, CCIS 86, pp. 199–205, 2011.

design, we think much of comparative deformation in the inner of the tunnel. As a matter of fact, inner and outer deformations of the tunnel are a uniform organism, none but considering and analyzing both inner and outer deformation of the tunnel conjunctly, it is much better to ensure the quality of tunnel construction. Therefore, the writer and research team put forward the thought concerning inner and outer deformation integrated monitoring for the tunnel construction as well as offer corresponding monitoring method and technical instrument.

Hereinafter combining with the 6th contract section of No.1 Huancheng subway, we talk about monitoring system design and practice of soft-soil tunnel construction with inner and outer integrated deformation.

2 Integral Monitoring System of External and Internal Deformation of Tunnel Underground Soft Soil in Construction

The 6th contract section of No.1 Huancheng subway belongs to the section of Mining Method Construction, where groundwater is abundant with water table depth of 1.2~4m, and the average depth between the tunnel and ground is 18.1m, and down the earth's surface in turn, there are loose-slightly-dense miscellaneous filling, hard—malleable sticky clay, soft flowing malleable sticky clay, malleable sticky clay, soft flowing malleable sticky clay mixing into thin sticky soil, malleable sticky clay, soft flowing malleable sticky clay mixing into clay, and hard-- malleable sticky clay, soft-flowing malleable sticky clay with features of high compressibility, high water-content, high flexibility, and low intensity, which easily generate vermiculate phenomenon. The intensity of soil lowers obviously after interruption, then the soil can take concretion and secondary concretion as well as sedimentation for long time. This section of stratum wall rock has worse self-stability ability, which easily results in collapse after starting digging, and mud-outburst is possible to happen when serious, then construction can't go on wheels, and it is hard to control sedimentation of the ground; extraordinary sedimentation shall cause ground's crack even sink, affecting traffic security, further easily cause surrounding underground pipeline broken, endangering security of the building. Therefore, the construction adopts freezing method.

Soft-soil Tunnel Construction with Inner and Outer Integrated Deformation Monitoring System includes uniform ground GPS(Global Positioning System) three-dimensional benchmark net (consisting of GPS basic point, GPS liaison point, see Fig.1)of entire tunnel system, ground GPS monitoring point(including deformation monitoring points of earth's surface and whole building, see Fig.1), and underground monitoring system. The whole system adopts ground reference frame and elevation system based on WGS-84, and the three-dimensional coordinates benchmark of monitoring system underground is connected by GPS contact point located in entrances (flat hollow, inclined well or vertical well)of each district and section. GPS three-dimensional benchmark net on the ground adopts static GPS to observe for long time for accomplishment, while GPS monitoring point on the ground is fulfilled by

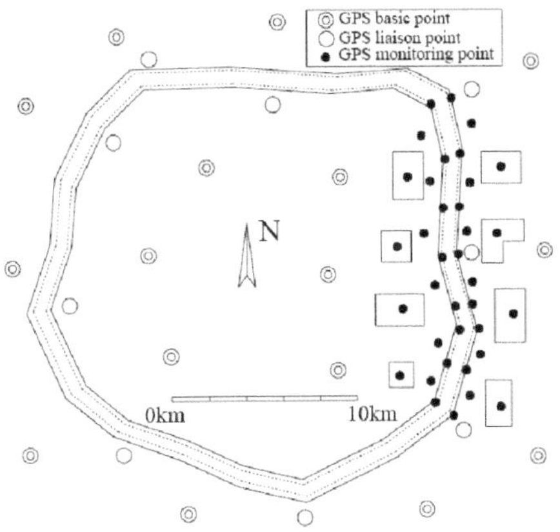

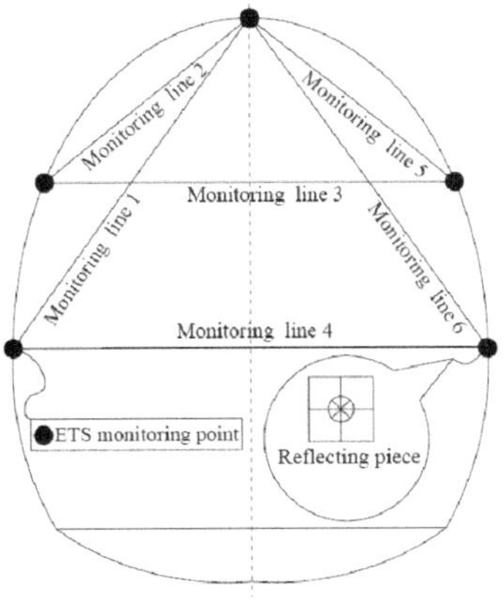

Fig. 1. Ground control and monitoring **Fig. 2.** Cross-section measure point layout

GPS-RTK technology, and monitoring underground is in virtue of automatic ETS(Electronic Total Station)(Surveying Robot). To ensure accuracy of construction space position underground, peg-top theodolite orientation is adopted as well [1]-[3].

3 Soft-Soil Tunnel Construction Underground Deformation Monitoring

Wall rock headroom displacement measurement is the important basis to judge stability of wall rock as well as guide construction. Traditional wall rock headroom displacement measurement adopts mechanism or machine-electron convergence meter for measure. The traditional method has some features including low-cost, simple and reliable, able to adapt to bad environment, great interference to construction, human-factor's large influence on measure accuracy, low speed of measure, which can't meet construction requirements (rapidness, security and high effectiveness)of modern soft soil tunnel. The emergence of automatic ETS lays good foundation for modern wall rock headroom displacement measurement.

A set of TCRA1102 ETS of Leica, many Leica reflection disks(4cm×4cm), Leica Deformation software and processing software after wall rock convergence analysis.

According to criterion, relevant requirements and wall rock condition as well as digging manner and other factors, fix on position and interval of wall rock headroom displacement measurement cross-section, and baseline deployment form of each cross-section. Firstly join angle steel on the anchor pole, then stick slice on the angle steel (or sticking slice by anchoring steel plate using inflation bolt). The surface normal direction of slice is best to be vertical with axis direction of the tunnel, so as to make ETS receive strongest reflection signal. After palm surface spouts, there should be 5 slices stuck on cross-section of each tunnel, constituting 6 baselines(see Figure2).

The minimum show value of distance measurement should be 0.1mm,the minimum show value of angle measurement should be 0.1″,the minimum show value of position

(X,Y,H)should be 0.1mm,the tolerance of distance measurement should be 0.5mm, the tolerance of angle return to zero should be 5″, the tolerance of 2C should be 15″,the tolerance of angle i should be 30″,the mutual tolerance of 2C should be 6″,the mutual tolerance of angle i should be 10″. The observation mode of six rounds is adopted.

The measure content of ETS is mainly comparative distance of each measure point. Bury metal sign point on the ground, obtain three-dimensional coordinates of this sign point using ETS through previous three-dimensional coordinates points in the hole, then set ETS frame on the top of the sign point with leveling in the middle; turn on power supply of the instrument, enter into Deformation measure procedure by pressing the Key Program of the instrument, and make adjustment to each parameter of the instrument as per wall rock measure parameter standards.

After finishing setting of the instrument, confirm each slice point of the instrument on measure cross-section, use laser guiding function of the instrument for collimation with laser point pointing at central point of the slice(see Fig.2), so as to prevent actual shape error of the wall rock occurring due to different collimation when making measure every time. After pressing the Key DIST, the instrument can make electronic record for position information of this point (angle, horizontal distance, X, Y, H etc.) kept into EMS memory (or PC card) of the instrument, then input point number which has been set(such as 1, 2…and so on). Make target identification to other points as per the sequence. The function of ATR(Automatic Target Recognition) of the instrument could distinguish target, aim at the target, follow the target and measure automatically. Make use of the procedures to accomplish manual intelligent data collection, see Fig.3.To ensure accuracy of measure to cross-section of the tunnel, so as to inspect and compare with each other on the same baseline, there are left and right stations set on both left and right sides of axes of the tunnel for measure. The distance between left and right measure station can be set freely. The detailed observing course is as below:

1) Left-station measure: The instrument shall be set on the left measure station, and measure shall start after selecting slice point. Start Left Measure function of the instrument, servo motor of the instrument can measure information at each point for 6 rounds automatically in the sequence of aim (obverse-reverse lens full-round observation method).

2) Right-station measure: After left station measure is over, move the instrument to right station, and take right station measure in the same sequence with the left station measure.

3) Checking and processing of data: After both left and right stations measure are finished, adjust Checking Function of the instrument, examine the length of the same baseline of left and right measure stations; when the length tolerance measured on the baseline of left and right measure stations goes beyond the limit(0.5mm), take a measurement over again to this baseline.

4) Observation key point: The measure at both left and right stations could eliminate the influence on distance measurement by inclining of the slice, and measure station position should be fixed when taking measurement each time, at the same time, in order to improve reflection effect and measure accuracy of reflection slice to laser of the instrument, set the instrument within 5m before and after the measure cross-section.

Take observation for 5 times in 1, 4, 9, 18 and 48 hours after digging, and each observation takes cross-section number as name of measure station, thus, the instrument can make number to each observation data document automatically in terms of different time set in the instrument.

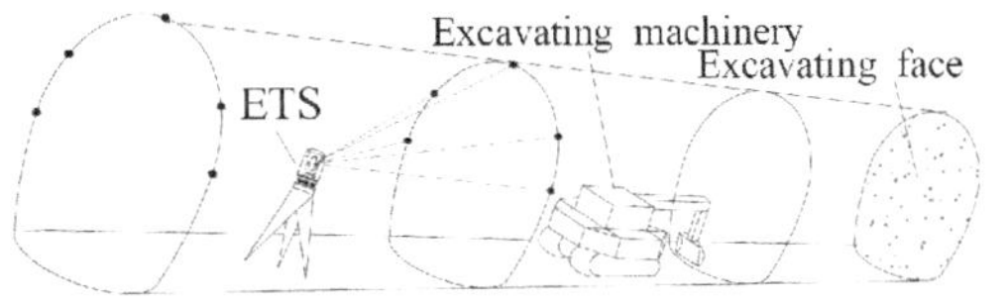

Fig. 3. Inner deformation monitoring process of the tunnel

After finishing measure, connect the instrument with the computer using data cable(field data are memorized in EMS memory of the instrument), or insert PCMCIA card in the instrument into PC Express of the computer(field data are memorized in PC card),lead original data measured each time in the computer, at the same time as per reminders of the procedure, fill out cross-section shape, digging method, covering thickness, geological description, wall rock category and tunnel support design etc. Wall rock convergence analysis software of the computer shall make automatic processing and analysis on the data measured. Output wall rock displacement outcome(including measure line displacement speed-time change curve and measure line displacement trend chart as well as regress analysis figure, see Fig.4, Fig.5, as well as each measure point's displacement value in different time section). In the meantime, make analysis on the outcome automatically so as to judge stability of wall rock(there is show like"how long have this measure line been stable at least" in the computer), supplying basic basis for tunnel support.

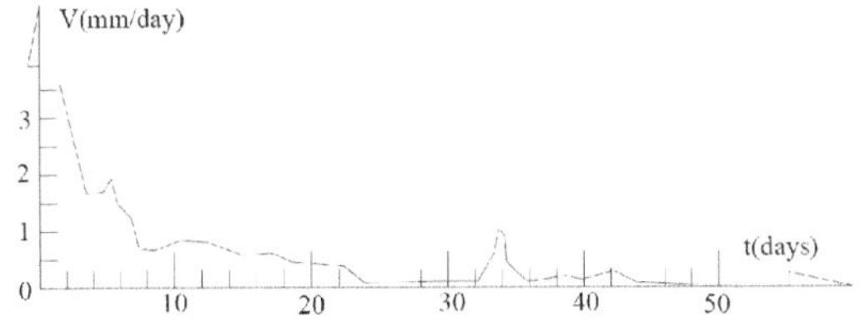

Fig. 4. Movement curve of replacement speed of measure line 6 accompanying the time

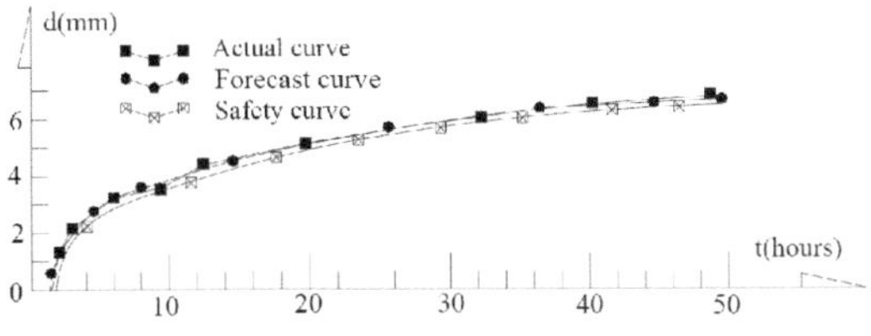

Fig. 5. Movement curve of replacement of measure line 6 accompanying the time

The processing and analysis of cross-section survey data mainly adopt Regress Analysis Approach (including exponent, logarithm and power function). By comprehensive comparison and analysis on several regress methods, combining with monitoring measure managerial expert system, make comprehensive analysis and judgment to the displacement quantity and rate.

4 Other Attentive Problems

To ensure construction security of the soft soil tunnel, it is a must to pay more attention to geological prediction in advance, the purpose of which is to discover, avoid and deal with various geological disasters in time. It is available to adopt TSP (Tunnel Seismic Prediction)technology(see Fig.6), geological radar, infrared water finder, pre-boring technology, and pay attention to gas inspection in the meantime.

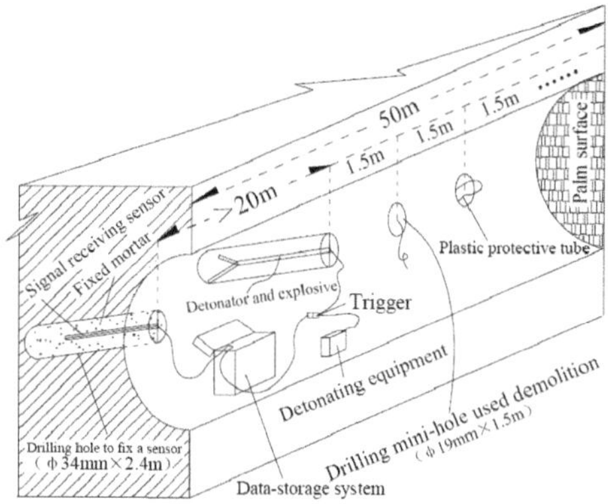

Fig. 6. The composition of tunnel seismic prediction system

When the tunnel is dug per 300m, the overall re-observation of construction control system and three-dimensional coordinates correction should be carried out once. The tunnel is dug towards single direction, taking 1km as the unit, divide the measure section, and there should be 1-3 peg-top side(s) measured additionally in each measure section (namely, determine peg-top azimuth angle of these sides).

When the tunnel is dug bi-directionally and multi-directionally coming up against horizontal-hole connection, put through the construction control system of the tunnel connected, accordingly form closed control system(closed lead, closed leveling line),and measure the closed system all around over again, thus get three-dimensional coordinates with high precision of each control point of closed control system, and make this coordinates in place of original coordinates.

5 Conclusion

Soft-soil Tunnel Construction with Inner and Outer Integrated Deformation Monitoring System can reflect inner and outer deformation situation of soft-soil tunnel entirely, accurately, directly, rapidly and timely, supplying reliable security for safe construction of soft-soil tunnel, offering trustable technical support for stability and support effect judgment of soft-soil tunnel, providing soft-soil tunnel construction technology optimization with scientific basis, and furnishing accurate judgment for security of various engineering structures of the earth's surface. Soft-soil Tunnel Construction with Inner and Outer Integrated Deformation Monitoring System is not only applied to soft soil tunnel, but also applied to other various tunnels with significant function and actual meanings. Limited by our cognition level, the opinions in the context have some deflections and errors unavoidably, welcome experts and comrades in this field to offer your comments.

Acknowledgment

The research project discussed in this paper has been funded by the National Natural Science Foundation of China (No. 79160173), 211Construction Fund released by Jiangnan University (No. 2004012). Sincere gratitude is presented to them.

References

1. Broch, E.: Support of large rock caverns in norway. Tunneling and Underground Space Technology 11(1), 11–19 (2006)
2. Shalabi, F.I.: FE analysis of time-dependent behavior of tunneling in squeezing ground using two different creep models. Tunnelling and Underground Space Technology 20(3), 271–279 (2005)
3. Parkinson, W.B., Spilker, J., Enge, P.: Global Positioning System: Theory and Applications. In: AIAA, Washington, DC (2003)

Empirical Research on Financial Expenditure Policy to the Effect of Inflation

Wen-jun Chen and Lin Hu

Central South University of Forestry and Technology
410004 Changsha, Hunan China
cwj1008@126.com

Abstract. By the global financial crisis, early 2009 China's economic continued last year's decline trends, in response to this situation, the government adopted a series of financial expenditure policy. In this paper, it adopts monthly data of financial expenditure and consumer price index which is from 1996 1 month - December 2008 in China, and analyzes impact which is from the government's expenditure to inflation. The results show that the explanatory power which is about financial expenditure to inflation is very low.

Keywords: Financial expenditure; Inflation; Analysis.

1 Introduction

By the global financial crisis, early 2009 China's economic continued last year's decline trends, in order to promote economic recovery, the government adopted a series of financial expenditure policy.

In the related to financial policy and inflation, most theoretical and empirical studies are analyzed from the perspective of the deficit. Barro said that if the debt stock's growth rate exceed output growth rate, sustained deficits would lead to inflation. Sargentand and Wal-lace said because the bonds have an upper limit, Persistent deficit bond financing will eventually be monetized. Many Chinese scholars make empirical and theoretical researches on fiscal deficit and inflation effects in China. Xiong qi Xu and Zong yi Zhang used the data from 1978-2002 in China to test financial deficit, financial deepening and inflation, The results show that in the addition that the monetary policy variables affect inflation , deficit levels in China lead to inflation each other.

The direct effect relationship on total amount of financial expenditure and inflation lacks adequate research and empirical evidence. In this paper, it used monthly data which is about the financial expenditure and the consumer price index from January 1996 to December 2008 in China, and ADF testing and variance decomposition, and observes the impact of financial expenditure to inflation from the data.

L. Qi (Ed.): ISIA 2010, CCIS 86, pp. 206–212, 2011.

2 Paper Preparation

2.1 Target Selection and Data Sources

Time-series data often has a trend, and performances non-stationary. To avoid non-stationary of macro-economic variables produces perverse regression, this paper intended to judge smooth of the data with the method of unit root test, and according testing the result is that the variables have non-stationary and co-integration relationship between variables, further, review the degree of contribution from shocking of each variable which exist the changing of Inflation by variance decomposition.

In this paper, GSA indicates the financial expenditure, Pindicates consumer price index which represents the rate of inflation level. Financial expenditure's (GSA) data is from the PRC Ministry of Finance website and Economic Information Network Database. Consumer price index's data (P) is from "China Statistical Yearbook" and Bureau of Statistics of China website.

2.2 Analysis Process

2.2.1 Stationary Test

Firstly, the time series' data of financial expenditure (GSA) and consumer price index (P) is adopted with ADF method to test the existence of the sequence to be stationary testing.

Table 1. Statistics of ADF Method and Quantity of P

	Trend of type c, t, k	Statistics of ADF test and quantity of P	Critical value 0.01	0.05	0.10
GSA	0, 0, 0	7.362095 (1.0000)	-2.580264	-1.942938	-1.615316
	c, 0, 0	4.566167 (1.0000)	-3.473672	-2.880463	-2.576939
	c, t, 0	0.962501 (0.9999)	-4.019151	-3.439461	-3.144113
P	0, 0, 0	-0.652674 (0.4329)	-2.581233	-1.943074	-1 615231
	c, 0, 0	-2.687604 (0.0786)	-3.476472	-2.881685	-2.577591
	c, t, 0	-2.986886 (0.1396)	-4.023506	-3.441552	-3.145341

As can be seen from table 1, in the ADF test significant levels at 5% of time series of each variable all have unit root, so all variables are non-stationary series. Because these series have the same single whole order number, we can analyze whether it exists the co-integration between them. If there is co-integration, it means that there is a long-term equilibrium relationship between sequences; otherwise, we need to further analyze the Possibility of one-way Granger.

2.2.2 Co-integration

After testing the series of inflation and economic growth rate by unit root, we judge the Possibility of co-integration between them.

Assuming that the two-dimensional random vector is Xt=(GSAt, Pt) ', for the sake of simplicity, the following analyses ignore the impact of exogenous variables, and it can set its p-order autoregressive models as:

$$\triangle X_{t=} A_0+\Pi X_{t-1}+ \sum_{i=1}^{p-1}\Gamma i \triangle X_{t-i}+\Sigma_t$$

$$\Pi= \sum_{i=1}^{p} Ai -I \qquad \Gamma_i=- \sum_{j=i+1}^{p} Aj$$

If matrix Π is reduced rank (0 < rank (Π) = r < 2), there is column vector α and β, and make Π=α β'(β is a Co-integration vector, and can be Standardization). It makes co-integration vector as combinatorial coefficient, and it can make t β 'X is a stationary series, which is the co-integration portfolio looking for. Table 2 gives the test results of co-integration. In the 5% significance level, track statistics 47.11009 is greater than the corresponding critical value 42.91525, so inspection refuses the original hypothesis which is that there is no co-integration.

In macroeconomics, co-integration between variables can serve as evidence of long-term equilibrium between them.

Co-integration test results show as in table 2 and table 3, which is about time series data of financial expenses (GSA) and consumer price index (P)

Table 2. The Test Results of Co-integration

Number Assumptions of Co-integration	The characteristic root	Track test statistic	5% Key value	Probability
none	0.144856	47.11009	42.91525	0.0180
At most 2	0.049324	7.2589	13.51798	0.3259

Table 3. The Test Results of Co-integration

Number Assumptions of Co-integration	The characteristic root	Maximum eigenvalue test statistics	5% Key value	Probability
none	0.144856	22.22101	25.82321	0.1394
At most 2	0.049324	7.182598	12.51798	0.3259

2.2.3 Pulse Response

(1) Pulse response

Pulse response characters that the response on the errors plus standard deviation affects the current and future value of the endogenous variables. We present a Co-integration ECM model as a basis, and set the model of impulse response. Figure 1 and Figure 2 are the function curves which are impulse response about financial expenditure and the price against inflation.

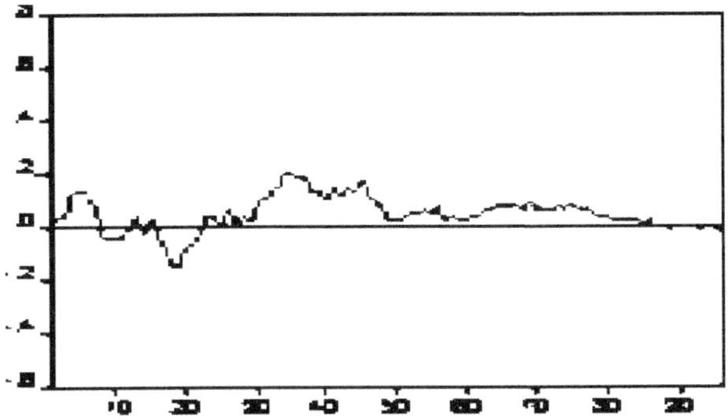

Fig. 1. The Response of P Caused by GSA

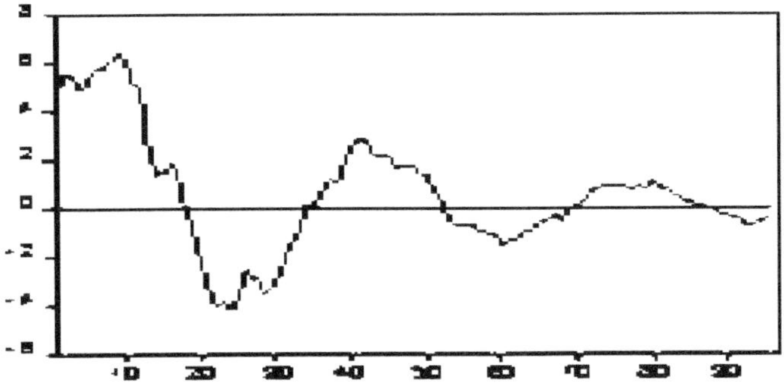

Fig. 2. The Response of P Caused by P

(2) Cumulative impulse response

The above is that the impact of the errors in a moment by a unit affects directly the each future moment of each variable, but in fact that, the direct impact of a response from a moment to the next variable value affects indirectly variable values from the future time. The cumulative impulse response reflects the cumulative impact which is about that direct and indirect impact in a moment is against the each subsequent period. Figure 3 and Figure 4 are the cumulative impulse response plans respectively about financial expenditure and the price against inflation.

From the impulse response of GSA, the positive impact of a standard deviation of GSA in the lagged 5-6 months will lead to a small increasing of P, but then it decreased gradually, and is absorbed gradually at last. Viewed from the cumulative effect, the changes of P in response amplitude caused by the impact of GSA is very small.

From the impulse response graph of the price itself, the positive impact of one standard deviation of P will lead to the rapid response of P, in the lagged 10 months will result that P increases in high rise, and after 10 months it will gradually decline. This response form of price shocks to reflect the Interaction between expectation itself and price, and current price has impact on future prices through expectations and the self-realization mechanism of expectations. Shock resistance patterns of response plan from P seem to reflect this feature that the price has viscosity.

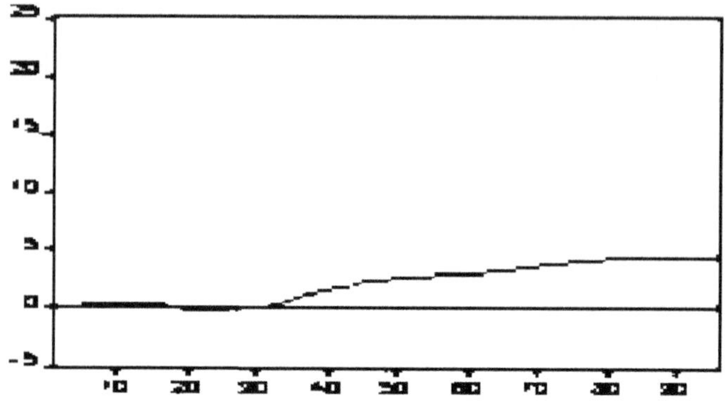

Fig. 3. The Cumulative Response of P Caused by GSA

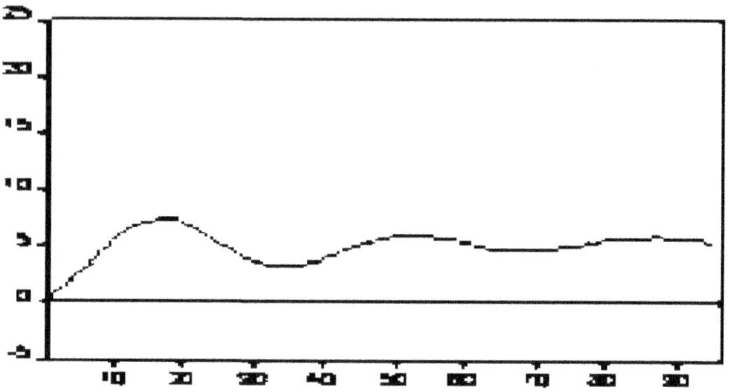

Fig. 4. The Cumulative Response of P Caused by P

2.2.4 Variance Decomposition

Co-integration provides long-term static relationship information between variables. After analyzing the impulse response of all variables, we set variance decomposition further. Variance decomposition can provide dynamic effect between variables for further information, and the changes between variables are what we wished to visit. Variance decomposition is used to analyze the dynamic effects which are about random disturbance to Variable system; the changes of endogenous variable decompose into component impact on VAR, and observe the relative importance of random disturbance. It can be investigated the contributing degree from GSA and P in changing of inflation according to variance decomposition.

As can be seen from the variance decomposition diagram, the explanatory power of financial expenditure is only about 5% for inflation, but the price is very powerful for inflation.

2.3 Empirical Results

Overall, financial expenditure has very small for the effects of inflation, but it will exert positive effect which is led by public expectations. Therefore, the government takes into account the delay factors in policy decisions, and improves the future capacity in making policy and evaluation and predictive ability after making policy, on this basis it can control scale intervention and increase the policy stability.

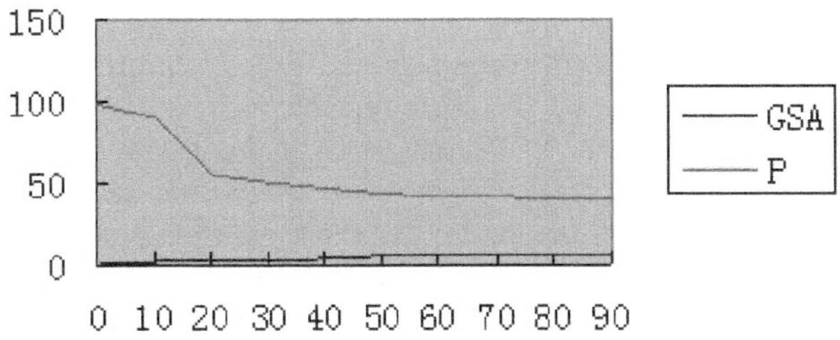

Fig. 4. GSA and The P Possess Percentage in Predicted Variance on P

3 Conclusions and Recommendations

Our empirical results show that the explanatory power of financial policy for inflation is very low, but considering the actual situation in China, this conclusion seems to have not applied in other countries .The reason is that our financial policy has applied the principle which is within your income and prudent financial policy over the years, even when implementing active financial policy, it is also in the premise of balance which is maintained by deficit level.

Therefore, under the existing condition in china, the government should take into account the policy delay factors in policy decisions, and improve the future capacity in making policy and evaluation and predictive ability after making policy, on this basis it can control scale intervention and increase the policy stability. Regarding to financial policy has less affected for the amount of inflation, we should adopt more active financial policy, not only solve the short-term problems through fiscal stimulus plans promote economic recovery, but also consider the long-term development, adjust the Structure and focus of financial policy and real the more active policy to promote the society transforms domestic demand-based economy.

References

1. Barro, R.J.: Reply to Feldstein and Buchanan. Journal of Political Economy (82), 1095–1117 (1976)
2. Sargent, T.J., Wallace, N.: Some Unpleas ant Mone-tarist Arithmetic. Federal Reserve Bank of Minne-apolis Quarterly Review (5), 1–17 (1981)
3. Xu, X.q., Zhang, Z.y.: Financial Deficit, Financial Deepening and Inflation——— theoretical analysis and empirical of Chinese experience (1978—2002). Management World (9) (2004)
4. Yin, G.: Financial Stimulus Plans, Money Supply, Public Expectations and Inflation. Research on Financial and Economic Issues 2, 8–16 (2012)

5. Zhang, Y.: Inflation Rate Volatility, Uncertainty and The Macroeconomic Research in China, vol. 6 (2009)
6. Yan, K.: Inflation's Risk Analysis of Active Financial Policy Tax Research
7. Li, J.l.: The Effect of The Positive Financial Policy and Realistic Options in China. Trade on Fiscal and Monetary 10, 15–17 (2008)
8. Wang, M.j.: Analysis of Inflation in China—Econometric Methods and Applications, vol. 136 (2001)

A Face Detection Method Based on Color Image

Wencheng Wang

School of Information and Control Engineering, Weifang University,
Weifang 261061 China
wwchpaper@126.com

Abstract. To the problem of face detection, a new method based on color image is proposed, which is used for face detection in complex scene. Firstly, the rough face region is detected by establishing the skin color model. Then, the human face candidate area is processed for further selecting based on facial features, and the selected areas with the original face images were fused. Finally, the gray projection algorithm was adopted to determine the location of the human eyes and the face was verified using template matching method. The experiment shows that the proposed method can reduce the error rate, shorten the operation time and has strong robustness.

Keywords: face detection; skin segmentation; template matching; skin model.

1 Introduction

Over the past decades, the problem of human face detection has been thoroughly studied in the computer vision community for its fundamental challenges and interesting applications, such as video surveillance, human computer interaction, face recognition, video indexing, face data management, and so on[1]. Numerous methods have been proposed for face detection such as the color-based method[2], the template-based method[3], the Eigenfaces method[4] and the ANN method[5], etc.. In which the template matching method can obtain good results for the face detection in simple background image. But to the complex background image, it is easy to appear the phenomenon that the non-human face is treated as a face. To this problem, a new face detection method is presented on the combination of facial skin color and gray projection in this paper. Firstly, on the basis of skin color clustering, it realized the image segmentation by using the skin model, and the skin region is obtained. Then, the human face candidate areas are processed for further selecting based on facial features, and the selected areas with the original face images were fused. Finally, the gray projection algorithm was adopted to determine the location of the human eyes and the face was verified using the template matching method.

The rest of this paper is organized as follows. Section 2 presented the segmentation method of facial regions based on skin model, and realized the rough segmentation of pure face through skin model and morphology. Section 3 described the eye location approach and the specific steps. The implementation of template matching is discussed in section 4. Section 5 showed the experimental results of the proposed method and section 6 gave some conclusions.

L. Qi (Ed.): ISIA 2010, CCIS 86, pp. 213–219, 2011.

2 The Rough Segmentation of Face Region

2.1 The Skin Model

Skin color is one of the most important features in the human face. In order to separate the face region with non-face regions, it needs a reliable skin color model to suit for different skin colors and different lighting conditions. The popular color spaces includes YCrCb, HSV, YIQ et al.. Based on Terrillon et al.'s [6] comparison of the nine color spaces for face detection, we use the YCbCr space since it is widely used in video compression standards. Since the skin-tone color depends on luminance, we nonlinearly transform the YCbCr color space to make the skin cluster independent. This also enables robust detection of dark and light skin tone colors. Through face color sample chart, it was found that the chrominance component of the skin color falls into a relatively stable range. If the skin region is denoted as 1 and the non-region is denoted 0, then, the discriminant function will be:

$$\begin{cases} 1 & (100 \le C_b \le 128) \& (138 \le C_r \le 168) \\ 0 & otherwise \end{cases} \tag{1}$$

The image segmentation using above function is shown in Fig.1.

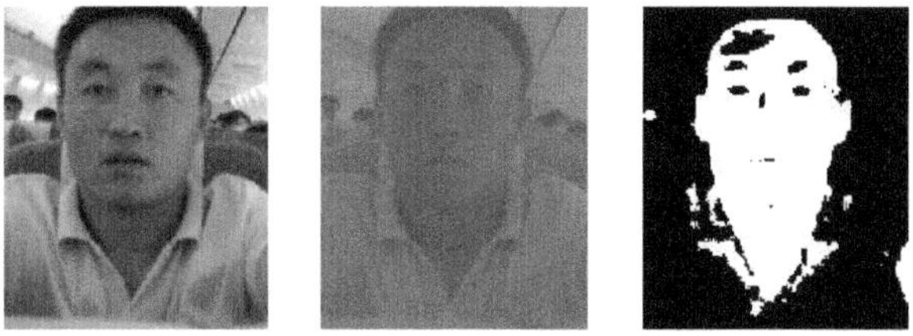

(a) Original image (b) YCrCr space image (c) Binary image

Fig. 1. Image segmentation

2.2 Morphological Processing

As only the Cb and Cr components are processed in YCbCr color space region, so it is inevitable to appear the situations after the segmentation. One is the introduction of noise which will result in the connection of face region and the around non-face regions, or emerge some isolated points. Another is the 'non-completely filled' cased by the non-skin regions such as eyebrows, eyes, etc. In order to remove the interference, morphological processing algorithm can be used. Morphology is a technique of image processing based on shapes. By using erosion operations repeatedly, the internal hollows in pipe image and the background noise will be removed, and it can back to original size if it is conducted corresponding dilation. That can make good preparation for further face region analysis and the candidate face region is obvious. The image is shown in Fig.2.

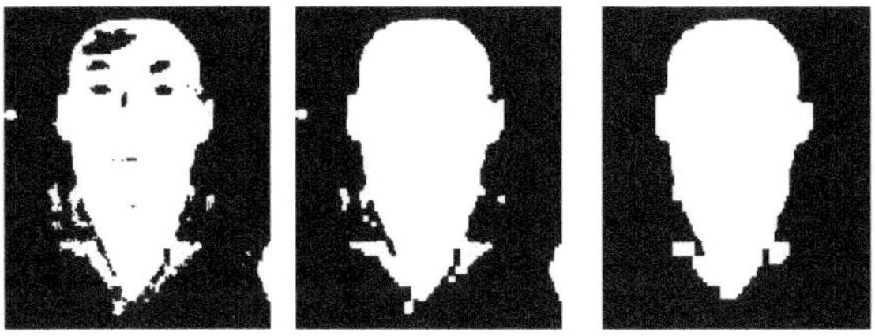

(a) Binary image (b) Morphological processing (c) Region selection

Fig. 2. Face candidate extraction

2.3 Region Selection

The connected face candidate region can obtain by morphological processing. But in the complex background, there may be exposed arms or legs are misjudged as face candidate regions, so it needs reprocessing. The specific method is based on some prior knowledge such as shape and size, aspect ratio, the length of the ellipse axis ratio and pixel validation prior knowledge of such share, which can eliminate the non-face regions obviously. In experiments, it is found that it can filter out most of the non-face region by using the following method to adjust individual parameters.

(1) Region selection based on sizes. It is not the face area if the area is too large or too small by calculating the number of pixels in the region. There are two thresholds, one is the higher limitation and the other is the lower limitation, only the area value is in the range can be regard as the face region.

(2) Region selection based on ratio of height and width. It is used to exclude a number of irregular objects close to skin tones. This approach is effective to exclude arms and legs obviously.

The combination of the above methods can be used to filter out most of the non-face areas, and providing convenience for the follow-up treatment, the processed image is shown in Figure 2 (c).

3 Eye Localization

In order to verify the face region, it needs to find the candidate eyes in this region.

3.1 Eye Candidate Region Cutting

The purpose of image fusion is to filter non-skin background information as possible and to retain the face candidate region. Suppose the image of skin segmentation is $F(x, y)$, the original image is $G(x, y)$, the fused image is $M(x, y)$, then the fusion strategy is:

$$I(x, y) = \begin{cases} F(x, y) & (F(x, y) = 0) \\ G(x, y) + 1 & (F(x, y) = 1) \end{cases} \tag{2}$$

The reason why source image pixel value adding 1 is mainly to avoid confusion with the mask, the fused image which may include a face is shown in Fig.3(a). In order to locate the eye more accurately, it needs to cut out the region including the human eyes. With the distribution of various facial parts, the candidate region of eyes is cut out with the same width and height. The region is cut from the top, and the cutting area does not include the mouth, but must include the human eye. The face model and eye candidate region is shown in Fig.3(b) and Fig.3(c), respectively.

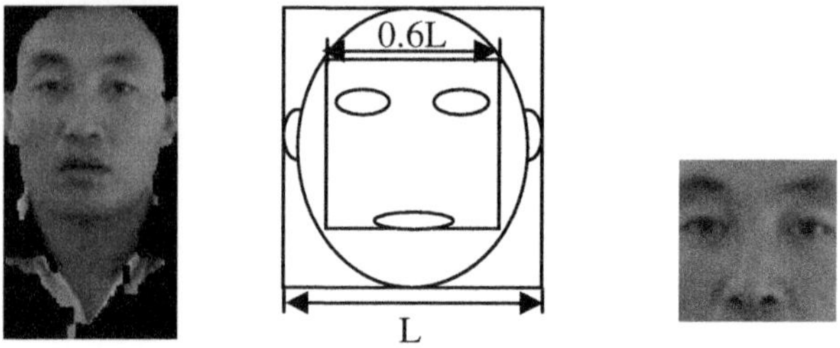

(a) Fused image (b) Face model (c) Eye candidate region

Fig. 3. Eye candidate region cutting

3.2 Gray Projection

In the facial image, the eye region has two salient features. One is the eye region has lower gray compared with its around. The other is the intensity presence of eye white and pupil has greater changes. The gray projection method is one of the popular eye localization methods[7]. Set $I(x, y)$ stands for the gray scale value of pixel (x, y), then, the vertical integral projection function $M_v(x_0)$ and the horizontal integral projection function $M_h(y_0)$ can be described as:

$$M_v(x_0) = \frac{1}{y_2 - y_1} \sum_{y=y_1}^{y_2} I(x_0, y) \quad M_h(y_0) = \frac{1}{x_2 - x_1} \sum_{x=x_1}^{x_2} I(x, y_0) \tag{3}$$

It is easy to find that integral projection considers only the mean of intensity. In the case of the same means, integral projection function will fail, so we use the variance in practice the projection method.

We attempt to mark the eye with cross Star "+". If the scope of cross-star label is in a black eye, it means the accurate detection. The detection steps are as follows:

(1) Conduct preprocessing to the candidate region.

(2) Separate the region to S_L and S_R.

(3) Calculate the average of each row and column in S_L and S_R, respectively.

(4) Realize the horizontal and vertical projection in S_L and S_R, respectively, and obtain the coordinates $E_L(x_l, y_l)$ and $E_R(x_r, y_r)$ of two eyes.

(5)Save the candidates of two eyes and label the symbol '+'.

The following images of eye location are shown in Fig.4. In the horizontal projection curve, ① Corresponds to eyebrows, ② corresponds to eyes, and ③ corresponds nose. In the vertical projection curve, ④ and ⑤ correspond to the left eye and right, respectively.

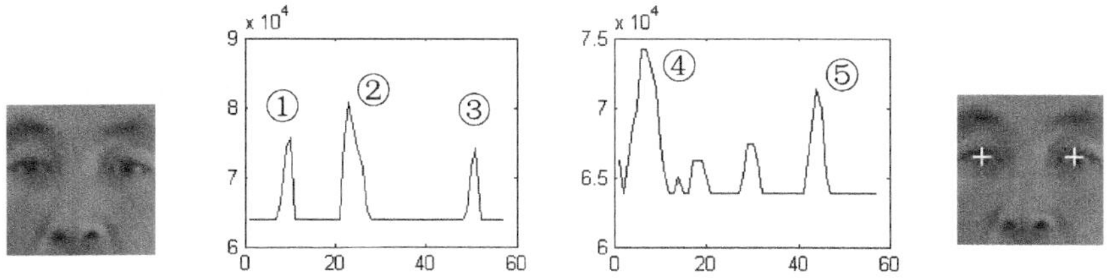

(a) The eye candidate (b) Horizontal projection (c) Vertical projection (d) Located eyes

Fig. 4. The course of eye location

4 Template Matching

After the eye localization, the template matching method is used to verify the face. We prepared 60 human face images and selected the eye centers manual, and then get the average template. The template is with the size of 32×32 pixels. It is shown in Fig.5.

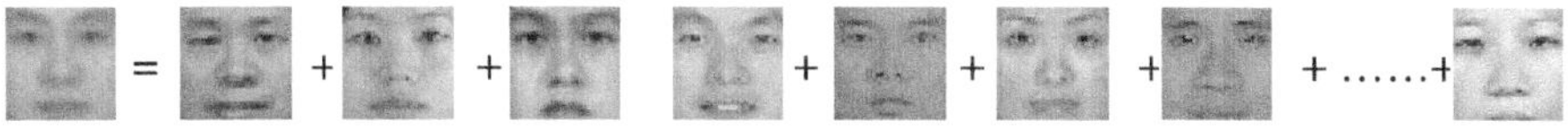

Fig. 5. The formation of the template

In order to de template matching, it needs the normalization. The expressions are as follows:

(1) Angle normalization. Generally, there will be a certain degree of face rotation. So, we should do the angle normalization to the matching sample and make the eyes on the same horizontal line. Assuming that the coordinates of eyes $E_L(x_{l,}y_l)$ and $E_R(x_{r,}y_r)$ are obtained, the distance of them is D, the angle between connection of the eyes and the horizontal line is θ, then:

$$D=\sqrt{(x_l-x_r)^2+(y_l-y_r)^2}$$
$$\theta=\arg\tan(y_r-y_l)/(x_r-x_l) \tag{4}$$

(2) Size normalization. Because the size of template is 32×32 pixels and the distance of two eye centers is 18 pixels. Therefore, the zoom ratio of the image to be detected is $ratio = D/18(0.1 \le ratio \le 10)$. The Scaling transformation matrix is:

$$[x_0, y_0, 1] = [x, y, 1]\begin{bmatrix} 1/ratio & 0 & 0 \\ 0 & 1/ratio & 0 \\ 0 & 0 & 1 \end{bmatrix} \quad (5)$$

(3) Correlation coefficient. Suppose $T[M][N]$, μ_T, σ_T are gray matrix, mean and variance of template image, respectively. $R[M][N]$, μ_R, σ_R corresponding values of test image. Then, the correlation coefficient can be defined as:

$$r(T,R) = \frac{\sum_{i=0}^{M-1}\sum_{j=0}^{N-1}(T(i,j)-\mu_T)(R(i,j)-\mu_R)}{M \bullet N \bullet \sigma_T \bullet \sigma_R} \quad (6)$$

To each testing window, we use face template to verify, the actual similarity coefficient is defined as r. If it is greater than the threshold value of 0.3, then, the region will be regard as a human face area.

5 Experimental Analysis

In order to test the performance of this method, we realized the algorithm with Matlab, and experiments are conducted on color face database. Some images of this database are from Internet and the others are collected by out laboratory. Those images have different background complexity and different illumination. The experiment result is shown in Table 1.

Table 1. Comparison of experiment results

Method	Accuracy(%)	Error detection(%)	Missed rate(%)
Skin+template	83.9	10.8	16.1
Our method	87.5	8.8	12.5

From Tab.1, it can be seen that our method is superior to the combination of skin color detection and template matching method, and can improve the detection accuracy and reduce the false positive rate. By analyzing missed images, we can find that the distortion of the color, background with similar skin color, hair over eyes and face rotation are the main reasons. Especially, when the face is in "looking up" posture, the distance between nose and eyes is small. So, it is easily to identify the nostril as an eye, which will result in the failure of face detection. Other failures were due to very small size of the face components that were eroded during opening by reconstruction performed using a single structuring element for all images irrespective of face sizes. Our future work will focus on these two aspects for further research. The goal is to design a system that detects faces and facial features, allows users to edit detected faces, and uses the facial features as indices for retrieval from image and video databases.

6 Conclusions

An algorithm has been developed to detect human face in a color image, which is no required a priori knowledge of the number of faces or the size of the faces in a given image. The algorithm starts with human skin color modeling and then generates face candidates based on the spatial arrangement of these skin patches. Comparing with other existing template matching methods, our method is on the basis of rough face detection and eye localization. Experiment shows that the method can be robust to light interference and has strong adaptability.

References

1. Brunelli, R., Poggio, T.: Face recognition:Features versus templates. IEEE Transactions on Pattern Analysis and Machine Intelligenee 15(10), 1042–1052 (1993)
2. Cai, J., Goshtasby, A.: Detecting human faces in color images. Image and Vision Computing 18(1), 63–75 (1999)
3. Liang, L.-H., Ai, H.-Z., He, K.-Z., et al.: Single rotated face location based on affine template matching. Chinese Journal of Computers 23(6), 640–645 (2002)
4. Moghaddam, B., Pentland, A.: Probabilistic visual learning for object representation. IEEE Transactions on Pattern Analysis and Machine Intelligence 19(7), 696–710 (1997)
5. Rowley, H.A., Baluja, S., Kanade, T.: Neural network-based face detection. IEEE Transactions on Pattern Analysis and Machine Intelligence 20(1), 23–38 (1998)
6. Terrillon, J.C., Shirazi, M.N., Fukamachi, H., Akamatsu, S.L.: Comparative performance of different skin chrominance models and chrominance spaces for the automatic detection of human faces in color images. In: Proc. IEEE Int'l Conf. on Face and Gesture Recognition, pp. 54–61 (2000)
7. Feng, G.C., Yuen, P.C.: Variance projection function and its application to eye detection for human face recognition. Pattern Reeognition Letters 19(9), 899–906 (1998)

Design and Evaluation of Variable Stages Pipeline Processor Chip

Tomoyuki Nakabayashi, Takahiro Sasaki, Kazuhiko Ohno, and Toshio Kondo

Graduate School of Engineering Mie University
1577 Kurimamachiya, Tsu, Mie, 514-8507 Japan
{tomoyuki,sasaki,ohno,kondo}@arch.info.mie-u.ac.jp

Abstract. In order to reduce the energy consumption in high performance computing, variable stages pipeline architecture (VSP) is proposed, which improves execution time by dynamically unifying the pipeline stages. The VSP processor adopts a special pipeline register called an LDS-cell (Latch D-flip flop Selector - cell) that unifies the pipeline stages and prevents glitch propagation caused by stage unification under low energy mode. The design and fabrication of the VSP processor chip on a Rohm $0.18\mu m$ CMOS process is described and the energy consumption is evaluated. The result indicates the VSP processor can achieve 13% less energy consumption than the conventional approach.

Keywords: VLSI, Low energy processor, Variable stages pipeline, Glitch.

1 Introduction

Recently, mobile computers are required to achieve both low energy consumption and high performance. To reduce energy consumption, dynamic valtage and frequency scaling (DVFS) [1] that lowers the supply voltage and clock frequency dynamically is currently employed in many applications. However, with prospective low supply voltage CMOS devices, the effectiveness of DVFS will be reduced. Futhermore, the performance decreases proportionally with the decrease of the clock frequency. As an alternative technique that is independent of the supply voltage and device technologies, we propose a variable stages pipeline (VSP) technique [2,3] that dynamically varies the depth of the pipeline stages to achieve both low energy and high performance computing. To vary the depth of the pipeline stages, a special cell called a latch D-flip flop (D-FF) selector-cell (LDS-cell) is introduced as a pipeline register instead of a general D-FF. The LDS-cell can not only vary the pipeline depth, but can also reduce the energy consumption by preventing glitch propagation caused by unification of the pipeline stages under low energy mode. To date, we have demonstrated the effectiveness of VSP by SPICE simulation; however due to computational complexity, the accuracy of the evaluation result is not sufficient. Therefore, to conduct a more accurate evaluation, we have designed and fabricated a VSP processor on a Rohm $0.18\mu m$ CMOS process, and measured the energy consumption of the fabricated chip.

L. Qi (Ed.): ISIA 2010, CCIS 86, pp. 220–226, 2011.

In this paper, we show the energy consumption of fabricated chip. According to the evaluation results, VSP can achieve 13% less energy consumption than the conventional approach.

2 Related Work

This section summarizes the research directed toward achieving low energy and high performance computing. One of the current major techniques is DVFS, which is used to dynamically control the supply voltage and clock frequency. The energy consumption is proportional to the square of the supply voltage; therefore, lowering of the supply voltage is effective in reducing energy consumption. However, the effectiveness of DVFS will be gradually reduced in the future, because the supply voltage is being lowered every year. Furthermore performance decreases proportionally with a decrease in the clock frequency.

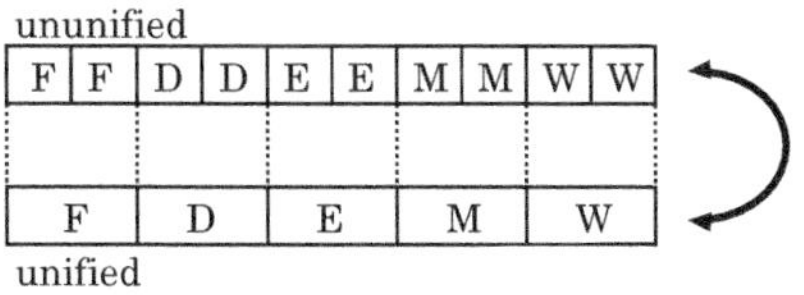

Fig. 1. Variable pipeline stages

Therefore, certain architectural approaches that are independent of device technologies and supply voltage, such as pipeline stage unification (PSU) [4,5] are proposed. Figure 1 shows the basic concept of PSU. PSU unifies the pipeline stages under low energy mode, and has the following two advantages:

- The depths of the pipeline stages under low energy mode become shorter, so that PSU can reduce penalty cycles caused by branch prediction miss and data/control dependency, compared with DVFS. Execution time can be reduced with a small penalty cycle.
- Under low energy mode, PSU can reduce energy consumption by stopping unused pipeline registers and units.

Unfortunately, with conventional approaches such as PSU, the number of glitches increases under low energy mode, because the scale of the combinational circuit increases by unifying the pipeline stages. In the next section, we propose a novel mechanism to solve this problem.

3 Variable Stages Pipeline

The VSP processor dynamically varies the depth of the pipeline stages, similar to PSU, to achieve low energy and high performance computing. Figure 2 shows

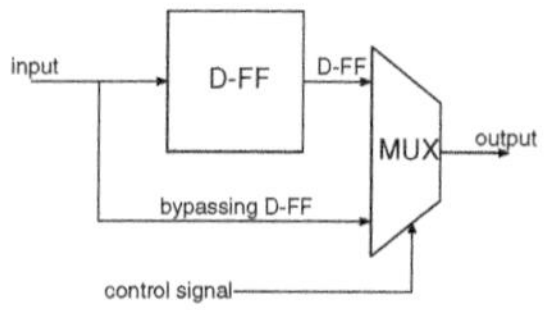

Fig. 2. Circuit diagram of pipeline register with D-FF + MUX

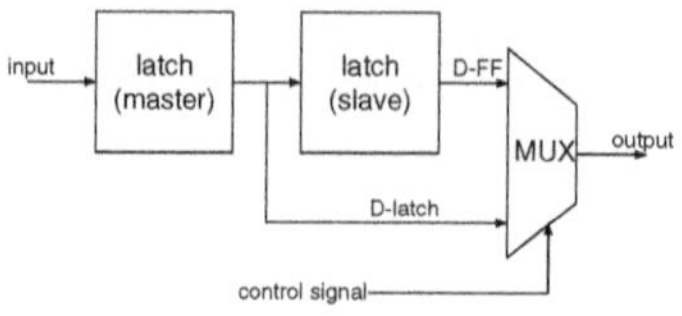

Fig. 3. Circuit diagram of pipeline register with LDS-cell

the pipeline register that is used on the PSU processor that can vary the depth of the pipeline stages. Under low energy mode, the D-FF+MUX (Fig. 2) unifies the pipeline stages by selecting a lower path to directly connect the input port to the output port.

Generally, glitch propagation is caused by irregular delays of logic gates and wires, and the frequency of glitch propagation increases in proportion to the scale of the combinational circuits. Because plural pipeline stages are unified and they have a large combinational circuit, energy dissipation caused by glitch propagation is increased.

In order to prevent glitch propagation, the LDS-cell is introduced into the pipeline registers between unified pipeline stages. Figure 3 shows the circuit schematic of the LDS-cell, and Fig. 4 schematically illustrates the function of the LDS-cell. Under low energy mode, the LDS-cell outputs a master D-latch signal to prevent glitches. The function of the LDS-cell under low energy mode is given as follows.

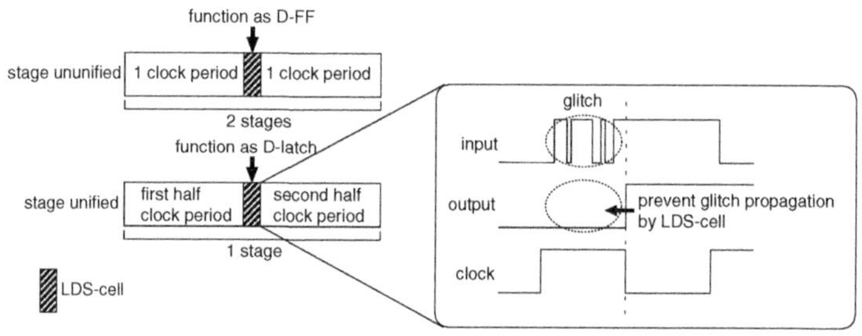

Fig. 4. Function of the LDS-cell

- In the first half clock period, the master latch keeps a data to prevent the glitch from propagating to the next stage.
- In the second half clock period, the master latch passes the input data to start the operations of the second half stages. Therefore, the LDS-cell functions as a D-latch.

The LDS-cell uses a master latch included in the D-FF, so that the number of transistors in an LDS-cell is the same as a pair of D-FF and MUX.

The VSP processor has the following two modes.

High speed (HS) mode

- Nine-stage pipeline for high clock frequency to achieve the highest performance. The LDS-cell functions as a pipeline register.

Low energy (LE) mode

- Three-stage pipeline for low energy. The LDS-cell functions as a D-latch to prevent glitch propagation.
- No interlock occurs by branch delay or data dependency between arithmetical operations.
- Clock lines for unused pipeline registers between unified stages are gated and the branch prediction unit is stopped.

We have previously reported the effectiveness of VSP by SPICE simulation [2,3]; however, due to computational complexity, the accuracy of the evaluation result was not sufficient. Therefore, a VSP processor was designed and fabricated on a Rohm $0.18\mu m$ CMOS process to conduct a more accurate evaluation. Before showing the measured results, we briefly show the SPICE simulation result in the next section.

4 Result of SPICE Simulation

To evaluate the VSP processor, two processors were designed; one adopting the PSU technique and the other using the VSP technique. The two processors were designed based on the MIPS R3000 compatible processor with a nine-stage pipeline under HS mode and a three-stage pipeline under LE mode.

The processors were design using Verilog HDL, and synthesized with a Synopsys Design Compiler on Rohm $0.18\mu m$ CMOS technology. The energy consumption was estimated using the Synopsys Nanosim tool. Both processors are designed to operate at 100MHz for HS mode and 25MHz (a quarter of 100MHz) for LE mode.

Figure 5 shows the energy consumption under HS and LE modes. with each value normalized using the PSU result. Figure 5(A) shows that the energy consumptions of both processors is almost same, because PSU and VSP are the same under HS mode except a part of the clock tree. Figure 5(B) shows that VSP achieves approximately 5% less energy consumption than PSU under LE mode, due to the decrease in glitch propagation by use of the LDS-cell. According to the above results, the VSP can achieve the lower energy consumption than PSU under LE mode.

Wiring capacitance and wiring resistance are not accounted for in this simulation, due to the computational complexity. For this reason, this simulation can evaluate only glitches caused by gate delay. If the VSP is evaluated including wiring capacitance and wiring resistance, then we surmise that VSP can prevent the propagation of glitches caused by wire delay and crosstalk noise, and it is expected that VSP can achieve higher efficiency than the present simulation results. Therefore, both VSP and PSU chip were fabricated and measured. The next section presents the evaluation results for the fabricated chip.

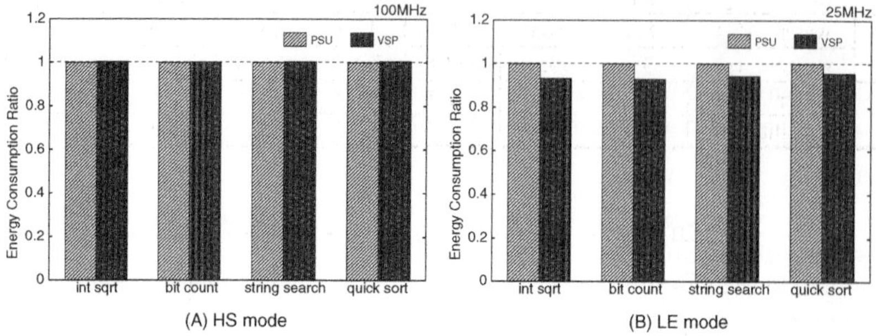

Fig. 5. Results of SPICE simulation

5 Evaluation of Fabricated Chips

Both VSP and PSU chips were fabricated on a Rohm $0.18\mu m$ CMOS process. Figure 6 shows a micrograph of a VSP processor on a chip. Figure 7 shows a shmoo plot of the VSP chip under LE mode measured using a HP 83000 LSI tester. The measured supply voltage is plotted against the clock period, and indicates that the VSP chip can operate at 47.6MHz when the supply voltage is 1.8V, and the lowest supply voltage is 1.55V at a 20MHz clock frequency.

In the SPICE simulation, VSP can operate at 100MHz under HS mode, and 25MHz under LE mode, because VSP unifies 4 stages into 1 stage under LE mode. However, the shmoo plot shows that VSP can operate at 47.6MHz under LE mode. The shmoo plot of the HS mode cannot be graphed, due to the electrical characteristic limitations of the subboard that connects the LSI tester and VSP chip.

Figures 8 and 9 show the energy consumption under the HS and LE modes respectively, with each value normalized according to the PSU results. Figure 8 shows that energy consumption of VSP in HS mode is approximately 20% less than that for PSU. However, the energy consumption of VSP and PSU are almost the same in the HS mode. According to the simulation result, the difference in energy consumption between VSP and PSU is approximately 1%, as shown Fig. 5(A). Figure 8 show that the difference in energy consumption has no relation to the type of benchmark program and the ratio is fixed. For this reasons, we consider this phenomenon to be caused by manufacturing variability.

Therefore, in evaluation of energy consumption in the LE mode, the difference under HS mode was corrected and evaluated according to the corrected value. According to Fig. 8, the difference in energy consumption for the two approaches is 20%; therefore, the error corrected by reducing the measured value of PSU by 20%.

Figure 9 shows that energy consumption of VSP is approximately 13% less than that of PSU. Compared with the simulation results shown in Fig. 5(B), it is obvious that the effectiveness based on actual measurement with the fabricated VSP chip is larger. As mentioned above, because of computational complexity, wiring capacitance and wiring resistance are not considered and the only

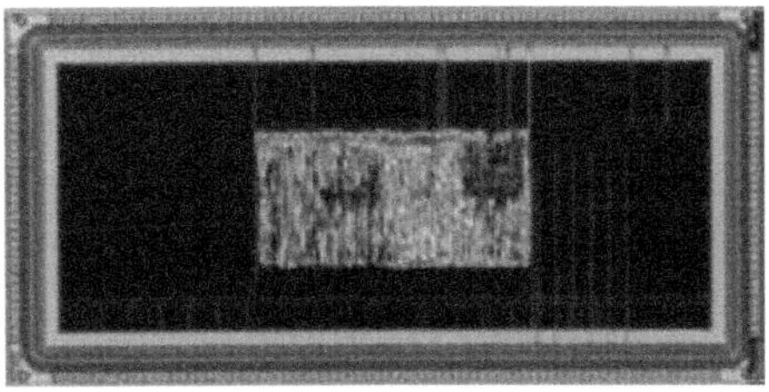

Fig. 6. Chip micrograph

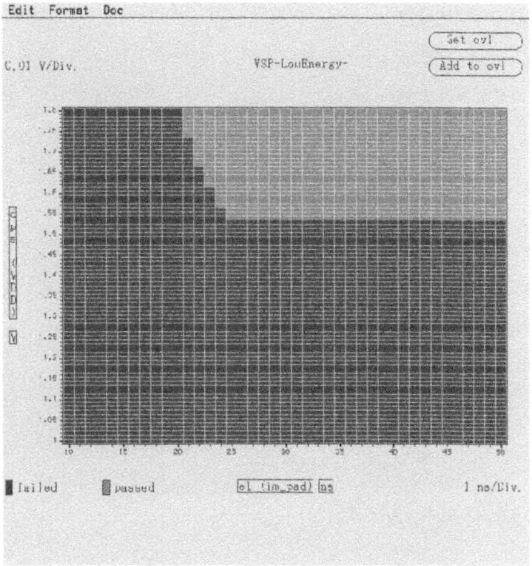

Fig. 7. Shmoo plot of VSP

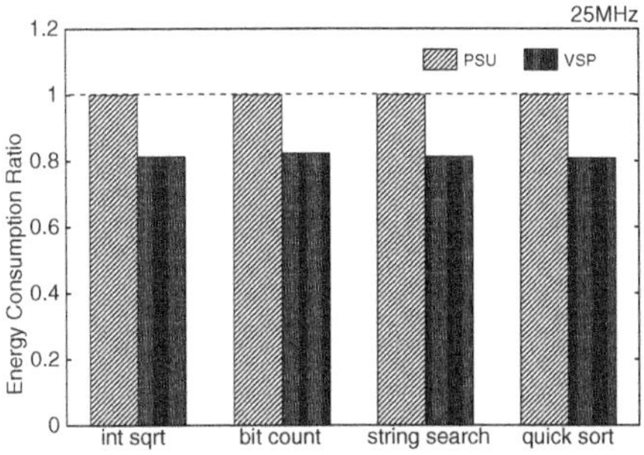

Fig. 8. Energy consumption on HS mode (actual value)

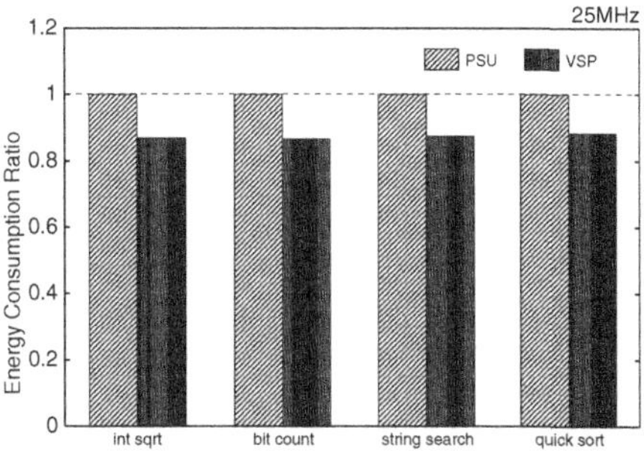

Fig. 9. Energy consumption on LE mode (corrected value)

glitches caused by gate delay are evaluated by the SPICE simulation. On the other hand, evaluation of the fabricated chip demonstrates that VSP can prevent glitches caused by wire delay and crosstalk noise. Therefore, the LDS-cell can reduce energy consumption further than that shown by the simulation results. Therefore, VSP is superior to the PSU according to results for the fabricated chip.

6 Conclusions and Future Work

A VSP chip was implemented and the energy consumption of the fabricated chip was evaluated to be 13% less energy consumption than that for the PSU processor.

As future research, we plan to introduce the VSP technique into more complex circuits, such as the superscalar processor and 64bit processor. In these processors, the scale of the combinational circuit increases and the amount of glitch becomes larger. Therefore, we consider that the LDS-cell can further reduce energy consumption caused by glitches.

Acknowledgments. This work is supported by VLSI Design and Education Center (VDEC), the University of Tokyo in collaboration with Synopsys, Inc., Cadence Design Systems, Inc., Mentor Graphics, Inc. The VLSI chip in this study has been fabricated in the chip fabrication program of VDEC in collaboration with Rohm Corporation and Toppan Printing Corporation. This work is supported by a Grant-in-Aid Young Scientists(19700042) from the Ministry of Education, Culture, Sports, Science and Technology of Japan.

References

1. Pouwelse, J., Langendoen, K., Sips, H.: Dynamic voltage scaling on a low power microprocessor. In: 7th ACM Int. Conf. on Mobile Computing and Networking (Mobicom), pp. 251–259 (2001)
2. Ichikawa, Y., Sasaki, T., Hironaka, T., Kitamura, T., Kondo, T.: A Design of Prototype Low Energy Processor by Variable Stages Pipeline Technique. In: International Technical Conference on Circuits/Systems Computers and Communications, vol. 2, pp. 561–562 (2005)
3. Sasaki, T., Ichikawa, T., Hironaka, T., Kitamura, T., Kondo, T.: Evaluation of low energy and high performance processor using variable stages pipeline technique. IET Journal of Computer and Digital Techniques 2(3), 230–238 (2008)
4. Shimada, H., Ando, H., Shimada, T.: Pipeline Stage Unification: A Low-Energy Consumption Technique for Future Mobile Processors. In: The International Symposium on Low Power Electronics and Design 2003, pp. 326–329 (2003)
5. Shimada, H., Ando, H., Shimada, T.: A Hybrid Power Reduction Mechanism Using Pipeline Stage Unification and Dynamic Voltage Scaling. In: Symposium on Advanced Computing Systems and Infrastructures, pp. 11–18 (2005) (Japanese)

Evaluation of Variable Level Cache

Nobuyuki Matsubara, Takahiro Sasaki, Kazuhiko Ohno, and Toshio Kondo

Department of Information Engineering, Mie University
1577 Kurimamachiya-Cho, Tsu City, Mie 514-8507, Japan
{nobu,sasaki,ohno,kondo}@arch.info.mie-u.ac.jp

Abstract. This paper proposes a variable level cache (VLC) to reduce leakage power of cache memory. Generally, required cache size depends on a program feature and data set. Therefore, VLC estimates the required cache size dynamically, and if the required size is small, it divides cache memory in half. The upper half memory functions as normal cache memory and the lower half memory shifts into stand-by mode to reduce leakage current and performs as lower level cache. According to our simulation results, the VLC is approximately 36% improvement of the energy-delay product to the conventional approach.

Keywords: cache system, variable level cache, low leakage.

1 Introduction

Recently, the developments of mobile computers, such as cellular phones, laptops and PDAs, have resulted in increased energy consumption, which has reduced the battery operation time. Therefore, a reduction in energy consumption without a decrease in performance is required.

Leakage power is one of the major factors that increases power consumption of above devices. So reduction of leakage energy in cache system is very important because leakage energy is in proportion to the number of transistors and high performance processor has large cache system using many transistors. The approach proposed by Yang et al. (referred to as the DRI cache)[1] is one of the representatives as low energy cache system. The DRI cache estimates required cache size and resizes cache capacity dynamically. To reduce cache capacity, it completely turns off the power supply for unused cache part. So data in the part is lost. Hence, the number of cache miss may increase and it causes to decrease performance. In order to solve the problem, we propose a variable level cache (VLC) to improve the DRI cache.

2 DRI Cache

DRI cache can reduce leakage energy; however, it has a serious drawback in performance degradation. In this section, we summarize some of the features of DRI cache and identify drawbacks.

L. Qi (Ed.): ISIA 2010, CCIS 86, pp. 227–233, 2011.

2.1 Basic DRI Cache Design

Figure 1 shows a block diagram of the DRI cache. The DRI cache counts the number of cache misses with a miss counter at fixed length intervals. At the end of each interval, the cache system expands/reduces the cache capacity depending on whether the miss counter is lower/higher than the statically defined threshold. Resizing the cache capacity requires the number of index bits to be dynamically changed depending on the cache size. Every time reduction of the cache size is required, the size mask shifts to the right to use a smaller number of index bits and vice versa. Therefore, reduction of the cache size removes the highest numbered sets in the cache in groups of powers of two. The DRI cache manages the used/unused part of the cache as one bank to control the power supply. Therefore, it can turn off the power supply not only for a memory cell, but also for a sense amplifier because the power supply for all the bank of the unused cache part is turned off.

2.2 Disadvantages of DRI Cache

Before turning off the voltage of the unused cache part, data in this part must be written back to the lower level storage, because the power supply for that part is turned off completely and the data is lost. Moreover, when the cache system expands the cache capacity, the data that exists in the cache is written back to lower level storage because the location of the data depends on the cache size. Such DRI cache processes decrease performance. To avoid such drawbacks, The DRI cache applied this approach dedicated only to the instruction cache.

In order to solve the problem, we propose a VLC that improves the DRI cache by reducing the penalty to write back datum on switching to the stand-by mode. Our approach can apply not only a dedicated instruction cache but also a data cache including unified L2 and/or L3 caches.

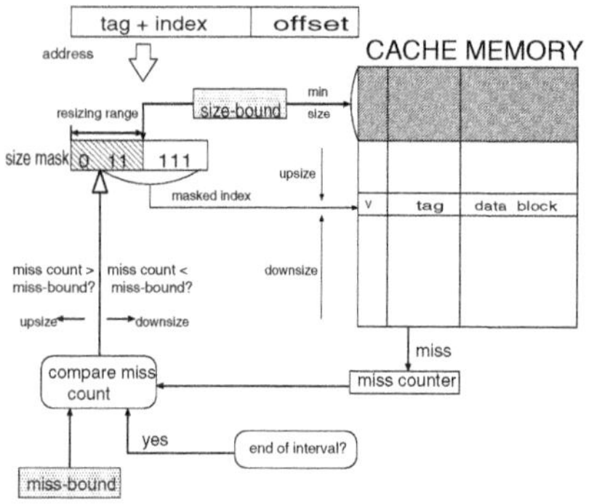

Fig. 1. Block Diagram of DRI cache

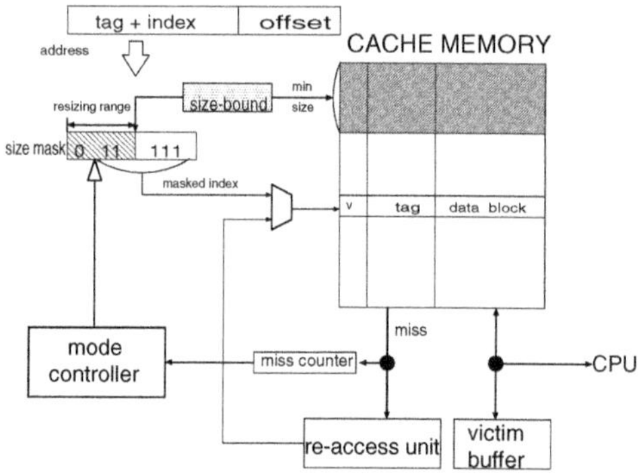

Fig. 2. Block Diagram of Variable Level Cache

3 Variable Level Cache

In this section, we discuss the basic concept and provide a detailed description of the proposed VLC. In this paper, we assume to introduce our approach into unified L2 cache.

3.1 Basic Design of the VLC

Similar to DRI cache, the VLC dynamically estimates the required cache size by counting the number of cache misses at fixed length intervals. However, the cache miss ratio increases if the cache capacity is simply reduced such as for a DRI cache. In order to solve this problem, we propose that the power supply for the unused cache part is not completely turned off, but shifted into stand-by mode, the power supply is reduced to the lowest voltage to maintain the memory content[2]. However, access to the memory part when in stand-by mode takes longer than in normal mode. Therefore, if memory access is concentrated in the memory part where is in stand-by mode, then this causes performance degradation. Hence, the cache operated in stand-by mode performs as a lower level cache(reffered to as the pseudo-L3 cache). We adapt an exclusive cache[3] into the lower level cache to reduce the access frequency to the stand-by part as much as possible. Figure 2 shows a block diagram of the VLC. The VLC has a re-access unit and victim buffer added to the circuit of the DRI cache. The re-access unit and the victim buffer are used only in stand-by mode. In stand-by mode, if the cache miss occurs in the L2 cache, then the pseudo-L3 cache is switched from stand-by mode to normal operating mode, and the pseudo-L3 cache is accessed using the re-access unit. The victim buffer is used to data swap between the L2 cache and pseudo-L3 cache. The state where all lines are activated is referred to as the "normal mode". The state where the lower level cache operates in stand-by mode is reffered to as the "low energy mode". When switching from normal mode to low energy mode in the case of a DRI cache, writing back to the lower level storage is required, because the data present is destroyed to turn off the power supply. In contrast, the VLC does not destroy the data, so that writing back is not necessary.

3.2 Detailed Operation

Our approach is explained using a simple example as follows. The cache system is first operated in normal mode when the program starts. It then switches to low energy mode, and switches back to normal mode again. When in normal mode, the VLC behaves as a conventional cache. Now we assume that a program is started on normal mode. When executing the program, the required cache size decreases, because the cache access frequency is low or the working set is small, and the condition to shift into the low energy mode is satisfied. Generally, index bits in the cache system are determined by the number of sets. The index bits are calculated as $\log_2(n)$, where n is the number of sets. The VLC varies the number of sets, so that the index bits are also varied. In low energy mode, the required number of index bits is changed; therefore the mask shifts to the right in order to use a smaller number of index bits, and the cache memory is divided in half.

In low energy mode, when the CPU executes load or store instructions, VLC firstly accesses the upper half L2 cache. If a cache miss occurs, then the pseudo-L3 cache, which is the half lower L2 cache, is activated and accessed by the re-access unit shown in Fig. 2. If the pseudo-L3 cache hits, then this data moves to L2 cache. On the contrary, if the pseudo-L3 cache also misses, then the main memory is accessed. After cache access is completed, the data stored in the victim buffer is written to the pseudo-L3 cache during program execution. As the program executes further, the required cache size increases, because the program behavior changes, and the condition to shift into the normal mode is satisfied. At this time, the data in the cache memory is written back to the main memory, the bits of the mask are all adjusted to '1', and the limitation to index is released, because the location of data in the cache memory changes. The VLC operates in this way to achieves high performance with low energy cost.

3.3 The Problem and Solution of Mode Switching

It is important to decide a suitable energy mode for the VLC. In the case of the DRI cache, the number of cache misses is counted at fixed length intervals. If the number of cache miss is higher/lower than the threshold, then it is switched to the normal/low energy mode. This approach is applied for the VLC, and is referred to as the "Base(baseline)-VLC". However, this approach may cause frequent mode switching due to the fixed threshold. When switching to the normal mode from low energy mode, data is lost. Therefore, frequent mode switching leads to an increase in the execution time. In addition, the number of cache misses cannot be used to estimate the required cache size, because the number of cache misses generally increases if the number of cache accesses increase. Therefore, we propose two mode switching approaches.

The first approach is referred to as "Dyn(dynamic)-VLC". The Base-VLC applies a fixed threshold, but if the number of cache misses is continuously taken near the threshold in some applications, then the mode may switch frequently. Frequent mode switching leads to an increase in the execution time. Therefore, this approach lowers the threshold when the mode switches frequently, and prevents performance degradation.

The second approach applies a cache miss ratio at fixed length intervals as the threshold. If the cache hit ratio is higher/lower than the threshold, then switching to low energy/normal mode occurs. This approach is referred to as "Hitrat(hit ratio)-VLC". Base-VLC and Dyn-VLC are applied to the number of cache misses as the threshold. However, the number of cache misses generally increases if the number of cache accesses increase. By using the cache hit ratio, the required cache size can be estimated more accurately. On the other hand, to obtain the cache miss ratio, a divide unit is required, which makes the hardware cost the largest of three approaches.

4 Performance Evaluations

The execution time and energy consumption in the DRI cache and VLC were evaluated. The energy-delay product was also evaluated, because the goal of this research is to yield high performance with low energy cost.

To estimate energy consumption, we use the following energy consumption model used in reference[5].

$DE_{total} = DE_{line} \times Access$ (1)
$LE_{total} = CC \times LE_{line} \times CSize$ (2)
$LE_{line} = SR \times LE_{sline} + (1 - SR) \times LE_{aline}$ (3)
$CE_{total} = CE_{line} \times BSize \times Access_{sline}$ (4)
$E_{total} = DE_{total} + LE_{total} + CE_{total}$ (5)

The total dynamic energy consumption (DE_{total}) is calculated according to Eq. (1), where DE_{line} is the dynamic energy per line and $Access$ is the number of cache accesses. The total leakage energy consumption (LE_{total}) is calculated according to Eq. (2), where CC is the number of execution clock cycles, LE_{line} is the average leakage energy per line, and $CSize$ is the number of total lines. In Eq. (3), SR represents the fraction of lines in the entire cache memory operating in stand-by mode. LE_{sline} is the leakage energy per line in stand-by mode, and LE_{aline} is the leakage energy in normal mode. In Eq. (4), CE_{total} implies the energy to dissipate when switching from stand-by mode to normal mode to access. CE_{line} is the energy caused by switching the mode per line, $BSize$ is the number of lines in which the mode is switched, and $Access_{sline}$ is the number of switches to normal mode; the number of accesses to pseudo-L3 cache. Thus, the total energy consumption (E_{total}) is calculated using Eq. (5).

In ordinary cache and DRI cache, $CE_{total} = 0$, because stand-by mode is not available. The value in which the energy of the cache is evaluated assumes a 32 nm proccess and 64B-line size using CACTI[6]. The obtained value is indicated as follows.

$DE_{line} = 2.20E - 10$ (J) $\qquad LE_{aline} = 5.39E - 15$(J)
$LE_{sline} = 8.61E - 16$(J) $\qquad CE_{line} = 1.92E - 16$(J)

In order to estimate the evaluation parameters used in above model, SimpleScalar [4], which is a widely used microprocessor simulator, was modified. We implemented DRI cache and VLC on unified L2 cache.

In the evaluation, VLC dynamically switches between normal mode and low energy mode; where the cache configuration for the normal mode is a conventional 512 KB L2 cache and the cache configurations for low energy mode are 256 KB L2 cache and 256 KB pseudo-L3 cache operated as exclusive cache. The DRI cache dynamically switches between normal and low energy modes; the cache configuration for the normal mode is a conventional 512 KB L2 cache and the that for the low energy mode is a 256 KB L2 cache by turning off the power supply for the lower half part. In addition, the DRI cache, Base-VLC and Dyn-VLC count the number of cache misses per million cycles, and estimate whether the threshold is exceeded. If the counter value is above the threshold,

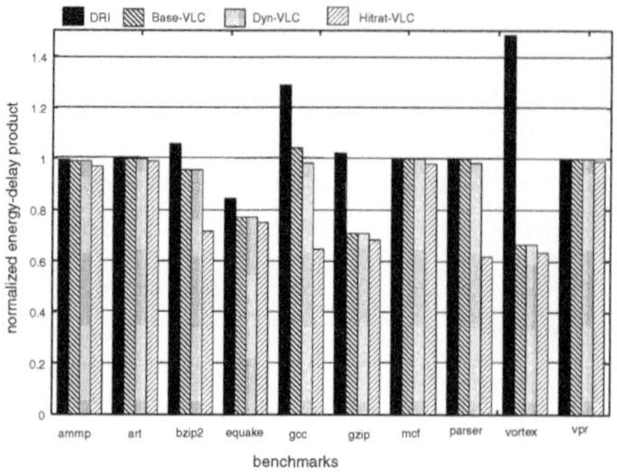

Fig. 3. Energy-Delay Product

a shift into normal mode is made. Otherwise, a shift into low energy mode is made. Hitrat-VLC evaluates the cache miss ratio per million cycles, and whether the threshold is exceeded. If the miss ratio is above the threshold, then a shift into low energy mode is made, and vice versa. The thresholds for DRI cache, Base-VLC and Dyn-VLC are 1000, and that for Hitrat-VLC is 25%. In Dyn-VLC, the threshold immediately after changing the mode is 500. These values are the optimized values that are obtained experimentally for each approach. In VLC, the experimentally assumed penalties for switching from stand-by mode to normal mode are 1, 3 and 5cycles. The influence of the switching penalty is very small (ca. 0.3%) compared with 1 cycle; therefore, only 5cycles are shown in this work. A benchmark program was executed using the SimpleScalar out-of-order simulation to evaluate the performance of each approach. The experiment was conducted using 10 programs from the SPEC2000[7] benchmark suite. For each program, the first two billion instructions were skipped to avoid the initial startup behavior of the benchmarks, and the next two billion instructions were simulated for the evaluations.

4.1 Evaluation Results

The proposed approaches were evaluated according to the energy-delay product, because the goal of this research is to yield high performance with low energy cost.

Figure 3 shows the experimental results of the energy-delay product. The horizontal axis shows the benchmark program, and the vertical axis is normalized execution time, energy consumption and energy-delay product by the result of the conventional cache. According to Fig. 3, all methods based on VLC (Base-VLC, Dyn-VLC and Hitrat-VLC) are superior to those based on the DRI cache for all benchmarks. One reason for this is the difference of the execution time. Although the DRI cache does not use the lower half L2 cache in low energy mode, VLC can use it effectively and reduce the execution time. In bzip2, Hitrat-VLC takes less time in low energy mode, but indicates the highest performance, which implies that Hitrat-VLC can estimate the required cache size more accurately and does not incur waste in mode switching.

For gcc, although the energy-delay product for Base-VLC is superior to that for DRI cache, the energy consumption is inferior to the DRI cache. The number of accesses to the pseudo-L3 cache is much larger than other benchmarks compared with the time executed in low energy mode, because cache misses frequently occur for the DRI cache and the number of main memory accesses is increased, with an apparent difference in execution times resulting. The Base-VLC can reduce the problem of the DRI cache by introducing an exclusive pseudo-L3 cache. However, accessing the pseudo-L3 cache cause mode switching of the memory block from low energy mode to normal mode. Dyn-VLC dynamically determines the threshold and prevents consecutive mode switching. Therefore, it is superior in both execution time and energy consumption compared with Base-VLC. Moreover, Fig.3 shows that Hitrat-VLC is superior to conventional cache and DRI caches by approximately 29% and 36% respectively.

Therefore, Hitrat-VLC is the best of three approaches to reduce leakage energy.

5 Conclusions

In this paper, VLC was proposed as a cache leakage reduction approach, and three implementation methods were presented. The evaluation results indicated high performance was achieved with low energy cost compared with that for a conventional cache. The energy-delay product of the proposed Hitrat-VLC can be decreased approximately 36% compared with the DRI cache. Hitrat-VLC applies the cache miss ratio as threshold to more accurately estimate the required cache size. In future work, we will design Hitrat-VLC VLSI chip and evaluate with fabricated chip.

References

1. Yang, S.H., et al.: An Integrated Circuit / Architecture Approach to Reducing Leakage in Deep-Submicron High-Performance I-Caches. In: Proc. of the 7th Int. Symp. on High-Performance Computer Architecture, pp. 147–157 (2001)
2. Qin, H., et al.: SRAM Leakage Suppression by Minimizing Standby Supply Voltage. In: Proceedings on the International Symposium on Quality Electronic Design. IEEE, Los Alamitos (2004)
3. Zheng, Y., et al.: Performance Evaluation of Exclusive Cache Hierarchies. In: IEEE International Symposium of Performance Analysis of Systems and Software, IS-PASS, pp. 89–96 (2004)
4. SimpleScalar Simulation Tools for Microprocessor and System Evaluation, `http://www.simplescalar.org/`
5. Zushi, J., et al.: Evaluation of Algorithms to Change Cache Line Mode in Drowsy Caches. IPSJ SIG Technical Reports, 2006-ARC-170, pp. 37–41 (2006) [in Japanease]
6. Thoziyoor, S., et al.: CACTI 5.1. In: HPL 2008, HP Laboratories, Palo Alto, 20 (2008)
7. SPEC -Standard Performance Evaluation Corporation, `http://www.spec.org/`

Intelligent Control System of BOF Steelmaking

Gongfa Li, Jianyi Kong, Guozhang Jiang, Jintang Yang, and Liangxi Xie

College of Machinery and Automation, Wuhan University of Science and Technology, 430081 Wuhan, China
ligongfa@yahoo.com.cn

Abstract. On the basis of analyzing BOF steelmaking characteristics, a prediction model and control model based on neural network were established. Taking endpoint temperature and endpoint carbon content as control target, oxygen-blown volume and added coolant were calculated, then endpoint control of convert steelmaking was realized. Simulation results show that the simulation result of endpoint temperature and carbon content are satisfying and control strategy is very effective. The application results demonstrate the efficiency of the method.

Keywords: BOF steelmaking; endpoint control; endpoint temperature; endpoint carbon content; intelligent control.

1 Introduction

At present, BOF steelmaking has such questions as low automatic degree, on-stability of quality of steel product at random and high accident pilosity. According to this kind of current situation, BOF steelmaking complex automatic system has been developed. This system can monitor the whole production course, automatically control and manage with the modernization, all parameters can be monitored and recorded in real time, report form can be typed automatically. The entering amount and proportions of different raw materials can be held accurately, the oxygen and nitrogen pressure can be adjusted in time, and then the requisite measure can be taken in time to prevent great accident from occurring. Though the above-mentioned comprehensive automation function has been realized, the automatic level of the steelmaking has been improved greatly, but steelmaking quality and production efficiency should be improved and the energy consumption can be reduced in an all-round way. For this reason, on the basis of automated system of the original foundation, intelligent control strategy of steelmaking endpoint and steelmaking intelligent control system are developed.

2 Intelligent Control of BOF Steelmaking

2.1 Structure of Intelligent Control

The BOF steelmaking intelligent control system can be divided into three modules, namely weight monitoring module of abolishing steel and molten iron, instrument monitoring and control module and instruction of electrically control module.

L. Qi (Ed.): ISIA 2010, CCIS 86, pp. 234–239, 2011.

Weight monitoring module of abolishing steel and molten iron includes pressing-type weigh sensor installed on the overhead traveling crane, batching controller, wireless data starting station, wireless data receiving station, monitoring computer of abolishing steel and molten iron. Instrument monitoring and control module includes such sensors as weight, flow, pressure and temperature, regulation valve, instrument PLC, communication adapter, communication dividing case, monitoring computers of the sub picture of the craft, the material in bulk, coal gas and workshop production scheduling. Instruction of electrically control module includes electric PLC, communication adapter and instruction computer of electrically control. Every above-mentioned computer makes up the computer NT network, realizes the data share.

2.2 Strategy of Intelligent Control

The endpoint control of BOF steelmaking is control of endpoint temperature and composition mainly, which is shown as Fig.1.Endpoint control not only guarantee accuracy of carbon content and temperature, meeting the demand of sulphur and phosphorus composition and reaching the content of oxygen of molten steel as low as possible.

BOF steelmaking system is a great complicated system with big-delay, non-linear, strong coupling, various physical-chemical reactions, alternate the material flows and energy flows.It's a difficulty in realizing multivariable goal function with high yield, low consumption, high-quality and longevity. In order to realize the optimization goal of high yield and low consumption, multivariable goal function is set up as following.

$$\int_{\tau}^{\tau+T} F_1\left(Y,\frac{\partial Y}{\partial t},Z,\frac{\partial Z}{\partial t},X,\frac{\partial X}{\partial t},H,\frac{\partial H}{\partial t}\right)dt = U_t \rightarrow Max . \tag{1}$$

$$\int_{\tau}^{\tau+T} F_2\left(Y,\frac{\partial Y}{\partial t},Z,\frac{\partial Z}{\partial t},X,\frac{\partial X}{\partial t},H,\frac{\partial H}{\partial t}\right)dt = K \rightarrow Min . \tag{2}$$

Where Y is a raw materials parameter, Z is control parameter, X is state parameter and H is equipment parameter. Each partial derivative is variation function, which changes with time. Integral region $[\tau,\tau+T]$ represents the initial time and end time of steelmaking, namely T is the duration of heating. U_t is output, K is energy consumption. Formula (1) represents output function, namely maximization of usage factor. Formula (2) represents energy consumption function, namely minimization of energy consumption.

BOF steelmaking is generally divided into three stages, namely initial stage, middle stage and later stage. It was finished on later stage, measured result according to the 1st time and one pair of sublance that endpoint is controlled, through blowing into right amount oxygen and a certain amount coolant joining, make molten steel temperature and carbon content in the stove reach the goal range of the demand. At BOF steelmaking later stage, most of impurity in the molten steel has already been got rid of, the response course is not as violent as mainly blowing stage, tending towards stability gradually, the content of carbon and temperature in the molten steel demonstrate that compares the law confirmed.

BOF steelmaking intelligent control system adopt an advanced control method in the world, namely control strategy combining static control and dynamical control. At the beginning of every stove steel blows, entering amount of stove iron, abolished steel and various kinds of material in bulk are put into systematic databases automatically, according to these value, through the calculation of optimization expert system, the best endpoint temperature and endpoint carbon with high yield and low consumption are produced. The steelmaking of the converter produces and uses one pair of sublance, the content of carbon and oxygen in molten steel, measure temperature online, taking a sample, can be carried on in the course of blowing, then these data are put into the real-time database of the computer. When the converter is close to endpoint, temperature and content of carbon examining are put into optimization computer, the necessary oxygen amount and joining amount of coolant are calculated reaching the goal temperature and content of carbon, and the actual number value examined is as initial value, dynamic model is started once every oxygen blast 3s afterwards, the temperature and content of goal carbon in the molten bath are predicted. When all entering the goal range in temperature and content of carbon, blowing is stopped.

In order to make BOF steelmaking endpoint hit the target, a certain amounts of oxygen and right amount coolant is needed during the course of mending and blowing. At the same time in order to enable smelting course smoothly, some auxiliary raw materials will join, usually include lime, compounding, ore and dolomite. Because RBF neural network is base function of the radial from inputting layer to implying layer, it is linear relations from implying layer to outputting layer, so linear adjustment technology is adopted, therefore the goal curve restrains the speed fast, and some extreme point difficult to fall into. So, measured information from one pair of sublance as the foundation, RBF neural network of the endpoint temperature and carbon content prediction is established. The structure of two networks is self-same, input nodes is 7, the first 6 nodes of two networks are the same, correspondent to the amount of blown oxygen, iron sheet a, lime blown, mixed material, ore and dolomite separately. The 7th input node is correspondent molten steel temperature in endpoint temperature prediction RBF model, the 7th input node is carbon content of molten steel in endpoint carbon content prediction model. Amount of nodes in hidden layer is confirmed by train result, one output node is correspondent to the molten steel temperature or endpoint carbon content respectively.

Goals of BOF steelmaking endpoint control are endpoint temperature and carbon content, dynamical control model according to molten steel state measured and goal blowing, the controlling amount is confirmed, namely amount of blown oxygen and cooling dosage. Because BP network has very strong nonlinearity shining upon ability, BP network is adopted to set up the controlling model. A target's straight model is set up at first, difference between expectation introduction going against the neural network and outputs with the straight model of controlled system is utilized to adjust the right value going against the neural network, therefore the difference that it makes the straight model of system output and expect to be outputted is minimum, but not it is minimum to enable outputting with the difference that is input systematically against the neural network. BOF steelmaking dynamic endpoint control system is shown as Fig. 1.

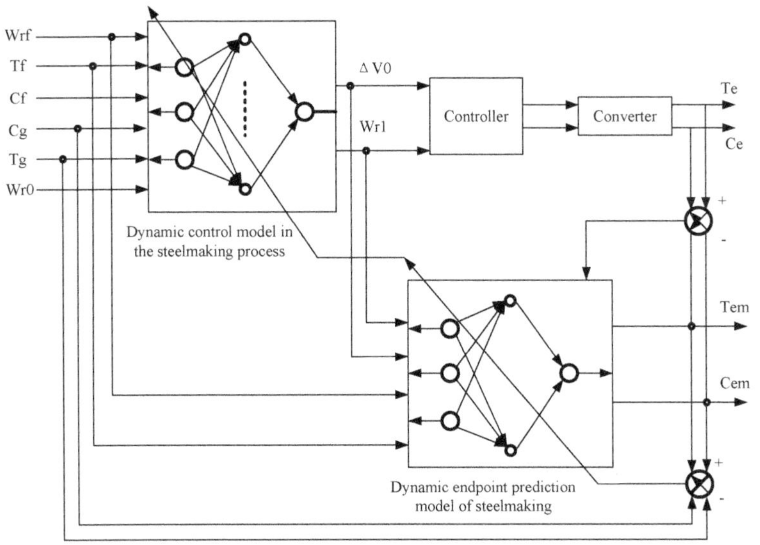

Fig. 1. Dynamic endpoint control system

Wrf represents the amount of cooling dosage joined, Cf represents the carbon content measured by sublance, Cg represents the goal carbon content Tg represents the goal temperature, Wr0 represents the amount of cooling dosage joined at the stage of first blowing, ΔV0 represents the amount of blown oxygen at the stage of first blowing, Tem represents the endpoint temperature of neural network prediction model, Cem represents the endpoint carbon content of neural network prediction model.

The neural network control model has two parts, a part is produced amount of blown oxygen during the reblowing, another part is produced amount of the cooling dosage joined during the reblowing, network structure is BP network of three layers. Amount of blown oxygen during the reblowing has relationship with carbon content measured by sublance,goal carbon content and amount of the cooling dosage joined during the reblowing.Among actual steelmaking, for being easy to operate, usually only a kind of coolant is joined, and other auxiliary raw materials are few in joining amount, neglect its impact on oxygen quantity are neglected, so, input nodes of controller of amount of oxygen blown is 3, correspond to above-mentioned three variables separately, the network outputting is correspondent to amount of blown oxygen, but amount of the cooling dosage joined during the reblowing has relationship with molten steel temperature and carbon content measured by sublance, goal molten steel temperature and carbon content, so the inputs nodes of neural network controller of amount of cooling dosage is 4, the network outputs is correspondent to the cooling dosage joined, Amount of nodes in hidden layer is confirmed by train result.

3 Simulation of Intelligent Control Strategy

With some real measurements data of steel factory 250t converter, the training of the neural network prediction model and control mode is carry on, hidden nodes of prediction model are 12, the learning rate is 0.995, the error rules is 0.001, hidden nodes of control model are 6, the learning rate is 0.05, the error rules is 0.001, simulation course is corresponds to the real steelmaking course. The corresponding endpoint temperature and carbon content are shown as Fig.2 and Fig.3.When the prediction error limits are $|\Delta t| \leq 10°C$, $|\Delta C| \leq 0.05\%$, then the correction rates of endpoint temperature and carbon content are 92.8% and 92.5% respectively. From the figures, it is known that the simulation of endpoint carbon content and temperature are closed to the practical value in most heats, its result is very ideal. This system has already been in some application o of steel factory 250t converter for two years, it reacted well, reached the anticipated index.

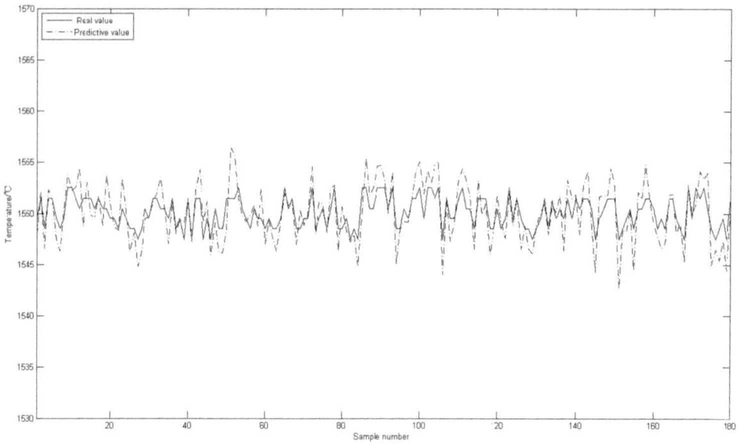

Fig. 2. Endpoint Temperature of BOF Steelmaking

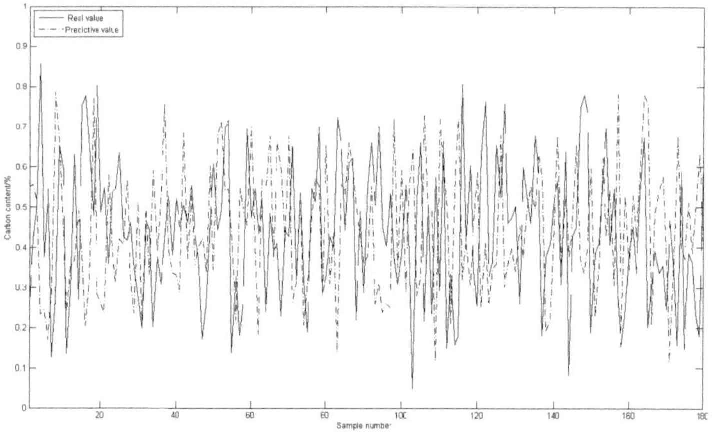

Fig. 3. Endpoint Carbon Content of BOF Steelmaking

4 Conclusion

BOF steelmaking is a very complicated, high temperature and many phase physical-chemical process. There are many factors to influence the endpoint, and some factors are difficult to be described quantitatively.Simultaneously, the temperature is very high, and the environment is abominable in the smelting process. As a result, it brings enormous difficulty for BOF endpoint control. An intelligent control system of BOF steelmaking based on neural networks was proposed. The accuracy of the endpoint prediction model is perfectly improved. The simulation and application results demonstrate the efficiency of the method.

Acknowledgement

This research reported in the paper is supported by Research Center of Green manufacturing and Energy-Saving &Emission Reduction Technology in Wuhan University of Science and Technology (B0911) and National Natural Science Foundation of China (70971102). This support is greatly acknowledged.

References

1. Li, G., Kong, J., Jiang, G.: Research and Application on Compound Intelligent Control System for Coke Oven Heating. Chinese Journal of Iron and Steel 43, 89–92 (2008)
2. Merriman, D.: Mass spectrometry for oxygen steelmaking control. Steel Times 225, 439–440 (1997)

Intelligent Diagnosis of Abnormal Work Condition in Coke Oven Heating Process by Case-Based Reasoning

Gongfa Li, Jianyi Kong, Guozhang Jiang, and Liangxi Xie

College of Machinery and Automation, Wuhan University of Science and Technology, 430081 Wuhan, China
ligongfa@yahoo.com.cn

Abstract. For reducing the fault ratio of coke oven heating process, based on the analysis of the fault mechanism and case-based reasoning (CBR), an intelligent abnormal work condition diagnosis model is proposed for the coke oven heating process. The probability of the typical fault and their operation guidance with the help of case-based reasoning technology is obtained. The knowledge representation of case and the structure of case-based are studied, and the algorithms of retrieval, learning and reuse are discussed. The proposed abnormal work condition diagnosis system is successfully applied to the coke oven heating process, the fault ratios during production process is decreased, and the proved benefit is achieved.

Keywords: Case-Based Reasoning; intelligent diagnosis; abnormal work condition; coke oven heating process.

1 Introduction

Coke oven is an important thermal equipment in the metallurgical industry, disposed by several char rooms and combustion chambers alternatively, coal gas (blast furnace gas, coke oven coal gas) and the atmosphere is mixed, spread and burnt in the combustion chamber, the heat that produces is given to char room through the form transmission of the radiation convection, the coal material is isolated the air and heated (high-temperature dry distillation) and formed the coke in the char room. The coke oven offers the coke for blast furnace, and offer the coal gas to the stove kiln of industry civilly too. Coke is important raw materials of metallurgy and chemical industry trade, quality and output of it concern stability that follow-up produces directly. If the coke oven breaks down, will influence the coke oven production seriously, will make the working state of the coke oven in the unstable state of even staying cool seriously, cause the coke oven to heat the situation not controllable of the course, so it's essential to carry on fault diagnosis of coke oven heating process.

The artificial intelligent technology has permeated through fault diagnosis field in recent years, for example: intelligent monitoring, intelligent control of automatic processing course, intelligent fault inspection and diagnosis of production process, aircraft intelligent control of flying and landing course and medical course intelligent control [1-5], have made the remarkable result. The research used in this field in the mixing of intelligent technology has been paid close attention to too [6]. Especially

L. Qi (Ed.): ISIA 2010, CCIS 86, pp. 240–244, 2011.

case-based reasoning technology is applied to fault diagnosis, a new way is offered to the research of fault diagnosis[7]. However, in the coke oven heating process, the effective method that fault diagnosis or prediction of fault trend in order to prevent the emergence of fault is not carried on yet.

Aiming at frequent fault in the coke oven heating process and limitation of tradition intelligent method in fault diagnosis field, abnormal work condition prediction model of coke oven heating process based on case-based reasoning is proposed.

2 Fault Description in Coke Oven Heating Process

According to the structure of the coke oven and characteristic of heating, there are 5 most common production faults in the course of coke oven heating, namely over-heating, premature coke, difficulty in coke pushing, bunch leaks and jam. For a long time, the coke oven operator judges and makes policy to avoid the craft equipment fault according to basis experience knowledge .Once operate the improper, will cause fault emergence, influence the personal security of operator and production process seriously, and influence coke quality. So a set of complete abnormal work condition diagnosis model need to set up to replace patrolling the work of examining of people, improve working efficiency, continuity and stability guaranteeing to be produced.

3 Intelligent Diagnosis of Abnormal Work Condition by Case-Based Reasoning

There is not very good theory model at present to summarize fault information in the coke oven heating process, knowledge of fault diagnosis is not complete or difficult definition,but experienced technologist can rely on abundant experience to take decision of coke oven heating course, judge the present stove condition variation tendency. Case-based reasoning technology can simulate such treatment course, as to traditional expert system, its reasoning ability is strong, knowledge acquisition is easy, it can be used in more complicated system. Case-based reasoning is applied to fault diagnosis, a new way is offered to fault diagnosis. Based on above-mentioned analysis, Case-based reasoning method is proposed to predict potential fault in coke oven heating process and offer operating guidance in order to avoid the emergence of the fault.

3.1 Case Representation and Storage

The case-base is composed of cases that are in pre-defined structure and organization. Case representations the foundation of CBR. The case representation needs to determine the content stored in case, find out a right structure for the content description, and decide how to organize and index for effective retrieving and reusing. The case features are the important factors for constructing case, they define the index information of case, the CBR system carries on retrieving and matching according to the case features. In this paper case can be defined as:

$$Case(F,W) = case((f_1,f_2,f_3,\ldots,f_n),(w_1,w_2,w_3,\ldots,w_n)). \quad (1)$$

where $F=(f_1,f_2,f_3,\ldots,f_n)$, f_i is the descriptive feature, it may be qualitative or quantitative; $W=(w_1,w_2,w_3,\ldots,w_n)$,$W_i$ is the solution of problem.

In this paper, because material composing, different material ratio and Jingtieshan ore grade which has larger proportion have obvious influence on rate of waste index, these parameters are taken as case features. In this paper, the case representation structure is as followed:

$$\text{Case}(F,W) = \text{case}((f_1,f_2,f_3,\ldots,f_8),(w_1,w_2,w_3,\ldots,w_5)). \quad (2)$$

Where the description feature is F= $(f_1,f_2,f_3,\ldots,f_8)$, f_1 denotes amount of coal injection, f_2 denotes electric current of coke pushing, f_3 denotes coke guide vehicle state, f_4 denotes coke-quenching vehicle state, f_5 denotes heating gas flow, f_6 denotes furnace genitive pressure, f_7 denotes temperature, f_8 denotes coke quality predicted; The solution of the problem is W=$(w_1,w_2,w_3,\ldots,w_5)$, w_1 denotes over-heating, w_2 denotes premature coke, w_3 denotes difficulty in coke pushing, w_4 denotes bunch leaks and w_5 denotes jam.

3.2 Case Index and Retrieval

CBR is a reasoning mode which obtains the solution of present problem by accessing past similar problem solution in case base. Obviously, how to complete case retrieval effectively and quickly is significant, and it has much influence on efficiency of CBR. So far, there are general several case retrieval strategies such as nearest-neighbor approach, inductive algorithm, Knowledge-guided and module-retrieval .In this paper, the nearest-neighbor approach is adopted, and the similarity between the input case and old cases is calculated, the key of this approach is how to determine the important weight of each features.

The case index is carried on according to problem description features of each case, namely, amount of coal injection, electric current of coke pushing, coke guide vehicle state, coke-quenching vehicle state, heating gas flow, furnace genitive pressure, temperature and coke quality predicted.

After the matching by main index, the similarity is calculated, and all the cases which meet the matching threshold are been searched. The similarity between new case and retrieved eases is given as equation (3):

$$\text{SIM}(c_i, c_R)=(w_i \times \text{sim}(f_i^I, f_i^R))/(w_i) . \quad (3)$$

Where, c_i is the input case, c_R is the retrieved case, SIM(c_i, c_R) is the similarity function, w_i is the significance weight of feature i,f_i^I and f_i^R respectively denotes the value of feature i in input case and retrieved case,i=8.sim(f_i^I, f_i^R) is the similarity function on feature i,and it is defined as followed:

$$\text{sim}(f_i^I, f_i^R) = 1 - |(f_i^I, f_i^R)| / \max(f_i^I, f_i^R) . \quad (4)$$

The threshold of similarity is given by experts, it is an adjustable parameter, if there are many matched cases, the plan maker may enlarge the threshold; If there are little matched cases, the threshold may be reduce, thus ore or several most similar cases to target case can be selected.

3.3 Case Reuse

Once the matched cases are obtained, CBR system will adapt the solutions of selected case according to input case so as to obtain the solution of input case, this process is named as case reuse. It is supposed that k similar cases in case-base are retrieved, then the case adaptation may be carried on according to equation (5):

$$W_{in,i} = \lambda_j W_{j,i} \ . \tag{5}$$

Where λ_j =e $SIM(C_{in},C_{Rj})$ / $SIM(C_{in},C_{Rj})$, $W_{in,i}$ denotes the ith variable of input case solution, $SIM(C_{in},C_{Rj})$ denotes the similarity between input case and retrieved case j, λ_j denotes the influence degree of retrieved case j to input case.

3.4 Case Learning

The quantity of initial cases in case-base is limited, it is necessary to generate new cases and add them to case-base during plan making process, so the experience is accumulated. Because there are certain rules during plan making, the new plan may be completely same or very similar to the retrieved cases from case base, if all the cases are added to the case-base, the case-base will become lager and larger, thus case learning must be carried on. The main case learning strategies are as followed:

a) When the case matching fails, the plan maker has the right to filter the new case, define and add the case considered to be valuable into the case-base;

b) If the all the similarities between input case and cases in case-base are all less than certain value (for example 0.8), then the case is added to the case base.

4 Industrial Application

In the process of coke oven heating, basic reason influences product and frequent fault is that coke oven production totally depend on experience knowledge of operator to judge stove condition and adopt corresponding operation, not only labor intensity is very great, but also dependability, security and economic validity are all difficult to ensure. Combine the reality of heating course of some coke oven plant, abnormal work condition diagnosis system based on human-computer interaction is developed and applied to production using the method put forward in this paper. The system is applied to some coke ovens No. 10 in Coking and Chemistry Plant, the characteristic weights of 8 variables based on case-based reasoning will be confirmed, described and accords with other relevant knowledge according to the corresponding case, through such operation as the search, matching of the case in the case storehouse of coke oven heating fault, get the diagnosis result in conformity with actual conditions, provide the treatment method. The actual operation result of more than two years indicates, this system can realize correctly the abnormal work condition diagnosis, the accuracy rate of fault diagnosis is up to about 90%, the operating rate of the equipment improves by 2.5%, coke quality is improved a lot, stability coefficient of the coke oven and even coefficient are improved.

5 Conclusion

Case-based reasoning technology is proposed to realize intelligent abnormal work condition diagnosis of coke oven heating process. Have avoided setting up the mechanism model difficult and obtaining the defect with difficult knowledge in use of the case-based reasoning technology, prediction in real time of key craft parameter that difficult measurement or lagging behind has offered the characteristic parameter for fault reasoning. Applying to the production practices of coke oven heating, has reduced the trouble incidence by a wide margin, have obviously improve production target, has played an important role in steadily control and stability run of optimization control system of whole course.

Acknowledgement

This research reported in the paper is supported by Key Laboratory of Metallurgical Equipment and Control of Ministry of Education in Wuhan University of Science and Technology (2009A04), Research Center of Green manufacturing and Energy-Saving &Emission Reduction Technology in Wuhan University of Science and Technology (A0902), National Natural Science Foundation of Hubei (2008CDB298) and National Natural Science Foundation of China (70971102). This support is greatly acknowledged.

References

1. Yang, Q., Xue, W., Lan, Z.: Development of an artificial intelligent diagnosis system for transformer fault. In: IEEE/PES Transmission and Distribution Conference & Exhibition: Asia and Pacific, pp. 1–5 (2005)
2. Grant, P.W., Harrism, P.M., Moseley, L.G.: Fault diagnosis for industrial printers using case-based reasoning. Engineering Application of Artificial Intelligence 9, 163–173 (1996)
3. Awadallah, M.A., Morcos, M.M.: Application of AI tools in fault diagnosis of electrical machines and drives-an overview. IEEE Transactions on Energy Conversion 18, 245–251 (2003)
4. Yang, B.S., Han, T., Kim, Y.S.: Integration of ART-Kohonen neural networks and case-based reasoning for intelligent fault diagnosis. Expert Systems with Application 26, 387–395 (2004)
5. Wu, M., She, J.H., Nakano, M.: Expert control and fault diagnosis of the leaching process in zinc hydrometallurgy plant. Control Engineering Practice 10, 433–442 (2002)
6. Chen, K.Y., Lim, C.P., Lai, W.K.: Application of a neural fuzzy system with rule extraction to fault detection and diagnosis. Journal of Intelligent Manufacturing 16, 679–691 (2005)
7. Lee, S.G., Ng, Y.C.: Hybrid case-based reasoning for online product fault diagnosis. International Journal of Advanced Manufacturing Technology 27, 833–840 (2006)
8. Li, G., Kong, J., Jiang, G.: Research and Application on Compound Intelligent Control System for Coke Oven Heating. Chinese Journal of Iron and Steel 43, 89–92 (2008)

FPGA Chip Optimization Based on Small-World Network Theory

Hai-ping Zhou[1,2] and Shao-hong Cai[2]

[1] Department of Computer Science, Guiyang College, Guiyang 550005, Guizhou, China
[2] College of Science, Guizhou University, Guiyang 550025, Guizhou, China

Abstract. With the development of semiconductor technology, the devices integrated in chips are more and more dense. As a result, the delay of circuit has become a bottleneck problem that impact on the efficiency of chip. In this paper, we proposed a new method to optimize the circuit. In the optimization, we added a few random edges in the regular circuit of FPGA chip. The result shows that the average path length of FPGA chip is significantly reduced after the optimization.

Keywords: small-world network; delay of circuit; circuit optimization.

1 Introduction

With the progress of micro electronic technology, the size of micro electron device is scaled and the number of device integrated in a chip with the unit size is increased. In order to make the devices work cooperatively and improve the operating efficiency of the chip, the optimization of the chip should be carried on. One of the goals of chip optimization is to decrease the delay of circuit. Generally, there are two kinds of delay in circuit, which are separately wiring delay and switch delay. In the classical circuits, the delay on the switch is far beyond that on wires so that the wiring delay is usually ignored and the optimization of circuit is concerned on switch delay. Along with the progress of processing technology, the size of electron device is smaller and smaller, meanwhile, the integration density of the chip is higher and higher. This process on the one hand lead to the decline of switch delay, on the other hand the wiring delay is increased with the growth of integration density. Nowadays, with the progress of the technology of integrated circuit the wiring delay has surpassed the switch delay. How to decline the wiring delay has become an important problem, which is call "wiring-crisis" [1].

Table 1 shows the data of wiring delay and switch delay in different submicron process, which is released by ITRS2002. From the table, we can see that when technology is advanced from 130nm to 22nm, the wiring delay is increased by 22 times and the switch delay is cut by 6/7, thus, the rapid increase of wiring delay has become a vital problem of Very Large Scale Integrated Circuits (VLSI).

L. Qi (Ed.): ISIA 2010, CCIS 86, pp. 245–251, 2011.

Table 1. Wiring delay and switch delay in different technology

Process technology (nm)	1mm wiring delay (ps)	Gate delay (ps)
130	21	2.55
90	37	1.84
65	79	1.14
45	131	0.85
32	248	0.56
22	452	0.35

For traditional programmable logic devices, wiring delay is far less than switching delay, so that the main way to improve the operating efficiency of circuit is to minimize the number or switches. However during the deep submicron process, the wiring delay has surpassed the switch delay, so we have to consider the effect on efficiency of circuit caused by wiring delay. These years, researchers have been studying all kinds of ways to deal with the "wiring-crisis", the common method is to use the low resistivity metal wires instead of the high resistivity wires. However, this process usually brings higher costs. So people have to look for other methods to solve this problem. The optimization of the circuits' structure is one of the most effective methods to improve the efficient of circuit. In the post years, complex network has been developed rapidly and now is being applied in many fields [2-6]. In this paper, we apply the small-world network theory to solve the optimization problem of FPGA chip and obtained satisfied effect. In the following sections, we introduced the basic theory of small-world network in the first and then apply this theory to optimize the FPGA circuit.

2 Small-World Network Model

The "small-world" character of small-world network means the average path length of the network is smaller than that of other kinds of network. Watts published the paper of "Collective dynamics of 'small-world' networks" in *Nature* in 1998 [2]. In the paper, Watts found many real networks' path length is very short. Soon he proposed a network model which is got a short average path length and is called WS network model [7]. This model revealed the small-world character of complex network and demonstrate the mechanism that small-world network generate by.

The small-world network is constructed by the following steps:

(1) Given an initial network which is made up by N nodes. These nodes are connected by some lines and form a ring. Each node connects to $K/2$ right neighbors and $K/2$ left neighbors. In order to make the network to be sparse, $N>>K$ should be meted. This network is of high clustering coefficient.

(2) Reconnect each edge of the network with probability p. When one edge is selected to be reconnected, one node of the edge is fixed, then the edge is cut off and the other node is randomly selected. There is no more one edge set between each pair nodes and each node can not link to itself. This step greatly reduces the average path length of networks. When p is small, the clustering coefficient changes faintly.

According to above evolving rules, when $p=0$ the network will keep in the regular structure, and when $p=1$, the model will evolve to random network. When p is turned from 0 to 1, the network will change from regular structure to random structure. Figure 1 shows the changing procedure of the network when p is turned from 0 to 1.

Figure 2 shows the relation between the average path length and reconnection probability in WS network. From which, we can see that as long as p is turned from 0 to 0.01, the average path length will reduce greatly.

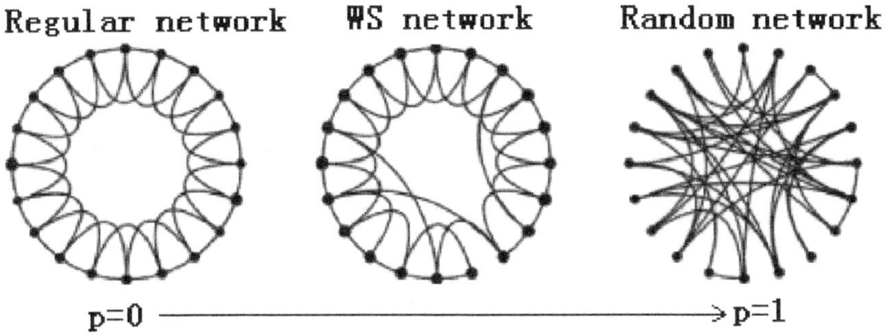

Fig. 1. The evolving procedure of WS small-world network

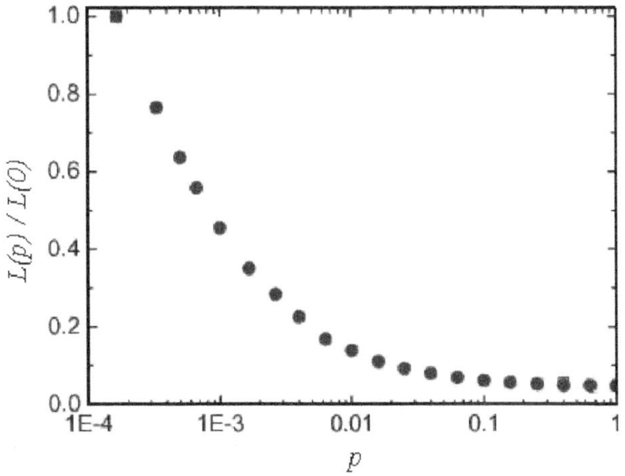

Fig. 2. The relation between the average path length and reconnection probability in WS network

In WS model, the reconnection of the network may break down the connectivity of the network. In order to avoid this defect Newman and Watts improved the WS model in 1999 [8]. In the new model, the old edges are not cut off, and the new edges are added into the network randomly with probability p. This model is called NW network model. In this model, when p is turned from 0 to 1, the network will change from regular network to complete network. The studies show that when p is small, WS network and NW network are blessed with similar characters.

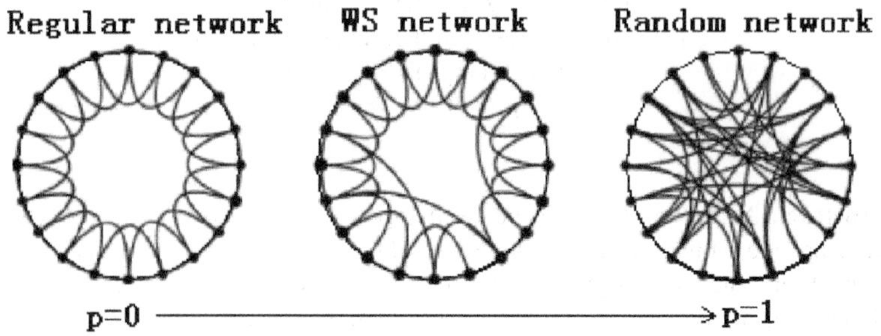

Fig. 3. The evolving procedure of NW small-world network

Enlightened by the idea of NW network, we try to add a few random edges in regular circuit such as FPGA so as to reduce the average path length and decrease the delay of the circuit. In the following section we are going to optimize the FPGA chip by the method of NW network.

3 Optimization of FPGA

The circuit of FPGA chip is often designed to an island structure shown as figure 4. The circuit is composed by logic block, switch block and channel. The connection mode is regular and the circuit looks like a two dimensional network. Each of the logic blocks is programmable so as to implement some functions. All logic blocks are connected by switches and tracks and forming a network with complete functions. The channel is composed by a series of metal wires of graded lengths. The magnified channel is shown as figure 5. The channel consists of a single length wire, double length wire, quad length wire and long line.

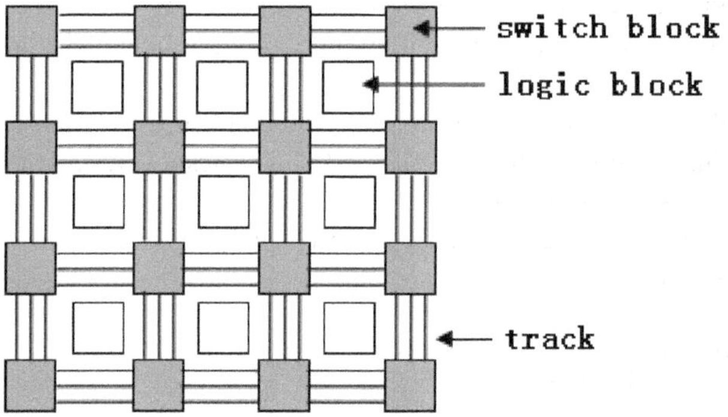

Fig. 4. The circuit structure of FPGA chip

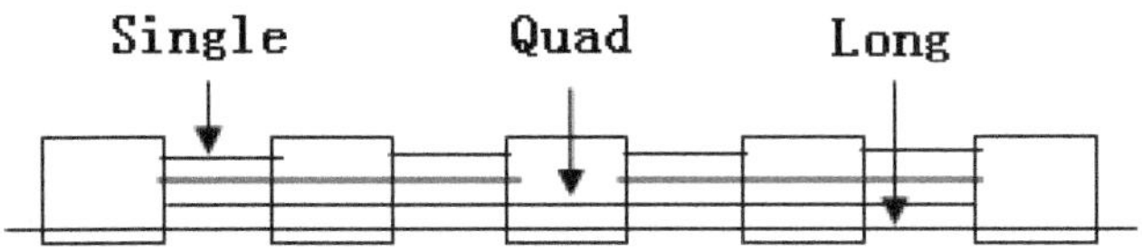

Fig. 5. The inner connections of channel

According NW small-world network model, we optimized the circuit structure of FPGA. The optimization steps are as follows:

(1) Select two blocks randomly in the circuit.
(2) Add a line between the selected blocks.
(3) Repeat the steps above until the average path length is short enough.

Figure 6 shows the circuit structure of FPGA chip. In the previous circuit, the information transferring from block A to block B has to pass two double length wires and two single length wires. However, in the optimized circuit it should pass only two single length wires and one small-world wires. Compared with the previous circuit, the optimized path not only is shorter than before but also cut off a switch block. Obviously, the efficient of the optimized circuit is enhanced greatly.

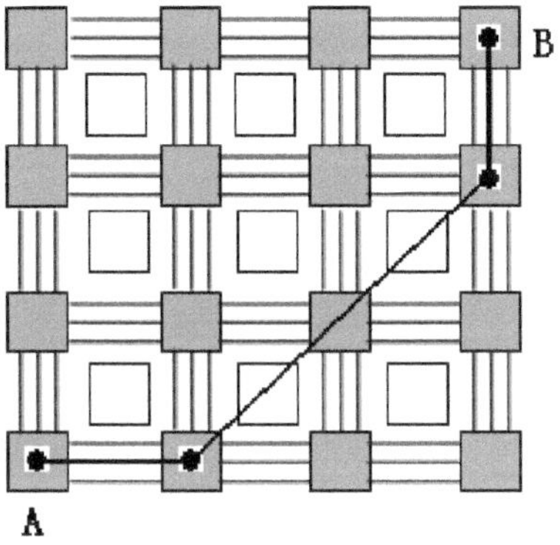

Fig. 6. The schematic diagram of chip optimization

The average path length of the network with $N \times N$ switches can be calculated by formula (1) , where $l_{i,j}$ denote the path length of block i and block j .

$$L = 2\sum_{i=1}^{N}\sum_{j=i+1}^{N} l_{i,j} \Big/ N(N-1) \quad (1)$$

Figure 7 shows the contrast of average path length between previous circuit and optimized circuit, from which we can see that with the increase of circuit's scale the average path length is increased deeply, meanwhile, the average path length of optimized circuit is much shorter than that of original circuit, which means the circuit

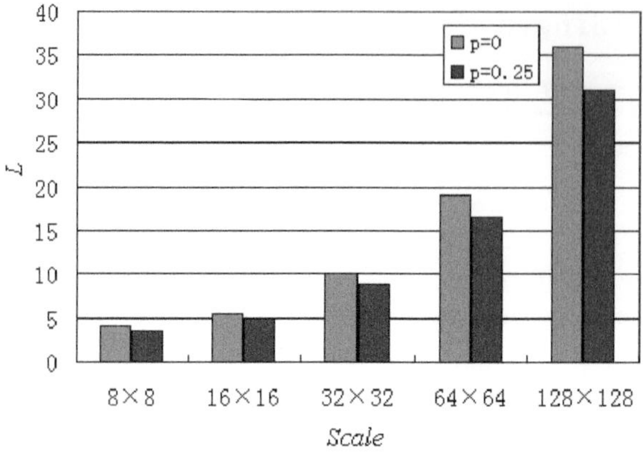

Fig. 7. The contrast of average path length between previous circuit and optimized circuit

delay will be decreased greatly. In fact, besides the reduction of average path length, the number of blocks information passing by is also decreased, so the operating efficiency of the circuit is further enhanced.

Figure 8 shows the relation between the scaled average path length and reconnecting probability, from which we can see that the average path length reduces deeply with the increase of reconnecting probability. The size of the chip is larger the optimizing effect is more remarkable.

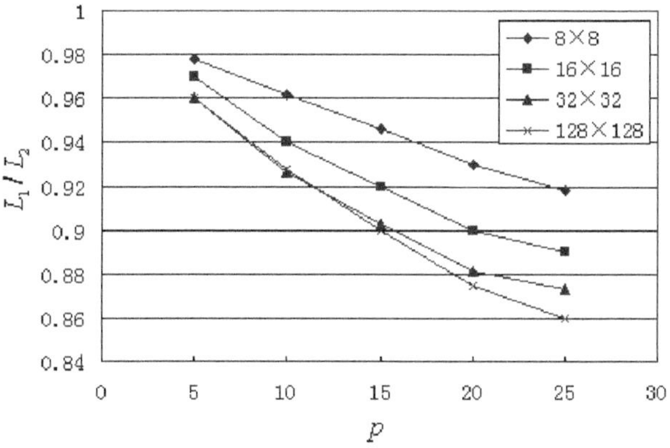

Fig. 8. The relation between the scaled average path length and reconnecting probability

4 Conclusions

Recently, under the driven of business needs, the integration density is getting higher and higher, how to reduce the delay of circuit has been a key problem. In this paper, we optimized the regular circuit such as FPGA chip with the theory of small-world network. The average path length of optimized circuit is shorter and the number of blocks in the paths is decreased, which is helpful to enhance the operating efficiency of circuit.

References

1. Ho, R., Mai, K.W., Horowitz, M.A.: The future of wires. Proc. IEEE 89(4), 490–504 (2001)
2. Watts, D.J., Strogatz Steven, H.: Collective dynamics of "small-world" networks. Nature 393(6684), 440–442 (1998)
3. Jeong, H., Mason, S.P., Barabási, A.L., et al.: Lethality and centrality in protein networks. Nature 411, 41–42 (2001)
4. Stragotz, S.H.: Exploring complex networks. Nature 410(8), 268–276 (2001)
5. Milo, R., Itzkovitz, S., Kashtan, N., et al.: Superfamilies of Evolved and Designed Networks. Science 303, 1538–1542 (2004)
6. Zhou, H.P., Cai, S.H., Jia, X., Long, Y.: Random-scale free unified evolving network model. Journal of University of Shanghai for Science and Technology 30(3), 283–286 (2008)
7. Watts, D.J., Strogatz Steven, H.: Collective dynamics of "small-world" networks. Nature 393(6684), 440–442 (1998)
8. Newman, M.E.J., Watts, D.J.: Scaling and percolation in the small-world network model. Phys. Rev. E 60, 7332–7342 (1999)

Virtual Exhibition and Customization Based on Web3D

Yanfang Wu, Kun Chen, Lei Yang, and Junfen Wang

Manufacture Engineering Institute, Southwest University of Science and Technology, 621010 MianYang, China
stonewyf@163.com

Abstract. The purpose of this paper is to present a method for virtual exhibition and product customization. The essential relations between the characteristics of virtual exhibition and customization is analyzed, the paper surveys some used technologies and design tools, presents a detailed method that includes virtual exhibition, color customization, form customization, and so on, by researching on action, event and script. At last, a system about automobile exhibition and customization is developed based on these methods.

1 Introduction

With the development of market economy, the competition is more and more intense, the market is subdivided, and the diversification of consumers' requirements will be the coming trend. The virtual reality technology can establish a realistic system of virtual exhibition and customization, during product development, particularly, in the early conceptual design stage, the enterprise can find out user demands and evaluation of design solutions in time by this system, and adjust corresponding product development strategy, which helps the enterprise to design and produce product meets the market requirements more closely, then to improve competitive power.

2 Virtual Exhibition and Customization

2.1 Virtual Exhibition

Differing from traditional static display system based on the Web, virtual exhibition emphasizes to give user more realistic experience, the virtual scene as realistic as photo rendering can be created by Web3D technology, so, users can have a all-dimensional observation to product shown like real product. At present, virtual exhibition mainly can create a sense of reality by the action and sound, in the future, user can acquire more such as taking up object in virtual exhibition system through digital equipment, and feeling its weight, texture and so on.

L. Qi (Ed.): ISIA 2010, CCIS 86, pp. 252–258, 2011.

2.2 Customization

In environment of the virtual exhibition, perfect interaction design can make user become a participant rather than merely a viewer, in the virtual platform user can customize product in accordance with individual tastes such as function, form, decorative pattern, texture and so on. The virtual platform opens not only to user, enterprises also can take advantage of it to know the needs and feedbacks of users, then improve the product, and adjust product design strategy timely, finally develop the product that will please and satisfy users.

3 Main Web3D Technologies

Web3D is the generic term commonly used to refer to any three dimensional graphics technology supported by the World Wide Web. Recently it has been recognized as the important technology for the new Internet age. The first ISO standard for Web3D was the Virtual Reality Modeling Language (VRML), currently Wed3D technologies have a rapid development, there are some to be applied wildly such as Cult3D, VRML, Java 3D, Viewpoint, and so on, they have respective advantage and are widely applied in manufacturing, e-commerce, real estate, urban planning, etc.. The paper focus on the Cult3D, Cult3D has the advantages of small file size, good picture and mutual display, moreover, its kernel bases on Java and supports embedded Java classes which user developed. For narrow-band network, the Cult3D is one of best solutions to product exhibition, so, a system about virtual exhibition and customization of automobile based on the Cult3D is researched in this paper.

4 Virtual Exhibition and Customization of Automobile

With the Cult3D Designer, you get some pre-made actions and JAS. With these you can control your scene and object to your desire. You connect an action to an event, and then connect the selected (i.e. object, sound, etc.) to the action you want the data to perform.

4.1 Virtual Exhibition System of Automobile

Create automobile model with material, lights and the camera in 3ds max, then export the model to the Cult3D intermediate file format, .c3d.

Open the Cult3D Designer, in the Actions window, you can find Rotation, Translation and Scale action from the Object motion section and Arcball action under the category interactivity. When those actions are triggered in Event map windows the object will be given special features like rotate, zoom and translate. In this way, users can achieve rotate, zoom and translate the object or the part, experience really functions and form of the object. If need sound, the actions under the category Sound can achieve the purpose. In addition, you can use the Select camera action in the Camera section to set different viewpoints for choose.

Fig. 1. A virtual exhibition of the automobile

The concrete example is shown as follow: drag the car object and a World start event to the Event map windows. Then add the Arcball action to World start event, and link it to the car object, so that when the user manipulate the mouse button, will have a all-dimensional observation to the car object. If you want look into the interior decoration of the car, you may drag the Left mouse click object event to the Event map windows, connect Rotation action to the car door object, and double click the Rotation graph icon to set details which can control the object to your desire. Then use Activate/Deactivate event action from the Event section of the Actions window to control the mouse button action, in this way, when the user click the door, the door will open around its own pivot point, if click again, the door will close, the concrete operating process is shown in figure1.

4.2 Color Customization of Automobile

The color scheme of automobile is relatively simple, the car body commonly has a single color, the fittings adopt achromatic colour. Compared with form and function customization, color customization is relatively easy to implement. In the Cult3D Designer color customization is implemented in two ways. One is by two Cult3D features, hotspots and toggle hotspot's alter ego. A hotspot is a rectangular area defined on the surface of the texture, you can control the color of the object by changing the texture, A hotspot alter ego is a piece of bitmap, which can be cropped out of any existing texture or from a externally loaded image file, which can be used later on to replace the original texture pixels in the area of the hotspot through the use of the "Set hotspot's alter ego" and "Toggle hotspot's alter ego" grouped under the Texture section in the Action dialog.

The concrete operation as shown in figure2: clicking the Textures node of Scene Graph window, you will find the red texture map assigned to the automobile model in 3dsmax, double click it, this should open the texture details window, in the texture details window add new texture map and the new hotspot, then resize the new hotspot to fill the space at the end of the hall in the middle of the texture, so a new Hotspot should be visible under the Textures in the scene graph. You can add other color you want in the same way. Next, from the

Textures section of the Actions window, select toggle hotspot's alter ego and drag it into the Event map and onto the new Manual event or Keyboard. From the texture section of the scene graph, drag the new hotspot into the Event map and onto the toggle hotspot's alter ego, so you can change the color of the object through the Keyboard or Manual event. In this example, choice Manual event, because it is more convenient than Keyboard key, and needs be triggered by script in other applications.

Fig. 2. The process of color customization

The other is by JavaScript. In the Cult3D Designer, we can execute JAS script for changing the color of objects, there are two ways to execute it, one is changing the color of texture, an other is changing the texture directly. The following script programs are written for changing the color, by this way, it becomes very simple when you need the different color, just to add the corresponding color data into the script program.

```
JavaActionscript(1,0,0){
MapExecute("yellow",1); }
actoinsList("yellow","right"){
setTexture("red",[255,0,0],[255,255,0],40,40);
mapExecute("red",1); }
actoinsList("red","left") {
setTexture("red",[255,0,0],[255,255,0],40,40);
mapExecute("yellow",1); }
```

4.3 Form Customization of Automobile

Because the automobile appearance design emphasizes the whole, the outline has little change and the varieties of the details are abundant. This facilitates the form customization of automobile.Product customization commonly demands universalization and standardization of parts or components. In this way, form customization is achieved by exchanging the form components in same series products. For form customization of automobile, the form components for choice commonly include the car body, rim, lights, the hood and the rearview mirror.

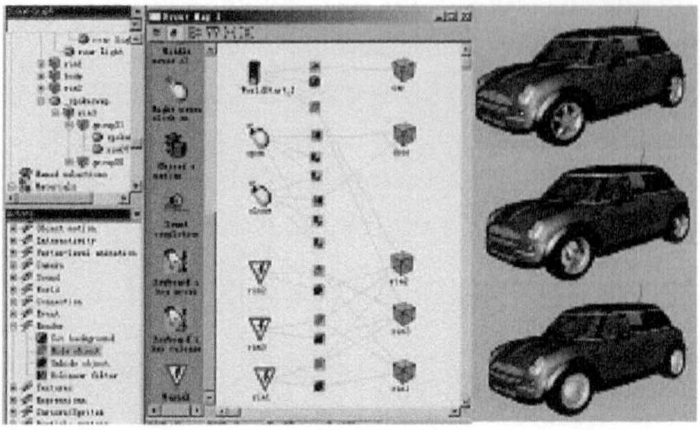

Fig. 3. The process of form customization

Form customization also can be implemented in two ways. One is by hide and unhide action. When modeling in 3D Software, in same position, put several different components for choice. After the model file is opened in the Cult3D Designer, at first, hide superfluous components and only keep one component for showing, then display them in turn by hide/unhide action. For rim customization in the example, there are three types of rims for customer to select. Drag three rims in the Scene graph to the Event map, drag the Hide object action and Unhide object action from the category Render in the Action window onto the new World start event in Event map, link the Hide object action to rim2 and rim3. This way, you may view only rim1 shown in the Stage window. Next, drag three Manual icons to the Event map, drag the Hide object action and Unhide object action on them, and link to the object which you want to show or hide in the Scene graph. Because the rim is composed of different parts, you must double click on the hide object icon and unhide object icon in the Event map and select the Apply to child option in Hide action details window. The concrete operating process is shown in figure3. The other is to apply the features of hotspots and toggle hotspot's alter ego. The method has been introduced above, it not only can achieve the color customization, but is effective to customize plane or plate, such as keystrokes in control panel, windows and decorative pattern. Because their forms depend on the patterns of the surfaces, it is easily to achieve the form change of the object by changing the hotspot.

4.4 Function Customization of Automobile

Because all functions of product are fulfilled by components, function customization actually is a part of form customization. A function customization of automobile can be achieved by combining diverse form components.

5 Media Mixing

After edit the Cult3D object in the Cult3D Designer, we save the file as Internet file. This .co file can be published in html, MS Office, Adobe Acrobat and Adobe Director.

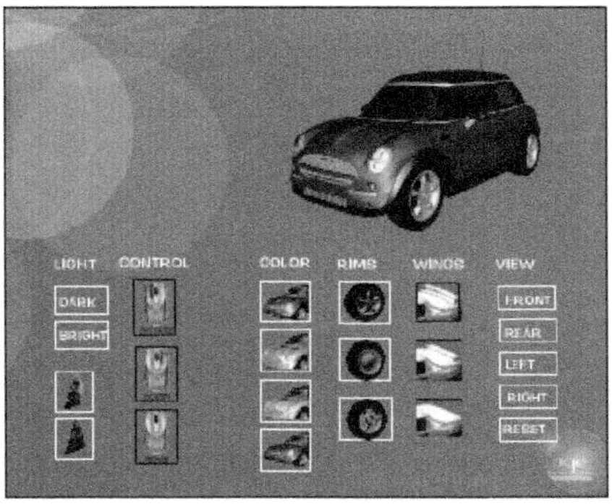

Fig. 4. The final interface

To embed a Cult3D object without requiring the user to activate the control inside the Web page, Cycore Systems has a created a Java script that embeds a Cult3D file dynamically. If the Cult3D object with events need be triggered, you have to write the JS code on your HTML page, and load the corresponding Cult3D object for each button. This is correspondingly complicated. Therefore, we choice Adobe Director to encapsulate the Cult3D object. By its convenient script language, you can easily trigger the event defined in the Cult3D Designer. In the addition, it has the ability of beautifying the interface, which helps to provide a comfort operation environment for user.

In this example, most of actions in the Cult3D Designer are linked to manual events, manual events can be triggered from lingo in Adobe Director, user can browse and customize only by clicking graph icon. At the same time, Director supports local demonstration which can publish in executable file for embedding into HTML or local Multimedia demonstration.

The concrete operating procedures include the following steps:

First, import the Cult3D object. Start Director, open the Property Inspector and choose the size and background color for your presentation. After choose the Control and ActiveX command from the insert-menu, the Select ActiveX Control dialog box will open. Choose the Cult3D Active Player in this dialog. When the ActiveX Control Properties dialog opens, click on the Custom button, the Properties dialog provides two ways to link the Cult3D object: embedding and typing a URL. In this example, choose the embed button, and select the car.co file, then the Cult3D object is placed in the Internal Cast, drag and drop it onto the stage.

Then, insert buttons to communicate with the events into this Cult3D object, change the name for each button. It is shown as the figure 4. Only keep the first frame the Cult3D object located in the Score window. Create buttons for the viewpoints. Create buttons for changing car color, this example offers red, blue, gray and yellow to choose. Create buttons for customizing form, for example, rim1, rim2 and rim3 as shown in the figure 4. For practical application, you can adjust the objects customized, operational motions and the viewpoints according to your need.

Next, trigger manual event. Right-click on any button and choose the Script, the Behavior Script window will open. In this window, trigger the different manual event defined in the Cult3D Designers by entering corresponding script. For example, to active the red button, use the following script:

```
on mouseUp
      TriggerEvent sprite (1), "red"
end
```

"1" is the number of the channel the Cult3D object located in, "red" is the name of the manual event defined in Cult3D Designer. Other scripts about forms and viewpoints can be created by modifying the name of the corresponding manual event.

Finally, optimize interface by graph and text to help user operation and understand. After finishing it, play the presentation by clicking on the play button. At this point, you can operate car model, also can customize a car with your favor and select a viewpoint you need by clicking on a corresponding graph icon.

6 Conclusion

The application of virtual reality technology will make the methods of future product design and product promotion have a radical change. The virtual exhibition of product gives user a more convenient access to get information about products, and more realistic experience. The customization system based on Web can provide individualized products and services for consumers, and offer design principles for product development. With the development of Web3D technology, in the future, the user will get more abundant of experience on browsing Web and selecting products, virtual design platform will become an important tool to promote enterprise competitive power.

References

1. Li, X.L.: Research on Interactive Virtual Presentation technology based on Web. J. Computer Engineering and Application 43, 90–92 (2007)
2. Liu, S.: Interactive 3D Virtual Exhibition on Web Environment. J. Journal of Engineering Graphics 4, 20–25 (2009)
3. Wei, S.Z.: Cult3D Application Guide. MTsinghua University Press, Peking (2001)
4. Cycore Inc. Cult3D Homepage, `http://www.cult3d.com`

Sensorless Pressure Control for Special Industrial Applications

Tianshu Peng[1], Craig Struthers[2], Jianwu Zhe[1], Guangming Liu[1], Yulin Shen[1], and Yitong Sun[1]

[1] Computing Centre of Gansu Provincial Department of Science & Technology
Lanzhou, Gansu Province 730030, China
pts22@yahoo.cn, {zhejw,Liugm}@mail.gspcc.com
[2] Faculty of Engineering and Surveying, University of Southern Queensland
Toowoomba, QLD 4350, Australia
Craig.Struthers@usq.edu.au

Abstract. An automated pumping controller, which is able to maintain constant pressure in a hydraulics system and will not have any physical contact with the medium, is addressed in this paper. The reason to do some research on the automated pumping controller is that it will surely address the lack of products currently available within the industrial control field for pump pressure control in a hydraulics system. Constant pressure control within a hydraulic pumping system is conventionally performed by the use of an electric pump with a variable speed controlling of the pump. In the traditional control system, a pressure transmitter is used as a feedback of the pump which in turn, also controls the pressure in the system. This research proposes to establish that control of the pumping system without the use of a pressure transmitter. Along with the control system itself, a user interface was developed to operate over the Ethernet ensuring the ability to utilize current WEB server interfaces such as windows Explorer and the like.

Keywords: Sensorless, pressure control and industrial pump.

1 Introduction

Pumps are utilized in almost all aspects of industry and engineering with an almost endless assortment of pumping equipment available ranging in size, type and material of construction. At the same time, each year there are many more pumps in use than are actually being supplied new [1].

With today's focus on energy efficiency and sustainability, one way of achieving this goal is by reducing the number of components within the pumping system itself. Not withstanding the environmental advantages that this would provide, the added benefits of removing a component from a conventional pump system include ease of installation, reduction in the labour content, cost savings, and improved reliability of the system which will limit the components prone to failure.

The specific aim of this project is to develop and test an automated pumping controller, which is able to maintain constant pressure in a hydraulics system. In

L. Qi (Ed.): ISIA 2010, CCIS 86, pp. 259–267, 2011.

completing the functionality of this project, a real-time monitoring, configuration and control system software package is also being developed.

This project was chosen in order to find a practical solution to achieving pressure feedback in pumping applications with highly corrosive or dangerous mediums. In some situations, pressure transmitters are available for mediums that are considered dangerous and corrosive the cost of this component is highly prohibitive. Furthermore, replacement or calibration of the pressure transmitter component can be difficult and again not cost effective under such severe conditions.

For the purposes of this project pumped material is described in terms of fluid. It must be acknowledged that some pumps can manage solids, however, the material must demonstrate an overall liquid behaviour to do so [3]. Arguably the most fundamental means of categorizing pumps is by the way in which energy is conveyed to the pumped fluid [3]. By this method all pumps can be separated into two major categories, either kinetic or positive displacement.

The following figure 1 illustrates a typical pump curve. The shape of the curve varies depending on the type of pump used. Pump curves such as this will be utilized within this project to provide information as to how the pump performs with respect to speed and pressure within the system.

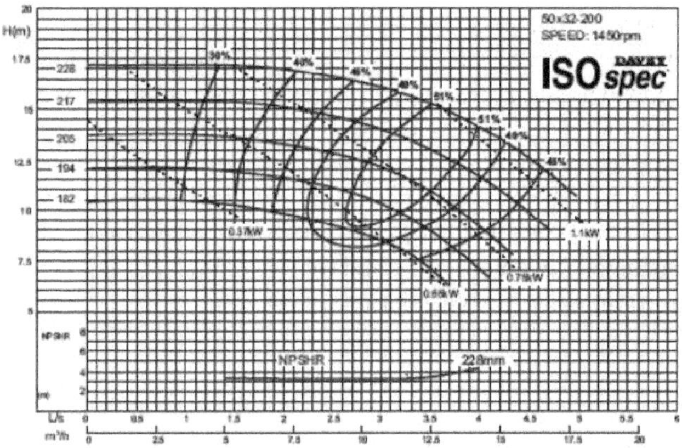

Fig. 1. Illustration of a typical Pump Curve

Not withstanding the fact that the centrifugal pump is one of the most widely used pumps for transferring liquids, it also has excellent ability to control pressure, is quiet in comparison to other pumps, has relatively low operating and maintenance costs, takes up minimal floor space and can create a uniform, non pulsating flow [3,10]. For these reasons the centrifugal pump has been chosen for the purposes of this project.

The primary reason for this appears to be the need to avoid invasive extra components such as pressure transmitters within the human body [2,10]. One example that is of interest to this project is a ventricular assist device which can be permanently implanted within the human body. The computer modeling of the interaction of the electric motor and the blood pump within the circulatory system has parallels with objectives of this project [2]. It must be noted however, that this development is on a much smaller scale (i.e. physical flow and pressure rates) of the pump than is anticipated within this project. In addition, the application is highly specialized. Along similar lines is a project based upon non-invasive measurements of blood pressure and flow utilizing a centrifugal pump [6].

2 Methodology

This research proposes to remove the pressure transducer component of pump pressure control in order to achieve an efficient process [1]. Following is the equipment that has been utilised for this project. Each item has been carefully chosen for its suitability for the purpose of the project. Meanwhile, some technical specifications and the reasoning behind the selection of the specific items utilized will be provided.

2.1 Resource Planning

In this test a centrifugal pump has been chosen for this project. This form of pump consists of a shaft mounted impeller(s) rotating unidirectionally within a casing. Usually there are 3 main types of impellers: radial glow impellers, mixed flow impellers and axial flow impellers. Here the type of pump chosen is a centrifugal pump with a radial flow impeller driven by a suitably sized squirrel cage induction motor.

Various brands of VSDs were considered, with final selection being a PDL Micro Drive Elite. The other VSDs considered include Danfoss, Telemechanique, Moeller and ABB, all of which were capable of controlling the speed of the motor but the PDL drive had the added feature of being able to write user software within the VSD thus eliminating the need for an additional controller. The PDL drive uses an icon function block based programming language called VYSTA which has been developed by PDL.

The current method of control utilises VVVF which stands for Variable Voltage Variable Frequency. This method of control also varies the voltage in proportion with the Frequency so that the V/HZ ratio is kept constant. [4]Also, a Lantronix Din Rail mounted Serial to Ethernet convertor / Web server was selected. The unit can also be programmed to perform calculations and control functions for the VSD system as well as be able to be configured as an HMI (Human Machine Interface) from the VSD system to the real world. And The Test Tank used is to be a 500 litre poly tank with a 2 inch outlet valve which will feed the pump and a return line which will be fed back into the top of the tank so that the system can maintain constant circulation.

2.2 Construction of Test Equipment

The final test product can be seen from figure 2. The figure shows a 500 litre water tank with the outlet of this tank feeding into the pump. On the output of the pump is installed a pressure transmitter. This pressure transmitter has then been connected back into the VSD and the VSD is then monitored for pressure, speed and motor current readings.

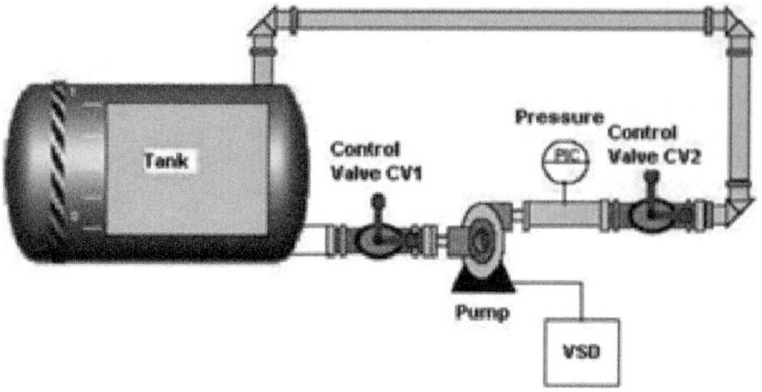

Fig. 2. Closed Loop Testing System PI Diagram

2.3 Programming of the VSD

The VSD has a user programming software which was developed by PDL called Vysta. This software is a graphical icon based connection software. The VSD has 30 registers that can be used for a user application written in Vysta. The PDL Microdrive Elite Series are primarily motor controllers. Therefore any programming that is done in Vysta must not interfere with their ability to control the motor [4]. The inputs and outputs of each of the function blocks are updated each cycle. As a result it is imperative that the input of a function block must not in any way be dependent on its own output [4]. In order to program Vysta, the use of the Schematic Editor is required in addition to the screen lists and it is relevant to the objectives of this project to possess a broad understanding of both.

The Schematic Editor enables a function block based control configuration to be assembled as can be seen in figure 3. Function blocks are selected from Vysta's Menu and interconnected using click and drag. Each function block has its own configuration dialog boxes for the various parameters associated with that function block as can be seen in figure 4.

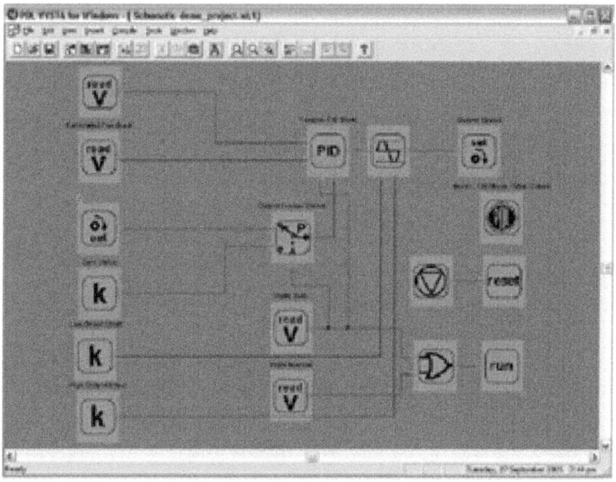

Fig. 3. An example of a VYSTA Schematic Connection Diagram

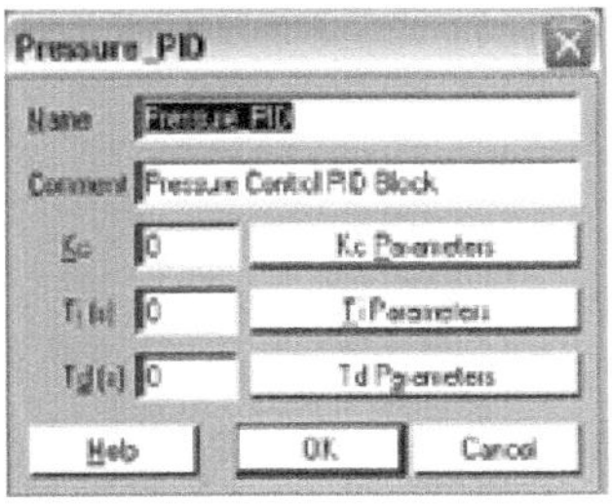

Fig. 4. An example of a Read Variable Dialog Box

It is possible to access Standard Elite Series system variables, however, custom variables can be created for control purposes. Variable values are both displayed and entered by way of the motor controller's Display Unit. When writing a Vysta program that will run on a PDL AC motor controller, the Schematic Program is required to interface with some of the standard motor controller functions [4]. The Standard Program function block is used once in each Vysta program's Schematic to select the control source for the various functions. As mentioned previously the programming of Vysta requires not only the use of a Schematic Editor but the Screen Lists as well.

2.4 Java Applications

In today's world nearly all PCs are connected to the Internet. As part of this transformation a new way to program was developed which is known as JAVA [7]. JAVA is the superior language of the Internet and is a critical tool for programmers worldwide. JAVA has had a profound effect on programming. In a network, there are two categories of objects that can be transmitted between the server and a PC, which is passive information and active programs. JAVA enables both types of objects to be transmitted. To perform the required programming a browser-supported language like JAVA is needed. In order to connect to the Device Server with a JAVA applet communicating with a serial device attached to the Device Server requires familiarity with JAVA programming as well as a JAVA compiler [5].

As a result of JAVA's programming abilities as well as its compatibility with the Lantronix embedded Web Server it is the programming tool chosen for this project. In order to compile and run JAVA programs it is essential to acquire a JAVA development system. The one chosen for this project is a JAVA Development Kit available from Sun Microsystems.

3 Results and Discussion

3.1 Open Head and Closed Head System Tests

The first test that was conducted was the Open Head System Test. It shows that the motor current at 3.75 amps at full speed results in a pressure of approximately 155 Kpa. This indicates that the motor current is linearly proportional to both speed and pressure in the Open Head System.

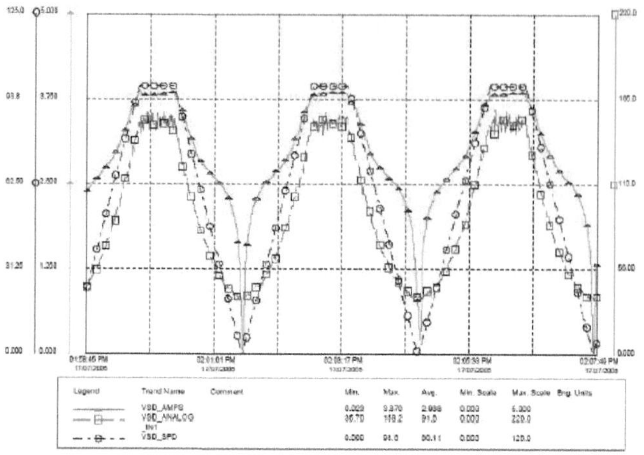

Fig. 5. Plot of Pressure and Motor Current with speed being varied for 0% -100% with the pump outlet valve fully open

Figure 6 illustrates the closed head pressure. This graph shows that the motor current drops down to approximately 3 amps whilst the pressure increases up to 200 Kpa or thereabouts. This data indicates that there are pressure limitations within the system. These limitations are that once the pressure of approximately 165 Kpa is reached, then the motor current begins to decrease until such time that a minimum current of approximately 3 Amps is reached and maintained. This current of 3 Amps will be held at this level as long as the motor pressure is greater than the 165 Kpa.

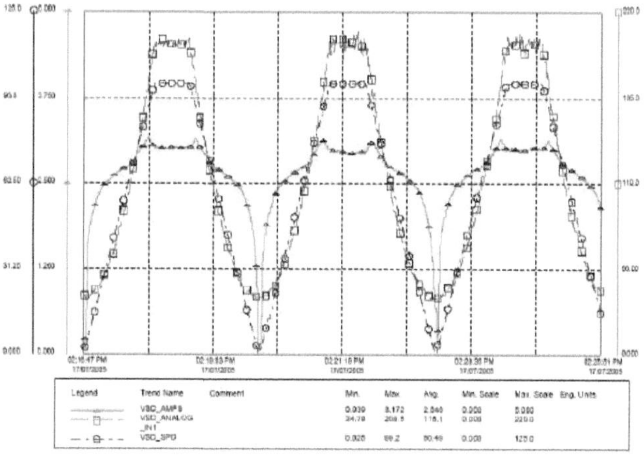

Fig. 6. Plot of Pressure and Motor Current with speed being varied for 0% -100% with the pump outlet valve fully closed

3.2 Closed Loop System Tests with Pressure Transducer

It can been seen that the pressure is controlled constantly about the required setpoint of 65Kpa in figure 7.

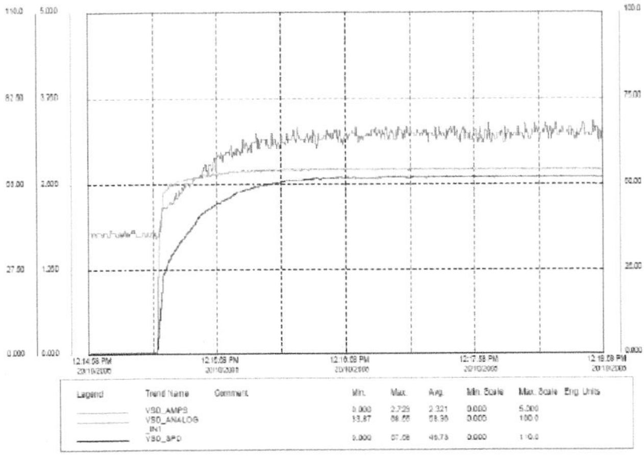

Fig. 7. Plot showing closed loop system Pressure - with pressure transmitter feedback

Utilising this data a program was created to test the conventional system. This program can be seen in figure 8. The schematic shown in figure 9 provides the foundationfor the final control algorithm VSD program.

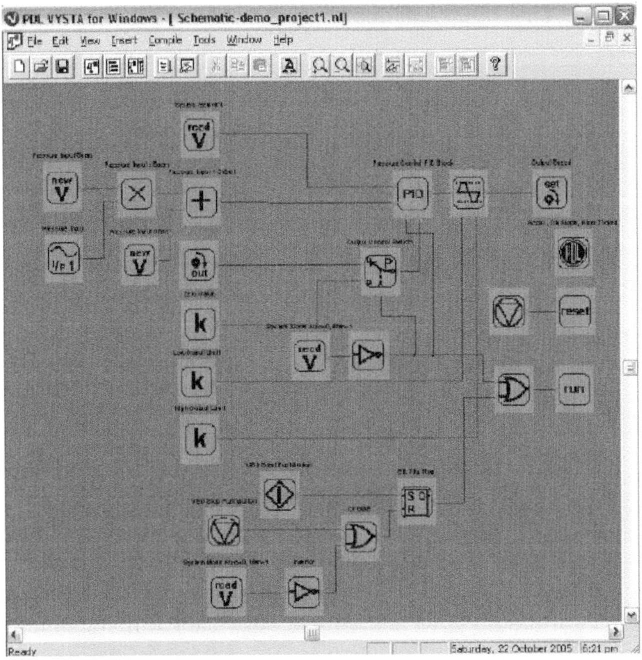

Fig. 8. Closed loop system Vysta Program - with pressure transmitter feedback

This program makes use of the built in PID Function Block within Vysta. The PID function block has two input variables which are connected to the left of the function block. The top variable is the setpoint that is being controlled to and the bottom variable is the feedback reference. The feedback reference is from the pressure transmitter and is connected into the analog input of the VSD. During this test the setting of the setpoint and controlling of the drive was preformed by entering and controlling from the VSD local display screen.

3.3 Closed Loop System Tests Using Algorithm

From the graph obtained in figure 9, it can be shown that control around the setpoint was achieved.

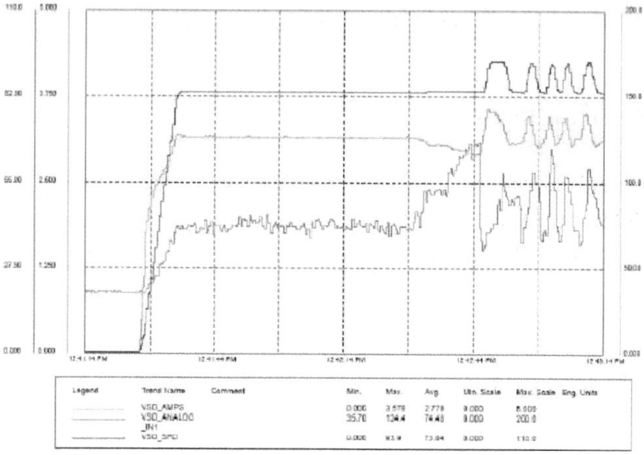

Fig. 9. Plot showing closed loop system Pressure - with pressure transmitter feedback

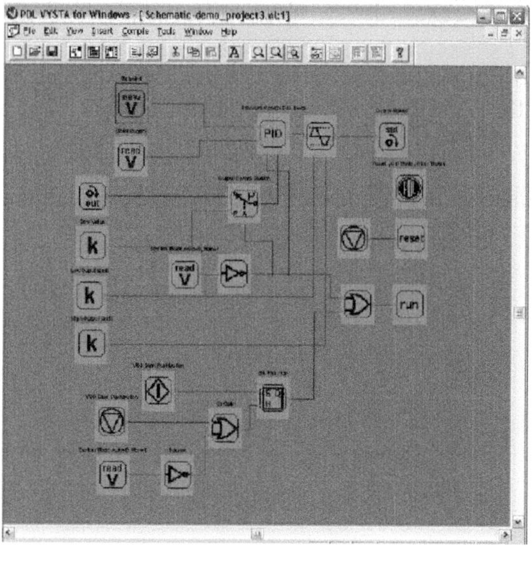

Fig. 10. Closed loop system Vysta Program - with pressure transmitter feedback

The program used to achieve the control with the control Algorithm is shown in figure 10. This has been programmed in Vysta, with ppendix F providing the screen listof the user control screen. The VSD is then placed into Automatic mode via the user control screen and the previous data was obtained.

4 Conclusion

There are a number of instances where applications have required a pressure feedback but because the medium is so corrosive and/or dangerous, the pressure sensor becomes prohibitively expensive. This research proposes to establish the control of that pumping system without the use of a pressure transmitter/sensor. The results obtained from our simulation and experiments confirm the control of that pumping system can be achieved. However, the performance is not yet as good as that of a pressure control system with sensor. Therefore, further refinement is necessary to obtain a more accurate and smoother control response. As most of these applications are implemented through SCADA (supervision control and data acquisition) system, we also developed a real-time monitoring, configuration and control system software package to interface with SCADA systems.

It is ultimately envisaged that the pressure control system identified in this research will, after further development, refinement and thorough testing, provide immense benefits for the engineering and industry sectors.

References

1. Hydraulic Institute, Europump & the U.S. Department of Energys (DOE) Industrial Technologies Program: Variable Speed Pumping – A Guide to Successful Applications, `http://www.bpma.org.uk/Executive/Summary/-vsp.pdf`
2. Minghua, F., Longya, X.: Computer Modeling of Interactions of an Electric Motor, Circulatory System, and Rotary Blood Pump, vol. 46. Lippincott Williams and Wilkins, Inc. (2000)
3. Nelik, L.: Centrifugal and Rotary Pumps – Fundamentals with Applications. CRC Press, Boca Raton (1999)
4. Vysta Virtual Automation Programming Platform Version 2.0 – Help File: PDL Electronics. Napier, New Zealand (2002)
5. Web Enabling Your Serial Device: Lantronix, Irvine, California (2002)
6. Kitamura, T., Matsushima, Y., Tokuyama, T., Kono, S., Nishimura, K., Komeda, M., Yanai, M., Kijma, T., Nojin, C.: Physical Model-Based Indirect Measurements of Blood Pressure and Flow Using a Centrifugal Pump, vol. 24 (2000)
7. Schildt, H.: Java - A Beginner's Guide, 3rd edn. McGraw Hill Osborne, Emeryville (2005)
8. Darby, R.: Chemical Engineering Fluid Mechanics, 2nd edn. Marcel Dekker Incorporated, New York (2001)
9. Davidson, G.: Centrifugal Pump: Parallel & Series Operation. University of Pittsburgh School of Engineering (2002)
10. Trinkl, J., Mesana, T., Havlik, P., Mitsui, N., Demunck, J., Dion, I., Candelon, B., Monties, J.: Control of Pulsatile Rotary Pumps Without Pressure Sensors. Compendex Database 37 (1991)

Balanced Orthogonal Multi-Wavelet Blind Equalization Algorithm Based on Coordinate Transformation

Yecai Guo[1,2] and Xueqing Zhao[1]

[1] School of Electrical and Information Engineering,
Anhui University of Science and Technology,
Huainan 232001, Chnia
[2] College of Electronic & Information Engineering,
Nanjing University of Information Science & Technology,
Nanjing 21004, Chnia
{guo-yecai,1212_zxq}@163.com

Abstract. Aiming at the defect of not suitable for using Constant Modulus Algorithm(CMA) to equalize nonconstant modulus higher-order QAM signals, on the basis of analyzing coordinate transformation(CT) and multi-wavelet transformation, a balanced orthogonal Multi-Wavelet Transformation blind equalization algorithm based on Coordinate Transformation and CMA(CT-MWTCMA) is proposed. In this proposed algorithm, first, nonconstant modulus 16QAM signals are transformed into constant modulus 4QAM via coordinate transformation method to be equalized via using blind equalization algorithm based on CMA, second, input signals of equalizer is transformed via multi-wavelet function to reduce its autocorrelation. Accordingly, the proposed algorithm has fast convergence rate and small mean square error comparison with orthogonal wavelet transformation blind equalization algorithm based on Coordinate Transformation and CMA(CT-WTCMA), and blind equalization algorithm based on Coordinate Transformation and CMA(CT-CMA).The advantages of CT-MWTCMA were presented by underwater acoustic channel simulation.

Keywords: Coordinate transformation; balanced orthogonal multi-wavelet; 16QAM signals; underwater acoustic channel.

1 Introduction

Constant modulus algorithm (CMA) with the simple structure, conservatism property and the remarkable ability of opening eye diagram only is suitable for equalizing constant modulus signals[1], whereas its ability to equalize nonconstant modulus higher order QAM signals is relatively poor. Presently, the constant modulus blind equalization algorithms equalizing nonconstant modulus higher order QAM signals, such as mode adaptive blind equalization algorithm, weighted multi-mode blind equalization algorithm, and switch blind equalization algorithm based on the decision domain method,etc., can improve the convergence rate and reduce mean square error(MSE) to a degree, however, the nonconstant modulus property of higher order

L. Qi (Ed.): ISIA 2010, CCIS 86, pp. 268–274, 2011.

QAM signals and the reduction of the autocorrelation of input signals to equalizer doesn't considered[2,3,4], so their convergence rates still are low and their mean square errors big. The researches show that the ratio of maximum to minimum eigenvalue of the signal's auto-correlation sparse matrix obtained by multi-wavelet or wavelet packets transformation with good decorrelation property can be reduced, so it can improve the convergence rate and reduce mean ssquare error[5,6,7,8,9], and that when the nonconstant modulus higher-order QAM signals are turned into constant modulus 4QAM signals via coordinate transformation, the constant modulus bind equalization algorithm(CMA) is suitable for equalizing higher-order QAM signals[10].

In this paper, according to the advantages of multi-wavelet transformation and coordinate transformation, a balanced orthogonal Multi-Wavelet Transformation blind equalization algorithm based on Coordinate Transformation and CMA(CT-MWTCMA) is proposed, and its effectiveness is verified via underwater acoustic channel simulation.

2 Coordinate Transformation Principle

When the traditional CMA is used to equalize the nonconstant modulus higher-order QAM signals, the output signals of equalizer is equalized to a circle with radius R and $R = \mathrm{E}\{|a(k)|^2\}/\mathrm{E}\{|a(k)|\}$. In fact, the blind equalization property is influenced since nonconstant modulus signals are regarded as constant modulus signals in CMA algorithm, whereas the principle of coordinate transformation is to turn the nonconstant modulus higher order QAM signals into constant modulus 4QAM signals. For example, the process that the 16QAM signals are turned into 4QAM signals is shown in Fig.1.

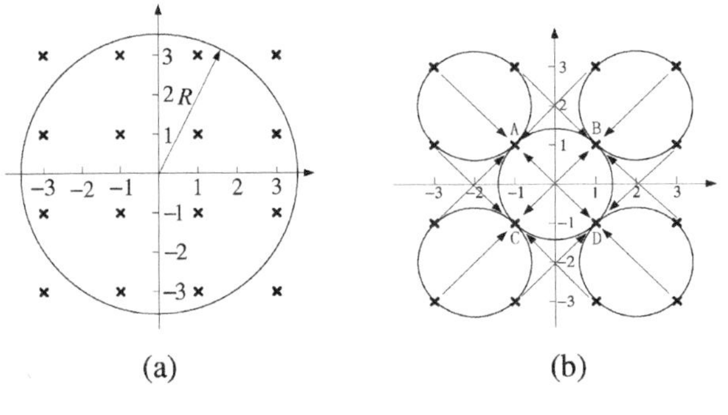

Fig. 1. Constellation graph of 16QAM signals. (a) Constellations of 16QAM signals, (b) Process of constellation transformation.

In CMA, the error function can't become zero after equalization completely, so we introduce the theory of coordinate transformation. Constellation points of 16QAM signals are $\{\pm1\pm i, \pm1\pm3i, \pm3\pm i, \pm3\pm3i\}$, and distribute in the three circles with the different radius, whose center coordinates are in $\{2+2i, -2+2i, -2-2i, 2-2i\}$. In Fig.1, the input signals are adjusted along the arrow, then the coordinates are turned into $\{1+i, -1+i, -1-i, 1-i\}$ .The coordinate transformation of 16QAM signals is shown in Table 1.

Table 1. Coordinate transformation of 16QAM signals

Original coordinates	New coordinates	Original coordinates	New coordinates
1+i	-1-i	-1-i	1+i
1+3i	-1+i	-1-3i	1-i
3+i	1-i	-3-i	-1+i
3+3i	1+i	-3-3i	-1-i
-1+i	1-i	1-i	-1+i
-1+3i	1+i	1-3i	-1-i
-3+i	-1-i	3-i	1+i
-3+3i	-1+i	3-3i	1-i

The relation between the original coordinate X of the signals and the new coordinate Y is given by

$$Y = [X_r - 2\text{sign}(X_r)] + i[X_i - 2\text{sign}(X_i)]. \tag{1}$$

where X_r is the real part of X and the imaginary part of X is X_i. $\text{sign}(\cdot)$ is signum function.

The signals distribute in the same circle after turning 16QAM signals into 4QAM signals and their modulus is turned into $\sqrt{2}$. In case, the error function reduces to zero.

3 Balanced MWT Blind Equalization Algorithm Based on CT

3.1 Multi-Wavelet Equalizer

Assume that $\boldsymbol{w}(k)$, $\boldsymbol{y}(k)$,and $z(k)$ are the weight vector of equalizer, the input signal and the output, respectively. $w(k)$ is expressed by wavelet function and scale function of multi-wavelet. According to the multi-resolution analysis of multi-wavelet transformation, $z(k)$ is written as

$$z(k) = \sum_{i=0}^{N_f-1} w_i(k) y(k-i) = \sum_{l=1}^{r} \left(\sum_{j=1}^{J} \sum_{n} u_{j,n}^{l}(k) r_{j,n}^{u,l}(k) + \sum_{n} v_{J,n}^{l}(k) r_{J,n}^{v,l}(k) \right). \tag{2}$$

where

$$r_{j,n}^{u,l}(k) = \sum_{i=0}^{N_f-1} \psi_{j,n}^{l}(i) y(k-i), \quad r_{J,n}^{v,l}(k) = \sum_{i=0}^{N_f-1} \varphi_{J,n}^{l}(i) y(k-i). \tag{3}$$

In order to implement and calculate easily, the binary translation relation between wavelet function and scale function is adopted and given by

$$r_{j,n}^{u,l}(k) = r_{j,0}^{u,l}(k - 2^j n), \quad r_{J,n}^{v,l}(k) = r_{J,0}^{v,l}(k - 2^J n). \tag{4}$$

Based on the above analyses, we can obtain balanced orthogonal multi-wavelet transformation equalizer. In fact, the equalizer based on multi-wavelet transformation is a kind of transformation domain equalizer[6,7] and the multi-wavelet transformation matrix corresponding to this transformation is written as

$$\boldsymbol{T}=[\boldsymbol{Q}_1;\boldsymbol{Q}_2\boldsymbol{P}_1;\cdots;\boldsymbol{Q}_J\boldsymbol{P}_{J-1}\cdots\boldsymbol{P}_2\boldsymbol{P}_1;\boldsymbol{P}_J\boldsymbol{P}_{J-1}\cdots\boldsymbol{P}_2\boldsymbol{P}_1]. \tag{5}$$

In blind equalization algorithm, after employing multi-wavelet transformation for the input signals of equalizer, the convergence rate of the blind equalization algorithm has a great improvement duo to decorrelation performance of multi-wavelet transformation.

3.2 Description of Algorithm

The diagram of blind equalization algorithm based on coordinate transformation (CT) and MWTCMA is shown in Fig.2.

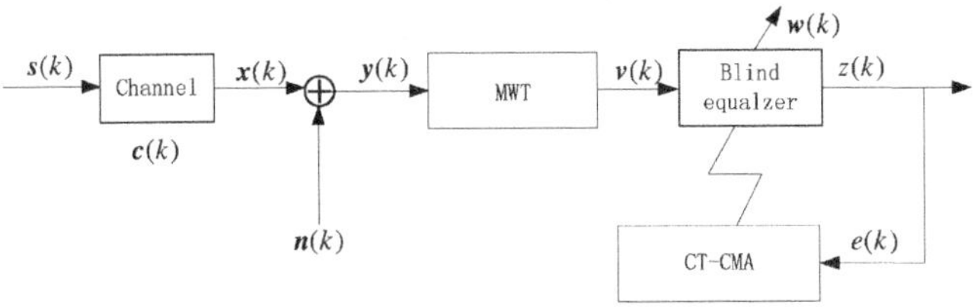

Fig. 2. The diagram of CT-MWTCMA

The input signal $\boldsymbol{y}(k)$ of equalizer is turned into $\boldsymbol{v}(k)$ via the orthogonal multi-wavelet transformation matrix $\boldsymbol{T}$, $\boldsymbol{y}(k)=[y(k),y(k-1),\cdots,y(k-N+1)]^T$, $\boldsymbol{v}(k)=[v(k),v(k-1),v(k-1),\cdots,v(k-N+1)]^T$, and

$$\boldsymbol{v}(k)=\boldsymbol{T}\boldsymbol{y}(k). \tag{6}$$

The output of equalizer is written as

$$z(k)=\boldsymbol{w}^T(k)\boldsymbol{v}(k). \tag{7}$$

The weight vector of MWTCMA is updated by

$$\boldsymbol{w}_{\text{MWTCMA}}(k+1)=\boldsymbol{w}_{\text{MWTCMA}}(k)+\mu\hat{\boldsymbol{v}}^{-1}(k)e_{\text{MWTCMA}}(k)\boldsymbol{v}^*(k). \tag{8}$$

where μ is iteration step-size and a constant, $e_{\text{MWTCMA}}(k)$ is constant modulus error.

$$z(k)=\boldsymbol{w}^T(k)\boldsymbol{v}(k). \tag{9}$$

$$\hat{\boldsymbol{v}}^{-1}(k)=\text{diag}[\sigma^2_{J,n,0}(k),\sigma^2_{J,n,1}(k),\cdots,\sigma^2_{J,n,m}(k),\sigma^2_{J+1,n,0}(k),\cdots,\sigma^2_{J,n,m}(k)]. \tag{10}$$

where $\sigma^2_{J,n,m}(k)$ is average power estimation and a constant. Its iteration formula is given by

$$\sigma_{J,n,m}^2(k+1) = \beta\sigma_{J,n,m}^2(k) + (1-\beta)|z_{j,n,m}(k)|^2 . \tag{11}$$

Accordingly, blind equalization algorithm based on MWTCMA and coordinate transformation is obtained. Its weight vector $w_{\text{CT-MWTCMA}}(k)$ is updated as follows

$$w_{\text{CT-MWTCMA}}(k+1) = w_{\text{CT-MWTCMA}}(k) + \mu\hat{v}^{-1}(k)e_{\text{CT-MWTCMA}}(k)v^*(k) . \tag{12}$$

where

$$e_{\text{CT-MWTCMA}}(k) = R_{new}^2 - |z_{new}(k)|^2 . \tag{13}$$

$$z_{new}(k) = \{z_R(k) - 2\text{sign}[z_R(k)]\} + i\{z_I(k) - 2\text{sign}[z_I(k)]\} . \tag{14}$$

$$R_{new}^2 = \frac{E\{|[s_R(k) - 2\text{sign}[s_R(k)]] + i[s_I(k) - 2\text{sign}[s_I(k)]]|^4\}}{E\{|[s_R(k) - 2\text{sign}[s_R(k)]] + i[s_I(k) - 2\text{sign}[s_I(k)]]|^2\}} . \tag{15}$$

4 Simulation Tests

To present the effectiveness of the proposed algorithm,the simulation tests with underwater acoustic channel were carried out and compared with CT-WTCMA and CTCMA. In tests,the transfer function of the mixed-phase underwater channel was given by $c = [0.3132\ -0.1040\ 0.8908\ 0.3134]$, the transmitted sequence was 16QAM signals, SNR was 25dB, and the weight length of equalizer was 16. Simulation parameters were shown in Table 2 and simulation results were shown in Fig. 3.

Table 2. Simulation parameters

Algorithm	Step-size	Wavelet	Decomposition layer	Original power	β value	Original weight
CTCMA	0.0004					
CT-WTCMA	0.003	DB2	2	10	0.99	The fifth tap is 1
CT-MWTCMA	0.008	GHM	2			

Fig. 3(a) shows that the convergence rate of CT-MWTCMA has an improvement of about 5500 steps comparison with CT-WTCMA, and 10000 steps comparison with CT-WTCMA, and that its steady-state error has a drop of about 4dB comparison with that of CT-WTCMA, and about 7dB comparison with that of CTCMA. Fig.3(b) shows that the steady-state error of CT-MWTCMA was smaller than that of CT-WTCMA or CTCMA at the same SNR. From Fig.3(c) to Fig.3(f), we can know that the output constellations of CT-MWTCMA are closest and clearest in all algorithms. From

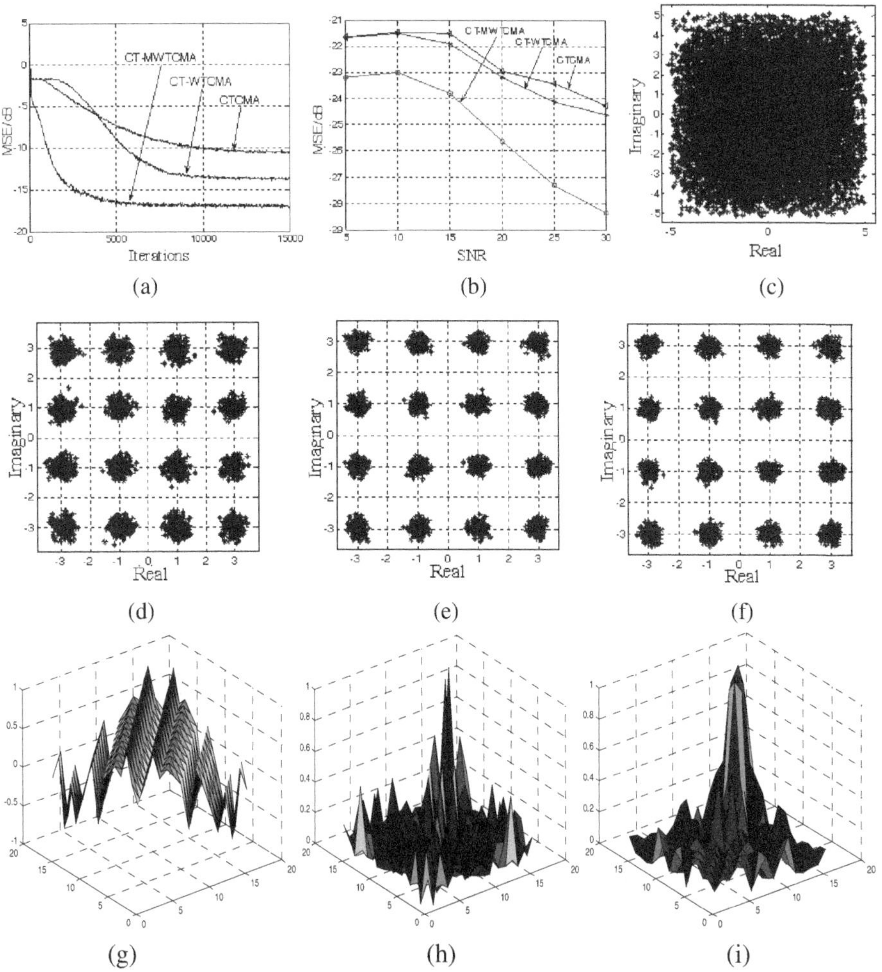

Fig. 3. Simulation results. (b) Curves of MSE, (c) Relation between MSE and SNR, (d) Output of CTCMA,(e) Output of CT-WTCMA ,(f) Output of CT-MWT CMA, (g) Autocorrelation matrix $\boldsymbol{R}_1$ of $y(k)$, (h) Auto-correlation matrix $\boldsymbol{R}_2$ of $y(k)$ after orthogonal wavelet transformation,(i) Auto-correlation matrix $\boldsymbol{R}_3$ of $y(k)$ after orthogonal multi-wavelet transformation

Fig.3(g) to Fig.3(i), we know that the energy of $\boldsymbol{R}_3$ is most focused on the main diagonal line. Accordingly, balanced orthogonal multi-wavelet transformation has the best ability to reduce the autocorrelation of input signals of equalizer.

5 Conclusions

CT-MWTCMA is proposed based on the analysis of 16QAM signals characteristic, coordinate transformation, and multi-wavelet property. The convergence rate of the

CT-MWTCMA is improved via the multi-wavelet decomposition of input signals to equalizer, its mean square error is reduced via turning higher-order QAM signals into constant modulus 4QAM signals. The advantages of CT-MWTCMA were presented by underwater acoustic channel simulation.

Acknowledgment

This paper is supported by Specialized Fund for the Author of National Excellent Doctoral Dissertation of China (200753), Natural Science Foundation of Higher Education Institution of Jiangsu Province (08KJB510010) and "the peak of six major talent" cultivate projects of Jiangsu Province(2008026), Natural Science Foundation of Higher Education Institution of Anhui Province (KJ2010A096), Natural Science Foundation of Jiangsu Province(BK2009410).

References

1. Xu, H., Zheng, H.: A Simple Initialization Method for Bussgang Class Blind Equalization. Acta Simulata Systematica Sinica 17(1), 217–219 (2005)
2. Yan, X., Wang, Q., Li, G., et al.: Adaptive Blind Equalization Algorithm with Multi-Module For High-Order QAM Real-Time Multi-Domain Measurement. Journal of Electronic Measurement and Instrument 23(5), 22–28 (2009)
3. Xu, X.-d., Dai, X.-c., Xu, P.-x.: Weighted Multimodulus Blind Equalization Algorithm for High-Order QAM Signals. Journal of Electronics & Information Technology 29(6), 1852–1855 (2007)
4. Guo, Y.-c., Zhang, Y.-p.: Dual-mode Multi-modulus Blind Equalization Algorithm for High-order QAM Signals. Journal of System Simulation 20(6), 1423–1426 (2008)
5. Wang, J.-f., Song, G.-x.: Adaptive Linear Equalization Algorithm Based On Wavelet Packets Transform. Jounal of Xidian University (School of Science) 28(4), 516–519 (2001)
6. Lebrun, J., Vetterli, M.: High-Order Balanced Multiwavelets: Theory, Factorization, and Design. IEEE Trans. Signal Process. 49(9), 1918–1930 (2001)
7. Lian, J.A., Chui, C.K.: Balanced Multiwavelets With Short Filters. IEEE Trans. Signal Process. 11(2), 75–78 (2004)
8. Han, Y., Guo, Y., Wu, Z., et al.: Design and Algorithm Simulation of Orthogonal Wavelet Transform Based Multi Modulus Blind Equalizer. Chinese Journal of Scientific Instrument. 29(7), 1441–1445 (2008)
9. Han, Y.-g., Guo, Y.-c., Li, B.-k., et al.: Momentum Term and Orthogonal Wavelet-based Blind Equalization Algorithm. Journal of System Simulation 20(6), 1559–1562 (2008)
10. Rao, W., Yuan, K.-m., Guo, Y.-c., et al.: A Simple Constant Modulus Algorithm For Blind Equalizer Suitable for 16-QAM Signal. In: International Conference on Signal Processing Proceedings, pp. 1963–1966 (2008)

A Combined Time Diversity Blind Equalization Algorithm Based on Orthogonal Wavelet Transform

Yecai Guo and Xuejie Ding

College of Electronic & Information Engineering,
Nanjing University of Information Science & Technology,
Nanjing 21004, Chnia
guo-yecai@163.com

Abstract. To overcome the defects of multi-path underwater acoustic channel and Constant Modulus Algorithm (CMA), time diversity blind equalization algorithm based on orthogonal wavelet transform (WT-CTDE) was proposed. In this proposed algorithm, time diversity technique is applied to hyperbolic tangent error function blind equalizer based on constant modulus algorithm (HCMA) and combined with Decision Directed (DD) algorithm and Phase Lock Loop(PLL), and orthogonal wavelet transformation is employed for a transformation to input signals. The proposed WT-CTDE algorithm can not only overcome phase rotation and multi-path propagation but also get small Mean Square Error (MSE) and fast convergence rate. Its performance was proved by the simulations in underwater acoustic channels.

Keywords: time diversity technique; orthogonal wavelet transformation; hyperbolic tangent function; decision directed; phase-locked loop.

1 Introduction

Inter-Symbol Interference (ISI) caused by the multi-path propagation and channel distortion decreases transmission speed and reliability of information in underwater acoustic communication system. Blind equalization technique without training sequences is very suitable for the bandwidth limited underwater acoustic channel[1]. The traditional blind equalization algorithm can't reduce the influence of multi-path propagation on communication quality, but the blind equalization algorithms based on diversity techniques such as spaced diversity, time diversity, frequency diversity,etc., has good performance in compensating the influence of multi-path fading channel on communication signals[2][3].

The hyperbolic tangent error function blind equalizer based constant modulus algorithm (HCMA) is of small Mean Square Error(MSE) caused by asymmetry curve of error function, but its convergent rate is still slow[4][5]. Decision Directed (DD) algorithm can speed up the convergent rate and decrease the MSE[6][7], but it can't reduce the auto-correlation of the input signals. However, in blind equalization algorithm, orthogonal wavelet transformation to input signals of equalizer can greatly improve its convergence rate via reducing the auto-correlation of the input signals[6].

L. Qi (Ed.): ISIA 2010, CCIS 86, pp. 275–281, 2011.

But all these blind equalization algorithms can't overcome the phase rotation caused by Doppler frequency shift, the research shows that one-order Phase Lock Loop(PLL) can correct phase rotation and realize the carrier restoration effectively[6].

Therefore, in this paper, we proposed a new combined time diversity blind equalization algorithm based on orthogonal wavelet transformation(WT-CTDE). Firstly, time diversity technique is applied to the HCMA to get a time diversity blind equalization algorithm based on HCMA (TDE-HCMA). Secondly, the TDE-HCMA is combined with DD algorithm and Phase Lock Loop(PLL). Finally, the orthogonal wavelet transformation is used to make a transformation to the input signals of equalizer and the WT-CTDE algorithm is simulated by using underwater acoustic channel.

2 Time Diversity Blind Equalization Algorithm(TDE)

Time diversity technique is to transmit the same signals repeatedly at the interval of a period,which exceeds the coherence time and obtains some signals with independent fading. At the same time, we use an appropriate combing technique to amalgamate these signals to improve output signal-to-noise ratio(SNR) and decrease Bit Error Rate (BER). The structure of *D* braches time diversity blind equalization algorithm is shown in Fig.1(a).

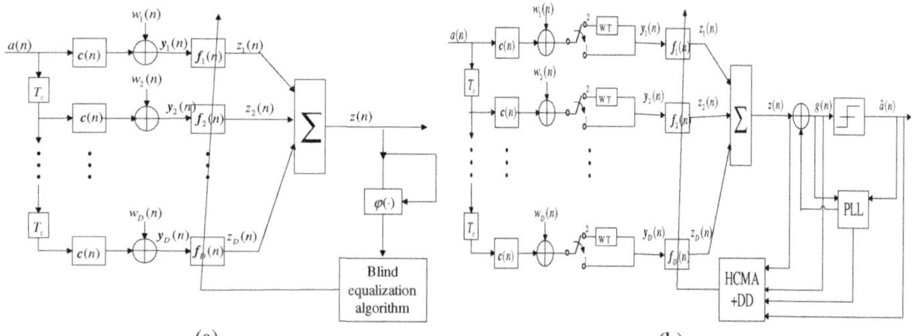

Fig. 1. Blind equalization algorithms. (a) Structure of time diversity blind equalization algorithm, (b) Structure of time diversity combined blind equalization algorithm.

In Fig.1, $\{a(n)\}$ is the channel input sequence at discrete time instants *n*, $c(n)$ presents the impulse response vector of every channel, and T_c is time interval. $\{w_l(n)\}$ denotes an i.i.d. additive white Gaussian noise of the *l*th channel, and $f_l(n)$ denotes the weight vector of the *l*th equalizer and is given by $f_l(n) = [f_l(n), f_l(n+1), \cdots, f_l(n+M_f-1)]$ (M_f is the length of equalizer). $z_l(n)$ is the output sequence of the *l*th equalizer $(l = 1, 2 \cdots D)$ and $z(n)$ is the output signal after combining.

In Fig.1, each branch is made up of a channel and an equalizer, and the channels of all breaches are the same and the equalizers of each breach are different. The outputs of every branch are merged in the combiner. In all combing techniques, maximal ratio combining(MRC) is best when D is larger. The combined gain of equal gain combining (EGC) is almost the same as that of the MRC, but the EGC is relatively easy achieved. So, the EGC method is chosen in this paper.

3 Combined Time Diversity Blind Equalization Algorithm(CTDE)

In this section, time diversity technique is applied to HCMA, whose performance is better than that of traditional CMA , to obtain small MSEs and combined with DD algorithm via soft-decision model to get a fast convergence rate, and PLL is used to correct phase rotation caused by the Doppler effect.

3.1 TDE-HCMA Algorithm

Traditional CMA has large MSEs at end convergence because of asymmetric error function. For improving the performance of traditional CMA, the error function of the lth branch is modified and defined as

$$e_l(n) = \tanh(|z_l(n)| - R). \tag{1}$$

where $R^2 = E\{|a(n)|^4\} / E\{|a(n)|^2\}$, the output of the lth branch equalizer is written as

$$z_l(n) = \sum_{i=0}^{M_f} f_l(i) y_l(n-i). \tag{2}$$

Then the output combined signal is given by

$$z(n) = \sum_{l=0}^{D} g_l z_l(n). \tag{3}$$

In Eq.(3), when $g_l = 1$, equal gain combining technique is adopted, and the weight vector of the lth breach equalizer is updated by the following equation

$$f_l(n+1) = f_l(n) - \mu_l \frac{e_l(n)}{\cosh^2(|z_l(n)| - R)} y_l^*(n) \text{sign}[z_l(n)]. \tag{4}$$

Eq.(1)~Eq.(4) are called as time diversity based on hyperbolic tangent error function constant modulus blind equalization algorithm (TDE-HCMA).

3.2 CDTE Algorithm

When TDE-HCMA is combined with DD algorithm in a soft-decision mode and PLL, a combined time diversity blind equalization algorithm(CTDE) is obtained. Its structure is shown in Fig.1(b)(all the switches turn to 1). In Fig.1(b), $g(n) = z(n)e^{-j\hat{\theta}(n)}$. $\hat{\theta}(n)$ is an estimated value of the constant phase rotation and initialized into zero, and $\hat{a}(n)$ is the decision output of $g(n)$.The weight vector of the lth breach equalizer is given by

$$f_l(n) = f_{l(\mathrm{HCMA})}(n)e^{j\hat{\theta}(n)} + f_{(\mathrm{DD})}(n)e^{j\hat{\theta}(n)}. \tag{5}$$

where $f_{l(\mathrm{HCMA})}(n)$ is the lth branch weight vector of the HCMA algorithm, $f_{l(\mathrm{DD})}(n)$ is the lth branch weight vector of the DD algorithm, and they are updated by following equations

$$f_{l(\mathrm{HCMA})}(n+1) = f_{l(\mathrm{HCMA})}(n) - \mu_{l(\mathrm{HCMA})}\frac{\tanh(|z_l(n)-R|)}{\cosh^2(|z_l(n)|-R)}y_l^*(n)\mathrm{sign}[z_l(n)]. \tag{6}$$

$$f_{l(\mathrm{DD})}(n+1) = f_{l(\mathrm{DD})}(n) + \mu_{l(\mathrm{DD})}\delta[\hat{\hat{a}}(n) - \hat{a}(n)]\cdot[\hat{a}(n) - g(n)]^* y_l(n)e^{j\hat{\theta}(n)}. \tag{7}$$

where $\hat{\hat{a}}(n)$ is the decision output of $\hat{z}(n)$, and $\delta(n)$ is defined as

$$\delta(n) = \begin{cases} 1, & n = 0 + j0 \\ 0, & n \neq 0 + j0 \end{cases}.$$

$$\hat{z}(n) = \sum_{l=1}^{D} y_l^H(n) f_{l(\mathrm{HCMA})}(n+1) + y_l^H(n) f_{l(\mathrm{DD})}(n). \tag{8}$$

When the decision output of $z(n)$ is as the same as that of $\hat{z}(n)$, the weight vector of the DD algorithm is updated. We call Eq.(2),Eq.(3),Eq.(5)~Eq.(7) as combined time diversity blind equalization algorithm(CTDE).

4 CTDE Based on Orthogonal Wavelet Transformation

In Fig.1(b), when all the switches turn to 2, the structure of a new combined time diversity blind equalization algorithm based on orthogonal wavelet transform (WT-CTDE) is obtained.

The research shows that the essence of orthogonal wavelet transformation is to make a transformation to the input signals, i.e.,

$$\begin{cases} r_{jk}(n) = \sum_i y_l(n-i)\varphi_{jk}(i) \\ s_{jk}(n) = \sum_i y_l(n-i)\phi_{jk}(i) \end{cases}. \tag{9}$$

Assume that $R_l(n) = [r_{1,0}(n), r_{1,1}(n), \cdots r_{J,k_J}(n),\ s_{J,0}(n), \cdots s_{J,k_J}(n)]^{\mathrm{T}}$ and the unknown weight vector is $D_l(n) = [d_{1,0}n),\ d_{1,1}(n),\ \cdots,\ d_{J,k_J}(n), v_{J,0}(n)\ ,\cdots,\ v_{J,k_J}(n)]$. Q presents the orthogonal wavelet transform matrix and is given by $Q = [G_0;\ G_1H_0; G_2H_1H_0;\ G_{J-1}G_{J-2}\cdots\ H_1H_0\ ;\ H_{J-1}H_{J-2}\cdots H_1H_0]$, where

$\boldsymbol{H}_j$ and $\boldsymbol{G}_j$ are made of the coefficients of wavelet filter and the coefficients of scale filters, respectively, then

$$\boldsymbol{R}_l(n) = \boldsymbol{Q}\boldsymbol{y}_l(n). \tag{10}$$

The output of the lth breach equalizer is given by

$$z_l(n) = \boldsymbol{D}_l(n)\boldsymbol{R}_l(n). \tag{11}$$

The weight vectors of the lth breach equalizer are updated by the following equations

$$\begin{aligned} \boldsymbol{f}_{l(\mathrm{HCMA})}(n+1) &= \boldsymbol{f}_{l(\mathrm{HCMA})}(n) - \mu_{l(\mathrm{HCMA})}\hat{\boldsymbol{R}}_l^{-1}(n) \\ &\cdot \frac{\tanh(|z_l(n) - R|)}{\cosh^2(|z_l(n)| - R)}\boldsymbol{R}_l^*(n)\mathrm{sign}[z_l(n)]. \end{aligned} \tag{12}$$

$$\begin{aligned} \boldsymbol{f}_{l(\mathrm{DD})}(n+1) &= \boldsymbol{f}_{l(\mathrm{DD})}(n) + \mu_{l(\mathrm{DD})}\hat{\boldsymbol{R}}_l^{-1}(n) \\ &\cdot \delta[\hat{\hat{a}}(n) - \hat{a}(n)][\hat{a}(n) - g(n)]^* \boldsymbol{R}_l^*(n)e^{j\hat{\theta}(n)}. \end{aligned} \tag{13}$$

where $\boldsymbol{R}_l^*(n)$ is the conjugate of $\boldsymbol{R}_l(n)$, $\hat{\boldsymbol{R}}^{-1}(n) = \mathrm{diag}[\sigma_{j,0}^2(n), \sigma_{j,1}^2(n), \cdots, \sigma_{J,k_J}^2(n), \sigma_{J+1,0}^2(n), \cdots \sigma_{J+1,k_J}^2(n)]$, $\sigma_{j,k}^2(n)$ and $\sigma_{j+1,k}^2(n)$ are the average power estimate of $r_{j,k}(n)$ and $s_{j,k}(n)$, respectively, and updated by the following equations

$$\begin{aligned} \hat{\sigma}_{j,k}^2(n+1) &= \beta\hat{\sigma}_{j,k}^2(n) + (1-\beta)|r_{jk}(n)|^2, \\ \hat{\sigma}_{j+1,k}^2(n+1) &= \beta\hat{\sigma}_{j+1,k}^2(n) + (1-\beta)|s_{jk}(n)|^2, \end{aligned} \tag{14}$$

According to Eq.(3), Eq.(9) to Eq.(14), a new combined time diversity blind equalization algorithm using orthogonal wavelet transform (WT-CTDE) has been established.

5 Simulation Tests

For testing the performance of the proposed WT-CTDE algorithm, 16QAM data symbols were transmitted to an underwater acoustics channel with the Doppler phase rotation and its impulse response $c = [e^{-0.7j},\ 0,\ 0,\ 0.3e^{-1.8j}]$.The SNR was set to 20dB. All the equalizers had 16 taps and their center taps were initialized to 1. In Fig.1, we assumed that there were two branches, i.e., D=2 and all branches had the same parameters. Simulations were carried out via employing the Db2 wavelet function, the wavelet layer was 2, the initial power was 4, $\beta = 0.999$,the step-size of TDE-CMA was set to 0.001, the step-size of TDE-HCMA was set to 0.005,the step-size of HCMA algorithm in the CTDE algorithm was set to 0.005, the step-size of DD algorithm was set to 0.0185, the step-size of HCAM algorithm in WT-CTDE algorithm was set to 0.0195, the step-size of DD algorithm was set to 0.0225. The simulation results of Monte Carlo 500 times were shown in Fig.2.

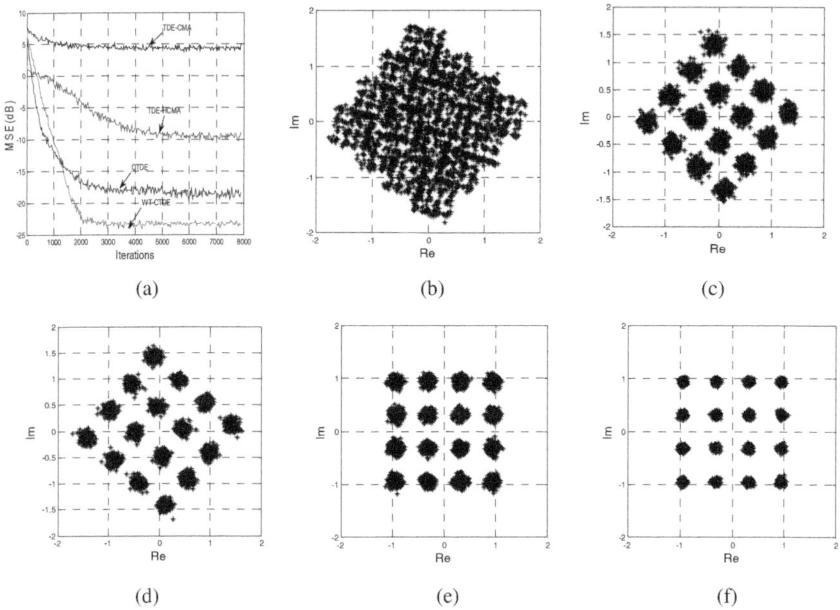

Fig. 2. Simulation results. (a) The MSE curves of all algorithms, (b) Input of equalizer, (c) Output of TDE-CMA, (d) Output of TDE-HCMA, (e) Output of CTDE, (f) Output of WT-CTDE.

Figs.2(a) shows that when the TDE-CMA and TDE-HCMA have the same convergence rates, the MSE of TDE-HCMA has a drop of about 14dB comparison with the CTDE, whereas the convergence rate of WT-CTDE has an improvement of about 4000 steps,1700 steps comparison with the TDE-HCMA and the CTDE, respectively. The MSE of WT-CTDE performs a drop of about 27dB, 13dB and 5dB comparison with the TDE-CMA, the TDE-HCMA and the CTDE, respectively. From Figs.3(c, d), it is seen that the TDE-HCMA's constellations are clearer and more focus than the TDE-CMA's, but they can't modify carrier phase rotation. From Figs.3 (e, f), we know that the CTDE and the WT-CTDE can modify carrier phase rotation, and the WT-CTDE's constellations are clearest and closest.

6 Conclusions

The proposed algorithm, which is called a new combined time diversity blind equalization algorithm based on orthogonal wavelet transform(WT-CTDE), makes full use of the advantages of time diversity, the HCMA, DD algorithm, PLL technique, and orthogonal wavelet transformation. This novel blind equalization algorithm can reduce the multi-path propagation via time diversity, is of the characteristic of small MSE , and has a fast convergence rate and it can modify carrier phase rotation by PLL technique. The simulation results with underwater acoustics channel show that the proposed WT-CTDE algorithm has favorable performance.

Acknowledgment

This paper is supported by Specialized Fund for the Author of National Excellent Doctoral Dissertation of China (200753), Natural Science Foundation of Higher Education Institution of Jiangsu Province (08KJB510010) and "the peak of six major talent" cultivate projects of Jiangsu Province(2008026), Natural Science Foundation of Jiangsu Province (BK2009410), Natural Science Foundation of Higher Education Institution of Anhui Province (2010A096).

References

1. Guo, Y., Han, Y.-g., Rao, W.: Blind Equalization Algorithms Based On Different Error Equations With Exponential Variable Step Size. In: The First International Symposium on Test Automation & Instrumentation (ISTAI), pp. 497–501. World Publishing Corporation, Xi'an (2006)
2. Kibangou, A.Y., Favier, G.: Blind Equalization Of Nonlinear Channels Using A Tensor Decomposition With Code, Space, Time Diversities. Signal Processing 89, 133–143 (2009)
3. Guo, Y., Zhang, Y.: Decision Circle Based Dual-Mode Constant Blind Modulus Equalization Algorithm. Journal of Data Acquisition & Processing 22(3), 278–281 (2007)
4. Hadef, M., Weiss, S.: Concurrent Constant Modulus Algorithm and Decision Directed Scheme for Synchronous DS-CDMA Equalization. IEEE Statistical Signal Processing 17(20), 203–205 (2005)
5. Cooklev, T.: An efficient architecture for orthogonal wavelet transforms. IEEE Signal Processing Letters 13(2), 77–79 (2006)
6. Bae, H.-M., Ashbrook, J.B., et al.: An MLSE Receiver for Electronic Dispersion Compensation of OC-192 Fiber Links. IEEE Journal of Solid-State Circuits 41(11), 2541–2554 (2006)
7. Yuan, J.-T., Tsai, K.-D.: Analysis of the Multi-modulus Blind Equalization Algorithm in QAM Communication Systems. IEEE Transactions on Communcations 53(9), 1427–1431 (2005)
8. Lucky, R.W.: The Adaptive Equalizer. IEEE Signal Processing, Magazine, 104–107 (2006)

Variable Momentum Factor Decision Feedback Blind Equalization Algorithm Based on Constant Parameter Error Function

Yecai Guo[1] and Juanjuan Ji[1,2]

[1] School of Electrical & Information Engineering,
Anhui University of Science and Technology,
Huainan 232001, China
guo-yecai@163.com
[2] Anhui Xinhua College,
Hefei 231000, China

Abstract. According to disadvantages of low convergence rate, big steady-state and poor ability to track time-varying channel of traditional constant modulus blind equalization algorithm(CMA), the variable momentum factor decision feedback blind equalization algorithm based on the constant parameter error function(VMCDFECMA) is proposed by introducing the variable momentum factor into the momentum decision feedback blind equalization algorithm based on the constant parameter error function(MDFECMA). The proposed VMCDFECMA can improve convergence rate and tracking per-formance, and reduce steady-state error via making full use of the features of the decision feedback equalizer, the momentum term,and the error functions. The performance of the proposed VMCDFECMA algorithm is simulated with underwater acoustic channels.

Keywords: variable momentum factor; constant parameter error function; decision feedback equalizer; constant modulus algorithm.

1 Introduction

In underwater acoustic communication systems, ISI(inter-symbol interference) caused by the multi-path effect and the limited bandwidth is an important factor on affecting the communication quality. To overcome ISI, blind equalization algorithm without training sequences is a research hot spot. In the blind equalization algorithm based constant modulus algorithm(CMA), its error function is asymmetrical. The research results show that the error function has very important influence on the performance of blind equalization algorithm[1],[2],[3],[4], decision feedback equalizer(DFE) with nonlinear structure outperforms linear blind equalizer in reducing the length of weight vector[5],[6],[7],[8],and the convergence rate of blind equalization algorithm can be improved by introducing the momentum term into the adaptive algorithm[9] [10] [11], as well as, variable step-size can improve the tracking performance of algorithm.

L. Qi (Ed.): ISIA 2010, CCIS 86, pp. 282–288, 2011.

In this paper, aiming at the advantages of DFE ,momentum term, and the features of error functions, a variable momentum factor constant parameter error function based on decision feedback blind equalization algorithm and CMA(VMCDFECMA) was proposed. The proposed algorithm has good ability to track time-varying channel, fast convergence rate, and low mean square error. Its performance is proved by computer simulation with underwater acoustic channels.

2 Description of VMCDFECMA

The constant parameter error functions are defined as

$$e_1(n) = |z_1(n)|^2 - R^2 . \tag{1}$$

$$e_1(n) = |z_1(n)|^2 - R^2 . \tag{2}$$

$$e_3(n) = \begin{cases} \sqrt{|z_3(n)|^2 - R^2} & |z_3(n)|^2 \geq R^2 \\ -\sqrt{R^2 - |z_3(n)|^2} & |z_3(n)|^2 < R^2 \end{cases} . \tag{3}$$

where $R = E[a^2(n)]/E[a(n)]$ and $R^2 = E[a^4(n)]/E[a^2(n)]$. R and R^2 are called as the modules of the transmitted signal $a(n)$ .The output signal of the equalizer is written as $z(n)$. The error function $e_1(n)$ is a quadratic function, its value is equal to zero when $|z_1(n)| = R$, and it is asymmetrical. Eq.(2) is a linear function and symmetrical near the $|z_2(n)|^2 = R^2$. From Eq.(3), we can know that $e_3(n)$ is piecewise function and asymmetrical near the $|z_3(n)|^2 = R^2$. So, when momentum term, decision feedback structure, and different error functions are introduced into traditional blind equalization algorithm, a variable momentum factor constant parameter error function based on decision feedback blind equalization algorithm and CMA(VMCDFECMA) can be obtained. Its structure is shown in Fig.1.

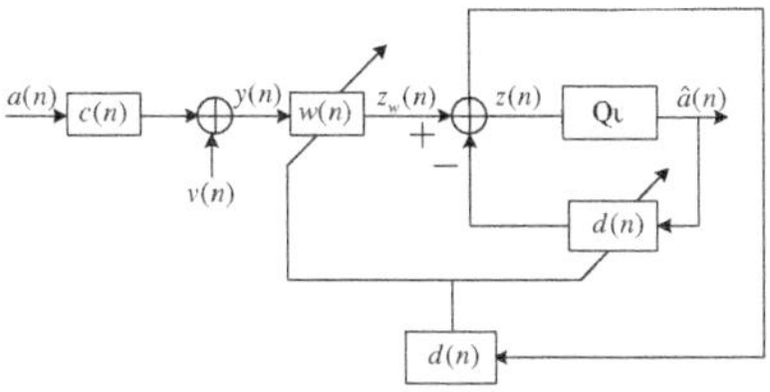

Fig. 1. Basic structure of VMCDFECMA blind equalization algorithm

In Fig.1, $\boldsymbol{a}(n)$ is a transmitted sequence vector. $\boldsymbol{c}(n)$ is an impulse response vector of the channel. $\boldsymbol{v}(n)$ is an additive white Gaussian noise vector. $\boldsymbol{y}(n)$ is the equalizer input sequence. $\boldsymbol{w}(n)$ denotes the weight vector of the feed-forward filter. $\boldsymbol{d}(n)$ is the weight vector of the feed-back filter. $z_w(n)$ is the feed-forward filter output sequence. $z(n)$ is the equalizer output sequence. $\hat{\boldsymbol{a}}(n)$ is the estimation of $\boldsymbol{a}(n)$. The iterative formulas of the weight vector corresponding to different error functions are written as

$$\begin{aligned} \boldsymbol{w}_1(n+1) &= \boldsymbol{w}_1(n) + \mu_{w1} z_1(n) \boldsymbol{y}^*(n) e_1(n) + \alpha_1[\boldsymbol{w}_1(n) - \boldsymbol{w}_1(n-1)], \\ \boldsymbol{d}_1(n+1) &= \boldsymbol{d}_1(n) - \mu_{d_1} z_1(n) \boldsymbol{A}(n) e_1(n). \end{aligned} \tag{4}$$

$$\begin{aligned} \boldsymbol{w}_2(n+1) &= \boldsymbol{w}_2(n) + \mu_{w2} z_2(n) \boldsymbol{y}^*(n) e_2(n) + \alpha_2[\boldsymbol{w}_2(n) - \boldsymbol{w}_2(n-1)], \\ \boldsymbol{d}_2(n+1) &= \boldsymbol{d}_2(n) - \mu_{d_2} z_2(n) \boldsymbol{A}(n) e_2(n). \end{aligned} \tag{5}$$

$$\begin{aligned} \boldsymbol{w}_3(n+1) &= \boldsymbol{w}_3(n) + \mu_{w3} z_3(n) \boldsymbol{y}^*(n) e_3(n) + \alpha_3[\boldsymbol{w}_3(n) - \boldsymbol{w}_3(n-1)], \\ \boldsymbol{d}_3(n+1) &= \boldsymbol{d}_3(n) - \mu_{d_3} z_3(n) \boldsymbol{A}(n) e_3(n). \end{aligned} \tag{6}$$

where μ_{w1}, μ_{w2} and μ_{w3} are the step-sizes of feed-forward filter corresponding to error function $e_1(n)$, $e_2(n)$,and $e_3(n)$. μ_{d_1}, μ_{d_2} ,and μ_{d3} denote the step-sizes of feedback filter. α_1,α_2, and α_3 are the momentum factors of the momentum terms and constants. When $\alpha_1 = 0$ in Eq.(4), $\alpha_2 = 0$ in Eq.(5), and $\alpha_3 = 0$ in Eq.(6), Eq.(4), Eq.(5),and Eq.(6) are called as constant parameter error function decision feedback blind equalization algorithm based on CMA (CDFECMA) . Otherwise, Eq.(4), Eq.(5),and Eq.(6) are called as momentum term based CDFECMA (MCDFECMA). Because α_1 in Eq.4, α_2 in Eq.(5), and α_3 in Eq.(6) are constants, so the MCDFECMA has lower performance in tacking time-varying channel, but outperforms the CDEFCMA and CCMA(constant parameter error function blind equalization algorithm based on CMA) in improving convergence rate and reducing mean square error(MSE). In order to improve the performance of the MCDFECMA better, α_1 in Eq.(4), α_2 in Eq.(5), and α_3 in Eq.(6) may be defined as variables. Accordingly, Eq.(4), Eq.(5), and Eq.(6) are modified into

$$\begin{aligned} \boldsymbol{w}_1(n+1) &= \boldsymbol{w}_1(n) + \mu_{w1} z_1(n) \boldsymbol{y}^*(n) e_1(n) + \alpha_{M1}(n)[\boldsymbol{w}_1(n) \text{-} \boldsymbol{w}_1(n-1)], \\ \alpha_{M1}(n) &= \beta_1[1 - e^{\rho_1 - MSE_1(n)}], \\ \boldsymbol{d}_1(n+1) &= \boldsymbol{d}_1(n) - u_{d_1} z_1(n) \boldsymbol{A}(n) e_1(n). \end{aligned} \tag{7}$$

$$\begin{aligned} \boldsymbol{w}_2(n+1) &= \boldsymbol{w}_2(n) + \mu_{w2} z_2(n) \boldsymbol{y}^*(n) e_2(n) + \alpha_{M2}(n)[\boldsymbol{w}_2(n) \text{-} \boldsymbol{w}_2(n-1)], \\ \alpha_{M2}(n) &= \beta_2[1 - e^{\rho_2 - MSE_2(n)}], \\ \boldsymbol{d}_2(n+1) &= \boldsymbol{d}_2(n) - u_{d_2} z_2(n) \boldsymbol{A}(n) e_2(n). \end{aligned} \tag{8}$$

$$\boldsymbol{w}_3(n+1) = \boldsymbol{w}_3(n) + \mu_{w3} z_3(n) \boldsymbol{y}^*(n) e_3(n) + \alpha_{M3}(n)[\boldsymbol{w}_3(n) - \boldsymbol{w}_3(n-1)],$$
$$\alpha_{M3}(n) = \beta_3[1 - e^{\rho_3 - MSE_3(n)}], \tag{9}$$
$$\boldsymbol{d}_3(n+1) = \boldsymbol{d}_3(n) - u_{d_3} z_3(n) \boldsymbol{A}(n) e_3(n).$$

where $\alpha_{M1}(n)$ in Eq.(7), $\alpha_{M2}(n)$ in Eq.(8), and $\alpha_{M3}(n)$ in Eq.(9) are called as variable momentum factors. Form Eq.(7) to Eq.(9), a variable momentum factor constant parameter error function decision feedback blind equalization algorithm based on CMA(VMCDFECMA) has be established.

3 Simulation Results

To present the effectiveness of the proposed VMCDFECMA algorithm, simulation tests with underwater acoustic channel were carried out and compared with MCDFECMA, CDFECMA and CCMA. In simulation tests, 8PSK signals were transmitted to mixed-phase underwater acoustic channel and its impulse response was given by $c_1 = [0.9656\ \ -0.0906\ \ 0.0578\ \ \ 0.2368]$, SNR was set to 20dB. For CMA, the weight length $L = 16$, and its center tap was initialized to one. For VMCDFECMA, MCDFECMA and CDFECMA, the weight lengths of feed-forward filters and feed-back filters were 16 ,the 8th tap of the weight vector of feed-forward filter was initialized to one ,and all taps of the weight vector of feed-back filter were initialized to zero. Simulation parameters were as follows:

Blind equalization algorithm based on error signal $e_1(n)$. For CCMA, μ_{CCMA} =0.002. For CDEECMA, $\mu_w = 0.001$ and $\mu_d = 0.0002$ .For MCDEECMA, $\mu_w = 0.002$, μ_d = 0.001.For VMCDFECMA, $\mu_w = 0.003$, μ_d=0.006, α_1=0.001, $\beta_1 = 0.2$, and $\rho_1 = 0.04$.

Blind equalization algorithm based on error signal $e_2(n)$. For CCMA, μ_{CCMA} =0.008. For CDEECMA, $\mu_w = 0.001$ and $\mu_d = 0.0006$ .For MCDEECMA, $\mu_w = 0.002$, μ_d = 0.0015. For VMCDFECMA, $\mu_w = 0.003$, μ_d=0.008, α_1=0.001, $\beta_1 = 0.2$,and $\rho_1 = 0.04$.

Blind equalization algorithm based on error signal $e_3(n)$. For CCMA, μ_{CCMA} =0.002. For CDEECMA, $\mu_w = 0.005$ and $\mu_d = 0.00001$ .For MCDEECMA, $\mu_w = 0.004$, μ_d = 0.004. For VMCDFECMA, $\mu_w = 0.001$, μ_d=0.006, α_1=0.001, $\beta_1 = 0.2$, and $\rho_1 = 0.04$.

Simulation results were shown in Fig.2,Fig3, and Fig.4.

From Fig.2,Fig.3, and Fig.4, we can know that the VMCDFECMA has the fastest convergence rate, the smallest mean square error(MSE), and the clearest and closest constellations in all algorithms. In Fig.2, Fig.3,and Fig.4, it has been seen that the performances of blind equalization algorithms based on different constant parameter error functions are different. Accordingly, it is very obvious that the performance of blind equalization algorithm depend on variable step-size, momentum term, and the features of error functions.

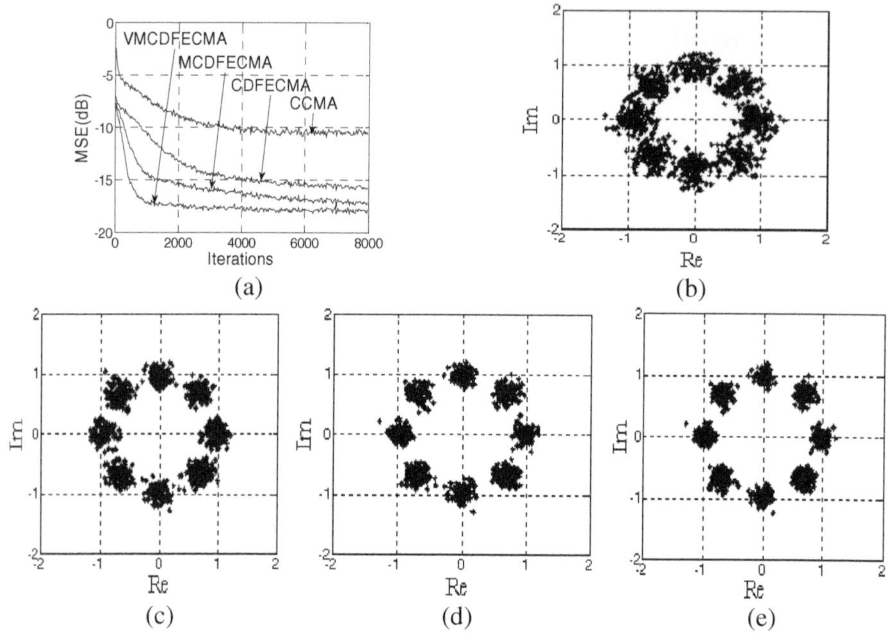

Fig. 2. Simulation results for $e_1(n)$. (a) Curves of MSE, (b) Input of equalizer, (c) Output of CDFECMA, (d) Output of MCDFECMA, (e) Output of VMCDFECMA.

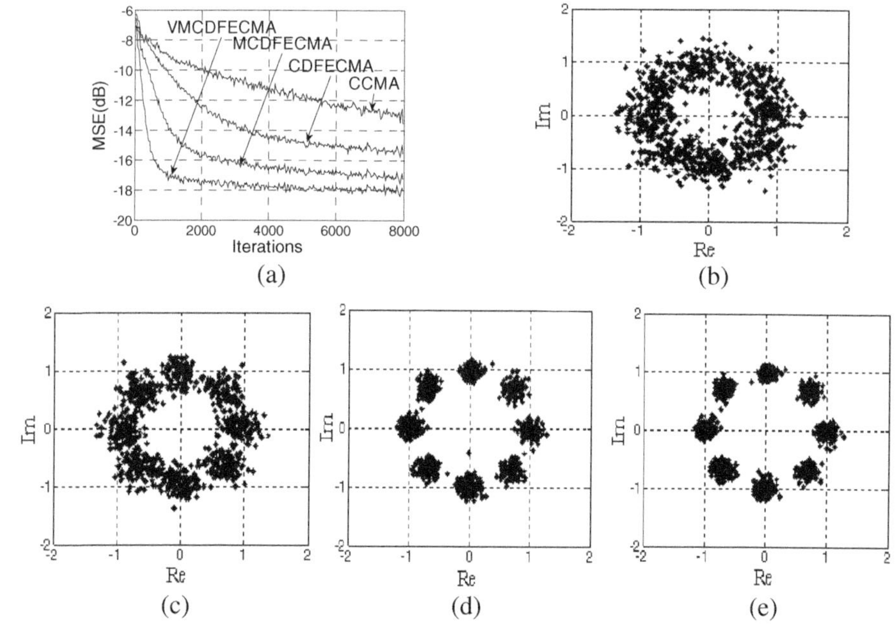

Fig. 3. Simulation results for $e_2(n)$.(a) Curves of MSE, (b) Input of equalizer, (c) Output of CDFECMA, (d) Output of MCDFECMA, (e) Output of VMCDFECMA.

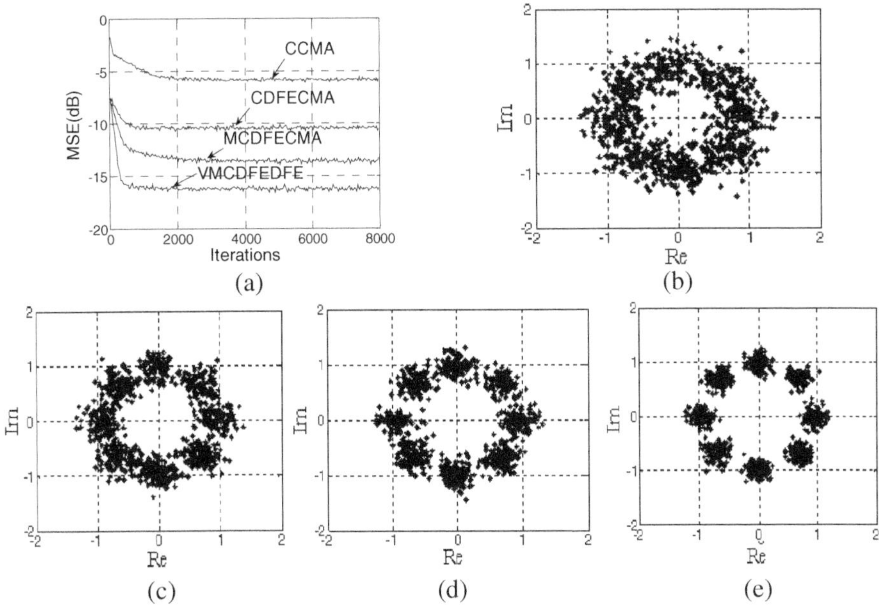

Fig. 4. Simulation results for $e_3(n)$. (a) Curves of MSE, (b) Iutput of equalizer, (c) Output of CDFECMA, (d) output of MCMADFE, (e) output of VMFCMADFE

4 Conclusions

In this paper, a variable momentum factor decision feedback blind equalization algorithm based on constant parameter error function(MCDFECMA) is proposed by introducing the momentum factor into the momentum term decision feedback blind equalization algorithm based on constant parameter error function(MCDFECMA). The proposed VMCDFECMA algorithm outperforms the MCDFECMA, CDFECMA (constant parameter error function decision feedback blind equalization algorithm based on CMA),and CCMA(constant parameter error function blind equalization algorithm based on CMA) in improving convergence rate ,tracking time-varying channel, and reducing mean square error. The performance of the proposed VMC-DFECMA algorithm is proved by simulation tests with underwater acoustic channel.

Acknowledgment

This paper is supported by Specialized Fund for the Author of National Excellent Doctoral Dissertation of China (200753), Natural Science Foundation of Higher Education Institution of Jiangsu Province (08KJB510010) and "the peak of six major talent" cultivate projects of Jiangsu Province(2008026), Natural Science Foundation of Higher Education Institution of Anhui Province (KJ2010A096), Natural Science Foundation of Jiangsu Province (BK2009410).

References

1. Scott, C.D., Meng, T.H.Y.: Stochastic Gradient Adaptation Under General Error Criteria. IEEE Transactions on Signal Processing 42(6), 1335–1351 (1994)
2. Roy, P., Beex, A.A.(Louis): Blind Equalization Schemes With Different Error Equations. In: IEEE International Symposium on Intelligent Signal Processing and Communication Systems, ISPACS 2000, pp. 835–840 (2000)
3. Abrar, S.: A New Cost Function for the Blind Equalization of Cross-QAM Signals. In: The 17th International Conference on Microelectronics, ICM 2005, pp. 290–295 (2005)
4. Guo, Y.-c., Hang, Y.-g.: Novel Variable Step Size Blind Equalization Algorithm Based On Lognormal Error Function. Journal of System Simulation 19(6), 1224–1226 (2007)
5. Wei, R.: The Extended Research On Bussgang Blind Equalization Algorithm. Anhui University Of Science And Technology (2007)
6. Zhang, H.-b., Zhao, J.-w.: Decision Feedback Blind Equalization Algorithm Based On RENYI Entropy For Underwater Acoustic Channels. Journal Of Electronics & Information Technology 31(4), 911–915 (2009)
7. Zhu, J., Guo, Y.-c.: Decision Feedback Blind Equalization Algorithm Based On Maximal Ratio Combining Spatial Diversity. Journal Of System Simulation 20(11), 2843–2845 (2008)
8. Guo, Y.-c., Lin, R.-g.: Blind Equalization Algorithm Based On T/4 Fractionally Spaced Decision Feedback Equalizer. Journal of Date Acquisition & Processing 23(3), 284–287 (2008)
9. Han, Y.-g., Guo, Y.-c.: Momentum Term And Orthogonal Wavelet-Based Blind Equalization Algorithm. Journal of System Simulation 20(6), 1559–1562 (2008)
10. Guo, Y.-C.: The Information Processing Technology Based On Higher Order Statistics. Press of Hefei University of Technology (2005)
11. Guo, Y.-C., Zhao, J.-W.: An Adaptive Filtering Algorithm Of Higher-Order Cumulant-Based Signed Coherent Integration. Journal of System Simulation 10(4), 1280–1283 (2002)

Fuzzy PID Control and Simulation Analysis of Cruise Control System*

Meilan Zhou, Jing Sun, Hanying Gao, and Xudong Wang

College of Electrical & Electronic Engineering , Harbin University of Science and Technology, XueFu Road52, 150080 Harbin, China
zhoumeilan001@163.com

Abstract. Considering the running situation of automobile being complex and variable frequently, and cruise control system(CCS) having high nonlinearity and nondeterminacy, it will not obtain a good effect in all the conditions by using the method of traditional PID control. A new kind of CCS is designed based on fuzzy PID. The simulation model of dynamics system is established. By dealing with the subject functions and adjusting rules of parameters, the table of fuzzy matrix of PID parameters had been got. Adjusted the control parameters, we made a simulation contrast of the system based on fuzzy PID control and traditional PID control. The simulation results indicate the fuzzy PID controller has a better effect in keeping the speed steady.

Keywords: Cruise Control System (CCS); Fuzzy PID; Simulation.

1 Introduction

The automobile cruise control system, which called CCS for short is also called cruise driving equipment, speed control system, auto-drive system and so on. The driver could set a cruise speed by using the cruise control switches, when the speed of the automobile equipped with CCS exceeds a level (commonly 40km/h). In the course of cruising control, the automobile will alter the opening range of the throttle or shift automatically along with the change of the road gradient and the resistances during the automobile runs. And the automobile could run steadily with the optimal fuel economy or power rule in the storage of microcomputer.

Cruise control system could lessen the drivers' oppressiveness, reduce unnecessary change of speed, economize fuel farthest, reduce the pollution of exhaust gas and increase the efficiency of using engine. And CCS could also improve the dynamical performance and driving comfort to some extent.

At present, CCS has been the equipment set or chosen by many vehicles especially on advanced cars. The research on CCS begins late at home, and its technology is backward comparatively. So the research is mainly about keeping the speed changeless.

* Supported by the Key Project of Chinese Ministry of Education (No.: 208037); Scientific Research Fund of Heilongjiang Provincial Education Department (No.:11551072).

L. Qi (Ed.): ISIA 2010, CCIS 86, pp. 289–295, 2011.

Although the research on the electronic cruise control system has already begun at home, it is not mature on the whole. And a suitable control method is very important for the research on CCS [1].

2 Principle of CCS

A cruise control system comprises a controller, an executive unit of throttle, an engine and a gearbox, a speed sensor and so on. The principle structure of cruise control system is shown as Fig.1.

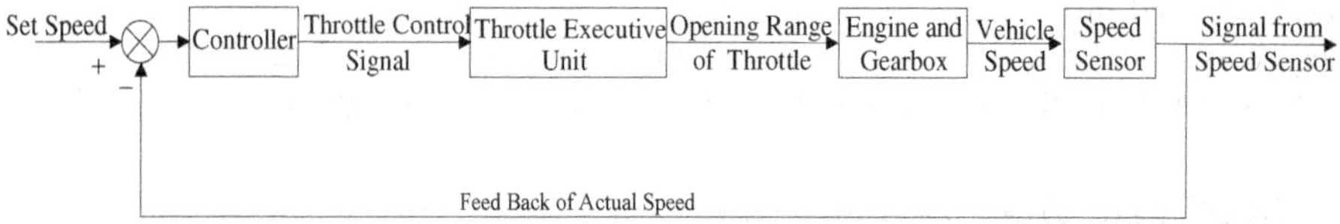

Fig. 1. Principle structure of CCS

The controller has two input signals. One is the set speed signal which is set by the driver, and the other is the feed back signal of actual speed. After the electro-controller detects the error between the two inputs, it will produce a throttle control signal and send the signal to the executive unit of throttle. For correcting the error that the electro-controller detects, the opening range of the throttle will be changed by the throttle executive unit based on the signal it receives. Then the speed will be kept changeless.

3 Modeling and Simulation

3.1 Model of Automobile

While the vehicle is running, it will be affected by driving force F_c, driving resistance F_f, air resistance F_w, gradient resistance F_h and accelerated resistance F_δ. The model of the vehicle is divided into three parts including the driven wheel, the driving wheel and the body [2]. Suppose the vehicle is running on the ramp, the driving equation of the vehicle is shown as the follow:

$$F_c = F_f + F_w + F_h + \delta ma \,. \tag{1}$$

Where δ is the mass conversion coefficient which reckons in the inertial moment of revolving mass, and δ=1.05.

The simulation model of automobile dynamics system is built up in the Simulink of MATLAB (Fig.2). In the figure, max thrust and max brake are the maximal driving force and the maximal braking force respectively. Suppose the simulation vehicle model is a car, and the mass of the vehicle is 1100kg.

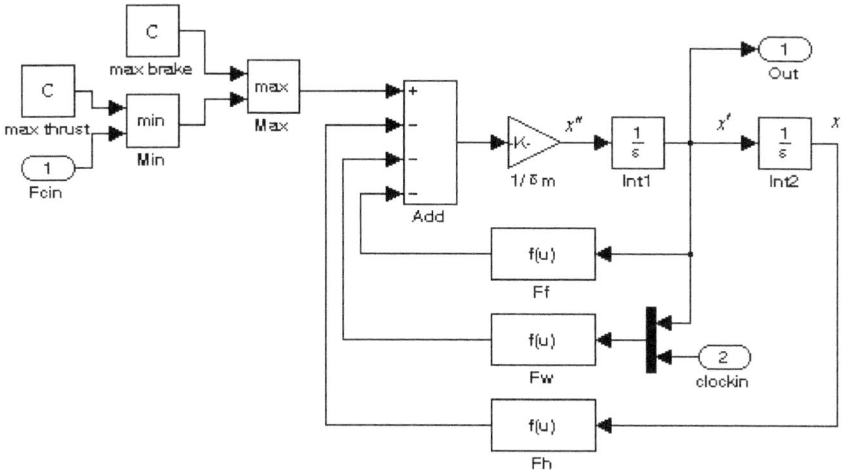

Fig. 2. Simulation model of automobile dynamics system

where

$$F_f = mg \times 0.014 \times (1 + x'^2/19440)\ . \tag{2}$$

$$F_w = 0.01 \times [x' + 20\sin(0.01t)]^2\ . \tag{3}$$

$$F_h = mg \times 0.01 \times \sin(0.0001x)\ . \tag{4}$$

3.2 Design of Fuzzy PID Controller

During the vehicle is running, it could affected by factors such as disturbance from external load, the indeterminacy of the vehicle's mass and transmission unit's efficiency, and high nonlinearity of the object under control. The process parameters will be changed, so it will not promise to have a content effect in all the conditions by using the method of traditional PID control for CCS. To content the request of timely control, the parameters of PID control are needed to adjust on line during the control process.

Based on PID, fuzzy PID checks the fuzzy matrix to adjust the parameters according to the fuzzy consequence result after calculating the error and the error change rate of current system. Fuzzy PID absorbs the advantages of fuzzy control and traditional PID control. Fuzzy PID has adapting ability, and it not only could recognize and adjust the process parameters automatically and be adapted to the change of process parameters, but also has the excellences of traditional PID controller such as simple configuration, high robustness and high reliability [3]. The principle of fuzzy PID control is shown as Fig.3.

The change of the absolute value of error $/E/$ and the absolute value of error change rate $/EC/$ are defined as the discourse universe of fuzzy set $/E/$, $/EC/=\{0,1,2,3,4,5\}$, and the term set of the linguistic values are defined as $/E/$, $/EC/=$ {zero(Z), small(S),

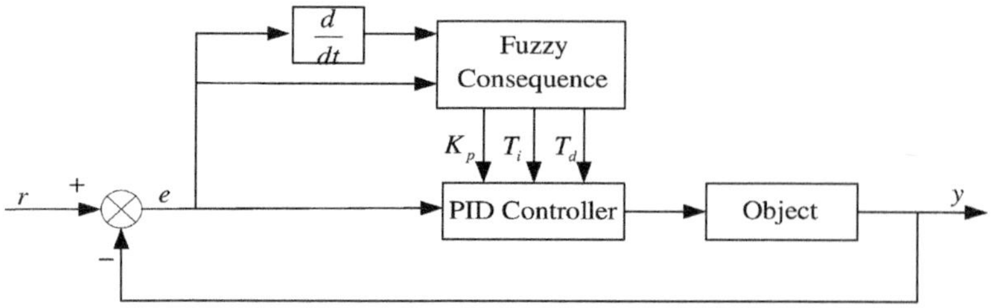

Fig. 3. Structure of fuzzy PID control

medium(M), big(B)}. The subject functions are shown as Fig.4. The linguistic values of proportional compensation factor K_p', integral compensation factor T_i' and differential compensation factor T_d' are defined as zero(Z), small(S), medium(M), big(B) , and the subject functions are shown as Fig.5.

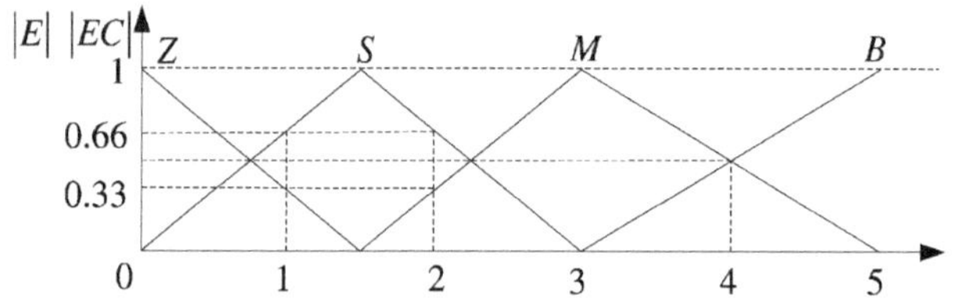

Fig. 4. Subject function of */E/* (*error*) and */EC/* (*error change rate)*

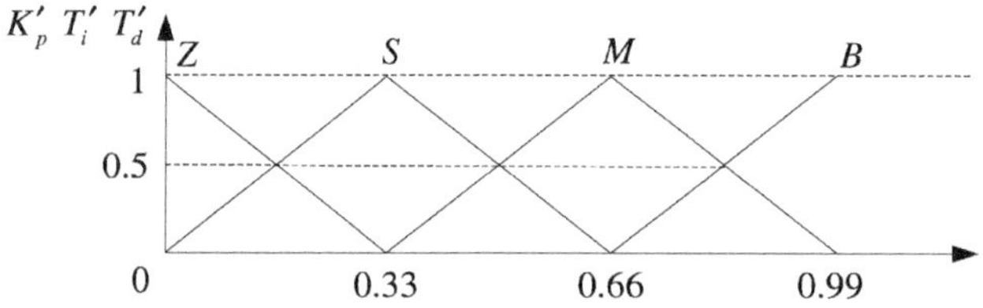

Fig. 5. Subject function of K_p'(*proportional compensation factor*), T_i'(*integral compensation factor*)and T_d'(*differential compensation factor)*

Considering the stability, the response speed, the overshoot and the stable state precision of the system, we may get the following constraints [4]:

1) When */E/* is bigger, the system should have faster response speed and K_p should be bigger; to prevent the bigger overshoot and the differential supersaturation caused by */E/* becoming big suddenly at the beginning, T_i should be bigger and T_d should be smaller.

2) When */E/* and */EC/* are medium, K_p should be smaller for a smaller system overshoot; T_i should be medium to avoid the effect of dynamic stability; and T_d should be bigger because the adjustive character is sensitive to the change of T_d.

3) When /*E*/ is smaller, for having a good stable state stability, reducing the static error and enhancing the restraining ability of disturbance, the system should have a bigger K_p, a smaller T_i and a smaller T_d.

For the actual automobile cruise control system, the driver will feel uncomfortable when the speed error is zero. So the speed error should not be zero, but be kept in a definite error range. So when we design the fuzzy rules of CCS, the followings are also need to be considered: for allowing an error, the effect of integral element should be weakened and the effect of proportional element should be enhanced when the speed error is smaller correspondingly.

The adjusting rules of K_p, T_i and T_d could be educed according to the rules above and the control requests. The subject functions and the adjusting rules of parameters are input into the FIS editor of MATLAB, and then the table of fuzzy matrix will be got. When the system is running on line, the control system will correct the PID parameters through dealing with the result of fuzzy control, looking up the table and operating [5].

3.3 Simulation and Result

The model of CCS will be built up after connecting the automobile dynamics system with the fuzzy PID controller or the PID controller in the Simulink of MATLAB. The initial values of PID control are: K_p =200, T_i =200, T_d =0.15. And the three parameters will be adjusted according to the speed error and the error change rate on line [6].

The following capability of the automobile speed is the ability that the actual speed is changed to the set speed. The simulation is based on the comparison between the fuzzy PID control and the traditional PID control. When the simulation speed is 70km/h and 100km/h with the phase step 30km/h, the results of simulation are shown as Fig.6 and Fig.7. And the y-axis is the ratio of the actual speed to the set speed.

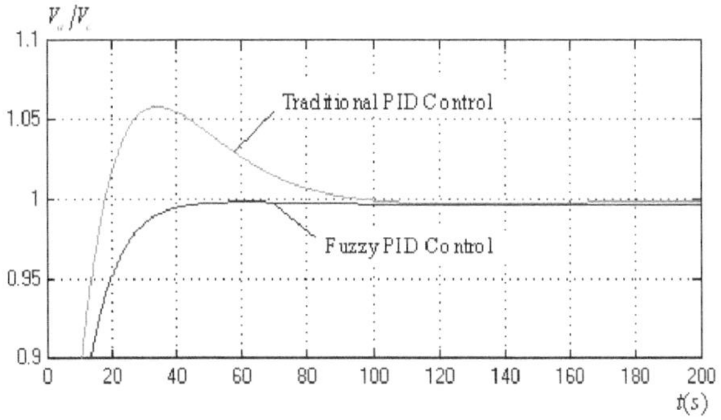

Fig. 6. Control curve of 70km/h cruise speed

The keeping capability of the automobile is the ability of the automobile to keep the cruise speed changeless under the disturbers outside. When the simulation speed is 60km/h and 100km/h with the phase step 20km/h, the results of simulation are shown as Fig.8 and Fig.9.

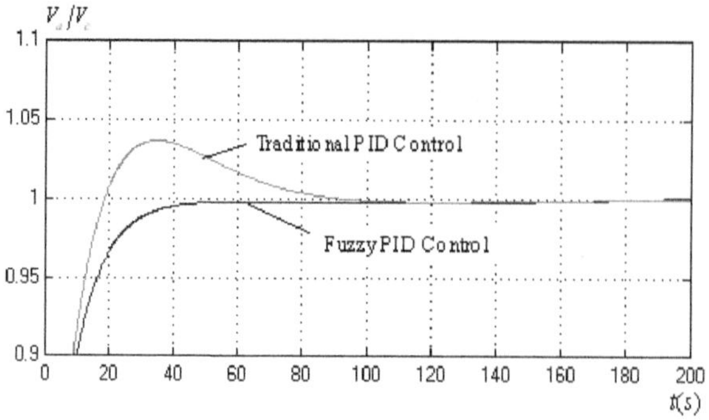

Fig. 7. Control curve of 100km/h cruise speed

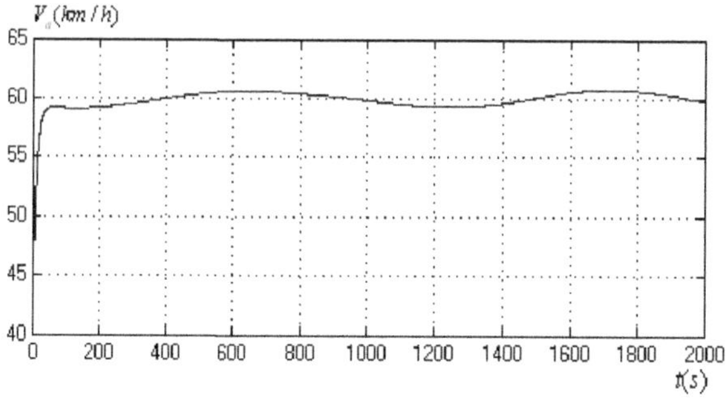

Fig. 8. Fuzzy PID simulation result of 60km/h cruise speed

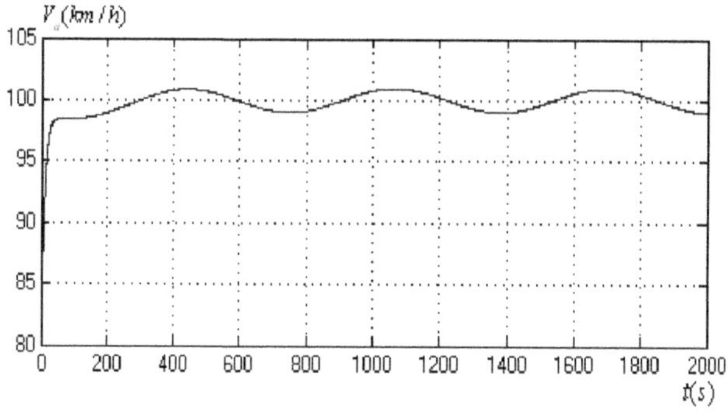

Fig. 9. Fuzzy PID simulation result of100km/h cruise speed

Through the result of the simulation we may know that fuzzy PID controller could make the overshoot smaller and the reaction time shorter comparing with the traditional PID control. Fuzzy PID control could keep the driving speed at the cruise speed well. The fluctuate as sine wave in Fig.8 and Fig.9 is caused by using disturbers with sine wave instead of the actual disturbers when building up the automobile dynamics system. After analysis the results of simulation we may know that fuzzy PID control is better than the traditional PID control with changeless parameters. It could have a good effect at different cruise speeds, and it is a suitable control method for CCS.

After the simulation, the system is debugged on the automobile control system dais based on the accomplishment of its hardware circuit board and software programming. The result of the test shows that the system could execute some simple orders ideally.

4 Conclusions

Through the result of the simulation we may know that the system have the anticipative control effect with small fluctuate of speed and good stability when using fuzzy PID control. And the test shows that there is a good effect in keeping the speed steady.

References

1. Bifulco, G.N.: Experiments toward a Human-like Adaptive Cruise Control. In: 2008 Intelligent Vehicles Symposium, pp. 919–924. IEEE Press, Eindhoven (2008)
2. Ko, S., Lee, J.: Fuzzy Logic Based Adaptive Cruise Control with Guaranteed String Stability. In: 2007 International Conference on Control, Automation and System, pp. 15–20. IEEE Press, Seoul (2007)
3. Khan, A.A., Papal, N.: Fuzzy-PID controller: Design, Tuning and Comparison with Conventional PID Controller (2006)
4. Li, H.-X., Zhang, L., Cai, K.-Y., Chen, G.: An Improved Robust Fuzzy-PID Controller with Optimal Fuzzy Reasoning. IEEE Trans. Syst. 35 (2006)
5. Echegaray, S.: The Modular Design and Implementation of an Intelligent Cruise Control System. In: 19th International Conference on Systems Engineering, pp. 1–6. IEEE Press, Las Vegas (2008)
6. Kljuno, E., Williams II, R.L.: Vehicle Simulation System: Controls and Virtual- Reality-Based Dynamics Simulation, pp. 79–99 (2008)

An Improved FastSLAM Algorithm Based on Genetic Algorithms

Yi-min Xia and Yi-min Yang

College of Automation, Guangdong University of Technology,
Guangzhou 510006, China
echosun80@163.com

Abstract. In order to mend the problem of particle filter's sample depletion, the paper introduces the adaptive algorithms into FastSLAM. Besides using selection, crossover and mutation operation of genetic algorithm to improve the diversity of samples, this algorithm imports adaptive controlling parameters to overcome the premature convergence at the same time protect the excellent individual. Theoretical analysis and simulation experiments show that the algorithm can effectively improve the accuracy of simultaneous localization and localization.

Keywords: Particle Filter, FastSLAM, Genetic Algorithms.

1 Introduction

The Simultaneous Localization and Mapping (SLAM) problem of mobile robot can be described as: robot starts moving from an unknown position in the unknown environment, locating itself according to state estimation and sensor observation, and building environmental map at the same time [1]. Since its important theoretical and application value, many scholars believe that SLAM is the key of realize really autonomous mobile robot, and that's why SLAM becomes the hotspot in mobile robot field[2].

In recent years, because of the ability of trailing multiple maps and estimating robot's pose, Rao-Blackwellized Particle Filter (RBPF) is believed to be an effective mean to solve SLAM problem by a lot of scholars[3-5]. Murphy was the first one to introduce RBPF to SLAM and solve SLAM problem of 10x10 grids environment successfully [4]. Montemerlo put forward FastSLAM algorithm based on it by decomposing SLAM problem to localization problem and mapping problem, which is a mixed algorithm of Particle Filter (PF) algorithm and Extended Kalman Filter (EKF). Localization problem can be realized by PF with Gauss approaching of new pose's posterior probability by EKF [5].

Sample degradation problems will inevitably exist in FastSLAM algorithm for using PF algorithm[6,7]. Since genetic algorithm can improve solution quality by choose solutions with higher fitness than average through iterated calculation, it's a kind of widely used evolutionary algorithm.

L. Qi (Ed.): ISIA 2010, CCIS 86, pp. 296–302, 2011.

This paper import evolution mechanism into FastSLAM algorithm, optimize samples by choose, crossover and mutation operators in genetic algorithm, and adjusting control parameters are set by forcing samples moving to larger posterior probability density area, which can effectively reduce sample numbers, then diversity of samples can be improved and sample degradation can be reduced.

2 FastSLAM Algorithm

FastSLAM algorithm represents motion model $p(s_t \mid u_t, s_{t-1})$ and observation model $p(z_t \mid s_t, \Theta, n_t)$ with probabilistic method, in which s_t means pose of time t, u_t means robot controlled variable, $\Theta = (\theta_1, \theta_2, \cdots, \theta_k)$ means environment landmarks, z_t means observed value of time t, n_t means index number of observed landmark in time t. Then SLAM problem is a process of confirming all landmarks Θ and pose s_t based on observed value $z^t = z_1, \cdots, z_t$ and controlled variable $u^t = u_1, \cdots, u_t$. If robot's moving path s^t and data association variable is known, estimation of all landmarks are relatively independent, then FastSLAM can be represented by product of two independent posterior probability[6] as follows:

$$p(s^t, \Theta \mid z^t, u^t, n^t) = p(s^t \mid z^t, u^t, n^t) \prod_{i=1}^{N} p(\theta_i \mid s^t, z^t, u^t, n^t) \tag{1}$$

In formula (1), first factor is posterior probability of robot's path estimated by PF, and second factor means posterior probability of N landmarks in given path estimated by EKF.

FastSLAM samples path with PF, each particle maintains single map composed of n EKFs. The mth particle $S_t^{[m]}$ means path $s^{t,[m]}$ composed of N Gauss landmark estimation which can be represented by mean $\mu_{n,t}^{[m]}$ and covariance $\Sigma_{n,t}^{[m]}$, that is:

$$S_t^{[m]} = \{s^{t,[m]}, \mu_{1,t}^{[m]}, \Sigma_{1,t}^{[m]}, \cdots, \mu_{N,t}^{[m]}, \Sigma_{N,t}^{[m]}\} \tag{2}$$

FastSLAM algorithm can be described as following steps:

1) Sampling new pose. Landmark estimation depends on robot' motion path, and this path is composed of robot's pose from time 1 to time t, so the sample steps of particle is important. New pose s_t^i will be sampled from posterior probability $p(s_t \mid s^{t-1,i}, z^t, u^t, n^t)$.

2) Updating the Observed Landmark Estimate. When landmark is not been observed, its posterior probability remain unchanged; otherwise, updating the estimate.

3) Calculate importance weight and resample.

3 Improved FastSLAM Algorithm with Genetic Algorithms

In order to solve sample degradation problems in FastSLAM algorithm, this paper import genetic mechanism, diversity of samples can be improved by choose, crossover and mutation operators in genetic algorithm, and adjusting control parameters are set to improve performance of the algorithm.

3.1 Genetic Operator

Evolution operator in genetic algorithms [9] can optimize the system state represented by particles, and Metropolis algorithm is used to make sure those evolved particles will accurately approximate posterior. As the weight of particle reflect the quality of path estimation, this paper define it as fitness function f. To avoid trouble brought by binary coding and decoding, crossover and mutation will operated on floating-point directly.

1 Operator choosing : use roulette select operator

(1) Calculate fitness value $eval(U_k) = p(z_k \mid x_k)$ of each chromosome U_k

(2) Calculate sum of group's fitness value $F = \sum_{k=1}^{m} eval(U_k)$, m is chromosome populations

(3) Calculate choice probability P_k : $P_k = \frac{eval(U_k)}{F}$ corresponds to each chromosome U_k

(4) Calculate cumulative probability Q_k : $Q_k = \sum_{k=1}^{m} P_j$ of each chromosome U_k

(5) Generate a random number r between [0,1]

(6) Compare Q_k with r in turn , the first individual k whose Q_k is not less than r will be choose as replication object

(7) Repeat (5)- (6) , until the required number of replication individual be satisfied

2 Crossover operator: random choose two parents particles $\{x_k^{(a)}, x_k^{(b)}\}$ from particle collection, and generate two new particles defined as followed[9]:

$$\begin{cases} x_k^{(an)} = \xi x_k^{(a)} + (1-\xi) x_k^{(b)} \\ x_k^{(bn)} = \xi x_k^{(b)} + (1-\xi) x_k^{(a)} \end{cases} \tag{3}$$

$\xi \in [0,1]$ is a random number uniform distributed. parents particles will be replaced by new particles according to Metropolis algorithm: new particle $x_k^{(an)}$ meet $p(z_k \mid x_k^{(an)}) > \max\{p(z_k \mid x_k^{(a)}), \{p(z_k \mid x_k^{(b)})\}$ will be accepted, otherwise it will be accepted with probability $p(z_k \mid x_k^{(an)}) / \max\{p(z_k \mid x_k^{(a)}), \{p(z_k \mid x_k^{(b)})\}$; and operate new particle $x_k^{(bn)}$ with the same rule.

3 Mutation operator: random choose parents particl $\{x_k^{(c)}\}$ from particle collection, and generate new particle defined as followed:

$$x_k^{(cn)} = x_k^{(c)} + \gamma \tag{4}$$

γ is a random number gaussian distributed, with mean value 0 and covariance σ^2. New particle $x_k^{(cn)}$ meet $p(z_k \mid x_k^{(cn)}) > p(z_k \mid x_k^{(c)})$ will be accepted, otherwise it will be accepted with probability $p(z_k \mid x_k^{(cn)}) / p(z_k \mid x_k^{(c)})$.

Crossover and mutation operators operate adaptively with probability Pc and Pm, increase Pc and Pm when individual fitness tends to agree or tend to local optimum, decrease Pc and Pm when fitness is dispersed.

3.2 Parameter Adjustment

Improper selection of crossover probability P_c and mutation probability P_m will cause premature convergence. This problem can be solved by using adaptive parameters adjustment. Define fmin as the fitness value of the most optimal individual in a generation, and f_{ave} as average fitness value of this generation. Convergence of particle group can be reflected by the gap between average fitness and optimal fitness approximately, define $F = f_{ave} - f_{\min}$, then P_c and P_m will be decided by F . To avoid premature convergence, increase P_c and P_m when F is small, and decrease P_c and P_m when F is big [11].

In order to protect excellent individuals while avoiding premature convergence, different individuals in the same generation should have different P_c and P_m. Increase P_c and P_m of individual with high fitness, and decrease P_c and P_m of individual with low fitness,. So P_c and P_m will not only related to ,but also related to $f_{\max} - f_m$,and $f_{\max} - f_c$, f_m is fitness value of mutated individual, and f_c is the bigger fitness value of two crossover individuals:

$$P_c = k_1(f_c - f_{\min})/(f_{ave} - f_{\min}), f_c \le f_{ave} \tag{5}$$

$$P_c = k_3, f_c > f_{ave} \tag{6}$$

$$P_m = k_2(f_m - f_{\min})/(f_{ave} - f_{\min}), f_m \le f_{ave} \tag{7}$$

$$P_m = k_4, f_m > f_{ave} \tag{8}$$

$0 \le k_1, k_2, k_3, k_4 \le 1$ .Experiments test that : $k_1 = 0.75, k_2 = 0.1, k_3 = 0.45, k_4 = 0.8$.

3.3 Algorithm Flow

Steps of improved FastSLAM algorithm based on genetic algorithms:
Input: initial robot pose S_0, particles number N, evolution number M;
Output: built map and robot's path
Step 1 initialization : t=1 , map of each particle Mi=Φ.

Step 2 prediction of robot pose: calculate possible pose s(i)t at time t of each s(i)t-1 at time t-1 according to movement model.

Step 3 evolution based on genetic algorithms and repeat for M times: ① calculate fitness of every particles according toobserved information at time t, get Pc and Pm; ② crossover among particles with probability Pc; ③ mutation of particles with probability Pm.

Step 4 refreshing map: refresh each particle's features coordinates with EKF according to observed information at time t, if the observing landmark is fresh observed value then refresh it's location with EKF and add it into landmark map.

Step 5 calculate each particles' importance weight, resample.

Step 6 map building: if the map is built , output the map corresponding to the biggest weight value and back to step 1; otherwise back to step 2.

4 Simulation Experiments

Mobile robot can make arbitrary rotation by controlling its two wheels. When we put control value u_k included with forward speed and angular velocity, motion model of robot can be described as [11,12]:

$$\begin{bmatrix} x_r(t) = x_r(t-1) + v\Delta T\cos(\phi_r(t-1) + \gamma(t)\Delta T) + w_x \\ y_r(t) = y_r(t-1) + v\Delta T\sin(\phi_r(t-1) + \gamma(t)\Delta T) + w_y \\ \phi_r(t) = \phi_r(t-1) + \gamma(t)\Delta T + w_\phi \end{bmatrix} \tag{9}$$

In the formula (9), $(x(k), y(k))$ and $\phi(k)$ means position and azimuth angle of robot in time k ; v means moving speed , γ means angular velocity, $w_{xy\phi} \in N(0,\sigma_{xy\phi})$ means Gauss noise with covariance of 0, used to describe influence of unknown characteristic such as wheel skidding.

As environment is stable, state of landmark keeps the same all the time, we get landmark motion model:

$$\begin{bmatrix} x_\theta(t) = x_\theta(t-1) \\ y_\theta(t) = y_\theta(t-1) \end{bmatrix} \tag{10}$$

Motion model of system is composed of robot motion model and landmark motion model. In FastSLAM algorithm, each PF means a path, each particle has it's own independent map composed of N landmarks, so state variable can be shown as $\{x_r, y_r, \phi_r; x_{\theta_1}, y_{\theta_1}; \cdots; x_{\theta_N}, y_{\theta_N}\}$.

Robot gets distance and azimuth angle of landmarks with its ranging sensors. The observation model can be represented as[12]:

$$\begin{bmatrix} R_i = \sqrt{(x_r - x_{\theta_i})^2 + (y_r - y_{\theta_i})^2} + w_R \\ \theta_i = \arctan(\dfrac{x_r - x_{\theta_i}}{y_r - y_{\theta_i}}) - \phi_r + w_\theta \end{bmatrix} \tag{11}$$

In the upper formula, (x_r, y_r) means position coordination of robot, ϕ_r means orientation, $(x_{\theta_i}, y_{\theta_i})$ means position coordination of landmark i, w_R and w_θ means noise series of distance and angle, and these two series match Gauss distribution of $N(0,\sigma_R)$ and $N(0,\sigma_\theta)$.

Covariance of observation model is :

$$R_t = \begin{bmatrix} \sigma_R^2 & 0 \\ 0 & \sigma_\theta^2 \end{bmatrix} \tag{12}$$

We use Matlab to simulate the experiment. Suppose data association is known, moving speed of the robot is 1m/s, maximum striking angle is 30π/180, time interval of control signal is 0.05s, speed noise is 0.1m/s, farthest distance of observation is

30m, time interval of observation is 0.2s, distance noise of observation is 0.1m, angle noise of observation is $\pi/180$, and we set 64 landmarks in simulation environment of 100m×100m.

Fig. 1 shows the comparison of robot path estimation and landmark position estimation between normal FastSLAM algorithm and improved algorithm when sample particle number is 30. In the figure, * shows real landmark position in the environment, • shows estimated landmark position. Final landmark estimated position is mean value of all particles. Real line represents robot's real motion path, dotted line represents estimated path. From the figure, we can see that improved algorithm has higher precision than normal FastSLAM algorithm in estimating robot's motion path and landmark position.

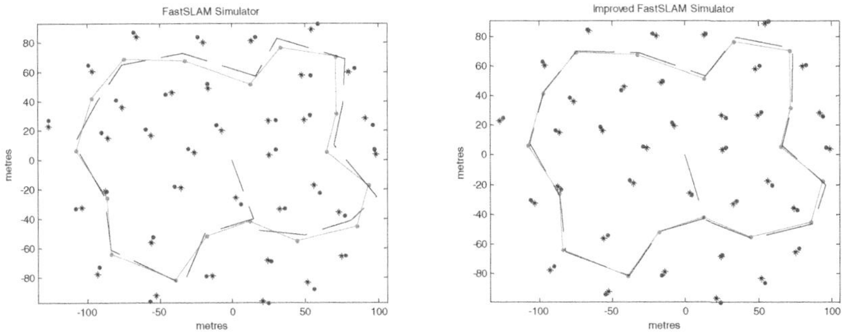

Fig. 1. Comparison of robot pose estimation and landmark position estimation of two algorithms

Fig. 2 and Fig. 3 show when particle number changing from 0 to 100, root mean square (RMS) error of estimation of robot's pose and landmark position of both algorithms reduces with increase of particle number. And we can see that improved algorithm has smaller RMS error and higher precision than normal FastSLAM algorithm.

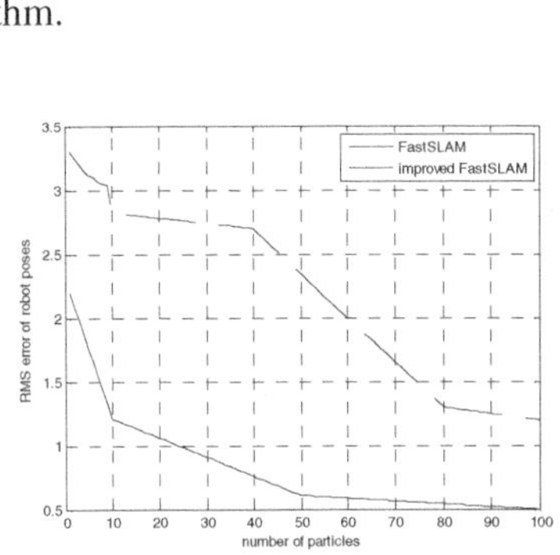

Fig. 2. Comparison of RMS error of robot pose estimation of two algorithms

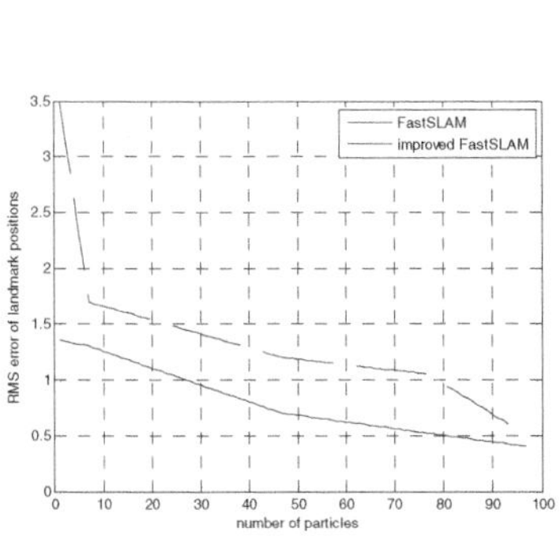

Fig. 3. Comparison of RMS error of landmark position estimation of two algorithms

5 Conclusion

To solve the sample degradation problems and improve performance, this paper import evolution mechanism into FastSLAM algorithm, optimize samples by choose, crossover and mutation operators in genetic algorithm, set adjusting control parameters. Theoretical analysis and simulation results show that compared to FastSLAM algorithm, improved algorithm with adaptive genetic algorithm has higher estimation precision and lower RMS error.

References

1. Wang, L., Cai, Z.-x.: Progress of CML for Mobile Robots in Unknown Environments. Robot 26, 380–384 (2004)
2. Dissanayake, G., Newman, P.M.: A Solution to the Simultaneous Localization and Map Building (SLAM) problem. IEEE Transactions on Robotics and Automation 17, 229–241 (2001)
3. Chi, J.-n., Xu, X.-h.: Research on Simultaneous Localization and Mapping of Mobile Robot. Robot 26, 92–96 (2004)
4. Murphy, K.P.: Bayesian Map Learning in Dynamic Environments. In: Advances in Neural Information Processing System, vol. 12, pp. 1015–1021 (2000)
5. Montemerlo, M., Thrun, S.: FastSLAM: a Factored Solution to the Simultaneous Localization and Mapping Problem. In: Proceeding of the Eighteenth National Conference on Artificial Intelligence, pp. 593–598. AAAT Press, Edmonton (2002)
6. Bailey, T., Nieto, J., Nebot, E.: Consistency of the FastSlam Algorithm. In: Proceedings of the IEEE International Conference on Robotics and Automation, pp. 424–427 (2006)
7. van der Merwe, R., Doucet, A., de Freitas, N., Wan, E.: The Unscented Particle Filter. In: Technical Report CUED/FINFENG/TR 380, Cambridge University, Department of Engineering (2000)
8. Montemerl, M., Thrun, S., Koller, D., et al.: FastSLAM 2.0: an Improved Particle Filtering Algorithm for Simultaneous Localization and Mapping that Provably Converges. In: Proceedings of the 18th International Joint Conference on Artificial Intelligence, Acapulco, Mexico, pp. 1151–1156 (2003)
9. Wang, X., Cao, L.: Genetic Algorithm, pp. 202–250. Xi'an jiao tong university press, Xi'an (2002)
10. Li, M.-h., Hong, B.-r., Luo, R.-h.: Improved Rao-Blackwellized Particle Filters for Mobile Robot Simultaneous Localization and Mapping. Journal of Jilin University (Engineering and Technology Edition) 37, 401–406 (2007)
11. Tan, B.-c., Lian, C.-y., Xu, A., Zhang, H.-g.: A Method of Improved Genetic Algorithm for Robotic Path Planning. Journal of Xi'an Technological University 28, 456–459 (2008)
12. Thrun, S., Burgard, W., Fox, D.: Probabilistic Robotics, pp. 189–279. MIT Press, London (2005)

A Study on the Protection of Consumers' Rights and Interests in the C2C Mode of Network Transaction ——Taking www.taobao.com as an Example

Qinghua Zhang

School of Information Management & Engineering,
Shanghai University of Finance & Economics, Shanghai 200433, China
zhangqh@mail.shufe.edu.cn

Abstract. The mode of network transaction creates objective conditions for undesirable businessmen to accuse rights and interests of consumers because of its virtualization and uncertainty while it brings convenience to the numerous consumers. This paper analyzes the challenges for sake of protecting consumers' rights and interests under environment of network transaction as well as makes researches, by taking www.taobao.com as an example, on the strategy of protecting consumers' rights and interests in the C2C mode of network transaction. Via application on the protective service to consumers and on Alipay, the paper puts forward an idea of establishing a risk fund for improvement aiming at their defects with consideration of affirming the positive effect they have on the protection of consumers' rights and interests, which also be of valuable reference to other network transaction platform of C2C mode.

Keywords: E-Commerce, Protection of consumers' rights and interests, www.taobao.com, Fund of risk management.

1 Influence on the Protection of Consumers' Rights and Interests by Network Transaction

The start of network transaction has increased ways of obtaining information for consumers, lowered consumption cost as well as has offered more convenient ways of consumption in the meanwhile of that information content of commodities and transparency of market have been increased significantly, which enables overall consumers to benefit. However, due to the network virtuality and hysteresis of related laws and regulations, consumers are frequently harmed in transactions. Infringing act under network transaction not only hurts the consumers' rights and interests, but also greatly restricts the development of network transaction itself, hence, the protection of consumers' rights and interests in network transaction has become a pressing issue.

Methods hurting consumers' rights and interests in network transaction appear more complicated and diversified than those in the traditional ways. More and more fake and exaggerated information are seen over the net than in reality. It becomes easier for illegal operators to reach their purpose of cheating consumers by network and network

L. Qi (Ed.): ISIA 2010, CCIS 86, pp. 303–308, 2011.

technology could also be used to create updated deceiving manners in an easier way while the anonymity of Internet has make it more difficult to exactly find out the law-breakers.

2 Analysis on the Existing Policy of Protection of Consumers' Rights and Interests in the C2C Mode

Being a new economic form, the industry of network transaction has so far not yet been covered by the unified management of the state, in which we still have various vacancies in supervision and the industry normalization seems harder compared with that of the traditional trading. Typically, for the disputes in network transaction, there is still a way to go to the legal level in spite of certain industry standards. Since the fact that there are no existence of mature laws and regulations like those from the traditional industry, it is only the individually initiated action between network businessmen and network transaction platform in terms of protecting the online shopping consumers. We still have a distance to go in perfecting the normalization of trading environment.

For the time being, *Taobao* is alone in a big situation in China's shopping websites in concern with C2C. According to a research report about China's online shopping market recently released by CNNIC, online shop of C2C occupies 85% of market share and *Taobao* 76.5% from the point of users' initial choice of online shops. From the loyalty of users to online shops, *Taobao* is the highest, meaning that 94.6% users keep using *Taobao* after they choose *Taobao* half a year ago. From the eyes of many consumers, it has become a habit if they have the desire of "Doing online shopping in *Taobao*". That's why we take *Taobao* as an example in this paper to analyze its policy in protecting consumers' rights and interests.

There are altogether six rules in protecting consumers in *Taobao* as following:

(1) Commodities truthfully described: this is the essential service in Consumer Protection Service, i.e. the default service, meaning that commodity should be truthfully described by sellers and be matched with commodity itself without any description that does not match the fact or any exaggerated descriptions. The most concerns in online shopping is to buy products which do not match the description and all sellers of *Taobao* promise to prove service of truthful description. If what bought is not as the same as the descriptions, buyers may ask for refund from sellers. If rejected by sellers, buyers may start complaints and apply for compensation from *Taobao*.

(2) Non-reason return and refund within 7 days: Within 7 days upon receipt of goods, if buyers dissatisfy at or dislike what they have bought, they may apply for return and refund from the sellers under the condition that the second time sales of products are not affected. If rejected by sellers, buyers may start complaints and apply for compensation from *Taobao*.

(3) Three-time compensation for one fake found: In concern with sellers who have joined Consumer Protection Service and are ready to provide service of "Three-time compensation for one fake found", buyers may start complaints against sellers from whom the fake is bought. Meanwhile, buyers may apply for "Three-time compensation for one fake found" from *Taobao*.

(4) 30-day maintenance for digital and household appliances: For sellers who have joined Consumer Protection Service and are ready to provide 30-day maintenance, they will be unconditionally responsible for one time free maintenance no matter what problems occur. If not fixed at willingness or not ready within the appointed time, buyers may start complaints against sellers and apply for compensation from *Taobao*.

(5) Guarantee of authentic commodities: Dealers of *Taobao* sell authentic commodities and offer regular invoices and they are strictly forbidden to sell fake goods and the unauthentic items of non-original. If what the buyers buy is not authentic, *Taobao* reserves right to terminate agreement with dealers all at once. Buyers have right to start complaints and may apply for compensation from *Taobao*.

(6) Prompt delivery of virtual items: For those sellers who promise to provide service of Prompt delivery should make deliveries within 2 hours. If they fail to meet this responsibility, buyers may start complaints and ask for over-time compensation from *Taobao*.

3 Limitations of Protective Service to Consumers

From the above analysis, we can find that the six rules to protecting consumers may in deed protect the consumers' rights and interests to certain extent, helping consumers escaping from being violated by undesirable businessmen even though their limitations including:

(1) Except that the "Commodities truthfully described" is the default service in the item of consumer service protection, which requires to be followed by those no matter they are individual online businessmen or *Taobao* Stores while the other five parts of service has no such kind of regulations that allows businessmen to do at their own willing. Because of the fact that *Taobao* never makes mandatory requirement for all businessmen to provide all the service, this brings forth hidden troubles to result in trading disputes when the two sides conclude transactions between each other, which also lays basis for the suffering of rights and interests from consumer side.

(2) When buyers put forward to retreat goods from sellers, the referred sellers should have joined "Consumer Protection Service" and promised to supply the corresponding service, otherwise, the buyers shall not have application access. For instance, the precondition to start a complaint of "Three-time compensation for one fake found" is that the sellers have promised to supply the service of "Three-time compensation for one fake found". Except for "Commodities truthfully described", sellers are not requested to join other five service items, which is obviously quite unfavorable to consumers and which may directly lead consumers to the passive situation when they apply for compensation. In this case, the undesirable businessmen may realize their purpose of violating consumers' rights and interests by means of thoroughly rejecting to provide the related service.

(3) *Taobao* regulates that transactions not paid by *Alipay* are not entitled to enjoy parts of the service concerning consumer protection, for example, transactions not paid by *Alipay* have no access to apply for "Non-reason return and refund in 7 days service". For the time being, in spite of that the majority of *Taobao* registered members choose *Alipay* in the first place to conclude their transactions, there still are other possibilities

for certain malicious sellers who try to persuade buyers without knowledge of the truth in escaping *Alipay* in the transactions so that their purpose of cheating buyers be realized.

4 Countermeasures of Improvement: Establish the Fund of Risk Management

Virtuality of network transaction and the universality of internet fraud have greatly sapped consumer confidence for network transaction. When serious deficiencies are found by buyers upon their confirmation and acceptance of goods they order, sellers will try to escape and play hide-and -seek when buyers lay hopes for negotiations with sellers. In most cases, consumers only get a taste of their own medicine. If this kind of phenomenon occurs frequently, it will badly block the development of network transaction. Hence, the most efficient way to enhance consumers' confidence in network transaction is to have consumers feel that network transaction is safe and reassuring and particularly that they can get compensation via certain channels if fraud happens.

The existing compensation form of *Taobao* in regard with the above situations mainly lies in the system of "Advance compensation", meaning that when the transaction is concluded via *Alipay* between *Taobao* buyers and sellers who have already signed Consumer Protection Agreement, if such transaction leads to a loss of buyers' rights and interests and buyers have failed in getting a direct solution from the sellers, buyers shall have right to start complaints against *Taobao* even after conclusion of the transaction and to apply for compensation at the same time. *Taobao* makes judgment on whether the buyers' application is reasonable or not in accordance with the related rules and it also has right to inform *Alipay* to directly deduct the corresponding amount from the sellers' account to compensate buyers' loss.

In another sense, *Taobao* regulates that sellers who apply to join Consumer Protection Service should deposit corresponding amount in their *Alipay* accounts according to the difference of their identities or the commodity category they deal with. For example, sellers who are in business of ladies' wear or boutique are required to deposit a guarantee of RMB1000 in advance in their *Alipay* accounts while sellers dealing with daily necessities or cooking utensils are only required a deposit of RMB500.

By analyzing the above two clauses, it becomes easier to find out they can protect the consumers' rights and interests to certain extent while their limitations are obviously noted. All these two clauses are pre-conditioned by sellers' participation of Consumer Protection Service and the arrangement of security deposit is not so reasonable. Security deposit of *Taobao* is categorized according to commodity types and arrangement of deposit lacks of flexibility and changeability in terms of amount, for instance, sellers who deals with shirts and fur coats in the category of ladies' wear are all required to deposit RMB1000 under *Taobao* regulations. When Infringement occurs, the RMB1000 deposit will never means a deterrent force to fur traders.

In conclusion, the author of this paper holds that *Taobao* may establish a fund of risk management on the basis of *Alipay*. And the fund of risk management is a pre-charged amount by sellers in *Alipay*, which is similar to security deposit in nature and which plays the function of preventing malicious sellers from cheating via *Taobao* platform as well as makes some economic compensation on the buyers who have been violated in rights and interests.

Calculation method for recharge amount: From the first day to run shops, sellers should, within one week, recharge RMB2000 as fund of risk management in *Alipay* and then RMB500 will be returned to sellers on a one-month basis until the 3rd month that the frozen amount in sellers' *Alipay* will be RMB500 which will be then compared with variable Y and the maximum amount will be chosen to serve as the fund of risk management for the 4th month.

Calculation formula of Y value:

$$Y = \frac{x1 + x2}{2} \times (1 - g) \times \frac{a1 + a2}{2} \tag{1}$$

In formula (1):

x1: Largest turnovers of orders within recent three months
x2: Smallest turnovers of orders within recent three months
g: Positive rate
a1: Most expensive items in the shop
a2: Most cheapest items in the shop
Therefore,

$\frac{x1 + x2}{2}$: Average turnovers within recent three months

1-g: Percentage of all dissatisfying buyers in the transaction

$\frac{a1 + a2}{2}$: Average price of items in the shop

$\frac{x1 + x2}{2} \times (1 - g) \times \frac{a1 + a2}{2}$: Maximum amount that shops pay for compensation

For example, if x1=1752, x2=827, g=99.5%, a1=238, a2=45, then $Y = \frac{x1 + x2}{2} \times (1 - g) \times \frac{a1 + a2}{2} = \frac{1752+827}{2} \times (1-99.5\%) \times \frac{238+45}{2} = 912.3 > 500$. Therefore, the amount of the fund of risk management in *Alipay* is RMB 912.3.

It should be noted that there is need to make a monthly rolling adjustment in respect of Y and the related amount will be sent to sellers' Email addresses in bill way. If sellers withdraw from online operation three months ahead of opening their shops, the related fund of risk management will be automatically refunded after a three-month time.

Fund of risk management established on the basis of *Alipay* is in favor of effective right protection by consumers in case of occurrence of infringing act. Due to virtuality of network transaction, this kind of paperless trading only leaves consumers the complicated and confusing identity of sellers. Particularly, when the consumers' rights and interests are violated, they are often unable to express bitter feelings because of the unrecognized identity of sellers. From this sense, we can say that the establishment of fund of risk management will efficiently help to safeguard the road of right protection by consumers, in creasing safety of network transaction. Advantage of fund of risk management on the basis of *Alipay* lies in the function of warning to illegal businessmen, stopping acts violating consumers' rights and interests as well as protecting consumers' rights and interests when sellers run away from their cheats so that the buyers could still receive a certain percentage of compensation.

5 Conclusions

This paper analyzes the challenges for sake of protecting consumers' rights and interests under environment of network transaction as well as makes researches, by taking www.taobao.com as an example, on the strategy of protecting consumers' rights and interests in the C2C mode of network transaction. Via application on the protective service to consumers and on *Alipay*, the paper puts forward an idea of establishing a risk fund for improvement aiming at their defects with consideration of affirming the positive effect they have on the protection of consumers' rights and interests, which also be of valuable reference to other network transaction platform of C2C mode.

Frankly speaking, the fund of risk management is only targeted at guarding against the unscrupulous businessmen who aim to cheat consumers, which can not cover all the cases of incidental infringing acts. As for the calculation of fund of risk management, this paper only suggests one method and the author believes that there will be more objective and detailed solutions in the future.

Acknowledgments. This work was supported by the Humanities and Social Sciences Foundation from the Ministry of Education of China (09YJCZH073) (Research on Government Regulation for Bad Faith of Auto Bidding for Search Engine Keywords), Shanghai Philosophy and Social Science Foundation from Shanghai Planning Office of Philosophy and Social Science (2009EZH001), and Leading Academic Discipline Program, 211 Project for Shanghai University of Finance and Economics (the 3rd phase).

References

1. Turban, E., King, D., Lee, J., Viehland, D.: Electronic Commerce——A Managerial Perspective, 4th edn. China Machine Press, Beijing (2007)
2. Murray, B.H.: Defending the Brand: Aggressive Strategies for Protecting your Brand in the Online Arena, American Management Association (2004)
3. Roussos, G.: Ubiquitous and Pervasive Commerce: New Frontiers for Electronic Business. Springer, Heidelberg (2006)
4. CNNIC: Research Report about China's Online Shopping Market (2009), http://www.cnnic.cn/index/0E/00/11/index.htm
5. http://www.taobao.com

Application Research on WebGIS Index System Based on Fractal Theory

Jiancun Li[1,2], Mingguang Diao[1], and Tao Xue[1]

[1] China University of Geosciences, Beijing 100083, China
[2] China Aero Geophysical Survey & Remote Sensing Center for Land and Resources, Beijing 100083, China

Abstract. To solve the problem of information island. Established the cluster spatial database in a distributed network environment. Built a unified spatial index system based on Hilbert space-filling curve scan matrix, to get one-dimensional mapping between Hilbert space arrangement code and spatial index information. Realized spatial data rapidly index across spatial databases. Experiments show that the system directly to a relational database table through the scales of operation and maintenance, as much as possible to avoid the call to the system relies on the development of the ArcGIS platform package ArcObjects correlation function, built a key spatial index system and a good support for data sharing.

Keywords: spatial index; metadata; Hilbert; spatial ordering; data sharing; ArcGIS.

1 Introduction

With the development of network technology and Web GIS. GIS-related sector generates huge amounts of spatial data are all stored in the distributed database servers, sharing of spatial data has become a bottleneck in the GIS applications. Therefore, how to respond to the relevant researchers and policy makers concerned about data services. Han Xinchun [1] proposed a set of metadata and data to achieve integrated management of the effective ways in the metadata and data sets distributed management, while coordinating metadata services and data service consistency and synchronization. Xue Tao [2] designed metadata management platform based on the geographic information metadata standard, platform application effectively solves the integration and sharing of spatial data between departments. Chen Zhanlong et al [3] for distributed spatial data index mechanism, based on R-tree and hashing hash table, a massive distributed environment for spatial data distributed index tree structure DR-H tree effectively improve spatial data retrieval performance.

For the same purpose, the spatial index system proposed in this paper is based on B / S architecture, designed application code of spatial index algorithm with filling Hilbert space curve in the two-dimensional space region of spatial index, can efficiently and accurately search spatial information to present to the user. Although the index precision maybe not absolutely accurate, but has done a good balance between

L. Qi (Ed.): ISIA 2010, CCIS 86, pp. 309–316, 2011.

accuracy and speed of index. Provided a good data supporting platform for the applications such as spatial data warehouse and spatial data mining, and effectively addressed the problem of spatial information island. It has played a important role in Remote Sensing Geological Survey Spatial Information Platform(RSUSIP).

2 RSUSIP Index Optimization

The goal of the RSGSIP is data services and data sharing, to provide data resources for requirement in the relevant fields. So, the key element of the RSGSIP is a efficient spatial index system, can be quickly and easily find data of interest and not produce unrelated data. The system have a spatial index database that store spatial index metadata, basic information about data such as index layer name, index layer minimum coordinate range, Hilbert space arrangement of coding data. Through make a unified mapping between spatial data and spatial index metadata in the cluster spatial database, user can quickly retrieve target data from vast amounts of spatial data in the distributed network environment.

2.1 Spatial Index Architecture

RSUSIP organize spatial data using ESRI's geographic database (Geodatabase) data model. The data model is implemented by ArcSDE application server in a standard relational database such as Oracle. ArcSDE is an open user-defined database interface, it can directly access spatial data and spatial index data with SQL in distributed relational database. Spatial Index System structure chart shown in Figure 1.

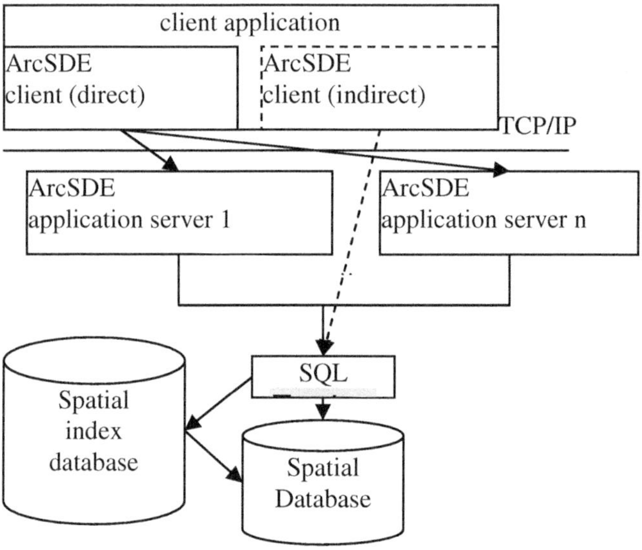

Fig. 1. Spatial Index System structure chart

2.2 Spatial Index Algorithm

The spatial index utilize based layer (such as standard maps, administrative division layers, into the mine with layers, etc.) in spatial index database as index condition to search data deployed RSUSIP spatial database. The data has topology associated with region of the index layer will be found rapidly and accurately. Due to index target layer deployed in different distributed spatial database, in order to reduce the burden, the local spatial index database server only stores relevant spatial index information of the target layer. Then, in the RSUSIP, how to create a relation between index region spatial information and spatial data information in distributed spatial database, such as GIS spatial analysis software to the same association, as key of spatial index system.

The system fill Hilbert space curve into two-dimensional region of the spatial index. The grids of curve make arrangement coding of Hilbert space, shown in Figure 2. Arrangement coding of Hilbert space has many to one mapping with index basic layer element in the local server (shown in Figure 2, surface elements of A corresponding Hilbert code: 1, 2, 3, 4, 8, 9, 14 , 15; point elements of B corresponding Hilbert code: 13). The mapping of all the Hilbert codes corresponding to the grid arrangement of the scope of information extracted from the target layer with the index information in the scope of the minimum coordinates of the associated topology. If the result consistent with default topologic relationship in spatial index system, the distributed index target layer will be extracted from the spatial database, to facilitate the users to re-use of geospatial data.

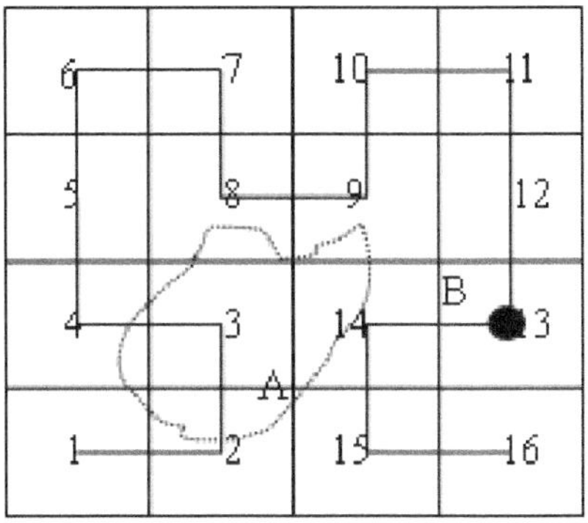

Fig. 2. Hilbert index elements arranged in coding and mapping

2.2.1 Hilbert Space Filling Curve

Hilbert curve derived from the classical Peano curves, is a FASS curve [5]. Peano curves is a closed unit interval I=[0,1] To the closed rectangular cell $S=[0,1]^2$ Continuous mapping, it is general of all continuous fractal curves be able to fill two-dimensional or higher dimensional space, it is also known as space-filling curve. In this paper only study two-dimensional Hilbert space filling curve. Hilbert space filling curve is described as follows: four equal to a square region, connecting the center of each small square, curve shown in Figure 3 (a) shall be the first order Hilbert curve;

In accordance with the first-order Hilbert curve generation method 4 iterations, the first order Hilbert curve do array in order in second order Hilbert curve generated four quadrants: in the first and second quadrant curve hold prototype of the first order Hilbert curve, in third and fourth quadrant the original first-order Hilbert curve clockwise and counterclockwise rotation were 90 °, so each small square would be subdivided into four identical small square, and connect the small square center, in accordance with the continued subdivision of this iteration go, and according to certain rules one by one connection, you can get Figure3 (b), Figure 3 (c) second and third order of the Hilbert curve. Hilbert filling curve is a good mapping sorted method of two-dimensional to one-dimensional, also achieve mapping of the grid array coordinates to the one-dimensional sorted serial number $n=f(x,y)$. Through the mapping function, it can determine the grid cell place in the one-dimensional sort, to find space mapping, to prepare the necessary conditions for spatial index.

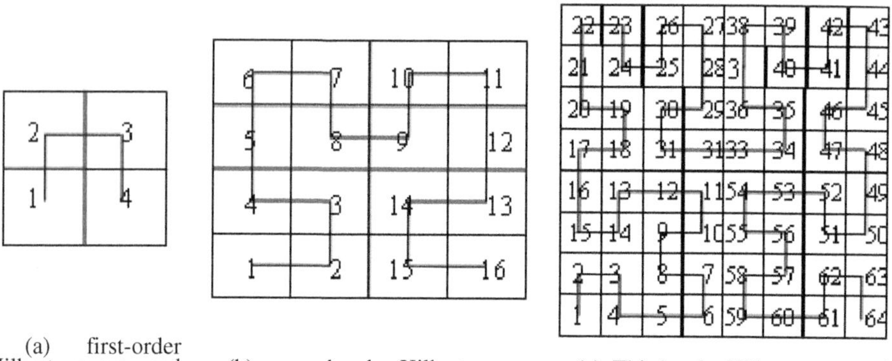

(a) first-order Hilbert curve and the arrangement of coding

(b) second order Hilbert curve and the arrangement of coding

(c) Third-order Hilbert curve and the arrangement of coding

Fig. 3. Hilbert curves and the grid partition diagram

2.2.2 Hilbert Space Arrangement Coding

In spatial index system, Hilbert space filling curve corresponds to the grid code that is Hilbert space arrangement code, it is realized by Hilbert scan matrix generation algorithm based matrix operation [6]. The algorithm in literature 6 made a detailed proof, the following specific description.

Set:

$$A=\begin{bmatrix} a_{1,1} & a_{1,2} & \cdots & a_{1,m} \\ a_{2,1} & a_{2,2} & \cdots & a_{2,m} \\ \vdots & \vdots & \vdots & \vdots \\ a_{n,1} & a_{n,2} & \cdots & a_{n,m} \end{bmatrix}_{n\times m},\ \tilde{A}=\begin{bmatrix} a_{1,m} & \cdots & a_{1,2} & a_{1,1} \\ a_{2,m} & \cdots & a_{2,2} & a_{2,1} \\ \vdots & \vdots & \vdots & \vdots \\ a_{n,m} & \cdots & a_{n,2} & a_{n,1} \end{bmatrix}_{n\times m},\ A^{\wedge}=\begin{bmatrix} a_{n,1} & a_{n,2} & \cdots & a_{n,m} \\ \vdots & \vdots & \vdots & \vdots \\ a_{2,1} & a_{2,2} & \cdots & a_{2,m} \\ a_{1,1} & a_{1,2} & \cdots & a_{1,m} \end{bmatrix}_{n\times m} \tag{1}$$

Set 2^n Order Hilbert curve scan matrix as $\boldsymbol{H}_{2^n}$, And when $n=1$,

as $\boldsymbol{H}_2=\begin{bmatrix} 2 & 3 \\ 1 & 4 \end{bmatrix}$, The structure of recursive Hilbert curve scan matrix algorithm:

$$H_{2^{k+1}} = \begin{cases} \begin{bmatrix} H_{2^k} & 2^{2k}E_{2^k} + H_{2^k}^T \\ (2^{2(k+1)}+1)E_{2^k} - \hat{H}_{2^k} & (3\times 2^{2k}+1)E_{2^k} - (\tilde{H}_{2^k})^T \end{bmatrix} & k \text{ as odd number} \\ \begin{bmatrix} H_{2^k} & (2^{2(k+1)}+1)E_{2^k} - \tilde{H}_{2^k} \\ 2^{2k}E_{2^k} + H_{2^k}^T & (3\times 2^{2k}+1)E_{2^k} - H_{2^k}^T \end{bmatrix} & k \text{ as even number} \end{cases} \quad (2)$$

In formula (2) , E_{2^k} corresponds orders on behalf of the unit matrix.

3 Spatial Analysis Research

In this paper, spatial analysis using ArcGIS Server9.3 and ArcSDE9.3 for Oracle10.2 technology. ArcSDE Geodatabase for Oracle provided ST_Geometry type to store geometric data. ST_Geometry is a follow ISO and OGC specifications, can be directly read storage type of spatial information by SQL. Using this storage method can make better use of resources compatible with Oracle features, the business data and spatial data stored in a table. ST_Geometry table structures for storage spatial data shown in Table 1:

Table 1. ST_Geometry table structures for storage spatial data

Name	Type	Memo
ENTITY	NUMBER (38)	Elements of the type of space
NUMPTS	NUMBER (38)	The number of coordinates
MINX	FLOAT (64)	The scope of outsourcing rectangular geometry
MINY	FLOAT (64)	
MAXX	FLOAT (64)	
MAXY	FLOAT (64)	
MINZ	FLOAT (64)	Elevation values
MAXZ	FLOAT (64)	
MINM	FLOAT (64)	Measurements
MAXM	FLOAT (64)	
AREA	FLOAT (64)	Space element area
LEN	FLOAT (64)	Perimeter space elements
SRID	NUMBER (38)	Spatial Reference System ID
POINTS	BLOB	Coordinate sequence

In distributed Spatial database, you can simply import the ST_Geometry data in SDE to the local spatial index database. Using SQL statement with a record constructed ST_Geometry object , then set of ST_Geometry objects take the place of spatial index layer, and perform spatial analysis with the index base maps in the local server that can greatly improve the system efficiency.

4 Application of Spatial Index System

The spatial index system has been used for data service in RSUSIP [4,7,8]. Usually access to data resources with a name as an identifier, the data resources maybe a spatial database or many spatial database, and find the resources. the system generates the designated order (generally no more than 9 bands) Hilbert space arrangement code according to basic index layer, then traverse the minimum coordinates of distributed target spatial data range compared to the Hilbert space arrangement code in the spatial index database. If the target layer shall meet the requirements, the purpose of achieving a spatial index.

4.1 Operation of Spatial Index System

Implementation of spatial index shown in Figure 4 to determine the spatial index method, after select the index layer, click on the element in the display option layer, based on user needs can choose one or more elements to determine the spatial index range.

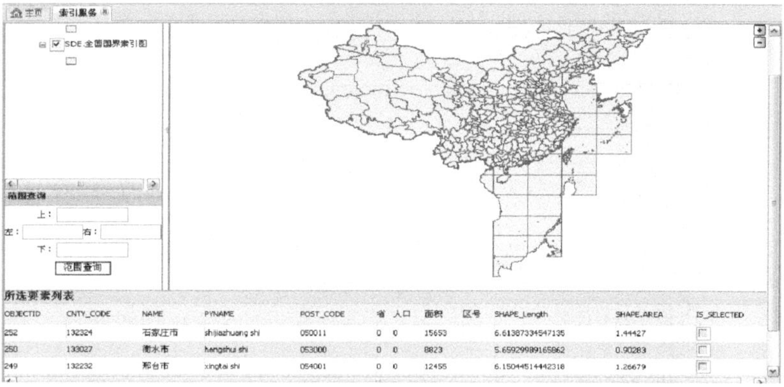

Fig. 4. Factors determining spatial index

Selected elements, the basic information of the elements is also displayed in the interface, click the checkbox of "inquiry intersection elements" under the list of the elements, to query all layers in database those have intersection topology relationship with the spatial index range.

The system designed two index model of based on Hilbert curve and ArcObjects, to promote the system user-friendly and the spatial index efficiency. Shown in Figure 5, the elements of the list below, if do not check the "fine Search" (Figure 6 (a) show) checkbox, then the index based on Hilbert space. Otherwose the system is precise spatial index based on ArcObjects (Figure 6 (b) show).

Figure 6 (a), Figure 6 (b) comparison showed that the "Beijing grass" layer only is found with ordinary search model in Figure 6(a). This indicate that MBR corner boundary of the Beijing grass" layer intersect with Hilbert values of MBR corner boundary of the selected elements.

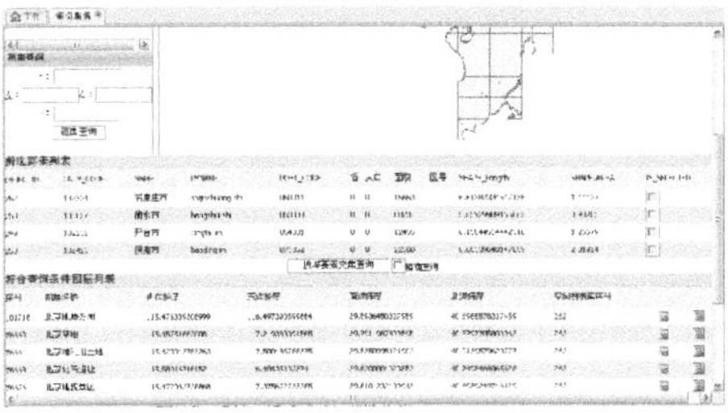

Fig. 6. (a) Index based on Hilbert space

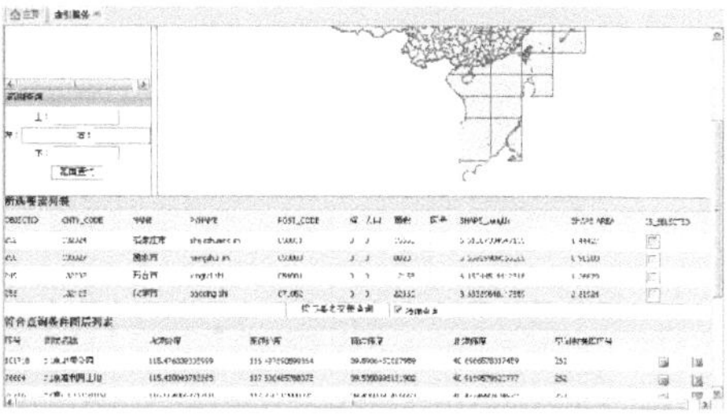

Fig. 6. (b) results based on ArcObjects Spatial Index

In Figure 6 (b), select the absolute query, the "Beijing grass " layer was filtered. This point that the selected index elements and all elements of the "Beijing grass " layer are not intersect, spatial index based on ArcObjects is better accurate than spatial index based on Hilbert space.

4.2 Analysis of Experimental Data

Test data: maps of Remote Sensing colligation survey for land and resources in China, the data volume is 8GB, include 432 layers, the total elements number is 628423.

By comparison, the index algorithm based on Hilbert space as in the process of index need to calculate the Hilbert curve values and the decomposition of leaf nodes in the minimum bounding rectangle of spatial objects, time of index than the basic package with ArcObjects index function to establish a long time, but due to Hilbert value simplifies the search path, so search for a shorter time; in the index accuracy, because of the Hilbert order of the threshold value is set, will have some redundant data, while the spatial index based on ArcObjects function does not produce redundant data. So you can change the threshold Hilbert decomposed under the premise of an index based on user requirements and the accuracy of the index of time, select the spatial index algorithm integrated. Here are two kinds of spatial index algorithm efficiency comparison table, shown in Table 2.

Table 2. Index based on Hilbert space compared with the spatial index based on ArcObjects

Test Project	Index based on Hilbert space	Spatial index function based on ArcOjects
Time to establish a spatial index structure	16.43s	8.27s
Spatial index time	21.41s	50.57s
The number of redundant data	3	0

5 Conclusion

Under WebGIS development platform, the spatial index speed depends on efficiency of the GIS spatial analysis function and operational volume. However in spatial index system based on Hilbert transform the main influence fact is order number of Hilbert space filling curve, because the order is a Hilbert system can be controlled, so the index speed of spatial index system also can be controlled. If the order is more high then the index speed is more slow, but the index precision is better. While Run the spatial index, the system generate Hilbert space ordered code, build one-dimensional mapping table between spatial index data and index terms, the records stored in a relational database. Thus greatly improve the efficiency of the spatial index when dealing with records.

However, due to Hilbert filling curve is a rectangular two-dimensional filling curve in the index system. So if the Hilbert order is too low, will produce some redundant data when establish the mapping relationship between the space-filling curve and spatial index layers. The accuracy of spatial index also is reduced while the speed of spatial index is increased. Therefore, users need to weigh the relationship of efficiency and accuracy based on the actual needs.

References

1. Han, X.: Application research of management model based on spatial metadata and dataset. Chengdu University of Technology, College Information Management, Applied Mathematics, Chengdu, pp. 10–15 (2005)
2. Xue, T., Diao, M.: Unified platform about information issuing, integration and management of meta-database and spatial-datbase based on geology metadata standatd. Earth Science Frontiers 16(77), 288 (2009)
3. Chen, Z., Wu, X., Xie, Z., et al.: Microelectronics & Computer 24(10), 54–57 (2007)
4. GB/T 19710-2005. National Standard of the People's Republic of China -Geographic Information-Metadata. China Standards Press, Beijing (2005-2008)
5. Chen, N., Wang, N., Chen, Y.: Mini-micro Systems 26(10), 1754–1756 (2005)
6. Wang, S., Xu, X.: Journal of Image and Graphics 11(1), 119–122 (2006)
7. Diao, M., Xue, T., Pan, W.: Design of universal authority management model of online teaching system. In: Proceedings of International Conference on Computer Science and Software Engineering, CSSE 2008, vol. 3, pp. 610–613 (2008)
8. Zhao, P., et al.: Remote Sensing for Land & Resources 20(4) 101–104 (2009)

Fault Diagnosis of Automobile Based on CAN Bus*

Meilan Zhou, Xue Ao, and Jian Wang

College of Electrical & Electronic Engineering, Harbin University of Science and Technology, Harbin150040, China
zhoumeilan001@163.com

Abstract. Aimed at the CAN technology utilized on automobile currently, and the complexity of the automobile fault information and the difficulty of diagnosis, CAN bus adapter is designed .Microsoft Visual C++ 6.0 is utilized to build the Kalman digital filter and automobile fault diagnosis system based on BP network. Incepting the signal from the CAN bus, filtering and removal of noise, and online fault diagnosis and forecast to the main systems of automobile. Experiments show that Kalman filtering plays good on removal of noise from the automobile fault signals, and the BP network trainings of the systems are effective to implement non-linear mapping from the fault phenomenon of automobile to the fault position.

Keywords: Fault diagnosis of automobile; CAN bus; interface design; Kalman filtering; Back-propagation Network.

1 Introduction

The fast development of computer technology provides new vitality for automobile fault diagnosis. If judging automobile's current mode of operation, and determining cause or position of the fault timely and accurately, on condition that automobile is non-separate, it will be more convenient for automotive performance testing and maintenance. Aimed at the CAN bus technology applied to automobile[1], and the complicated trait of the fault information of automobile[2] [3], according to the strongpoint of the nonlinear function approaching ability, the self-learning ability and the self-adapting ability of the artificial neural network. This paper designs fault diagnosis of automobile based on CAN bus.

2 System Functions and Design

Fig. 1 shows that automobile fault diagnosis system is made up of CAN bus adapter based on USB technology, BP neural network fault diagnosis system, and fault show interface. Data about automobile's mode of operation are uploaded from child nodes to CAN bus, and accepted by CAN bus adapter, then transmitted to the BP network part of

* Supported by the Key Project of Chinese Ministry of Education (No.: 208037); Scientific Research Fund of Heilongjiang Provincial Education Department (No.:11551072).

L. Qi (Ed.): ISIA 2010, CCIS 86, pp. 317–323, 2011.

upper monitor. Then they are calculated by the BP network which has been already trained, and judged that whether there is a fault or not and the fault position. Finally, results are showed on the fault show interface to be made reference by the driver and maintainer.

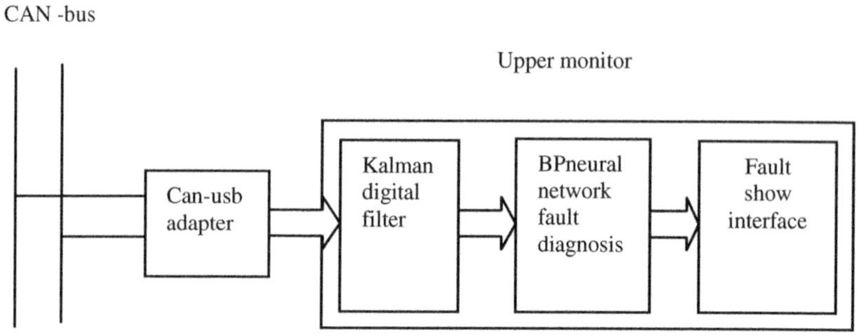

Fig. 1. System structure

3 CAN -USB Adapter

In Fig.2, CAN bus controller SJA1000 and CAN transceiver 82C250 produced by Philips corporation compose CAN bus interface mostly. Interface between CAN controller and physical bus adopts 82C250, which mainly applied to translation of electrical level, enhance of systemic drive capability, and offer for differential transmitting and acceptable capability, to meet the requirement of CAN bus communications. It can adjust speed of transition and countercheck interference too. SJA1000 is taken for CAN systemic communication control unit. Microcontroller P89C52 affords frame data error detection and automatic address recognizes function. USB device PDIUSBD12 is used as universal high-speed parallel interface communicated with microcontroller.

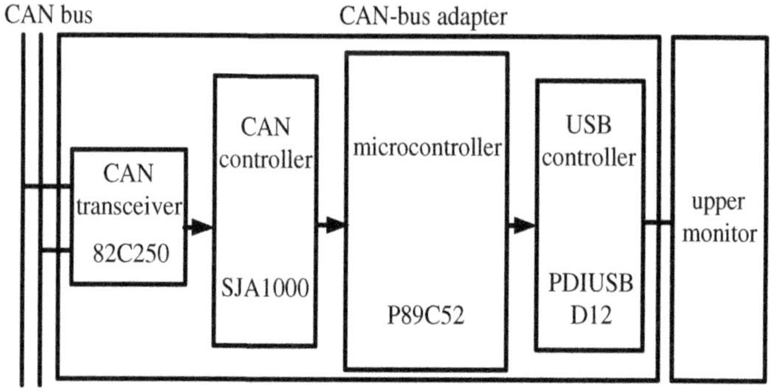

Fig. 2. CAN-USB adapter

4 The Design of Kalman Digtal Filter

Kalman filter is an efficient recursive filter,That is optimal from the regression data-processing algorithms. It can select measurements from a series of incomplete noise, and estimate the status of the dynamic system [4]. Using the process model of system to predict next State system. The supposition of present's system mode is k, according to system's model, it may forecast current condition based on system's previous status

$$X(k|k-1)=AX(k-1|k-1)+BU(k)\ . \quad (1)$$

In Type (1), $X(k|k-1)$ is prediction of using the results of the previous state. $X(k-1|k-1)$ is the result of a state of optimal. $U(k)$ is the control volume for the current state.

$$P(k|k-1)=AP(k-1|k-1)A'+Q\ . \quad (2)$$

In Type (2) $P(k|k-1)$ is the corresponding covariance of $X(k|k-1)$ $P(k-1|k-1)$ is the corresponding covariance of $X(k-1|k-1)$, A' Represents the transpose of A matrix , Q is a system process covariance.

Now we have the prediction of the present state , then we will collect the measurement of current state. Combining with predicted value and measured value, we can obtain optimized estimate $X(k|k)$ in the condition.

$$X(k|k)=X(k|k-1)+Kg(k)(Z(k)-HX(k|k-1))\ . \quad (3)$$

In Type (3), Kg is the Kalman gain:

$$Kg(k)=P(k|k-1)H'/(HP(k|k-1)H'+R)\ . \quad (4)$$

In order to make the Kalman filter continuously run until the end of the system, it also need to update the covariance of $X(k|k)$ in the state of k :

$$P(k|k)=(I-Kg(k)H)P(k|k-1)\ . \quad (5)$$

When the system enters state of $k+1$, $P(k|k)$ is $P(k-1|k-1)$ in the type (2). This algorithm can go from the return operation.

5 BP Network Adopted for Automobile Fault Diagnosis

The first stage is network train stage as Fig.4 shown, which performs the problem field by adjusting the network weights and thresholds. The second stage called work stage, has fixed weights and thresholds. As experimental data is input into the network, the network can classify them [5] [6].

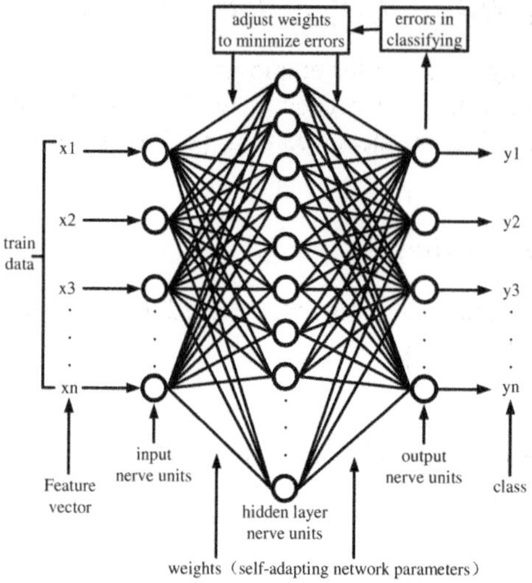

Fig. 4. Training stage

BP network adopts nonlinear seriate derivable S-type (Sigmoid) function as activation function

$$f(Net_{kj}) = \frac{1}{1+e^{-Net_{kj}}} . \tag{6}$$

In formula (1), Net_{kj} is the state of network unit u_j, $Net_{kj} = \sum_i \omega_{ji} o_{ki} + \theta_j$

So to the unit output is

$$o_{kj} = \frac{1}{1+\exp(-\sum_i w_{ji} o_{ki} - \theta_j)} = \frac{1}{1+e^{-Net_{kj}}} . \tag{7}$$

In which, θ_j is field value of the unit u_j. In the circumstance of the kind of activation function like formula (7), $f_j'(Net_{kj})$ is

$$f_j'(Net_{kj}) = \frac{\partial o_{kj}}{\partial Net_{kj}} = o_{kj}(1-o_{kj}) . \tag{8}$$

So to the output layer units

$$\delta_{kj} = (t_{kj} - o_{kj}) \cdot o_{kj}(1-o_{kj}) . \tag{9}$$

To the hidden layer units

$$\delta_{kj} = o_{kj}(1-o_{kj}) \sum_m \delta_{km} \omega_{mj} . \tag{10}$$

The weights adjustment is

$$\Delta\omega_{ji}(t+1)=\eta\delta_{kj}o_{ki} \ . \tag{11}$$

Learning rate η has heavy impact on the training process in the actual process of training. η is step length searching according to the gradient. η grows larger with the weight changing more acutely. In practical application the value should be maximum regularly in the premise of not leading to vibration. Add a momentum term to the δ formula in order to make learning speed fast enough to not to produce oscillations.

$$\Delta\omega_{ji}(t+1)=\eta\delta_{kj}o_{ki}+\alpha\Delta\omega_{ji}(t) \ . \tag{12}$$

In formula (12), α is a constant, which decides the impact degree on variety of the by-pass weight from the variety of the current weight.

Selection of the number of hidden layer units depends on the Kolmogorov theorem $n_1=2n+1$,in which n is number of input layer units, and in this system $n_1=2*4+1=9$.

Learning rate η, which is also called learning step length grows larger with the weight changing more acutely, but this will cause vibration; if η is min, learning rate would change slowly, but learning process would be smooth [7]. So a constant should be adopted fulfilling that $0<\eta<1$, and in this system is 0.5.

Momentum term correction factor α should be adopted coordinately with learning rate η. If α is biggish, it could improve convergence rate, but would not actively process on improving precision [8]. α should be a constant, and in this system is 0.5.

6 Software Design of Automobile Fault Diagnosis System

The system designs software program of automobile fault diagnosis based on programming environment of Microsoft Visual C++ 6.0, which mainly includes three parts of program, such as communication between USB and upper monitor, training and forecast program of BP neural network as well as design and show interface of fault diagnosis system.

Fig.4 shows main interface of automobile fault diagnosis system. To push the four system buttons (*such as veer system, brake system, engine, and transmission*) enters the training interface to train all of the four BP networks in the system; then to push the button of Demo Program enters Detection Interface to input data and verify the validity of BP network; finally to push the OK button diagnoses and forecasts faults of the four systems online.

Take transmission system as an example, because this system meet the requirements of any double-data as input sample and target output, set input sample data as automobile fault data, and outside the scope of data as normal condition; set target output data as automobile fault location code, corresponding automobile fault position.

After 180 times of training, as the data shown in table.1, effective output achieves the accuracy range, so the training is effective. BP network train interface based on transmission fault is shown in Fig.6.

After the actual test, the main interface in Fig.4 shows the sub-node data received from CAN bus system. Basing on the BP neural network fault diagnosis system, we can determine an automobile transmission system being fault, and that is clutch sump pressure is too low, fault location or cause of the malfunction is the liquid pressure controller oil route.

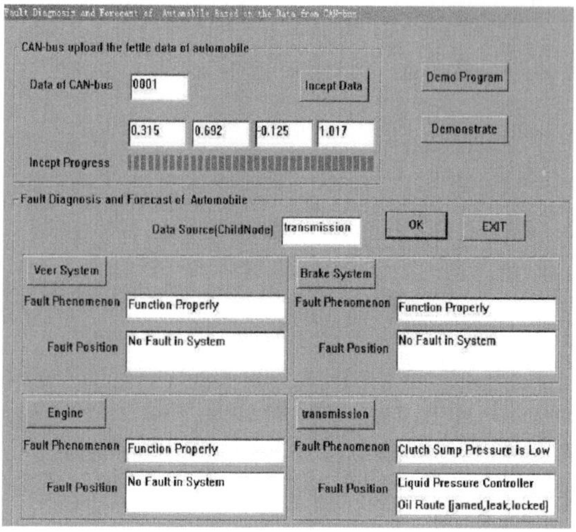

Fig. 5. Automobile fault diagnosis main interface

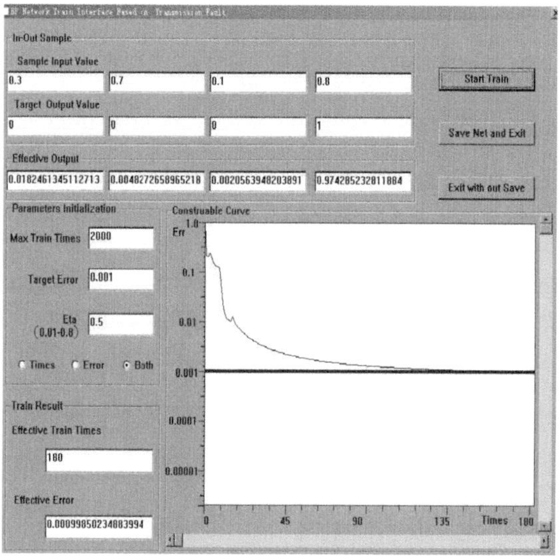

Fig. 6. BP train interface

7 Conclusions

This paper presents Automotive fault analysis system based on CAN bus .CAN-USB interface card, signal denoising of Kalman digital filter, BP neural network fault diagnosis are the three parts of the idea. The text sets forth BP network theoretics, analyses the structure of BP network, the training process and complete the software design basedon VC + +.

The results show that the CAN-USB intelligent interface card can receive the high-speed signals from sub- node of CAN bus, Kalman digital filter is suitable for automobile signal complexity and filtering effect is obvious, the use of BP neural network can carry on effective to diagnose and forecast the fault of the car effectively and accurately.

In addition, because of complex structures and car parts, it is inevitable that in the future, you can improve the software upgrade further to make it to have more comprehensive functionality.

References

1. Chen, Y., Hu, Y.: Failure diagnose research for engine of basic BP NN. CAT (1), 81–83 (2008)
2. GmbH, R.B.: BOSCH'S Control ler Area Network [EB/OL], `http://www.can.bosch.com`
3. Szabo, S., Oplustil, v.: Distributed CAN based control system for robotic and airbormapplications. J. IEEE on Control, Automation, Robotics and Vision 3(11), 1233–1238 (2007)
4. Haykin, S.: Adaptive Filter Theory, 4th edn., vol. 11, pp. 369–402. Publishing House of Electronics Industry (2006)
5. Zhang, X.D., Lu, M.: Research on neural network integration fusion method and application on the fault diagnosis of automotive engine. In: 2007 2nd IEEE IEA, pp. 480–483 (2007)
6. Wang, S.W., Yu, D.L.: Neural network in stable adaptive control law for automotive engines. In: Liu, D., Fei, S., Hou, Z.-G., Zhang, H., Sun, C. (eds.) ISNN 2007. LNCS, vol. 4491, pp. 122–131. Springer, Heidelberg (2007)
7. Yang, D.Z., Wang, J.: Estimation and control of hybrid electric vehicle using artificial neural networks. In: 2007 2nd IEEE IEA, pp. 35–40 (2007)
8. Chen, X.J., Gao, Z.F.: Applications of ANNs in geotechnical engineering. In: EI CEMI, Papers 3, pp. 656–659 (2007)

Heuristic Evolutionary Approach for Weighted Circles Layout

Zi-qiang Li, Hong-liang Zhang, Jin-hua Zheng, Meng-juan Dong, Yan-fang Xie, and Zhuo-jun Tian

School of Information and Engineering, Xiangtan University
Xiangtan, P.R. China 411105
xtulzq@hotmail.com

Abstract. In order to improve lack of GA, we combined it with case-based reasoning(CBR) and constructing non-isomorphic layout pattern (NILP) to solve layout problem in this paper. Its basic idea includes 2 components: (1) initial population of GA consists of reasoning solutions of current layout problem attained from case database, their NILP solutions and random solutions; (2) For every generation of optimization, a NILP solution of the current best individual is used in replacing the worse one in the population. The experimental results demonstrate good performance of approach in the paper on quality of solution and computational efficiency.

Keywords: Layout algorithm; Case-based reasoning; Non isomorphic layout pattern; Heuristic; Genetic algorithm.

1 Introduction

The layout problems such as the layout scheme design of printed circuit board and placement of machine tool equipments for factory[1] etc can be formulated as weighted circles layout problems which are NP-hard. Because the problems are very difficult to solve, they have been being studied for long time. So far, the solving algorithms proposed mainly include: heuristic algorithms, evolutionary algorithms (EAs), graph theory etc. EAs(for example GA) are a class of effective algorithms[2,3].

Because GA is easy to occur premature phenomenon and has slow convergence speed[4], its optimal solution often fails to engineering requirements for the high precision layout problem. What's more, GA doesn't possess human intelligence. So scholars probe out many strategies to improve the performance of GA. Combining case-based reasoning (CBR) with GA is one of them. For example, Sanja et al[5] used it to solve examination scheduling problem in 2007; Jin et al[6] combined CBR with GA to solve layout problem of spacecraft modules in 2007. Combining GA with human intelligence or other layout algorithm is another strategy. For example, Cho[7] proposed an interactive genetic algorithm for modeling design and image retrieval in 2002; Qian et al[8] proposed a human computer interactive genetic algorithm (HCIGA) and applied it to layout problems with performance constraints in 2001. Thus, both human and computer can exert their respective advantages to the utmost.

L. Qi (Ed.): ISIA 2010, CCIS 86, pp. 324–331, 2011.

This paper presents a heuristic evolutionary approach for the weighted circles layout, called HELA. It is combining GA with CBR and constructing non-isomorphic layout pattern (NILP). It has two combining points: (1) By CBR and constructing NILP[9][10][11], its initial population can involve larger solution space and include elite solutions; (2) combining evolutionary operation of GA with constructing NILP effectively avoid falling into the local optimal in the process of evolution.

This paper is organized as follows: CBR rules and constructing NILP are described in section 2 and section 3. HELA is proposed in section 4. A test example is given in section 5. The summary is showed in section 6. The final part is acknowledgment.

2 Layout Case Based Reasoning

CBR possesses many advantages such as fast reasoning, remembering, easy explanation and so on[12]. It is becoming a research focus in the artificial intelligence field and is widely applied [5][6][9]. In the viewpoint of application, CBR system can be classified into two kind types of explanation and resolution. Their difference is that the former is to take directly the solution of case satisfied the reasoning rules as that of the current problem by judging, but the latter is to reconstruct the solution of the problem by amending or modifying the solution of case. CBR in this paper is to take solutions of cases, with the same model as the current layout problem, as its reasoning solutions and to generate its reasoning solutions by modifying solutions of cases with the approximated model. It is hybrid of two kind of reasoning.

2.1 Case Storage and Case-Based Reasoning

2.1.1 Case Storage

All cases are stored in base according to model.[1].

2.1.2 Related Definition

To describe CBR in this paper, we give several related definitions at first.

Definition 1. Let D_C denote the distance between two containers of the current problem and case, and if their shape is the same then D_C=0; otherwise D_C=1.

Definition 2. Let N_P and N_C denote the number of circles of the current problem and that of case respectively, D_{Obj} denote the distance between two sets of circles, then D_{Obj} can be calculated by formula (1).

$$D_{Obj} = \frac{|N_P - N_C|}{\max(N_P, N_C)} + \frac{a_1 + a_2 + \ldots + a_p}{\max(N_P, N_C)} \tag{1}$$

Where, if the ith circle in problem can be found in case, then $a_i = 0$; else $a_i = 1$.

---

[1] The layout model of case: Area (double), weighted distance (double), weighted factor (double); The shape (char) and size (double) of container, The number (int) of circles. Size (double) of circles, layout scheme (solution) and weighted matrix is stored in the same text file, its name corresponds the number of case in case base.

Definition 3. Let λ and W denote the weighted factor and the matrix respectively (see formula (4) and (5)), let D_I denote the distance between two sets of the constraint indices. Then D_I can be calculated by formula (2).

$$D_I = \frac{2\eta_1}{L(L-1)} \sum_{i=1}^{L} \sum_{j=i+1}^{L} \frac{|a_{P(i,j)} - a_{C(i,j)}|}{\max(a_{P(i,j)}, a_{C(i,j)})} + \frac{\eta_2 |\lambda_P - \lambda_C|}{\max(\lambda_P, \lambda_C)} \tag{2}$$

Where η_1 and η_2 are the weight factors, $W_P=[a_{P(i,j)}]_{N\times N}$ and $W_C=[a_{C(i,j)}]_{M\times M}$ are the weight matrix of the problem and that of case respectively, L=max(M, N). If $M<N$, then extend W_P so that their orders are the same and the values of the increased elements are 0. If $M>N$, extend W_C so that their orders are the same and the values of the increased elements are 0.

Definition 4 Let D_{LC} denote the weighted distance between problem and case, then D_{LC} can be calculated by formula (3), and three weighted factors $W_C+W_{Obj}+W_I=1$.

$$D_{LC} = W_C D_C + W_{Obj} D_{Obj} + W_I D_I \tag{3}$$

2.1.3 Reasoning Rules

For each case, according to formula (3) we calculate D_{LC} between the case and the current problem respectively. Because $0 \le D_{LC} \le 1$ for all cases, we can select less threshold $\varepsilon(\varepsilon>0)$, and retrieval all cases satisfied $D_{LC} \le \varepsilon$, which consist of case subset ***C***. Reasoning rules we present are as follows.

(i) for a case in ***C***, if D_C=0 and D_{Obj}=0, then take its layout scheme as the reasoning solution of the current problem.

(ii) for a case in ***C***, if D_C=0 but $D_{Obj} \neq 0$ then construct the reasoning solution as following rule: After two sets of circles of case and the current problem are sort as radius from small to large, approximate circle mapping between them is done. If $N_P>N_C$, center coordinates of the circles do lack generate randomly.

(iii) If none of the 2 cases above is satisfied, after executing (ii), use GA in this paper to obtain layout schemes and take it as the reasoning solution.

3 Non-Isomorphic Layout Pattern (NILP)

For constructing NILP, The current main methods include mass-center exchanging [10], Quasi-boundary line approach (QBLA) [11] and so on. In this paper, on the one hand, we use QBLA to construct NILP solutions and take it as a part of the initial population individual; on the other hand, in process of iteration, we use mass-center exchange method to construct NILP for avoiding premature. This reason is that when trapping into local optimization, the circles of the current best individual are the dense state, overlapping amount of NILP solution constructed by exchanging centers of its two circle can't increase sharply compared with evolutionary operator of GA. The detail description of QBLA sees [11].

4 Heuristic Evolutionary Layout Approach

The optimal maintaining strategy make GA possesses the global convergence in theory, but when practically solving the layout problem it easily appears premature phenomenon and falls into the local optimum. So, domestic and foreign scholars do future research on it. Many results show that obtaining the layout global optimal solution by GA related closely to the initial population distribution and the population diversity in process of iteration and so on.

For the former the current main strategies are (i) human-computer cooperation [8], (ii) uniformly distributed strategy [13] and (iii) combining random with NILP [10] or CBR [6]. In this paper, the initial population constructed by combining CBR and NILP consists of reasoning solutions of current layout problem attained from case database, their NILP solutions and random solutions. Let M denote population size, the concrete method is as follows: firstly m reasoning solutions ($0<m<<M$) are got by CBR; for each reasoning, solution $[(M\text{-}m)/m_1]$ NILP solutions are constructed through QBLA, where $[(M\text{-}m)/m_1]$ is the integer part of $(M\text{-}m)/m_1$ and m_1 is an integer with $m< m_1<3m/2$; finally, $M\text{-}m\text{-}[(M\text{-}m)/m_1]\times m$ solutions are generated randomly.

For the later, the current main strategies are (i) hybrid method [14] and (ii) human-computer cooperation [8]. In this paper, a NILP solution of the current best individual is constructed by exchanging center of its two circles and used in replacing the worse one in the population in each of iterations. Where their numbers i and j are generated randomly ($1\leq i,j\leq n$ and $j\neq i$), n denotes the number of circles of layout problem.

After introducing the idea of HELA, its algorithm steps are described as follows:

Step 1. Definite and initiate variable M: population size, P_c and P_m: crossover and mutation probability, k iteration number counter: maximum iteration number, Δf: the fitness threshold, $\Delta f>0$ and it is determined by many time test

Step 2. According to definition 4, calculate D_{LC} for all cases retrieved. Store m cases satisfied $0\leq D_{LC}\leq\varepsilon$. If $m>0$, then go to Step3; otherwise go to Step4;

Step 3. According in section 2.1.3 do CBR and construct the initial population p(k);

Step 4. Randomly generate the initial population p(k);

Step 5. Calculate the fitness $f_{i,k}$ of individual $x_{i,k}$ for i=1,2,...M and the average fitness f_k ,save the optimal solution into $x_{opt,\,k}$. If $x_{opt,k}$ satisfies the requirements, then after output it and go to Step9; otherwise go to Step 6.

Step 6. Construct NILP of the best individual to replace one of worse fitness in $p(k)$.

Step 7. Execute evolutionary operations to generate population $p(k+1)$. So $k = k + 1$;

Step 8. Calculate the fitness $f_{i,k}$ for i=1,2,...M and the average fitness f_k and retain the optimal solution $x_{opt,\,k}$. If $x_{opt,\,k}$ satisfies the requirements, or $f_{k\text{-}1} - f_k > \Delta f$, or $k>k_{\max}$, the output $x_{opt,\,k}$ and store the problem model and $x_{opt,k}$ into case library and corresponding documents, go to Step9; Otherwise go to Step 6;

Step 9. Algorithm end.

5 Experiments

The problem is cited from [10], and its background is the printed circuit board design problem. Let d_{ij} and w_{ij} denote the weighed factor and distance between circle A_i and A_j respectively, Try to place 15 circles in a rectangular container with following conditions: (i) there is non-overlapping between circles and between circles and container; (ii) circles with larger weight, should gather each other to make

$C=\sum_{i=1}^{n-1}\sum_{j=i+1}^{n} d_{ij}w_{ij}$ get the minimum value; (iii) Area S of envelope rectangle is the smallest. Its mathematical model is showed in formula (4). Formula (5) is the weight matrix.

$$\begin{aligned}&\text{Find } X=(x_i,y_i)\ \ i=1,2,\ldots,n\\&\min F(X)=S+\lambda C\ (\lambda \text{ is the weight of } C \text{ relative to } S)\\&\text{s t. int } A_i\cap A_j=\Phi\ \ i,j=1,2,\ldots,n \text{ and } i\neq j\end{aligned} \tag{4}$$

$$W=\begin{bmatrix}0&0&0&98&98&0&81&0&92&93&45&61&99&84&27\\0&0&34&0&0&0&93&44&0&0&33&60&0&0&56\\0&34&0&0&0&0&0&0&0&85&0&65&39&0&50\\98&0&0&0&91&50&5&24&73&0&4&0&0&31&23\\98&0&0&91&0&37&0&16&78&95&0&0&73&32&0\\0&0&0&50&37&0&0&35&0&31&0&0&0&48&0\\81&93&0&5&0&0&0&94&33&34&26&61&0&87&87\\0&44&0&24&16&35&94&0&91&0&0&0&59&39&0\\92&0&0&73&78&0&33&91&0&0&30&0&0&0&0\\93&0&85&0&95&31&34&0&0&0&0&0&0&0&0\\45&33&0&4&0&0&26&0&30&0&0&0&21&35&2\\61&60&65&0&0&0&61&0&0&0&0&0&56&0&43\\99&0&39&0&73&0&0&59&0&0&21&56&0&1&0\\84&0&0&31&32&48&87&39&0&0&35&0&1&0&0\\27&56&50&23&0&0&87&0&0&0&2&43&0&0&0\end{bmatrix} \tag{5}$$

Solving: According to the layout model of this problem weighted factor:0.8; weight matrix: W; container type: rectangle; number of circles: 15. The matrix w, radius of circles and layout schema are stored in the text file. Using CBR in section 2,we obtain 6 reasoning solutions $X_1(S=5241\ mm^2, C=75301\ mm)$, $X_2(S=5700\ mm^2, C=91707\ mm)$, $X_3(S=5737\ mm^2, C=85123\ mm)$, $X_4(S=5155mm^2, C=77774mm)$, $X_5(S=5309mm^2, C=84143\ mm)$, $X_6(S=5193mm^2, C=76351mm)$. The layout schemes of the reasoning solutions $X_1 \sim X_6$ are shown in the figue1 (a)~(f).

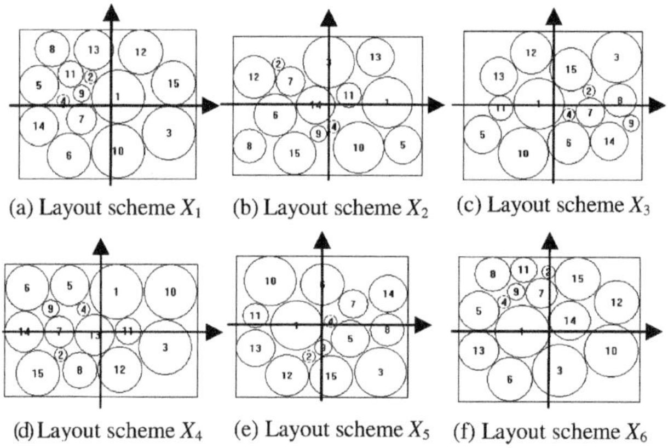

(a) Layout scheme X_1 (b) Layout scheme X_2 (c) Layout scheme X_3

(d) Layout scheme X_4 (e) Layout scheme X_5 (f) Layout scheme X_6

Fig. 1. Layout schemes of six reasoning solutions for example

Parameters of genetic algorithm are as follows: Selection operator in GA is roulette wheel method, two point cross and basic bit mutation. Probability of crossover and mutation is 0.99 and 0.30 respectively, the size of population is 60 and the maximal number of iterations is 10000. By exchanging circle center, we construct the 30 NILP solutions of 6 reasoning solutions. The initial population of HELA consists of 30 NILP solutions, 6 reasoning solutions and 24 random initial solutions. The initial populations(size 60) of HCICA and GA are generated randomly. We use penalty-function method to run HCIGA and HELA to solve the problem 50 times respectively. The optimal layout schemes of PHAIA [10], HCIGA and HELA are shown in Figure 2(a)~ (c), which three sets of circle center coordinates are shown in table 1. Table 2 shows the performance comparison of the optimal solution of them and GA.

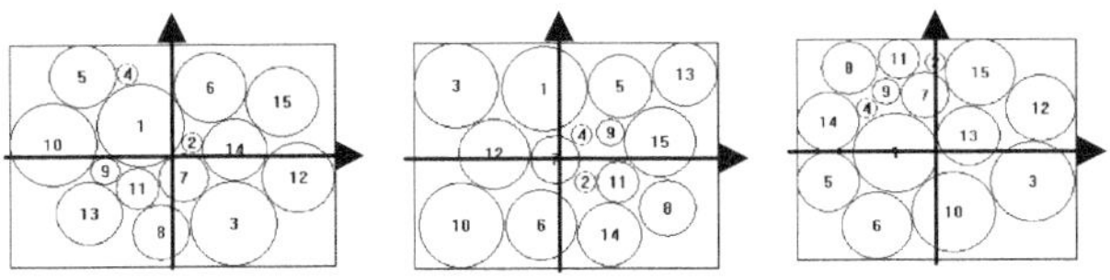

(a) Layout of PHAIA (b) Layout of HCIGA (c) Layout of HELA

Fig. 2. Corresponding results of layout optimization of PHAIA, HCIGA and HELA for example

Table 1. Circle's radius and optimized circle center coordinates of example

Objects num	r_i /mm	HELA x_i /mm	HELA y_i /mm	PHAIA x_i /mm	PHAIA y_i /mm	HCIGA x_i /mm	HCIGA y_i /mm
1	12	-11.49	-1.02	-8. 80	8.49	-5.82	8.45
2	3	-0.10	25.75	5.23	3.38	5.44	-7.89
3	12	27.08	-9.84	16.61	-19.63	-30.27	19.60
4	3	-19.36	12.30	-12.37	23.06	4.67	5.55
5	9	-30.59	-9.73	-24.36	22.63	15.12	19.29
6	10	-16.81	-22.67	10.39	19.18	-6.65	-19.57
7	7	-3.16	16.05	3.01	-6.36	-2.63	-1.45
8	8	-24.43	24.06	-3.26	-21.94	28.29	-15.07
9	4	-14.11	17.06	-18.24	-4.45	12.51	6.23
10	12	4.78	-18.71	-32.29	3.19	-28.90	-19.60
11	6	-10.63	26.67	-9.61	-9.48	14.74	-7.63
12	10	29.53	12.27	34.29	-6.54	-19.59	0.37
13	9	9.03	3.99	-22.79	-16.62	33.27	22.46
14	9	-30.62	8.23	17.02	1.370	12.15	-22.38
15	10	12.54	22.68	30.00	15.24	26.43	3.76

Table 2. Optimized performance indexes of example

method	S/mm^2	C/mm	$\triangle$S/mm^2	t/s
GA	5718	79967	0	789
HCIGA	5246	81023	0	789(Human200)
PHAIA[10]	5603	77841	0	213.9
HELA	5173	69497	0	90.0

GA, HCICA and Our HELA are run on PC with 933M CPU/512M memories. PHAIA, which layout scheme and performance indexes are from [10], is run on computer with 566M CPU/128M memories. From Table 1 and Table 2, the solution quality and computation efficient of our HELA are superior to that of HCIGA and PHAIA. Although the machine ran PHAIA is inferior to one run HELA, A big gap between 90.0s and 213.9s in table 2 shows our HELA is better than PHAIA in performance. In addition, the three methods are superior to GA.

6 Conclusion

This paper suggests heuristic evolutionary layout algorithm by combining CBR and construction of NILP with GA. In the initialization stage, reasoning solutions obtained through CBR and their NILP solutions compose a part of initial population of GA, the rest are randomly generated, so that its search space is as large as possible. At iteration of the evolutionary stage, each time constructing NILP solutions of optimal individual and replacing population individual with worse fitness, it could updates the current optimal solution with maximum probability to avoid the local optimal. In this way, we improve the quality of the solution and efficiency for every generation. Because of no manual intervention, the solving automation degree is further increased. Numerical experiments show that this algorithm HELA has high accuracy and efficiency, compared with HCIGA, parallel hybrid ant immune algorithm (PHAIA) and GA.

Acknowledgments. This work is supported by the Doctor Startup Foundation of Xiangtan University of China (Grant No. 09QDZ18) and the National Natural Science Foundation of China (Grant No.60773047) and 2th Xiangtan university and 3th Hunan province innovation experiment program for university students.

References

1. Holland, J.H.: Adaptation in nature and artificial systems. MIT Press, Cambridge (1992)
2. Pierre, M.G., Georges, M.F.: A GA based configuration design optimization Method. Journal of Mechanical Design 126(1), 6–14 (2004)
3. Yin, J., Li, M.: An optimization layout solution based genetic algorithm. Chinese Journal of Computer Research and Development 39(10), 1273–1296 (2002)
4. Rudolph, G.: Convergence analysis of canonical genetic algorithms. IEEE Transactions on Neural Networks 5(1), 96–100 (1994)
5. Sanja, P., Yong, Y., Moshe, D.: Case-based selection of initialization heuristics for metaheuristic examination timetabling. Expert Systems with Applications 33(2), 772–785 (2007)
6. Jin, B., Teng, H.F.: Case-based evolutionary design approach for satellite module layout. Journal of Scientific and Industrial Research 66(12), 989–994 (2007)
7. Cho, S.B.: Towards creative evolutionary systems with interactive genetic algorithm. Application Intelligent 16(2), 129–138 (2002)

8. Qian, Z.Q., Teng, H.F., Sun, Z.G.: Human-computer interactive genetic algorithm and its application to constrained layout optimization. Chinese Journal of Computers 24(5), 553–559 (2001)
9. Park, Y.-J., Kim, B.-C.: An interactive case-based reasoning method considering proximity from the cut-off point. Expert Systems with Application 33(4), 903–915 (2007)
10. Li, G.Q.: Research on the Theory and Methods of Layout Design and Their Applications [Doctoral Dissertation]. Dalian University of Technology, Dalian (2003)
11. Teng, H.F., Li, Z.Q., Shi, Y.J., et al.: An approach to constructing isomorphic or non-isomorphic layout pattern. Chinese Journal of Computers 29(6), 987–991 (2006)
12. Earltta, B.: An introduction to Case Based Reasoning. AI Expert (8), 43–49 (1991)
13. Wang, Y.S., Shi, Y.J., Teng, H.F.: An improved scatter search for circles pack-ing problem with the equilibrium constraint. Chinese Journal of Computers 32(6), 1214–1221 (2009)
14. Young, H.L., Moon, H.L.: A shape black layout approach to facility layout problems using hybrid genetic algorithm. Computing & Industrial Engineering 42(2-4), 237–248 (2002)

Robust H∞ Reliable Guaranteed Cost Control for Delta Operator Uncertain Systems

Lili Guan, Shan Meng, and Duanjin Zhang

School of Information Engineering
Zhengzhou University
Zhengzhou, 450001, China
djzhang@zzu.edu.cn, duanjin.zhang@uni-due.de

Abstract. This paper is concerned with the problem of guaranteed cost robust H-infinity reliable control for delta operator formulated uncertain systems with actuator failures. Based on the pattern of actuator continuous failures, a sufficient condition of the existence of guaranteed cost controller of delta operator system with H-infinity performance constraints is derived in terms of linear matrix inequality (LMI) approach. When this LMI is feasible, the parameterized representation of the desired state feedback controller is also obtained.

Keywords: Delta operator system; Guaranteed cost control; LMI; Actuator failures; H-infinity Control.

1 Introduction

To improve the reliability and safety, the design method of fault tolerant control has played an important role in the modern control systems. In recent years, the research on fault tolerant control has attracted much attention, and a lot of results have been reported. Considering the continuous actuator failures and sensor failures, the reliable control has become one of the active fields of fault tolerant control. Furthermore, as performance constraints of control systems, the guaranteed cost control design was firstly introduced in 1972. Since then, many results on robust reliable guaranteed cost control have been obtained [1-4]. In [5], an optimal guaranteed cost state feedback controller was given for linear uncertain systems with actuator failure. In fact, it is desirable to design a control system which not only possesses guaranteed cost constraints, but also has some level of performance such as H∞ performance. This motivates the multi-objective reliable control design for fault tolerant control systems. For example, robust H∞ guaranteed cost control for network systems has been studied in [6].

On the other hand, delta operator method proposed by Middleton and Goodwin [7] has been attracting much attention. By using the delta operator, ill-conditioned problems can be avoided when fast sampling, but some control problems of continuous-time and discrete-time systems can be investigated in the unified form. For instance, [8] examined the non-fragile H∞ filter design with pole placement

L. Qi (Ed.): ISIA 2010, CCIS 86, pp. 332–339, 2011.

constraints for delta operator systems via LMI optimization. Ref. [9] studied the problem of robust D-stable fault tolerant control for the delta operator time-delay systems, but did not consider the H∞ performance and the guaranteed cost constraints. As for the delta operator guaranteed cost control, a few results have been reported in the literature. In [10] the problem of guaranteed cost control for delta operator uncertain system was addressed. Xiao [11] studied the problem of reliable guaranteed cost control for delta operator uncertain systems with sensor failure.

This paper deals with the problem of robust H∞ guaranteed cost control for delta operator uncertain systems with actuator failures. The purpose of the paper is to design a state feedback controller such that the closed-loop systems can simultaneously satisfy the guaranteed cost and H∞ performance constraints. A sufficient condition for guaranteed cost control with H∞ norm bound is obtained, and a suitable state feedback control law is also given in terms of the solution of a certain LMI.

2 Problem Formulation

Consider the delta operator formulated linear uncertain system

$$\begin{cases} \delta x(t) = (A+\Delta A)x(t) + (B+\Delta B)u^f(t) + G\omega(t) \\ z(t) = Cx(t) \end{cases} \tag{1}$$

where $\delta = (q-1)/T$ is called the delta operator , T is the sampling interval, q is the shift operator, $x(t) \in R^n$ is the state vector, $u^f(t) \in R^p$ is the control input vector, $z(t) \in R^k$ is the controlled output vector, $\omega(t) \in R^q$ is the external disturbance vector, $\|\omega(t)\|_2 \le \beta$, A , B , C , G are known real constant matrices with appropriate dimensions, ΔA and ΔB are the uncertainties in the systems and defined as follows

$$[\Delta A \quad \Delta B] = DF[E_1 \quad E_2] \tag{2}$$

where D , E_1 and E_2 are known constant real matrices with appropriate dimensions, which represent the structure of uncertainties, and F is unknown matrices with Lebesgue measurable elements and satisfies

$$F^T F \le I \tag{3}$$

In which I denotes the identity matrix of appropriate dimension.

The state feedback control law is as follows

$$u(t) = Kx(t) \tag{4}$$

The failure is adopted as

$$u^f(t) = MKx(t) \tag{5}$$

where $M = diag(m_1 \ldots m_i \ldots m_p)$ is the actuator failure model, and satisfies $0 \le m_{il} \le m_i \le m_{iu}$. When $0 \le m_{il} < m_i < m_{iu}$ and $m_i \ne 1$, it corresponds to the partial failure.

Denote

$$M_0 = diag\ (m_{01}, m_{02}, \ldots, m_{0p}),\ J = diag\ (j_1, j_2, \ldots, j_p),$$

$$L = diag\ (l_1, l_2, \ldots, l_p), \quad |L| = diag\ (|l_1|, |l_2|, \ldots, |l_p|). \tag{6}$$

with

$$m_{0i} = \frac{1}{2}(m_{li} + m_{ui}), \quad j_i = \frac{m_{ui} - m_{li}}{m_{ui} + m_{li}}, \quad l_i = \frac{m_i - m_{0i}}{m_{0i}} \ (i = 1,2,\ldots, p).$$

Then we can have

$$M = M_0(I + L), \quad |L| \le J \le I \tag{7}$$

The quadratic cost function associated with system (1) is

$$J = S_{t=0}^{\infty}[x^T(t)Qx(t) + (u^f(t))^T Ru^f(t)]dt \tag{8}$$

where S_t^{∞} denotes the Riemann integral in a continuous time and the Riemann sum in the discrete time. $Q > 0$ and $R > 0$ are symmetric positive-definite matrices.

Now the delta operator closed-loop system can be written as

$$\begin{cases} \delta x(t) = ((A + BMK + DF(E_1 + E_2MK))x(t) + G\omega(t) \\ z(t) = Cx(t) \end{cases} \tag{9}$$

The purpose of the present work is to design a state feedback $u(t) = Kx(t)$ such that for the following performance requirements are simultaneously satisfied:

(i)The closed-loop systems (9) is asymptotically stable, and the value of the cost function (8) satisfies $J < J^*$, where J^* is said to be a guaranteed cost upper bound.

(ii)The H∞ norm of the transfer function $T_{wz}(\gamma)$ from disturbances $w(t)$ to control output $z(t)$ satisfies the constraint $\|T_{wz}(\gamma)\|_{\infty} = \|(C(\gamma I - A)^{-1}G\| < \gamma_1$, where $\gamma_1 > 0$ is a given positive constant. Here $\gamma = (z-1)/T$ is the transform variable for the delta operator, where z refers to the transform variable for the shift operator.

Definition 1 : For delta operator uncertain system (1), if there exists a matrix P such that $A_b^T P + PA_b + TA_b^T PA_b < 0$, then the closed -loop system (9) is quadratic stable, where $A_b = A + BMK + DF(E_1 + E_2MK)$.

Definition 2: If there exists a symmetric positive definite matrix $P \in R^{n \times n}$, for all admissible uncertainties and M, such that

$$A_b^T P + PA_b + TA_b^T PA_b + Q + C^T C + K^T MRMK + (I + TA_b)^T PG(\gamma_1^2 I - TG^T PG)^{-1} G^T P(I + TA_b) < 0 \tag{10a}$$

$$\gamma_1^{-2} I - TG^T PG > 0 \tag{10b}$$

then $u(t) = Kx(t)$ is said to be the robust H∞ reliable guaranteed cost control law.

Remark 1. Definition 2 can deal with the problem of H∞ reliable guaranteed cost control for continuous time systems and discrete time systems in delta unified form.

Let $T = 0$, the condition (10) is suitable for continuous systems. And when $T = 1$, $A = TA + I$, the condition (10) is suitable for discrete systems.

Here we give some lemmas which are very useful in the poof of theorems.

Lemma 1(Schur complement): Given symmetrical matrices,

$$S = \begin{bmatrix} S_{11} & S_{12} \\ S_{21} & S_{22} \end{bmatrix} < 0 \text{ if and only if}$$

1) $S_{11} < 0, S_{22} - S_{12}^T S_{11}^{-1} S_{12} < 0$

2) $S_{22} < 0, S_{11} - S_{12} S_{22}^{-1} S_{12}^T < 0$

Lemma 2[12]: Given matrices D , E and Y with appropriate dimensions and with Y symmetrical, then

$Y + DFE + E^T F^T D^T < 0$

For all satisfying $F^T F \le I$, if and only if there exists scalar $\varepsilon > 0$ such that

$Y + \varepsilon DD^T + \varepsilon^{-1} E^T E < 0$

Lemma 3[13]: Let $T_{wz}(\gamma)$ be a matrix transfer function with $[C \quad A_b]$ detectable realization of the following delta operator system

$$\begin{cases} \delta x(t) = A_b x(t) + G\omega(t) \\ z(t) = Cx(t) \end{cases}$$

If there exist a positive semi-definite matrix $P \ge 0$ and a real constant $\gamma_1 > 0$ such that

(a) $A_b^T P + PA_b + TA_b^T PA_b + (I + TA_b)^T PG(\gamma_1^2 I - TG^T PG)^{-1} G^T P(I + TA_b) + C^T C < 0$

(b) $\gamma_1^2 I - TG^T PG > 0$

Then A_b is stable, and $\|T_{wz}(\gamma)\|_\infty = \|C(\gamma I - A)^{-1} G\| < \gamma_1$.

Lemma 4 : For delta operator uncertain systems (1) and cost function (8) and the external disturbance $\omega(t)$, If there exist a positive definite matrix P and a matrix K, for all admissible uncertainties and all possible M, such that

$$\begin{bmatrix} A_b^T P + PA_b^T + TA_b{}^T PA_b{}^T + Q + C^T C & G^T P(I + TA_b) & (MK)^T \\ (I + TA_b)^T PG & -\gamma_1{}^2 I + TG^T PG & 0 \\ MK & 0 & -R^{-1} \end{bmatrix} < 0 \tag{11}$$

Then the closed-loop system is asymptotically stable, the cost function (8) satisfies

$J \le x_0{}^T P x_0 + \gamma_1{}^2 \beta^2$

and $\|T_{wz}(\gamma)\|_\infty < \gamma_1$,where x_0 is the initial state.

Proof: Construct a Lyapunov function as $V(x) = x^T Px$, taking the time derivative of $V(x)$, by using the Lemma 3 and Definition 2, we have

$$\begin{aligned} \dot V(x) &\le x^T (A_b{}^T P + PA_b + TA_b{}^T PA_b + C^T C + (I + TA_b)^T PG(\gamma_1{}^2 I - TG^T PG)^{-1} G^T P(I + TA_b)x + \gamma_1{}^2 \omega^T \omega \\ &\le -x^T (Q + K^T MRMK)x + \gamma_1{}^2 \omega^T \omega \end{aligned} \tag{12}$$

When $\omega(t)=0$, then $\hat{V}(x)<0$ the closed-loop system is asymptotically stable.
If $\omega(t)\neq 0, \lim_{t\to\infty}(t)=0$, as $t\to\infty$, then $V(x_0)=0$, $V(x_\infty)=0$.
We obtain
$J_{zw}=\int_0^\infty [z^T(t)z(t)-\gamma^2\omega^T(t)\omega(t)]dt<0$, then $\|z(t)\|_2<\gamma\|\omega(t)\|_2$
Considering the upper bound of cost function (8), we have

$$J=S_{t=0}^{\infty}[x^T(Q+K^TMRMK)x]dt<V(0)+\gamma_1^2\beta^2=x_0{}^TPx_0+\gamma_1^2\beta^2=J^* \tag{13}$$

This completes the proof of Lemma 4.

3 Main Results

Theorem 1: For the delta operator closed-loop system (9) and the cost function (8), given a level of noise attenuation $\gamma_1>0$, if exist a symmetric positive definite matrix P and a scalar $\varepsilon_1>0$, for all admissible uncertainties and all possible M, such that the following inequality holds

$$\begin{bmatrix} -TP^{-1}+\gamma_1^{-2}T^2GG^T+\varepsilon_1D_1D_1^T & I+TA+TBMK & 0 & 0 & 0 \\ (I+TA+TBMK)^T & -T^{-1}P+CC^T & (E_1+E_2MK)^T & I & (MK)^T \\ 0 & (E_1+E_2MK) & -\varepsilon_1 I & 0 & 0 \\ 0 & I & 0 & -Q^{-1} & 0 \\ 0 & (MK) & 0 & 0 & -R^{-1} \end{bmatrix}<0 \tag{14}$$

$$\gamma_1^{-2}TGG^T-P^{-1}<0 \tag{15}$$

Then there exist the control gain matrix K such that the closed-loop system is asymptotically stable, and the guaranteed cost function has upper bound, and satisfies $\|T_{wz}(\gamma)\|_\infty\leq\gamma_1$.

Proof: Let us introduce the following signs

$$P_1=P/T, D_1=TD.$$

Then denote

$$V=\begin{bmatrix} -P_1^{-1}+\gamma_1^{-2}T^2GG^T & [I+TA_b] \\ [I+TA_b]^T & -P_1+C^TC+Q+K^TMRMK \end{bmatrix}<0 \tag{16}$$

where $\gamma_1^{-2}I-TG^TPG>0$ is equivalent to (15)

By using Lemma 1 and Lemma 2, we can complete the proof of the theorem.

Theorem 2: For the delta operator formulated linear uncertain system (1) and the cost function (8), given H∞ performance index $\gamma_1>0$, if exist constants $\varepsilon_1>0, \varepsilon_2>0$, a symmetric positive definite matrix X and a matrix W, such that for all admissible uncertainties and all possible M the following LMI holds

$$\begin{bmatrix} -TX+\gamma_1^{-2}T^2GG^T+\varepsilon_1 D_1 D_1^T+\varepsilon_2 TBM_0J(TBM_0)^T & (I+TA)X+TBM_0W & * & * & * & * & * \\ X(I+TA)^T+(TBM_0W)^T & -T^{-1}X & * & * & * & * & * \\ \varepsilon_2 E_2M_0J(TBM_0)^T & E_1X+E_2M_0W & -\varepsilon_1 I+\varepsilon_2 E_2M_0J(E_2M_0)^T & * & * & * & * \\ 0 & X & 0 & -Q^{-1} & * & * & * \\ \varepsilon_2 M_0J(TBM_0)^T & M_0W & \varepsilon_2 M_0J(E_2M_0)^T & 0 & -R^{-1}+\varepsilon_2 M_0JM_0^T & * & * \\ 0 & \sqrt{J}W & 0 & 0 & 0 & -\varepsilon_2 I & * \\ 0 & CX & 0 & 0 & 0 & 0 & -I \end{bmatrix}<0 \quad (17)$$

$$\gamma_1^{-2}TGG^T-X<0 \quad (18)$$

where * denotes the symmetric block in symmetric matrices, then the controller $u(t)=WX^{-1}x(t)$ is said to be a reliable H∞ guaranteed cost controller which satisfies the two inequalities (17) and (18) simultaneously and the quadratic cost function satisfies $J \le x_0^T Px_0+\gamma_1^2\beta^2$.

Proof: Let $M=M_0(I+L)$ replace M in (14), then we obtain

$$Y_2=\begin{bmatrix} -TP^{-1}+\gamma_1^{-2}T^2GG^T+\varepsilon_1 D_1D_1^T & I+TA+TBM_0K & 0 & 0 & 0 \\ (I+TA+TBM_0K)^T & -T^{-1}P+CC^T & (E_1+E_2M_0K)^T & I & (M_0K)^T \\ 0 & (E_1+E_2M_0K) & -\varepsilon_1 I & 0 & 0 \\ 0 & I & 0 & -Q^{-1} & 0 \\ 0 & (M_0K) & 0 & 0 & -R^{-1} \end{bmatrix}$$

Consider $J>0$, according to Lemma 1 and Lemma 2, we complete the proof of Theorem 2. Here $X=P^{-1}$, $W=KX$.

Theorem 3: For delta operator system (1) and cost function (8), if the following optimization problem has a feasible solution, then the following optimization problem is meaningful

$$\min(\gamma_1):(X, W, \gamma_1, \varepsilon_1, \varepsilon_2) \quad (19)$$
$$\text{s. t.1) } (17) \sim (18)$$

Proof: If $(X_0, W_0, \gamma_0, \varepsilon_{10}, \varepsilon_{20})$ is a feasible solution of the LMI (19), if given $\gamma_1>\gamma_0$, the solution γ_1 of the above optimization problem is the minimum value of reliable H∞ guaranteed cost, then the LMI (19) is feasible.

Theorem 4: Consider delta operator system (1) and cost function (8), if the following optimization problem

$$\begin{aligned} &\min \operatorname{Trace}(S_1) + \gamma_1^2 \beta^2 \\ &\text{s. t.} 1)\ (17) \sim (18) \\ &\quad 2) \begin{bmatrix} S_1 & I \\ I & X \end{bmatrix} > 0 \end{aligned} \tag{20}$$

has a solution $(\tilde{\varepsilon}_2, \tilde{\varepsilon}_1, \tilde{W}, \tilde{X}, \tilde{S}_1)$, then $u(t) = \tilde{W}\tilde{X}^{-1}x(t)$ is the optimal reliable robust H∞ guaranteed cost controller, which ensures the minimization of the upper bound for the closed-loop quadratic cost function.

Proof: From Theorem 2, we obtain the result of the theorem. If $(\tilde{\varepsilon}_2, \tilde{\varepsilon}_1, \tilde{W}, \tilde{X}, \tilde{S}_1)$ is a feasible solution of the LMI (20), then $(\tilde{\varepsilon}_2, \tilde{\varepsilon}_1, \tilde{W}, \tilde{X}, \tilde{S}_1)$ is also a feasible solution of the constraint condition 1) in (20), then $u(t) = \tilde{W}\tilde{X}^{-1}x(t)$ is the optimal reliable robust H∞ guaranteed cost controller.

By using the Lemma 1, the constraint condition 2) in (20) is equivalent to

$$S_1 > X^{-1} > 0,\ \operatorname{Trace}(S_1) > \operatorname{Trace}(X^{-1}).$$

Then the minimal of $\operatorname{Trace}(S_1)$ can ensure the minimal of $\operatorname{Trace}(X^{-1})$, so the $J \le x_0^T P x_0 + \gamma_1^2 \beta^2$ is the minimal upper bound of the cost function. The problem (20) is a convex optimization process, thus reaching a global optimum. Therefore the problem with LMI constraints can be solved in terms of the MATLAB-LMI toolbox.

4 Conclusion

Based on LMI method, this paper presents the design of robust H∞ guaranteed cost control for delta operator uncertain systems with actuator failure. Considering the model of actuator continuous failure, we have constructed the consistency indices of H∞ performance constraint and the quadratic stable guaranteed cost index. The sufficient condition for robust H∞ guaranteed cost reliable control is obtained, and a suitable state feedback controller is also given.

Acknowledgments. This work was supported by the Natural Science Foundation of Educational Department of Henan of China under Grant 2010A510017.

References

[1] Zhang, Y.L., Tang, Z.Y., Zhou, H.B.: Robust guaranteed cost control for jump stochastic systems subject to actuator saturation. In: 2010 International Conference on Measuring Technology and Mechatronics Automation, Changsha, China, pp. 270–273 (2010)

[2] Gong, Y.M., Wo, S.L., Yan, P.: Guaranteed cost H (control for stochastic systems. In: 2009 Chinese Control and Decision Conference, Guilin, China, pp. 4117–4121 (2009)

[3] Wang, L., Shi, H.B.: An LMI approach to robust guaranteed cost fault-tolerant control for a class of uncertain parameter systems. In: 7th World Congress on Intelligent Control and Automation, Chongqing, China, pp. 955–959 (2008)

[4] Han, X.D., Xie, D.X., Ge, L., Wang, Z.Q.: Robust H∞ guaranteed cost D-stable satisfactory fault-tolerant control against actuator failures. In: 2008 Chinese Control and Decision Conference, Yantai, China, pp. 2696–2699 (2008)
[5] He, G.N., Liu, Z.D.: Guaranteed cost control for uncertain linear systems under actuator failures. J. Techniques of Automation & Applications 24, 18–21 (2005)
[6] Lun, S.X., Wang, S.Q.: Robust H∞ guaranteed cost control for networked control systems based on T-S model. In: 2010 International Conference on Networking Sensing and Control, Chicago, pp. 447–451 (2010)
[7] Goodwin, G.C., Leal, R.L., Middleton, R.H.: Rapprochement between continuous and discrete model reference adaptive control. Automatica 22, 199–207 (1986)
[8] Guo, X.G., Yang, G.H.: Non-fragile H∞ filter design with pole placement constraints for delta operator formulated systems via LMI optimization. In: 2009 American Control Conference, St. Louis, MO, pp. 3188–3193 (2009)
[9] Zhang, D.J., Song, Q.H., Zhang, L.H., Ding, S.X.: Robust D-stable fault tolerant control for delta operator formulated uncertain systems with state delays. In: 2009 ISECS Int. Colloquium on Computing, Communication, Control and Management, Sanya, China, pp. 505–508 (2009)
[10] Hu, G., Ren, J.C., Xie, X.S.: Optimal guaranteed cost control for delta operator uncertain linear systems. J. Electric Machines and Control 7, 139–142 (2003)
[11] Xiao, M.Q.: Reliable robust guaranteed cost control of delta operator linear uncertain systems with sensor failure. In: 2009 Second International Conference on Intelligent Computation Technology and Automation, Changsha, China, pp. 834–837 (2009)
[12] Xie, L.: Output feedback H∞ control of systems with parameter uncertainty. International Journal of Control 63, 741–750 (1996)
[13] Shor, M.H., Perkins, W.R.: Reliable control in the presence of sensor / actuator failures: a unified discrete / continuous approach. In: 30th IEEE Conference on Decision and Control, Brighton, pp. 1601–1606 (1991)

The Operational Efficiency Evaluation of China's Mobile Payment Enterprises

Xiao-liang Zhao, Bin Qiao, and Bao-zhi Zhang

School of Management and Economics, Taiyuan University of Science and Technology, 030024, P.R. China

Abstract. Firstly, this paper analyzed the factors which affect our mobile payment by analytic hierarchy process and entropy value method. Secondly, we use the fuzzy comprehensive evaluation method to evaluate operating performance of the three companies in mobile payment. From the results of the evaluation, we know that the consumer level and economic level are the most important two factors. The operational performance of relevant enterprises in China is between middle and good. Mobile payment is still has much developing space in China.

Keywords: mobile payment, analytic hierarchy process, entropy value method, fuzzy comprehensive evaluation.

1 Introduction

In general, mobile payment is to allow mobile users to use their mobile devices to pay for the goods or services. Mobile payment enterprises combined with the characteristics of a variety of industries are beneficial to themselves. This attracts more attention of most mobile operators.

Mobile payment has much developing space in China. First, China has the world's largest mobile phone user market, which made it has natural advantages compared with other countries. Second, China's current mobile payment market is from warm-up period into the early start. Compared to the global development pace of mobile payment services, the Warm-up period of China's mobile payment market is relatively late for a few years.

With the development of mobile payment industry, domestic and foreign scholars have begun to move into the area and kept up with the same pace. In recent years the research achievement is increasing step by step. Domestic scholar's research mainly focuses on the industrial chain of mobile payment and the business models of mobile payment. Mainly as follows, description of existing means of payment, the opportunities and advantages of mobile payment; describes the development status of mobile payments abroad; introduced the Payment environment in China and the relationship between the industrial chains; list the possible business model of mobile payment, compare the advantages and disadvantages of various business models, and suggest effective measures.

By studying the existing literature, we found that domestic scholars have not yet analyze the operating performance of China's mobile payment industry, and the

L. Qi (Ed.): ISIA 2010, CCIS 86, pp. 340–346, 2011.

existing studies are mainly qualitative analysis which is short of quantitative analysis. So this paper adopted the method of quantitative analysis to evaluate the operational performance of mobile payment companies in China.

2 Building the Index System of Performance Evaluation Index System

The accuracy and effectiveness of performance evaluation are developing a reasonable evaluation index system. Because mobile payment has many influencing factors and complex structures, so only from various angles and dimensions to design index system can we make a reasonable assessment of mobile payments. Therefore, the method to establish evaluation index system and principles are as follows:

(1)Select a key role index set in the more complete set of selected indicators.
(2)The select set should cover all aspects of assessment process.
(3)Pertinence and comparability principle.
(4)Combined with qualitative indicators and quantitative indicators.

According to these principles, the relevant index sets of mobile payment are as follows:

Table 1. Assessment index system

consumer level (customer loyalty)	Brand value
	Culture of Social radiation
	Business and service portfolio
economic level	State financial benefits (ROE)
	Assets operating conditions (Total assets turnover)
	Solvency position (Asset-liability ratio)
	Development capacity (Growth rate of total assets)
technical level	Transaction security
	Transaction convenience
	Technological Innovation
social level	the privacy of trading participants
	Consuming Habits
legal and policy level	The perfection of supporting laws
	Government's support

3 Using AHP Method to Determine the Subjective Weights

The analytic hierarchy process was first formally extracted by the U.S. operations research Thomas • sati in the mid-20th century. AHP is a method which has characteristics of qualitative, quantitative, systematic and hierarchical analysis. Because it has advantages in complex decision-making in dealing with the issue of practicality and effectiveness, it is widely used in the worldwide. The characteristics of AHP are the use of less quantitative information to make decisions thinking mathematical, so as to provide an easy method of decision-making for multi-objective, multi-criteria or no structural characteristics of complex issues. The analytic hierarchy process is especially suitable for indirect and inaccurate measurement situations.

3.1 Hierarchy Model Established

According to the index in table 1, we establish the hierarchy model as follows:

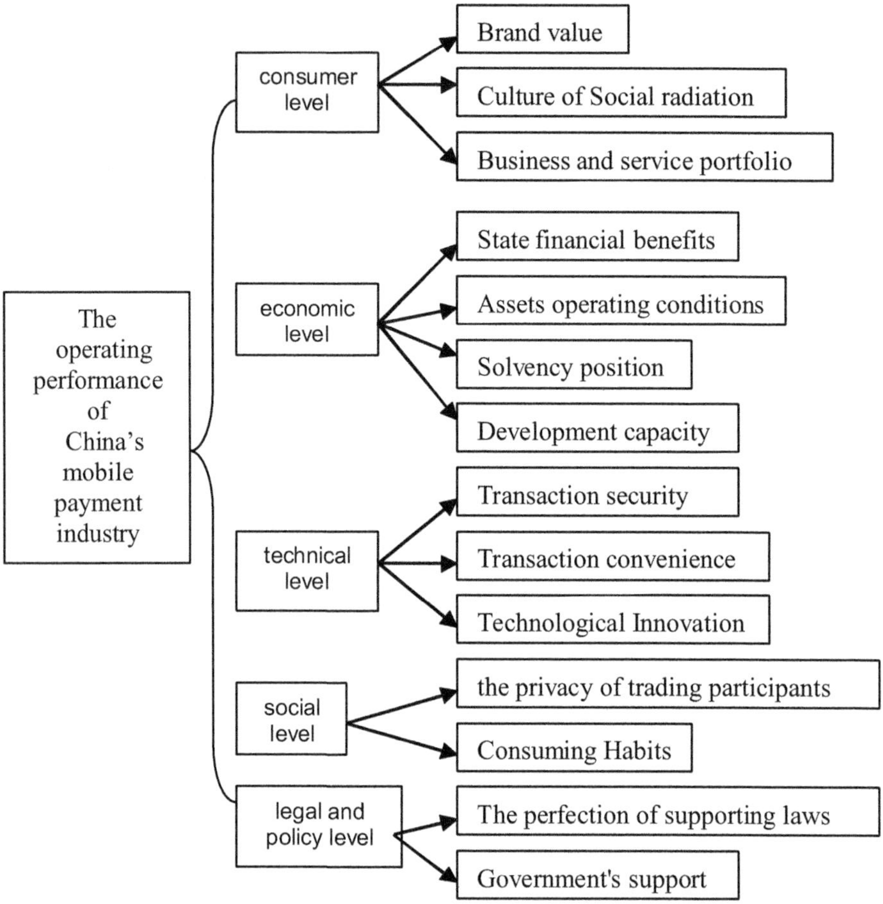

3.2 By Using the Software of Yaaph5.1 We Can Calculate the Index Weights of Various Secondary Indicators. The Results Are as Follows:

Table 2. The subjective target weight

evaluation index		weights
consumer level (0.3078)	Brand value	0. 1744
	Culture of Social radiation	0. 0795
	Business and service portfolio	0. 0539
economic level (0.3061)	State financial benefits	0. 1162
	Assets operating conditions	0. 0651
	Solvency position	0. 0598
	Development capacity	0. 0650
technical level (0.1078)	Transaction security	0. 0610
	Transaction convenience	0. 0230
	Technological Innovation	0. 0238
social level (0.1981)	privacy of trading participants	0. 1580
	Consuming Habits	0. 0401
legal and policy level (0.0802)	perfection of supporting laws	0. 0511
	Government's support	0. 0291

4 Using Entropy Method to Determine the Objective Weights

From the principle of AHP we can know that the weights calculated by AHP are mainly obtained based on practical experience of experts. The weights are subjective and don't taking into account the target data itself on the impact of weight. According to information theory, entropy can be used to measure the size of the amount of information. The concept of entropy first been proposed by Clausius in 1864 and soon applied in thermodynamics. Later in 1948 it was introduced to information theory. Now it has widely used in the engineering, socio-economic and other fields. Entropy is a measurement of the uncertainty of a system state, which quantitative describes the changes of the economic system complexity, organizational state of uncertainty and degree of order. Entropy method is constituted by the evaluation matrix to determine a method of index weight on objective conditions. It shows that the data more scattered distribution, the greater the uncertainty. It can weight as much as possible to eliminate the subjective factors so as to make the results of evaluation more realistic.

4.1 Dimensionless and Standardized of Index

Because the indexes' dimensions are different in entropy method, so in order to facilitate comparison, the raw data needed to do first dimensionless. As the selected indicators in this paper are positive indicators, we use the following indicators of fuzzy quantitative models. The formulas are as follows:

$$R_j(x) = \frac{1}{2} + \frac{1}{2}\sin[\frac{\pi}{x_{j\max} - x_{j\min}}(x_j - \frac{x_{j\max} + x_{j\min}}{2})] \tag{1}$$

We assume that x_{ij} is the data that which is dimensionless, than using the following formula to standardize it.

$$y_{ij} = \frac{x_{ij}}{\sum_{i=1}^{m} x_{ij}} \tag{2}$$

4.2 Determine the Objective Weight

When the data are standardized, we will use the formula of entropy to calculate the weight of every evaluation index. The results are as follows:

Table 3. The objective target weight

evaluation index		weights
consumer level (0.4803)	Brand value	0.106226
	Culture of Social radiation	0.267889
	Business and service portfolio	0.106226
Economic level (0.2127)	State financial benefits	0.106226
	Assets operating conditions	0.106226
	Solvency position	0.000122
	Development capacity	0.000122
technical level (0.1004)	Transaction security	0.050122
	Transaction convenience	0.050122
	Technological Innovation	0.000122
social level (0.1063)	privacy of trading participants	0.000127
	Consuming Habits	0.106226
legal and policy level (0.1003)	perfection of supporting laws	0.050122
	Government's support	0.050122

5 AHP and Entropy Method to Determine Comprehensive Weight

The weights determined by AHP reflect the importance of experts' idea on the related indexes, while the weights determined by entropy theory mainly reflect the objective relationship between numerical which based on the effectiveness of mobile payment operators. Because evaluation should be objective and fair assessment, we should take full account of both subjective and objective factors. In order to fully reflect the importance of evaluation, we first combine with the experts' subjective weight s and the objective weights, than ultimately determine the weights as follows:

$$w_j^* = \frac{w_j^1 \times w_j^2}{\sum_{j=1}^{n} w_j^1 \times w_j^2} \text{ ,where } 0 \le w_j^* \le 1 \text{ and } \sum_{j=1}^{5} w_j^* = 1 \tag{3}$$

(w_1^i is subjective weight; w_2^i is Objective weight)

6 Evaluate the Mobile Payment Companies in China Using the Method of Fuzzy Comprehensive Evaluation

Fuzzy comprehensive evaluation method is a kind of method which based on fuzzy mathematics. Fuzzy comprehensive evaluation method changes the qualitative evaluation into quantitative evaluation according to the degree of membership of fuzzy mathematics theory. That is, we use fuzzy mathematics to make an overall evaluation of the object which is restricted by many factors.

By Part 5 we can conclude that the comprehensive weights which relate to the factors of China's mobile payment. The results are as follows:

Table 4. The subjective and objective weights

evaluation index		weights
consumer level (0.584636)	Brand value	0.239552053
	Culture of Social radiation	0.275387738
	Business and service portfolio	0.07403587
Economic level (0.257475)	State financial benefits	0.159609797
	Assets operating conditions	0.089419946
	Solvency position	9.43373E-05
	Development capacity	0.000102541
technical level (0.042801)	Transaction security	0.039534916
	Transaction convenience	0.014906608
	Technological Innovation	3.75456E-05
social level (0.083276)	privacy of trading participants	0.000259468
	Consuming Habits	0.055080489
legal and policy level (0.031811)	perfection of supporting laws	0.033118593
	Government's support	0.018860099

According to the relevant calculations, we invited 10 experts to score the relevant indexes. The score is divided into five, that is V= (excellent, good, middle, qualified, unqualified). As a result, we can determine the membership degree of each factor and the evaluation matrix. In the end, we calculated the three companies' Fuzzy evaluate matrixes. The results are as follows:

China Mobile: B_1=(0.159609797, 0.159609797, 0.089419946, 0.055080489, 0)

China Unicom: B_2=(0.239552053, 0.275387738, 0.275387738, 0.07403587, 0)

China telecom: B_3=(0.1, 0.275387738, 0.275387738, 0.2, 0)

Ultimately using fuzzy comprehensive evaluation evaluates the three companies who carry out mobile payment (the amount of evaluation grades assigned as follows).

$$V = \{V_1, V_2, V_3, V_4, V_5\} = \{2,\ 1,\ 0,\ -1,\ -2\}$$

Note : V_1 : excellent; V_2 : good; V_3 : middle; V_4 : qualified; V_5 : unqualified

China Mobile's fuzzy comprehensive evaluation is: $G_1 = B_0 V^T$ =0.423749

China Unicom's fuzzy comprehensive evaluation is: $G_2 = B_0 V^T$ =0.680456

China Telecom's fuzzy comprehensive evaluation is: $G_3 = B_0 V^T$ =0.275388

7 Conclusion

From the evaluation results we can seen that the operating performances of China's three companies are between "middle" and "good" .The results of this evaluation is consistent with the actual situation. Because 2009 is the first year of 3G in China, while mobile payment is a specific application of 3G services. In China, mobile payment is still at the initial stage and has a long way to go. In 2010, various operators are seeking to develop mobile payment services. Such as China Mobile will include mobile payment into the Group's KPI evaluation system and develop it vigorously. While China Unicom and China Telecom will also pay mobile payment as the first important business of 3G value-added services. This shows that the three major domestic carriers are active in the layout for mobile payments.

In addition, we can see from the results of this evaluation that the most important factors affecting China mobile payment are consumer level and economic level. So the operational decision of making mobile payment enterprises should continue to optimize their asset structure and optimize portfolio. So as to taking a step forward to enhance brand value and community radiation and to make enterprises to become stronger.

References

1. Liu, D., Fang, X.: The review of mobile payment at home and abroad. Business Times (2009)
2. Luo, Y.: The obstacles and prospects of China's mobile payment. Technology and Market (2009)
3. Xu, P., Zhang, X.: Mobile Payment factors analysis. Journal of BUPT (2009)
4. Studying the typical case of mobile payment at home and abroad. Communication World (2009)
5. Guo, Q.: Research on user's acceptance behavior on mobile payment. Master's thesis of Beijing University of Posts and Telecommunications (2009)
6. Li, J.: The situation of mobile value-added business situation and development trend. Master's thesis of Beijing University of Posts and Telecommunications (2009)
7. Zhao, Y.: The research and implementation of secure mobile payment system. Master's thesis of Zhejiang University of Technology (2009)

JDEL: Differential Evolution with Local Search Mechanism for High-Dimensional Optimization Problems

Xingbao Liu, Liangwu Shi, and Rongyuan Chen

Educational Center of Modern Technology
Hunan University of Commerce
Changsha, Hunan Province 410205, China
Liuxb0608@gmail.com

Abstract. JDE, proposed by J. Brest and et al. is an efficient variant of differential evolution algorithm. JDE algorithm is focused on global search ability. However, its local search ability also need further improvement. Therefore a novel variant of JDE is proposed, which combines JDE and a local search operator simplex crossover operator aiming to improve the local search ability of JDE. The experimental results show that the novel hybrid algorithm improves the performance of JDE in term of precision and efficiency.

Keywords: Evolutionary computation, Differential evolution, Simplex crossover, Global optimization problems.

1 Introduction

Differential evolutionary (DE) proposed by Storn and Price [1], is a simple and efficient intelligent heuristic algorithm for real optimization, and has obtained some outstanding achievements in real-world applications. However, some practical experiments shows that DE may occasionally stop proceeding toward the global optimum even though the evolutionary population could not converged to a local optimum or any other point [2]. The above situation is usually called stagnation or premature convergence.

Over the past few ten years researches have been investigating ways of improving the performance of DE algorithm through tunning control parameters mutation step F, crossover rate CR, and population size NP, or strategy selection for different mutation operators and crossover operators. Gamperle et al. [4] evaluated different parameter settings for DE, and their experiments revealed that the global searching ability and convergence are very sensitive to the choice of control parameters F and CR. As can be seen from the literature, some claims on the setting of control parameters are contradictory. In this content, Qin proposed a self adaptive DE algorithm, in which both the trial vector generation strategies and their associated parameters F and CR are gradually self-adapted by learning from their previous experience of generating promising solution. In another hand, Zaharie[5] investigated the relationship between

L. Qi (Ed.): ISIA 2010, CCIS 86, pp. 347–352, 2011.

the population and the algorithmic performance. He proposed a parameter adaptation strategy for DE (ADE) based on the idea of controlling the population diversity, and implemented a multi-population approach.

The mentioned variants of DE employ all or at least one mutation strategies, which could provide more diversity information during the evolutionary process. However, how to play the best performance of every strategy is a difficulty problem. Brest [3] presented a unique perspective, called JDE, to solve the dilemma. He selected rand/1 mutation operator as only trial vector generation strategy, and encoded control parameters F and CR into every individual and adjusted their values according to two new probabilities τ_1 and τ_2, which are set to constant 0.1. He also evaluated the performance of his novel algorithm on a benchmark suit[9], and the experiments showed that the global searching ability of DE is improved. However the local searching ability is neglected by most researches.

In the paper we present a novel variant JDEL based on Brest's work in order to improve the local search ability for global high-dimensional optimization problems. JDEL generates new trial vector using rand/1 mutation strategy, meanwhile the control parameters F and CR are set based on JDE. In another hand, a simplex crossover operator is introduced to JDEL, which is selected according to a probability τ. A serial of experiments on a benchmark suits reveal that JDEL outperforms in term of time cost and solution precision.

2 Prelimary of DE

A set of *D*-dimensional vectors is called population P containing *NP* vectors, where *NP* is the number of vectors. The vectors with *D* dimension is call individual, and it can be noted as x_i, $x_i = (x_{i1}, x_{i2}, \cdots, x_{iD})$. The classic DE algorithm can be summarized as follows:

(1) Mutation operator:

$$v_i = x_{r_1} + F * (x_{i_2} - x_{i_3}),\ r_1 \neq r_2 \neq r_3 \tag{1}$$

where $x_{r_i}, i = 1,2,3$ are individuals taken from **P** randomly. F is a scale factor controlling the step size of the difference vector $(x_{i_2} - x_{i_3})$.

(2) Crossover operator

The operator combines v_i and x_{r_1}, then gets the target vector $u_i = (u_{i1}, u_{i2}, \cdots, u_{iD})$,where

$$u_{ij} = \begin{cases} v_{ij} & \textit{if } U(0,1) < CR \text{ or } j = j\_rand \\ x_{ij} & \text{otherwise} \end{cases} \tag{2}$$

In (2), $U(0,1)$ stand for random distribution, j_rand is a random number chosen from $\{1, 2, \cdots, NP\}$ to ensure target vector u_i getting at least one component from x_{r_1}.

(3)Selection operator.

$$x_i^{'} = \begin{cases} u_i & \text{if } u_i \text{ is superior } x_i \\ x_i & \text{otherwise} \end{cases} \quad (3)$$

where $x_i^{'}$ enters the next generation.

According to Storn et al[6][7], DE algorithm is much more sensitive to the choice of F than it is to the choice of CR. In JDE, Brest [3] used a self-adaptive control mechanism to change the control parameters F and CR during the evolutionary process. Each individual in the population is associated with two parameters F and CR, which is self adjusted by means of the evolution process. The better values of F and CR are more likely to survive and produce offspring, and propagate these better parameter values. The control parameters $F_{i,G+1}$ and $CR_{i,G+1}$ are calculated as

$$F_{i,G+1} = \begin{cases} F_l + rand1 * F_u & \text{if } rand2 < \tau_1 \\ F_{i,G} & \text{otherwise} \end{cases} \quad (4)$$

$$CR_{i,G+1} = \begin{cases} rand3 & \text{if } rand4 < \tau_2 \\ CR_{i,G} & \text{otherwise} \end{cases} \quad (5)$$

and generate new parameters F and CR. $rand_j$ j=1,2,3,4 are uniform random values $\in [0,1]$, τ_1 and τ_2 are adjust probabilities, and both set to 0.1 in their experiments. F_l and F_u are set to 0.1 and 0.9 respectively.

3 JDEL: JDE with Neighborhood Search

From serial experiments we conclude that JDE is efficient to global search, and outperforms DE algorithm in term of efficiency. Practical experience, however, shows that the neighborhood searching capability of JDE is not improved correspondingly. Therefore, in the paper we propose a neighborhood searching mechanism that utilizes an explorative crossover operator, with an objective of balancing their effects.

3.1 Local Searching Mechanism

In the paper, a simplex crossover operator is introduced to the JDE algorithm. The simplex crossover generates offspring individuals based on uniform probability distribution and does not need any fitness information of individuals. In the domain **D**, n+1 mutually independent antibodies vector $\mathbf{x}_i$,$i = 1, 2, \cdots, n+1$, form a simplex. The process of generating offspring individuals consists of two steps [8]. (1)Employing a certain ratio to expand the original simplex in each direction ($\mathbf{x}_i$-$\mathbf{O}$), where the vector **O** is the center of original simplex. Therefore a new simplex is formed. (2)Choose one point within the new simplex as an offspring individual.

In order to illustrating the above process, we consider three vector $\mathbf{x}_1$, $\mathbf{x}_2$ and $\mathbf{x}_3$, they form a simplex. When the ratio of expansion is given, the vertexes of new simplex are generated by $\mathbf{y}_i = (1+\lambda)(\mathbf{x}_i - \mathbf{O}), i = 1,2,3$. Next, we choose a point $\mathbf{z}$ randomly within the new simplex. i. e. $\mathbf{z} = \sum_{i=1}^{3} k_i \mathbf{y}_i$, where k_1, k_2, and k_3 are randomly selected within [0, 1], and they satisfy $k_1 + k_2 + k_3 = 1$. Control parameter λ affects the distribution of offspring individuals generated through SPX operator,. However, the parameter is fixed to 3.5 according to reference [8].

3.2 Framework of JDEL

In the proposed JDEL algorithm, both simplex crossover operator and rand1 mutation operator are selected based on a probability *p*, which is set to a constant 0.15 according to our precious experiments. A detailed description of JDEL is given in table 1. The terminal conditions may be Function evaluations (FEs), the number of generation, solution tolerance, and other conditions.

4 Experiments and Discussion

Experimental validation for the proposed JDEL is conducted on a benchmark suits[9]. Functions f_1- f_7 are high dimensional, unimodal problems. And Function f_6 is a step function which has one minimum and is discontinuous. Function f_7 is a noisy quadratic function. Functions f_8- f_{13} are high-dimensional multimodal benchmarks, where the number of local minima increases exponentially with the number of problem dimension.

Table 1. The description of JDEL

Algorithm JDEL. JDE algorithm with local search
1. Initialize population **P** randomly within its domain **D**
2. Initialized parameters F_l and CR
3. Evaluation fitness function for all individuals
4. While the terminal conditions are not met
1. If U(0,1)<*p* //generate offspring using SPX operator.
2. V=SPX(P, alpha)
3. Else //generate offspring using rand/1 operator.
4. V=rand_1(P, F)
5. End_If
6. U=crossover(V,P,CR)
7. Evaluate fitness function for all individuals
8. Update the evolutionary population
5. End_While
6. Output experimental results.

4.1 Comparison with JDE Based on FEs

In the experiment study, we set the maximum evolutionary generations as it is given in ref [3]. The population size *NP* is set to 100, while it is 30 in proposed JDEL, therefore we set FEs=1E+5 as the terminal condition for fair comparison. The selection probability p is set to 0.15 according to our precious experiments. The initial mutation step F_l=0.1, and F_u=0.9, which means F is within [0.1, 1]. Then no more parameters are to set in JDEL. The average results of 25 independent runs are listed in table 2.

For unimodal functions f_1-f_7, JDEL achieved much better experimental results than JDE, except on the simple step function f_6, where all two optimization algorithms found the optima. Apparently, two algorithms performed almost same for function f_5 although the solution quality are all poor.

For multimodal functions f_8-f_{13}, JDEL outperformed JDE in term of solution quality obviously except f_8. With further observation of experimental results, JDEL is sometimes able to find the optima, and sometimes fall into local trap. Therefore the average value of experimental results of f_8 is worse than that of JDE.

Table 2. The comparison between JDE and JDEL. "Mean" stands the mean value, and "Std" represents standard deviation of optimization results of 25 independent runs.

Functions	JDEL		JDE	
	Mean	Std	Mean	Std
f_1	2.28e-294⁺	0.00	1.95e-11	8.28e-12
f_2	1.15e-167⁺	0.00	1.16e-7	4.62e-8
f_3	3.89e-52⁺	7.11e-52	1.34e+4	2.52e+3
f_4	5.16e-65⁺	1.15e-64	7.03e-1	1.07e-1
f_5	2.40e+1	5.52e-1	2.47e+1	1.52e-1
f_6	0.00	0.00	0.00	0.00
f_7	2.66e-4⁺	1.40e-4	1.69e-2	5.43e-3
f_8	2.34e+1	5.30e+1	1.08e-3	1.08e-3
f_9	0.00⁺	0.00	3.57e+1	3.34
f_{10}	8.88e-16⁺	0.00	1.06e-6	3.31e-7
f_{11}	0.00⁺	0.00	5.84e-9	1.28e-8
f_{12}	1.57e-32⁺	0.00	1.79e-12	1.02e-12
f_{13}	1.37e-32⁺	5.51e-34	9.66e-12	2.50e-12
	+10		**+2**	

Table 3. The comparison between JDE and JDEL. "Mean" stands the mean value, and "Std" represents standard deviation of optimization results of 25 independent runs.

	JDEL		JDE	
	Mean	Std	Mean	Std
f_1	2.15e-021	4.74e-021	8.07e-1	2.84e-1
f_2	1.76e-13	3.64e-13	1.78e-1	1.75e-2
f_3	5.10e-1	6.34e-1	2.20e+4	2.69e+3
f_4	4.13e-4	6.67e-4	1.66e+1	1.44
f_5	2.73e+1	5.26e-1	3.00e+1	1.26
f_6	0.00	0.00	0.00	0.00
f_7	2.50e-3	.70e-3	6.84e-2	1.51e-2
f_8	2.68e+3	4.01e+2	3.25e+3	3.39e+2
f_9	0.00⁺	0.00	9.52e+1	7.56
f_{10}	9.28e-13	1.39e-12	3.70e-1	5.04e-2
f_{11}	0.00	0.00	8.06e-1	1.21e-1
f_{12}	1.91e-4	1.62e-4	1.50e-1	7.13e-2
f_{13}	5.80e-1	1.98e-1	6.51e-1	2.17e-1
	+13		**0**	

4.2 Comparison with JDE Based on the Iteration Number

The above observation of comparison results are based on a fixed FEs, which indicates the efficiency of two algorithms. On the other hand, the terminal condition of the generation number, #Gen, is able to indicates the individual progress during every generation. Consequently, two algorithms JDE and JDEL are run at same terminal condition of the generation number, that is #Gen=300. The corresponding experimental results are listed in table 3.

It is clear that, our proposed JDEL algorithm performs much better than JDE for the whole benchmark suits. The reason behind the experimental results is the contribution of local searching mechanism.

5 Conclusions

JDE proposed by Brest is an outstanding variant of DE, and it works better than most intelligent optimization algorithm utilizing to the global optimization problems. However, JDE focuses on global searching mechanism, and has little work on its local searching mechanism. Based on the observation, we introduce a local searching operator, the simplex crossover operator, to JDE and form a new algorithm JDEL. A serial experiments show that the local searching operator improves the performance of JDE largely.

References

1. Storn, R., Price, K.V., Lampinen, J.: Differential Evolution - A Practical Approach to Global Optimization. Springer, Berlin (2005)
2. Lampinen, J., Zelinka, I.: On stagnation of the differential evolution algorithm. In: Ošmera, P. (ed.) Proc. of MENDEL 2000, 6th International Mendel Conference on Soft Computing, Brno, Czech Republic, June 7-9, pp. 76–83 (2000)
3. Brest, J., Greiner, S., Bošković, B., Mernik, M., Žumer, V.: Self-adapting Control parameters in differential evolution: a comparative study on numerical benchmark problems. IEEE Transactions on Evolutionary Computation 10(6), 646–657 (2006)
4. Gamperle, R., Muller, S.D., Koumoutsakos, A.: Parameter study for differential evolution. In: WSEAS NNA-FSFS-EC 2002, Interlaken, Switzerland, February 11-15 (2002)
5. Zaharie, D.: Control of population diversity and adaptation in differential evolution algorithms. In: Matousek, D., Osmera, P. (eds.) Proc. of MENDEL 2003, 9th International Conference on Soft Computing, Brno, Czech Republic, pp. 41–46 (June 2003)
6. Storn, R., Price, K.: Differential evolution—A simple and efficient heuristic for global optimization over continuous spaces. J. Global Optimiz. 11, 341–359 (1997)
7. Storn, R., Price, K.: Differential Evolution—A Simple and efficient adaptive scheme for global optimization over continuous spaces, Berkeley, CA, Tech. Rep. TR-95-012 (1995), `http://citeseer.ist.psu.edu/article/storn95differential.html`
8. Yong, W., Zi Xing, C., Guan Qi, G., Yu Ren, Z.: Multi-objective optimization and hybrid evolutionary algorithm to solve constrained optimization problems. IEEE Transactions on Systems Man and Cybernetics Part B-Cybernetics 37(3), 560–575 (2007)
9. Xin, Y., Liu, Y., Lin, G.: Evolutionary Programming Made Faster. IEEE Transaction on Evolutionary Computation 3, 82–102 (1999)

Model of Fuzzy Optimizations about the Proposed Plans and Its Application

Ke Lihua and Ye Yicheng

College of Resources and Environmental Engineering,
Wuhan University of Science and Technology, Wuhan, Hubei, 430081
Hubei Key Laboratory for Efficient Utilization and Agglomeration of Metallurgic Mineral Resources, Hubei, Wuhan 430081
klh518@sina.com

Abstract. It is important to evaluate comprehensively the proposed mining plans from many perspectives in view of the condition and mining present situation about the deposited iron ore in western hubei.The opinions about all the experts in the decision-group have been dealed with based on triangle fuzzy numbers.The proposed mining plans have been evaluated by using the the fuzzy borda number and the superior functions based on interval numbers in relative comparing space. The fuzzy uncertainty of qualitative indexes and the tiny differences of experts' opinions have been sloved well to avoid the disadvantageous influences on the appraising results and make the evaluation be more comprehensively and systematically.

Keywords: Fuzzy Optimizations; Triangle Fuzzy Number; Interval Numbers; Priority Degree of Function.

1 Introduction

The deposited iron ore in western hubei is gentle dip thin orebody in which ore layers strike eighty degrees from north to east, fourteen kilometers in length and three meters in average thickness. The ore is composed of hematite, limonite, oolitic chlorite and manganese carbonate one etc.There are rich groundwater in the range of mining district and iron shale,calcareous shale or shale around the orebody. Its thickness and grade vary largely along alignment and inclination.The orebody grade is rich and thick from three meters to four meters in the west of No.40 exploration line, poor and thick from one meter to three meters in the other partion.The orebody is rich, thick from three to seven meters in the north, poor and thin in the south in which there are many dissections and its grade is too low to meet the industry requirements.The structure of ore is tuberculosis, oolitic and lenticularis in which there are one to four layers of oolitic hematite granule existing mainly in zone centralization. The core is mainly composed of hematite and the main components of oolitic zone is collophanite ore and clay one.

In summary, the deposited iron ore is inclining gentlely and thin. Its thickness and grade vary largely along alignment and inclination. There are iron shale,calcareous shale or shale around it.So the collapsed ore can't completely arrive the predetermined places

L. Qi (Ed.): ISIA 2010, CCIS 86, pp. 353–358, 2011.

by gravitational force.It's too difficult to obtain the collapsed ore by blasting power because of the blasting restriction.It is easy to be no safety, low efficiency, high cost and bigger dilution ratio and loss one. It is important to determine the suitable mining method and the best implemented scheme in the prophase research from society, economy and technical aspects. The indexes are fuzzy because of the differences of all the expersts'opinions who are from different fields. Thus the triangle fuzzy numbers have been selected to discribe objectively the indexes and the the expersts'opinions by which the three mining plans have been analysed comprehensively in the paper in order to make a scientific decision.Thus the best one will be selected to carry out.

2 The Model of Evaluation Based on Triangle Fuzzy Numbers

2.1 The Determination of Basic Information

We define the decision group as $P=\{P_1,P_2,\ldots,P_k\}(k\geq 2)$, the collection of proposed plans as $A=\{A_i\}(i=1,2\ldots,m)$ and the collection of appraising indexes as $X=\{X_j\}(j=1,2\ldots,n)$.

We get the fuzzy appraising value S_i^r about mining plan A_i.which is evaluated by P_r.based on the triangle fuzzy number x_{ij}^r and W_j^r.The expression is followed as:

$$S_i^r = \frac{1}{n} \otimes [(x_{i1}^r \otimes W_1^r) \oplus \cdots \oplus (x_{in}^r \otimes W_n^r)], \quad (i = 1,2,\cdots,m; r = 1,2,\cdots,k) \tag{1}$$

$$x_{ij}^r = (c_{ij}^r, a_{ij}^r, b_{ij}^r) \tag{2}$$

$$W_j^r = (e_j^r, f_j^r, g_j^r) \tag{3}$$

Where xijr is the fuzzy appraising value about the opinions of expert P_r on the index X_j of mining plan A_i and W_j ris the weight coefficients about the opinions of expert P_r on index X_j.

We define S_i^r with the expansion theory in order to simplify the calculation.The expression is followed as:

$$S_i^r = (T_i^r, Q_i^r, H_i^r) \ (i = 1,2,\cdots,m; r = 1,2,\cdots,k) \tag{4}$$

Where

$$T_i^r = \sum_{j=1}^{n} c_{ij}^r e_j^r / n \tag{5}$$

$$Q_i^r = \sum_{j=1}^{n} a_{ij}^r f_j^r / n \tag{6}$$

$$H_i^r = \sum_{j=1}^{n} b_{ij}^r g_j^r / n \tag{7}$$

2.2 The Evaluation about the Plans

2.2.1 The Analysis Based on Fuzzy Borda Numbers

The fuzzy borda numbers have been selected to analyse the plans in order to use fully the representative information from Q_i^r which means the greatest possible value.

Set

$$B(i)=\sum_{r=1}^{k}B_r(i) \tag{8}$$

$$B_r(i)=m-n(i) \tag{9}$$

Where $B(i)$is the opinions of the decision-group when $B_r(i)$ is the ordinal number of the opinion P_r on A_i and $n(i)$ means the ordinal number of A_i in accordance with $Q_i{}^r$.

The plan A_i is better when its $B(i)$ is bigger.

2.2.2 The Analysis Based on Interval Numbers

Set

$$a=\langle a^-,a^+\rangle\,,\ b=\langle b^-,b^+\rangle$$

We get

$$l(a)=a^+-a^-,l(b)=b^+-b^-,l(a,b)=\max(a^+,b^+)-\min(a^-,b^-) \tag{10}$$

We set up the superior function $p(a\geq b)$ expressing the possible superior degree of a to b in complete comparison space which is followed as:

$$p(a\geq b)=\begin{cases}0 & a^+\leq b^- \\ \dfrac{l(a,b)}{l(a)+l(b)} & a^-\leq b^+\leq a^+ \\ 1-\dfrac{l(a,b)}{l(a)+l(b)} & b^-\leq a^+\leq b^+ \\ 1 & b^+\leq a^-\end{cases} \tag{11}$$

Set

$$\mu_D(i)=\frac{1}{k}\sum_{r=1}^{k}\mu_D^r(i)\,,(i=1,2,\ldots,m) \tag{12}$$

$$\mu_D^r(i)=\frac{1}{m-1}\sum_{j=1,\,j\neq i}^{m}p(S_i^r\geq S_j^r)\,,(i=1,2,\ldots,m;r=1,2,\ldots,k) \tag{13}$$

Where$\mu_D(i)$ is the opinions of the decision-group when $\mu_D^r(i)$ is the ordinal number of the opinion P_r on Ai in accordance with $G_i{}^r$=(T_i^r , H_i^r).

The plan Ai is better when its$\mu_D(i)$ is bigger.

2.2.3 The Comprehensive Analysis

We can obtain the comprehensive value $B_z(i)$ based on $B(i)$ and$\mu_D(i)$ which is expressed as:

$$B_z(i) = w_1 B(i) + w_2 \mu_D(i) \tag{14}$$

Where w_1 and w_2 are the wegiht coefficients about the two analysis while the sum of w_1 and w_2 is one. Thus the plan A_i is better when its $B_z(i)$ is bigger[1].

3 Fuzzy Optimizations and Its Application about Mining Plans of the Deposited Iron Ore in Western Hubei

3.1 The Determination of the Index Value and the Wegiht Coefficients

The appraising indexes $X=\{X_j\}$(j=1,2,...n) have been set up in view of the present information about the deposited iron ore in western hubei.The indexes are mining cost X_1, cutting ratio X_2,manufacturability of mineral block X_3, dilution ratio X_4, loss rate X_5; the mining process X_6,mechanization X_7,security X_8, ventilation X_9, labor intensity X_{10} surface damage X_{11}. The quantitative indexes have been forecasted by the design of schemes such as cutting ratio X_2,manufacturability of mineral block X_3, dilution ratio X_4, loss rate X_5 while the qualitative indexes have been determined by using the fuzzy collection of language such as$\{\mu_1,\mu_2,\mu_3\}$={better grade,good grade,general grade}. The grades of different indexes are different because of the different meaning of qualitative indexes. So the corresponding grades about the indexes is defined as Table 1 by which all the experts determine the indexes values followed by Table 2.

The wegiht coefficients about eleven indexes have been determined by all the experts in the.group based on SPA[2] and gathering iterative method [3][4][5][6][7] on which the triangle fuzzy weight coefficients is followed as Table 2.

Table 1. The Grades of the Qualitative Indexes

Indexes	Grades of The Qualitative Indexes		
	μ_1	μ_2	μ_3
Mining Cost	(0.80, 0.92, 1.00)	(0.50, 0.65, 0.80)	(0.00, 0.35, 0.50)
Mining Process	(0.80, 0.90, 1.00)	(0.50, 0.65, 0.80)	(0.00, 0.30, 0.50)
Mechanization	(0.80, 0.88, 1.00)	(0.50, 0.63, 0.80)	(0.00, 0.28, 0.50)
Safety	(0.90, 0.96, 1.00)	(0.50, 0.72, 0.90)	(0.00, 0.38, 0.50)
Ventilation	(0.80, 0.90, 1.00)	(0.50, 0.62, 0.80)	(0.00, 0.35, 0.50)
Labor Intensity	(0.90, 0.94, 1.00)	(0.50, 0.72, 0.90)	(0.00, 0.28, 0.50)
Surface Damage	(0.90, 0.96, 1.00)	(0.50, 0.72, 0.90)	(0.00, 0.25, 0.50)

Table 2. The Parameters of Indexes

X_j	W_j	A_1	A_2	A_3
X_1	(0.10,0.12,0.14)	$(\mu_1,\mu_1,\mu_1,\mu_1,\mu_1)$	$(\mu_2,\mu_2,\mu_1,\mu_1,\mu_2)$	$(\mu_2,\mu_2,\mu_1,\mu_2,\mu_2)$
X_2	(0.10,0.12,0.14)	2-12	3-5	10-12
X_3	(0.08,0.10,0.12)	200-400	150-300	200-300
X_4	(0.03,0.03,0.05)	10-15	10-15	10-15
X_5	(0.03,0.03,0.05)	20-25	10-15	10-15
X_6	(0.08,0.10,0.10)	$(\mu_1,\mu_1,\mu_1,\mu_1,\mu_1)$	$(\mu_2,\mu_2,\mu_2,\mu_2,\mu_2)$	$(\mu_2,\mu_2,\mu_2\mu_2,\mu_2)$
X_7	(0.12,0.15,0.15)	$(\mu_1,\mu_1,\mu_1,\mu_1,\mu_1)$	$(\mu_1,\mu_1,\mu_1,\mu_1,\mu_1)$	$(\mu_1,\mu_2,\mu_1,\mu_1,\mu_1)$
X_8	(0.08,0.10,0.10)	$(\mu_1,\mu_1,\mu_1,\mu_1,\mu_1)$	$(\mu_2,\mu_2,\mu_2,\mu_2,\mu_2)$	$(\mu_2,\mu_3,\mu_2,\mu_3,\mu_2)$
X_9	(0.03,0.06,0.06)	$(\mu_2,\mu_2,\mu_2,\mu_2,\mu_1)$	$(\mu_2,\mu_2,\mu_1,\mu_1\mu_2)$	$(\mu_2,\mu_2,\mu_2,\mu_2,\mu_2)$
X_{10}	(0.08,0.13,0.13)	$(\mu_1,\mu_1,\mu_1,\mu_1,\mu_1)$	$(\mu_2,\mu_2,\mu_1,\mu_2,\mu_2)$	$(\mu_2,\mu_2,\mu_1,\mu_2,\mu_2)$
X_{11}	(0.10,0.13,0.13)	$(\mu_3,\mu_2,\mu_3,\mu_2,\mu_2)$	$(\mu_2,\mu_3,\mu_3,\mu_3,\mu_2)$	$(\mu_3,\mu_3,\mu_3,\mu_3,\mu_2)$

3.2 The Analysis of Mining Plans

The parameters in Table 2 have been dealed with the the formula(1)-formula(7) and we obtain the fuzzy appraising value $S_i^{\ r}$ followed as Table 3.The analysis based on fuzzy borda numbers and interval ones have been dealed with he formula(8)- formula(14) followed as Table 3, Table 4 and Table 5.

Table 3. The Fuzzy Appraising Value S_i^r

P_r	A_1	A_2	A_3
P_1	(0.0406,0.0617,0.0936)	(0.0367,0.0628,0.0931)	(0.0307,0.0533,0.0788)
P_2	(0.0452,0.0672,0.0983)	(0.0321,0.0572,0.0884)	(0.0238,0.0470,0.0714)
P_3	(0.0452,0.0672,0.0983)	(0.0389,0.0633,0.0932)	(0.0366,0.0578,0.0815)
P_4	(0.0406,0.0617,0.0936)	(0.0359,0.0617,0.0920)	(0.0271,0.0502,0.0742)
P_5	(0.0460,0.0687,0.0994)	(0.0367,0.0628,0.0931)	(0.0353,0.0588,0.0825)

Table 4. The Analysis Based on Fuzzy Borda Numbers

A_i	B_1	B_2	B_3	B_4	B_5	$B(i)$	$n_1(i)$
A_1	1	2	2	2	2	9	1
A_2	2	1	1	2	1	7	2
A_3	0	0	0	1	0	1	3

Table 5. The Analysis Based on Interval Numbers

A_i	P_1	P_2	P_3	P_4	P_5	$\mu_D(i)$	$n_2(i)$
A_1	**0.5743**	**0.6729**	**0.5914**	**0.5970**	**0.6044**	**0.6080**	**1**
A_2	**0.5413**	**0.5079**	**0.5088**	**0.5496**	**0.4932**	**0.5202**	**2**
A_3	**0.3843**	**0.3192**	**0.3999**	**0.3534**	**0.4025**	**0.3718**	**3**

We can know that the comprehesive ordinal number is same as that of the analysis based on fuzzy borda numbers and the analysis based on fuzzy interval numbers because the two results are identic. Thus the mining plan A_1 has been selected to carry out.

4 Conclusion

In summary,the triangle fuzzy numbers have been introduced to solve the opinions about all the experts in the decision-group in view of the condition and mining present situation about the deposited iron ore in western hubei. The analysis based on fuzzy borda numbers and interval ones have been carried out so that the comprehesive ordinal number can be obtained. The fuzzy uncertainty of qualitative indexes and the tiny differences of experts' opinions have been sloved well to avoid the disadvantageous influences on the appraising results and make the evaluation be more comprehensively and systematically.

References

1. Ye, Y., Hu, W., Ke, L.: The Model and Its Application Based on Fuzzy Priority Degree about the Projects. J. Technoeconmics & Management Research, 37–38 (2003)
2. Ke, L., Ye, Y.: The Model of Index Weights in SPA About Comprehensive Evaluation. J. Dynamics of Continuous Discrete and Impulsive Sytems-Series, 202–209 (February 2006)
3. Guo, Y.: The theories and methods about comprehensive evaluation. Science and Technology Press, Beijing (2000)
4. Ye, Y., Ke, L., Huang, D.: The Comprehensive Evaluation Technology and its Application. Metallurgical Industry Press, Beijing (2006)
5. Ye, Y., Huang, Y.: Study on the value function of MADMin mining system. J. Wuhan Yejin Univ. of Sci. & Tech. (Nature Science Edition), 18–20 (1999)
6. Ye, Y., Zhao, Y.: The Study about Model of Mine Comprehensive Evaluation Based on Interval Number. Dynamics of Continuous Discrete and Impulsive Sytems-Series, 210–215 (February 2006)
7. Ye, Y., Ke, L.: The Analysis of Sensitivity about the Selection of Mining Methods Based on Multi-attribute Appraising. In: Proceeding of 2nd International Conference on Biomedical Engineering and Informatics, pp. 2245–2248 (2009)

GPRS-Based Electric Power Remote Monitoring System

LiPing Wang

ChangChun Institute of Technology, ChangChun, JiLin, China

Abstract. With the help of the GPRS (General Packet Radio Service) information transmission technology; the proposed system can remote monitor electric power system's devices (i.e., electricity meter box). When devices were opened, the system can identify them location in time and transmit information to the monitoring center through GPRS. By software's process, we can realize remote monitor electric power system's devices and prevent electricoty-stolen.

Keywords: GPRS; Remote monitoring; Power consumption; Power line transmission.

1 Introduction

In our society, electric power supply plays an important role; people's work and life are inseparable from it. But the power supply is facing a very serious situation in complex environment; the most important challenge is high power consumption, lead to great part of power generation capacity was loss in the transmission process. Stealing power constantly repeated, despite prohibitions, also contributed to the power system the main reason for the high power consumption. In this situation, how to prevent steal power and not change the existing power equipment based on the realization is a serious problem. GPRS-based remote monitoring system can effectively prevent the stealing occurs, thereby reducing power consumption.

Current power system is also used for many remote anti-theft alarm systems, which had always been used in long-range fixed point wireless alarm technology to prevent the extension stealing. Traditional methods have many drawbacks: coverage area is small, high emission power, handicap by wireless committee management constraints and standby time is short when power-down. For these reasons, the development of GPRS-based remote monitoring system can solve the shortcomings of traditional methods, effective solve the power consumption problem.

For the electricity supply department, a good alarm system should be based on or rely on the use of existing power facilities, and should not impose extra burdens and pressures, and can improve network security and reliability. At present lots of departments from aboard adopted soft wares: anti-theft device or software, such as power system alarms and security systems, but the software or systems are independent, not dependent on electric power transmission line, which caused the electricity sector a lot of pressure, it also brings a lot of insecurity[1]. The GPRS-based remote monitoring system of power aims to solve these shortcomings, in the meter box an anti-theft alarm device is utilized, through real-time transmission power lines the

L. Qi (Ed.): ISIA 2010, CCIS 86, pp. 359–364, 2011.

current state of the device was acquired, and then the state was transmitted to the centralized control position via GPRS. The software identifies the location of meter boxes and monitor meter boxes; by the help of this function, we can prevent the electronic power stealing effectively.

GPRS-based remote monitoring system can be utilized in areas which coverage by GSM network; it adapt to self-service transformer substation, storage, finance room, parking lot etc, and so on, and convenient in the Electric Power Industry Bureau or electric manage place are on the low-level power supply facilities, security, anti-theft and anti-theft for large users[3].

2 System Design

In order to prevent stealing and reducing power consumption, GPRS-based remote monitoring system of power meter box's key detection method is set the alarm in switch. If the software grand open permission privileges to the meter box, normal operation can not trigger alarm, if not authorized to open meter box, the alarm system will send non-real time through the normal power line to open the meter box identification number, at the same time through GPRS signal transmission to the special duty personnel office or related service center, and then further processing.

Remote monitoring system is based on traditional monitoring systems, combined with the current wireless communications technology and information processing technology, further developed a new control system. Generally, the system consists with the "control center - monitoring stations" mode. Control Center is the core of the entire system is responsible for collecting information that monitoring stations uploaded. Monitor stations were deployed at the areas away from the control center, responsible for the completion of information collection and initiate transmission. Control center and apply on general computer system, workstation, or industrial control, software development can be based on an existing Windows or UNIX operating systems. The design and implementation of applications of stations according to different purposes and application environment, using single chip or Intel X86 series of microprocessors.

The network mode adopted by remote monitoring system is also very flexible, it can make use of existing wireless communications network, such as GSM / GPRS network, CDMA mobile networks, and can also use wireless LAN. The microcontroller-based design proposal is generally applicable to less demand data processing, less computational system, based on remote monitoring and switching technology to packet based data network by using IP protocol [4].

GPRS-based electric power remote monitoring system can be constructed by following steps:

(1) Design meter box security hardware module;
(2) Set the identification for each meter;
(3) Transmission identification information to the centralized control of the scene;
(4) The information transmitted via GPRS to a wireless control terminal, and further transmitted to the control center host through Internet;
(5) Accept TCP / IP data packets by VB environment WinSocket control, and restore the original data analysis for further processing.

3 Hardware Design

3.1 Hardware Module Design

In order to remote monitor meter box by alarm switch and transmit signal, the system adopted the 8051 microcontroller used here as the trigger of the components, simply open the meter box, it will automatically alert the logo has been well set to return to the control center, the software identifies the location of alarm occurred, and then processed. The meter identification, before the program was under the location number, for example, "xy1234", where "x" can be expressed as the partition, "y" that belongs to a certain line, while the 1234 is that the number of specific order; as long as the date in the software to add information indicating this number represents a specific location, you can quickly lock the target, and then processed.

3.2 The Selection of Multi-function Digital Power Meter Box

The DTSD188S multi-function three-phase electronic energy meter was selected by the system. It uses the up-to-date split case, composes of sub-metering of the main unit and communication unit in two parts. Measurement of the main unit has all the features of multi-functional table can accurately measure and record large amounts of data. DTSD188S three-phase electronic energy meter is an independent 485 communication port, GPRS remote communication interface and other functions [6]. This hardware design of the microcontroller set meter can identification information corresponding to the communication unit, through the digital code modulated analog signal transmission through power lines.

3.3 The Selection of the Modem

Aims at transmit information through power line, the modem identify the signal (digital signal) modulated high-frequency analog signals. Based on the above reason, use an external modem, external modem and need extra power cord.

3.4 The Adoption of Control Center Computer

As the central monitoring station deploy in the call center or dedicated computer room, and only used to transmit the meter box's identification, and then analyzed by software to identify their location, so the functions and work through the realization of the environmental decision to choose Pentium V serious or compatible computers, and can access Internet terminal for receiving the transmitted control over GPRS data.

3.5 GPRS Terminal Setting

A GPRS data terminal was settled in transformer region for different meter box to send a message or receive message. It receives power line to the high-frequency analog signals to demodulation into a digital signal, through the hiring GPRS service, GPRS wireless communication transmission to the control terminal. GPRS wireless data communication control terminal first packaged into TCP / IP data packets, and then transform GPRS data packets transmitted via a wireless link to the wireless data exchange center, then peel GPRS data packets through the gateway GPRS support node TCP / IP packets sent to the Internet, then sent to the Internet communication control computer, communications control computer through the Socket to accept TCP / IP data packets, and restore it to original data and further processing data [7]. System design as shown in Figure 1:

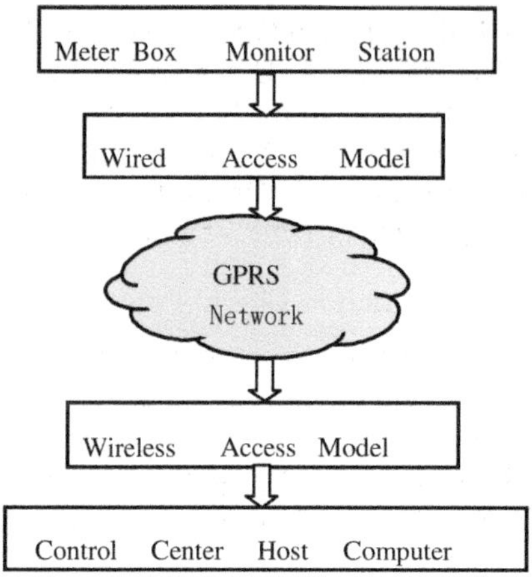

Fig. 1. System design scheme

4 Software Design

Based on current software and hardware development situation, the design aimed how to design rational software architecture to achieve high quality and efficient electric power remote monitoring. In the programming method, we use the structured object-oriented programming method, and for modular design. Software architecture and programming method to determine, based on the function of the system, first draw a rough diagram of procedures, through the block diagram for expansion and specific, are detailed program flow chart, and then the preparation of specific procedures. Program language can be choose from VB, VC + + or. NET, and integrated development environment which design interface and process data. This system designed by VB6.0.

Software design has three main parts. First, the location of the user identification alarm software interface design, interface, concise, crisp, clear, includes general management information system to add, delete, change, and search and other basic functions. Secondly, the use of VB controls under the WinSocket accepted Internet, TCP/IP packet transmission, and restores the pre-packaged raw data. Finally, enter the meter box in the database identification number and its specific location in which, due to large amount of data using SQL Server to do the background database for rapid and accurate positioning the alarm meter box location.

Require special note is that because meter box is opened it will return the warning message, and for the power sector because of their normal maintenance and inspection should not open the meter box alarm. Therefore, in the interface design should also include an area or a certain part of the list box to set an invalid received TPC alarm signals, so you can open without any affect.

The main interface of GPRS-based electric power remote monitoring system can be seen in Figure 2:

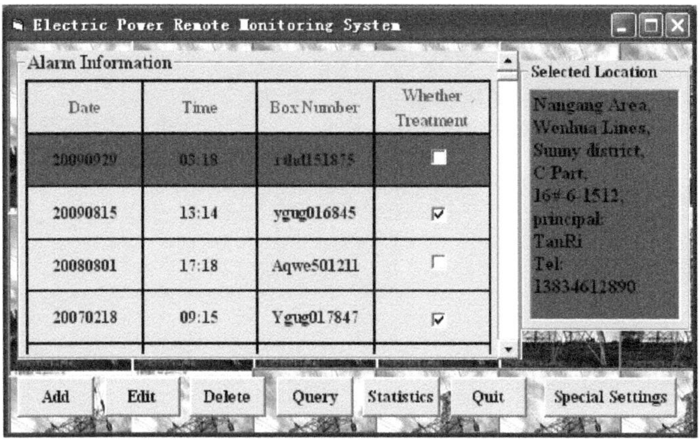

Fig. 2. Main interface of system

5 Conclusion

To the GPRS-based electric power remote monitoring system remote monitoring meter box prevent stealing power is an important means of reducing power consumption. The research of wireless communication and software-based technology is the integration of GPRS data transmission technology, computer and communication related fields such as mutual integrated new technologies.

The monitoring system can be used in huge electricity consumption areas such as large industrial and mining enterprises and distribution of concentrated residential areas. To the system itself, if the success operated, we can archive double profit: power system would reduce huge economic losses than before, while the modern enterprise management of power enterprise would be enhanced.

References

[1] Han, W.: The GPRS Wireless Communication System Applies to the Field of Operational Monitoring of Electric Power System. Power System Technology, 276–279 (2007)

[2] Ge, Y., Ma, X., Zang, K.: Control and Simulation of Continuously Variable Transmission for Electric Transmission System. Journal of Beijing Institute of Technology, 426–429 (2004)

[3] Xu, H.: The Design Of Reading Meter System Based On 10KV Electric-Power Line Communication. Mechanical and Electrical Engineering 33(10), 31–33 (2004)
[4] The Power Industry Of the People's Republic of China, Electric Power System Management Procedures (DL/T544–94), pp. 504–519 (1994)
[5] Huang, Y., Huang, Q., Sun, F.: Research and Development of Engine-Generator Set Control System for Tracked Vehicle Electric Transmission System. Journal of China Ordnance (2007)
[6] Zhang, Y., Li, H.: Analysis on Thermal Current Field in Powertrain Cabin of Tracked Vehicles With Electric Transmission System Configuration-2. Journal of China Ordnance (2007)
[7] Zhao, Z., Chao, L.: VB-Based Modbus Communication in the Distributed Control System. Techniques of Automation and Application, 91–93 (2006)
[8] Sun, F., Chen, S., Zhang, C.: Steering Dynamic Performance of an Electric Transmission Tracked Vehicle Based on Rotating Speed Control. Journal of China Ordnance (2006)
[9] Zhang, L., Long, F.: Distribution Equipments Security & Power Theft Monitoring System. China Electric Power, 101–104 (2007)

Identification of Memristor-Based Chaotic Systems Using Support Vector Machine Regression

Xiao-Dong Wang[1] and Mei-Ying Ye[2]

[1] Department of Electronic Engineering, Zhejiang Normal University, Jinhua 321004, P.R. China
wxd@zjnu.cn
[2] Department of Physics, Zhejiang Normal University, Jinhua 321004, P.R. China
ymy@zjnu.cn

Abstract. The memristor-based Chua's chaotic system provides a new framework for secure communications with chaos. This paper deals with an identification technique of memristor-based chaotic systems via support vector machine (SVM) regression. An illustrative numerical example, where the memristor was characterized by a smooth continuous cubic monotone-increasing nonlinearity, was given to verify the effectiveness of the identification scheme. The identification results of memristor-based chaotic systems via the SVM technique has been found to give excellent performance even in the presence of additive noise.

Keywords: memristor, identification, chaos, support vector machines.

1 Introduction

The memristor, characterized by a relation of the type $f(\varphi, q) = 0$, is a missing circuit element studied by Chua in 1971 [1] and realized by Williams's group of HP Labs in 2008 [2]. Memristor-based chaotic systems have attracted much attention recently. Itoh and Chua derived several chaotic systems from Chua's chaotic oscillator by replacing Chua's diodes with memristors characterized by a monotone-increasing and piecewise-linear nonlinearity [3]. Bao, Liu and Xu presented the memristor characterized by a smooth continuous cubic monotone-increasing nonlinearity was assumed and has been used in Chua's chaotic system [4-6]. These chaotic systems present opportunities for developing applications under the constraints of scalability and low power. They also provide a memristor-based framework for secure communications with chaos [7].

Nowadays, Zhong and Yu discussed a fuzzy identification method of a memristor-based chaotic system [8]. Recent some results show that support vector machine (SVM) technique seems to be very effective to identify a broad category of complex nonlinear systems when complete model information cannot be obtained. In this work, we focus on the identification of memristor-based chaotic systems using SVM regression.

L. Qi (Ed.): ISIA 2010, CCIS 86, pp. 365–371, 2011.

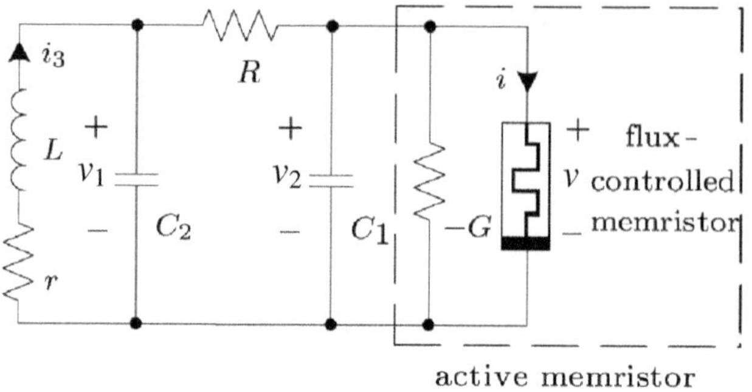

Fig. 1. Chua's oscillator with a flux-controlled memristor

2 Memristor-Based Chaotic Systems

A memristor Chua's oscillator with an active memristive circuit is shown in Fig. 1, which is directly extended from Chua's oscillator with smooth equation [9] by replacing the Chua's diode with a smooth flux-controlled memristor and a negative conductance. The smooth flux-controlled memristor shown in Fig. 1 is a passive two-terminal electronic device described by

$$q(\varphi) = a\varphi + b\varphi^3 , \tag{1}$$

where $a, b > 0$. In this case, the memductance $W(\varphi)$ is given by

$$W(\varphi) = \mathrm{d}q(\varphi)/\mathrm{d}\varphi = a + 3b\varphi^2 . \tag{2}$$

As shown in Fig. 1, if $r = 0$ and $C_2 = 1$ we can obtain a set of four first-order differential equations, in which there are four circuit variables (v_1, v_2, i_3, φ)[3]:

$$\begin{cases} \mathrm{d}v_1/\mathrm{d}t = (1/(RC_1))(v_2 - v_1 + GRv_1 - RW(\varphi)v_1), \\ \mathrm{d}v_2/\mathrm{d}t = (1/(RC_2))(v_1 - v_2 + Ri_3), \\ \mathrm{d}i_3/\mathrm{d}t = -(1/L)v_2 - (r/L)i_3, \\ \mathrm{d}\varphi/\mathrm{d}t = v_1, \end{cases} \tag{3}$$

where the $\varphi - q$ characteristic curve of the flux-controlled memristor is given by Eq. (1) and $W(\varphi) = \mathrm{d}q(\varphi)/\mathrm{d}\varphi$.

Let $y_1 = v_1$, $y_2 = v_2$, $y_3 = i_3$, $y_4 = \varphi$, $\alpha = 1/C_1$, $\beta = 1/L$, $\gamma = r/L$, $\xi = G$, $C = 1$ and $R = 1$. Then Eq. (3) can be written in dimensionless form as follows:

$$\begin{cases} \dot{y}_1 = k\alpha(y_2 - y_1 + \xi\, y_1 - W(y_4)y_1), \\ \dot{y}_2 = k(y_1 - y_2 + y_3), \\ \dot{y}_3 = -k(\beta y_2 + \gamma y_3), \\ \dot{y}_4 = ky_1, \end{cases} \tag{4}$$

where k is a time scale factor.

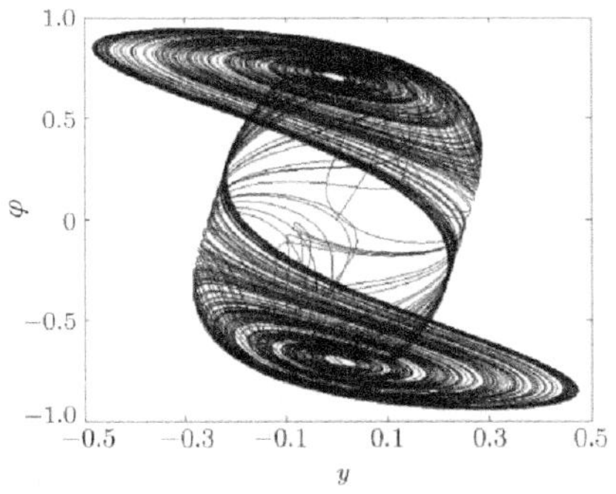

Fig. 2. Chaotic attractor of smooth memristor oscillator: y - φ

For $\alpha = 9.8$, $\beta = 100/7$, $\gamma = 0$, $\xi = 9/7$, $a = 1/7$, $b = 2/7$, $k = 20$, and initial conditions (0, 10^{-10}, 0, 0), the system (4) is chaotic [6] and displays a symmetrical 2-scroll chaotic attractor as shown in Fig. 2.

3 Identification Problem of Chaotic Systems

A chaotic systems with an output y can be described in discrete time by the NARX (nonlinear autoregressive with exogenous input) an input–output model

$$y(k+1) = f(\mathbf{x}(k)), \tag{5}$$

where $f(\cdot)$ is some nonlinear function, $y(k+1)$ denotes the output predicted at the future time instant $k+1$ and $\mathbf{x}(k)$ is the regression vector, consisting of a finite number of past inputs and outputs:

$$\mathbf{x}(k) = [y(k),\cdots,y(k-n_y+1),u(k),\cdots,u(k-n_u+1)]^T. \tag{6}$$

where the dynamic order of the system is represented by the number of lags n_u and n_y.

The task of system identification is essentially to find suitable mappings, which can approximate the mappings implied in a chaotic system. The function $f(\cdot)$ can be approximated by some general function approximators such as neural networks, neuro-fuzzy systems, splines, interpolated look-up tables, etc. The aim of system identification is only to obtain an accurate approximator for y. In this work, we use the SVM for chaotic system identification.

4 Support Vector Machines Regression

The SVM [10], which have become very popular as methods for learning from examples and have been recently introduced as a general alternative to neural networks, is established based on the structural risk minimization principle rather than minimize the empirical error commonly implemented in the neural networks, SVM achieves higher generalization performance than the neural networks in solving these machine learning problems. Another key property is that unlike neural network' training that requires non-linear optimization with the danger of getting stuck into local minima, training SVM is equivalent to solving a linearly constrained quadratic programming problem. Consequently, the solution of SVM is always unique and globally optimal. A simple description of the ν-SVM regression is provided here, for more details please refer to Ref. [11].

Given a set of training data points $\{(\mathbf{x}_1, y_1), \cdots, (\mathbf{x}_L, y_L)\}$ ($\mathbf{x}_i \in R^n$ an input and $y_i \in R$ a desired output). SVM approximate the function in the following form:

$$y = \mathbf{w}^T \phi(\mathbf{x}) + b, \tag{7}$$

where $\phi(\mathbf{x})$ represents the high dimensional feature spaces mapped from the input space $\mathbf{x}$. The $\mathbf{w}$ and b are estimated by minimizing the regularized risk function

$$R(C) = (1/2)\mathbf{w}^T\mathbf{w} + CR_{emp}^{\varepsilon}, \tag{8}$$

$$R_{emp}^{\varepsilon} = (1/L)\sum_{i=1}^{L} |d_i - y_i|_{\varepsilon}, \tag{9}$$

$$|d - y|_{\varepsilon} = \begin{cases} |d - y| - \varepsilon, & |d - y| \geq \varepsilon, \\ 0, & \text{otherwise}. \end{cases} \tag{10}$$

The first term $(1/2)\mathbf{w}^T\mathbf{w}$ is called regularized term. The second term CR_{emp}^{ε} is empirical error (risk) measured by the ε-insensitive loss function given in Eq. (10). C is referred as regularized constant determining the trade off between the empirical error and the regularized term. ε is called the tube size of SVM. Both C and ε are user-prescribed parameters and selected empirically.

The parameter ε can be useful if the desired accuracy of the approximation can be specified beforehand. In some case, however, we just want the estimate to be as accurate as possible, without having to commit ourselves to a specific level of accuracy. Hence, Ref. [11] presented a modification of the SVM that automatically minimizes ε, thus adjusting the accuracy level to the data at hand. A new parameter ν ($0 \leq \nu \leq 1$) was introduced, which lets one control the number of support vectors and training errors. To get the estimations of $\mathbf{w}$ and b, Eq. (7) is transformed to the primal problem of ν-SVM regression:

$$\begin{aligned} \text{minimize} \quad & (1/2)\mathbf{w}^T\mathbf{w} + C(\nu\varepsilon + (1/L)\sum_{i=1}^{L}(\xi_i + \xi_i^*)), \\ \text{subject to} \quad & (\mathbf{w}^T\phi(\mathbf{x}_i) + b) - y_i \leq \varepsilon + \xi_i, \\ & y_i - (\mathbf{w}^T\phi(\mathbf{x}_i) + b) \leq \varepsilon + \xi_i^*, \\ & \xi_i, \xi_i^* \geq 0, \ \mathrm{i} = 1, \cdots, L, \ \varepsilon \geq 0. \end{aligned} \tag{11}$$

Here, the slack variables ξ and ξ^* are introduced. The ξ is the upper training error (ξ^* is the lower) subject to the ε-insensitive tube $|y - (\mathbf{w}^T\phi(\mathbf{x}) + b)| \leq \varepsilon$.

By introducing Lagrange multipliers and exploiting the optimality constraints, solving Eq. (12) is equivalent to finding

$$\begin{aligned} \text{minimize} \quad & (1/2)(\alpha - \alpha^*)^T\mathbf{Q}(\alpha - \alpha^*) + \mathbf{y}^T(\alpha - \alpha^*), \\ \text{subject to} \quad & \mathbf{e}^T(\alpha - \alpha^*) = 0, \ \mathbf{e}^T(\alpha - \alpha^*) \leq C\nu, \\ & 0 \leq \alpha_i, \alpha_i^* \leq C/L, \ \mathrm{i} = 1, \cdots, L, \end{aligned} \tag{12}$$

where $Q_{ij} = K(\mathbf{x}_i, \mathbf{x}_j) = \phi(\mathbf{x}_i)^T \phi(\mathbf{x}_j)$ is the kernel and $\mathbf{e}$ is the vector of all ones. α, α^* are the introduced Lagrange multipliers. Thus, the regression estimative function given by Eq. (7) can be takes the following form:

$$y = \sum_{i=1}^{L} (\alpha_i - \alpha_i^*)\ K(\mathbf{x}_i, \mathbf{x}_j) + b\,. \tag{13}$$

where $K(\mathbf{x}_i, \mathbf{x}_j) = \phi(\mathbf{x}_i)^T \phi(\mathbf{x}_j)$ is the kernel function. Any function that satisfies Mercer's condition [10] can be used as the kernel function. Common kernel function is the radial basis function (RBF) kernel:

$$K(\mathbf{x}_i, \mathbf{x}_j) = \exp(-\gamma \|\mathbf{x}_i - \mathbf{x}_j\|^2)\,, \tag{14}$$

where γ is the parameter of the RBF kernel.

5 Numerical Example

The mean squared error metric (MSE) metric is used to evaluate the performance of SVM identification for the examples. The MSE of the testing set is calculated as follows:

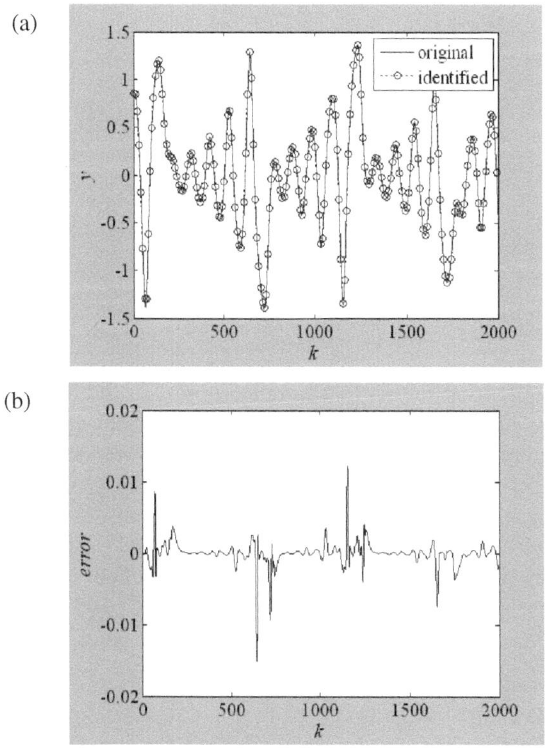

Fig. 3. (a) The testing result and (b) the testing error $error = y - y'$ obtained by using the SVM with noise-free training samples

(a)

(b)

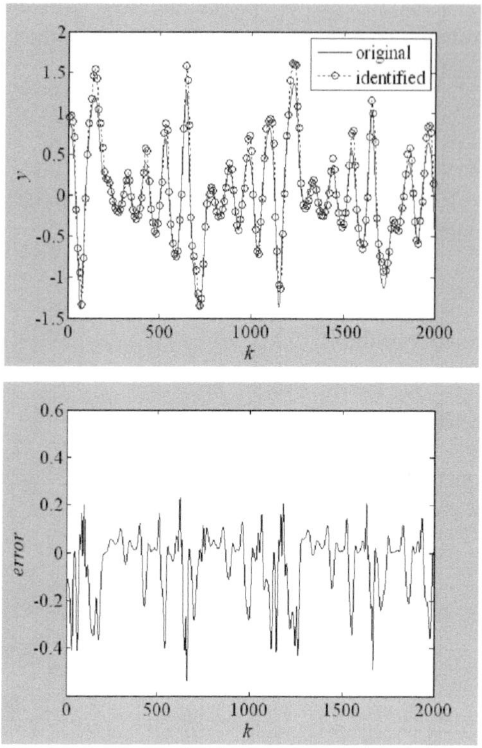

Fig. 4. (a) The testing result and (b) the testing error $error = y - y'$ obtained by using the SVM with noisy training samples

$$\mathrm{MSE} = \frac{1}{N}\sum_{k=1}^{N}(y_k - y_k')^2, \tag{15}$$

where N denotes the total number of samples points in the testing set. y represents an output component of the original memristor-based chaotic system and y' represents an estimation output component of the ν -SVM model.

Consider a memristor-based chaotic system described by Eq. (4). The simulated training and testing samples are obtained by applying the conventional fourth-order Runge-Kutta algorithm with the sampling time of 0.001s to determine the numerical solution to Eq. (4). In our work, the number of simulated samples from Eq. (4) is 7000. The samples are divided into two sets. The first 5000 samples are used for training to build the SVM model, and the rest 2000 samples are used for testing. In addition, the lags n_u and n_y in Eq. (6) are 4 and 0 respectively. The SVM parameters ν and γ are 0.8 and 1 respectively.

In real identification problems, samples are usually corrupted by noise. Uncertainty can arise from measurement instruments, system noise, or un-modeled dynamics. In practice, we should take into account the effect of noise. Hence, we have shown that the proposed method operates very well in the presence of noise. In this work, the noisy training samples are derived by adding the normal distribution noise to the simulated samples. This normal distribution noise is denoted by $\mathcal{N}(\mu,\sigma^2)$, here mean $\mu = 0$ and variance $\sigma^2 = 5$.

The testing results and testing errors obtained by using the SVM with noise-free and noisy training samples are plotted in Figs. 3 and 4 respectively. Correspondingly, the MSE values are 3.135×10^{-6} and 0.0215 respectively. Obviously, if the training samples are noise-free, the proposed method allows one to find the original chaotic dynamics with a high precision. The outputs of the original system and the SVM are seen to be indistinguishable. It still gives acceptable performance even in the presence of additive noise.

6 Conclusion

In this paper, we investigate the identification problem of memristor-based Chua's chaotic system using the SVM regression. We also consider the influence of noise on the simulated training samples. The testing results indicate that excellent performance can be obtained by the SVM technique even in the presence of additive noise. The proposed method can be extended to the identification of other memristor-based dynamic systems.

References

1. Chua, L.: Memristor-the Missing Circuit Element. IEEE Transactions on Circuit Theory 18(5), 507–519 (1971)
2. Strukov, D.B., Snider, G.S., Stewart, D.R., Williams, R.S.: The Missing Memristor Found. Nature 453(7191), 80–83 (2008)
3. Itoh, M., Chua, L.O.: Memristor Oscillators. International Journal of Bifurcation and Chaos 18(11), 3183–3206 (2008)
4. Bao, B.C., Liu, Z., Xu, B.P.: Dynamical Analysis of Memristor Chaotic Oscillator. Acta Physica Sinica 59(6), 3785–3793 (2010)
5. Bao, B.C., Liu, Z., Xu, J.P.: Steady Periodic Memristor Oscillator with Transient Chaotic Behaviours. Electronics Letters 46(3), 228–229 (2010)
6. Bao, B.C., Liu, Z., Xu, J.P.: Transient Chaos in Smooth Memristor Oscillator. Chinese Physics B 19(3), 030510 (2010)
7. Muthuswamy, B., Kokate, P.P.: Memristor-Based Chaotic Circuits. IETE Technical Review 26(6), 417–429 (2009)
8. Zhong, Q.S., Yu, Y.B., Yu, J.B.: Fuzzy Modeling and Impulsive Control of a Memristor-Based Chaotic System. Chinese Physics Letters 27(2) (2010)
9. Cafagna, D., Grassi, G.: Decomposition Method for Studying Smooth Chua's Equation with Application to Hyperchaotic Multiscroll Attractors. International Journal of Bifurcation and Chaos 17(1), 209–226 (2007)
10. Vapnik, V.N.: Statistical Learning Theory. John Wiley & Sons, Inc., New York (1998)
11. Scholkopf, B., Smola, A.J., Williamson, R.C., Bartlett, P.L.: New Support Vector Algorithms. Neural Computation 12(5), 1207–1245 (2000)

Intrusion Detection System Based on Immune Algorithm and Support Vector Machine in Wireless Sensor Network

Yu Sheng Chen[1,*], Yu Sheng Qin[2], Yu Gui Xiang[3],
Jing Xi Zhong[4], and Xu Long Jiao[4]

[1] Computer Science Department of North China University of Science & Technology, Beijing, 101601, China
yusc5291@sina.com
[2] Gu Yuan Earthquake Platform of Earthquake Bureau of Ning Xia Province, Gu Yuan 756000
[3] Beijing Zhao Fang Investment Trust Co., Ltd. Beijing 100028
[4] Kongzhuang mine, Shanghai Energy Co., Ltd. Xuzhou, Jianshu province 221000

Abstract. In order to improve the reality and whole performance of network intrusion detection system (IDS), after the characteristics of data used in IDS were analyzed, an approach, in which intruders are recognized, was presented in the paper, which was based on immune algorithm (IA) and support vector machine (SVM). In this method, immune algorithm is used to preprocess the network data, SVM is adopted to classify the optimization data, and recognize intruders. Experimental results showed that the method was feasible and efficient.

Keywords: Intrusion Detection System (IDS); immunity algorithm (IA); support vector machine (SVM); Feature selection.

1 Introduction

Wireless Sensor Network (WSN) is a Mobile Ad-hoc Network. It is applied widely in military and common domain. An Intrusion Detection System (IDS) in WSN generally detects unwanted manipulations to computer systems, mainly through the WSN .The manipulations may take the form of attacks by skilled malicious hackers, or script kiddies using automated tools[1]. An Intrusion Detection System is used to detect all types of malicious network traffic and computer usage that can't be detected by a conventional firewall[2]. This includes network attacks against vulnerable services, data driven attacks on applications, host based attacks such as privilege escalation, unauthorized logins and access to sensitive files, and malaria (viruses, Trojan horses, and worms) [3].

An IDS is a large and complicated project. Now, there are yet many weakness in IDS, it should be improved. IDS includes choosing a set of features, building up a set of attack models or normal models, matching or comparing the models with activation of now executed program, obtaining the difference of the built models and activation of now executed program, and alarming. In IDS, to choose a set of features

[*] Corresponding author.

L. Qi (Ed.): ISIA 2010, CCIS 86, pp. 372–376, 2011.

is very important. If the set of features used in IDS are chosen wrongly, no matter how perfect next later works are, the final results of IDS are wrong. In addition, action classification in IDS is completed through mathematics method, which requires that number of features is small. How to determine a good group of features and how to classify and recognize intruders are important in IDS. The solutions to the problems are provided in the paper. In this paper, immune algorithm (IA) is used to preprocess the network data, to determine a good group of features, and support vector machine (SVM) is adopted to classify and recognize the intruders. Experimental results show that the method is feasible and efficient. The approach is introduced in follow sections.

2 Support Vector Machines

In the practical application, the data used to classification in IDS is extremely complex. The data is frequently high in dimension, small in number of sample, and hard in divisibility and classification. So-called high-dimension means that the data set used in IDS have very many attributions in the normal condition. For example, the dimension of the data in KDDCUP99 is 41. The attribuion of data is called as feature. The so-called small-sample means that the information that the data set can manifest is limited. At sometime, the sample of a specific type of attack is very deficient in number. For example, only several samples of some type of attacks are found in about 5000000 samples. Under such sample structure, that each kind of attack is recognized accurately is very hard. The so-called indivisibility means that the classified samples are possibly overlapped so highly that they are not classified and recognized by simple linear sorter. Support vector machine (SVM) was developed from the theory of structural risk minimization, proposed by Vapnik et al, which is sorter design method based on the small sample study, which is suitable to the classification of small sample data. Therefore, the SVM method is suited to classify the high-dimension data in IDS. Maintaining the Integrity of the Specifications.

3 Immunity Algorithm

Immunity algorithm work in the same way the biological immunity principle does. According to biological immunity principle, the biology immune system produces automatically the corresponding immune body (antibody) against the antigen through the cell fission and the disintegration, to resist the antigen that invades the life body. This process is called the immunity reply. In the immunity reply process, part of antibodies is preserved as memory cell. When the similar antigen invades the life body once more, the memory cell is activated and produces rapidly the massive immune bodies. This causes that 2nd reply is more intense than the primary reply, which manifests memory function of the immune system. A kind of immune body and another kind of immune body are also promoted and suppressed mutually, to maintain the multiplicity of the immune body and the balance of immunity.

This kind of balance is carries on according to the density mechanism, namely the density of the immune body is higher, the immune body is suppressed more; The density of antibody is lower, the antibody is promoted more. This manifests self-adjustment function of the immune system.

The immunity algorithm is a new method developed recently in the artificial intelligence, which is a new algorithm that is designed under enlightenment of the biology immune system.

In the immunity algorithm, the antigen is corresponding the objective function and variant kind of constraint conditions; and the immune body (antibody) corresponds with the optimal solution. The kissing degree between the antibody and antigen corresponds with the match degree of solution and the objective function.

3.1 Encoding Methods

In the immunity algorithm, encoding method of the immune body and antigen mainly has three kinds, that is, the binary code, the real number code and the character code, and in minor case, the grey code (gradation) is used, and so on. Binary code has merits of strong search ability, simple crossover and variation operation. Therefore, the binary code method is selected in this article. As to P attributes of the network data, each attribute is expressed by the binary string which is E in the length. The decimal base value in correspondence with this binary value represents the weight value of the attribute. Therefore, each immune body contains $P \times E$ binary code.

3.2 Methods for Computing Immune Body Density

Immune body density is mainly used to maintain multiplicity of the immune body, and raise the partial search ability. The methods for computing immune body density are mainly ones based on Euclidean space distance and information entropy. The latter is used in this article, to compute immune body density. Information entropy is used in indicating multiplicity of the immune body in the community. The immune body density is computed and obtained by average information entropy of a group of the immune bodies.

3.3 Method for Computing the Affinity between Immune Body and Antigen

The affinity between immune body and antigen is used in indicating the degree in which the immune body recognizes the antigen. In this article, SVM, which is a classification method, is applied to obtain the rate of alarm missing and the rate of false alarm, then compute the affinity between immune body and antigen.

3.4 Steps of Immunity Algorithm

The immunity algorithm step proposed in this article is as follows:

① Analysis question: Analysis and understand the characteristic of solution, and design the appropriate coded form of the solution.

② Initialization: Initialize immune body group as well as memory storehouse (base). Set the initial values of immune body group size n and memory storehouse size mnum.

③ Evaluate each immune body in the immune body group: Expectation reproduction rate of the immune body is taken as the standard evaluate index of immune body in this algorithm. The index is determined together, according to immune body density and affinity between the immune body and the antigen.

④ Save the superior immune body to memory storehouse (base): The immune bodies are ordered by descent, according to expectation reproduction rate. Save the former mnum immune bodies with higher expected reproduction rate into memory storehouse (base).

⑤ Judge whether the stop condition satisfies: Yes, then algorithm stops; otherwise, continues the operation of the next step.

⑥ Production new community: Based on the fourth step computed result, the selection, crossover and variation operations of each immune body in each immune body group are carried on, to obtain new community, according to its expectation reproduction rate. The new generation of community is constituted of the new community obtained just about and mnum individuals in the memory storehouse together.

⑦ Change to the step③, and continues to carry out.

4 The Intrusion Detection System Model Based on the AI-SVM

The essential idea of SVM classification is to use a kernel function to map the initial input data into a high dimensional space (Hibert space), so that the two classes of data become linearly separable, as far as possible. The pretreatment of the data used in intrusion detection system is carried on with the immunity algorithm, to withdraw main attribute characteristics related with intrusion detection. The classification of the data processed is carried on with SVM. The outline of the concrete intrusion detection system model based on the AI-SVM is: ①network data set, ② immunity pretreatment, ③ SVM training, ④ SVM sorter, ⑤policy-making response (react).

5 Experiments and Result Analysis

The data used in the experiment is KDD CUP 99 data sets. This data set provides 9 week's network connection data that gathered from on a simulation local area network. Each record in the data set has contains 41 characteristics and 1 marking of kind. The record in the data set has been divided into 5 kinds, namely normal connection (Normal) denial of service (DoS), the unauthorized remote service to login (R2L), the unauthorized access to the local super user's privilege, scanning and probing (Probing).

Because primitive data sets of KDD CUP 99 are too huge, in order to carry on the confirmation to the algorithm, the representative data sets had chosen in this article, to product stochastically training set and test set, which have contained 5072 and 5280 records separately. In the practical test, the mean value of 10 test results is taken as the test finishing result. The practical immunity algorithm parameters are as follows:

The community size is 100, alternate number is 50 generations, the memory storehouse size is 5, crossover rate is 0.6, and the variation rate is 0.01. Experimental result is as shown in Table 1.

Table 1. Comparisons of correct identification rates for intruders in some different IDS

Method for IDS	the rate of alarm missing	Correct recognition rate
SVM	5.2%	81.4%
Immunity algorithm (IA) +SVM	5.7%	95.8%

The experimental results show that correct recognition rate is enhanced greatly by method developed in the paper, while the rate of false alarm rises lightly.

5 Conclusion

In data used in intrusion Detection System (IDS), there are non-correlated or the redundancy characteristics which causes performance of IDS to drop. Other, in data used in IDS, the dimension of characteristics is higher. Thus, the extraction of characteristic was carried on in the paper. Considering the global restraining of immunity algorithm as well as the immune body multiple characteristics, it was proposed to carries on pretreatment of the data and extraction of essential attributes with the immunity algorithm, then classify and recognize intruders with SVM. IDS model based on the AI-SVM was established. The experiments indicated that this algorithm was feasible and effective.

References

1. Vapnik, V.N.: Statistical study theory essential. Qinghua University publishing press, Beijing (1995)
2. Yu, S.-c.: Method for Getting Inter-Independent Features Used to Intrusion Detection System In Controllable and Trusted Networks. In: GCIS 2009, pp. 461–465 (2009)
3. Fan, Y.-t., Yu, S.-c.: System for Performing Resource Management and Control and Task Scheduling in Grid Computing. In: ISCSCT 2008, vol. 12, pp. 648–650 (2008)

A Haptic Interface for Virtual Reality Based Teleoperation System

Zhao Di*, Li Shiqi, Zhu Wenge, and Wang Mingming

Department of Industrial & Manufacturing System Engineering,
Huazhong University of Science & Technology, 430074, Wuhan, China
vbscripzt@126.com, {sqli,wgzhu2000}@mail.hust.edu.cn,
kahadeswmm@gmail.com

Abstract. Virtual Reality technologies are widely used in teleoperation system. Haptics feedback can provide environmental interactions. A virtual force feedback method is studied in this paper. A virtual force extrapolation method is used to provide smooth force-feedback. Finally, an on-orbit satellite service experiment is achieved with this system and provides a convenient immersion, transparency and interaction.

Keywords: Haptic, Virtual Reality, Teleoperation, Time delay.

1 Introduction

The teleoperation system can drive the telerobots working under a dangerous environment or a place unreachable to human being. The working environments lead to distribution of the master-slave parts are in distance. And the time delay is also an important problem. The experiments indicate that when the time delay is larger than 0.25s, it will sharply decrease the system stability and effect the judgment of the operator in common bilateral control teleoperation system with force feedback .This will reduce efficiency and safety of teleoperation .Scientists suggested a lot to reduce the efficiency such as transparency, caused by time delay, such as tele-program, predictive control and event based supervisory control .Traditional method to reduce that efficiency is using VR based predictive display to simulate the working environment [1], [2], [3]. The operator interacted with the virtual reality environment continually, and the results can be showed in VR environment, then the telerobot would act as the one in the VR environment by time delay [4].

The network transfer delay can not be ignored when there is long distance between the master and slave .For example, the transfer delay between ground station and on-orbit telerobot reach 3~6s [5]. When only traditional methods are used in the teleoperation system, the system will become instability because of the time delay .Under this situation, the force-feedback and the video-feedback will lag behind a lot.

* Zhao Di is a PHD candidate in the School of Mechanical Science and Engineering, Huazhong Universty of Science and Technology. His research interests include virtual reality, human-computer interface and teleoperation.

L. Qi (Ed.): ISIA 2010, CCIS 86, pp. 377–383, 2011.

This kind of delay will interfere the operator's judgement. If the operator will wait for the corresponding feedback before making the judgement, the operation will be move-wait-move. This is not efficient. In order to solve the problem above, we build a force-feedback virtual reality environment that is not only a graphic mode simulation but also with the force-feedback. Instead of the remote force-feedback the VR environment gives the operator real time force-feedback. This will help operator makes the right choice. This kind of feedback varies with the real, and we have to revise some parameters by comparing with the real force-feedback. After that we can gain better immersion, interaction, and transparency. In this paper we use CyberGrasp force-feedback glove as the haptic equipment.

In this paper, we study a ground research system which is used to simulate the ground-space teleoperation system. This system includes VR based teleoperation system with force-feedback and video based supervisory control.

2 System Descriptions

The hole system includes VR equipment (such as Flock Of Birds, CAS DataGlove and CyberGrasp force-feedback glove etc.), two servers located in master/slave side, cameras which provide remote video-feedback, the model of satellite, a 4DOF robot arm, a robot hand and sensors which provide remote force-feedback etc. The master and slave are connected through Internet; the purpose of this experiment is to simulate satellite on-orbit services, under large time delay, with video feedback through Internet. The detail works including to unfold the sun panels when they failed to do so automatically and to wipe the polluted camera lens on the satellite Fig1.

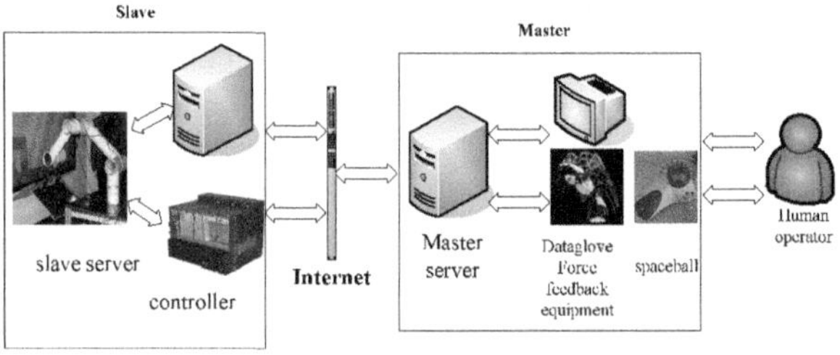

Fig. 1. Haptic master–slave teleoperation

The operator on the ground station acts as the master while the telerobot located in long distance acts as the slave. The bilateral control is used in this system.

3 Time Delay

1. In this teleoperation system, the time delay mainly include: the transfer delay d_t and the executancy delay d_τ. Transfer delay d_t mainly consist in the network transfers [6].It equals the command transfer delay add on the feedback information transfer delay, and this kind of delay are always very large and vary by random

when the distance is very long. executancy delay equals the sum of executancy delay in both master and slave sides. ($d_{\tau} = \tau_m + \tau_s$).Our study is to reduce the bad efficiency caused by transfer delay d_t , and to improve the stability and transparency of the system. The time delay block is show in Fig2 :

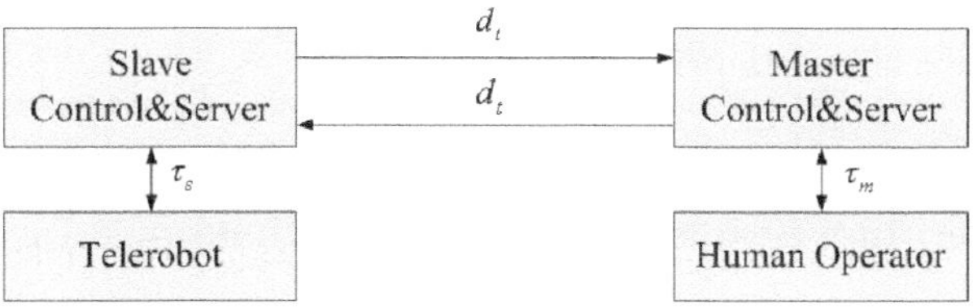

Fig. 2. The transfer time delay in Master-Slaver system

4 Building the Virtual Reality Environment

We use the OpenInventor 5.0 & Visual C++ 6.0 as our develop tools. And we build the 3D model of the telerobot and the satellites in Pro Engineer based on their real size, and then export the model as .iv format files. Then we assemble these models in OpenInventor view and constrain their degrees of freedom almost the same we do to the real environment in the slave side. These programs will run on the master server. The VR environment is shown as Fig 3:

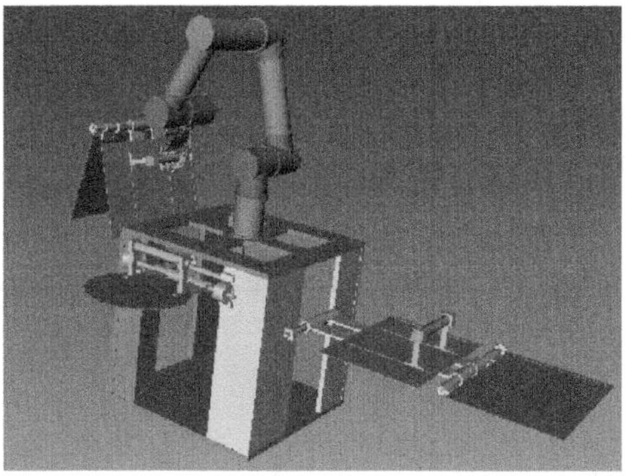

Fig. 3. On-orbit satellite service in Virtual Reality Environment

The telerobot is a kind of joint robot according to the mechanics. It is an open chain link machine in 3D space which is a machine make up of several components connected with several joints with multiple degrees of freedom [7]. The one we operate is a 4DOF open chain robot. In order to study the robotic kinematics we have to establish the relationship between joints movement and the coordinates of the manipulator. We describe these relationships in mathematic ways so that we can analyse and control the robot movement.

Building a robot model is not only to build the geomagnetic model but also to build the kinematic model. The kinematic model including the DOF and the constrain of the movement and the velocity and so on. Furthermore, we have to make the robot motion plan and the control parameters. At the end, we cannot avoid to model the environment.

We use the D-H method to solve the inverse-kinematics problem. The joint value can be worked out by the T matrixs. Then we map the 16 sensors in CAS glove to the manipulate joints because the manipulate act like a human hand [8]. After a demarcation, operator can operate the manipulator by easily bend one or several fingers.

5 Tele-Operation System with Force Feedback

According to the research, force feedback can enhance the immersion of the tele-operation system, and improve the efficiency and veracity of the operation. If people operate the master side only depends on the VR graphic simulation without force feedback, the modeling error will lead to the failure. Some scientists studied on the tele-operation system, the research indicate that force feedback can reduce 40% time to accomplish the task.

Human can feel force variations about 300~1000HZ [9]. The lower frequency will lead intermission to human's haptic feelings. Direct sensor force feedback frequency from remote is low. The force feedback depends on the calculating speed of the virtual force. In this paper, two methods are combined to calculate the virtual force in VR environment, and feedback to the human operator in time.

Based on the mapping relationship of the robot hand and the VR environment, using the angle displacement of fingers on robot hand to instead of the move of the robot hand, and calculate the virtual force by Hooke's law [10]. Then update the force feedback to the operator.

Use the insert-value method to extrapolate the virtual force based on time sequence, disperse the force value which is gotten in experiment and save to a database. Calculate the virtual force in VR environment as a symbol, and query from the database to extrapolate and feedback to the human operator. This will achieve smooth force feedback and conquer the impulse of the force feedback.

The force changes because of the move of the robot hand which made the distortion of face-contacted of the object. Considered the characteristics of the robot hand, the angle displacement of fingers R is referred to the position changes P of the fingers. While the angle displacement is easier to get than the position change, the angle displacement is used to instead the distortion of the face-contacted in virtual force calculation. This approximation will reduce the calculation of virtual force. The t time virtual force $F(t)$ changes from

$$F(t) = \begin{cases} f(P_n) & t = n \\ f(a_{n+i}) & n < t < n+1 \end{cases} \tag{1}$$

$$F(t) = \begin{cases} f(R_n) & t = n \\ f(a_{n+i}) & n < t < n+1 \end{cases} \tag{2}$$

Function f is referred to the mapping of the angle displacement and it can be query in the database, P_n referred to the position of the finger, R_n referred to the angle displacement of the finger, $f(a_{n+i})$ referred to the extrapolation of the virtual force.

5.1 Virtual Force Extrapolation Based on Material Characteristics

Calculating speed of the virtual force is not fast enough to catch up human operator's feeling. An extrapolation of virtual force based on material characteristics by testing in experiment is studied in this paper. The approaches are:

1. Operate the HIT/DLR robot hand to grasp object and save the force value and draw a curve.
2. Disperse the force value and save to the database
3. Using the calculated virtual force in the VR environment as a symbol, extrapolate the virtual force.

An angle displacement R_n is gotten in every refresh of the VR environment. The virtual force is calculated by function $F(t) = f(R_n)$. Query the most near force value $f(a_n)$ in database, feedback $f(a_n) \cdots f(a_{n+i})$ to human operator until next update of VR. The calculation should judge the grasp and release action to decide the orientation of the extrapolation. The extrapolation of $F(t_n)$ is

$$F(t) = \begin{cases} F(t_n) & t = t_n \\ F(a_{n+i}) & t_n < t < t_{n+1} \\ F(t_{n+1}) & t = t_{n+1} \end{cases} \tag{3}$$

This extrapolation of virtual force is based on material characteristics, so the extrapolation is very close to the real force feedback. The next update and re-query will avoid the accumulative errors. The process is shown as fig4

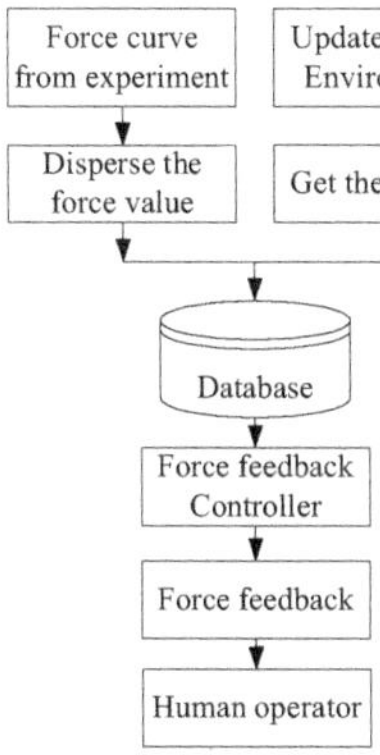

Fig. 4. The approach of virtual force extrapolate

6 The Experiment

The operator interacts with the virtual slave in the master side. The virtual slave provides VR environment to simulate the real slave side and graphic feedback to the operator. The tele-operation commands are examined in virtual slave to be inerrant before send to the robot in slave side by transmission network.

A 4DOF robot arm and a 4-finger robot hand (13DOF) are used as the slave robot system in this tele-operation experiment. Spaceball 5000 (6DOF) is used as a input equipment to operate the robot arm and CyberGlove dataglove is used to operate the robot hand, while CyberGrasp is used as the force feedback equipment. Through the internet, a tele-operation experiment between Wuhan and Harbin was held to push to unfolding the invalid sun panels and to wipe the polluted camera lens on the satellite. A time-vary delay was added in the transmission network to simulate the maintenance on the of airspace equipment.

When the tele-operation started, the contact was established between the master & slave side, initialized by the remote sensor feedback.

- With the help of local VR environment, the operator use the Spaceball to move remote robot system (including robot arm & hand) to approach the sun panel
- When reaches the panel, adjust the pose of robot arm and initialize the robot hand to grasp
- Grasp the handle on the sun panel with robot hand by using CyberGlove dataglove, the force feedback is provided by Cybergrasp equipment so that the operator could confirm a stable grasp to the handle
- Program a path and push the sun panel to a correct position
- Release the handle and take back the robot system in the slave side

The experiment shows that the force feedback becomes smooth by extrapolating virtual force (fig5).

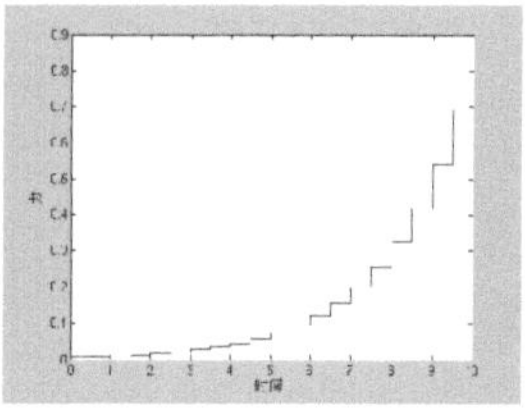

(a)

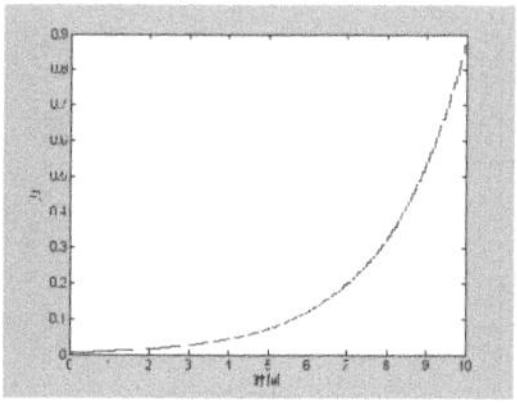

(b)

Fig. 5. The Force-feedback results with & without virtual force extrapolate, (a) is virtual force feedback without extrapolation, (b) is virtual force feedback with extrapolation

Obviously, the force feedback is smoother when virtual force extrapolation is used. The disadvantage of this method is the material characteristics should be tested beforehand.

7 Conclusion and Future Work

In this paper, we discuss a VR based teleoperation system adds on virtual force feedback. First, we build a VR environment and video feedback system in master server. Second, VR equipments are used to create a more natural operation. At the end, we unfold the folded sun panels successfully through Internet.

In the future, we will improve the virtual force feedback environment, and establish the module of multiple operator-virtual robot manners to approach the Immersion, Interaction and transparency.

References

1. Kim, W.S.: Computer Vision Assisted Virtual Reality Calibration. IEEE Transactions on Robotics And Automation 15(4), 450–464 (1999)
2. Hirzinger, G., Brunner, B.: ROTEX— The First Remotely Controlled Robot in Space. In: IEEE International Conference on Robotics and Automation, pp. 2604–2611 (1994)
3. Oda, M., Kibe, K., Yamagata, F.: ETS-VII, Space Robot In-Orbit Example Satellite. In: IEEE International Conference on Robotics and Automations, pp. 739–744 (1996)
4. Zhuang, J., Qiu, P., Sun, Z.: Distributed telerobotic system with large time delay. Journal of Tsinghua University 40(1), 80–83 (2000) (in Chinese)
5. Sheridan, T.B.: Space teleoperation through time delay: review and prognosis. IEEE Trans. on Robotic and Automation 9(5), 593–606 (1993)
6. Tavakoli, M., Patel, R.V., Moallem, M.: A haptic interface for computer-integrated endoscopic surgery and training. Virtual Reality 9, 160–176 (2006)
7. Jiao, E.Z.: Simulation of Robot Motion Based on AutoCAD. Journal of Computer-aided Design & Computer Graphics 13, 932–936 (2001) (in Chinese)
8. Xiong, Y., Li, S., Wang, W.: Operating Technology of Virtual Robot Based on Data Glove Drive. Mechanical Science And Technology 23 (2004)
9. Shimoga, K.B.: A survey of perceptual feedback issues in dextrous telemanipulation: part I. Finger force feedback. In: Proceedings of the IEEE Annual Virtual Reality International Symposium, pp. 271–279 (1993)
10. Song, A., Dan, M., Edward Colgate, J., Li, J.: Real Time Softness Haptic Display Device for Teleoperation and Virtual Operation. Chinese Journal of Scientific Instrument. 27 (2006) (in Chinese)

Comparison of Pagination Algorithms Based-on Large Data Sets

Junkuo Cao, Weihua Wang, and Yuanzhong Shu

Nanchanghangkong University, Information and engineer school, Fenghe south Road. 696, 330063 NanChang, China
wwh985176@163.com

Abstract. At present, many kinds of databases are widely used in Web applications. When Search Engines find tens of thousands of results for a keyword, it's very important to spit out and display the results in multiple pages quickly and efficiently instead of just putting them all in one long page. This paper firstly presents several kinds of pagination algorithms based-on Large Data Sets which belongs to the database-driven method. Then, we will experiment on the large data sets and discuss different pagination approaches. The experimental results show that proc_Rownumber() method can greatly improve the performance on the query speed of the pagination.

Keywords: pagination algorithm; Large Data Sets; web application.

1 Introduction

With the amount of data in the database increasing sharply, too many rows will be returned in the result set, which often reach millions of lines, sometimes even hundreds of millions of lines. This will lead to the increase of the query time and inconvenience for people to browse information. So pagination is practically required.

On the Internet, pagination is used for such things as displaying a limited number of results on search engine results pages, or showing a limited number of posts when viewing a forum thread[1]. In other words, Paging means showing your query result in multiple pages instead of just put them all in one long page.

This paper is organized as follows. In Section 2, we firstly introduce the general principle of pagination algorithm. Following that, we will describe several pagination algorithms in detail. Then in Section 3, experiments and analysis are conducted. Finally, we give our conclusion in Section 4.

2 Summary of Pagination Algorithms

Unless the returning result set is guaranteed to be very small, any web application with search capabilities must have pagination. For instance, if the result set is less then 30 rows, pagination may be optional. However, if it's bigger then 100 rows, pagination is highly recommended, and if it's bigger then 500 rows, pagination is practically

L. Qi (Ed.): ISIA 2010, CCIS 86, pp. 384–389, 2011.

required. Pagination can be handled client-side or server-side. Server-side pagination is more common. Client-side pagination can be used when there are very few records to be accessed, in which case all records can be returned, and the client can use Javascript to view the separate pages. By using AJAX, hybrid server/client-side pagination can be used, in which Javascript is used to request the subsequent page which is loaded and inserted into the Document Object Model via AJAX[2]. Server-side pagination is appropriate for large data sets providing faster initial page load, accessibility for those not running Javascript, and complex view business logic.

2.1 General Principle of Pagination Algorithm

Principle of pagination algorithm is as follows: Besides showing correctly records per page in the web application, we often consider the problem of links page: home page, previous page, next page, tail page. When clicking each time for its links, we will return the value of Page number. The detailed design is as follows:

1) Setup the initial parameters of this program. Through the database query, we can get the total records' number which can be expressed as allCount. Setup the size of page which can be expressed as pageNum to display rows of records per page. Then according to allCount and pageNum, we can get the total page number named pageCount.
2) Returned page number will be assigned to thePage which is on behalf of the current page number. At first, thePage will be established to 1. Then we show pageNum pieces of records from first record of the current page.
3) Finally, we design its links: first page, previous page, next page, tail page. Attention, first page will return 1 as the value of page number; Previous page will return the difference of the current page number and 1 as the page number; Next page will return the sum of the current page number and 1 as the page number; Tail page will return pageCount as the value of page number.

Correctly implementing pagination can be difficult[3]. There are many different usability questions such as should "previous" and "next" links be included, how many links to pages should be displayed, and should there be a link to the first and last pages[4].

2.2 Several Pagination Algorithms

At present, there are several pagination methods by using SQL query. This article will introduce several common paging stored procedures. Details are as follows:

1) Using Select Max and Select Top Statements to Page

```
create procedure proc_selectMax(@pageIndex int,
 @pageSiint)
 as
  begin
```

```
    set nocount on;
    declare @timediff datetime
    declare @sql nvarchar(500)
    select @timediff=Getdate()
    set @sql='select top '+str(@pageSize)+' * From
   TestTable where(ID>(select max(id) From (select top
   '+str(@pageSize*@pageIndex)+' id From TestTable order
   by ID) as TempTable)) order by ID'
    execute(@sql)
    select datediff(ms,@timediff,GetDate()) as
Time-consuming
   set nocount off;
 end
```

2) Using Select Not In and Select Top Statements to Page

```
create procedure proc_notin•@pageIndex int,@pageSize int)
 as
 begin
     set nocount on;
     declare @timediff datetime
     declare @sql nvarchar(500)
     select @timediff=Getdate()
     set @sql='select top '+str(@pageSize)+' * from
     TestTable where(ID not in(select top
     '+str(@pageSize*@pageIndex)+' id from TestTable order
     by ID ASC)) order by ID'
     execute(@sql)
     select datediff(ms,@timediff,GetDate()) as
     time-consuming
     set nocount off;
  end
```

3) Using startNo and endNo To Page

```
create procedure proc_endNo ( @pageIndex int,@pageSize
int)
```

```
 as
 begin
     set nocount on;
     declare @timediff datetime
     declare @sql nvarchar(500)
     select @timediff=Getdate()
     set @sql = 'select top '+str(@pageSize)+' * from
    (select top '+str(@pageSize)+' * from ( select top '
    + str(@pageIndex*@pageSize) +' * from (select * from
    TestTable)t1 order by id desc)t2 order by id asc )t3
    order by id desc'
     execute(@sql)
     select datediff(ms,@timediff,getdate()) as
Time-consuming
    set nocount off;
end
```

4) Using proc_Rownumber() To Page

```
create procedure proc_Rownumber( @pageIndex int,@pageSize
int )
  as
  begin
     set nocount on;
     declare @timediff datetime
     select @timediff=getdate()
     select * from (select *,proc_Rownumber() over(order by ID
     asc) as 'IDRank' from testTable) as IDWithRowNumber
     where IDRank>@pageSize*@pageIndex and
     IDRank<@pageSize*(@pageIndex+1)
      select datediff(ms,@timediff,getdate()) as
Time-consuming
      set nocount off;
 end
```

3 Experiments and Analysis

Our experiment includes 2000000 pieces of records. And we test the pagination algorithms in large data sets. To evaluate the effectiveness, we design some experiments on the data set of 2000000 records in the table named TestTable.

3.1 Test Environment

The test environment we adopt is made up of hardware and software.The section of hardware adopts personal computer that the single frequency of its Dual-Core CPU is 2.8GHz and the capacity of its memory is 2GB while the part of software adopts the XP SP3 operating system and the database of Microsoft SQL Server 2005.

3.2 Experiments Setup

Firstly, we create a test database named TestDatabase. Then we create a test table named TestTable on the base of the database. The pseudo-code is as follows:

```
create database TestDatabase  --create database
GO
use TestDatabase
GO
create table TestTable   --create table:TestTable
(    id bigint identity(1,) primary key,
     userName nvarchar(20) not null,
     userPWD nvarchar(20) not null,
     userEmail nvarchar(40) null )
GO
```

Then we insert 2000000 pieces of records into the TestTable. The pseudo-code is as follows:

```
set identity_insert TestTable on
declare @count int
set @count=1
while @count<=2000000
begin
  insert into TestTable(id,userName,userPWD,userEmail)
     value(@count,'wangwu',wangwu','wangwu@163.com.cn')
    set @count=@count+1
end
set identity_insert TestTable off
```

3.3 Test Result and Analysis

So far, experimental programs are created. We separately carry out the tests on page 1000, on page 10000, on page 100000, on page 199999 on the conditions of 10 rows records per page, time-consuming unit: ms. We test five times per page and use their average value as final test value. As Table2 shows, proc_Rownumber() method uses a fat lot time to execute a piece of query statement in such large data set which contains 2000000 pieces of records. Average executive time only has 406ms. So proc_Rownumber() approach is very high-efficient on the part of executive speed, especially in large data sets.

Table 1. We separately carry out the test on page 1000, on page 10000, on page 100000, on page 199999 on the conditions of 10 rows records per page. We test five times per page and use their average value as final test value. time-consuming unit: ms.

Procedure name	Page 1000	Page 10000	Page 100000	Page 199999	Rank
proc_notin	13	30.6	319	643.2	3
proc_selectMax	0	23.8	218.4	436.8	2
proc_endNo	9	79.4	777.2	1540.4	4
proc_Rownumber	0	14.2	205.6	406	1

4 Conclusion

In this paper, we have discussed several pagination algorithms based-on the large data sets. Experimental result indicates that proc_Rownumber() method can greatly improve the efficiency of queries compared with other methods. Above procedures are implemented by Microsoft SQL Server 2005 and applied equally to other relational databases such as MYSQL, ORACLE and so on. Now they are used widely by all kinds of web applications. So pagination technology discussed in this article has practical value.

References

1. Paging (Web), `http://en.wikipedia.org/wiki/Pagination_web`
2. Mikheevc, O.: Ajax programming with Struts 2. Network World (2009)
3. Baptiste, L.: Perfect PHP Pagination. SitePoint (2009)
4. Gervasio, A.: Previous or Next? Paginating Records with PHP-Part 3. Developer Shed (2009)

An Effective Conflict Management for Large Transactions in Hardware Transactional Memory System

Chen Fu*, Dongxin Wen, Xiaoqun Wang, and Xiaozong Yang

Harbin Institute of Technology,
Harbin 150001, China
fuchen2003@gmail.com

Abstract. In Transactional Memory (TM) for multicore systems, contention management refers to the mechanisms used to guarantee forward to avoid performance pathologies, and to promote throughput. The choice of contention management police impacts strongly on the performance of applications. In this paper, we study contention management policies for Hardware Transactional Memory (HTM). Although the results were obtained from a HTM framework, the conclusions and proposals are applicable to any type of TM: hardware, software or hybrid. We first characterize transactions' behaviors on execution time.We find that significant time is wasted when few transactions forward progress due to the data conflicts among different transactions. To reduce this kind of waste time, we propose a novel contention management(CM) scheme that reduces waste time and guarantees forwarding. We evaluate our techniques using a detailed execution-driven simulator. The results show that our scheme can effectively minimize the amount of waste time and, consequently, decrease execution time than previous approaches for large transactions.

Keywords: hardware transactional memory, contention management, parallel programming, multicore processors.

1 Introduction

Transactional memory (TM) [1] makes concurrent programming easy by providing atomic execution for a block of code. A programmer can replace locks with transactions in a multi-threaded application. TM systems get high performance by speculatively executing transactions concurrently and only committing transactions serially. Transactions can be concurrently committed if they do not conflict. A conflict occurs when two or more concurrent transactions access the

* This work is supported by the National High Technology Development 863 Program of China under grant #2008AA01A201. In the meanwhile, this work is also supported by the National Science Foundation of China under grant #60503015 and the ShanDong Province Science and Technology Development Foundation under grant #2007GG10001020.

L. Qi (Ed.): ISIA 2010, CCIS 86, pp. 390–396, 2011.

same location and at least one access is a write. TM systems may resolve some conflicts by stalling one or more transactions, but must be able to abort transactions with cyclic conflicts.While some TM systems operate completely in software transactional memory (STM) [2] or in software with hardware acceleration [3], this paper focuses on those implemented with hardware support.

Guaranteeing throughout has been the focus of contention management (CM). Currently, many CM policies are proposed. Scherer et al. [4] studies a set of arbitration heuristics on the STM framework. They use transactions gather information (e.g., priority, read/write set size, and number of aborts) to decide which transactions to abort. However they did not evaluate conflict resolution time as an important design choice available to the contention manager. As for hardware supported TM systems, main focus is on implementation tradeoff and contention management issues are ignored. Bobba et al. [5] firstly studied the performance pathologies due to specific conflict detection, management, and versioning policies in HTMs. They proposed enhanced HTMs to mitigate the impact of specific pathologies. Furthermore, they analyzed specific points in the design space, making it difficult to choose a universal policy which can be effective for TM applications with different kinds of characteristics. Most recently, Ramadan et al. [6] have proposed dependence-aware transactions which forward data between speculative transactions and tie their destiny together in order to uncover more concurrency. Whether the performance improvements promised by dependence-awareness can make up the hardware complexity is not yet clear. Their goals and approach is different than our work.

In this paper, we introduce a new CM scheme for large transactions. Our study across a wide set of applications makes the following contributions:

(1) We corroborate that stall, abort, and backoff are the main factors that hurt the TM system performance. Therefore, to reduce the amount of their time can cut down waste work.

(2) We show that scheduling the conflicted transactions from global perspective is helpful to avoid waste work and improve performance.

(3) We reveal that while using the right contention management policy can avoid pathological situations, the choice of conflict resolution time is also important to improve performance.

2 TMProf

Yen [7] proposes TMProf which is an online lightweight profiling framework to diagnose and analysis the performance a TM system. TMProf implements a set of hardware performance counters to track the execution cycles of the eight transactional related events in each processor core. They are defined as follows:

Useful transaction: Cycles executed by successful transactions that commit.

Wasted transactions: Cycles executed by lose transactions that eventually abort.

Non-transaction: Cycles spent by non-transaction instructions.

Commit: Cycles spent by committing operations.
Stall: Cycles spent by waiting for conflict to resolve.
Abort: Cycles spent by processing aborts.
Backoff: Cycles spent in backoff.
Barrier: Cycles spent in barrier.

3 Global Conflict State and Waste Work

We use eight TMProf measurements: useful transaction, wasted transaction, non-transaction, commit, barrier, stall, abort, backoff and barrier to obverse the whole system runtime state. In the runtime of benchmarks, considerate cycles are used to stall, abort or backoff transactions. In our experiment, we find one phenomenon. When the cycles of stall, abort backoff or wasted transaction increase rapidly, useful transaction cycles increase slowly. Considerate CPU cycles are spent in stall, abort and backoff events. To reduce these cycles is helpful to improve performance. So we exploit a dynamic scheme named SAB Contention Management (SABCM) to reduce the cycle of waste work in runtime.

In the traditional methods, the most CM policies only use some local conflict states to decide when and which transactions will stall or abort. They will get best performance only if there are few local conflicts and all the conflict transactions are considered, otherwise the result may be poor. At the worst case, plenty of conflicts may be produced and the system stops to do effective work.

For example, transaction A shares data with transaction B. At the same time, transaction B conflicts with transaction C. using local conflict state, stall B is the best method to resolve the problem. After stalling B, transactions D, E, F may Conflict with A, but can execute with B in parallel. At this moment, the D, E, and F will be stalled. At last, the number of stalled transactions will increase evidently. There are two reasons lead to this result. One is the local greedy method is not the most optimal scheme in global perspective. By using the global conflict state, the most optimal scheme may be gotten. Another reason is premature optimization. This is the problem that when to resolve conflicts is the most optimal time.

According to our experiments, immediate intervention is not good to conflict resolution. In our scheme, there are two steps to decide when the contention manager will intervene. The first step is to wait a period and to collect the global conflict state. The second step is that if the cycles of stall, abort and backoff increase rapidly, the global contention manager will schedule some transactions in stall state with aborting them.

4 Platforms and Methodology

In this section, we describe the HTM system and the benchmarks used in our simulation.

4.1 The HTM System

We evaluate our CM scheme using full-system execution-driven simulation based on the Wisconsin GEMS toolset [8] coupled with Virtutech Simics [9]. Table 1 shows the base 16-core CMP system parameters we use for our simulation.

Table 1. Simulated CMP system parameters

	System model
Processor cores	16 5GHz in-order single-issue
L1 cache	32 KB 4-way split, 64-byte blocks, writeback, 1-cycle latency
L2 Cache	8 MB 8-way unified, 64-byte blocks, writeback, 34-cycle latency
Memory	4 GB, 448-cycle latency
L2 Directory	Bit vector of sharers, 6-cycle latency
Internet	Point-to-point, 7-cycle link latency
Protocol	MESI
Signature	perfect

The base HTM system we simulate is based on LogTM-SE [10]. It has basic CM policies named Base [7].The Base policy make requestors stall on conflicts and repeatedly retry the conflicting memory access until potential deadlock occurs.Furthermore, we combinate our idea to the base policy and get an enhanced CM policy: SABCM-BG.

4.2 Workloads

In order to observe a range of program behavior, we used four benchmarks (Delaunay, Genome, Vacation and Kmeans) from the STAMP workload suite [11] and one SPLASH-2 benchmarks [12] (Barnes). Table 2 highlights the benchmarks' characteristics such as contention level, read/write set size, transaction length and transaction time.

5 Results

Figure 1 shows the speedup of our SABCM techniques compared to a system that only uses the Base policy.

From this figure, it is clear that our enhanced policy with SABCM scheme can speed up significantly than the baseline policy for large transaction benchmarks. For example, SABCM-BG separately achieves speedups of 1.42, 1.41, 1.12 and 1.02 relative to the baseline for the Delaunay, Vacation, Genome and Barnes.

Table 2. Transactional workload characteristics

Benchmark	Contention level	Read/Write set size	Tx length	Tx time ratio
Delaunay	Medium	Large	Long	High
Genome	Low	Medium	Medium	High
Vacation	Medium	Medium	Medium	High
Kmeans	Low	Small	short	Low
Barnes	High	Small	Short	Low

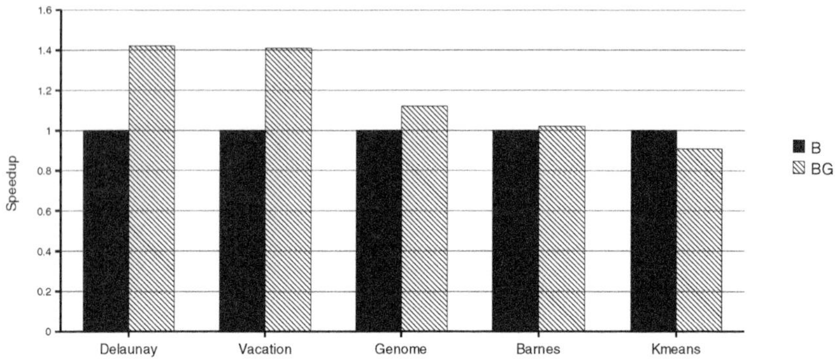

Fig. 1. Execution time speedup of TM systems with different CM policies

The reason is that SABCM scheme is very effective at directly reduce stall cycles when the numbers of stall, abort and backoff increase rapidly, thereby reducing the amount of waste time. Delaunay, Vacation, and Genome have more transaction length, transaction time or higher level of contention than Barnes. It leads to more conflicts than Barnes, therefore SABCM scheme can get more benefit on them than Barnes. Barns has short transaction length, low transaction time but high level of contention, therefore its conflict frequency is between large and small transaction benchmarks. It gets moderate benefit with SABCM.

Our enhanced policy with SABCM scheme get little lower speedup in small transaction benchmarks. For Kmeans, SABCM-BG, achieve 0.91 speedup only. It is because that SABCM's benefit can not merit its cost in the cases that there is few conflicts. On the other side, it illustrate that the cost by SABCM is little relative to its benefit.

Figure 2 shows the fraction of execution time of our SABCM techniques to the Base policy system. From this figure, it is clear that our SABCM-BG techniques can diminish even remove the backoff cycles than the baseline policy for all benchmarks. And it reduce the stall cycles for Delaunay and Vacation but it slightly increase the stall cycles for Genome and Barnes than the baseline policy. The wasted cycles is increased for most of the benchmarks except Kmeans with our SABCM-BG. From the positive and negative balance, the final result of our SABCM-BG is that the total cycles is reduced.

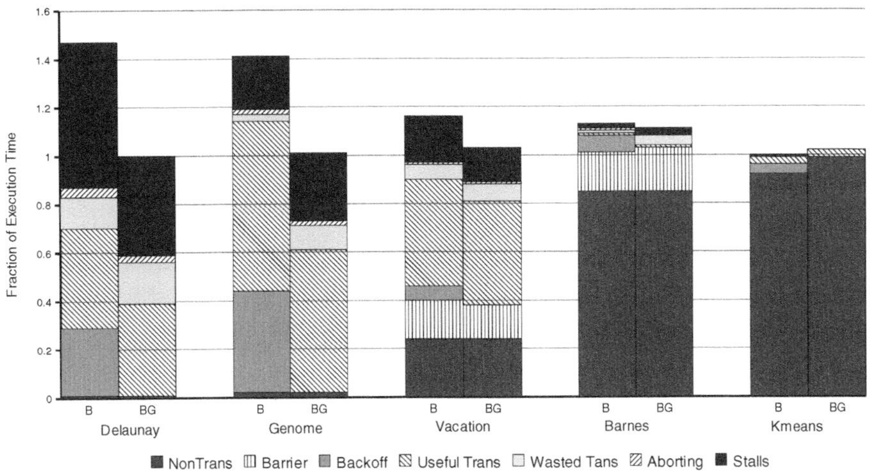

Fig. 2. Fraction of execution time with different CM policies

6 Conclusion

In this paper, we study of the behavior of large transactions in HTM systems. Our experiments corroborate recent studies that stallabort and backoff are the important facts of impact performance. In particular, reducing the amount of their time avoids wasted work. To reduce this waste, we proposed a new CM scheme using the global conflict state. We reveal that while the right choice of CM policy can reduce pathological situations, the choice of conflict resolution time is also important to improved performance. Finally, our results show that our CM scheme in HTMs provides a good compromise between exploiting concurrency, saving wasted work, and implementation complexity for large transactions.

Acknowledgments. We thank the Harbin Institute of Technology Fault Tolerant Computing Lab.

References

1. Herlihy, M., Moss, J.E.B.: Transactional Memory: architectural support for Lock-Free data structures. In: Proceedings of the 20th Annual International Symposium on Computer Architecture. Conference Proceedings - Annual Symposium on Computer Architecture, San Diego, CA, USA, pp. 289–300. IEEE Press, Los Alamitos (1993)
2. Larus, J.R., Rajwar, R.: Transactional Memory. Morgan & Claypool Publishers, San Rafael (2006)
3. Damron, P., Fedorova, A., Lev, Y., Luchangco, V., Moir, M., Nussbaum, D.: Hybrid transactional memory. In: Proceedings of the 12th International Conference on Architectural Support for Programming Languages and Operating Systems, ASPLOS-XII, pp. 336–346. ACM, New York (2006)

4. Scherer III, W.N., Scott, M.L.: Randomization in STM Contention Management (POSTER). In: Proceedings of the 24th ACM Symposium on Principles of Distributed Computing, Las Vegas, NV (2005)
5. Bobba, J., Moore, K.E., Volos, H., Yen, L., Hill, M.D., Swift, M.M., Wood, D.A.: Performance pathologies in hardware transactional memory. In: Proceedings of the International Symposium on Computer Architecture, San Diego, CA, USA, pp. 81–91. IEEE Press, Los Alamitos (2007)
6. Ramadan, H.E., Rossbach, C.J., Witchel, E.: Dependence-aware transactional memory for increased concurrency. In: Proceedings of the 2008 41st IEEE/ACM International Symposium on Microarchitecture, MICRO 2008, pp. 246–257. IEEE Computer Society Press, Los Alamitos (2008)
7. Yen, L.: Signatures in Transactional Memory systems. PhD thesis, Department of Computer Science, University of Wisconsin-Madison, WI, USA (2009)
8. Martin, M.M.K., Sorin, D.J., Beckmann, B.M., Marty, M.R., Xu, M., Alameldeen, A.R., Moore, K.E., Hill, M.D., Wood, D.A.: Multifacet's general execution-driven multiprocessor simulator (GEMS) toolset. SIGARCH Comput. Archit. News 33, 92–99 (2005)
9. Magnusson, P.S., Christensson, M., Eskilson, J., Forsgren, D., Hallberg, G., Hogberg, J., Larsson, F., Moestedt, A., Werner, B.: Simics: A full system simulation platform. Computer 35, 50–58 (2002)
10. Yen, L., Bobba, J., Marty, M.R., Moore, K.E., Volos, H., Hill, M.D., Swift, M.M., Wood, D.A.: LogTM-SE: decoupling Hardware Transactional Memory from caches. In: Proceedings of the IEEE 13th International Symposium on High Performance Computer Architecture, Scottsdale, AZ, USA, pp. 261–272. IEEE Press, Los Alamitos (2007)
11. Herlihy, M., Luchangco, V., Moir, M., Scherer III, W.N.: Software Transactional Memory for Dynamic-Sized data structures. In: Proceedings of the 22nd Annual Symposium on Principles of Distributed Computing, Boston, Massachusetts, USA, pp. 92–101. ACM Press, New York (2003)
12. Woo, S.C., Ohara, M., Torrie, E., Singh, J.P., Gupta, A.: The SPLASH-2 programs: characterization and methodological considerations. In: Proceedings of the 22nd Annual International Symposium on Computer Architecture, ISCA 1995, pp. 24–36. ACM, New York (1995)

A Comprehensive Scheme for Contention Management in Hardware Transactional Memory

Xiaoqun Wang*, Zhenzhou Ji, Chen Fu, and Mingzeng Hu

Harbin Institute of Technology,
Harbin 150001, China
wxiaoqun@gmail.com

Abstract. Transactional Memory (TM) is one kind of approach to maximize parallel performance for multicore systems. There are conflicts When two or more parallel transactions access the same location and at least one access is a write. Contention management(CM) refers to the mechanisms used to guarantee forward—to avoid performance pathology, and to promote throughput. In this paper, we introduce a new CM police. We remitted six of seven performance pathologies summered by Bobba. Our result shows high performance for large transactions, while get moderate improvement or little slowdown for small transactions. The performance of the systems used this policies combined with other policy are steady.

Keywords: hardware transactional memory, contention management, parallel programming, multicore processors.

1 Introduction

Transactional memory systems (TM) [1] are expected to be a technique for parallel programming for general-purpose computing. Especially for hardware transactional memory (HTM) systems, the programmers need not worry about the correctness of a multithreaded application. Generally, transactional memory systems can exceed the performance of lock based systems which serialize all the transactions when they share data. When the executing workload lacks sufficient parallelism, lock based systems do not hurt the performance since they were pessimistically serializing every transaction. For a transactional memory system, excessive transactions can actually result in considerate conflicts. A conflict is defined that when two or more concurrent transactions access the same location and at least one access is a write.TM systems resolve them by CM policies.

* This work is supported by the National High Technology Development 863 Program of China under grant #2006AA04A103. In the meanwhile, this work is also supported by the National Science Foundation of China under grant #60503015 and the ShanDong Province Science and Technology Development Foundation under grant #2007GG10001020.

L. Qi (Ed.): ISIA 2010, CCIS 86, pp. 397–403, 2011.

The concept of "contention management" is introduced by Herlihy et al. [2] in the context of software transactional memory (STM) up to date, many CM policies are proposed for resolve conflicts. Scherer et al. [3] studies a set of arbitration heuristics on the STM framework. They firstly use transactions gather information such as priority, read/write set size, number of and aborts. Then they it them decide which transactions will be aborted. However they did not evaluate conflict detection time as an important design choice. In most HTM systems,

The tradeoff of version management and conflict detection is the important research focus. Contention management is always ignored. Bobba et al. [4] firstly studied the performance pathologies due to specific conflict detection, contention management, and version management policies in HTMs. They proposed enhanced HTMs to remit specific pathologies. Furthermore, they analyzed specific points in the design space, making it difficult to choose a universal policy which can be effective for TM applications with different kinds of characteristics. TxLinux [5] exploits HTM and integrates transactions with the operating system scheduler. They focus on achieving synchronization in the kernel for future processors with TM hardware support. Their goals and approach is different than our work.

In this paper, we introduce a new CM police. Our study across a wide set of applications makes the following contributions:

(1)We remitted six of seven performance pathologies summered by Bobba.

(2) Our result shows that our techniques can effectively remit six performance pathologies and the tendency of deadlock, minimize the amount of waste time and consequently, decrease execution time than previous approaches.

(3) Our result shows that the performance of the systems used this policy combined with other policies are steady.

2 TMProf

Yen [6] proposes TMProf which is an online lightweight profiling framework to diagnose and analysis the performance a TM system. TMProf implements a set of hardware performance counters to track the execution cycles of the eight transactional related events in each processor core. They are defined as follows:

Useful transaction: Cycles executed by successful transactions that commit.

Wasted transactions: Cycles executed by lose transactions that eventually abort.

Non-transaction: Cycles spent by non-transaction instructions.

Commit: Cycles spent by committing operations.

Stall: Cycles spent by waiting for conflict to resolve.

Abort: Cycles spent by processing aborts.

Backoff: Cycles spent in backoff.

Barrier: Cycles spent in barrier.

3 Remitting Six Performance pathologies

We use eight TMProf measurements: useful transaction, wasted transaction, non-transaction, commit, barrier, stall, abort backoff and barrier to obverse the run-time state. We find three interesting phenomena. The first phenomenon is that all eight TMProf measurements are constant in a certain sampling period. As time is elapsed, no transaction is forwarded. This phenomenon is similar to dead-lock, so we defined it the tendency of deadlock. The second phenomenon is that only two TMProf measurements (useful transaction cycle and non-transaction cycle) are constant in a certain sampling period. When time is elapsed, the cycles of good transaction and nontransaction are not increased. However, the numbers of other six TMProf measurements are increased. All cycles are used in stall, barrier, abort, commit, backoff and non-transaction phase. No useful work is forwarded. This phenomenon is similar to livelock, therefore we define it the tendency of livelock. The third phenomenon is that when stall, abort backoff and wasted transaction cycles increase rapidly, useful transaction cycles increase slowly. Considerate cycles are spent in stall, abort and backoff events.

In the Bobba's paper [4], he summered that there are seven performance pathologies. When they occur, considerate cycles are wasted and programs are forwarded slowly or stopped. These pathologies are described as follows:

FriendlyFire: Concurrent transactions that conflict.

StarvingWriter: Transactions that modify a widely read shared variable.

SerializedCommit: Threads frequently use short, concurrent transactions.

FutileStall: Transactions that read and then later modify highly contended data.

StarvingElder: Conflicting accesses by a long transaction and a sequence of short transactions.

RestartConvoy: Repeated instances of a transaction that updates a contended memory location.

DuelingUpgrade: Concurrent transactions that first read a common set of blocks, and then update one or more of them.

I also find that some performance pathologies have relations with these three phenomena we observed. When pathology FriendlyFire or pathology FutileStall occurs, the system may lead to the tendency of livelock. When pathology Restart-Convoy occurs, the system shows that stall, abort backoff and wasted transaction cycles increase rapidly, useful transaction cycles increase slowly. When some of performance pathologies occur, the system may lead to the tendency of deadlock.

In our experiment, we find that there are considerable stall cycles when these three phenomena happen. When reduce the stall cycles, the waste cycles also decrease and the system is speedup finally. Therefore the system using this CM policy can get higher performance.

To remit the pathology DuelingUpgrade, StarvingElder and StarvingWriter, a retry queue is added to the policy to choice the transactions in stall state. We find that the queue is helpful to reduce waste time in these three pathologies. Up to now, we remitted six of seven performance pathologies which are summered in Bobba's paper. We called this policy SCM.

4 Platforms and Methodology

4.1 The HTM System

We evaluate our CM scheme using full-system execution-driven simulation based on the Wisconsin GEMS toolset [7] coupled with Virtutech Simics [8]. Table 1 shows the base 16-core CMP system parameters we use for our simulation.

Table 1. Simulated CMP system parameters

	System model
Processor cores	16 5GHz in-order single-issue
L1 cache	32 KB 4-way split, 64-byte blocks, writeback, 1-cycle latency
L2 Cache	8 MB 8-way unified, 64-byte blocks, writeback, 34-cycle latency
Memory	4 GB, 448-cycle latency
L2 Directory	Bit vector of sharers, 6-cycle latency
Internet	Point-to-point, 7-cycle link latency
Protocol	MESI
Signature	perfect

The base HTM system we simulate is based on LogTM-SE [9]. The Base CM policy makes requestors stall on conflicts and repeatedly retry the conflicting memory access until potential deadlock occurs[6]. Furthermore, we combine our idea with the base CM policy and get an enhanced CM policy: SCM-BG.

4.2 Workloads

In order to observe a range of program behavior, we used four benchmarks (Delaunay, Genome, Vacation and Kmeans) from the STAMP workload suite [2] and one SPLASH-2 benchmark [10] (Barnes). Table 2 highlights the benchmarks' characteristics.

5 Results

Figure 1 shows the speedup of our SCM techniques compared to a system that only uses the Base policy.

Table 2. Transactional workload characteristics

Benchmark	Contention level	Read/Write set size	Tx length	Tx time ratio
Delaunay	Medium	Large	Long	High
Genome	Low	Medium	Medium	High
Vacation	Medium	Medium	Medium	High
Kmeans	Low	Small	short	Low
Barnes	High	Small	Short	Low

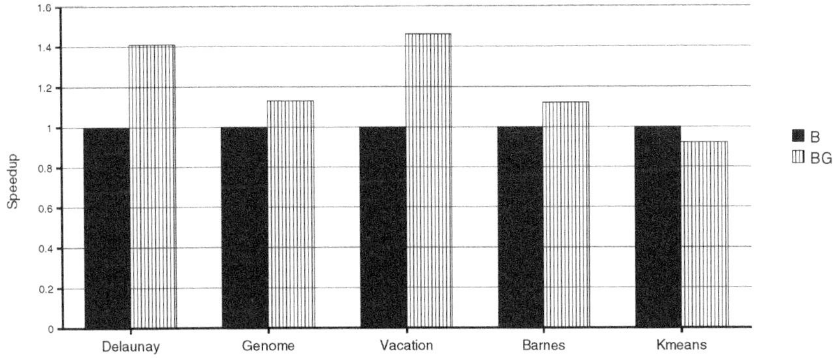

Fig. 1. Execution time speedup of TM systems with different CM policies

From this figure, it is clear that our enhanced policy with SCM scheme can speed up significantly than the baseline policy for large transaction benchmarks. For example, SCM-BG separately achieves speedups of 1.41, 1.46 and 1.13 relative to the baseline for the Delaunay, Vacation and Genome.

The reason is that SCM scheme is very effective at directly reduce stall cycles when the numbers of stall, abort and backoff increase rapidly or there appears the tendency of deadlock or deadlock, thereby reducing the amount of waste time. Delaunay, Vacation and Genome have more transaction length, transaction time or higher level of contention than Barnes. It leads to more conflicts than Barnes, therefore SCM scheme can get more benefit on them than Barnes. Barnes has short transaction length, low transaction time but high level of contention, therefore its conflict frequency is between large and small transaction benchmarks. It gets moderate benefit with SCM.

Our enhanced polices with SCM scheme get little slowdown in small transaction benchmarks for Kmeans. It is because that the cost of SCM scheme is slightly more than the waste time it saves. On the other side, as the slowdown of the policy with SCM scheme is limited in 0.08 than the correspond policy without SCM scheme, it illustrate that the cost by SCM is little relative to its benefit.

Figure 2 shows the fraction of execution time of our SCM techniques to the Base policy system. The figure presents results for SCM-BG, which is the combination of Base with SCM schemes.

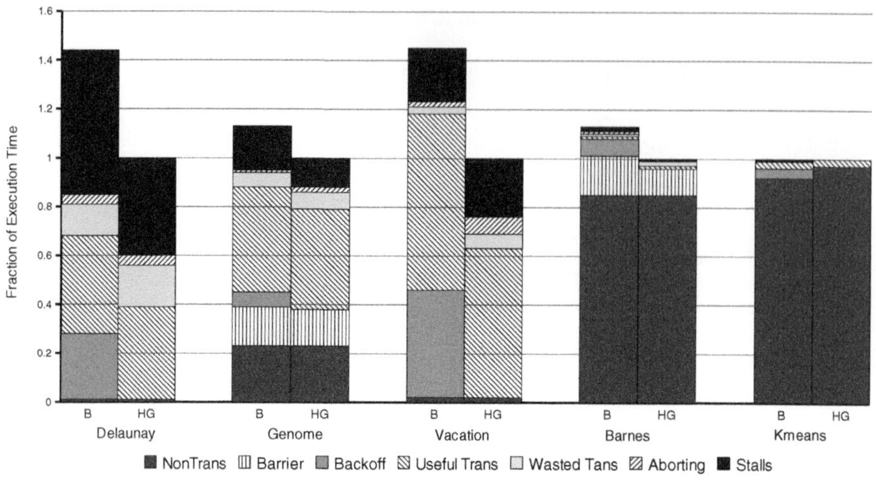

Fig. 2. Fraction of execution time with different CM policies

From this figure, it is clear that our SCM-BG techniques can reduce the backoff cycles and the stall cycles significantly than the baseline policy for all benchmarks. For Delaunay, Vacation, Genome and Barnes, it is interesting that SCM-BG reduces the backoff cycles and the stall cycles and increase the aborting cycles and the wasted transaction cycles.After positive and negative balance, the final result is that the total cycles is reduced.This figure shows clearly that aborting the stalled transactions increases the the wasted transaction cycles and the aborting cycles. On the whole, the SCM decreases the total cycles.

6 Conclusion

In this paper, we introduce a new CM police. We remitted six of seven performance pathologies summered by Bobba. The results show that our techniques can effectively remit most performance pathologies, minimize the amount of waste time and, consequently, decrease execution time than previous approaches. Our result shows high performance for large transactions, while get moderate improvement or little slowdown for small transactions. It also shows that the performance of the systems used this policy combined with other policy are steady.

Acknowledgments. We thank the Harbin Institute of Technology PACT group and Fault Tolerant Computing Lab for their help and support.

References

1. Herlihy, M., Moss, J.E.B.: Transactional Memory: architectural support for Lock-Free data structures. In: Proceedings of the 20th Annual International Symposium on Computer Architecture. Conference Proceedings - Annual Symposium on Computer Architecture, San Diego, CA, USA, pp. 289–300. IEEE Press, Los Alamitos (1993)

2. Herlihy, M., Luchangco, V., Moir, M., Scherer III, W.N.: Software Transactional Memory for Dynamic-Sized data structures. In: Proceedings of the 22nd Annual Symposium on Principles of Distributed Computing, Boston, Massachusetts, USA, pp. 92–101. ACM Press, New York (2003)
3. Scherer III, W.N., Scott, M.: Contention Management in Dynamic Software Transactiona Memory. In: Proceedings of the PODC Workshop on Concurrency and Synchronization in Java Programs, St John's, Newfoundland, Canada, pp. 128–140 (2004)
4. Bobba, J., Moore, K.E., Volos, H., Yen, L., Hill, M.D., Swift, M.M., Wood, D.A.: Performance pathologies in hardware transactional memory. In: Proceedings of the International Symposium on Computer Architecture, San Diego, CA, USA, pp. 81–91. IEEE Press, Los Alamitos (2007)
5. Rossbach, C.J., Hofmann, O.S., Porter, D.E., Ramadan, H.E., Aditya, B., Witchel, E.: TxLinux: using and managing hardware transactional memory in an operating system. In: Proceedings of Twenty-First ACM SIGOPS Symposium on Operating Systems Principles, SOSP 2007, pp. 87–102. ACM, New York (2007)
6. Yen, L.: Signatures in Transactional Memory systems. PhD thesis, Department of Computer Science, University of Wisconsin-Madison, WI, USA (2009)
7. Martin, M.M.K., Sorin, D.J., Beckmann, B.M., Marty, M.R., Xu, M., Alameldeen, A.R., Moore, K.E., Hill, M.D., Wood, D.A.: Multifacet's general execution-driven multiprocessor simulator (GEMS) toolset. SIGARCH Comput. Archit. News 33, 92–99 (2005)
8. Magnusson, P.S., Christensson, M., Eskilson, J., Forsgren, D., Hallberg, G., Hogberg, J., Larsson, F., Moestedt, A., Werner, B.: Simics: A full system simulation platform. Computer 35, 50–58 (2002)
9. Yen, L., Bobba, J., Marty, M.R., Moore, K.E., Volos, H., Hill, M.D., Swift, M.M., Wood, D.A.: LogTM-SE: decoupling Hardware Transactional Memory from caches. In: Proceedings of the IEEE 13th International Symposium on High Performance Computer Architecture, Scottsdale, AZ, USA, pp. 261–272. IEEE Press, Los Alamitos (2007)
10. Woo, S.C., Ohara, M., Torrie, E., Singh, J.P., Gupta, A.: The SPLASH-2 programs: characterization and methodological considerations. In: Proceedings of the 22nd Annual International Symposium on Computer Architecture, ISCA 1995, pp. 24–36. ACM, New York (1995)

Locomotive Driving Simulator for Multi-objective Train Operation and Movement

Yong Ding

School of Traffic and Transportation, Beijing Jiaotong University,
Beijing 100044, China
Yding@bjtu.edu.cn

Abstract. This paper explores the simulation problem on multi-objective train operation and proposes the calculation model of train performance. A simulation system has been developed suitable for different running conditions and train parameters. As an emphasis, the paper discusses the evaluation model of train operation under fixed running time between given stations and its elicitation. Finally, the paper introduces the system structure, function and interface. The simulator can help train drivers improve their operation skill and speed up learning process of train driving.

Keywords: railway transportation, train operation, multi-objective, simulation.

1 Introduction

Driver behaviors have a significant influence on train movement, which includes security, energy consumption, punctuality, smooth control and parking accuracy. Up to now, train operation is mainly determined by the operation skill of train drivers in Chinese railway system. According to the external information observed and the understanding of the railway system and locomotive equipments, train drivers manage to make the corresponding strategy on train operation with their driving experiences.

Teaching multi-objective train operation such as energy efficiency, runtime and stop precision etc. is part of the basic education of train drivers. The drivers need to learn the optimal regimes, find the optimal coasting point and get taught that switching off the traction engine early or late in train control process. As Train driving is a very complex control task, new driver must expend a long time to study the driving skill and enrich his experience under all kinds of circumstances. If a new driver wants to achieve multi-objective train operation, the only way maybe that they repeatedly drive the same train with the same train timetable on same track. However, multi-objective train operation is a difficult task for train drivers and repeating driving practice is not realistic in real railway environment. One possibility to speed up learning is driving under supervision in train simulator.

Locomotive driving simulator for multi-objective train operation and movement aims at improving the driver's operation skill for the train moving on the uneven rail from one given point to another within a given time and assisting train driver in multi-objective train driving. Some locomotive driving simulators have been developed at

L. Qi (Ed.): ISIA 2010, CCIS 86, pp. 404–410, 2011.

home and abroad [1-6]. The simulation model always needs to be simplified for avoiding the complexity of railway transportation system. As a result, the simulation didn't better reflect the reality. Other classical numerical operation methods are too slow to be utilized in an on-board computer for real-time calculations.

This paper probes into calculation model and assessment model for multi-objective train operation on an actual track scale. We also describe the structure, function, and interface of the simulator.

2 Calculation Model of Train Performance

In the theory of railway traffic, the motion of a train can be described by the following equation:

$$dv/dt = f(v) - b(v) - w(v) - g(s). \tag{1}$$

Where, v is train speed, s is train location, $f(v)$ is the specific traction force, $b(v)$ is the specific braking force, $w(v)$ is the specific resistance to motion, and $g(s)$ is the specific external force caused by track grade and curve resistance.

The train has three control modes: motoring, coasting and braking. Air braking and electric braking are two types of braking of the locomotive. Some certain control settings exist on the motoring and braking control. In reality, train traction force and braking force have directly relationship with train speed and discrete control handles in motoring and braking process.

To describe the control mechanism, we introduce a control variable C. Let ϕ $=\{q_1,q_2,\ldots,q_i,\ldots,q_k,0,w_1,w_2,\ldots,w_j,\ldots,w_r,zw,z_1,z_2,\ldots,z_l,\ldots,z_p\}$denotes the set of all possible value for C_i. If the number of traction control settings is k, q_i is the i-th notch of the locomotive. Coasting is zero and w_j determines the j-th handle of electronic braking. zw is the electro-pneumatic braking. z_l delegates different air braking with the different pressure of train pipe, and z_p is emergency braking.

Based on the analysis mentioned above, we revise the equation of the train movement, i.e. Eq. (1). The revised equation is as follows:

$$dv/dt = f(C,v) - b(C,v) - w(v) - g(s),\ C \in \phi \tag{2}$$

Where, $f(C,v)$ is the specific traction force based on train speed and the control handle of locomotive, $b(C,v)$ is the specific baking force based on train speed and the pressure of train pipe.

As the time-based simulation system, the simulator needs to calculate speed, distance and energy consumption at each time step according to track geometry and traction equipment. Eq. (2) can be further simplified to a discrete model by changing the time derivative to the difference:

$$v(t+\Delta t) = v(t)+(f(C,v) - b(C,v) - w(v) - g(s))*\Delta t,\ C \in \phi. \tag{3}$$

Train running distance can be calculated by Eq. (4).

$$s(t+\Delta t) = s(t)+v(t)*\Delta t.+0.5*(f(C,v) - b(C,v) - w(v) - g(s))*\Delta t^2,\ C \in \phi. \tag{4}$$

At each time step, the main task of the system is to determine the control modes and handles. Once the control modes and handle is established, the position, speed and energy consumption of the train can be calculated easily. The model of determining the train's control modes and handles takes the general form:

$$C(t+\Delta t)=f(C(t),s(t), v(t), v_a,Q(s)) , C \in \phi. \tag{5}$$

Where, Q(s) is the advance wayside signal of the train, v_a is the target train speed according to train schedule running time, subjects to constraints:

$$v(0)=0, v(T)=0, 0 \leqq v(t) \leqq v_{max} \tag{6}$$

$$s(0)=0, s(T)=S \tag{7}$$

Where, v_{max} is train limited speed which including the civil speed and temporary limit speed, T is the total running time and S is the distance of the section.

The calculation method of train energy consumption is similar to traction force, which is determined by train speed and control handles and modes. Energy consumption calculation model is given as:

$$E =\sum e(C,v)* \Delta t, C \in \phi. \tag{8}$$

Where, $e(C,v)$ is the energy consumption per unit time.

3 Evaluation Model of Train Operation

In order to ensure the quality of train movement (safety, punctuality, energy-efficient, comfort and so on), the assessment model for driver's operation is designed and produced based on experts advice and excellent drivers' experiences. The following shows the multi-objective function used to monitor the energy consumed, punctuality, security, comfort, and park accuracy of the train throughout the journey.

3.1 Security

During the whole process of train movement, the train speed must not exceed the train limited speed. When the train stops at the station, the train's head must not get across the position of the departure signal and the train's tail must enter into the inner part of the fouling post. The evaluation function of train movement safety $\mu_s \in [0,1]$ is given as:

$$\mu_s = \begin{cases} 0 & v(t) > v_{\max} \\ 0 & S_h(t) \geq Q(s) \\ 0 & S_t(t) \leq fp(s) \\ 1 & 0 \leq \mathrm{v(t)} \leq \mathrm{v}_{\max} \end{cases} \tag{9}$$

Where, $S_h(t)$ is position of train's head, $S_t(t)$ is position of train's tail, $fp(s)$ is the position of fouling post in arrival track.

3.2 Punctuality

The punctuality of train movement is determined by the difference between scheduled running time T_s and practical running time T. t_{lamx} is the maximum promised late time in this section. The evaluation function of train movement punctuality $\mu_p \in [0,1]$ is expressed by:

$$\mu_p = \begin{cases} 1 & T_s \geq T \\ 1 - \dfrac{T - T_s}{t_{l\max}} & T - t_{l\max} \leq T_s < T \\ 0 & T_s < T - t_{l\max} \end{cases} \tag{10}$$

3.3 Energy Consumption

Comparison is made between the practical energy consumption E which is calculated by Eq. (8) and the desired energy consumption E_d which is deduced according to excellent drivers' experience. The evaluation function of train energy consumption $\mu_e \in [0,1]$ is expressed by:

$$\mu_e = \begin{cases} 1 & E \leq E_d \\ e^{-\frac{(E-E_d)}{E_d}} & E > E_d \end{cases} \tag{11}$$

3.4 Comfort

If the value and variance ratio of train accelerate are bigger than standard, the passengers might feel uncomfortable, or even fall over themselves. The evaluation function of comfort $\mu_c \in [0,1]$ is given as:

$$\mu_c = \alpha \cdot \mu_{c1} + \beta \cdot \mu_{c2}, \ \alpha + \beta = 1 \tag{12}$$

$$\mu_{c1} = \begin{cases} 1 & |a| \leq a_s \\ 1 - \dfrac{a - a_s}{a_{\max} - a_s} & a_s < |a| \leq a_{\max} \\ 0 & |a| > a_{\max} \end{cases} \tag{13}$$

$$\mu_{c2} = \begin{cases} 1 & |\Delta a| \leq \Delta a_s \\ 1 - \dfrac{\Delta a - \Delta a_s}{\Delta a_{\max} - \Delta a_s} & \Delta a_s < |\Delta a| \leq \Delta a_{\max} \\ 0 & |\Delta a| > \Delta a_{\max} \end{cases} \tag{14}$$

Where, a_s and Δa_s are the value and variance ratio of train accelerate according to comfort criteria, a_{max} and Δa_{max} are the limits of train accelerate and it's variance ratio, α is the weight factor for accelerate, and β is the weight factor for variance ratio of accelerate.

3.5 Park Accuracy

When the train stops at arrival and departure track of the station, the park accuracy is determined by the difference between practical stop position and target stop position. The evaluation function of park accuracy $\mu_k \in [0,1]$ is expressed by:

$$\mu_k = \begin{cases} 1 & K \le K_s \\ e^{\frac{(K-K_s)}{K_s}} & K > K_s \end{cases} \tag{15}$$

Where, K is practical stop error and K_s is standard stop error which is promised based on the station transportation management rules.

3.6 Evaluation

On the basis of train control indexes involving security, comfort, punctuality, energy consumption, and park accuracy, the multi-objective optimality criterion takes the general form:

$$P = (\mu_s \cdot \mu_p \cdot \mu_e \cdot \mu_c \cdot \mu_k) \cdot p_{\max} \tag{16}$$

Where, p_{max} is maximum score of driving simulation. The simulator will give score after train driver finish the driving simulation and some consequent advices are proposed to improve his driving skill.

4 Structure and Interface

The simulator for multi-objective train operation and movement has been developed by use of object-oriented programming technology and it is mainly divided into two components. The first component is train operation simulator, which is used to calculate the train performance. The other is evaluation for driver's operation. A detailed configuration of the simulator is shown in Fig. 1.

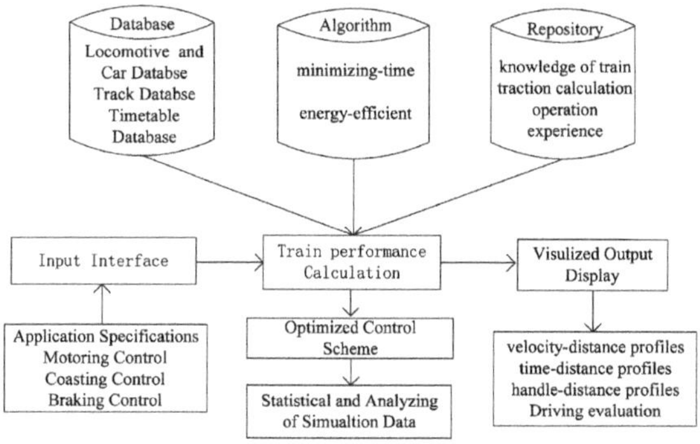

Fig. 1. Structure of Multi-objective Train Operation simulator

The simulator can provide the driver with the velocity-distance profile, time-distance profile, and the handle-distance profiles of optimization operation for train movement. The interface of the simulator is shown in Fig. 2.

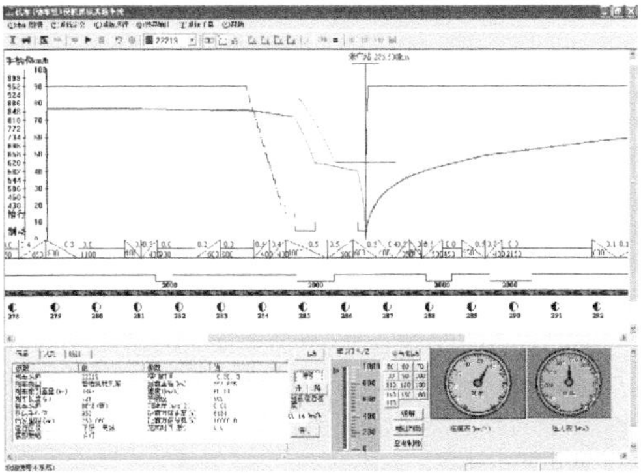

Fig. 2. Interface of Multi-objective Train Operation simulator

The simulator has two train operation control methods which are automatic control and manual control. If the driver chooses automatic control, the simulator will give an automatic simulation for train performance with excellent drivers' experience. If the driver chooses manual control, he can control the train by himself under simulated railroad environment and repeatedly practice to improve his driving skill.

The minimizing running time of rail operation and the energy-efficient train control are two operation methods in the simulator. The train will be controlled with minimizing-time operation if it is late and with energy-efficient operation if it is punctuality.

5 Conclusions

In this paper, we discuss the simulation problem for multi-objective train operation from one given point to another in fixed time and present the simulation models for train performance calculation. We have explored the key technique for evaluating driver driving behavior on train movement and developed the simulator which is suitable for different running conditions and parameters. The simulator is designed and produced based on experts, knowledge about train traction calculations, operation experiences of excellent drivers, longitudinal profiles of running lines, train compositions as swell as performance data of locomotives and rolling stocks. It can give a scientific and reasonable assessment on the driver's operation, and is easy to learn and practical at the same time. The simulator can realize the repeating driving to help train drivers master some optimal operation methods, for example, when it is optimal to start coasting without the risk of driving late.

Acknowledgements

The research described in this paper was partly supported by the National Natural Science Foundation of China (70173014).

References

1. Hamilton, W.I., Clarke, T.: Driver Performance Modeling and its Practical Application to Railway Safety. Applied Ergonomics 36, 661–670 (2005)
2. Butler, K.L., Ehsani, M., Kamath, P.: A Matlab-Based Modeling and Simulation Package for Electric and Hybrid Electric Vehicle Design. IEEE Transactions on Vehicular Technology 48, 1170–1778 (1999)
3. Fisher, D.L., Laurie, N.E., Glaser, R., Connerney, K., Pollatsek, A., Duffy, S.A.: Use of a Fixed-Base Driving Simulator to Evaluate the Effects of Experience and PC-Based Risk Awareness Training on Drivers' Decisions. Human Factors 44, 287–302 (2002)
4. Albrecht, T.: Energy-efficient train operation. In: Hansen, I.A., Pachl, J. (eds.) Railway Timetable & Traffic, Eurailpress/ DVV Media, pp. 83–105. Eurail Press, Germany (2008)
5. Dukkipati, R.V., Amyot, J.R.: Computer-aided simulation in railway dynamics. Marcel Dekker, Inc., New York (1988)
6. Feng, X.Y., Gui, X., Zhu, J.L.: Development of assessment system software for locomotive driver's operation. Electric Drive for Locomotives, 51–55 (2002)

A Network-Centric Architecture for Combat System-of-Systems Model

Hua Tian and Zhi-chun Gan

Department 2, Commanding Communications Academy,
Wuhan, China
tianh@126.com

Abstract. In the development of a system-of-systems combat simulation, how to design the architecture of the combat system-of-systems model determines if the final simulation result is satisfying or not. Here, a fresh network-centric architecture is put forward to support the modeling and simulation for combat system-of-systems. This architecture takes the communications network model as the basis platform, enable various combat entity models being flexibly plugged to the network model platform. Then, as the core component in the this architecture, the communications network model is researched weightily, about the resolution of the model, hot swapping model design, network scenarios generation, and so on.

Keywords: Network-Centric; Architecture; Combat Simulation; Modeling; Simulation; OPNET.

1 Introduction

System-of-systems combat is an important form in the information-based warfare. It emphasizes on constructing a system of systems, which suppresses, breaks, destroys or controls enemies' system of systems. Combat simulations and military training simulations have attached too much importance to modeling for firepower countermine, but little importance to modeling for communications network, so it can not meet with the demands of the system-of-systems combat. Different to the traditional combat simulation, system-of-systems combat simulation is a series of theories and methods, which regards the system-of-systems countermine as the basis, mainly researches modeling and simulation of information-based warfare and combat [1]. Moreover, it emphasizes the effects as the all and the one, especially thinks much of communications networking modeling.

As the most popular architecture, HLA (High Level Architecture) is a kind of outstanding distributed interactive architecture, which is presented by DoD, and mainly used for simulation. Almost all simulation applications all over the world adopt this architecture, however, HLA is only a type of bottom technological architecture, and handle mainly with the following problems: bad inter-manipulation and weak reuse level among simulators, high cost in developing, maintaining and using, bad verification and validity, and so on [2], so this architecture can not resolve

L. Qi (Ed.): ISIA 2010, CCIS 86, pp. 411–417, 2011.

the following problems: what components the target system is composed of, what are the relationship of these components, and so on. So, aiming to the System-of-systems combat simulations based on HLA/RTI, a kind of network-centric architecture are put forward here. In the following content, the network-centric architecture used for system-of-systems combat simulations is researched mainly, and some issues about constructing, implementing and applying the simulative network models.

2 Network-Centric Architecture for Combat System-of-Systems Model

Combat System-of-Systems is composed of various nodes, such as combat entities, computer information systems, and so on, all of that are linked by communication networks (including computer networks). So, it can be deemed that communication networks functions as the nerves of the combat system-of-systems. The communication networks control the transformation of nearly all information in the battlefield, and consequently lead to the situation that all kinds of combat powers form the target system-of-systems eventually. However, just because of the performance of the communication networks and other factors in the battlefield background, much information would be lost, error, and delayed, and accordingly the integrality, veracity and delay of the information transformation would be influenced every much, so that the result of the combat would have to be influenced [1]. Therefore, the same as that the modern joint operations take the communication networks as the center, combat system-of-systems should take the simulative network as the base platform, and carry on researching and designing based on the network. The network here is the simulative network in the simulation system. It could be the pure simulative network or the seamless integration of the physical network and simulative network.

Combat system-of-systems can be researched through three domains: physical domain, information domain and cognitive domain. According to the U.S. military viewpoints, the physical domain includes personnel, equipments, environments, and so on, and is the domain that the strikes are expanded in the circumstances of land, sea, sky and outer space, that is the domain composed of the physical platforms, and the communication networks linking these physical platforms. The information domain is the domain that the information exists, in which the information is generated, collected, processed, and shared. The information existed in the information domain perhaps reflects the real things, also maybe not. The cognitive domain exists in the consciousness of man, in which senses, recognitions, comprehensions, persuasion and worthiness are placed, and is the domain that the decision making must be made through reasoning [3]. According to the partition of the three domains, in the combat system-of-systems simulation, all kinds of entity models including combat ones and communication ones are located in the physical domain; the information models, information flow models and information system models are located in the information domain; the command and control models and decision models are located in the cognitive domain. In fact, all types of models, including entity ones, information system ones, command and control ones, decision ones, exchange information through information flow, that is, they exchange by the models of communication networks. And it reflects on other than the idea of "network-centric". Moreover, based on the distributed interactive architecture of HLA/RTI, all models (platform level models or aggressive level models) are taken as the federate participating in the simulation federation. Now, it can be considered that the communication network models are actually acting as the common base platform, and all other models are hot swapping based on it through HLA/RTI. This architecture is shown as Fig. 1. In Fig. 1, all other members (federate) are interacting with HLA/RTI, whereas, from the application point of view, which is higher than HLA/RTI in layers, their information about communications has to pass through the simulative networks. After the handling course is simulated in the simulative networks, it would be decided if the information should be transmitted or not. Thus, looking from the aspect of communication information flow, all members (except for the simulative network) are plugged directly in the simulative network logically, that is, the simulative network is regarded as the base platform in the simulation system-of-systems.

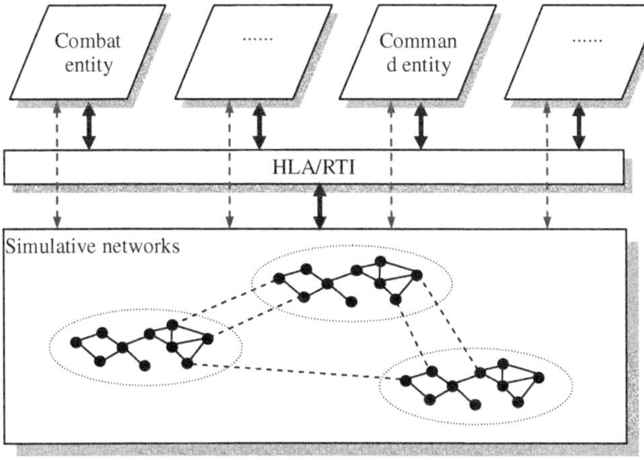

Fig. 1. The network-centric architecture for combat system-of-systems model

When the combat system-of-systems models are constructed based on the simulative networks, the following advantages could be anticipated:

- The weightiness that the simulative networks are regarded as the base platform is ascertained, so that the simulative networks are not slave to other members. Thus, the simulation for the networks is much more vivid, and the power of the architecture is also reflected.
- All communication factors are concentrated in the consideration for the simulative networks. Thus, designs and implementations of all other members are more independent and modularized.
- When the combat system-of-systems is changed, the importance of modifications is focused on the models other than the simulative networks. The structure of the simulative networks is relatively stable, so the total architecture is flexible and expansible.

3 Several Issues about the Construction and Application of the Network Models

As the base platform of the combat system-of-systems model, simulative networks are the core modeling issues with on doubt. In general modeling methods of communication networks, the network models' characters can be unfolded to the following two aspects:

- The network model is regarded as the passive model. It is driven by combat model, and so on, and modifies itself according to others' simulation results.
- The simulative network is absolute, thus the simulation results about success ratio and delay are mainly got through simple computing [5].

Just because of these two characters, simulative networks are slave to other models, accordingly can not support the model system-of-systems.

In order to construct the network-centric combat system-of-systems model, the research must be begun from the network models, and make simulative network drive the combat system-of-systems model. Thus, it is required to discover how to simulate and model from the inner networks. Now, we can adopt the advantages of the mechanism and platform of the commercial network simulation. OPNET Modeler platform is chosen here. There are several key issues needing consideration in the following, when we construct the base platform of combat system-of-systems model based on OPNET Modeler.

3.1 Resolution of Simulative Networks

Though OPNET Modeler provided perfect resolution scheme for the network performance simulation and modeling, it could not suit the need of combat system-of-systems model, so the traditional methods can not be copied directly, should be modified adaptively for the new modeling requirements. Most models in OPNET Modeler are equipment level models, with high resolution. Modeling in OPNET follows a type of three layers' mechanism: network, node, and process, is enough to construct subtle models of network equipments. These models can reflect the actual working status of most network equipments realistically. But, in system-of-systems models, the number of the network equipments in the battlefield is too large. Even if detailed network equipment models could be constructed, they can not be applied into the system-of-systems models or simulations, because of at least three aspects:

- It would influence the simulation largely, so that the simulation time runs badly behind the physical time. This is not allowed for the system-of-systems simulation.
- The resolution of the models is too high to reflect some monolithic specialties.
- In simulation, vast number of data, such as equipment parameters, initial data, and so on, exceeds the bound people in the combat/training simulation could endure, and restricts the scale of the system-of-systems simulation greatly.

In order to resolve these problems, the first issue to be solved is the resolution of nodes in simulative networks. Equipment level model accounts for the running speed of the simulation system, so a direct way to solve this question is to make the granularity of models larger, or debase the resolution of models, especially node models. That is, the models should adopt aggressive models other than platform level models. For example, in the simulation of field battle communication networks, many types of equipment are filled in one vehicle, so it is a good idea that each vehicle is taken as a fundamental node [6]. According this method of modeling, the communication networks are composed of vehicle nodes. Except for actual vehicles, other entities, such as command posts, fighters, and so on, can use the vehicle node, but some modifications should be made. The detailed method is adding or deleting one or more internal modules. And this requires the vehicle model should be flexible.

3.2 Hot Swapping Module Design

In OPNET Modeler, one node is composed of several modules, and each module can be denoted by a process. When it is determined that a node is made up of all equipments in the vehicle, the structure about these modules should be analyzed. According the given simulation requirements, the modules in the vehicle node may be just equipment, also may include several equipments, even made up of other factors than actual equipments. The first method is the same to the traditional network performance simulation, so the modeling difficulty and simulation speed can't be accepted; the third method is too abstract, so the coherence of model structure is absent. Thus, the second method is determined, that is, module is composed of several equipments. Its advantage is the complexity is depressed, and the total structure is clear. In this structure, all equipments in a vehicle can be divided to three parts, and each part denotes a layer. At first, the route equipments are regarded as the core of a vehicle node, hence the node becomes a subnet. Then, a part of other equipments can be viewed as the application layer (two ends of the information flow), and another equipments can be viewed as the transmitting layer (two peers of the information flow) [6].

The module structure is very important, because it should consider if the structure can suit most simulation applications. And another important aspect is that models will probably be changed more or less, such as upgraded, modified, maintained, and so on, but if each change is directly acted on the module, the workload is large, and the correctness will be incredible. To avoid this risk, a method of hot swapping module design is put forward. Hot swapping indicates that changing a module implementation outside OPNET Modeler. Of course, you don't want to construct models repeatedly after the simulation is completed. Now, according to the hot swapping method, you can only rewrite the code of module outside OPNET Modeler without any modification in OPNET Modeler, when some modifications about the original module require made. Even, the action can be taken place when simulation system is running [7]. This hot swapping method is based on the external system module provided by OPNET. Modules in OPNET Modeler include three kinds: processor, queue, and external system module. When an external system module works, it generates an interrupt to Esys interface, then all process is operated outside OPNET Modeler, at last it achieves the result of those operations. In fact the *hla_interface* module, usually used as the interface between HLA and OPNET Modeler, is an external system module. It is noticed that processor and queue modules could be got through simplifying external system module.

3.3 Generation of Network Scenario

Before the simulative network takes part in the system-of-systems simulation, the network topology, parameters of all network equipments, etc. should be configured. OPNET Modeler has provided a set of graphical interface, so that you can configure your network topology through dragging your mouse, and input your data in graphical windows. It is very convenient, but when the network scale becomes larger and larger, the data about topology and parameters also becomes huge. Now, if you insist on configuring topology and parameters following the old method, it will waste plenty of time, moreover, many mistakes will take place. To settle this problem, the EMA (External Model Access) should be adopted. EMA is similar with the TXT file, and its syntax is the same as the C language, so it is convenient for fast generating large scale of network scenario, including networks, nodes, modules, etc.

The procedure of generating network scenario using EMA is shown as the following. At first, xx.em.c file should be achieved by writing EMA codes, shown as the following. Then, the EMA file can be compiled to an executable file. At last, the network scenario is generated successfully by execute this file. The scenario can be used directly, and can also be added to another project. The EMA file is made up of five parts: entrancing main function, initializing library functions, constructing the model object, setting the attributes, generating scenario, and the it is very flexible to write the code.

Example of EMA codes

```
#include <opnet.h>
#include <ema.h>
#include <opnet_emadefs.h>
#include <opnet_constants.h>

/* array for all textlist attributes in model */
Prg_List*    prg_lptr [7];

/* array for all objects in model */
EmaT_Object_Id   obj [78];

int
main (int argc, char* argv [])
    {
    EmaT_Model_Id model_id;
    int              i;

    /* initialize EMA package */
    Ema_Init (EMAC_MODE_ERR_PRINT, argc, argv);

    /* create an empty model */
    model_id = Ema_Model_Create (MOD_NETWORK);

    /* create all objects */
    obj [0] = Ema_Object_Create (model_id,
OBJ_NT_SUBNET_FIX);
    ……
```

```
/* write the model to application-readable form */
Ema_Model_Write (model_id, "ethcoax_net-scenario1");

return 0;
}
```

4 Conclusions

The network-centric architecture was used in a certain network simulation training system, and now the system is running all right. As the base platform of combat system-of-systems model, the network model has satisfied some requirements of simulation system, however, from the viewpoint of the whole combat system-of-systems model, the network model has two main deficiencies: the lack of the common fitness, the shortcoming in integrality of the model parameters. So, the network-centric architecture for combat system-of-systems model needs to be improved and perfected step by step in future, and the integrated network-centric architecture also needs to be discovered. Moreover, some detailed issues yet need more research, such as how networks interact with HLA/RTI (especially when more than one federation exist) and how simulative networks interact with physical networks, also needs.

References

1. Hu, X.-f., Luo, P., et al.: Warfare complex systems modeling and simulation. National University of Defense Press, Beijing (2005)
2. The Institute of Electrical and Electronics Engineers, Inc. IEEE Standard for Modeling and Simulation (M&S) High Level Architecture (HLA) — Framework and Rules, IEEE Std1516-2000
3. Albert, D.S., Garstka, J.J., Stein, F.P.: Network centric warfare: developing and leveraging information superiority. CCRP Library of Congress Cataloging 2 in 2 Publication Data, Washington, DC (2000)
4. Yue, Q.-l., Bao, J., Zhang, M.-z., Shi, J.-j.: Issues of Network Model and Interative Model in Sos Countermine Simulaiton Modeling. Military Operations Research and Systems Engineering (2008)
5. Yang, R.-p., Guo, Q.-s., Zhao, H.-x., Zhang, Z.-r.: C3I System Modeling and Simulation. Defense Industry Press, Beijing (2006)
6. Gan, Z.-C., Tian, H., Shen, J.-J.: Networking Modeling for the Network-in-the-Loop Training Simulation. In: 2009 International Conference on Computer Technology and Development, ICCTD 2009, vol. 1, pp. 468–471 (2009)
7. Coyne, M.E.: 2d Lt, USAF. Hot Swapping Protocol Implementations in the OPNET Modeler Development Environment, Air Force Institute of Technology, Master's Thesis (2008)

A Handheld Testing System for Irrigation System Management

Jiachun Li[1,*], Wente Tu[2], Jian Fu[2], and Yongtao Wang[2]

[1] Associate professor, College of Mechanical Engineering, Guizhou University, China
Ljcgd211@tom.com
[2] Graduate students, College of Mechanical Engineering, Guizhou University, China
twentei@163.com

Abstract. Based on the C8051F microcontroller and the Rtx51tny operating system, a group of hand-held irrigation detection device is developed to meet the needs of the water-saving irrigation for the crops planted in karst slope area, in which there is quick changes and tremendous differences of the regional climate. The device has advantages such as simple operation, easy to carry, low cost, wireless control, and so on. By using the device, the complementarities can be integrated between manual control and automatic control. The management of personnel participating in irrigation control system according to the conditions of environment improves the execution efficiency of water-saving irrigation and achieves the crops yield. The test had proved that the device is stable and credible.

Keywords: irrigation, handheld, C8051F120, Rtx51tiny.

1 Introduction

From the viewpoint of modern control theory, soil-plant-water is concerned with a complicated system. Thus, precision irrigation needs all kinds of advanced techniques to archive high efficiency of water resource [1]. During the agricultural cultures' growing it's important to keep required water regime of soil. The objective of water-saving irrigation control is to gain maximum output or income and improve the quality of the crop in units of irrigation water. It can scientifically and timely irrigate according to the water requirement in different growth of crops[2]. At present computer technology, automation technology, geographic information systems, information technology, systems engineering are applied widely in irrigation automation of water management, to realize data acquisition, processing and feedback intelligent control. Shock et al. [3] used radio transmission for soil moisture data from data loggers to a central computer logging site. Pereira designed and simulated set sprinkler irrigation systems by using computer-aided design software that allowed the design of a simplified layout of the irrigation system [4].

The China national water-saving irrigation planning and development principles indicated that water-saving irrigation should suit local circumstances by means of

[*] Corresponding author.

L. Qi (Ed.): ISIA 2010, CCIS 86, pp. 418–423, 2011.

scientific and technological innovation in 2007. As the karst mountainous area is located in low latitude, the terrains with large differences in the vertical direction that result in three-dimensional climate significantly, and relatively large differences between temporal and spatial climate in this areas. Therefore, water-saving irrigation in karst fields requires not only expert system support, but also human factors involved, by which the optimal schema of water-saving irrigation can be implemented. The objective of this paper is to demonstrate the design, construction, and testing of a portable device for real-time in-field sensing and control of a variable irrigation system.

2 Handheld Wireless Control Testing Equipment and Technology Development

Hand-held device is the mobile device which can hold in hands used in normal . Because of the portability, simple operation and low power consumption of handheld devices, that has been widely used in testing, automation control industry, such as TV remote controllers, air conditioning controllers, millimeters, intelligent detectors, etc.

Wireless control technology has widely applied in data acquisition, information feedback and data communication due to its security, stability and remote control performances. At present , along with the development of wireless sensor meshwork based on ZigBee , GPS , the study on the wireless sensor in the intelligent control system of irrigation become dominant, integration application between wireless and meshwork is the trend of development of industry of irrigation control. So the research on the hand-held wireless control equipment in irrigation control system has a great significance, which can realize the measurement of irrigation environment variables, setting the parameters of irrigation system and sending irrigation control commands.

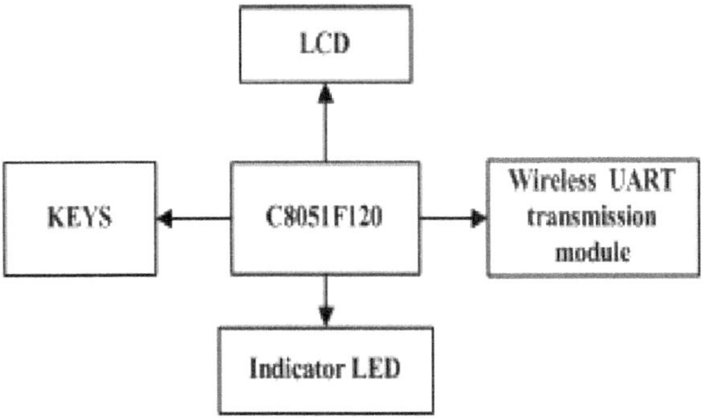

Fig. 1. Hand-held device

3 System Design

The general design structure of system consists of two parts, one is the control part, it consists of a series of on-site single-point controller. The other is handheld detection equipment in the figure 1 and figure 2 , by which management personnel can participate in the control of irrigation.

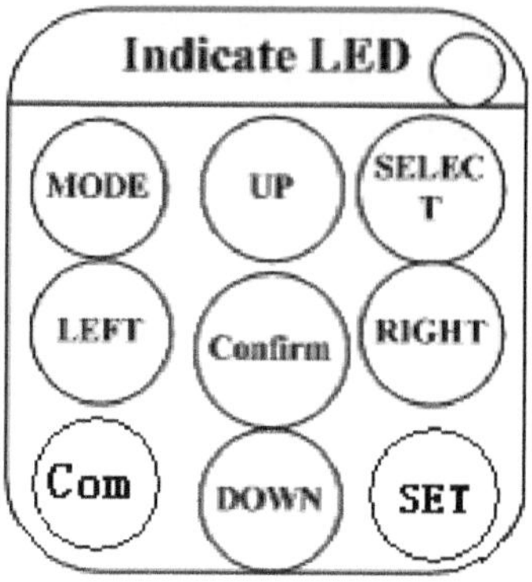

Fig. 2. Key graph

3.1 Micro Controller

The controller chip adopts single-chip microcomputer C8051F120, that is a industrial MCU chip with completely integration of mixed, and can work in the temperature range from - 45℃ to +85℃. Its maximum system clock frequency can reach 100MIPS. This chip is fully suitable for the climate conditions in irrigation system control for karst mountainous fields.

3.2 Embedded Operating System

By transplanting embedded operating system in micro controller, the real-time applications can be developed, and the program design and expansion become easier. We don't need big changes to add new functions. The application can be divided into a number of task modules to simplify the design process of applications, and deal with the real-time harsh events as fast and reliable as possible. Through effective system service, embedded real-time operating system makes the system resources better used.

Considering the equipment does not need execute too many tasks and the requirement of real-time, the embedded operating system adopts Rtx51tiny multitasking real-time kernel that is a multitask real-time operating system developed by KEIL Company for the 8051 series microcontroller. The RTX51 has two versions as RTX51FULL and Rtx51TINY. As a subset of Rtx51 FULL, Rtx51TINY support the maximum signal transmission and interrupt from 16 tasks in parallel according to time support circular task scheduling. Rtx51tiny is such a small kernel integrated fully on C51 KEIL compiler that only takes about 800 bytes of program storage space , and run in 8051 system without external memory, but the application still can access to external memory .

3.3 Hardware Design

The hardware system of the handheld device is composed of C8051F120control chip, wireless UART transmission module, LCD display, keyboard matrix and operation indicator.

In the panel of the handheld device, mode button , select button, the direction keys (up , down , right , left), command button , parameter settings button are designed, when operating the device, personnel press select key down firstly and to select the node number to communicate by direction keys, then press mode key to switch control object into manual control mode; secondly press the select button to choose which command to send by using the direction key to choose the commands in the LCD screen , then press confirm button to realize manual control. The control parameters can be set by pressing the parameters set button to enter parameter setting interface, in which the temperature and humidity upper can be configured through pressing the confirm button to communicate with single-point controller shown as figure 3. Finally through using the press select keys and mode button the device return automatic control mode. Commands include open and close the valve. The running process of the device can be demonstrated as: (1) setting irrigation parameters; (2) monitoring real-time soil temperature and humidity; (3)sending commands to control. Hand-held testing equipment use batteries as power supply.

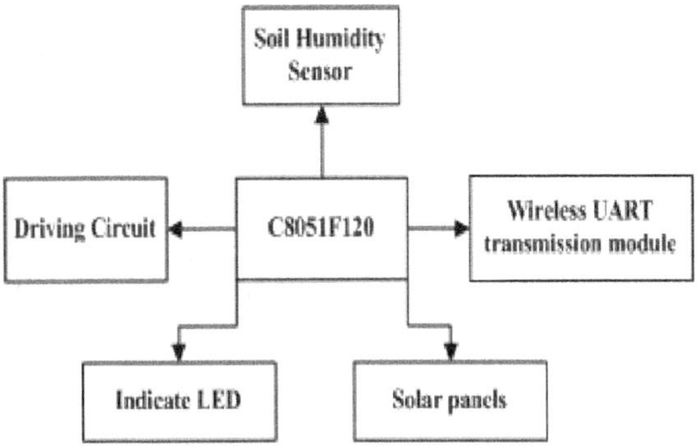

Fig. 3. Single-point controller

Single-point controller is composed by C8051F120 microcontroller, external wireless UART transmission module to communicate with the handheld devices, solar energy supply and storage. LED lights indicating whether the controller on work. The external solenoid valve driving circuit can increase the drive current to open or close the solenoid valves. The concrete schematic diagram is shown as figure 3.

The XL03-232AP2 series of wireless transmission of power micro module are selected as Wireless UART transmitter. The wireless module consists of a Wireless Transceiver IC and ATMEL industrial grade microcontroller PIC with anti-jamming well, stability and reliability, support UART interface in half duplex wireless transmission, and can be modify the settings online, The output power, serial port rate, RF rate, serial format and frequency can be set through serial debugging software. Moreover this module supports for sleep mode with transmission distance up to 600M.

3.4 Software Design

The real-time multitask control is realized through transplanting Rtx51tiny multitask real-time operating system into the hand-held devices and the single-point controller. The main task, send data tasks and switch valve task are established in various control points. In the non-handheld device's correspondence situation, the main task carries out the data acquisition of soil temperature and humidity according to the default time-gap internal recycling, and compares the monitoring data with the internal setting data to determine whether to open the irrigation valves. When the hand-held devices want to communicate with single-point controller, it send switch commands into manual control mode and cause the single-point controller serial interrupted, for which the interrupt function access the main task of the single-point controller through isr_send_signal ( ). Be inputted commands, the handheld device judges if the commands correctly and implement corresponding operations. If the received command is on-off valves, then open and close the valves. If the received command is for sensors data, the data frames from the single-point controllers would be judge whether the error, if not the corresponding data shows on the LCD screen of the handheld device. The software control diagram is shown as bellow.

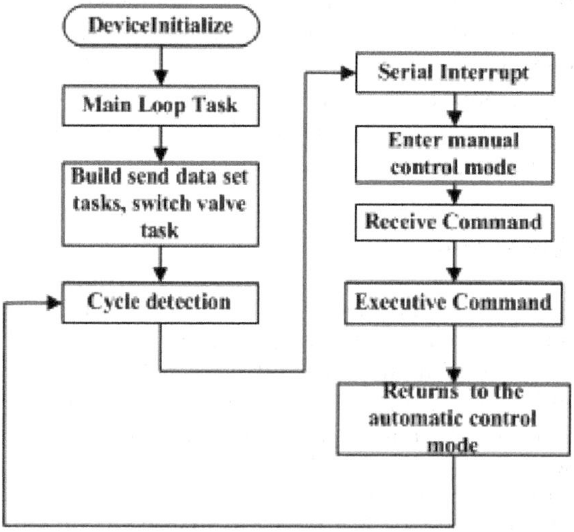

Fig. 4. Diagram of control flow

4 Conclusion

Through the development of the handheld device in irrigation control system, it is convenient for management personnel to participate in the management of irrigation system, and really realizing the complement between human and machine control. With the expert system, intelligent, wireless networked controls have been induced gradually into the irrigation control system., however some factors such as regional differences, climate anomalies in karst slope fields that cause the irrigation system not achieving the effect of precision irrigation agriculture, even there may be mistakenly irrigation and irrigation abnormal phenomenon in sometimes. One of the most available means is to induce human participation to analyze conditions of the district climate and the irrigation object's properties and realize the intellectualized irrigation in the true sense.

Acknowledgement

This work is supported by GuiZhou Key Research Projects and Research Foundation under Grant No. GDRJH[2007]024, QKH NY Z[2008]3054, and QKH GY Z[2008]3013. The support is gratefully acknowledged.

References

1. Zhao, Y., Bai, C.: An Automatic Control System of Precision Irrigation for City Greenbelt. In: 2007 Second IEEE Conference on Industrial Electronics and Applications, pp. 2013–2017 (2007)
2. Pastushenko, V., Stetsenko, A.: Development, modeling and technical implementation of automated control system of soil's moistness by underground irrigation. In: TCSET 2010, Lviv-Slavske, Ukraine, p. 33 (2010)
3. Shock, C.C., David, R.J., Shock, C.A., Kimberling, C.A.: Innovative, automatic, low-cost reading of Watermark soil moisture sensors. In: Proc. Irrig. Assoc. Tech. Conf., Falls Church, VA, pp. 147–152 (1999)
4. Abreu, V.M., Pereira, L.S.: Sprinkler irrigation systems design using ISADim. Presented at the ASAE Annu. Int. Meeting, Chicago, IL, pp. 27–31, Paper No. 022254 (July 2002)

Colonial Algorithm: A Quick, Controllable and Visible One for Gerrymandering

Hongjiang Chu[1], Yue Wu[2], Qiang Zhang[1], and Yuehua Wan[3,*]

[1] School of Science, Zhejiang University of Technology,Hangzhou 310032, Zhejiang, China
[2] Jianxing College, Zhejiang University of Technology, Hangzhou 310032, Zhejiang, China
[3] The MOE Key Laboratory of Special Purpose Equipment and Advanced Processing Technology, Zhejiang University of Technology, Hangzhou 310032, China
wanyuehua@zjut.edu.cn

Abstract. Gerrymandering is an issue of boundary delimitation for legislative purposes. In this paper, we presented a novel approach called Colonial Algorithm(CA) to draw legislative boundaries. Based on Voronoi Diagram, CA was developed originally for simulating colonies' behavior in a culture dish. We focused on the expanding speed and the competitive strategies on edges to create our algorithm and used Cellular Automaton to validate it. CA can be well applied in gerrymandering. When generated from places with big population density, every region undergoes a process of expansion and competition to finally reach a partition. In optimization, population equality and regional compactness were adopt to form the value function. To illustrate the process, a simplified example was taken, whose result indicates that regions drawn by Colonial Algorithm attain continuity, compactness, small population variances and relative geometric simplicity within a short time. And the whole process is controllable and visible.

Keywords: Colonial Algorithm, Gerrymandering, Growing speed, Competition strategies, Cellular Automaton, Voronoi Diagram.

1 Introduction

Districting is well known to be a critical determinant of the representation of population in legislatures.[1] In the case of redistricting electoral districts, since the district-drawers are chosen by those currently in power, the boundaries are often created to influence future elections by grouping an unfavorable minority demographic with a favorable majority; this process is called Gerrymandering. It is common for districts to take on bizarre shapes, spanning slim sections of multiple cities and criss-crossing the countryside in a haphazard fashion. The only lawful restrictions on legislative boundaries stipulate that districts must contain equal populations, but the makeup of the districts is left entirely to the district-drawers.[2] The process of how the shape and structure of districts is brought about has therefore received considerable interest both from political scientists and economists.

* Corresponding author.

L. Qi (Ed.): ISIA 2010, CCIS 86, pp. 424–430, 2011.

There are actually mathematical and numerical approaches that exist to eliminate gerrymandering. Such methods provides well-defined steps and constraints. These models always consider equal population, preservation of county integrity, and district area compactness to constrain partisan gerrymanderers. Gudgin and Taylor [3] used explicit enumeration ('brute force' search) method literally evaluate every possible plan. However, as the number of population blocs increases, the number of potential plans grows so rapidly that no computer can evaluate all of them. Many turn to heuristic methods. These methods, by definition, do not guarantee an optimal solution, but often perform well in optimization.[4] Browdy [5] recommended simulated annealing, which is based on a mathematical analogy to the slow cooling of metal. If the value function being optimized is sufficiently well behaved, simulated annealing asymptotically converges to the optimum value.[4] Chandrasekham et al.[6] demonstrated the effectiveness of genetic algorithm for graph-partitioning problems. Genetic algorithms are based on natural selection and genetic combination.

In this paper, we create a novel heuristic method called Colonial Algorithm for redistricting a state by treating the state continuously. Cellular automaton was applied to implement the algorithm. By utilizing the method, we obtained the final districts contiguously and compactly with almost equal population. Significantly, the method is quick with a controllable and visible process.

2 Method

2.1 Voronoi Diagram (VD)

The VD is normally defined as the 'region of influence'. Every location in the plane is assigned to the closest member in the point set.[7] The mathematical description is given as follows.

Let $S=\{P_1,P_2,...,P_n\}$ be a set of n points in the plane. For each P_i, let $R(S;P_i)$ be the set of points that are nearest to P_i than to other $P_j s,(i \neq j)$, that is

$$R(S;Pi)=\{P\|P-Pi\|<\|P-Pj\|, j\neq i\}$$

Where $\|P-Q\|$ denotes the Euclidean distance between two point P and Q. The plane is partitioned into $R(S;P_1),R(S;P_2),...,R(S;P_n)$ and their boundaries. This partition is called Voronoi diagram for S, and the elements of S are called the generators of Voronoi diagram. The region $R(S;P_i)$ is called the Voronoi region of P_i, and the boundary lines of Voronoi regions are called Voronoi edges.[8]

2.2 A Possible VD Algorithm

One possible algorithm of VD is given below. VD is generated over a finite subset of nodes in an integer plane, which is divided into $n\times m$ mesh grids. [9]

For $P_k \in S$, a neighborhood set of P_k denoted $N(P_k)$ is a collection of mesh points such that their convex hull. We define the neighborhood set of grid $P_k = (x_i, y_j)$ with $1 < i < m$ and $1 < j < n$ to be:

$$N(P_k) = \{(x_i, y_{j-1}), (x_i, y_{j+1}), (x_{i-1}, y_j)(x_{i+1}, y_j)\}$$

Notice that $P_k \notin S$ .The four tessellations centered at a generator, see Fig.1 below. Every time a Voronoi region expands, it swallows its neighborhood set.

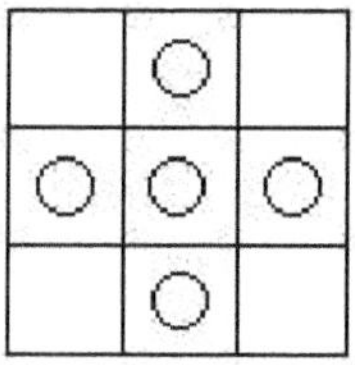

Fig. 1. Neighborhood set

A plane is partitioned in a VD way if generators expand at same speed, and stop when they meet.

Now we raise two considerations. what if the expending speed are not the same? Additionally, Voronoi regions may undergo a competition for edge grids to lower the system energy when they meet. Based on these considerations, we give a definition of Colonial Algorithm.

2.3 Colonial Algorithm (CA)

Firstly, let us pay attention to the following phenomenum. If bacteria are seeded separately in a culture dish, colonies will form around the generators. Colonies of different bacteria have different expanding capacities, i.e, "Hungry" colonies expand intensely while "full" colonies behave indolently. The expanding speed is determined by the physical conditions, the "appetite", of the colony. Meanwhile, there are competitions between colonies as resource scarcity. For simulating the growth of colonies, we created Colonial Algorithm as a development of VD with the two specific characteristics below:

① **Expanding Speed.** The expanding speed is a function of "starvation". Here "starvation"is an analog of a colony that is far from its equilibrium states. In the light of the famous Logistic curve, the initial stage of expansion is approximately exponential; then, as saturation begins, the expansion slows; and at maturity, expansion nearly stops. The speed function can have various forms in different cases.

② **Competition.** In VD, if two Voronoi regions meet at grid A, the expansions at A stop. However, in CA, the expansion may not stop at the edge grids and the regions keep competing for them. The competitions stop only when all the Voronoi regions reach a equilibrium state, or quite close to, with the lowest system energy. And we give one possible definition of system energy below.

2.3.1 System Energy (The Object Function)

In CA, every colony has an preset expected capacity of domain area, which symbolizes the growing potential of the colony. Efforts of expansion and competition are taken for better resource allocation to reach the lowest system energy, which is essentially the object function of optimization. If a colony is far from its expectation, it has a high degree of starvation and a sharp desire to expand; on the other hand, if it is over-expanded, the competition within population will diminish its domain area as well as the competition form the outside. Finally, the system will reach an equilibrium state and the system energy reached its lowest value, or quite close to.

We give the definition of system energy (namely, the object function):

$$E = \sum_{k=1}^{n} (P_k - P_{ke})^2 + wC_k$$

Where E is the system energy, n is the number of generators(colonies) in set S, P_k is the population of colony k and P_{ke} is the expected population , C_k is the compactness indicator, and w is a weight factor. Here the definition of system energy synthetically combines the compactness and deviation from expectation on population, thus being proper in evaluate the quality of gerrymandering results.

So the decrease of system energy transformed to the optimal question below:

$$P: \quad \min_{\sum P_k = const} \quad E = \sum_{k=1}^{n} (P_k - P_{ke})^2 + wC_k$$

Compactness is defined as ratio of region area and the square of the perimeter according to Sam Burden[2], $C_k = A_k / P_{erk}{}^2$, where P_{erk} is the perimeter of CA region k. As we know that the area of circle is the smallest among all the graphes with the same perimeter, so $C_k \geq \frac{1}{4\pi}$ and if C_k close to $\frac{1}{4\pi} \approx 0.0265$, it means the result is compact. In our algorithm, to calculate A_k and P_{erk}, we firstly give the definition of inner grids and edge grids. Inner grids are grids within the edge, with all its neighbors belong to the region, while edge grids are on the edge and have neighbor that not belong to the region. By calculating the sum of inner and edge grids, we generates the area of the region. By calculating the number of edge grids, we can obtain the perimeter. C_k is then generated.

According to the optimization theory, we know that in question (P) above, there must exist at least one optimal solution. But in this optimization question, the state space is huge, such as in the gerrymandering problems(see part 4), which will cost lots of time or even impossible to be solved.

3 Details of Colonial Algorithm

3.2 Three Steps of the Algorithm

① **Selection of Initial Generators.** Most of the time, the generators are not randomly selected, but chosen according to practical considerations. Take gerrymandering as an example, the generators are usually big cities with large population density. However, the choosing of generators does have something in common, to obtain well-proportioned regions we should seed them as separate as possible.
② **Period of Growth Step.** In the algorithm, we do not directly give the expanding speed, but fix the step size and change the expanding frequency. Every time Voronoi region k grows, it swallows its neighborhood set. This movement is called a growth step. If the frequency of growth step is intensified, the expansion of k is accelerated. By adjusting the expanding frequency, we can control the growing speed.
③ **Search Pattern.** In the algorithm, the Voronoi regions don't expand synchronously, but one by one. That's to say, during expansion region A firstly expand and annex its neighborhood set, followed by B,then C... recirculation... This method avoids the case that more than two colonies competing for one grid and reduces computation.

4 Result and Discussion

The problem is simplified in a 100*100 square with stochastic initial weight of each grid (we give the whole process a vivid look in fig.2.A) and solved by our CA algorithm. The process and the result are described in Fig. 2 below (process: fig.2.B-G and result: fig.2.H, where we stop at time 1930).

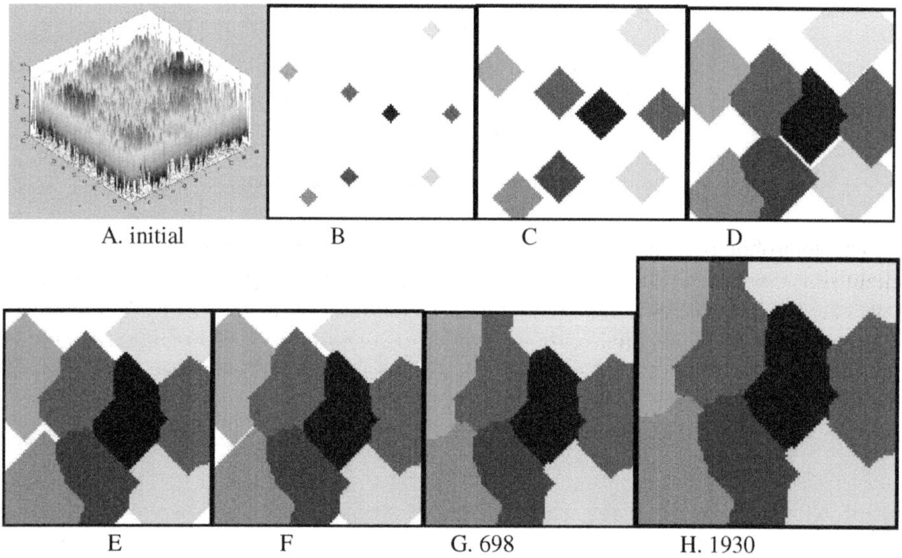

Fig. 2. The process of gerrymandering

The generators of CA in the gerrymandering case are big cities with large population density. Suppose we have n candidates in a state, we'd better choose n generators or the submultiples of n. To obtain population equality, we give every Voronoi region the same expected capacity and let them grow. As the figure depicts, the Voronoi regions expand, meet with each other and undergo competitions. The process is likely to stop at a equilibrium state, where all the regions keeps still.

However, there are cases the final result may not stop at one state, instead, it may keep shifting between two or more states, see Fig.2.G and H above. But as the differences of the final states are so small that could be ignored(see table.1 below, which is compact with small population variance). Empirically we can choose either of the states as final result in gerrymandering with admissible result.

Table 1. Resultant datas when time are 698 and 1930

T-698			T-1930		
region	population	compactness	region	population	compactness
1	5270.974841	0.037837	1	5172.188258	0.037535
2	5097.185035	0.042561	2	5243.699535	0.042760
3	5214.521795	0.035746	3	5260.863758	0.034676
4	5269.290996	0.038920	4	5063.033338	0.040496
5	5045.619865	0.035174	5	5122.707679	0.033133
6	5030.106970	0.034774	6	5224.951804	0.036175
7	5103.426960	0.037194	7	5250.448758	0.035763
8	5273.471151	0.039133	8	5269.141682	0.039547

From the above table, we can see the algorithm perform excellent in districting, with an almost convergent reslut.

Gerrymandering can be solved by other approaches, such as the optimization of combination results, which means that if we conduct gerrymandering at a 100×100 plane and want 8 electoral regions, we should try 8^100 times to obtain the optimized one. How time-consuming! So the utilization of CA greatly reduced the time consumption. Another big advantage of CA method is that we can obtain "compact"and "simple" regions, because of simple connectedness of CA results.

However, the algorithm has its shortcomings. We are not sure whether the result is convergent or not, neither can we know the convergence speed, and the distance to the optimized result. Although such a meta-heuristic algorithm can find excellent solutions fast, there is no proof that it won't generate a bad one. But luckily, though it has such shortcomings, CA method can be well used to solve practical problems, as bad answers only come from some abnormal or extreme conditions.

CA can be developed to partition there-dimentional space. It can be used in the simulation of crystal growth. And it is also an improvement of the population competition dynamical model by adding geographical factors, which we do not discuss in this paper.

5 Conclusion

We extend the definition of Voronoi Diagram and introduce a meta-heuristic algorithm called Colonial Algorithm (CA) in the light of expanding and competition behaviors of colonies in a culture dish. As an application, we use it in gerrymandering. Result shows that the algorithm is simple and convenient. The regions partitioned are "compact"and "simple".

Colonial Algorithm can be further used in many fields such as crystallization and regional monopoly. However, though we can find a quite admissible solution by CA, we cannot make sure whether the result is convergent or not, neither can we know the convergence speed, and the distance to the optimized result. They are the limitations of the algorithm.

Acknowledgments. This work is supported by the Zhejiang Provincial Natural Science Foundation of China (No.Y1080627).

References

1. Puppea, C., Tasnádi, A.: A computational approach to unbiased districting. Mathematical and Computer Modeling 48(9-10), 1455–1460 (2008)
2. Burden, S., et al.: Applying Voronoi Diagrams to the Redistricting Problem, MCM-Outstanding paper 07-A (2008)
3. Gudgin, G., Taylor, P.J.: Seats, Votes and the Spatial Organization of Elections. Pion Limited, London (1979)
4. Altman, M.: Modeling the effect of mandatory district compactness on partisan gerrymanders. Political Geography 17(8), 989–1012 (1998)
5. Browdy, M.H.: Simulated annealing:an improved computer model for political redistricting. Yale Law and Policy Review (8), 163–179 (1990)
6. Chandrasekham, et al.: Genetic algorithm for node partitioning problem and application in VLSI design. IEE Proceedings Series E 140(5), 255–260 (1993)
7. Gahegan, M., Lee, I.: Data structures and algorithms to support interactive spatial analysis using dynamic Voronoi diagrams. Computers, Environment and Urban Systems (24), 509–537 (2000)
8. Kobayashi, K., Sugihara, K.: Crystal Voronoi diagram and its applications. Future Generation Computer Systems (18), 681–692 (2002)
9. Schueller, A.: A nearest neighbor sweep circle algorithm for computing discrete Voronoi tessellations. J. Math. Anal. Appl. (336), 1018–1025 (2007)

An Improved New Event Detection Model

HongXiang Diao[1,2], Ge Xu[3], and Jian Xiao[1]

[1] Hunan agricultural university, ChangSha, China
[2] National University of Defense Technology, ChangSha, China
[3] Central South University, ChangSha, China
{dhxpaper,banitaa_xiao}@163.com, pigeonxu@sina.com

Abstract. New Event Detection (NED) aims to recognize the first story for a new event that had not been discussed before. The traditional event detection model basically adopts incremental TF-IDF weights for the terms, not considered the effect of part of speech and named entities in the document. The paper explore the application of weighting the part of speech and generates document theme terms based on the document named entity to detect new event, which can improve performance comparing with the traditional model.

Keywords: New Event Detection, Part of Speech, Named Entity.

1 Introduction

Topic detection and tracking (TDT) is a research work for information identification, extraction and organization, which provides a new platform of multi-language test for information retrieval, data mining and natural language processing (NLP) technology. New Event Detection (NED) is one of the five tasks in the Topic Detection and Tracking. The target of NED is to detect the first story on topics of interest. In the era of information explosion, the new topic is often submerged in the daily information flow, which greatly limits the people to grasp the important news in time. People who need the latest news when it happened, such as government situation, financial analysis and stock market, can use NED to more quickly identify new events. Therefore, the research of NED has important practical value.

The rest of the paper is organized as follows. Section 2 gives a short review of the related studies in NED. Section 3 presents the basic model for NED that most current systems use. Section 4 presents our new extensions and improved model. Section 5 presents experimental results and performance analysis. Section 7 gives our conclusions and future work.

2 Previous Research

NED models usually use a single pass incremental clustering algorithm. For a newly arrived document the similarity between the document and previous events is computed and the maximum similarity will be selected. If the similarity is more than the predefined threshold, the document will be assigned to a old event, otherwise it will be considered a new event.

L. Qi (Ed.): ISIA 2010, CCIS 86, pp. 431–437, 2011.

The prevailing method of NED is from James Allan [1] and Yiming Yang[2], Allen et al. used a modified version of TF-IDF and also penalized the threshold by the time distance between the document and the event; Yang et al. proposed an incremental IDF factor. They considered a time window and also a decay factor for the similarity between documents and events based on the time difference. Later related NED research involves two improvements based on the framework of statistical method, that is, a better document representation and full use of the event terms of corpus.

TF-IDF term weighting is the prevailing technique in document representation for the NED task. Yang firstly divided categories on the basis of previous stories, and then only selected the best category to describe the relevant stories on the topic. Brants[3] improved weight calculation method of incremental TF-IDF to include source-specific models, and represented the text using vector space model, then used Hellinger distance to match text correlation. Stocks[4] utilized a combination of evidence which was derived from two distinct representation. One of the representations was the usual text vector, the other made use of lexical chains (created using WordNet) to obtain the most prevalent topics discussed in the document.

The mainly method using corpus terms is expected to integrate into the Named Entity (NE) recognition technology. Papka[5] excluded from the NEs which frequently occur in the complete corpus, and endowed NE of sites with four times weight than other terms. Giridhar[6] described the reports into three vector space, namely, that contained all the term vector, only contained the NE term vector and excluded NE term vector. He compared the impact of the NED system based on three kind of vector space. Experiment verified that NE has promoted NED performance about some stories, but some stories in the absence of NE involvement is better. Based on this phenomenon, Giridhar classified the stories in advance, and distinguished the role of NE which stories categories helped NED system recognize a new topic. Another method using the stories corpus terms were adopted with time characteristics of the stories. The application of NED on time characteristics has two ways: one is based on the time sequence of input document using the KNN classification; the other is using the time decay function as the parameter to improve content-based correlation method. To certain extent, these studies raise the NED system performance.

TDT which is regard as a novel branch area of information processing gradually becomes an important research focus in China. The domestic research more focuses on own characteristics of TDT comparing to the main trend of statistical probability model in foreign studies. In addition, event generation and follow-up development include the relations between the time sequence, which implies that the TDT system can not be established by the single content-based topic model, but integrate the timing sequence characteristics and tracking trends in the evolution of topics. On this basis, the domestic-related research organizations[7][8][9] have put up preliminary attempts for the establishment of structured and hierarchical topic model.

3 Basic Model

This section presents the basic NED model that is similar to what most current systems use. We use it as a base for our extensions. General a NED system consists of three parts: story representation, the similarity calculation, the NED process. Story representation is building up the vector representation of stories by preprocessed operation; Similarity calculation is calculating the similarity between the stories based on vector of news representation. NED process is comparing the latter story and historical ones to determine whether it is the new event in chronological order.

3.1 Story Representation

The story representation firstly preprocesses the stories. We tokenize the stories, recognize abbreviations, normalize abbreviations, segment word, remove stop words, add part of speech tags, calculate word frequency and so on, then select the text representation model. Vector space model is one of the simple and efficient text representation models. Firstly, we extract term to represent the text as a collection of items, and then construct the weight vector according to the weight of items. In the traditional vector space model, we consider a single word as the term extraction.

In this paper, we adopt incremental TF-IDF model to calculate similarity. In a TF-IDF model, the frequency of a term in a document(TF) is weighted by the inverse document frequency(IDF). In the incremental model, document frequencies df (w) are not static but changeable in time steps t. At time t, a new set of test documents Ct is added to the model by updating the frequencies.

Incremental TF-IDF model:

$$df_t(w) = df_{t-1}(w) + df_{Ct}(w) \tag{1}$$

Where $df_t(w)$ is the document frequency of the word w at time t , $df_{Ct}(w)$ is the document frequency of the word w in the additional document corpus C_t.

The document frequency is used to calculate weights for the terms w in the documents d. At time t, we use

$$weight_t(d,w) = \frac{1}{Z_t(d)} f(d,w) \cdot \log \frac{N_t}{df_t(w)} \tag{2}$$

Where $weight_t(d,w)$ is a weight of the word w at time t in document d, N_t is the total number of documents at time t. $Z_t(d)$ is the normalization value with

$$Z_t(d) = \sum_w f(d,w) \cdot \log \frac{N_t}{df_t(w)} \tag{3}$$

3.2 Similarity Calculation

This paper uses cosine distance to calculate similarity between two documents. The document similarity between d and q is expressed as

$$Sim(d,q) = \frac{\sum_w weight(w,d) * weight(w,q)}{\sqrt{\sum_w weight(w,d)^2} * \sqrt{\sum_w weight(w,q)^2}} \tag{4}$$

Other possible similarity metrics include the Hellinger distance, the Kullback-Leibler divergence, Jensen-Shannon distance, or Clarity-based distance, which have been found useful in other work.

3.3 New Event Detection

The new document q at the time t will be compared to all the previous documents d. We identify the document d^* with the greatest similarity to q:

$$d^* = \arg\max_d sim_t(q,d) \tag{5}$$

The value

$$score(q) = 1 - sim_t(q,d^*) \tag{6}$$

is used to determine whether a document q is a new event and at the same time is an indication of the confidence in our decision. If the score (q) exceeds the threshold θ, we consider document q as a new event, on the contrary it describes an old event. Also the greater the difference between similarity and the threshold value, the higher the credibility of decision. The threshold θ can be determined by using labeled training data and calculate similarity scores for document pairs on the same or different events. This paper adopts θ=2.

4 Improved Model

We consider that documents often focus on specialized key words in different domain based on the term of the part of speech, as well as time, place, person, organization and other named entities have a major impact to event detection. So we calculate term weight taking into account of the term of the part of speech, and generate a document theme term set according to the document named entity in order to improve the performance of NED.

4.1 Feature Selection

The term weighting calculation of this model is based on the part of speech and the value of incremental TF-IDF. Typically, the only noun, verbs, adjectives, adverb and numbers in the document contain semantic terms. The paper[8] validated that different types of documents have different preferences for different types of entities, and part of speech. This paper considers the impact of the parts of speech in the calculation of the term weight value based on the incremental TF-IDF values.

$$weight_t(d,w) = \frac{c(w)}{Z_t(d)} f(d,w) \cdot \log \frac{N_t}{df_t(w)} \tag{7}$$

c (w) is a weight value for the word w what is affiliated with the part of speech. The value of c (w) is confirmed by statistical methods that different types of topics were given a different c (w) value.

4.2 Named Entity Extraction

NEs are divided into seven categories: person, organization, location, date, time, money, percentage. These phrases are the most basic information elements, which usually indicate the main content of the documents. In the understanding of the text, the role of NE is more important than common term. In this model, we extract person, organization, location and time appearing in the News as the subject term set.

4.3 Algorithm Model

(1) We firstly preprocess the document including part of speech recognition and named entity recognition, and then single out named entity to construct document theme term set.
(2) Firstly, we calculate the weights of document terms, construct terms of vector, and then calculate the document similarity by Hellinger distance comparing to all the stories that have been added. If the maximum similarity is below the threshold A, the story is considered a new Event; if the similarity with other documents above the threshold A, into (3).
(3) We compare the overlap of named entity between the story and each history documents that the similarity is above the threshold. If overlap value is above the threshold B, then these two documents are combined as one event set; on the contrary, the story is a new event.

5 Experimental Result and Performance Analysis

We use the NED system described in the paper to participate in the Topic Detection and Tracking evaluation. In the following, we shortly describe the data sets and then present improvement and evaluation results.

5.1 Experimental Data and Test Method

Experimental data sets are crawled about 2000 documents between January to June 2010 from the Internet portal (Sina, Sohu, Netease) . Firstly we select 20 training sample set about 500 stories from the corpus, then randomly select 1000 document as the test set from the remaining corpus. All the data set was artificially classified into 11 subject categories: (1) elections; (2) scandal/hearing; (3) crime; (4) natural disasters; (5) accidents; (6) military conflicts; (7) scientific discoveries; (8) finance; (9) laws; (10) Sports; (11) entertainment.

We designed and implemented three experiment programs to validate improvement in this paper. Experiment 1 is a benchmark system with the basic model; experiment 2 uses improved strategy of terms weighting based on benchmark system; Experiment 3 introduces the document theme term of named entity based on experiment 2.

5.2 Evaluation Methodology

To quantify the comparison of different systems, TDT meeting developed a set of evaluation norms.

$$C_{Det} = C_{Miss} \cdot P_{Miss} \cdot P_{target} + C_{FA} \cdot P_{FA} \cdot P_{non-target} \tag{8}$$

$$P_{non-target} = 1 - P_{target} \tag{9}$$

C_{Det} is a cost function that combines the probability of missing a new story P_{Miss}, the probability of seeing a new story in the data P_{target}, the cost of missing a new story C_{Miss}, the probability of a false alarm P_{FA}, the probability of seeing an old story $P_{non-target}$, and the cost of a false alarm C_{FA}. C_{Miss}, C_{FA} and P_{target} is the default value, as adjusted proportion coefficient of missing rate and false alarm rate for evaluation results. The cost is usually normalized between 0 and 1:

$$(C_{Det})_{Norm} = C_{Det} \Big/ \min\{C_{Miss} \cdot P_{target}, C_{FA} \cdot P_{non-target}\} \tag{10}$$

We directly use $(C_{Det})_{Norm}$ as a performance evaluation scores, $(C_{Det})_{Norm}$ is equal to 0 for a fully judicious system, $(C_{Det})_{Norm}$ is equal to 1 when all stories are judged as new event or not new events.

5.3 Experimental Result and Performance Analysis

The following table lists the experimental results of three new event detection system.

Table 1. Comparison table of test results

Experiment	Miss/(%)	FA/(%)	Norm/(%)
Experiment 1	41.38	4.85	0.5885
Experiment 2	41.12	4.53	0.5665
Experiment 3	36.45	4.95	0.5324

The experiment results show that the introduction of part of speech in term weight calculation makes better construction weight vector of the document, reduces the Miss and FA. After the introduction of the document theme term set, detection of new events have a more stringent standards, decrease Miss, and increase FA, but compared to the decrease degree of Miss, only a small increase of FA is tolerable.

6 Conclusions

This paper firstly analyzes the defect of traditional NED methods and the different weighting impacts in connection with different parts of speech and different types of named entity, proposes weighted method based on characteristics of part of speech, and in accordance with the named entities we generate document theme term set for event detection, which significantly decreases the system missing rate comparing with the traditional model.

Current research of NED mainly focuses on information retrieval, information filtering, classification, clustering and other technology based on traditional statistics strategy, ignores itself terms of the corpus, such as the suddenness and jump topic, continuity and succession of relevance documents, hierarchy and time-ordered of document content and so on. Based on this issue, the current research trend is to integrate many methods and embed corpus terms to achieve recognition and tracking topics, such as the topic model representation combining named entities, the weights and the threshold value estimation with the time parameter, etc. To some extent, the method can increase TDT system performance, but it just is a kind of supplementary and amendment for traditional statistical strategy, and not form its own research formwork and model for the TDT area. Therefore, future research direction in TDT will focus on construction of representation model and fusion of some lingual processing methods.

References

1. Allan, J., Papka, R., Lavrenko, V.: On-line New Event Detection and Tracking. In: Proceedings of 21st Annual International ACM SIGIR Conference on Research and Development in Information Retrieval, pp. 37–45. ACM Press, New York (1998)
2. Yang, Y., Pierce, T., Carbonell, J.: A Study on Retrospective and on-line Event Detection. In: Proceedings of 21st Annual International ACM SIGIR Conference on Research and Development in Information Retrieval, pp. 28–36. ACM, CMU (1998)
3. Thorsten, B., Francine, C., Ayman, F.: A System for New Event Detection. In: Proceedings of the 26th Annual International ACM SIGIR Conference, pp. 330–337. ACM Press, New York (2003)
4. Nicola, S., Joe, C.: Combining Semantic and Syntactic Document Classifiers to Improve First Story Detection. In: Proceeding of the 24th Annual International ACM SIGIR Conference, pp. 424–425. ACM Press, New York (2001)
5. Yang, Y., Carbonell, J., Brown, R., Pierce, T., Archibald, B.T., Liu, X.: Learning Approaches for Detecting and Tracking News Events. IEEE Intelligent Systems Special Issue on Applications of Intelligent Information Retrieval 14(4), 32–43 (1999)
6. Giridhar, K., Allan, J.: Text Classification and Named Entities for New Event Detection. In: Proceedings of the 27th Annual International ACM SIGIR Conference, pp. 297–304. ACM Press, New York (2004)
7. Hong, Y., Zhang, Y., Fan, J.-L.: New Event Detection Based on Division Comparison of Subtopic. Chinese Journal of Computers 31(4), 687–689 (2008)
8. Zhang, K., Li, J.-Z., Wu, G.: A New Event Detection Model Based on Term Reweighting. Journal of Software 15(4), 817–825 (2008)
9. Fan, X.-Q., Zhang, Y.-K.: New Event Detection Method Based on Word Pairs Vector Space Model. Computer Engineering and Applications 46(12), 123–125 (2010)

Calibration Model for Electrical Capacitance Tomography Sensor with Thin Radial Guards

Xiangyuan Dong and Shuqing Guo

Zhongyuan University of Technology, Zhengzhou 450007, China

Abstract. Electrical capacitance tomography (ECT) is one of process tomography technique which developed rapidly in recent years. There are three types of widely used ECT sensor: the sensor without radial guards, the sensor with thin radial guards and the sensor with radial guards. The sensor with thin radial guards is different from the other two. A calibration model for the sensor with thin radial guards was discussed in this paper. The simulation experiments were carried out to evaluate the method. The results show that the method can produce good results.

Keywords: Calibration model, Electrical capacitance tomography, Radial Guards.

1 Introduction

Electrical capacitance tomography (ECT) is a technique to obtain permittivity distribution in the pipe by using capacitance measurements. Due to its low cost, fast response, no radiation, non-intrusive and non-invasive characteristics, ECT has been widely used in oil pipeline flows, gas/solids flows and fluidization processes in recent years[1-3].

In ECT, it is often necessary to use two sets of reference capacitances and calculate the normalized capacitance

$$\boldsymbol{C}_{\mathrm{n}}=\frac{\boldsymbol{C}_{\mathrm{m}}-\boldsymbol{C}_{\mathrm{l}}}{\boldsymbol{C}_{\mathrm{h}}-\boldsymbol{C}_{\mathrm{l}}} \quad . \tag{1}$$

where $\boldsymbol{C}_{\mathrm{h}}$ and $\boldsymbol{C}_{\mathrm{l}}$ are the reference capacitances when the ECT sensor is filled with the low permittivity material and the high permittivity material, respectively. Then the normalized capacitance can be used for image reconstruction. It is known as calibration.

However, in some cases, it is difficult to fill the sensor with a material with a suitable high permittivity. So it is difficult to reconstruct good images. Yang [4, 5] proposed four mathematical models to describe the four types of ECT sensor. These models can be used to adjust the measurement range of the ECT system according to the defined permittivity value of a third material to be imaged.

L. Qi (Ed.): ISIA 2010, CCIS 86, pp. 438–443, 2011.

In this paper, another widely used ECT sensor, the sensor with thin radial guards is discussed. The mathematical model was described. Simulation and experimental data were used to demonstrate the effectiveness of the model.

2 Mathematical Model

According to the different radial guards, the ECT sensor can be classified into three types: the sensor without radial guards, the sensor with thin radial guards and the sensor with radial guards. They are shown in Fig. 1. The potential distributions for the three sensors are illustrated in Fig. 1 when the sensing region is filled with air. As can be seen, the potential distributions for the three sensors are slightly different. The external field lines are confined by the radial guards(or thin radial guards). Consequently, the capacitances measured between these electrodes with radial guards(or thin radial guards) mainly depend on the field lines that cross the interior of the pipe. It results in lower capacitances, as shown in Figure 2. From Fig. 2, we can also see that the capacitances between these electrodes with thin radial guards are larger than those with radial guards, so their external capacitances can not be negligible. The simplified model for the sensor with thin radial guards is shown in Fig. 3.

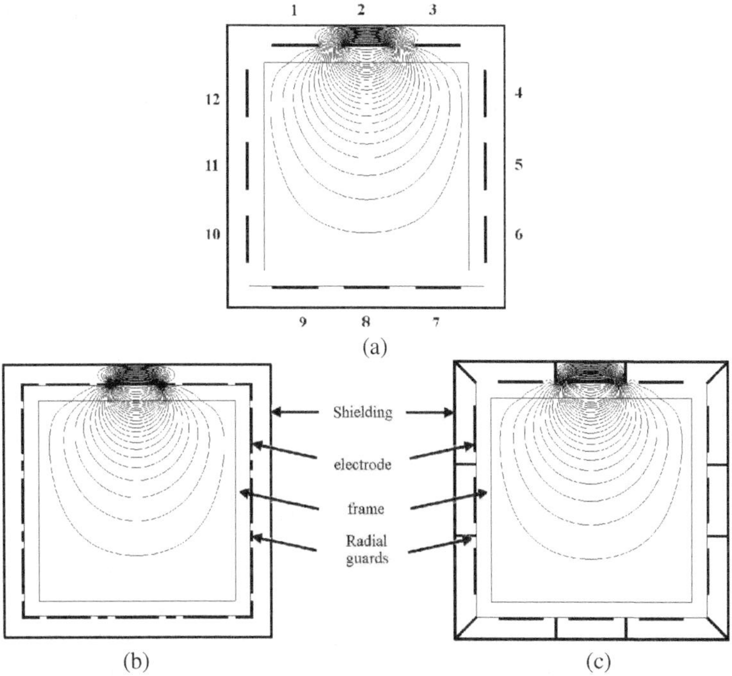

Fig. 1. Equi-potential lines for: (a) sensor without guards, (b) sensor with thin radial guards, and (c) sensor with radial guards

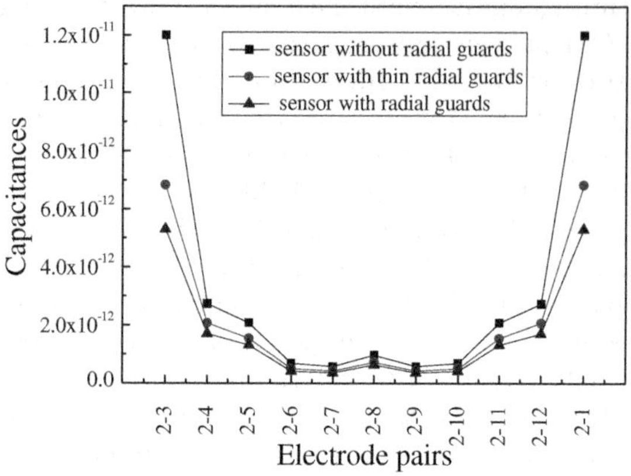

Fig. 2. Capacitances for 3 ECT sensors. Source electrode is number 2.

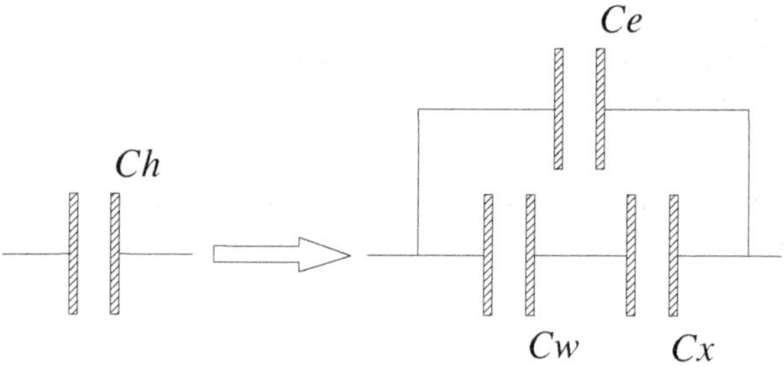

Fig. 3. Model of ECT sensor with thin radial guards

From Fig.3, the measured capacitance, C_h, can be written as

$$C_h = C_e + \frac{C_w C_x}{C_w + C_x} . \tag{2}$$

where C_w is the pipe wall capacitance, which is a constant for a given sensor. C_x is the material capacitance, which is proportional to a relative permittivity ε_r of the homogenous material in the sensor. It can be expressed by the following equation:

$$C_x = \varepsilon_r C_0 . \tag{3}$$

where C_0 is the material capacitance when the sensor is filled with air ($\varepsilon_r = 1$).

$$C_h = C_e + \frac{C_w \varepsilon_r C_0}{C_w + \varepsilon_r C_0} . \tag{4}$$

C_w, C_e and C_0 are the undetermined coefficients. In order to determine the three coefficients, three kinds of materials with known relative permittivity constant (i.e. ε_{r1} , ε_{r2} and ε_{r3}) can be used on the calibration respectively. And three measurements of measured inter-electrode capacitances (i.e. C_{h1} , C_{h2} and C_{h3}) can be obtained. Substituting ε_{r1} ε_{r2} and ε_{r3} in equation (4) gives three simultaneous equations:

$$C_{h1} = C_e + \frac{C_w \varepsilon_{r1} C_0}{C_w + \varepsilon_{r1} C_0} . \tag{5}$$

$$C_{h2} = C_e + \frac{C_w \varepsilon_{r2} C_0}{C_w + \varepsilon_{r2} C_0} . \tag{6}$$

$$C_{h3} = C_e + \frac{C_w \varepsilon_{r3} C_0}{C_w + \varepsilon_{r3} C_0} . \tag{7}$$

Solving these equations gives

$$A = \frac{\varepsilon_{r3}(\varepsilon_{r2}-\varepsilon_{r1})(C_{h3}-C_{h1})-\varepsilon_{r1}(\varepsilon_{r3}-\varepsilon_{r1})(C_{h2}-C_{h1})}{(\varepsilon_{r3}-\varepsilon_{r1})(C_{h2}-C_{h1})-(\varepsilon_{r2}-\varepsilon_{r1})(C_{h3}-C_{h1})} . \tag{8}$$

$$C_0 = \frac{(C_{h2}-C_{h1})(A+\varepsilon_{r1})(A+\varepsilon_{r2})}{A^2(\varepsilon_{r2}-\varepsilon_{r1})} . \tag{9}$$

$$C_w = \frac{(C_{h2}-C_{h1})(A+\varepsilon_{r1})(A+\varepsilon_{r2})}{A(\varepsilon_{r2}-\varepsilon_{r1})} . \tag{10}$$

$$C_e = C_{h1} - \frac{\varepsilon_{r1}(C_{h2}-C_{h1})(A+\varepsilon_{r2})}{A(\varepsilon_{r2}-\varepsilon_{r1})} . \tag{11}$$

3 Numerical Results

The finite element simulation method (FEM) has proven to be a useful tool for studying the ECT sensors [6, 7]. In this section, FEM was used to verify the model. The ECT sensors studied in this paper are shown in figure 2. The measuring zone is 80 mm by 80 mm, and the thickness of the frame is 5mm. A PC with a Pentium IV processor is used for simulation.

In order to determine the three coefficients C_w, C_e and C_0, three kinds of materials with the relative permittivity constant 1, 3, 6 are used. The Corresponding capacitances can be calculated by using the finite element method according to (12).

$$C = -\frac{1}{V} \oint_{\Gamma} \varepsilon(x, y) \nabla \phi(x, y) d\Gamma . \tag{12}$$

where V is the potential difference between the source and detector electrode. Γ is the electrode surface. $\varepsilon(x, y)$ and $\phi(x, y)$ are the dielectric constant and the electrical potential distributions, respectively. The electrical potential distributions $\phi(x, y)$ inside an ECT sensor can be obtained numerically by solving the following equation with the Dirichlet boundary conditions:

$$\nabla \cdot [\varepsilon(x, y) \nabla \phi(x, y)] = 0 \ . \tag{13}$$

After C_w, C_e and C_0 are calculated, the relationship between the measured inter-electrode capacitance and the relative permittivity constant can be determined. The other three relative permittivity constant 2, 5, 8 and the Corresponding capacitances calculated according to (12) are used to verify the model.

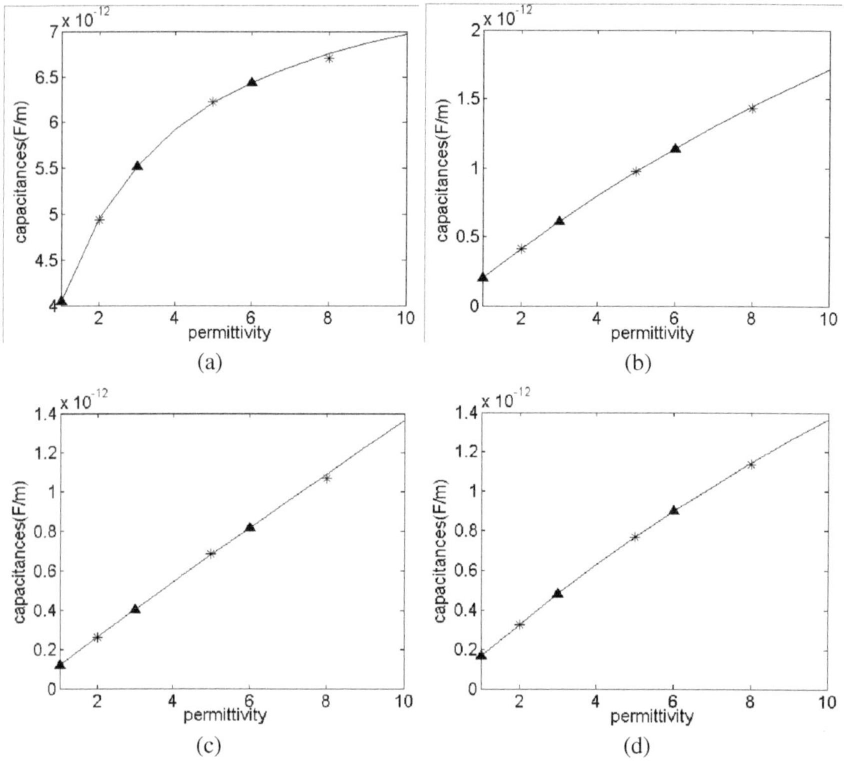

Fig. 4. Capacitances plotted against relative permittivity for: (a) Electrode pair 1–2, (b) Electrode pair 1–3, (c) Electrode pair 1–4 and (d) Electrode pair 1–5

Figure 4 and Table 1 show the results of the simulations. It can be seen that the model gives the small capacitance errors when compared with the simulation results according to (12).

Table 1. Relative capacitance errors for different electrode pairs (%)

permittivity	electrode pairs				
	1-2	1-3	1-4	1-5	1-6
2	0.328	0.718	1.119	0.292	1.058
5	-0.183	-0.302	-0. 564	-0.137	-0.606
8	0.717	1.05	2.222	0.52	2.647

4 Conclusion

In this paper, the sensor with thin radial guards is discussed. The mathematical model for the sensor with thin radial guards was described. The simulation experiments were carried out to evaluate the method. It shows that the predicted capacitances by this method are satisfactory.

References

1. Liu, S., Wang, H.G., Jiang, F., et al.: A new image reconstruction method for tomographic investigation of fluidized beds. AIChE J. 48, 1631–1638 (2002)
2. Liu, S., Li, J.T., Chen, Q.: Visualization of flow pattern in thermosyphon by ECT. Flow Measurement and Instrumentation 18, 216–222 (2007)
3. He, R., Xie, C.G., Waterfall, R.C.: Engine flame imaging using electrical capacitance tomography. Electronics Letters 30, 559–560 (1994)
4. Yang, W.Q.: Calibration of capacitance tomography systems: a new method for setting system measurement range. Measurement Science and Technology 7, 863–867 (1996)
5. Yang, W.Q.: Modelling of capacitance tomography sensors. IEE Proceedings: Science, Measurement and Technology 144, 203–208 (1997)
6. Khan, S.H., Abdullah, F.: Finite element modelling of multi electrode capacitive systems for flowing imaging. IEE Proc-G 140, 216–222 (1993)
7. Dong, X.Y., Guo, S.Q., Liu, S.: On-line calibration method for combustion visualization in porous media by using electrical capacitance tomography. Proceedings of the Chinese Society of Electrical Engineering 28, 44–48 (2008)

A New Traffic Data-Fusion Approach Based on Evidence Theory Coupled with Fuzzy Rough Sets

Hongzhao Dong[1], Min Zhou[1], and Ning Chen[2]

[1] ITS Joint Institute, Zhejiang University of Technology, 310014, Hang Zhou, China
[2] School of Mechanical and Automotive Engineering, Zhejiang University of Science and Technology, 310023, Hang Zhou, China
its@zjut.edu.cn

Abstract. The traffic detecting result is always short of accuracy by different kinds of individual sensors in urban China. To solve the issue, a new data fusion approach is raised. The algorithm combines fuzzy and rough set theory based on evidence theory. The method is improved to concise attribute rules and to measure fuzzy likelihood. Furthermore, a new combination rule is given to dissolve the confliction among the traffic evidence data collected by different individual sensors. Finally, the experiment to fuse the traffic data from an intersection in urban Hangzhou showed that the proposed approach could obtain a high accuracy.

Keywords: traffic data fusion, evidence theory, rough set method, attribute reduction, fuzzy probability measure.

1 Introduction

At present, there are several methods to detect traffic flow data in urban China, such as loops, video detector, and dynamic OD analyzer. Practically, the different methods often give the contradictory outcome. It is difficult to remark the credibility of the traffic data detected by the different methods individually. To solve the issue, the multi-source data fusion algorithm was researched widely such as Kalman filter [1], Bayes reasoning [2], Fuzzy set theory [3]. Unfortunately, these kinds of algorithm show poor capacity to figure out the conflict among the traffic flow data to be fused.

It is a relief that D-S evidence theory can fuse the uncertain message with unknown conditions by means of both trusted function and likelihood function coming from the essential probability function, and the fused outcome becomes more accurate[4,5,6]. However, to fuse the traffic data using D-S evidence theory, there are also several problems to be solved such as the traffic data redundancy need be reduced, the essential probability formula of each evidence group should be extracted to avoid depending too much on subjective experience. Meanwhile, the conflict of different evidence shows another serious problem to be figured out. There usually exists the deficiency or difference among the detected traffic data due to the failure of one or more sensors in a multi-sensor fusing system [7].

To solve these problems, a new approach is proposed in this paper. Firstly, the raw data should be preprocessed by means of the attribute reduction principle of rough set,

L. Qi (Ed.): ISIA 2010, CCIS 86, pp. 444–451, 2011.

and the principle is improved based on the classical reduction method considering the dependency degree between the existing attributes of the reduction set and the new ones. The fuzzy likelihood measure is used to obtain the essential probability formula of the traffic flow message. More over, on the basis of Yager improved combination rule of evidence theory [8], a new combination rule is raised to eliminate conflicts in the fusing process with a consideration of the conflict degree between evidence data.

2 Evidence Theory Based on Fuzzy Rough Set

The fusion parameters of traffic flow are defined as follows: vehicular flux, lane occupancy ratio (A_{flu}), average speed (B_{ocp}), queue length (C_{spe}), waiting time (D_{seq}), average traveling time (E_{tim}). These parameters can be formed as a vector $x=(A_{flu},B_{ocp},C_{spe},D_{seq},E_{tim},F_{dur},K)$, where K denotes the different methods to detect these parameters.

2.1 Data Preprocessing Based on Rough Set

For the classical attribute reduction method, the function *SGF(a,R,D)* to indicate the important degree, is employed to describe the influence to decision attribute *D* after new attribute '*a*' of condition attribute set *C* joined into reduction attribute set *R*. But it lacks consideration about the influence to set *R*. Using dependent degree k_{ab} to judge whether the addition of the new attribute makes the certain ones of set R become unimportant is proposed in this paper. The algorithm as follows:

Step 1: Select condition attribute set $C=(A_{flu},B_{ocp},C_{spe},D_{seq},E_{tim},F_{dur})$, and decision attribute *D=K*. The current collecting data, historical data and sensor characteristic constitute the decision attribute table expressed as formula 1:

$$P=\begin{pmatrix} y_1 \\ \vdots \\ y_i \\ \vdots \\ y_n \end{pmatrix}=\begin{pmatrix} 1 & \widehat{A_{flu}^{\ 1}} & \widehat{B_{ocp}^{\ 1}} & \widehat{C_{spe}^{\ 1}} & \widehat{D_{seq}^{\ 1}} & \widehat{E_{tim}^{\ 1}} & \widehat{F_{dur}^{\ 1}} & \widehat{K^1} \\ \vdots & \vdots & \vdots & \vdots & \vdots & \vdots & \vdots & \vdots \\ i & \widehat{A_{flu}^{\ i}} & \widehat{B_{ocp}^{\ i}} & \widehat{C_{spe}^{\ i}} & \widehat{D_{seq}^{\ i}} & \widehat{E_{tim}^{\ i}} & \widehat{F_{dur}^{\ i}} & \widehat{K^i} \\ \vdots & \vdots & \vdots & \vdots & \vdots & \vdots & \vdots & \vdots \\ n & \widehat{A_{flu}^{\ n}} & \widehat{B_{ocp}^{\ n}} & \widehat{C_{spe}^{\ n}} & \widehat{D_{seq}^{\ n}} & \widehat{E_{tim}^{\ n}} & \widehat{F_{dur}^{\ n}} & \widehat{K^n} \end{pmatrix}=F(X) \tag{1}$$

Where $X=\begin{pmatrix} x_1 \\ \vdots \\ x_i \\ \vdots \\ x_n \end{pmatrix}=\begin{pmatrix} A_{flu}^{\ 1} & B_{ocp}^{\ 1} & C_{spe}^{\ 1} & D_{seq}^{\ 1} & E_{tim}^{\ 1} & F_{dur}^{\ 1} & K^1 \\ \vdots & \vdots & \vdots & \vdots & \vdots & \vdots & \vdots \\ A_{flu}^{\ i} & B_{ocp}^{\ i} & C_{spe}^{\ i} & D_{seq}^{\ i} & E_{tim}^{\ i} & F_{dur}^{\ i} & K^i \\ \vdots & \vdots & \vdots & \vdots & \vdots & \vdots & \vdots \\ A_{flu}^{\ n} & B_{ocp}^{\ n} & C_{spe}^{\ n} & D_{seq}^{\ n} & E_{tim}^{\ n} & F_{dur}^{\ n} & K^n \end{pmatrix}$ is the original collected data set. In $F=(f_{A_{flu}},f_{B_{ocp}},f_{C_{spe}},f_{D_{seq}},f_{E_{tim}},f_{F_{dur}},E)$, each component expresses the mapping from x_i to y_i.

Step 2: In set C, if $\forall b \in C$, select it as the original element of set R.

Step 3: If $\forall a \in C \wedge a \notin R$, compute its importance degree of formulate 2.

$$SGF(a,R,D)=\frac{Card(pos_{R\cup\{a\}}(D))}{Card(U)}-\frac{Card(pos_R(D))}{Card(U)} \tag{2}$$

If $\exists a' \in C$ and $SGF(a',R,D)=\max(\{SGF(a,R,D)|a\in C\})$, denote $\overline{R}=R\cup\{a'\}$.

Step 4: To compute $k_{a'b}=r(a',b)=\frac{card(POS_{IND(a')}IND(b))}{card(U)}$ for the new attribute a' and $\forall b\in R$. If $\exists b' \in R$ and $k_{a'b'}=\max(\{k_{a'b}=r(a',b)\,|\,b\in R\})$, delete the attribute b' temporarily from reduction set $\overline{R}$. Then $\overline{R}$ is denoted as R' (if the dependent degrees are all equal, b' is selected as the element which has the longest survival time in the set). Compute $SGF(b',R',D)$. If $|SGF(b',R',D)-SGF(a,R,D)|<\delta$ (δ is given in advance as threshold value), do not delete attribute b from set $\overline{R}$, and denoted as $R=\overline{R}$.

Step 5: To compute $\gamma_R(D)$. If $\gamma_R(D)=\gamma_C(D)$, R satisfies the condition, then the calculation is over. Otherwise calculation turns to *step 2*.

2.2 To Evaluate the Essential Probability Base on Fuzzy Likelihood Measure

To avoid the subjectivity while obtaining the essential probability function, this method is proposed base on fuzzy likelihood measure (case study of three collecting methods). The algorithm as follows:

Step 1: Select identification frame $\Theta=\{X,Y,Z\}$, where X, Y, Z respectively denote the traffic flow parameters of the three collecting methods. The collected data set is denoted as $Q=\begin{pmatrix} t_{11} & & \cdots & & t_{16} \\ t_{21} & \cdots & t_{ij} & \cdots & t_{26} \\ t_{31} & & \cdots & & t_{36} \end{pmatrix}$ $(i\in[1,3], i\in Z; j\in[1,6], j\in Z)$ after attribute reduction. Where t_{ij} denotes the fuzzy membership function of the *j*-th traffic parameters which are collected by the j-th collecting method respectively. Historical fusion data in the same condition are selected to be the basic traffic flow data, which are expressed as the fuzzy membership function of relevant attribute, and denoted as $S=\left(\overline{t_1},\ \overline{t_2},\ \cdots,\ \overline{t_j}\right)$ $(j\in[1,6], j\in Z)$, where $\overline{t_i}$ is the fuzzy membership function of relevant attribute.

Step 2: Matrix multiplication is defined as the fuzzy likelihood calculation between two fuzzy membership functions.

$$\text{So}\ \overline{M} = Q \bullet S = \begin{pmatrix} \rho(t_{11},\overline{t_1}) & & \cdots & & \rho(t_{16},\overline{t_6}) \\ \rho(t_{21},\overline{t_1}) & \cdots & \rho(t_{ij},\overline{t_j}) & \cdots & \rho(t_{26},\overline{t_6}) \\ \rho(t_{31},\overline{t_1}) & & \cdots & & \rho(t_{36},\overline{t_6}) \end{pmatrix}_{3\times 6}$$

$(i \in [1,3], i \in Z;\ j \in [1,6],\ j \in Z)$ and

$\rho(t_{ij},\overline{t_j}) = p(t_{ij} \cap \overline{t_j} \neq \varnothing) = p(A \leq (M \wedge N)(x)) = \sup_x \min\{M(x), N(x)\}$.

Where $M(x)$ and $N(x)$ are the relevant membership functions of t_{ij} and $\overline{t_i}$.

Step 3: To evaluate the essential probability function of $\overline{M}$ divided by column to have normalization processing. The outcome is $M = \{(m_{i1}, m_{i2}, m_{i3}, m_{i4}) \mid i \in [1,6], i \in Z\}$, and each group as follows:

$$m_{i1} = \frac{\rho(t_{i1},\overline{t_i})}{(\rho(t_{i1},\overline{t_i}) + \rho(t_{i2},\overline{t_i}) + \rho(t_{i3},\overline{t_i}) + \rho_i(\Theta))} \tag{2}$$

$$m_{i2} = \frac{\rho(t_{i2},\overline{t_i})}{(\rho(t_{i1},\overline{t_i}) + \rho(t_{i2},\overline{t_i}) + \rho(t_{i3},\overline{t_i}) + \rho_i(\Theta))} \tag{3}$$

$$m_{i3} = \frac{\rho(t_{i3},\overline{t_i})}{(\rho(t_{i1},\overline{t_i}) + \rho(t_{i2},\overline{t_i}) + \rho(t_{i3},\overline{t_i}) + \rho_i(\Theta))} \tag{4}$$

$$m_{i4} = \frac{\rho_i(\Theta)}{(\rho(t_{i1},\overline{t_i}) + \rho(t_{i2},\overline{t_i}) + \rho(t_{i3},\overline{t_i}) + \rho_i(\Theta))} \tag{5}$$

($\rho_i(\Theta) = 1 - \max(\rho(t_{i1},\overline{t_i}), \rho(t_{i2},\overline{t_i}), \rho(t_{i3},\overline{t_i}))$, $i \in [1,6]$, $i \in Z$), respectively denote the essential probability functions of the message collected by three collecting methods and the uncertain message.

2.3 Conflict Solution of the Evidence Combination

The disaccord to the real traffic scene may be occurred by the fused outcome if there is high conflict evidence, namely the conflict coefficient $k \to 1$. Yager has improved the D-S composite formula. And the new formula is as follows (two evidence sources):

$$m(\varnothing) = 0 \tag{6}$$

$$m(A) = \sum_{A_i \cap B_j = A} m_1(A_i) \bullet m_2(B_j), A \neq \varnothing, X \tag{7}$$

$$m(X) = \sum_{A_i \cap B_j = X} m_1(A_i) \bullet m_2(B_j) + k \left(k = \sum_{A_i \cap B_j = \varnothing} m_1(A_i) \bullet m_2(B_j) \right) \tag{8}$$

Yager's formula shows that if the conflict evidence can't be resolved reasonably, it should be thrown into unknown field, but it will induce another issue. Although most of evidences have proved the conclusion is right, the combination outcome would be negative. On the basis of Yager's formula, Sun Quan proposed an evidence combination formula which transforms the conflict by the credibility [9]. But this method ignores the evidence contribution to the combination outcome when computing the credibility of each group of conflict evidence [10,11].

In this paper, we have improved the combination method of Yager evidence theory, with the consideration about the credibility of group conflict evidence. The evidence credibility is used as proportional coefficient of the probability of the conflict evidence in the combination formula. The new evidence combination formula is as follows:

$$m(\varnothing)=0 \tag{9}$$

$$m(A)=p(A)+k\bullet q(A), A\neq\varnothing, X \tag{10}$$

$$m(X)=p(X)+k\bullet q(X)+k\bullet\prod_{\substack{i=1\\ j\leq i}}^{n}(1-\varepsilon_{ij}) \tag{11}$$

Where $p(A)=\sum_{\substack{A_i\in F_i\\ \bigcap_{i=1}^{n}A_i=A}} m_1(A_1)m_2(A_2)\cdots m_n(A_n)$, and $q(A)=\sum_{i=1}^{n}\beta_i\bullet m_i(A)$.

The credibility between two evidences, m_i and m_j, is denoted as $\varepsilon_{ij}=e^{-k_{ij}}$, which is decreasing function. The conflict magnitude between the two evidences is denoted as $k_{ij}=\sum_{A_i\cap A_j=\varnothing} m_i(A_i)m_j(A_j)$. The average credibility between evidence m_i and other evidences is denoted as $\alpha_i=\frac{1}{n-1}\sum_{\substack{j=1\\ j\neq i}}^{n}\varepsilon_{ij}$. The weight value of normalization is denoted as $\beta_i=\frac{\alpha_i}{\alpha_1+\alpha_2+\cdots+\alpha_n}$.

It could be proved that $m(A)$ could be essential probability function, as there exists the below conditions:

1) $m(\varnothing)=0$; 2) $0\leq m(A)\leq 1$; 3) $\sum_{A\subset X} m(A)=1$.

It shows the normalization credibility β_i of the evidence in each group is used as the weight. It embodies fully the contribution degree of the evidence in each group to the combination outcome.

3 Application

In the urban area of Hangzhou City, the principle traffic data come from the sensors of loops, video detector, and dynamic OD analyzer. Here sets Qingchun Road and Yan'an Street intersection, one intersection in the transportation grid, as an example to testify the above algorithms. The time slice (t) is 12:00:00 to 12:05:00 on Jan 1st, 2007 and the traffic data are derived from one lane. The historical fused data of three periods before the time (the length of period is T) are shown as the table 1 and table 2. The number in

Table 1. Collected data of Qingchun-Yan'an intersection

Parameter method	Flux (vehicle /5minutes)	Traveling Time (second)	Queue Length (m)	Waiting Time (second)	Average Speed (km/h)	Lane Occupancy Ratio
Loops	16(1)	-	-	-	-	0.8(1)
Video Detector	25(2)	-	56(2)	90(2)	60(2)	-
OD Analyzer	18(3)	196(3)	-	-	53(3)	-

Table 2. Historical fusion data of Qingchun-Yan'an intersection at (t-T), (t-2T) and (t-3T)

Time	Flux (vehicle/ 5minutes)	Traveling Time (second)	Queue Length (m)	Waiting Time (second)	Average Speed (km/h)	Lane Occupanc y Ratio
t-T	20(1)	150(3)	60(2)	80(2)	56(3)	0.6(1)
t-2T	18(1)	160(3)	70(2)	150(2)	43(2)	0.7(1)
t-3T	16(3)	240(3)	40(2)	130(2)	30(3)	0.5(1)

Table 3. Important parameters in the process

Key Parameter	Value
Fuzzy Membership Matrix	$Q=\begin{pmatrix} e^{-(x-16)^2/18} & 0 & 0 \\ e^{-(x-25)^2/18} & e^{-(x-56)^2/18} & e^{-(x-60)^2/18} \\ e^{-(x-18)^2/18} & 0 & e^{-(x-53)^2/18} \end{pmatrix}$
Benchmark Conversion Matrix	$S=(e^{-(x-20)^2/18}, e^{-(x-60)^2/18}, e^{-(x-56)^2/18})$
Fuzzy Likelihood Matrix	$M=Q\bullet S=\begin{pmatrix} 0.8 & 0 & 0 \\ 0.9 & 0.8 & 0.8 \\ 0.7 & 0 & 0.9 \end{pmatrix}$
Vector of Essential Probability Function	$\overline{M}$ ={(0.320,0.360,0.280,0.040), (0,0.800,0, 0.200),(0,0.444,0.500,0.056)}

the bracket represents the mode of collecting data. (In the bracket '-'denotes no data; '1' denotes loops; '2' denotes video detector; '3' denotes OD analyzer.)

1) The identification frame is $\Theta = \{X, Y, Z\}$. (X, Y and Z denote the collected data of loops, video detector and OD analyzer). Compute the essential probability function. Table 3 shows the important parameters in this process.

2) Based on the essential probability function and the new combination formula of evidence proposed in this paper, we can achieve more credible fusion outcome. Table 4 shows the important parameters in the process.

In table 4 the combination outcome of evidence shows that the traffic flow data which are collected by the loops have the maximal credibility. Table 5 shows the fusion outcomes, the practical data of the traffic flow which are collected by manual work in the same condition, the relative error and the average value of relative error of each traffic flow parameter.

Table 4. Important parameters in the process

Key Parameter	Value
Conflict Coefficient $\overline{K}$	$\overline{K}$ =(0.7392, 0.7900, 0.9282)
Credibility ε	ε =(0.4775, 0.4538, 0.3953)
Conflict Coefficient k	0.3436
Average Credibility α	α =(0.4657, 0.4364, 0.4245)
Value of Normalization β	β =(0.3510, 0.3290, 0.3200)
Combination Outcomes of evidence	M=(0.4209, 0.2833, 0.1831, 0.1127)

Table 5. Fusion outcome of the traffic flow message

Detection Parameter	Vehicular Flux (vehicle/ 5minutes)	Traveling Time (second)	Queue Length (m)	Waiting Time (second)	Average Speed (km/h)	Lane Occupancy Ratio	Average Value Of Relative Error
Fusion Outcome	25	196	56	90	60	0.8	-
Practical Collecting Value	22.0	160	45.0	78	45	0.60	-
Relative Error	0.136	0.225	0.244	0.154	0.333	0.333	0.232

Table 5 shows that the approach can obtain the fusion outcome effectively. The errors occurred due to the below factors. The threshold value δ that affects the final outcome is obtained by human experience in table 2. Another one is that the variance of normal distribution is determined with the principle of 3σ.

4 Conclusion

The method of attribute reduction has been improved based on rough set and it could consider the relation of attributes during reduction process. The essential probability function is obtained by the maximum fuzzy likelihood function which helps to diminish the effect of subjective factor. Finally, a new combination formula has been raised based on the Yager's formulas. It can reduce the negative effect on fusion accuracy caused by the conflict of different evidences. The experiment demonstrates that the proposed method is effective and practical to cope with issues such as urban traffic data fusion in urban Hangzhou.

Acknowledgment

This work has been sponsored by the Important Project for Key Subject of Zhejiang Province, China (2009C03016-3).

References

1. Yang, Z.-s.: Fusion technology of basic traffic information and its application. Science Press, Beijing (2005)
2. Meng, X.-r., Bai, G.-l., San, B.-g., Han, X.-j.: Application of Bayes date fusion on intelligent fault diagnosis in engine room. Journal of Dalian Maritime University 28, 389–405 (2002)
3. Zadeh, L.A.: Fuzzy algorithm. Information and Control (1965)
4. Bogler, P.L.: Shafer-Dempster reasoning with applications to multisensor target identification system. IEEE Trans. System, Man and Cybernetics SMC-17, 968–977 (1987)
5. Wang, J.-s., Zhou, H.-s., Zhou, W.-g.: Application of information fusion technology on traffic information management of Hangzhou traffic police. China ITS Journal 6, 27–30 (2003)
6. He, X.-g.: Theory and Technology of Fussy Knowledge Processing. National Defence Industry Press, Beijing (1998)
7. Selzer, F., Gutfinger, D.: LADAR and FLIR based sensor fusion for automatic target classification. SPIE, 1003, 236–241 (1988)
8. Yager, R.R.: On the dempster-shafer framework and new combination rules. Information Sciences 41, 93–137 (1987)
9. Sun, Q., Ye, X.-q., Gu, W.-k.: A New Combination Rules of Evidence Theory. Acta Electronica Sinica 8, 706–739 (2000)
10. Dempster, A.P.: Upper and lower probabilities induced by a multi-valued mapping. Ann. Math. Statist. 38, 325–339 (1967)
11. Shafer, G.: A mathematical theory of evidence. Princeton U.P., Princeton (1976)

The Self-calibration of Varying Internal Camera Parameters Based on Image of Dual Absolute Quadric Transformation

Ze-tao Jiang[1] and Shan-chao Liu[2]

[1] School of Informatin Engineering, Nanchang HangKong University, Nanchang, 330063, China
[2] NanChang HangKong University Nondestructive Testing Laboratory of Ministry of education, Nanchang, 330063, China
zetaojiang@126.com, liusanchao611@163.com

Abstract. This paper presents an method of self-calibration of varying internal camera parameters that based on image of dual absolute quadric transformation. Absolute dual quadric elements have so large differences in magnitude that solutions are extremely sensitive to noise.Through the transformation, all the elements of the image of dual absolute quadric are transformed into the same magnitude, So the solutions become more stable. The theorical analysis and experiments with both simulated and real data demonstrate that this self-calibration method can lead to an enormous improvement on the stability and robustness of the results without increasing computation.

Keywords: absolute quadric varying internal camera parameters transformation self-calibration.

1 Introduction

Camera calibration is the key steps to obtain three-dimensional information from the two-dimensional image [1]. The traditional method is through the structure of known objects (such as the calibration block) in the image of the projection to calculate the camera intrinsic parameters. Defect of this approach requires a calibration block, for most practical applications is not appropriate.The early 90s of the 20th century, Faugeras, Luong, Maybank [2,3] etc proposed the concept of self-calibration causes the calibration is possibly under unknown scene and the camera random movement of the general situation. Faugeras,etc.From the perspective of projective geometry between two images show the existence of each of the two quadratic equations of the form Kruppa constraint, by directly solving the Kruppa equations can be solved the internal parameters. In order to solve conveniently, the people also proposed the lamination demarcates gradually thought that like the QR resolution of Hartley [4], the absolute quadratic of Triggs [5], the modulus constraint of Pollefeys [6] etc. However, these self-calibration methods are assumed to be constant internal reference camera, which target motion does not match the actual applications. Heyden,Ustrgm,Pollefeys etc [7,14] also proposed a camera calibration under the varying internal parameters and

L. Qi (Ed.): ISIA 2010, CCIS 86, pp. 452–461, 2011.

proved theoretically and Pollefeys etc [14] also presents a variable internal parameters of the camera self-calibration method. Present a more popular method of self-calibration is absolute dual quadric right way, however, this method is still orders of magnitude as the gap as the elements, resulting in poor robust.

The article propose a proved self-calibration method of varying internal parameters based on image of dual absolute quadric transformation in connection with above situation. The method uses the transformation and makes the various elements of dual quadric change to the same order of magnitude, so to solve the results of a more robust. The theorical analysis and experiments with both simulated and real data demonstrate that this self-calibration method can lead to an enormous improvement on the stability and robustness of the results without increasing computation.

2 Image of Dual Absolute Quadric Transformation

This article assumes that the classical pinhole camera model [2], if there exists a three-dimensional point M and the correspondingl image plane point position m, then the following relationship:

$$m \cong P_E M = K[R \quad t]M \tag{1}$$

And $P_E = K[R \quad t], K = \begin{bmatrix} f_u & s & u_0 \\ 0 & f_v & v_0 \\ 0 & 0 & 1 \end{bmatrix}$,R is the external parameters matrix,t is the translation vector, (u_0, v_0) is the coordinate of principal point, f_u is scale factor of u axe, f_v is scale factor of v axe,s is skew factor.

We only can obtain projective reconstruction P_{proj} from the point of the corresponding sequence of images under the conditions of the unknown camera position,unknown orientation and without calibration,but the purpose of self-calibration is to obtain the internal parameters and external parameters,and then obtain camera projection matrix P_E in the Euclid case. Suppose projective space to Euclid space transformation matrix H,that is:

$$P_E = P_{proj} H \tag{2}$$

We can obtain this:

$$\begin{aligned} P_E Q_\infty^* P_E^T &= P_{proj} H Q_\infty^* H^T P_{proj}^T \\ &= P_{proj} Q^* P_{proj}^T \cong \omega^* \end{aligned} \tag{3}$$

The absolute dual quadric in the projective space under the form of $Q^* = HQ_\infty^* H^T$.

If each piece of a view of the image sequence m_i through the appropriate transformation T_i ,then the image sequence into m_i' , from (1) we can obtain:

$$\begin{aligned} m_i' &\cong T_i m_i \cong T_i K_i [R_i \quad t_i] M_i \\ &= K_i' [R_i \quad t_i] M_i \end{aligned} \tag{4}$$

Here $K_i' = T_i K_i$, also from the absolute conic in the image on the dual $\omega_i^* = K_i K_i^T$, we can obtain:

$$\omega_i'^* = K_i' K_i'^T = T_i K_i K_i^T T_i^T = T_i \omega_i^* T_i^T \tag{5}$$

Here $\omega_i'^*$ is the image of absolute dual quadric when the internal parameters matrix is K_i'. From (5) can be seen, if through the transformation matrix T_i, so transformed various elements of the image of absolute dual quadric $\omega_i'^*$ in the same order of magnitude,then solve $\omega_i'^*$,we can obtain the unconverted image of absolute dual quadric ω_i^* through the inverse transform.If we set the matrix $T_i = K_i^{-1}$,then $\omega'^*_i = T_i \omega_i^* T_i^T = K_i^{-1} K_i K_i^T K_i^{-T} = I_{3*3}$,we can see, the transformed absolute dual quadric as the unit matrix,then using the constraint of unit matrix must be higher robustness than using the constraint of ω_i for solving the internal parameters.

3 Self-calibration

3.1 The Internal Camera Parameters Initialization

In this paper the method requires the initial value within the parameter matrix K_i ,used for assigning the transformation matrix T_i . A good initial value of the calibration accuracy and robustness of a key impact. We assume that the camera center of the first picture view and the center of the world coordinate system coincides, then the first image of the projective camera matrix $P_{proj}^1 = [I_{3*3} \quad 0]$, Euclidean camera matrix $P_E^1 = [K_1 \quad 0]$,use (2) can obtain:

$$\begin{aligned} & P_E^1 \cong P_{proj}^1 H \Leftrightarrow [K_1 \quad 0] \cong [I_{3*3} \quad 0] H \\ & \Leftrightarrow \exists (q, q_{44}) \mid H \cong \begin{bmatrix} K_1 & 0 \\ q^T & q_{44} \end{bmatrix} \end{aligned} \tag{6}$$

The assumption that skew factor is 0, aspect ratio is 1 and the principal point is constant conditions, H can be expressed as:

$$H=\begin{bmatrix} f_1 & 0 & u_0 & 0 \\ 0 & f_1 & v_0 & 0 \\ 0 & 0 & 1 & 0 \\ q_1 & q_2 & q_3 & 1 \end{bmatrix}=\begin{bmatrix} K_1 & 0 \\ q^T & 1 \end{bmatrix} \tag{7}$$

Here H has six unknowns.And from (3) can obtain:

$$\lambda_i\omega_i^* = P_{proj}^i HH^T P_{proj}^{iT} \Leftrightarrow \lambda_i \begin{bmatrix} f_i^2+u_0^2 & u_0v_0 & u_0 \\ u_0v_0 & f_i^2+v_0^2 & v_0 \\ u_0 & v_0 & 1 \end{bmatrix} = P_{proj}^i \begin{bmatrix} f_1^2+u_0^2 & u_0v_0 & u_0 & f_1q_1+u_0q_3 \\ u_0v_0 & f_1^2+v_0^2 & v_0 & f_2q_2+v_0q_3 \\ u_0 & v_0 & 1 & q_3 \\ f_1q_1+u_0q_3 & f_2q_2+v_0q_3 & q_3 & \|q\|^2 \end{bmatrix} P_{proj}^{iT} \tag{8}$$

We use Bougnoux [13] calculated the focal length of the two views,and through the transformation to get the coordinate of principal point and the coordinate of image center coincidence,it is $u_0 = v_0 = 0$,then we are left with four unknowns λ_i and (q_1, q_2, q_3) ,and equation (8) provides six independent equations.Thus,We avoid the false initial value by least squares by overconstrained equations. After the initial value of H obtained,we can obtain the initial value of metric reconstruction through (2), the initial value of K_i can be obtained by QR decomposition.

3.2 Calibration Use the Transformation

After the initial value of the internal parameters within each image obtained,we set the transformation matrix $T_i = K_i^{-1}$,then the transformation matrix acting on each image,in this case the camera projection matrix in the projective space as follows: $P'^i_{proj} = T_i P^i_{proj}$ .Use (3) we can obtain:

$$\lambda_i T_i \omega_i^* T_i^T = T_i P_{proj}^i Q^* P_{proj}^{iT} T_i^T = P'^i_{proj} Q^* P'^{iT}_{proj} \tag{9}$$

Then, equation (9) can be expanded:

$$\lambda_i K_i^{-1} K_i K_i^T K_i^{-T} = \lambda_i \begin{bmatrix} 1 & 0 & 0 \\ 0 & 1 & 0 \\ 0 & 0 & 1 \end{bmatrix} = K_i^{-1} P_{proj}^i \begin{bmatrix} f_1^2 & 0 & 0 & f_1q_1 \\ 0 & f_1^2 & 0 & f_1q_2 \\ 0 & 0 & 1 & q_3 \\ f_1q_1 & f_1q_2 & q_3 & \|q\|^2 \end{bmatrix} P_{proj}^{iT} K_i^{-T}$$

$$= K_i^{-1} P_{proj}^i \begin{bmatrix} a_1 & 0 & 0 & a_2 \\ 0 & a_1 & 0 & a_3 \\ 0 & 0 & 1 & a_4 \\ a_2 & a_3 & a_4 & a_5 \end{bmatrix} P_{proj}^{iT} K_i^{-T} = P_{proj}^{i}{}' \begin{bmatrix} a_1 & 0 & 0 & a_2 \\ 0 & a_1 & 0 & a_3 \\ 0 & 0 & 1 & a_4 \\ a_2 & a_3 & a_4 & a_5 \end{bmatrix} P_{proj}^{iT}{}' \tag{10}$$

Here, $P_{proj}^{i'} = K_i^{-1} P_{proj}^{i}$, $a_1 = f_1^2$, $a_2 = f_1 q_1$, $a_3 = f_1 q_2$, $a_4 = q_3$, $a_5 = \|q\|^2$ .So we can get the constraint equations from(10) as follows:

$$\omega_{11}^{*'} = \omega_{22}^{*'} = \omega_{33}^{*'} = 1 \tag{11}$$

$$\omega_{12}^{*'} = \omega_{13}^{*'} = \omega_{23}^{*'} = 0 \tag{12}$$

$$\omega_{21}^{*'} = \omega_{31}^{*'} = \omega_{32}^{*'} = 0 \tag{13}$$

These constraints can thus be imposed on the right-hand side of (10), yielding 5(n-1) independent linear equations in $(a_1, a_2, a_3, a_4, a_5)$:

$$P_{proj}^{i(1)'} Q^* P_{proj}^{i(1)T'} = P_{proj}^{i(2)'} Q^* P_{proj}^{i(2)T'} = P_{proj}^{i(3)'} Q^* P_{proj}^{i(3)T'}$$

$$2P_{proj}^{i(1)'} Q^* P_{proj}^{i(2)T'} = 0$$

$$2P_{proj}^{i(1)'} Q^* P_{proj}^{i(3)T'} = 0$$

$$2P_{proj}^{i(2)'} Q^* P_{proj}^{i(3)T'} = 0$$

With $P_{proj}^{i(j)'}$ representing row j of P_{proj}^{i} and Q^* parametrized as in (12). Substituted the expression of matrix into equation to calculate,we can obtain the linear equation group like AX=B,here

$$A=\begin{bmatrix} p_{11}^2 + p_{12}^2 - p_{21}^2 - p_{22}^2 & 2(p_{11}p_{14} - p_{21}p_{24}) & 2(p_{12}p_{14} - p_{22}p_{24}) & 2(p_{13}p_{14} - p_{23}p_{24}) & p_{14}^2 - p_{24}^2 \\ p_{11}^2 + p_{12}^2 - p_{31}^2 - p_{32}^2 & 2(p_{11}p_{14} - p_{31}p_{34}) & 2(p_{12}p_{14} - p_{32}p_{34}) & 2(p_{13}p_{14} - p_{33}p_{34}) & p_{14}^2 - p_{34}^2 \\ 2(p_{11}p_{21} + p_{12}p_{22}) & 2(p_{11}p_{24} + p_{14}p_{21}) & 2(p_{12}p_{24} + p_{22}p_{14}) & 2(p_{13}p_{24} + p_{23}p_{24}) & 2p_{14}p_{24} \\ 2(p_{11}p_{31} + p_{12}p_{32}) & 2(p_{11}p_{34} + p_{14}p_{31}) & 2(p_{11}p_{34} + p_{32}p_{14}) & 2(p_{13}p_{34} + p_{33}p_{14}) & 2p_{14}p_{34} \\ 2(p_{21}p_{31} + p_{22}p_{32}) & 2(p_{21}p_{34} + p_{31}p_{24}) & 2(p_{22}p_{34} + p_{32}p_{24}) & 2(p_{23}p_{34} + p_{33}p_{24}) & 2p_{24}p_{34} \end{bmatrix}$$

$B=\begin{bmatrix} p_{23}^2 - p_{13}^2 \\ p_{33}^2 - p_{13}^2 \\ -2p_{13}p_{23} \\ -2p_{13}p_{33} \\ -2p_{23}p_{33} \end{bmatrix}$, $X = (a_1, a_2, a_3, a_4, a_5)$, p_{ij} representing row i and line j of $P_{proj}^{i'}$.

Each line of matrix A are linearly independent, By the nature of linear equations, the solutions of equations are divided into the following three kinds of situations:

- Matrix A determinant is not equal to zero, then the equation has a unique solution;
- Matrix A of determinant equal to zero, but the rank (A)> = rank (B), there are infinitely many solutions;
- Matrix A of determinant equal to zero, but the rank (A) <rank (B), no solution;

When the matrix A of determinant equal to zero, there may be no solution or infinitely many solutions. If come out infinitely many solutions, it will randomly generate a solution, certainly the wrong solution,calibration failure; If no solution, calibration failure. Taken from the experiment can be seen when the two images are in the same plane and parallel , there will be the critical situation that the matrix A is equal to 0. Therefore, we must avoid this during the experiment.

When the matrix A is not equal to zero, the equation has a unique solution. We can see,only one view projective reconstruction of the linear equations can be obtained, Therefore, this method requires only two images can be obtained as the absolute quadric, then substituted into (5) can be obtained ω_i^* and using the Cholesky decomposition the matrix of intrinsic parameters from the equation $\omega^* = KK^T$.

4 Simulation and Real Experiment

4.1 Simulation Experiment

The purpose of simulation is to verify whether the increased calibration accuracy and robustness through image transform. Within the parameters of the camera is set to $f_u = f_v = 1200,\ \ u_0 = 320,\ \ v_0 = 240$. The simulated data consists of the set of 30 points are taken as random points with a unit sphere in front of the camera and a simulated series of rigid displacements of the camera. Thus the images of the set of points are mapped by the same camera at different view points and 6 views among the obtained image sequences are taken in our experiment.

As with noise, the error matching points may be extracted , which makes the determinant of matrix A is equal to 0, then the equation has no solution or wrong solution, can not to strike or to strike within the parameters of the imaginary. Therefore, if the determinant of matrix A is 0, that the calculation of failure. In this paper, 0.5 ~ 3 pixels gaussian noise, respectively, within the parameters of the initial assumption that the camera is f = 1, f = 120, f = 1200 and f = 6000,after calculating 100 times, results shown in Figure 1. As can be seen from Figure 1, there is the highest number of calculation failures when the assumption f = 1, because this time the transformation matrix is the unit matrix correspond to that have no image transformation treatment. the various elements of the transformed $w_i^{*\prime}$ of magnitude difference in the maximum, $w_i^{*\prime}$ not as a unit matrix, so the most frequent failure. There is the lowest number of calculation failures when the assumption f= 1. At this point the assumption value of the initial internal paremeters is equal to the true value, all elements of $w_i^{*\prime}$ in the smallest difference of magnitude, $w_i^{*\prime}$ is also closest to the unit matrix, therefore, the least number of failures. When the assumption f = 6000, the failures become more again. Assuming the initial value is greater than the true value of internal parameters, not all elements of the transformed $w_i^{*\prime}$ in the same order of magnitude,so failures added.

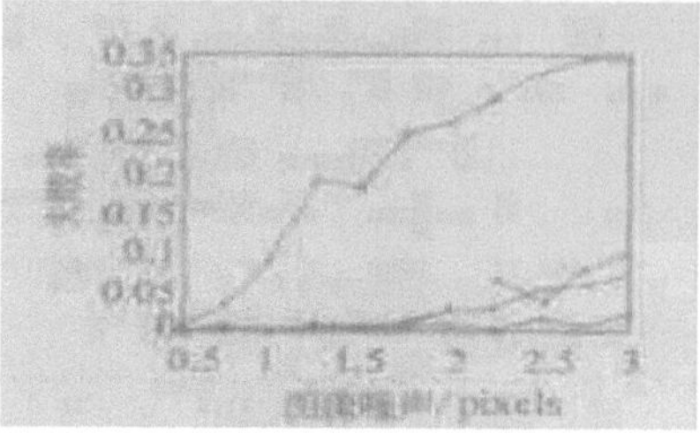

Fig. 1. Distribution of calculated noise failures: the highest curve is f = 1, the lowest f = 1200. The middle two cross curve is f = 120 and f = 6000.

In order to verify whether the transformation method can improve accuracy and stability of self-calibration. In this paper, 0 ~ 2 pixels gaussian noise, respectively, within the parameters of the initial assumption that the camera is f = 1, f = 120, f = 1200 and f = 6000,after calculating 100 times, results shown in Figure 2. As can be seen from Figure 2, when the assumption f = 1, there is the largest value of reprojection error and standard deviation that means the lowest stability and accuracy. When the initial assumption that within the parameters and within the parameters of almost real time, it means the smallest value of reprojection error and standard deviation and highest stability and accuracy. This shows that as the transformation method proposed in this paper can improve the self-calibration of precision androbustness.

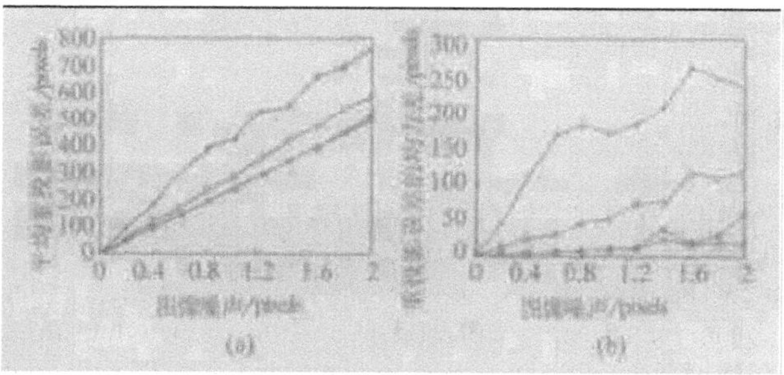

Fig. 2. Re-projection error and the standard deviation distribution with the noise

4.2 Real Experiment

In the real experiment, we use pictures taken by myself and from http://www.robots.ox.ac.uk/~vgg/data/data-mview.html download Valbonne church image sequences comparison calibration. We take three pictures of vase in the library for calibration,the size of these images is 640×480 pixels in Figure 3. Using our calibration method, only needs two views to calculate the internal and external camera parameters, so we calculated only the vase before the two images a) and b) internal parameters focal length of the initial value of $f_{ua} = f_{va}$=504.77, $f_{ub} = f_{vb}$=511.33, after using image transformation, the value of internal parameters focal length is

$f_{ua} = f_{va}$ =660.52, $f_{ub} = f_{vb}$ =670.48. Pollefeys [14] is the classical variational parameter calibration, but this method requires three images, so the three images to calculate focal length for the internal parameters is $f_{ua} = f_{va}$ =655.65, $f_{ub} = f_{vb}$ =670.26, $f_{uc} = f_{vc}$ =705.18.

a) b) c)

Fig. 3. Vase image sequences

Figure 4 is the Valbonne church image sequences that are downloaded from http://www.robots.ox.ac.uk/~vgg/data/data-mview.html, the size of this images is 512×768 pixels.

d) e) f)

Fig. 4. Valbonne church image sequences

Using our calibration algorithm obtain the initial value of internal parameters focal length of the image d) and e): $f_{ud} = f_{vd}$ =553.77, $f_{ue} = f_{ve}$=573.16, after using image transformation, the value of internal parameters focal length is $f_{ud} = f_{vd}$ =676.61, $f_{ue} = f_{ve}$=685.78.Using Pollefeys's [14] method to calculate the value of internal parameters focal length is: $f_{ud} = f_{vd}$ =679.57,

$f_{ue} = f_{ve}$=687.35, $f_{uf} = f_{vf}$=681.46. From the experimental results can be seen, the calibration accuracy og this transformation method is similar to the classical variational parameter calibration accuracy,so this article provide a high accuracy calibration and robustness method of self-calibration.

5 Conclusion

In this paper, we let the absolute dual quadric as the elements in the same order of magnitude through the transformation matrix to reduce the absolute dual quadric as sensitivity to noise, then can improve the robustness of calibration. We need to calculate the initial internal parameters in this paper,so we can obtain the transformation matrix,then get the the image of absolute dual conic to be a unit matrix and we can improve the calibration accuracy by using unit matrix. This calibration method can also be used for non-linear calibration, we can do the non-linear self-calibration through construct the object function of unit matrix.The non-linear self-calibration will be the priority of the calibration work in the future.

Acknowledgment

Sponsored by: Nature Science Foundation of China (60973096) and Nature Science Foundation of Jiangxi province in China (2008GZS0033).

References

[1] Meng, X.-q., Hu, Z.-y.: Recent Progress in Camera Self-Calibration. Acta Automatica Sinica 29(1), 110–124 (2003)

[2] Faugeras, O., Luong, Q.T., Maybank, S.: Camera self2calibration: Theory and experiments. In: Proceedings of the 2nd European Conference on Computer Vision, Italy, pp. 321–334 (1992)

[3] Maybank, S., Faugeras, O.: A theory of self-calibration of a moving camera. International Journal of Computer Vision 8(2), 123–151 (1992)

[4] Hartley, R.: Euclidean reconstruction and invariants from multiple images. IEEE Transactions on Pattern Analysis and Machine Intelligence 16(10), 1036–1041 (1994)

[5] Triggs, B.: Auto2calibration and the absolute quadric. In: Proceedings of Computer Vision and Pattern Recognition, pp. 609–614 (1997)

[6] Pollefeys, M., Van Gool, L., Oosterlinck, A.: The modulus constraint: A new constraint for self2calibration. In: Proceedings of International Conference of Pattern Recognition, Vienna, pp. 349–353 (1996)

[7] Heyden, A.: @str jm K. Euclidean reconstruction from image sequences with varying and unknown focal length and principal point. In: Proceedings of Computer Vision and Pattern Recognition, pp. 438–443 (1997)

[8] Liu, S.-g., Wu, C.-k., Tang, L., Jia, J.: Robust self-calibration method based on image of dual absolute quadric transformation. Systems Engineering and Electronics 27(2), 212–215 (2005)

[9] Li, H., Hu, Z.-y.: A Linear Camera Self-Calibration Technique Based on Projective Reconstruction. Journal of Software 13(12), 2286–2295 (2002)
[10] Hao, Y.-t., Zhou, W., Zhong, B.-t.: Self-calibration technique based on Kruppa equations obtaining initial solution by translational motion. Computer Engineering and Applications 45(22), 38–40 (2009)
[11] Hu, H.-l., Jiang, Z.-t.: Camera self-calibration method based on genetic algorithm. Computer Engineering and Desig. 2009 30(1), 204–206 (2009)
[12] Ha, J.-E., Kang, D.-J.: Initialization method for self-calibration using 2-views. Pattern Recognition, 143–150 (2004)
[13] Bougnoux, S.: From Projective to Euclidean Space under any practical situation, a criticism of self-calibration. In: Proc. ICCV, pp. 790–796 (1998)
[14] Pollefeys, M., Van Gool, L.: Self-calibration and metric reconstruction in spite of varying and unknown internal camera parameters. In: Proceedings of the IEEE International Conference on Computer Vision, Bombay, India, pp. 90–95 (1998)
[15] Lowe, D.G.: Distinctive image features from scale-invariant keypoints. International Journal of Computer Vision (60), 91–110 (2004)

Stereo Video Segmentation Used Disparity Estimation and Redundant Discrete Wavelet Transforms

Tao Gao[1,2], Jiantao Zhao[3], and Yongjiang Jia[3]

[1] Electronic Information Products Inspection Institute of Hebei Province (河北省电子信息产品监督检验院), Shijiazhuang 050071, Hebei, China
[2] Industry and Information Technology Department of Hebei Province, Shijiazhuang 050051, Hebei, China
taogao@its.la
[3] Hebei College of Administration(河北行政学院), Shijiazhuang 050031, Hebei, China

Abstract. Stereo video object segmentation is a critical technology of the new generation of video coding, video retrieval and other emerging interactive multimedia field. This paper presents a redundant discrete wavelet transforms based stereo video object segmentation algorithm. First, the redundant discrete wavelet transforms (RDWT) are used to obtain the parallax and then the stereo video parallax is used to do object segmentation. For the moving objects in the stereo video sequence, motion area is extracted form the redundant wavelet transform domain. Experimental results show that the algorithm can not only segment the overlapping objects, but also segment the stationary objects and moving objects at the same time with better accuracy and robustness.

Keywords: Stereo video object segmentation, disparity estimation, disparity correspondence, stereo images, redundant discrete wavelet.

1 Introduction

As human brain can deal with the weak difference between the left and right eye images, we can sense the external 3-D world; and this ability is called stereoscopic vision. Stereo image sequence is a three-dimensional visual image form, it uses left and right groups of images to describe a group of scenes, and the human eyes apperceive the 3-D depth information by addressing the relative position between the two images. Compared with traditional two-dimensional images, stereo images are more "realism" and the description of the scene is more natural. Currently, 3-D vision system has been widely applied to stereo video communications [1], robot vision [2], aviation navigation [3], and other fields. With ever-growing material and spiritual needs, the stereo images will gradually replace the traditional single vision images, and will be more used in television, online shopping, remote medical diagnosis and other civilian areas. To store or transmit the stereo images, a more efficient compression encoding program must be developed. Stereo image sequence compression method was put forward in the late 1980s; after more than 25 years of development, people have developed several comparatively mature algorithms. However, to the practical application point, there is still not a unified coding standard.

L. Qi (Ed.): ISIA 2010, CCIS 86, pp. 462–468, 2011.

The video object based stereo video coding method is to separate the video object from the scenes and to extract its borders, texture, movement and other parameters, then to code these parameters to achieve the purpose of coding the whole image [4]. This method uses the hidden 3-D depth information, through the creation of 3-D objects and coding model to improve the coding efficiency and to reduce the influence of the block. It provides a more natural scene interpretation. However, this approach requires sophisticated image analysis process, such as: object segmentation, object modeling, and all these are not ripe at present so it can only be applied to a single background image with simple motion; its widely using depends on better solving some of these key technologies.

This paper presents a redundant wavelet transform based stereo video object segmentation algorithm. First, we use the redundant wavelet transform to extract the feature points of stereo video images, then according to the feature points we do the disparity estimation, to form a disparity map. The stationary objects are segmented from the stereo images by the disparity map. For the moving objects, we use a redundant wavelet transform based moving object extraction algorithm to segment the moving target from the redundant wavelet domain. Experimental results show that our algorithm can segment video objects from stereo video images, including stationary objects and moving objects with good results, highlighted details, and simple calculation process; all these can help to the subsequent coding operation.

2 Stereo Vision Geometric Theories

Three-dimensional camera system generally can be classified as three-dimensional parallel camera system and three-dimensional clustering camera system. If the two cameras are installed with parallel optical axis, it is called three-dimensional parallel camera system; if the two optical axises intersect at the objects, it is called three-dimensional clustering camera system. This paper considers the 3D parallel camera system. Optical axis of two cameras parallel to each other and are perpendicular with the baseline. The space is determined by the plane known as the heart polarization plane. With a space (X, Y, Z), X is the horizontal direction and Y is the vertical direction, Z is the depth direction. The camera lens in the image plane is at point (x, y), where x parallel to the X, y parallel to the Y. In parallel camera system, the point in space around the image plane of the projector is at the same general location of the y coordinate, and the two set up their corresponding image brightness of the same point, that disparity search may be mainly concentrated in the horizontal direction, thus speeding up the searching and matching process.

Let f is the focal length, s is the image plane size, x_{l1}, x_{r1}, x_{l2}, x_{r2} are the positions of points: p_1 and p_2 in planar image around the projection image plane to the right position between the left and right images. The disparity definition for the same point in the space of two planar images is the differences of the positions. If left image is the reference image, then $d_1 = x_{l1} - x_{r1}$, $d_2 = x_{l2} - x_{r2}$. By triangular geometric principles, we can get:

$$\frac{x_{l2}-s/2}{f}=\frac{X_2+b/2}{Z_2},\ \frac{x_{r2}-s/2}{f}=\frac{X_2-b/2}{Z_2}. \tag{1}$$

$$X_2=\frac{b*(x_{l2}+x_{r2}-s)}{2*(x_{l2}-x_{r2})},\ Z_2=\frac{b*f}{x_{l2}-x_{r2}}. \tag{2}$$

Figure 1 shows the parallel camera system.

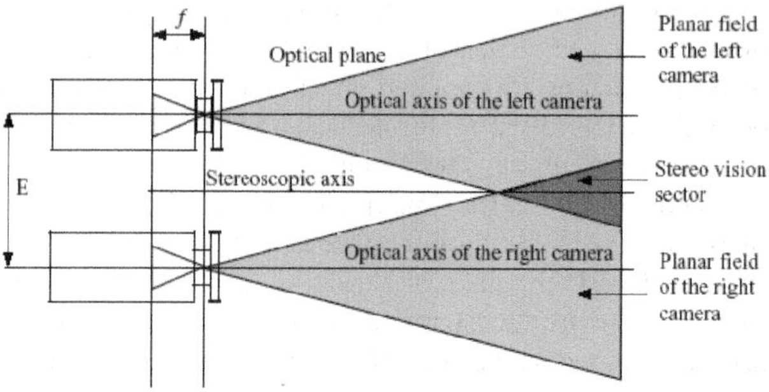

Fig. 1. Parallel camera system

3 Parallax Estimation Based on RDWT

The RDWT is an approximation to the continuous wavelet transform that removes the down-sampling operation from the traditional critically sampled DWT to produce an over-complete representation. The shift-variance characteristic of the DWT arises from its use of down-sampling; while the RDWT is shift invariant since the spatial sampling rate is fixed across scale. As a result, the size of each sub-band in an RDWT is the exactly the same as that of the input signal. It turns out that, by appropriately sub-sampling each sub-band of an RDWT, one can produce exactly the same coefficients as does a critically sampled DWT applied to the same input signal. We use the redundant wavelet to transform an image $P(i, j)$ in such ways:

$$PLL_j(x,y)=PLL_{j-1}(x,y)*([h]_{\uparrow 2^{j-1}},[h]_{\uparrow 2^{j-1}})(-x,-y). \tag{3}$$

$$PLH_j(x,y)=PLH_{j-1}(x,y)*([h]_{\uparrow 2^{j-1}},[g]_{\uparrow 2^{j-1}})(-x,-y). \tag{4}$$

$$PHL_j(x,y)=PHL_{j-1}(x,y)*([g]_{\uparrow 2^{j-1}},[h]_{\uparrow 2^{j-1}})(-x,-y). \tag{5}$$

$$PHH_j(x,y)=PHH_{j-1}(x,y)*([g]_{\uparrow 2^{j-1}},[g]_{\uparrow 2^{j-1}})(-x,-y). \tag{6}$$

The correlation mask consists of multiplying the high-low (HL) bands, the low-high (LH) bands, and the high-high (HH) bands together and combining the products [5]: $MS(x,y)=\left|\prod_{j=J_0}^{J_1} HL^{(j)}(x,y)\right|+\left|\prod_{j=J_0}^{J_1} LH^{(j)}(x,y)\right|+\left|\prod_{j=J_0}^{J_1} HH^{(j)}(x,y)\right|$. Where J_0 and J_1 are the starting and ending scales. We note that calculation of the correlation mask in this manner is possible due to the fact that each RDWT sub-band is the same size as the original image. We can obtain the feature points according to equation (7).

$$Point(x,y)=\begin{cases}1 & MS(x,y)\geq T\\ 0 & MS(x,y)<T\end{cases}. \tag{7}$$

T is the threshold. We transform $MS(x,y)$ into a gray picture, and then use the Iterative threshold method to obtain the T automatically. We choose the middle gray value of the gray picture as the initial threshold T_0, and do the iterative operation according to equation (8). h_k is the number of pixels with the k gray value, do the iterative operation until $T_{i+1}=T_i$, then T_i is the segmentation threshold.

$$T_{i+1}=\frac{1}{2}\left\{\frac{\sum_{k=0}^{T_i} h_k * k}{\sum_{k=0}^{T_i} h_k}+\frac{\sum_{k=T_i+1}^{L-1} h_k * k}{\sum_{k=T_i+1}^{L-1} h_k}\right\}. \tag{8}$$

Parallax estimation is based on the classic block-matching disparity estimation method [6]. That is, in the left image we make 16×16 or 8×8 blocks with feature points as the center, and then for each block we search the best-matching block in the right image from top to bottom about $\pm k$,from left to right about $\pm l$ ($l>k$), the offset is the disparity vector. The processes are as follows:

$$\nabla f(T)=\sum_{i=0}^{N-1}\sum_{j=0}^{N-1}\left|f_l(i,j)-f_r(i+m,j+n)\right|,\ \nabla f(T^*)=\min(\nabla f(T)),\ \left|T\right|=\sqrt{m^2+n^2}.$$

Let D=T^*, then D is the parallax. In parallax matching, we can dynamically change the search window size according to a prior estimate disparity vector and the change of the regional disparity map. Window expanding direction is the four side direction; to select the window size properly can get the optimal vector and then obtain the disparity vector [7]. Through obtaining the feature points in the redundant wavelet transform domain and doing the disparity estimation according to the feature points, the computation accuracy can be improved. But the disparity map is sparse; to have the dense disparity map we still need to match the remaining large number of non-feature points. We can think that, the feature points around the edge image segment the image into a number of regions; the points are matched in the region according to the feature points. The specific algorithm are as follows: First, for the different objects, we search from left to right, top to down to find the target marginal position, as there may be gaps in the feature points, when we find a edge point in the eight neighborhood and no other point, the point will be linked to the point below, then we can get the close regions of various objects, and obtain the mask map. According to the mask map, the dense disparity map can be obtained.

4 Objects Segmentation Based on the Disparity Map

Because disparity is in inversely proportional relationship with the depth, to segment the disparity map can obtain the objects at different depths. For the objects with larger proportion of the image area, we use the iteration threshold segmentation method which used before, and for the objects with smaller proportion of the image area, we use local adaptive segmentation method. Assuming gray values in the object are higher than the gray values in the background, the original image can be segmented by the Otsu method [8], and it can be divided into target region and background. The gray histogram of the original image which is higher than the threshold value can be considered as the regional gray statistics of the object. For the image composed by the objects and background, the gray histogram can be considered as the distribution of pixel gray mixed probability density, and we usually assume that the distribution of two-component mixture is the normal distribution.

Because the coefficients of the sub-bands of the redundant wavelet transform are highly correlated, and the direction and size are the same as the image, also, there is no translation in the sub-band; we put forward a new method which based on redundant wavelet transform to obtain the motion area [9].

$$MAS(x,y)=\sum_{j=J_0}^{J_1}\left(\begin{aligned}&\left|LL_1^{(j)}(x,y)-LL_2^{(j)}(x,y)\right|+\left|LH_1^{(j)}(x,y)-LH_2^{(j)}(x,y)\right|\\&+\left|HL_1^{(j)}(x,y)-HL_2^{(j)}(x,y)\right|+\left|HH_1^{(j)}(x,y)-HH_2^{(j)}(x,y)\right|\end{aligned}\right). \tag{9}$$

The binary motion mask obtained from the redundant wavelet transforms can be considered as the original mask of the moving video object. As the inner district of the object is usually flat, the characteristic is not obvious and there are some crevices in the mask, we use an assimilation method to fill the mask. The advantage of the assimilation operation is that it only fills the cragged crevice; the slow-changing edge will not be modified. However, some masks have big holes; they can not be completely filled, so we use the closure operation of the mathematical morphology to do the second filling [10].

5 Experimental Results

In this paper, 'puppy' stereo video is used for test. The video is sampled at a resolution of 768x576 and a rate of 25 frame/s. Once the frames are loaded into memory, the method averages 30 frame/s on a 1 400 MHz Celeron CPU. The accuracy measurement is used to quantify the degree that the algorithm matches the ground-truth. It is defined in this context as $E = N / M$, where N is the number of foreground pixels detected by the algorithm; and M is the number of foreground pixels in ground-truth. The frame numbers are 8 and the segmentation accuracy is showed in Table 1.

Table 1. The accuracy of segmentation

frame	1#	2#	3#	4#	5#	6#	7#	8#
bear	0.96	0.954	0.981	0.943	0.963	0.945	0.932	0.973
flower	0.912	0.864	0.867	0.916	0.832	0.791	0.843	0.907
box	0.746	0.731	0.743	0.774	0.702	0.744	0.801	0.764
vase	0.644	0.635	0.562	0.617	0.679	0.704	0.516	0.508

From Figure 2 we can see that the algorithm proposed in this paper can effectively segment the flowers, vase and carton. All these stationary objects are difficult for separation in the single-channel video. Form Figure 3 we can see that our method is better than the frame-subtraction method for moving object segmentation. It is advantageous for us to do the next operation to segment the moving object.

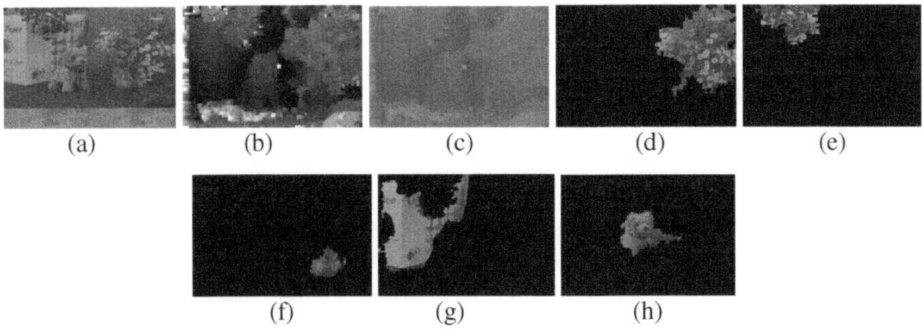

Fig. 2. Segmentation results based on the disparity map: (a) original image, (b) disparity map, (c) disparity map, (d&e) flower, (f) vase, (g) cartons, (h) bear

Fig. 3. Segmentation of the moving video object

6 Conclusions

This paper presents a new redundant discrete wavelet transforms based stereo video object segmentation algorithm. The algorithm segments the video objects by combining the redundant wavelet transform and the disparity map. The experimental results show that the algorithm can segment the overlapping objects, stationary objects and moving objects at the same time with better accuracy and robustness. It is obviously advantageous for us to do the video object-based stereo video coding in the future.

Acknowledgments. This research is funded by Project (60772080) supported by the National Natural Science Foundation of China, and funded by Project (08JCYBJC 13800) supported by the Science Foundation of Tianjin.

References

1. Waldowski, M.: A New Segmentation Algorithm for Videophone Applications Based on Stereo Image Pairs. IEEE Transactions on Communications 39(12), 1856–1868 (1991)
2. Nishihara, H.K., Poggio, T.: Stereo Vision for Robotics. In: Proc. First Int'l Symp. Robotics Research, pp. 489–505. MIT Press, Cambridge (1984)
3. Antonisse, H.: Active Stereo Vision Routines Using PRISM3. In: Proc. SPIE-Int. Soc. Opt. Eng., pp. 745–756 (1993)
4. Aizawa, K., Huang, T.S.: Model-based Image Coding: Advanced Video Coding Techniques for Very Low Bit-Rate Applications. Proc. IEEE 83(2), 259–271 (1995)
5. Gao, T., Liu, Z.G., Yue, S.H., Zhang, J., Mei, J.Q., Gao, W.C.: Robust Background Subtraction in Traffic Video Sequence. Journal of Central South University of Technology 17(1), 187–195 (2010)
6. Kanade, T., Okutomi, M.: A Stereo Matching Algorithm with an Adaptive Window: Theory and Experiment. IEEE Transactions on PAMI 16(9), 920–932 (1994)
7. Izquierdo, E.: Disparity/segmentation Analysis: Matching with an Adaptive Window and Depth- Driven Segmentation. IEEE Transactions on CSVT 9(4), 589–607 (1999)
8. Otsu, N.: A Threshold Selection Method from Gray-Level Histogram. IEEE Transactions on Systems, Man, and Cybernetics 9(1), 62–66 (1979)
9. Gao, T., Liu, Z.G., Zhang, J.: Redundant Discrete Wavelet Transforms based Moving Object Recognition and Tracking. Journal of Systems Engineering and Electronics 20(5), 1115–1123 (2009)
10. Gao, T., Liu, Z.G., Zhang, J.: BDWT based Moving Object Recognition and Mexico Wavelet Kernel Mean Shift Tracking. Journal of System Simulation 20(19), 5236–5239 (2008)

Study of Information Grid Structure Methods

YuChen Luo and Chenhan Wu

Wuhan Digital Engineering Institute, Wuhan 430074, China

Abstract. This paper firstly carried on the basic information resources level, the intelligence transmission level, the coordination computation level and the service polymerization level, respectively. Secondly reasonably planed to the sea battlefield information grid node according to the position property and the function property. Thirdly established the information description standard of physics resources for sea battlefield information grid, and proposed the description methods of the general kind, the information survey class, the information processing class and the weapon attack class metadata, and last described a typical metadata.

Keywords: sea battlefield, information grid, architecture, node, metadata.

1 Introduction

The sea battlefield information grid is a project which uses the military communication networks facility to support the tendency, distributed sensor system, physical resources of controlling system and armament system in the sea battlefield according to plan and reorganization, realizing the interconnection, intercommunication, interoperability among all nodes, and enhancing the highly effective use of transmission and demanded sharing for information acquisition, processing and distribution.

The fabrication of the sea battlefield information grid is a item of brand-new and complex engineering system. Without scientific practical top design plan, it is unable to carry on the standard and unification of the construction for the sea battlefield information grid. Without carrying on the standard and unification will produce many information isolated island which are unavailable interconnection and intercommunication, and also will be unable to realize the military grid which has integrated information supporting ability [1,2]. Therefore, this article will describe a typical metadata and research the sea battlefield information grids tectonic system according to establishing its hierarchical structure, defining its grid nodes and describing its typical metadata, based on its military demand and substantive characteristics.

2 Hierarchical Structure of Sea Battlefield Information Grid

The sea battlefield information is a complex, huge scale project. It may be composed of the foundation resources level, the intelligence transmission level, the coordination computation level and service polymerization level [3] according to the structure level. As shown in Fig. 1.

L. Qi (Ed.): ISIA 2010, CCIS 86, pp. 469–474, 2011.

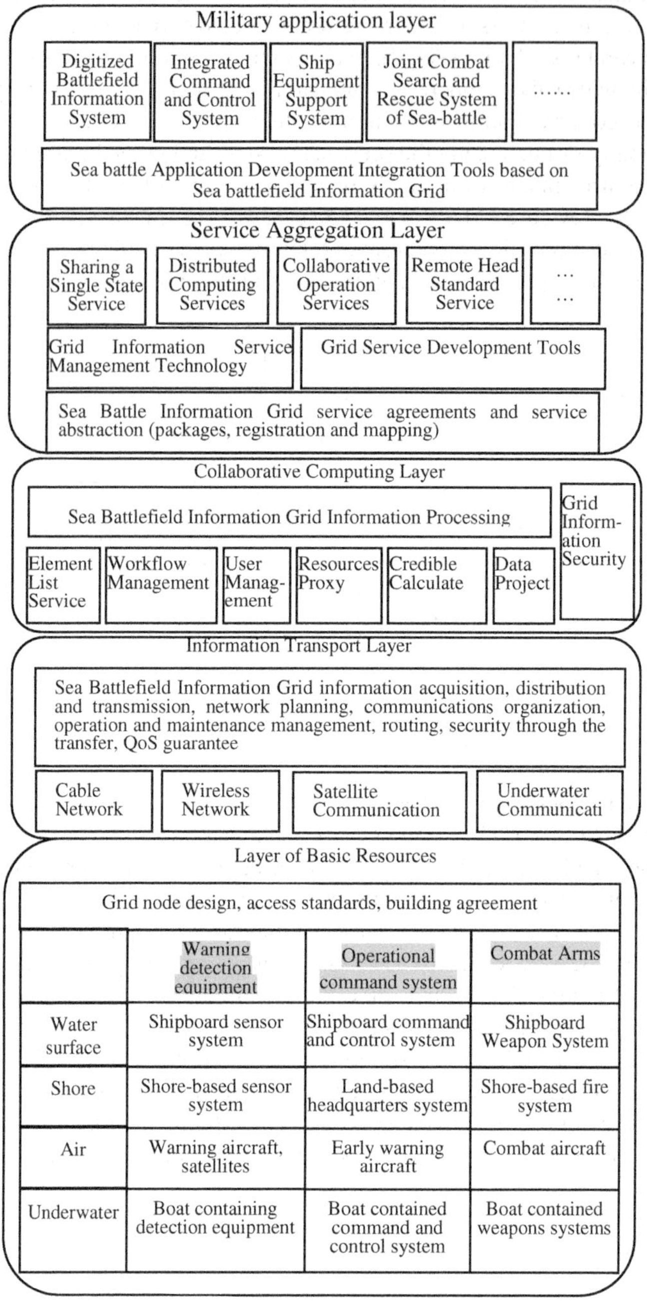

Fig. 1. Level structure of sea battlefield information grid

2.1 Foundation Resources Level

The sea battlefield physics resources are numerous, such as Warning detection equipment,Operational command system, Combat Arms,which constitute the sea battlefield information grid nodes. The grid nodes ensure the fast deployment and the integrated open style system standard system, integrates the grid system to realize the existing system to the in through the standard design and construction agreement of the sea battlefield information grid safeguard.

2.2 Intelligence Transmission Level

Through the network planning and organization management for the existing wired network, the wireless network, communication satellite system, we should build the sea battlefield information grid transmission facility, in order to realize the goal that sea battlefield moving platform can access anytime and anywhere nimbly, the dynamic network and the information can be reliable to transmit according to the operational mission.

2.3 Intelligence Transmission Level

Through the network planning and organization management for the existing wired network, the wireless network, communication satellite system, we should build the sea battlefield information grid transmission facility, in order to realize the goal that sea battlefield moving platform can access anytime and anywhere nimbly, the dynamic network and the information can be reliable to transmit according to the operational mission.

2.4 Coordination Computation Level

The sea battlefield information processing and the management interactive function is realized by the platform management such as each navy operational control systems, electronic information systems. Using the user management, the resources proxy, the credible computation and the data project and so on each kind of commercial mature tool, uses the grid information processing,We can establish coordination calculating facility for the sea battlefield information grid.

2.5 Service Polymerization Level

The establishment of a synthesized service frame of the electronic information system which is asked for a standard unification, perfect function, can realize the integration for the software platform and the information service, by means of abstracting each kind of combat application software into a complete service polymer, according to a kind of construction unification, a application conforming the development technology standard, and the service seal, registration and mapping methods.

3 Definition and Plan for the Sea Battlefield Information Grid Nodes

3.1 Definition Standard for Grid Nodes

The definition of the sea battlefield information grid nodes should follow a standard: Firstly, the nodes are certainly a computing power computer system, a software system, and a data-storage system, or a digitization instrumentation equipment, a control system. Secondly, it should also have the following characteristics [4,5]:(a) The nodes can be described simply. (b) The node can only be described by the attribute set. (c) The nodes may be long-standing. (d) The nodes value is effective to the entire grids.

3.2 Definition for Sea Battlefield Information Grid Nodes

Combining the node definition standard, we can comprehensive analysis the characteristic of the sea battlefield operating resource. According to the sea battlefield information grid structure need, we carry on classification for the grid nodes which is shown in Table 1.

Table 1. Definition for sea battlefield information grid nodes

Features / Geographic Distribution	Information Detection	Information Processing	Weapon strike
Surface Node	Shipboard sensor system, including radar, sonar and electronic detection equipment, infrared surveillance equipment,etc.	Shipboard command and control systems, including the ship command and control systems, formation command and control systems,etc.	Shipboard weapon systems, including the main / auxiliary guns, missile systems, anti-submarine weapon systems, electronic warfare systems, etc.
Shore-based Node	Shore-based observation facilities: including radar, observation, communication station, Technical Investigation stations, etc.	Shore-based command center: including the command post, command post and other superiors, etc.	Shore-based weapons: including the Coast fire systems.
Air Node	Aerial surveillance detection equipment: including warning aircraft, detection satellites, etc.	Air Command and Control System: early warning aircraft.	Air Weapons: combat aircraft.
Underwater Nodes	Boat contained sensor system	Boat contained command and control system	Boat contained weapons systems

3.3 Plan for Sea Battlefield Information Grid Nodes

Based on the nodes above, and profited from OGSA open style service standard, this paper introduces the plan way of Intra-grid ,Extra-grid and Inter-grid, and plans the sea battlefield information grid topologies by the three grid elements above. The relation of Intra-grid ,Extra-grid and Inter-grid is shown in Fig. 2. The Intra-grid is the smallest topology, the next is Extra-grid and finally is the Inter-grid.

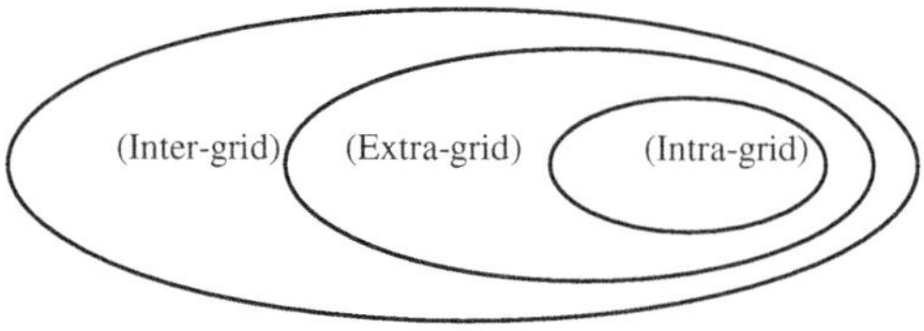

Fig. 2. The relation of Intra-grid, Extra-grid and Inter-grid

(1) Intra-grid

The characteristic of intra-grid is that it's a independent organization, has no cooperation with other nodes, and is an independent bunch of group.

(2) Extra-grid

The characteristic of extra-grid is composed of many dispersive intra-grids, has dispersive peaceful closed region, and connect by long-distance wireless..

(3) Inter-grid

The sea battlefield information grid is a inter-grid.

4 Description Method for Sea Battlefield Information Grid Information

Metadata is a data about data. It has abstracted the description of a data object, and is a intermediate level between data users and data. The metadata is abstracted from the sea battlefield physics resources data. The users can operate physical resources of the grid through a metadata definition. Therefore, the description of the metadata is very significant in the construction for sea battlefield information grid.

4.1 Classification Principle of Metadata

The classification principle of metadata is shown below:

(1) Using the multi-dimensional classification approach.
(2) For a certain classified plan, each kind is not accommodating
(3) For a certain classified plan, exhausting the whole resources.
(4) Not jumping ranks for the classification.
(5) It is unique for the classification under the identical classified element.

4.2 Classification for Metadata of Sea Battlefield Information Grid

Following the metadata classified principle, according to the different function of different data, the metadata is divided into the utility category, the information detection feature category, the information processing feature category and the Weapons striking feature category.

(1) Utility category
(2) Information detection feature category
(3) Information processing feature category
(4) Weapons striking feature category

4.3 Description of Metadata for Sea Battlefield Information Grid

The description for respective category of the metadata, needs to describe the element meaning specifically, point out whether the terminology is the element or the category, carries on the detailed annotation to the element, the description resources suggests the use in the system the technical name and the element data type. Take the general metadata for example, we carry on the description for the metadata.

Unique Identifier

Type: General category

Definition: Identifier that uniquely identifies the physical resources

Type of Term: Element

Comment: Uniquely identify in accordance with the water surface, shore, air and underwater

Label: Uni_Identity

Data Type: String

Value Domain: String of length X.

5 Conclusion

It is possible to establish a standard and open style sea battlefield information grid. Using sea battlefield information grid to provide the uniform information transmission service, may support the interconnection and information sharing among the sea battlefield physics resources. Using sea battlefield information grid to provide the uniform information processing service, may support the battlefield information reliable processing and the integrated management, causing the synthesis electronic information system of the sea battlefield to integrate by using "namely inserts namely uses" way. In order to construct a seal battlefield information grid which is adapt to the sea battlefield information service, it needs to explore many key questions and technologies, but the most importance are information grid architecture, grid node plan and the unification information description method. This article has carried on some beneficial explorations.

References

1. Foster, I., Kesselman, C., Nick, J.M., Tuecke, S.: The physiology of the Grid: An Open Grid Services Architecture for Distributed Systems Integration (June 2002)
2. Berman, F., Fox, G., Hey, T.: The grid: past, present, future. In: Berman, F., Fox, G.C., Hey, A.J.G. (eds.) Grid Computing: Making the Global Infrastructure a Reality, pp. 9–50. Wiley, Chichester (2003)
3. Bote-Lorenzo, M.L., Dimitriadis, Y.A., Gmez-Snchez, E.: Grid Characteristics and Users: a Grid Definition. In: Proc. European Across Grids Conference, Santiago, Spain (2003)
4. Tuecke, S., Czajkowski, K., et al.: Grid Service Specification. Open Grid Service Infrastructure WG, Global Grid Forum, Draft 2 (July 2002)
5. OGC Web Service Initiatives, `http://ip.opengis.org/ows2/`
6. Kreger, H.: Web Services Conceptual Architecture (WSCA 1.0), IBM Software Group (2001)

The New Grid Task Attemper Layer Model Based on Role

Zhou Xin Zhong

Shenzhen Housing Dedicated Fund Management Center for Public Facilities,
Shenzhen, China
zxzfriend@163.com

Abstract. The purpose of the grid system is to provide a set of service operations independent of the platforms through interface. This group of service operations can meet the needs of the caller, to accomplish the relative task.in the information grid system, there are a large number of running application tasks, which share the resources of the information grid system. The problem to be solved by the task scheduling is how to make the running application tasks get faster and more efficient access to spatial information services. a role Agent attemper layer model for task scheduling is proposed in this thesis. The model can segregate the huge amounts of data according to the interaction of Agent and the Agent sub-system on each layer can segregate again until it reaches the basic bottom sub-system.

Keywords: grid task, role, task attemper.

1 Introduction

Currently, the study is mainly on tasks as the scheduling objects while the specific application characteristics on the task execution time are less studied.

University of California[1,2], a study showed that: the use of the most busy times, an average of nearly 60% of the nodes in the idle state can use these idle nodes are mostly due to resource owners for various reasons such as temporary out result. Despite the different systems for "free to use the state definition of" different, but one thing is certain, that is less load on these nodes, it can take more tasks. Efficient task optimization model for resource owners, to maximize resource utilization, while the workload of resources would not be too high, so you can ensure that resources are adequate for most of the work load conditions, so as to resource owners with to more economic benefits.

Ian Foster and Carl Kesselman consider that the task scheduling is in a dynamic, with multiple departments or groups complex virtual organization and processing safe and flexible collaborative resource sharing and problem-solving[3,4]. Effective use of grid resources requires a powerful and flexible task scheduling mechanism. Whether the grid technology is success or not is largely determined by whether the grid platforms can automatically do the resource discovers, the task decomposition and the monitoring in processing through the inner link and needs of the user. In the grid system, there are a lot of running applications which share the various resources of the grid. The problem to be solved by the task scheduling is how to make the running

L. Qi (Ed.): ISIA 2010, CCIS 86, pp. 475–481, 2011.

application tasks get faster and more efficient information services. In the mean while it also needs to ensure the high utilization rate of the resource.As the resources nodes in the grid system are in low utilization state in most of the time.

A study of the University of California[5,6] shows that in even the busiest time an average of nearly 60% of the nodes are in the idle usable state, which result from owner of the resource temporary getting out and so on. Despite different systems define differently on "free usable state ", but one point is certain that these points load less and they can take more tasks. Efficient task optimization model can increase the resource utilization rate for resource owners. Meanwhile it doesn't make the workload of the resource too high in this way it can ensure the resource working in the appropriate status so as to bring more economic profit for the resource owners.

The agents within the organization interact and have the advantages of rapid evolution. This hierarchy is especially suitable to solve large-scale parallel and burst, the large-scale distribution in the physical space and large range of dispersion in high-dimensional space. When the scale of the problem increases, this hierarchical structure model is basically need no change accordingly. This article also builds experimental system to test and demonstrate to provide demonstration and basis for large-scale grid model application.

2 Role Theories

Role is accepted as the object for information receiving, information processing and the message sending[7,8]. In the role theory, role is the entia of responsibility and right. Responsibility stipulates code of conduct and constraints. In other words, the role is a characteristic collection of a particular object structure, properties, behavior, and functions etc. It is an essential characteristic general reflection of the object in its objection, ability, responsibility, licensing, constraints and agreements which can act as the rational classification standard. Collaborative activities are completed by the participants[9,10]. On the relationship between participants and roles, the concept of the role is to abstract the grouping of the participants through the factors such as their skills and abilities. One participant may play many roles and maybe more than one user play one role. During the collaborative process, the role is a proactive, independent abstract unit, has a certain goal and can complete a series of operations. A role has many factors such as activities, resources and states. Activities play the participant task roles of the role participants. Resources are the equipments, materials and information needed by the role. The contact of the roles is through information and information constitutes a precondition of the event. Activities are triggered by event trigger and change the state of the role. At different times, the roles can be in different states. The nature of the roles can be showed by the state relationship, role playing relationship, activities relationship and so on.

3 The Agent Role Level Task Processing

3.1 The Agent Role Level Task Processing Model

Task processing means have a great influence on the whole system tasks processing efficiency. In the past Multi-Agent System task processing works as job stream processing that means it considers the procedure as an ordered assemble of a series of logic relevant activities that aim at achieving certain objective. This kind of model is too simple to handle complex tasks.

This thesis divided the multi-Agent system into several layers according to the interaction of the Agents by improving the layer processing model. The multi-Agent systems are divided into many layers until the bottom subsystem .In any given layers, many Agent organizations work together to get their parents function of the system. Furthermore, within the organization, Agent works together to complete the whole function. And Agents within the organization interact and have the advantages of the rapid evolution. This hierarchy is especially suitable to solve large-scale parallel and unexpected, large-scale distribution in the physical space and wide range of dispersion problems in high-dimensional space.

3.2 The Agent Role Layer Task Processing

It's difficult to seek an efficient static task processing. Dynamic strategy can serve as an effective solution, because even though any machine is heterogeneous, the machine's load is kind of self-management and even self-balanced. The real enemy of dynamic strategy in heterogeneous platform is data dependent which may be dragged to the slowest speed, processor speed. Thus, as the heterogeneous platforms for grid system, the thesis comes up with a proposal that orients the combination of the role's static and dynamics. The processor data diversity achieves through the data reflects of each recognizable static item and calculation. And it applies the workflow task process and manages each Agent resource efficiently by role. When a certain Agent resource got a threshold it stopped distributing resource to the Agent. In addition, the division of tasks not only needs take the resource data specialty itself into account, but also think of the active information of corresponding task. And also it needs consider the relations of many role nodes and constraint of communication information. The task division should be moderate meeting the needs of the clients and the characteristic of cooperatives work of grid system. When carrying out the tasks division, the following principles should follow: (1) Independence: the divided task should have certain kind of independence, which would help each role node process independently tasks independently, reduce mutual coordination and communication work; (2) Levels: a task can be decomposed into many sub-tasks, the sub-tasks can be divides into several other lower sub-task. Complex task can divide into several simple, easy-to-handle tasks; (3) Combinations: it can complete a certain task by appropriate combination and can also complete another task through appropriate change; (4) Uniformity: The decomposition size, scale and level of difficulty should be as moderate as possible. And it also needs avoid a task execution time getting too long, resulting in uneven node burden, affecting the implementation efficiency of the overall system.

Role is also a level concept. Its granularity corresponds with the division granularity. According to workflow organization, from a work thread to a grid node, they all are responsible for a certain role. Shown in Figure below, from the perspective of the role, the decision of a role is dependent on the decomposition of the grid. High-level role is responsible for high-level goals.

The whole grid services can be seen as a target. In macro perspective, it can be seen as highly-leveled abstracted role. The service goal of the entire grid system is complete through several core processes. From the internal of the system, each process can be completed by roles taken on different nodes. Between the processes there are often predecessor and successor relationship. For example taking three-dimensional task workflow, including creating three-dimensional scene to the rendering of interactive scene, it involved there sub-process: create three-dimensional visualization interactive area, paint and texture treatment to the three-dimensional model. They are completed by different roles, holding customer and service relationships. Similarly, each process also includes different roles. When a user submits a request, the process task is allocated to the appropriate role of the grid nodes to complete. At this time the node and the role take on the Agent form, holding collaborative relationship between them. Seen from the analysis above, the customer - service provider relationship of the role was relative and it achieved the ultimate task processing target and value through the client - service provider chain.

4 Experiments and Analysis

The experiment uses 20 PC to carry out the Agent role grid task scheduling model test, common queue model and the buffer pool model test. Experimental environment is in the Local Area Network. We tested the simulation data simulation focusing on the three task scheduling model.

(1) Analysis on file conversion rate

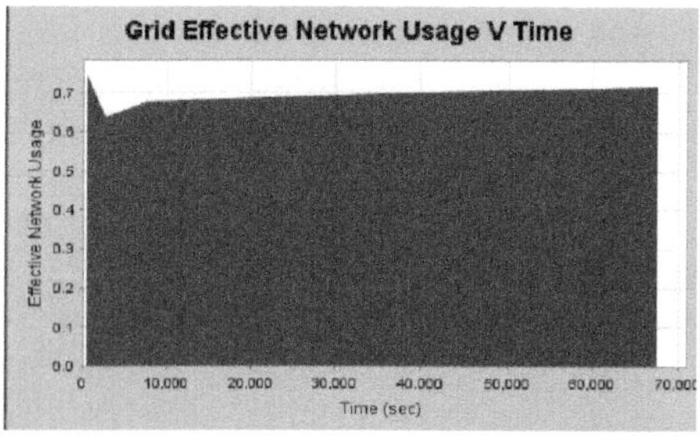

(a) File conversion rate of common queue model

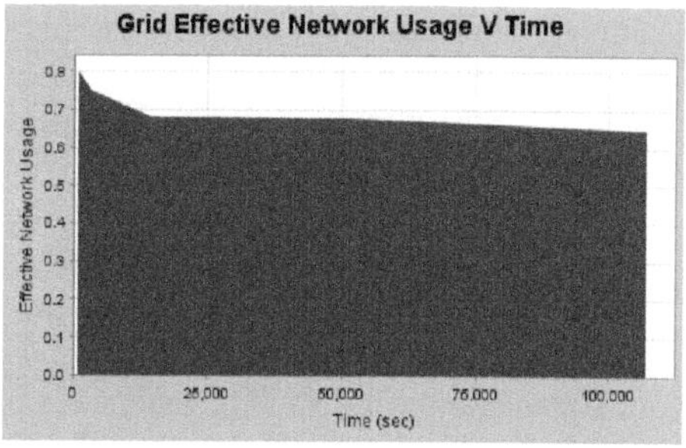

(b) File conversion rate of buffering pool model

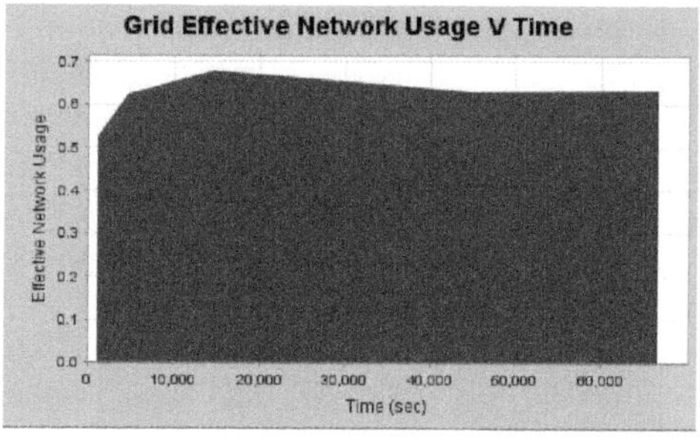

(c) File conversion rate of Agent role

Fig. 1. File conversion rate of the three scheduling models

The higher the network file conversion rate, the higher of the failure possibility of the file in grid scheduling (the lower of the conversion rate it means the documents have been decomposed appropriate and need no additional conversion). Therefore, it can be seen from Figure 2 that the conversion rate of Agent role scheduling model is relatively low, and the overall conversion rate is rather in balance with time passing by.

(2) The analysis on average consumption time of grid operation

It can be seen form Figure 3 that with time increasing, the average grid processing consumption time of the Agent role scheduling model is much lower than two models before.

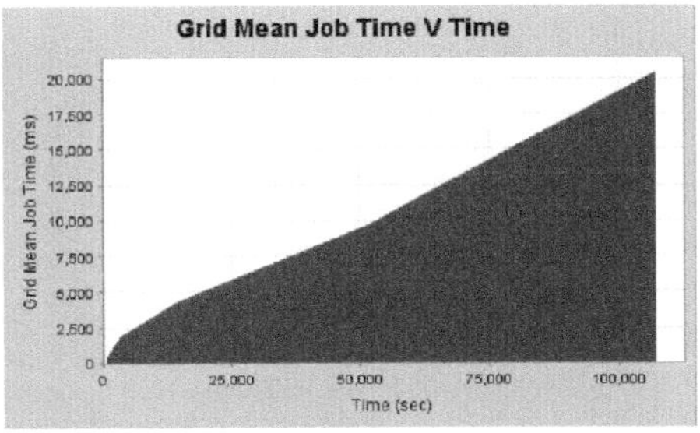

(a) Average consumption time of common queue model

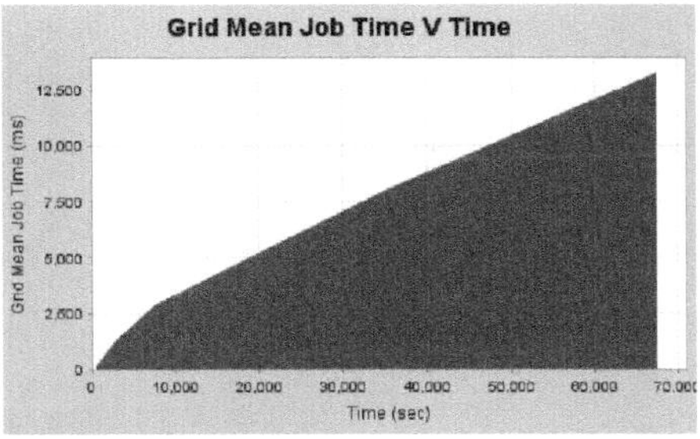

(b) Average consumption time of buffering pool model

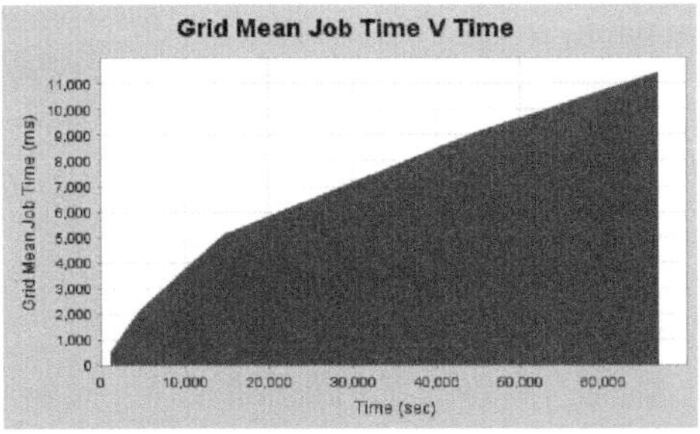

(c) Average consumption time of Agent role model

Fig. 2. Average consumption time of the three models

(3) Analysis on consumed grid workload

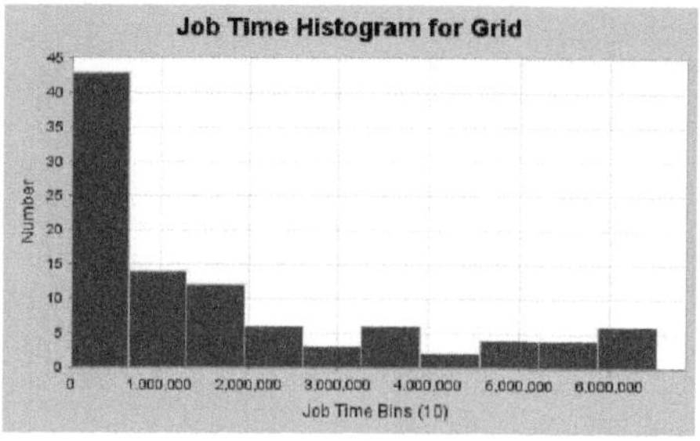

(a) Consumed grid workload of common queue model

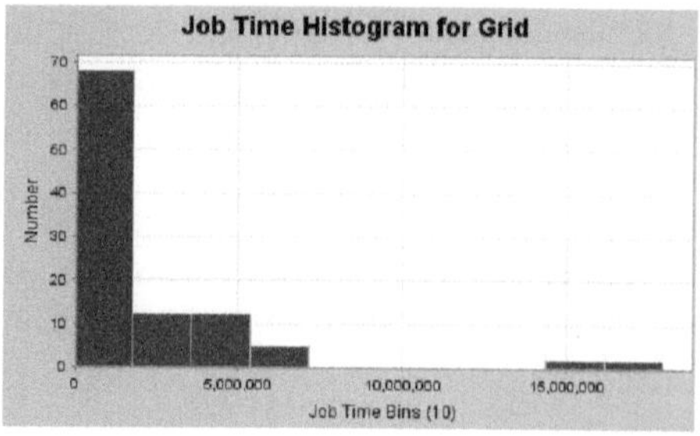

(b) Consumed grid workload of buffering pool model

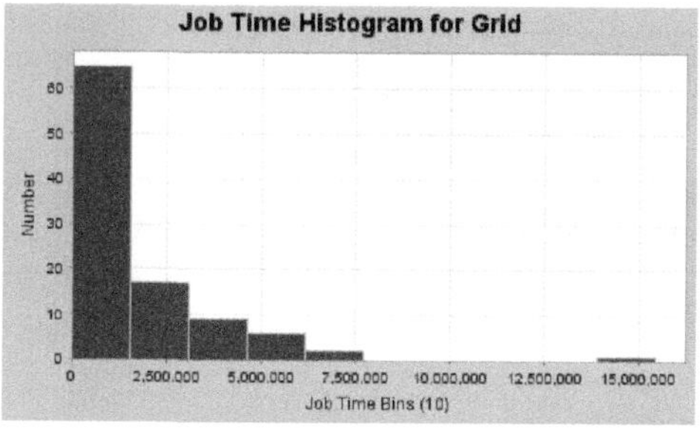

(c) Consumed grid workload of Agent role model

Fig. 3. Consumed grid workload of the three models

It can be seen from the figure that the workload consumption of the common queue model is the most, while Agent role consumption is the least. After the experiment, the grid model of the shows great speed advantage in large mount of task processing than common queue model and buffer pool model.

5 Conclusions

Tasks scheduling is building reasonable reflection between task and resource, thus it's some kind of resource distribution problem and the first goal of the resource distribution is efficiency. In this thesis, it introduced the role task distribution through Agent role making the task load rise or descent according to the supply and demand that is acting according to its ability. Finally it gets to a beneficial balance state to both provider and customer. Through the experiment the model has great advantages in task processing, but due to the limit node in the experiment, further experiment with more nodes needs to be done.

References

1. Smith, T.F., Waterman, M.S.: Identification of Common Molecular Subsequences. J. Mol. Biol. 147, 195–197 (1981)
2. May, P., Ehrlich, H.C., Steinke, T.: ZIB Structure Prediction Pipeline: Composing a Complex Biological Workflow through Web Services. In: Nagel, W.E., Walter, W.V., Lehner, W. (eds.) Euro-Par 2006. LNCS, vol. 4128, pp. 1148–1158. Springer, Heidelberg (2006)
3. Foster, I., Kesselman, C.: The Grid: Blueprint for a New Computing Infrastructure. Morgan Kaufmann, San Francisco (1999)
4. Czajkowski, K., Fitzgerald, S., Foster, I., Kesselman, C.: Grid Information Services for Distributed Resource Sharing. In: 10th IEEE International Symposium on High Performance Distributed Computing, pp. 181–184. IEEE Press, New York (2001)
5. Foster, I., Kesselman, C., Nick, J., Tuecke, S.: The Physiology of the Grid: an Open Grid Services Architecture for Distributed Systems Integration. Technical report, Global Grid Forum (2002)
6. National Center for Biotechnology Information, http://www.ncbi.nlm.nih.gov
7. Foster, I., Kesselman, C., Tuecke, S.: The Anatomy of the Grid: Enabling Scalable Virtual Organizations. International J., Supercomputer Applications 3(15) (2001)
8. Laszewski, G.V., Foster, I., Gawor, J., et al.: Peter Lane: A Java Commodity Grid Kit. Concurrency and Computation: Practice and Experience 13(8-9), 643–662 (2001)
9. Laszewski, G.V., Gawor, J., Lane, J., et al.: Features of the Java Commondity Grid Kit. Concurrency and Computation: Practice and Experience 14(13-15), 1045–1055 (2001)
10. Chue, H.N., Baxter, R.: Grid Data Transport Service Specification (in preparation), http://www.cs.man.ac.uk/grid-db/papers/GXDS-spec-1.0.pdf
11. http://www.globus.org/ [EB/OL] (2003)
12. Alexander, K., Sigrid, B.: 3D for urban purposes. Geoinfomatics 2(1), 79–103 (1998)
13. Xiao, L., Zhang, Y., Luo, J., et al.: A 3D four2level vectoredoctree structure integrating vector and raster features. In: The International Archives of Photogrammetry and Remote Sensing, vol. 32 (4W 12), pp. 265–272 (1999)
14. Lyster, P., Bergman, L., Li, P., et al.: CASA Gigabit Super computing Network CALCTUST Three-dimensional Real-time Mutlidata set Rendering. In: Proceeding of Supercomputing 1992 (1992)
15. Baker, M., Buyya, R., Laforenza, D.: The Grid: International Efforts in Global Computing. In: Intl. Conference on Advances in Infrastructure for Electronic Business, Science, and Education on the Internet (SSGRR 2000), Italy (2000)
16. Foster, I., Kesselman, C., Tsudik, G., et al.: A Security Architecture for Computational Grids. In: Fifth ACM Conference on Computers and Communications Security (November 1998)

The Research and Application of the Maritime Information Grid Service Technology Based on SOA

Hua Fang[1] and Hua Li[2]

[1] Wuhan Digital Engineering Institute, Wuhan 430074, Hubei, China
[2] Land Consolidation and Rehabilitation Center of Land Department, Beijing 100035, China
fanghwh@126.com, lh_vivian@163.com

Abstract. For meeting the need of real-time and seamless Information access, this paper researched the Web services and the grid services, proposed a service-centric maritime information grid, and gave its service architecture and the core services. It explained how various functional units should be transferred into the maritime information grid services through a concrete example, and how to recombine and converge the relevant services in order to accomplishing a specifically campaign task. The result shows that this bran-new method realizes a flexible sharing of information and enhances campaign efficiency.

Keywords: Grid Services; Web Services; SOA; WSRF.

1 Introduction

The military grid is an information infrastructure, which can provide all the facilities for the military information and interconnection, and holistically achieve the dynamic, responsive and cooperative engagement anytime or anywhere. The marine information grid is an important part of the military grid, it units control systems, information systems and combat cohesion a whole in the sea battlefield, carrying on the implementation of the entire maritime resource sharing.

The grid services is produced by the grid and Web services. In the grid services maritime information system, we can restructure and aggregate the services by changing a variety of business activities into a single standard interface military information service, so as to accomplish the specific tasks. The major purport of this article is the maritime information grid service technology based on the SOA.

2 The Web Services and the Grid Services

2.1 The Service-Oriented Architecture (SOA)

The service-oriented architecture (SOA) is an open, scalable, composable software architecture, is a solution for the design and build of the loose coupled software. The SOA architecture is shown in Fig. 1, which includes three roles: the service requesters, the service providers and the service registrars. The service requesters must be

L. Qi (Ed.): ISIA 2010, CCIS 86, pp. 482–487, 2011.

discovered by the service registrars who are required to use the services applications, software modules and other services. The service providers provides services which are published in the service registry. The service registrars store the information services, in order that the service requestors find them.

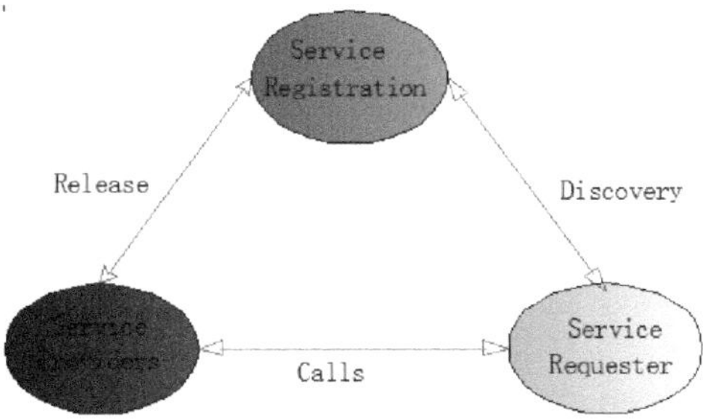

Fig. 1. The architecture of the SOA

2.2 The Web Services

The Web services is the main implementation of the SOA, which improves many problems of complexity and diversity that the SOA has to face to some extent. It has features of self-contained, self-describing, modular, you can publish, find and invoke the service through the Web. The Web service can perform various functions, including from the simple request response to the complex business processes.

2.3 The Grid Services

The OGSA is a service-centric service structure combined the Web services technologies. The Web services solve the problems of the discovery and stimulating eternal for services. But there is a large number of temporary services in the grid, so the OGSA Web services introduce the concept of grid services by extending the Web services. The grid services can be expressed as "The grid service = interface/action + service data" [1], which is defined in the OGSA, shown in Fig. 2. This service provides a set of interfaces, among which the Grid-Service interface is required, while the other interfaces are optional, including the handle maps, registration services, creating a temporary service, primary service, notification mechanism interface, upgrade, administration and so on, each service interface provides a appropriate action.

2.4 Relations between the Web Services and the Grid Services

The grid realizes on the interconnection and sharing for the resources all of the Internet. The grid service is a union of the grid and the Web service, the goal which is the resources conformity, the promoting the full sharing and the high effective utilization of the resources distributed in the network[2].

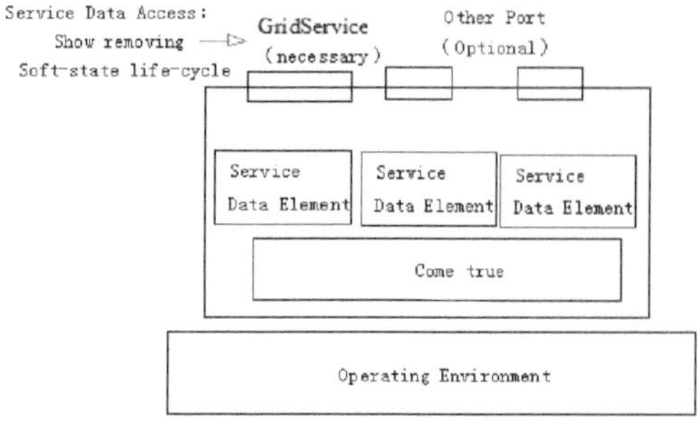

Fig. 2. The grid services schematic diagram

The difference between the grid services and the Web services is that: The grid service provides postural services and temporary Services, while the Web services provides a stateless service and permanent services. It can be said that the grid services is a postural Web service[3]. The state management is divided into two forms: The interactive perception the Web services used and the application-aware the state grid services used[4].

- The interaction perceived status: In the Web services world, the customers provide information and the service side is not responsible for the administration of any special information, while the service-side implementation is essentially scalable and stateless.
- The application-aware status: The grid service itself maintains some status information, and provide a set of standard interfaces for the customers, which allows customers to get or set the state information except the normal behavior of the service.

3 The Maritime Information Grid Services

Through the front research we can discover that the grid service locates on the top of the Web service, the function of which is stronger. Because the marine information grid services structure uses the technical system which faces the service, transforms kinds of military operational activities into the unification interface standard marine information grid service, carries on the reorganization and the polymerization through certain operational application operation flow to the corresponding service, will complete the specific operational missions. Two aspects will be discussed for the marine information grid service through the services architecture and the core services.

3.1 The Services Architecture

The division of the services architecture may be clearly indicated the organizational frame of the entire marine information grid, which includes three layers: the hosting environment layer, the core service layer and the application service layer.

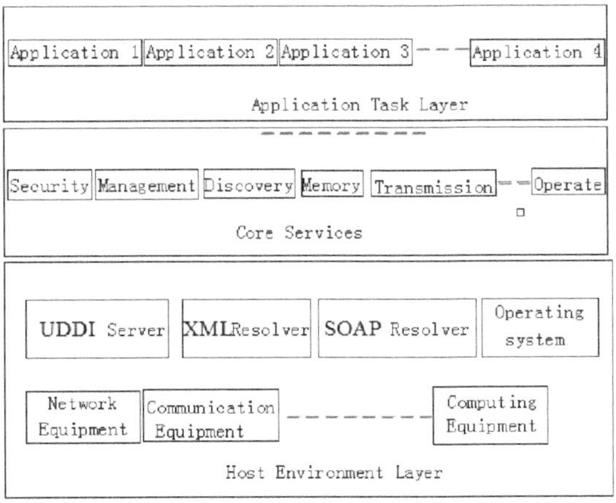

Fig. 3. The maritime information grid service system hierarchy

The bottom is the hosting environment layer, including the system environment and the equipment. The system environment layer mainly provides an operating environment, and shields the isomerism of the lower specific platforms and systems, in order to ensure the platform-independent for the services and applications. The intermediate layer a public core services layer, such as the security services, managed services, discovery services, and other basic services, or providing the application core services for the collaboration and interaction between the services and the applications. The top layer is the application service layer, the main task of which is providing better application service for the users, so the application service layer is the starting post for the maritime information grid construction, the value of the grid system is ultimately reflected in this layer.

3.2 The Core Services

The core services layer is located in the middle service architecture. This layer is very important, and should have the versatility and fundamental nature. Logically divided into two categories: first is the foundational core services, another is the applied core services. The foundational core services provide certain environmental support for the other applied core services, which includes:

- The store service: Providing the data integration and polymerization, and to the distributional data and the isomerism data a transparent visiting.
- The supervisory service: Used to monitor, control and manage all the core enterprise serves.
- The safety service: Guaranteeing each service to be safely transfered.

The role of the application core services is to guarantee the collaboration and interaction between the services and applications, and the sharing and interoperability among users and services, which includes:

- The integrated services: Implementing the integration of the existing data and resource through the access to the data and resources.
- The discovery service: Discovering the required service or resource through the retrieval.
- The information service: Accessing neatly to the static and dynamic information, accessing to the data of a expansive efficient way, and accessing multiple sources of information.
- The connection services: Supporting a variety of dynamic inter-working among the services.

4 The Application Analysis

The marine information grid changes from the system architecture to the services structure, causes each operational unit can obtain the service when carrying on the execution operational mission. Unified a operational plan faced a informational war following, we can understand how to realize the concrete application service if taking the marine information grid service as the center.

The red side executes the military exercise in the coast, while the blue side is the enemy. The red side investigates an unknown high-altitude surveillance aircraft launching of the blue side and flies to its exercise direction, so the red side starts to defense, the operational draft plan step is shown below:

Step 1: The early-warning aircraft and our warships firstly logins and authenticates on the maritime information grid system through the login authentication services, so as to access to the legal information resources.

Step 2: The early-warning aircraft preliminary obtains the enemy planes information, and publish them to the maritime information grid system. Achieving the objectives of the search service.

Step 3: The fleet command ships subscript and analyze the target information, and publish the target command. Implementing the given operational tasks services.

Step 4: Naval ships which belongs to the formation subscript the enemy information data and the target allocation order from the command ships, after the information processing, carrying on the fire distribution and the target interception. Achieving the target data service.

Step 5: The air missile launching system provides the missiles through the subscription of the target data and launches on the pre-defined trajectory. Realizing the firepower services.

Step 6: The missile will look for the targeted attacks when flight to the end, the early-warning aircraft radar systems continue to search for enemy target to determine the extent of the damage. Realizing the tracking of guidance services.

From the issues we can think that the customer demands are the most important in the context of a maritime information grid service-centric operational environment. the system offers various required maritime information grid services, and the users may access to the services and data anytime or anywhere, which is a new style of warfare, realizing the flexible sharing of information, and improving the operational effectiveness.

5 Conclusion

From system-centric to services-centric, is a major change for the implementation of the maritime information grid. This article researches a service-centric maritime information grid based on the understanding the Web services and the grid services, and describes how to transform the different functional units of the various types of military information systems into the maritime information grid services through the specific examples, in order to build a alternation in different systems through a unified manner, meeting the access to the real-time, secure and seamless information in the military applications.

References

1. Du, Z.: Grid computing, p. 10. Press of Qinghua University, Beijing (2002)
2. Hao, X., Zhou, F.: The research OGSA development grid services structurel. Shanxi Electronic Technology 1, 87–89 (2009)
3. Zhu, M., Zhang, D.: The grid architecture summary. Network Security Technology and Application 2, 69–71 (2009)
4. Sotomayor, B.: The Globus Toolkit 4 Programmer's Tutorial, pp. 14–17 (2005)

Application of Fuzzy PID Control in Marine Hydraulic Crane

Zhonghui Luo[1], Yuzhong Li[1], and Qijun Xiao[2]

[1] School of Mechatronics Engineering, Guangdong Polytechnic Normal University, 510635 Guangzhou, China
Lzh2382002@163.com
[2] School of Electronics Information Engineering, Zhaoqing University
400044 Zhaoqing, China
Shawqj@tom.com

Abstract. This paper has proposed the application of fuzzy PID control strategy in marine hydraulic crane. This crane has been applied in ships operating in the ocean, featuring the function of heave compensation. This paper presents the software and hardware structure of this control system, as well as the fuzzy PID control algorithm whose validity has been approved by emulation and simulation experiments.

Keywords: fuzzy PID; heave compensation; hydraulic crane.

1 Introduction

The 21st century is the century of ocean. Countries all over the world are under all-time high enthusiasm to develop and exploit marine resources, particularly the exploitation of marine oil and gas resources. Ships are the main carriers for ocean development. The cranes above the ocean, as influenced by oceanic wind, wave and stream, sway transversely and longitudinally, and heave vertically, which poses great difficulty for marine operations, such as transfer and transportation of goods between ships and between ships and docks, and can easily trigger collision accidents between goods and ships. In order to improve stability, safety and reliability of ship cranes, wave compensation crane is the preferred hoisting equipment, which means installing a set of electrohydraulic servo compensation device on the common crane to make it work like on the land above the ocean without being affected by the heaving waves. Structure of the wave compensation crane is shown in Figure 1. It compensates the ship's lateral movement through jib rotation, and detects the ship's relative heaving amount through the measurement encoder mounted on the jib.

The compensation hydraulic cylinder will automatically compensate the distance changes between the goods and the deck of the receiving ship caused by heaving, and thus avoids the losses caused by sudden collision between the goods and the deck. This paper researches the application of fuzzy PID control strategy in electrohydraulic servo heave compensation system.

L. Qi (Ed.): ISIA 2010, CCIS 86, pp. 488–493, 2011.

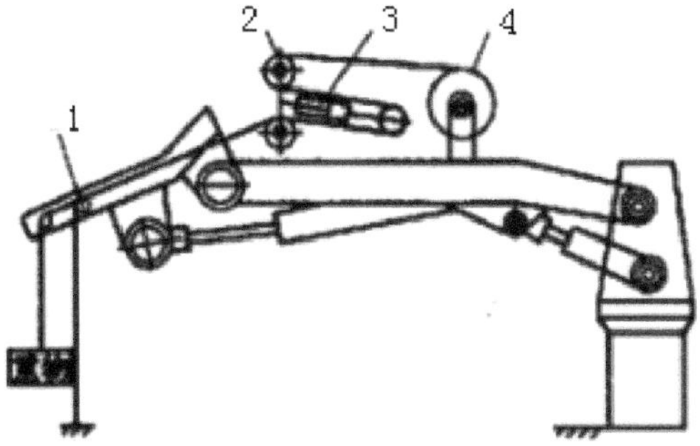

1. Sensor 2. Compensation mechanism

3. Hydraulic cylinder 4. Hoisting mechanism

Fig. 1. The Construction of heave compensation crane

2 Reasons for Adopting Fuzzy Control

The wave heave compensation crane uses hydraulic control. Its compensating cylinder uses electrohydraulic proportional closed-loop control. As the oceanic environment features nonlinear, time-varying and great random disturbance, etc., it is hard to build up a systematic mathematical model, but fuzzy control does not require any systematic and accurate mathematical model, so it has incomparable advantages compared with routine control as for time-varying, hysteretic and nonlinear systems; what's more, it also features simple control theory, easy implementation and low-cost development. Fuzzy control belongs to the intelligent control category. Its application degree represents the intelligent control level. Although fuzzy control has been widely used in hydraulic system, it also has defects of easy oscillation and poor stability. This paper adopts the method of combining fuzzy control and typical PID, so as to develop the advantages and avoid the weaknesses.

3 Reasons for Adopting Fuzzy Control

3.1 Discrete PID Controller

PID controller composes controlled variables with deviation proportion (P), integral (I) and differential (D) through linear combination to control the controlled objects. Its principle is shown in expression (1).

$$u(k)=K_pE(k)+K_I\sum_{j=0}^{k}E(j)+K_D[E(k)-E(k-1)] \tag{1}$$

In equation (1), K_p is proportion coefficient, K_I is integral coefficient, K_D is differential coefficient, $E(k)$ refers to the error at the k moment.

3.2 Fuzzy PID Controller

The fuzzy control system is mainly composed of the fuzzy controller, the input/output circuit, the controlled object and the sensor. The fuzzy PID control refers to a control method that combines fuzzy control and PID control. In actual control, as the fuzzy control passivates the sensitivity for input changes, the system has certain steady-state errors, and it is difficult to reach higher control accuracy. However, the integral part of the PID controller has great effect in eliminating steady-state errors, so the fuzzy control and PID control can be combined to improve steady-state control accuracy. Basic strategy of the control is: when the error is greater than some threshold value, fuzzy control is selected to accelerate its response; when the error is smaller than some threshold value, PID control is selected to improve its steady-state accuracy. For the time-varying and nonlinear hydraulic system, this method has incomparable superiorities compared with the typical PID control. It has both faster response speed and higher accuracy.

This fuzzy controller is a two-dimensional fuzzy controller. Error E and error change EC are the input variables of the fuzzy controller, the output variable is U, quantization discourse domain E=EC=U={-5,-4,-3,-2,-1, 0, 1, 2, 3, 4, 5}, and its fuzzy subset is {NB,NM,NS,ZE,PS,PM,PB}. The subset elements represent negative big, negative medium, negative small, zero, positive small, positive medium and positive big. Ei, ECi and Ui are the membership functions of the fuzzy set; the shape is triangular; the fuzzy control rules are the No. 45, which is following, If (E is NB) and (EC is NB) then (U is NB). The fuzzy implication uses minimization method, the composition uses maximum value method. The clear calculation uses weighted average method. U′ is obtained based on fuzzification and fuzzy inference.

$U' = \underset{i=1}{\overset{i=39}{U}} (E' \text{ and } EC') \, o \, /(E_i \text{ and } EC_i) \rightarrow U_i$. Then, output voltage U is obtained after clearing the U′. The control query table is obtained by putting the values of each calculation into a two-dimensional table.

The query control table method is often used in actual control. Obtain E and EC first, multiply them with respective quantification factor KE and KEC, and then quantify them; multiply the value calculated by fuzzy controller with a scale factor KU, and obtain the output quantity through clipping output. The following points shall be paid attention for the parameter selection:

(1) The more the discrete points of the discrete discourse domain is, the better the control function and control effects of the fuzzy controller and the higher the accuracy will be. However, if the discrete level is too high, response speed will be reduced, so it is suggested to consider using uneven grading method;

(2) Shape of the membership function is a steep triangle, which improves its sensitivity. Places with smaller error are steeper, which improves its accuracy. Places with greater error are more subdued, which enhances its speed;

(3) The fuzzy control is actually nonlinear PD control. KE is equivalent to the P parameter in PID control. Selecting a bigger KE is equal to narrowing the basic discourse domain of the system, enlarging the control function of the errors and shortening the system rising time. However, the overshoot will be enlarged, or even the system will oscillate and become unstable. If a bigger KEC is selected, the

overshoot will decrease, but the system response time is long. Reducing the KEC will cause a bigger overshoot or even cause oscillation. If the selected KU is too small, the system response time will be prolonged, the rapidity will deteriorate, and steady-state errors will increase; if the selected KU is too large, the system will oscillate. In the experiment, KE=500, KEC=40, KU=2.

4 Emulation Analysis of Fuzzy PID Control

Organically combine the FIS editor and Simulink in the fuzzy toolbox of the MATLAB, give full play to each other's advantages, will make the emulation design process simple and explicit. Emulation chart of the fuzzy PID control is shown in Figure 2, the emulation results are shown in Figure 3, and the emulation results of the typical PID control are shown in Figure 3. From the emulation results, we can see that the response time of fuzzy PID control is faster than that of the typical PID control.

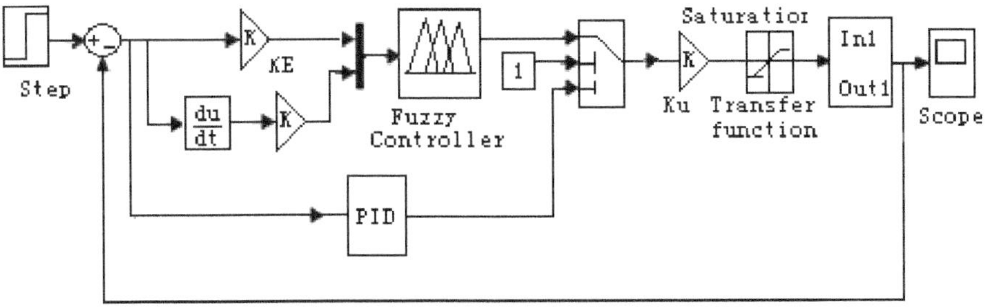

Fig. 2. Emulation chart of the fuzzy

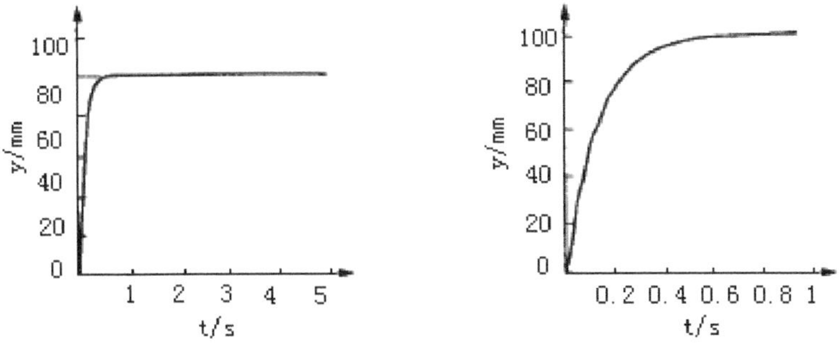

a) Response of traditional PID

b) Response of fuzzy control

Fig. 3. Results of simulation

5 Emulation Analysis of Fuzzy PID Control

To classify from functions, the simulation experiment system is composed of three parts: the heaving hydraulic cylinder that is used for simulating the heave movement of ships; the compensation hydraulic cylinder that compensates the heave movement with the aim to make the load keep still as much as possible; the last part is load. It is possible to simulate the lifting mass of the crane with heavy-duty mass block.

Closed-loop control system of the system hardware in heave-compensation simulation experiment mainly consists of industrial controlling computer, grating displacement sensor, grating sensor high-speed sampling interface card, D/A card, proportional controller, cylinder and proportional direction valve. Its composition structure is shown in Figure 4. The control principle is: the industrial personal computer receives the feedback signal from the grating displacement sensor; the signal is output through the controller to control the aperture and direction of the electro-hydraulic proportional valve, and thus controlling the displacement and speed of the hydraulic cylinder.

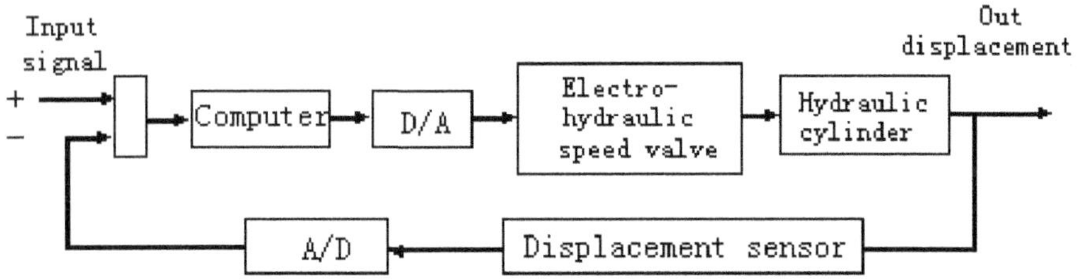

Fig. 4. The structure of simulation experiment system

The industrial personal computer is a fuzzy controller and the brain of the whole control system. It Pentium 4 Advantech industrial personal computer is used. The input/ output circuit is composed of grating displacement sensor, D/A and sampling card. Accuracy of the grating displacement sensor is 1μm. The sampling card converts the received pulse signal into digital signal and transfers it to the computer. This sensor mainly feeds back hydraulic cylinder displacement. The D/A card receives the digital signal calculated by computer and converts it into voltage signal.

The controlled objects include proportional controller, proportional direction valve and hydraulic cylinder. The output voltage is converted into current by proportional controller, and then transferred to the electromagnet of the proportional direction valve to control spool displacement and direction, and control the displacement and speed of the hydraulic cylinder by controlling the flow and direction of the oil.

This system adopts VB as programming tool and combining VC and MATLAB as auxiliary language for programming. The simulation experiment system has the following basic functions: data acquisition, parameter setting, data analysis and intelligent control algorithm. Software design mainly includes four parts: 1) data acquisition; 2) D/A output; 3) fuzzy control; 4) curve plotting.

The result of step experiment loading with 200kg is following: the step signal is 100mm, steady-state error of typical PID control is up to 3.26mm, while steady-state error of fuzzy PID control is only 0.49mm. The accuracy has been greatly improved; the response time of PID is 4.89s, while the response time of fuzzy PID control is 4.23s. The response speed has been improved.

It is supposed that the sinusoidal signal simulates the ship heaving signal (amplitude: 50mm; angular frequency: 1rad/s). If typical PID control is used, compensation displacement curve of the compensating cylinder is shown in Figure 5, the value of absolute errors is 3.46mm ,while fuzzy PID control's absolute errors value is 1.26mm Therefore, we can see fuzzy PID has greatly improved the system's stability and accuracy.

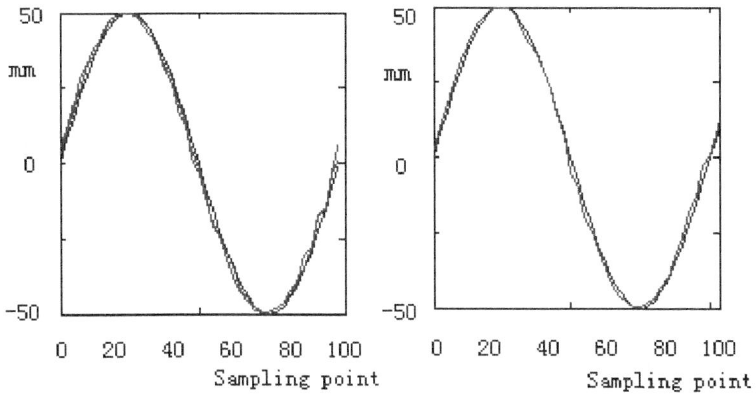

a) The sinusoid trace of PID b) The sinusoid of fuzzy PID control

Fig. 5. The compensation cure of the compensating hydraulic cylinder

6 Conclusion

This paper applies the fuzzy PID control in the wave heave compensation hydraulic crane, enabling the crane to compensate by electro-hydraulic proportional valve to control the hydraulic cylinder movement according to the detected wave heaving signal. As shown by computer emulation experiment and simulation experiment, fuzzy PID control can improve system stability, system response speed as well as anti-interference performance. It has vast application prospects in ship hydraulic crane.

References

1. Hong, S., Qingpu, Y.: The Application of Electro-hydraulic Proportional Control in the Wave-compensation Crane. Chinese Hydraulics & Pneumatics (7), 18–20 (2001)
2. Jiangfeng, P.: The Design of the Wave Compensating Device for Hydraulic Folding Type Crane. Ship & Boat (3), 39–41 (2000)
3. Zhicheng, Q., Mingyang, Z.: A Crane Applied in Ship With Heave Compensation. Engineering Machine, 12–13 (February 1999)
4. Qijun, X.: Research on Fuzzy Control Applied in the Deep Sea Mining Compensate Simulation System. The thesis for master degree of Guangdong University of Technology (May 2004)

Design of E-Commerce Platform Based on Supply Chain Management

Hui Tan

School of Economics & Management, Nanjing Institute of Industry Technology, Nanjing, Jiangsu, China
tanh@niit.edu.cn

Abstract. The e-commerce technology in the Internet time is a new generation driver of supply chain management. It is the key for increasing the core competitiveness of current enterprises to how to implement the process optimization of supply chain management and build a more complete and rapid e-commerce application platform under the environment of network. This article introduces the key technologies of design concept, functional structure and development of e-commerce platform based on supply chain management. This platform supports multiple business models, multi-site operations, and cross-channel selling, and can be integrated with the front-end trading systems, the back-end support system, and other external systems, such as partners and suppliers, which can meet e-commerce needs of enterprises with different scale and different development stages.

Keywords: B2C, B2B, E-Commerce, SCM.

1 Introduction

Until June 2009, the total number of Chinese netizens has reached 338 million, which has occupied the first place in the world and would be more hopeful to break through 400 million. A large number of netizens establish the great base for Chinese e-commerce. Since 2008, a large number of the traditional enterprises tried e-commerce and developed online B2C, B2B and other online channel business. Furthermore, the infrastructure like e-payment and logistics related to e-commerce developed rapidly, which established the base for large scale popularization of e-commerce. A powerful e-commerce system should be able to provide convenient operation to improve operational efficiency, strong marketing capabilities to increase the flow and trading volume, good integration capacity to inherit the front-end channel and the back-end systems, and excellent performance and security to ensure the reliability of the system, and the system has integrated with the industry's best practices. All this could not be met by a simple website or ERP foreground, which is a more complex system project. The e-commerce platform based on supply chain management studied in this article is a solution which can meet requirements of the parties involved, such as ultimate customers, enterprises, manufacturers and distributors, and can reply new emerging problems and challenges that enterprises face and orient to the next generation of e-commerce.

L. Qi (Ed.): ISIA 2010, CCIS 86, pp. 494–500, 2011.

2 Overall Structure Design of Platform

E-commerce platform design includes contents at three levels (hardware, software and services), which covers plan, build, deployment and operation and maintenance phase of e-commerce system, which is proved with the capability of providing single-stop-type services for customers. Figure 1 is overall structure design of e-commerce platform.

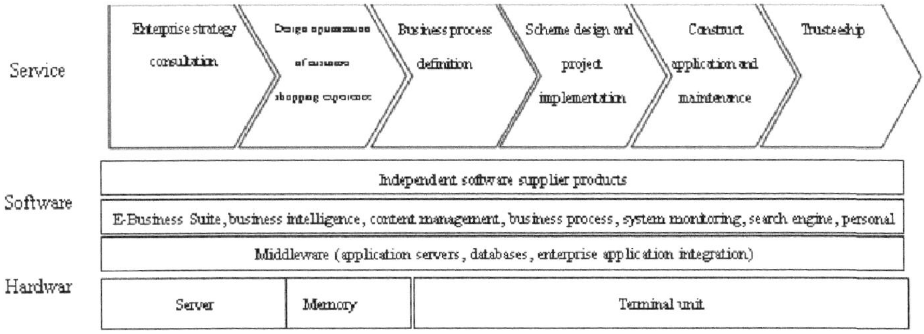

Fig. 1. Overall structure design of e-commerce platform

The platform is composed of IT infrastructures at hardware level, such as server, memory and other terminal equipments, which are the physical basis of e-commerce system operation. While the soft platform is composed of a series of system software, application software and independent third-party software, and this part is the main part of this program and the embodiment model of core function. At the service level, the platform includes business and technical supports provided for customers at every stage of system construction.

3 Business Function Design of Platform

Figure 2 is the main business functions provided by the e-commerce solutions based on supply chain management. These functions cover various aspects, such as application integration, business rules, product management, marketing management, online sales, customer service, and order management. Furthermore, a series of management, operation and maintenance and development tools are provided. Relying on these tools and features, enterprises can finish this complete end to end business operation loop from the commodity information distribution, definition of marketing, order capture, and customer service to order delivery, tracking, sales data collection, and customer behavior analysis. Configurable business functions.

In this e-commerce platform, the supply chain is composed of enterprise entities providing commercial services, and in the system, at a hosted manner, the buyers and supplies in this system can trade directly or through a middle trading market. The market provides a place of selling product and service for suppliers. The buyers enter into this private market, browse products, and select the products or services. The

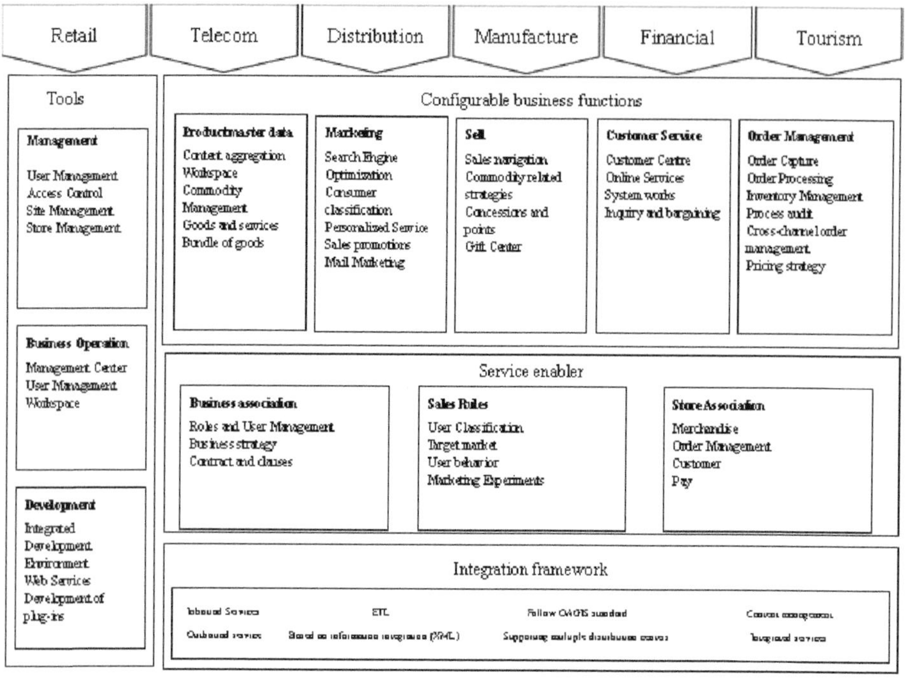

Fig. 2. Business function of e-commerce platform based supply chain management

buyers can establish contract with suppliers and also can inquiry interested suppliers. Under the environment, an organization can be a supplier, but also a buyer.

4 Expanding Design of Platform

4.1 Support to E-Payment

The platform not only provides payment support for upper order management system in various business application scenes, also provides integration with back-ground payment service to complete real-time payment transaction. Meanwhile, the platform also provides a strategy-based online payment solution and the payment strategy engine is the core of this program. In different e-commerce application scenes, the enterprises can configure or customize the appropriate strategy to accomplish their different business logic, according to different payment patterns, different payment service providers and different business needs.

4.2 Integration of SOA E-Commerce System

The key feature of this solution is integration with external systems. In the business logic of the platform, the default implementation of common integration points, adapters and interfaces are provided. The platform can be integrated with the following:

(1) multi-channel customers,
(2) Enterprise back-end systems,
(3) partners and suppliers.

In order to make integration easier, the system provides reference programs, support pages, documents, and code samples.

Use of information systems of e-commerce platform is interacted with external systems, which includes sending messages to external systems and receiving messages from the external systems. Meanwhile, the events within system (changes in order status, delivery notification, system failure, etc.) also can be sent to customers and site administrators through the messaging system.

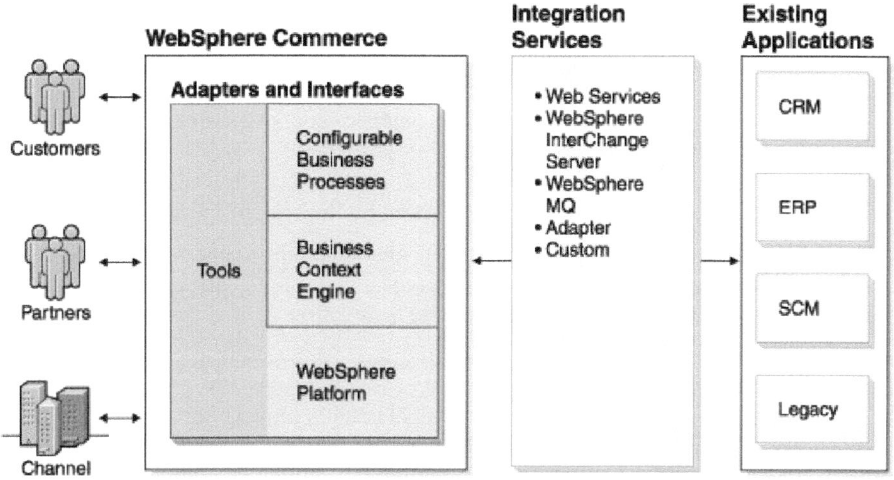

Fig. 3. Integration of e-commerce platform with external systems

Fig 3 describes the e-commerce platform how to integrate with external system. For example, establish a message system and send email to customers to notify the customer that the order has been shipped; the configuration system send message to the back-end system; when the front-end shop captures the order which contains required necessary information for back-end system carrying out the order, the back-end system will then send the order status to the platform for suggesting that the commodities of this order has been sent, the invoices printed and email sent to customers.

The platform runs corresponding orders according to the type of inbound messages from the back-end systems. The outbound message is made from the platform system and updated by the back-end systems, for example, when the customer orders, the order was sent to the back-end system to perform.

4.3 Integration with Purchasing System

The e-commerce platform is provided with integration capabilities with the purchasing system, which allows the system suppliers participate in the integration with the purchasing system from the customer of downstream supply chain as a supplier, expand their markets, increase sales and B2B businesses at the WEB end. When buyers of downstream supply chain purchase products, the order can directly generate in their own purchasing system and be sent to the platform to complete the transaction.

The platform supports two integration patterns with the purchasing system: local product catalog and request product catalog.

Under the local mode, the suppliers release their own product catalogs in the purchasing system from the customer of downstream supply chain, and the buyers can browse the product catalog without connecting the site of suppliers. The purchasing process is as follows:

(1) the buyers browse the product catalog in their own purchasing system and place an order;

(2) the order information are sent to the platform by the purchasing system, such as buyer, supplier login information, user ID, order, billing, and shipping information;

(3) the order request information are mapped into order creating command to generate order in the platform;

(4) the order requests are treated and approved, followed by the response massages are sent back to the purchasing system by the system.

Under the model of request product catalog, the suppliers maintain a single product catalog in the platform and the buyer purchasing system can access the product catalog from internet. The purchasing process is as follows:

(1) the buyers choose suppliers, and the request message of product catalog is sent to the platform;

(2) the message is mapped into the product catalog request command;

(3) the supplier certification is called by the product catalog request command followed by the user login of buyers and suppliers are authenticated;

(4) after successful authentication, check whether this user exists in the system, if no, how to create this user and the user is endowed with the role of purchaser;

(5) the product catalog is sent for URL request by the system and the session information is bound and sent to the purchasing system;

(6) the purchasing system opens a new browser window to display the product catalog;

(7) the buyers place a order, followed by production of this purchasing order, and then this order is sent to the purchasing system for approval;

(8) after approval completed, the purchasing system sends an order request to the platform to create order.

5 Key Technologies

5.1 Based on Latest Version of Middleware Platform

The platform is constructed on the latest version of the middleware platform, and the advantages of each function of the middleware platform are fully excavated, which ensures high reliability, scalability and access performance of the whole application and provides a robust and stable base platform for e-commerce of enterprises.

Furthermore, supports of various latest versions of technical specifications, such as Java EE 5, EJB3 and JSE6, and other enhanced features package, such as WEB2.0, SCA, are provided, thereby richer and more flexible e-commerce functionality will be designed and developed based on these new technical specifications and characteristics.

5.2 Perfect Integration

Each module of this platform all adopts the standard development languages and development platforms, and follows generally accepted agreement, standards and technical specifications. While the design of the whole solution is conducted under the guiding principle of SOA, the platform is provided with strong capabilities of business integration and conformity. Enterprises can integrate the internal and external business systems within this program at the same time. The integrated level can cover the presentation layer, business logic layer and data layer. a variety of integrated technology means are provided, which includes data transmission and conversion based on XML, Web services, adapters and message system.

5.3 Cloud Computing Platform Support

This e-commerce solution can be deployed in the cloud computing platform to provide business subscription in the form of service and IT resource allocation changed on demand. Therefore, enterprises can start e-commerce operations with low-cost and process better mercantile rate of return and lower cost of ownership.

6 Conclusions

Now, it is appropriate opportunity for Chinese manufacturing enterprises and retail enterprises to enter the e-commerce field and open up new sales channels. However, conventional enterprises have insufficient experiences on new internet channels, in addition to high IT investment risks. Many companies have to construct the solution meeting their own needs on a mature e-commerce platform. The design of e-commerce platform introduced in this paper can provide reference for enterprises in need. Now there are leading IT suppliers like IBM which has proposed an e-commerce solution, namely services. The customers do not need to purchase software licenses and the corresponding hardware devices, and can provide business subscription in the form of service, special customization and business process optimization, and provide IT resource allocation changed on demand and implementing schemes with low-cost and low risk.

References

1. Lu, J.: SCM-based Enterprise Resource Management Study. Chinese Market (2), 102–103 (2007)
2. He, Y., Yang, D.l., Huang, C.: Supply Chain Coordination under E - Commerce: A Literature Review. Computer Engineering and Applications 27(2), 12–15 (2006)

3. Yuan, H.-j., Zhang, Y.-n.: Design and Application of Supply Chain Management System. Logistics Technology 28(4), 106–108 (2009)
4. Fu, Q.-a., Wang, W.-b.: Realization of SCM System Based on Shortening Responding Time. Logistics Technology (10), 190–193 (2008)
5. Wang, Y.-q., Shi, X.-l.: Management of Supply Chain. Mechanical Industry Press, Beijing (2005)
6. Bowersox, D.J., Closs, D.J.: Logistical Management: The Integrated Supply Chain Process. Mc Graw-Hill, New York (1996)
7. Cachon, G.P.: Supply chain coordination with contracts. In: Graves, S.C., De Kok, T. (eds.) Handbooks. In operations Research and Management Science: Supply Chain Management, Elsevier, North-Holland (2004)
8. Chen, F.: Information sharing and supply chain coordination. In: Graves, S.C., De Kok, A.G. (eds.) Handbooks in Operations Research and Management Science: Supply Chain Management: Design, Coordination and operation. Elsevier Publishing Company, Amsterdam (2003)

An Adiabatic Content-Addressable Memory Based on Dual Threshold Leakage Reduction Technique

Jintao Jiang, Xiaolei Sheng, and Jianping Hu

Faculty of Information Science and Technology,
Ningbo University 315211 Ningbo City, China
nbhjp@yahoo.com.cn

Abstract. This paper presents a CAM (Content-Addressable Memory) using dual threshold leakage reduction technique. A 16×16 CAM is demonstrated using the proposed dual threshold technique based on CPAL (complementary pass-transistor adiabatic logic) circuits. All circuits are verified using HSPICE in different temperature, high-threshold voltage, and active ratios in 45nm technology. BSIM4 model is adopted to reflect the leakage currents. Simulation results show that leakage losses of the CPAL CAM using dual threshold leakage reduction technique are obviously reduced both in work mode and idle mode compared with basic CPAL one using single threshold transistors.

Keywords: Dual threshold CMOS, Leakage reduction, Adiabatic logic, CAM.

1 Introduction

In the past, dynamic energy loss has always dominated total power dissipation, while leakage power is small and can be ignored [1]. Adiabatic computing can recycle the energy stored in capacitance, and thus reduce dynamic dissipations [2]. When CMOS scales down into 100nm, leakage consumption increases faster than dynamic power and is expected to dominate total power consumption [3].

Dual threshold technology has proved to be a very effective method for reducing leakage [3]. A higher threshold voltage can be assigned to some transistors on non-critical paths to reduce leakage current, while the performance is maintained due to low threshold transistors in the critical paths. Therefore, no additional transistors are required, and both high performance and low power can be achieved simultaneously. This paper presents a CAM (Content-Addressable Memory) using dual threshold leakage reduction technique based on CPAL (complementary pass-transistor adiabatic logic) circuits. All circuits are verified using HSPICE in different temperature, high-threshold voltage, and active ratios in 45nm technology.

2 CPAL Circuits with Dual Threshold CMOS

The CPAL circuit using two-phase power-clock scheme is shown in Fig. 1 [4]. It is composed of two main parts: the logic function circuit and the load driven circuit. The

L. Qi (Ed.): ISIA 2010, CCIS 86, pp. 501–507, 2011.

logic circuit consists of four NMOS transistors (N5-N8) with complementary pass-transistor logic (CPL) function block. The load driven circuit consists of a pair of transmission gates (N1, P1 and N2, P2). The clamp transistors (N3 and N4) ensure stable operation by preventing from floating of output nodes. Its simulated waveforms and two-phase power clocks are also shown in Fig. 1. The detailed description on two-phase CPAL circuits can be found in [4].

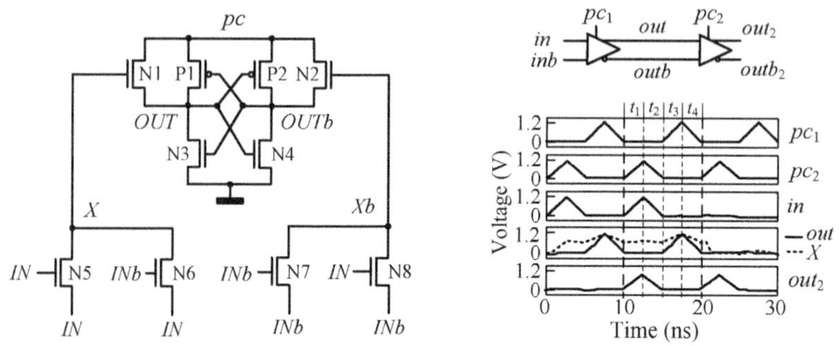

Fig. 1. CPAL buffer using two-phase scheme and its simulated waveforms

Leakage currents of the CPAL buffer are shown in Fig. 2 (b). When the output (Q) is "0", the transmission gate TG1 (N1, P1) and the transistor N4 are off, while the transmission gate TG2 (N2, P2) and the transistor N3 are turn on. TG1 and N4 dissipate sub-threshold leakage dissipations through the turn-on N3 and TG2. In order to reduce the sub-threshold leakage dissipations, TG1 and N4 are assigned to the high threshold, as shown in Fig. 2 (b) [5]. The sub-threshold leakage will be reduced when the output is "0" compared with the basic CPAL buffer.

However, the high threshold transistors will increase their turn-on resistance. When the output (Q) is "1", the transmission gate TG1 is on, and the output Q follows pc, so that the dynamic energy loss will become larger than the basic CPAL buffer. Therefore, we use DTCMOS structure for the logic gates in the register file, whose output (Q) often is "0".

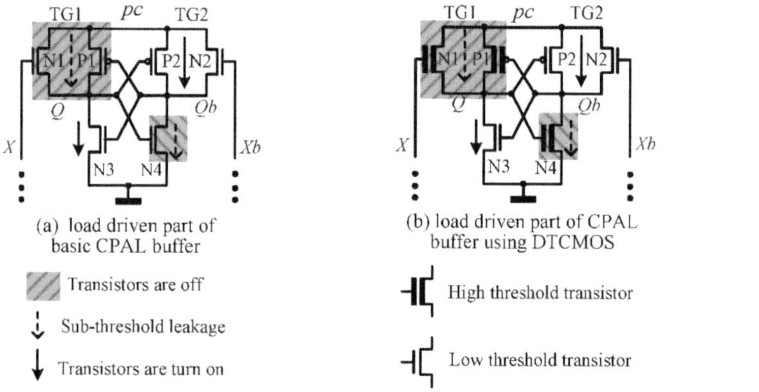

Fig. 2. Leakage currents in the basic CPAL circuit and the CPAL circuit with DTCMOS

3 CAM Based on CPAL Circuits Using DTCMOS Technique

A 16×16 CPAL CAM using the dual threshold technique is demonstrated.

3.1 The Structure of CPAL CAM

The two-phase CPAL CAM core consists of a CAM cell array, address decoders, write word-line drivers, match-line drivers, bit-lines drivers and write bit-line and read data-line drivers, as shown in Fig. 3.

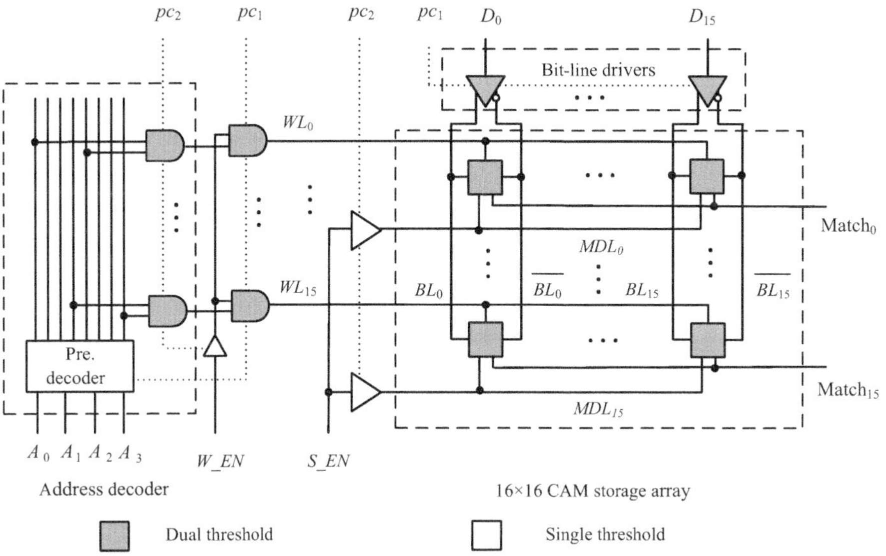

Fig. 3. The structure of the two-phase CPAL CAM

Address decoder with 4-bit addresses is used for selecting CAM cells by charging the word lines, as shown in Fig. 3. The 4-bit address is divided into the two-level address decoding. The word-line signals are produced by using AND gates with the two output of the pre-decoding and the write enable signals (W_EN). The AND gates in the address pre-decoding use the basic CPAL structure with low threshold transistors. In the second-decoder, the output of only one AND gate is "1", and the other 31 word-lines remain "0". Therefore, the AND gates in the second-decoder use the DTCMOS CPAL to reduce their leakage dissipations. The same analysis can also be carried out for the other circuits in the CAM.

In modern electronic devices, memory capacity can reach 1GB. This means the memory composed by about 2^{15} columns storage cells. The write bit-line and read data-line drivers are as much as the number of memory columns. But only several of them implement the read and write operations, most of read and write driver circuits are in idle state. Therefore, using dual-threshold structure can reduce leakage dissipations effectively.

3.2 CAM Storage Cell

The adiabatic CAM storage cell consists of a cross-coupled inverter pair (N4, P1, and N5, P2) for storing data, three transistors (N1, N2 and N3) for compare operation, and one pair of access transistors (N6 and N7) that is enabled by *WL* (word line) for write operation, as shown in Fig. 4. Its structure is similar to the conventional CAM cell except for the source port of the transistor N1, which is connected to GND in the conventional one. Here, it is connected to match-driving lines (*MDL*), which are driven by the power-clock pc_2, so that the energy of match lines (*ML*) can be recycled to the power-clock.

In the inactive state with no data inputs, *BL*/*BLb* holds 0/1, and N2 is off, so that sub-threshold leakage is dissipated. Therefore, N2 uses high threshold device. In the storing part of CAM cell, some transistors may dissipate large leakage power depending on the value stored in the cell. For example, when the cell stores '0,' as shown in Fig. 4, the transistors N5 and P1 dissipate sub-threshold leakage. Therefore, we can use double-threshold technique for storage cell. The transistors N5 and P1 are assigned to the high threshold, as shown in Fig. 4. Because about 70% of memory store data are "0" [5], the storage cell with double-threshold technique can reduce leakage in the inactive state.

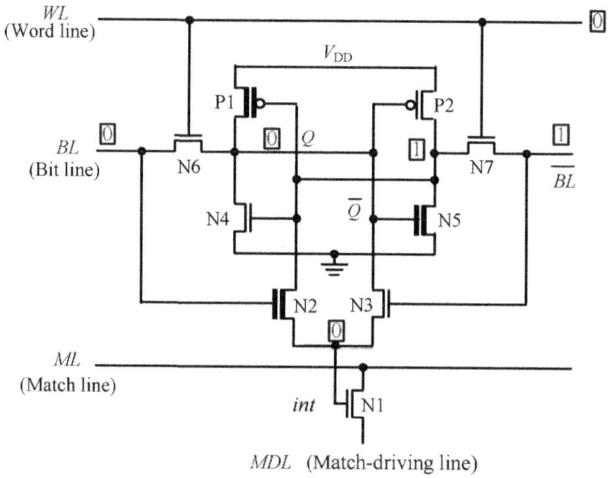

Fig. 4. Dual threshold CAM cell

4 HSPICE Simulations and Energy Dissipations

A single-phase adiabatic 16×16 CAM based on the CPAL circuits with the DTCMOS techniques is realized. The minimizations for the total energy overhead of the CAM have been carried out by choosing optimal sizes. The simulated waveforms for the 16×16 CAM with DTCMOS using a sinusoidal single-phase power clock are shown in Fig. 5.

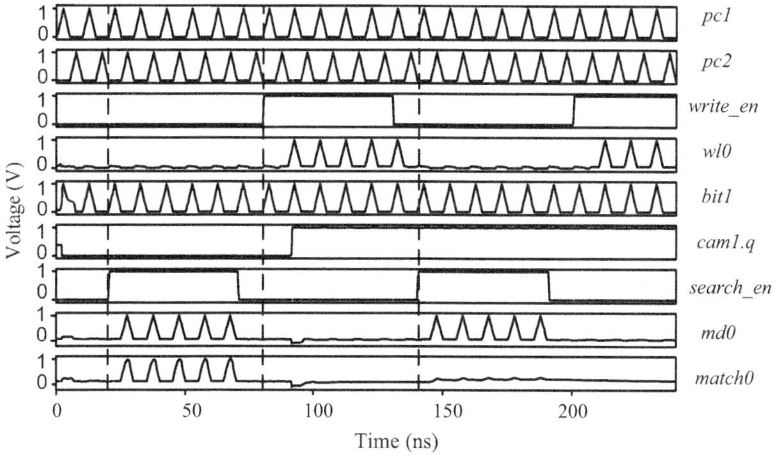

Fig. 5. Simulated waveforms of the 16×16 CAM based on DTCMOS CPAL circuits

4.1 Energy Dissipations

The function verifications and energy loss tests have been carried out both at standard temperature (27°C) and high temperature of 60°C in 45nm processes. Different threshold voltage of the devices and work modes are also taken into account. The total energy consumption and leakage dissipations have been tested respectively.

At 27°C and 60°C, the leakage loss per cycle of the adiabatic 16×16 CAM is shown in Table 1, and V_t is threshold voltage of the high threshold transistors.

For comparison, a 16×16 adiabatic CAM is also simulated, whose structure is the same as Fig. 3, but all circuits uses single threshold (low V_t). Fig. 6 and Fig. 7 shows the leakage loss saving rate of the adiabatic CAM using DTCMOS technique compared with single threshold voltage at 45nm processes.

Table 1. Leakage losses per cycle of the adiabatic CAM Based on CPAL with DTCMOS technique. Frequency is 50MHz, and V_t is threshold voltage of the high threshold transistors.

Temperature	State	V_t (V)		
		0.18	0.25	0.3
27°C	active	116fJ	79fJ	86.3fJ
	idle	82fJ	67.2fJ	60.4fJ
60°C	active	119fJ	103fJ	96.7fJ
	idle	101fJ	81.3fJ	74.7fJ

As is mentioned above, compared with single threshold CAM, the total energy loss of CAM using dual threshold are obviously reduced both in work mode and idle mode. When frequency is 100MHz, the optimal threshold voltage is 0.3V. Therefore, we can select a suitable high threshold voltage based on the temperature and the clock frequency.

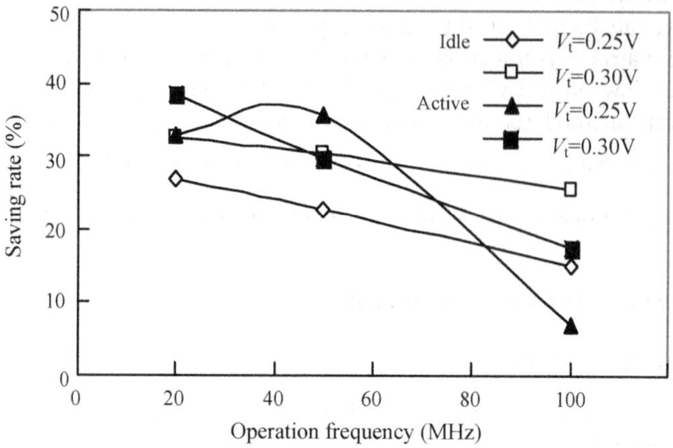

Fig. 6. Energy loss savings of the 16×16 CAM using DTCMOS technique compared with single threshold voltage in active time and idle time at 27°C. V_t is threshold voltage of the high threshold transistors.

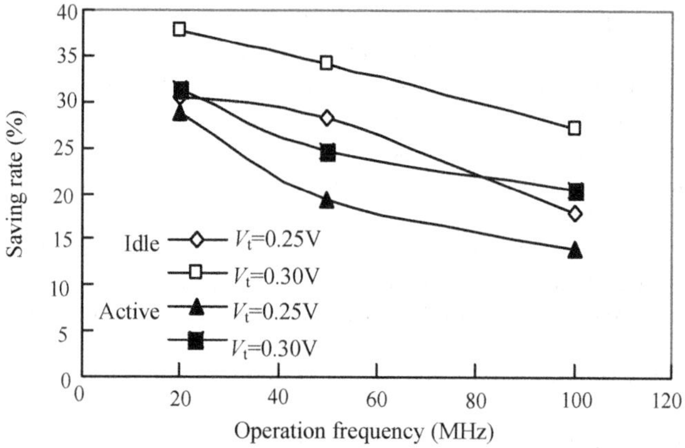

Fig. 7. Energy loss savings of the 16×16 CAM using DTCMOS technique compared with single threshold voltage in active time and idle time at 60°C. V_t is threshold voltage of the high threshold transistors.

4.2 Max Operation Frequency

The high threshold will influence the speed and stability of the circuits. Therefore, the ability of storage cell to retain data, and the driving ability of address decoder and bit drivers in the 16×16 CPAL CAM are changed when using DTCMOS technique. The maximum frequency of the 16×16 CPAL CAM which adopts various high-thresholds have been tested in 27°C and 60°C, as shown in Fig. 8. The CAM using DTCMOS technique with 0.18v high-threshold obtains the highest operation frequency.

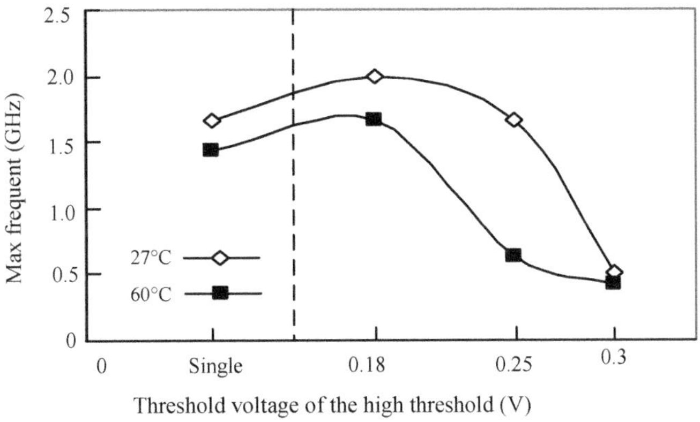

Fig. 8. The max operation frequency of 16×16 CPAL CAM using DTCMOS

5 Conclusion

This paper focuses on leakage reduction of the CAM based on CPAL circuits using DTCMOS schemes. The energy dissipations of the CAM have been investigated in different temperature, frequencies and operation modes. The results show that the leakage loss of the CAM using the DTCMOS technique can be reduced more effectively than the single-threshold one.

Acknowledgments. Project is supported by National Natural Science Foundation of China (No. 61071049) and Scientific Research Fund of Zhejiang Provincial Education Department (No. Z200908632), and supported by Ningbo Natural Science Foundation (No. 2009A610066).

References

1. Kim, N.S., Austin, T., et al.: Leakage Current: Moore's Law Meets Static Power. Computer 36(12), 68–75 (2003)
2. Hu, J.P., Xu, T.F., Li, H.: A Lower-Power Register File Based on Complementary Pass-Transistor Adiabatic Logic. IEICE Trans. on Inf. & and Sys. E88-D(7), 1479–1485 (2005)
3. Fallah, F., Pedram, M.: Standby and Active Leakage Current Control and Minimization in CMOS VLSI Circuits. IEICE Trans. on Electronics E88–C (4), 509–519 (2005)
4. Hu, J.P., Xu, T.F., Xia, Y.S.: Low-Power Adiabatic Sequential Circuits with Complementary Pass-Transistor Logic. In: 48th IEEE International Midwest Symposium on Circuits and Systems, vol. 2, pp. 1398–1401 (2008)
5. Hu, J.P., Yu, X.Y., Sheng, X.L.: An Adiabatic Register Based on Dual Threshold Leakage Reduction Technique. Advanced Materials Research 121-122, 148–153 (2010)

A New Integer Programming Model about Counterfeit Coin Problem Based on Information Processing Method and Its General Solution

Bai Xiaoping and Ke Rui

School of Management
Xi'an University of Architecture & Technology
Xi'an 710055, Shanxi, China
jingbei1010@163.com

Abstract. Based on the Counterfeit Coin Problem that is often met in ordinary life, this paper looks weighting coins as an information processing process, introduce some interrelated information processing knowledge in Information Theory and use it to establish a new Integer Programming model based on information processing method. Its detailed logic inference process, general solution is discussed, its universal laws and specific academic explainer are also presented, so this problem is solved, these can expand the application scope of information processing method.

Keywords: information processing; logic inference, Counterfeit Coin problem; Integer Programming model; general solution.

1 Introduction

The Counterfeit Coin Problem is a classic intelligence problem. Although it looks like very common, it can be related to Data Structure (such as the binary tree), computer program algorithm, Graph Theory, and etc, therefore researching this problem is very meaningful. The general Counterfeit Coin Problem can be described as: "given a set of same shape and size coins (or other objects) (such as 24 coins) among which there is a counterfeit coin that is lighter (or heavier) than other coins whose have same weights. Find this counterfeit coin using a balance without counterweights, ask how to determine the minimum number of weightings necessary and how to weight." For this problem, if the quantity of coins is less, it is not difficult to be solved. But if generalizing the quantity of coins to n (such as 345 coins), ask how to determine quickly the minimum number of weightings necessary and how to weight, and need look for its general law and theoretical basis, then it is not easy to be solved. Compared with other references, this paper considers this question from another point of view, looks weighting coins as an information processing process, and establishes a new Integer Programming model based on information processing method. In the other hand, this paper also presents detailed logic inference process and discusses its general solution. The minimum number of weightings necessary, its general law and theoretical basis can be quickly gotten according to presented model and method, so application ranges of the information processing method can be widened.

L. Qi (Ed.): ISIA 2010, CCIS 86, pp. 508–513, 2011.

2 Correlated Information Knowledge Description

Information theory is a branch of applied mathematics and electrical engineering involving the quantification of information. Information theory was developed by Claude E. Shannon to find fundamental limits on signal processing operations such as compressing data and on reliably storing and communicating data. Since its inception it has broadened to find applications in many other areas, including statistical inference, natural language processing, and cryptography generally, networks other than communication networks.

The information quantity firstly should be defined on quantitative studying information processing, in Shannon information theory, Information quantity I is defined as follows:

$$I(m) = \log(\frac{1}{p(m)}) = -\log(p(m)) \tag{1}$$

Where p (m) = Pr (M = m) is the probability that message m is chosen from all possible choices in the message space M. The base of the logarithm only affects a scaling factor and, consequently, the units in which the measured information content is expressed. If the logarithm is base 2, the measure of information is expressed in units of bits.

3 A New Integer Programming Model about Counterfeit Coin Problem Based on Information Processing Method

This paper looks weight coins as an information processing problem. Supposing there are n coins, among which there is a counterfeit coin that is lighter (or heavier) than other coins whose have same weights. The total information quantity must be satisfied to find out which coin is counterfeit one is

$$-\log_2(1/n) = \log_2 n \tag{2}$$

One-time weighting can be looked an information transmission process. All coins can be allocated to three groups, two number groups of coins are respectively set on two sides of a balance without counterweights, the rest group of coins is set balance outside (allocating to two groups can be looked as a special case , nothing but the number of third group coins is 0). Supposing it need to find the counterfeit coin from n1 coins on every time weighting, the value of n1 decreases continuously with weighting process. n1 coins can be allocated to three parts according to above allocation method, such as x_{max},x_1,x_2, here $x_{max} \geq \max(x_1, x_2)$, and among three values there are two equal ones. Results of one-time weighting have (1) if the weight of coins set on two sides of the balance is same, then counterfeit coin exists in among coins outside balance; (2) if left inclination of the balance occurs, light counterfeit coin exists in the right pan of the balance; (3) if right inclination of the balance occurs, light counterfeit coin exists in the left pan of the balance. Information quantities transmitted in three cases are respectively $l_{i1} = -\log_2 x_{max} / n_i$, $l_{i2} = -\log_2 x_1 / n_i$, $l_{i3} = -\log_2 x_2 / n_i$. In order to find general method, the worse case need be considered. Information quantities obtained in the worse case for one-time weighting can be expressed by

$$\begin{aligned} Qi &= \min\ \{l_{i1},\ l_{i2},\ l_{i3}\} \\ &= -\log_2 x_{max} / n_i \\ &= \log_2 n_i / x_{max} \end{aligned} \quad (3)$$

The problem of searching for the optimal weighting method can be transformed into an integer programming model presented in this paper, shown as follows.

Objective Function: Ti=max Qi=max $\{\log_2 n_i / x_{max}\}$ (4)

Constraint conditions:

$$\begin{cases} x_{max} + x_1 + x_2 = n_i \\ x_{max}, x_1, x_2, n_i \text{ are integers} \\ x_{max} \geq \max(x_1, x_2) \\ x_{max}, x_1, x_2, \text{among them there are two equal ones} \end{cases} \quad (5)$$

4 The Solution of the Integer Programming Model Presented in This Paper

According to formula (4), x_{max} must be the smallest in order to keeping Qi being the biggest, so

$$\begin{aligned} & x_{max} \geq x_1, x_{max} \geq x_2, x_{max} + x_1 + x_2 = n_i \\ & \Rightarrow 3x_{max} \geq x_{max} + x_1 + x_2 = n_i \\ & \Rightarrow x_{max} \geq n_i / 3 \end{aligned} \quad (6)$$

That is, the smallest value of x_{max} is $n_i / 3$, substitutes formula (6) into formula (4), obtains the maximum limit information quantities value gotten for per time weighting, shown as follows.

$$\log_2 n_i / x_{max} = \log_2 n_i / (n_i / 3) = \log_2 3 \quad (7)$$

According to formula (2), the total information quantity that must be satisfied to find out which coin is counterfeit one is $\log_2 n$, the maximum limit information quantities value gotten for per time weighting is $\log_2 3$, therefore, for the counterfeit coin problem including n coins, the minimum number of weightings necessary is

$$\frac{\log_2 n}{\log_2 3} \quad (8)$$

5 The General Weighting Method Based on the New Information Processing Model

In addition, further discussion on how to weight can be make according to new model. Supposing information quantities obtained on per time weighting is the maximum. So above integer programming problem can be discussed as follows.

(1) When $n_i=3k_i$, the smallest value of x_{max} is $n_i/3 = k_i$, the optimal weighting method is that allocating 3k1 coins to three parts (k_i, k_i, k_i), the maximum information quantities obtained in the worse case is

$$-\log_2 \frac{k_i}{3k_i} = \log_2 3 \tag{9}$$

(2) When $n_i=3k_i+1$,the smallest value of x_{max} is$[n_i/3]+1= k_i+1$, the symbol [] expresses integral function , the optimal weighting method is that allocating 3ki+1 coins to three parts (k_i+1, k_i+1, k_i-1) or (k_i+1, k_i, k_i), the maximum information quantities obtained in the worse case is

$$-\log_2 \frac{k_i+1}{3k_i+1} < -\log_2 \frac{k_i+1}{3k_i+3} = \log_2 3 \cdot \tag{10}$$

(3) When $n_i=3k_i+2$,the smallest value of x_{max} is$[n_i/3]+1= k_i+1$, the optimal weighting method is that allocating 3ki+2 coins to three parts (k_i+1, k_i+1, k_i) the maximum information quantities obtained in the worse case is

$$-\log_2 \frac{k_i+1}{3k_i+2} < -\log_2 \frac{k_i+1}{3k_i+3} = \log_2 3 \tag{11}$$

6 The General Weighting Method Based on the New Information Processing Model

According to above model and solution, supposing there are 25 coins, the minimum number of weightings necessary by formula (8) is

$$\frac{\log_2 25}{\log_2 3} = \log_3 25 = 2.93 \tag{12}$$

That is, the minimum number of weightings necessary is 3. In addition, the optimal weighting method can be described as follows.

Step 1: Because 25=3×8+1, ki=8, 25 coins can be allocated to three parts (9, 9, 7) or (9, 8, 8), the maximum information quantities obtained in the worse case is $-\log_2 9/25 = 1.465$ bits

Step 2: Because 9=3×3, ki=3, 9 coins can be allocated to three parts (3,3,3), the maximum information quantities obtained in the worse case is $-\log_2 3/9 = 1.585$ bits

Step 3: Because 3=3×1, ki=1, 3 coins can be allocated to three parts (1,1,1), the maximum information quantities obtained in the worse case is $-\log_2 1/3 = 1.585$ bits, so the result of which coin is counterfeit one can be gotten.

That is, for the counterfeit coin problem including 25 coins, the total information quantity must be satisfied to find out which coin is counterfeit one is $-\log_2 1/25 = 4.64$ bits, considering the error of rounding; this value equals total information quantity transmitted on three steps weighting method, shown as

$$-\log_2 9/25 + (-\log_2 3/9) + (-\log_2 1/3)$$
=1.465+1.585+1.585
=4.64 bits

7 Conclusions

The information processing method in Shannon information theory can be regarded as the new methods and means to study science and application problem. This paper integrated information processing method in Shannon information theory with operational research to study the classical intelligence problem, and establish a new Integer Programming model based on information processing method. Its detailed logic inference process, general solution is discussed; its universal laws and specific academic explainer are also presented. Although the counterfeit coin Problem looks like very common, it can be related to Data Structure (such as the binary tree), computer program algorithm, Graph Theory, and etc, so the model and method presented in this paper is meaningful aiming to these aspects.

Acknowledgment

This work is supported by Shanxi province natural science basic foundation (2010JM7004), Shanxi province key discipline" management science and engineering" construction special fund subsidized project and the national key discipline cultivation project. (No.200808265)

References

1. Born, A., Hurkens, C.A.J., Woeginger, G.J.: How to detect a counterfeit coin: Adaptive versus non-adaptive solutions. Information Processing Letters 86(3), 137–141 (2003)
2. Liu, W.-A., Zhang, Q.-M., Nie, Z.-K.: Searching for a counterfeit coin with two unreliable weightings. Discrete Applied Mathematics 150(1-3), 160–181 (2005)
3. Xiao, X.-p.: A Non-Adaptive Algorithm for the General Counterfeit Coin Problem. Journal of University of Electronic Science and Technology of China (S1), 451–453 (2007)
4. Qi, M.N., Li, W.: Unified search procedure with two devices for equilibrium models of two-counterfeit coin problem. System Engineering Theory and Practice 21(9), 73 (2001)

5. Alon, N., Kozlov, D.N.: Coins with Arbitrary Weights. Journal of Algorithms 25(1), 162–176 (1997)
6. Xiao, X.-P.: Non-adaptive Solutions to the General Counterfeit Coin Problem. Journal of Nanjing University (Natural Sciences) (5), 506–511 (2006)
7. Li, W., Pan, P.: Unified search procedure with two devices on equilibrium models of two-counterfeit-coin problem. Journal of Southeast University (Natural Science Edition) 3, 536–540 (2002)
8. Wang, X.-d.: Dynamic programming algorithm for counterfeit coin problem. Mini-micro Systems (12), 1301–1308 (2000)

Robot Remote Control Internet Architecture

R. Yu and X.G. Huang

School of Mechanical and Electrical Engineering,
North China University of Technology, Beijing, China
march08@126.com

Abstract. An internet architecture for robot remote control is proposed in order to have high availability, performance, and scalability. This system can be mainly divided into two major parts: the robot part (terminal) and the control software part (controller). A remote control protocol based on Internet for remote communication between users and robot is designed. The proposed architecture is comparatively concise and reduces the operation time because fewer data are transmitted on network.

Keywords: Internet, remote communication, remote control.

1 Introduction

Today's robot system has been broadly applied in factory automation, surgery, space exploration, military, and also in our daily life in the coming future. For example, an autonomous land vehicle in CMU can steer autonomously with high intelligence on the road under different conditions [1]. The intelligent autonomous mobile robot LOLA has the abilities of office delivery and trash cleaning [2, 3]. Although the robot behaviors for such applications have its autonomy but the owner still has to worry about the safety of the robot when the robot is working alone remotely. These kinds of problems lead to another research topic of remote control techniques.

The remote robot control technologies have been used in the fields like factory automation or space exploration, which it is difficult for human to treat. The exploration of the seabed and the planets are dramatic examples. Radio-controlled model aircraft and ships are more prosaic examples; lighthouses and power stations are more economically significant examples. In recent years with the popularity of the Internet, WWW based remote-control system is becoming an interesting and promising field of investigation in robotics, VR and visualization. Applications of these investigations in areas such as space and underwater robotics, remote manufacturing, operator training, remote education and entertainment are of great importance.

One of the first successful World Wide Web (WWW) based robotic projects was the Mercury project [4]. This later evolved in the Telegarden project [5], which used a similar system of a SCARA manipulator to uncover objects buried within a defined workspace. Users were able to control the position of the robot arm and view the scene as a series of periodically updated static images. In September 1994, Ken Talyor [6] at Western Australia University provides Internet control of an industrial ASEA IRB-6 robot arm through the WWW. Manipulator at remote place could

L. Qi (Ed.): ISIA 2010, CCIS 86, pp. 514–518, 2011.

operate the robot and control it to simulate assembly operation, at the same time robot could receive position input of remote workplace and control the motion of joint. The previously mentioned projects [4], [5], [6] rely on cameras to locate and distribute system data and the current environment to the user via the WWW. It is clear that such an approach needs a high-speed network to achieve on-line control of the mobile robot. Data transmission times across the WWW depend heavily on the transient loading of the network, making direct teleoperation (e.g. the use of cameras to obtain a mobile robot position feedback) unsuitable for time critical and dangerous (e.g. collision between the robot arm and the environment) interactions.

All of above system mostly regard browsers as control modules. In fact, distributed structure is often adopted. Three servers are connected to remote computer via Internet, web server provides information service of WEB, image server provides the function of image collection and storage, and robot control server provides control function. But there are some disadvantages, such as much more data are transmitted on network, the efficiency of transmission is low, and real time is not ideal and so on. In order to take full advantage of the Internet resources, A new internet architecture for robot remote control is proposed in the paper.

2 Communication Protocol

The robot remote communication protocol is designed according to the detailed conditions of power cube modular robot with 9-DOF in control process. The protocol is proposed to reduce network loading, improve network transport speed and insure the real time of system. Since client-server system is adopted in study, manipulators at client use remote communication protocol to connect with servers. After connecting, servers receive and comprehend commands sent by client accurately, and provide information that is relative to motion states of robot to client. Thus real states of motion are simulated when client receives these data exactly, and then we can real time control robot motion at client. The remote communication protocol accomplishes its objective of transmitting all data rapidly via the Internet connection. For client, sending command packets and receiving state packets are its main tasks in communication. For server, receiving command packets and sending state packets are tasks of it.

The data format of command packet is defined as table 1. Here indications are the characteristic of data packets and some command functions. There are many API command functions that relate to operating robot, such as Scan, PCube-syncModule, PCube-haltModule, PCube-moveCurrent, PCube-moveVel, and PCube-moveRamp and so on. Module ID is the identifier of module that command can apply to. Length is the length of this data packet. Parameter 1, 2, and 3 are some parameters related to certain commands indicated in Indication segment.

Table 1. The format of command packet

Indication
Module ID
Length
Para1
Para2
Para3

State packets divide into static packet and dynamic packet. Parameters in static packet are constant, such as numbers of robot modules and all parameters solidified in modules. For server, static packet is sent only when server begin to detect. Parameters in dynamic packet are variation, such as velocity and position and so on. The information of dynamic packet is only relates to robot motion and it contains few byte codes. So sending such information is more favorable than sending images in remote communication. In order to control the time of sending dynamic packet reasonably, we set threshold value at server. The feedback information is send to client when parameter value is higher than threshold value.

3 Remote Control System

Robot remote control system based on Internet can operate robot at remote place using existing network technology [7-9]. This system can accomplish some special functions. First, manipulator sends control command at client to server via Internet, then robot control system at server controls actual robot to perform corresponding operation and collects robot states at the same time. Secondly, the data information of robot motion states is transmitted to client and accomplishes real time dynamic simulation. So the effect of control is related to not only physics structure of robot but also control theory, even influenced by many problems of network, such as great data stream, low transport speed, delay and losing data packet caused by network overload and so on. Robot remote simulation and control system must have high real time capability and reliability. It is a crucial problem that how to send and receive data packets according to robot remote communication protocol between client and server. The remote control will be performed when manipulator at client sends command packets to server and could receive state packets accurately from server. Moreover, dynamic simulation is necessary for manipulator to know the control effect and get valuable information.

In order to accomplish robot remote operations, application software of client must be cooperating with that of server. The application software of client consists of data communication and graphics simulation. The functions of robot simulation and control are accomplished by their cooperation. Since server sends instructions to operate robot directly, its application is more complex than that of client. The function of real time control is added to perform the tasks of motion control and state monitoring. Both data communications of client and server must be abidance by same communication protocol to accomplish sending and receiving data, though they have different tasks and functions. Graphics simulation of client and server are same. Robot real time control of server is accomplished by API function provided by manufacturer. Figure 1 gives the principle of robot remote simulation and control system.

First, client connects to server by Internet and Send scan command to it to scan the numbers and states of robot modules on CAN bus. After checking, server packs static parameters and sends them to client according to robot remote communication protocol. Then client send control command to control the position, velocity and electric current of all robot joints. When server receives instructions from client, corresponding operations of robot are performed by API functions. And the motion

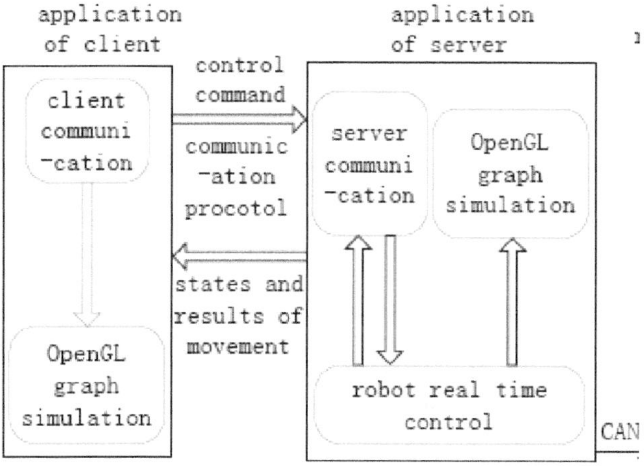

Fig. 1. Principle of remote simulation and control system

states and parameters of robot are detected in robot motion process. The feedback information is sent to client according to certain format stated in robot remote communication protocol. When receiving this information, client performs 3D simulation and displays the current state of robot. Thus manipulator can know the real state of remote robot in time and control it to perform operations accurately.

4 Conclusions

In this paper, internet architecture for robot remote control is proposed in order to have high availability, performance, and scalability. The robot remote communication protocol proposed in this paper could perform data transmission accurately in client-server system. The proposed architecture is comparatively concise and reduces the operation time because fewer data are transmitted on network.

Acknowledgment

The authors gratefully acknowledge supports from Beijing Undergraduate Research and Entrepreneurship Action Plan (BEIJ 1018), Funding Project for Academic Human Resources Development in Institutions of Higher Learning Under the Jurisdiction of Beijing Municipality and Research Foundation of North China University of Technology.

References

1. Jill, D.C., Jon, A.W.: The Warp Machine on Navlab. IEEE Transactions on Pattern Analysis and Machine Intelligence 13(5), 451–465 (1991)
2. Ricardo, G., Luo, R.C.: LOLA Probabilistic Navigation for Topological Maps. AI Magazine, 55–62 (1996)

3. Richard, L., Luo, R.C.: LOLA Object Manipulation in an Unstructured Environment. AI Magazine, 63–70 (1996)
4. Goldberg, K., Maschna, M., Gentner S., et al.: Desktop Tele operation Via the WWW. In: Proc. of the IEEE International Conf. on Robotics and Automation, pp. 654–659 (1995)
5. Goldberg, K., et al.: The mercury project: a feasibility study for Internet robots. IEEE Robotics & Automation Magazine 7(1), 35–40 (2000)
6. Taylor, K.: Australia's telerobot on the web. In: Proc. of 26th International Conference on Industrial Robotics, pp. 1235–1242 (2006)
7. Li, X.: Key technology and typical realization of robot remote control based on WWW. Industrial Control Computer 13(2), 51–53 (2000)
8. Nishimura, T., Takeno, J.: Remote-controlled humanoid robot system, robot and human interactive communication. In: Proceedings of 10th IEEE International Workshop on, pp. 551–555. IEEE Press, New York (2001)
9. Zhao, P.: Research of key technology of remote robot monitor system. Mini-micro Computer System 21(12), 1261–1263 (2000)

An Efficient Algorithm for an Industrial Robot

X.G. Huang

School of Mechanical and Electrical Engineering,
North China University of Technology, Beijing, China
march08@126.com

Abstract. The inverse kinematics problem of robot manipulators is to be solved analytically in order to have complete and simple solutions to the problem. In this paper, an efficient algorithm to compute all the closed-form inverse kinematics solutions of an industrial robot is proposed. Based on the method, we obtain all the closed-form solutions for the inverse kinematics of the robot.

Keywords: Industrial Robot, Algorithm, Closed-form solutions.

1 Introduction

An industrial robot is composed of a serial chain of several rigid links connected in series by revolute or prismatic joints driven by actuators. One end of the chain is attached to a supporting base while the other end is free and attached with a tool to manipulate objects or perform assembly tasks. The motion of the joints results in relative motion of the links. Since the robot servo system requires the reference inputs to be in joint coordinates and a task is generally stated in terms of the Cartesian coordinate system, controlling the position and orientation of the end effector of a robot arm to reach its object requires understanding of the kinematic relationship between these two coordinate systems.

The inverse kinematic control of an industry robot manipulator requires the transformation of end effector Cartesian task space coordinates into corresponding joint configuration space coordinates. The most common method is Denavit and Hartenberg notation for definition of special mechanism [1]. This method is based on point transformation approach and it is used 4×4 homogeneous transformation matrix which is introduced by Maxwell [2]. Maxwell used homogeneous coordinate systems to represent points and homogeneous transformation matrices to represent the transformation of points. The coordinate systems are described with respect to previous one. For the base point an arbitrary base coordinate system is used. Hence some singularity problems may occur because of this coordinate systems description. And also in this method 16 parameters are used to represent the transformation of rigid body while just 6 parameters are needed.

In this paper, an efficient algorithm procedure is presented to solve the inverse kinematics problem of an industry robot. Based on the method, we obtain all the closed-form solutions for the inverse kinematics of the robot.

L. Qi (Ed.): ISIA 2010, CCIS 86, pp. 519–523, 2011.

2 Homogeneous Transformations

As we all known, the 3x3 rotation matrix and the 3x1 translation vector can be assembled into the compact form of a 4x4 homogeneous transform. For instance a screw displacement along of the z-axis for rotation θ and slide s can be written as:

$$[Z(\theta,s)]=\begin{bmatrix}\cos\theta & -\sin\theta & 0 & 0\\ \sin\theta & \cos\theta & 0 & 0\\ 0 & 0 & 1 & s\\ 0 & 0 & 0 & 1\end{bmatrix} \tag{1}$$

According to [3], the matrix (1) can be expressed the product of two matrices as

$$[Z(\theta,\gamma)]=[G(\theta,\gamma)][H(\theta,\gamma)]^T \tag{2}$$

Where matrices $G(\theta,\gamma)$ and $H(\theta,\gamma)$ are

$$\begin{cases}\boldsymbol{G}(\theta,\gamma)=[\cos\frac{\theta+\gamma_i}{2},0,0,\sin\frac{\theta+\gamma_i}{2}]\\ \boldsymbol{H}(\theta,\gamma)=[\cos\frac{\theta-\gamma_i}{2},0,0,\sin\frac{\theta-\gamma_i}{2}]\end{cases} \tag{3}$$

According to [4], a screw displacement along z-axis can be denoted by

$$\tilde{Z}(\theta,\gamma)=\xi G(\theta,\gamma)+\eta H(\theta,\gamma) \tag{4}$$

Similarly a screw displacement along x-axis can be denoted as

$$\tilde{X}(\alpha,a)=\xi G(\alpha,a)+\eta H(\alpha,a) \tag{5}$$

where $G(\alpha,a)$ and $H(\alpha,a)$ can be written as:

$$\begin{cases}G(\alpha,\rho)=(\sin\frac{\alpha+\rho}{2},0,0,\cos\frac{\alpha+\rho}{2})\\ H(\alpha,\rho)=(\sin\frac{\alpha-\rho}{2},0,0,\cos\frac{\alpha-\rho}{2})\end{cases} \tag{6}$$

The rotation angle is calculated as r= a/R and $R=L/\delta_1/2$ [5].

3 Inverse Kinematics Using Double Quaternion

3.1 Inverse Kinematics Equations

Figure 1 shows an industry robot, the Denavit-Hartenberg convention allows us to write the kinematics equations in the standard form:

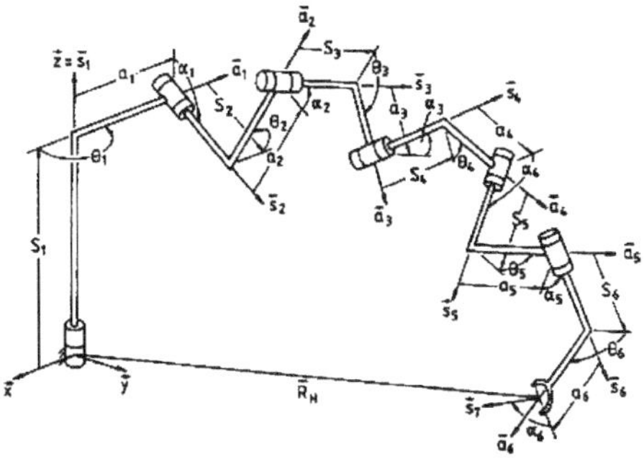

Fig. 1. An industry robot

$$[Z(\theta_1,s_1)][X(\alpha_1,a_1)][Z(\theta_2,s_2)][X(\alpha_2,a_2)]\cdots[Z(\theta_6,s_6)]=[M] \quad (7)$$

Converting each of these homogeneous matrices in (7) into the following equations:

$$\tilde{Z}(\theta_1,s_1)\tilde{X}(\alpha_1,a_1)\tilde{Z}(\theta_2,s_2)\tilde{X}(\alpha_2,a_2)\cdots\tilde{Z}(\theta_6,s_6)=\tilde{G} \quad (8)$$

Equation (8) can be written as

$$G(\theta_1,s_1)G(\alpha_1,a_1)G(\theta_2,s_2)G(\alpha_2,a_2)\cdots G(\theta_6,s_6)=G \quad (9)$$

$$H(\theta_1,s_1)H(\alpha_1,a_1)H(\theta_2,s_2)H(\alpha_2,a_2)\cdots H(\theta_6,s_6)=H \quad (10)$$

3.2 The Elimination Process

First moving terms associated with joint angles $\theta_4,\theta_5,\theta_6$ to the left side of (9) and (10) yields

$$G_{\theta1}G_{\alpha1}G_{\theta2}G_{\alpha2}G_{\theta3}G_{\alpha3}G_{\theta4}=GG^*_{\alpha6}G^*_{\theta6}G^*_{\alpha5}G^*_{\theta5}G^*_{\alpha4} \quad (11)$$

$$H_{\theta1}H_{\alpha1}H_{\theta2}H_{\alpha2}H_{\theta3}H_{\alpha3}H_{\theta4}=HH^*_{\alpha6}H^*_{\theta6}H^*_{\alpha5}H^*_{\theta5}H^*_{\alpha4} \quad (12)$$

Here $G_{\theta i}=G(\theta_i,s_i)$, $G_{\alpha i}=G(\alpha_i,a_i)$, $H_{\theta i}=H(\theta_i,s_i)$, $H_{\alpha i}=H(\alpha_i,a_i)$, and symbol "*" represents the quaternion conjugate.

Substituting (3) into (11) and (6) into (12) respectively, we get

$$\boldsymbol{E}_i\boldsymbol{M}^{\mathrm{T}} = \boldsymbol{D}_i[c\theta_1 c\theta_6, c\theta_1 s\theta_6, s\theta_1 c\theta_6, s\theta_1 s\theta_6]^{\mathrm{T}}, (i=1,2) \tag{13}$$

where $\boldsymbol{E}_i$ denotes 4×14 matrices, $\boldsymbol{D}_i$ denotes 4×4matrices, and $\boldsymbol{M}$ denotes 1×16 matrices as followings.

$$\boldsymbol{M} = [c\theta_2 c\theta_3 c\theta_4 c\theta_5, \cdots, c\theta_2 s\theta_2{}^{j_2-1} c\theta_3 s\theta_3{}^{j_3-1} c\theta_4, s\theta_4{}^{j_4-1} c\theta_5 s\theta_5{}^{j_5-1}, \cdots, s\theta_2 s\theta_3 s\theta_4 s\theta_5]$$
$$,(j_2, j_3, j_4, j_5 = 0,1)$$

where $c\theta_k = \cos(\theta_k/2), s\theta_k = \sin(\theta_k/2)$,k=1,2,3,4,5,6.

Solving straightforwardly the system (13) (when i=1) symbolically regarding $\cos(\theta_1/2)\cos(\theta_6/2)$, $\sin(\theta_1/2)\cos(\theta_6/2)$, $\cos(\theta_1/2)\cos(\theta_6/2)$,and $\sin(\theta_1/2)\sin(\theta_6/2)$ as linear unknowns, we obtain

$$c\theta_1 c\theta_6 = p_1(\theta_2, \theta_3, \theta_4, \theta_5) \tag{14}$$

$$c\theta_1 s\theta_6 = p_2(\theta_2, \theta_3, \theta_4, \theta_5) \tag{15}$$

$$s\theta_1 c\theta_6 = p_3(\theta_2, \theta_3, \theta_4, \theta_5) \tag{16}$$

$$s\theta_1 s\theta_6 = p_4(\theta_2, \theta_3, \theta_4, \theta_5) \tag{17}$$

Equation (13) (when i=2) can be expressed as

$$f_j(t_2, t_3, t_4, t_5) = 0, \qquad j=1\sim4 \tag{18}$$

According to [15][16], we can obtain

$$(\alpha^2\beta, \alpha\beta, \beta, \alpha^2, \alpha, 1) D_{6\times6}(t_5{}^2 t_4, t_4 t_5, t_4, t_5{}^2, t_5, 1)^T = 0 \tag{19}$$

where $D_{6\times6}$ is a 6x6 Matrix.

Equation (19) has a solution, if and only if

$$\det(D_{6\times6}) = 0 \tag{20}$$

The condition (20) can yield the following univariate polynomial

$$\sum_{i=-8}^{8} s_i t_2{}^i = 0 \tag{21}$$

where s_i $(i = 0, \ldots, 16)$ are real constants depending on input data only. Solving (21), we obtain 16 roots t_{2i} $(i = 1\ldots 16)$ for t_2 in the complex domain.

3.3 Back Substitution for Other Unknowns

Substitute t_{2i} into the following equation

$$D_{6\times6}(t_5^{\ 2}t_4, t_4t_5, t_4, t_5^{\ 2}, t_5, 1)^T = 0 \tag{22}$$

all solutions of t_4 and t_5 can be easily computed in the complex domain. For one solution of t_2 there will be one solution of t_4 and t_5.

Substituting t_2, t_4, t_5 into (18) solutions of t_3 can be gained easily. By the Euler equations, we can get values of θ_2, θ_3, θ_4, and θ_5, By (14)~(17), we can get the solutions of θ_1 , and θ_6 .

4 Conclusions

This paper introduces an efficient algorithm for the inverse kinematics of an industry robot. Compared with traditional homogenous matrix method, the procedure is more efficient. However it should be noted that a controllable error may be introduced to the solution process via conversion.

References

1. Pieper, D., Roth, B.: The kinematics of manipulators under computer control. In: Proc. of the 2nd International Congress on Theory of Machines and Mechanisms, pp. 159–169 (1969)
2. Maxwell, E.A.: General Homogeneous Coordinates in Space of Three Dimensions. Cambridge Univ. Press, Cambridge (1951)
3. Liao, Q.Z., Liang, C.G., Zhang, Q.X.: A Novel Approach to the Displacement Analysis of General Spatial 7R Mechanism. Chinese Journal of Mechanical Engineering 22(3), 1–9 (1986)
4. McCarthy, J.M.: An Introduction to Theoretical Kinematics. MIT Press, Cambridge (1990)

An Approach for Direct Kinematics of a Parallel Manipulator Robot

X.G. Huang

School of Mechanical and Electrical Engineering,
North China University of Technology, Beijing, China
march08@126.com

Abstract. Parallel platform robot's unique structure presents an interesting problem in its direct kinematics solution. It involves the solving of a series of simultaneous non-linear equation and, usually, non-unique, multiple sets of solutions are obtained from one set of data. In this paper, we present a concise approach for solving the direct kinematics problem of a parallel manipulator robot. The proposed method is comparatively concise and reduces the computational burden.

Keywords: Robot, parallel manipulator, univariate polynomial.

1 Introduction

In the last several years, there has been an increasing interest in applications of parallel manipulator in developing robotic devices. The parallel manipulator has been found to be suitable for situations where the requirements for accuracy and sturdiness outweigh those on workspace and maneuverability [1]. Parallel manipulators generally possess high positioning capability provided by the high structural rigidity and no serial accumulation of actuator errors. Furthermore, they also have higher strength-to-weight ratios as compared to conventional open-kinematic chain manipulators because payload is proportionally distributed to their links.

Although mechanical simplicity of parallel platforms provides potential for many engineering applications, their direct kinematics is very complex which limit their real-time applications. This is due to the requirements of solving a set of highly nonlinear equations or high degree polynomials for a general Stewart Platform. Broadly speaking, there are two approaches to the solution of the forward kinematics of the platform, namely analytical and numerical. In the first approach attempts are made to develop closed form solutions in special cases where some of the connection points at the platform or at the base are coalesced in groups of two or three. In this way the general 6-6 configuration, i.e. 6 separate joints at base and platform, is reduced to less than six at the base and/or the platform [2]-[4]. Another approach in the analytical category is to place restriction on the geometry of the platform or the base, or assume a certain relationship (e.g. linear) between the base and platform coordinates at attachment points [5]-[7].

L. Qi (Ed.): ISIA 2010, CCIS 86, pp. 524–528, 2011.

In this paper, we present a novel and relatively concise algebraic approach for solving the direct kinematics of a parallel manipulator robot. Based on the presented algebraic method, we derive a univariate polynomial. The proposed method is comparatively concise and reduces the computational burden.

2 Basic Equations for Direct Position Analysis

With reference to Fig. 1, which represents the GSM structure here considered, it can be observed that the position and orientation of the mobile platform are completely determined by the lengths of the 6 limbs L_i(i=1,…,6. In this paper, unless denoted specially, index i ranges from 1 to 6), i.e., the distance A_iB_i. Cartesian coordinate system $O_2x_2y_2z_2$ is fixed to the mobile platform, whereas $O_1x_1y_1z_1$ is fixed to the base platform. Let the coordinates of point A_i is $\boldsymbol{Q}_{ia}$ (a_{xi}, a_{yi}, a_{zi}) in $O_1x_1y_1z_1$, the coordinates of point B_i is $\boldsymbol{Q}_{ib}$ (b_{xi}, b_{yi}, b_{zi}) in $O_2x_2y_2z_2$, respectively.

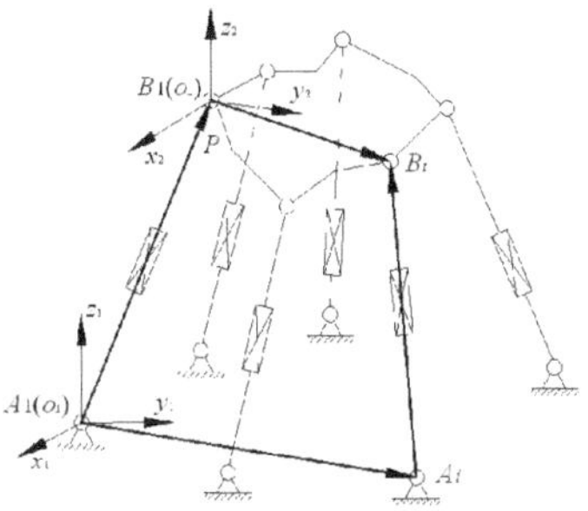

Fig. 1. The Generalize Stewart Manipulator

The DPA of the GSM can be stated as follows: Given L_i, $\boldsymbol{Q}_{ia}$ (a_{xi}, a_{yi}, a_{zi}) and $\boldsymbol{Q}_{ib}$ (b_{xi}, b_{yi}, b_{zi}), finding all possible values of the position vector $\boldsymbol{P}$ (x, y, z) between the two origin points $\boldsymbol{O}_1$ and $\boldsymbol{O}_2$, and the transformation matrix $\boldsymbol{R}$ between the two coordinate systems, which satisfy the kinematics constraints of the GSM.

The lengths of each limb can be calculated as follows:

$$L_i^{\,2} = [\boldsymbol{R}\boldsymbol{Q}_{ib} + \boldsymbol{P} - \boldsymbol{Q}_{ia}]^{\mathrm{T}}[\boldsymbol{R}\boldsymbol{Q}_{ib} + \boldsymbol{P} - \boldsymbol{Q}_{ia}] \qquad i=2,3,4,5,6. \tag{1}$$

$$L_i^2 = \boldsymbol{P}^{\mathrm{T}}\boldsymbol{P} \qquad i=1 \tag{2}$$

Substituting (2) into (1) and expanding, yields

$$\boldsymbol{Q}_{ib}^{\mathrm{T}}\boldsymbol{Q}_{ib} + L_1^2 - L_i^2 + \boldsymbol{Q}_{ia}^{\mathrm{T}}\boldsymbol{Q}_{ia} + 2\boldsymbol{P}^{\mathrm{T}}\boldsymbol{R}\boldsymbol{Q}_{ib} - 2\boldsymbol{Q}_{ia}^{\mathrm{T}}\boldsymbol{P} - 2\boldsymbol{Q}_{ib}^{\mathrm{T}}\boldsymbol{R}\boldsymbol{Q}_{ia} = 0 \qquad i=2,3,4,5,6 \tag{3}$$

In the three-dimensional coordinate system, a rigid body rotates an angle θ about an axis $\boldsymbol{K}$ (k_x, k_y, k_z), the rotation matrix $\boldsymbol{R}$ can be given as follow [7]:

$$\boldsymbol{R}_k(\theta)=\begin{bmatrix} k_x^2 v\theta+c\theta & k_x k_y v\theta-k_z s\theta & k_x k_z v\theta+k_y s\theta \\ k_x k_y v\theta+k_z s\theta & k_y^2 v\theta+c\theta & k_y k_z v\theta-k_x s\theta \\ k_x k_z v\theta-k_y s\theta & k_y k_z v\theta+k_x s\theta & k_z^2 v\theta+c\theta \end{bmatrix} \tag{4}$$

Where $c\theta = \cos\theta$, $s\theta = \sin\theta$, $v\theta = 1-\cos\theta$

Set

$$\begin{cases} c_1 = k_x tan(\theta/2) \\ c_2 = k_y tan(\theta/2) \\ c_3 = k_z tan(\theta/2) \\ \cos(\theta) = (1-\tan^2(\theta/2))/(1+\tan^2(\theta/2)) \\ \sin(\theta) = 2\tan(\theta/2)/(1+\tan^2(\theta/2)) \end{cases} \tag{5}$$

Substitute (5) into (4), we can obtain:

$$\boldsymbol{R}=\Delta^{-1}\begin{bmatrix} 1+c_1^2-c_2^2-c_3^2 & 2(c_1c_2-c_3) & 2(c_1c_3+c_2) \\ 2(c_1c_2+c_3) & 1-c_1^2+c_2^2-c_3^2 & 2(c_2c_3-c_1) \\ 2(c_1c_3-c_2) & 2(c_2c_3+c_1) & 1-c_1^2-c_2^2+c_3^2 \end{bmatrix} \tag{6}$$

Where $\Delta = 1+c_1^2+c_2^2+c_3^2$

Substitute (6) and $\boldsymbol{P}$ (x, y, z) into (3) we can get

$$f_i(c_1,c_2,c_3,x,y,z)=0 \qquad i=2,3,4,5,6 \tag{7}$$

Eq. 2 and (7), which represent 6 equations in 6 unknowns c_1, c_2, c_3, x, y and z, are the basic equations describing the DPA of the GSM.

3 An Algebraic Method

Eq. 2 and Eq. 7 form a polynomial system of 6 equations in the 6 unknowns. Since Eq. 2 is second-degree and Eq. 7 all are third-degree, the total degree (product of the degrees of the polynomials) of the basic equations equals $2\times3^5 = 486$. It is currently difficult and formidable to obtain the solutions using the traditional algebraic methods such as Dixon resultant method, Sylvester resultant method and Wu's method. Instead, we will use two step eliminating method to solve the problem. Firstly, we reduce the Groebner basis under lexicographic ordering of Eq. 2 and Eq. 7. Then, after analyzing the variants of the obtained basis and the terms of the different variants degrees, we construct Sylvester resultant by hiding proper variant such that the number of polynomials is equal to the number of monomials. At the same time, to reduce the time of symbol computing, we utilize the related principles in Linear Algebra to analysis the Sylvester resultant, minimize the degrees of the Sylvester resultant and deduce a 40th univariate equation.

3.1 The Groebner Basis under Lexicographic Ordering

The Groebner package available in Methematica-5.2 will be utilized to yield a reduced Groebner basis of Eq. 2 and Eq. 7 under the degree lexicographic term ordering with $x > y > z > c_1 > c_2 > c_3$. The reduced Groebner basis with 51 polynomials in 91 unknown monomials are not reported herein doe to space limitations.

Suppressing the unknown c_3, the obtained Groebner basis under degree lexicographic ordering can be viewed as a linear system of 51 equations in 51 unknown monomials and be expressed in the form of matrix

$$\boldsymbol{M}'_{51\times51}\boldsymbol{t} = 0 \tag{8}$$

where, $\boldsymbol{M}'_{51\times51}$ is the 51×51 coefficient matrix, $\boldsymbol{t}$ is the 51×1 column matrix of the unknown variables (the 51 monomials).

$\boldsymbol{t} = [1, c_1, c_1^2, c_1^3, c_2, c_1c_2, c_1^2c_2, c_2^2, c_1c_2^2, c_2^3, x, c_1x, c_1^2x, c_2x, c_1c_2x, c_2^2x, x^2, y, c_1y, c_1^2y, c_2y, c_1c_3y, c_1^2y, xy, c_1xy, c_2xy, y^2, c_1y^2, c_2y^2, xy^2, y^3, z, c_1z, c_1^2z, c_2z, c_1 c_2z, c_2^2z, xz, c_1xz, c_2xz, yz, c_1yz, c_2yz, xyz, y^2z, z^2, c_1z^2, c_2z^2, xz^2, yz^2, z^3]^T$

According to the Algebra, Eq. 8 has the solutions under the condition that the determinant of the matrix equate zero, i.e.

$$\det(\boldsymbol{M}'_{51\times51}) = 0 \tag{9}$$

Solving Eq. 9 will cost a large quantity of time and space of computer. So it is necessary to reduce the degree of the coefficient matrix to improve the efficiency.

3.2 Reducing the Degree of Resultant

For there is only one element in some columns of the matrix $\boldsymbol{M}'_{51\times51}$ is the constant (e.g. j^{th} column) and others are zero, i.e. only the ith equation has relation with j^{th} unknown variable, so removing the ith equation and decreasing an unknown variable will not affect solving the equations.

There remain 20 polynomial equations after similar removing above, given as g_i for i= 2, 3, 4, 5, 6, 7, 8, 39, 40, 41, 42, 43, 44, 45, 46, 47, 48, 49, 50, 51. With the unknown c_3, the 20 polynomial equations can be viewed as a linear system of 20 polynomials in 20 unknowns $1, c_1, c_1^2, c_2, c_1c_2, c_2^2, x, c_1x, c_2x, y, c_1y, c_2y, xy, y^2, z, c_1z, c_2z, xz, yz$, and z^2, with their coefficients expressed in terms of c_3.

$$\boldsymbol{M}_{20\times20}\boldsymbol{T} = 0 \tag{10}$$

where, $\boldsymbol{M}_{20\times20}$ is the 20×20 coefficient matrix of c_3, and $\boldsymbol{T}$ is the 20×1 column matrix in 20 unknown variables (the 20 monomials).

$\boldsymbol{T} = [1, c_1, c_1^2, c_2, c_1c_2, c_2^2, x, c_1x, c_2x, y, c_1y, c_2y, xy, y^2, z, c_1z, c_2z, xz, yz, z^2]^T$

3.3 Deriving a Univariate Ploynomial

According to Algebra, the condition of the Eq. 10 having solutions is the determinant of the coefficient is zero, i.e.

$$\det(\boldsymbol{M}_{20\times 20}) = 0 \tag{11}$$

Expanding directly the Eq. 11, we can obtain the following 40th degree univariate polynomial without factoring out or deriving any common divisor.

$$\sum_{i=0}^{40} s_i c_3^{\ i} = 0 \tag{12}$$

where, s_i (i=0,…,36) are real constants depending on input parameters only.

According to the analysis above, the maximum degree of the univariate coefficient determinant and the degree of the univariate polynomial derived from expanding the determinant is completely identical symbolically. 40 complex solutions will be obtained by solving Eq. 12.

3.4 Back Substitution for Other Unknowns

Solving the linear system, which is obtained by removing any one row from the matrix $\boldsymbol{M}_{20\times 20}$ in Eq. 11 with c_3 replaced by c_{3i} (i = 1, …, 40), solutions of c_1, c_2, x, y, and z can be easily computed in the complex domain.

4 Conclusion

This paper presents a novel and relatively concise algebraic method to solve the direct kinematics analysis of a parallel manipulator robot. Based on the presented algebraic method, we derive a univariate polynomial without factoring out or deriving the greatest common divisor. We also obtain a complete set of 40 solutions. Since the proposed method requires determinant calculation of smaller square matrices, the computational burden and computation time is greatly reduced comparing to the existing methods.

References

1. Dasgupta, B., Mruthyunjaya, T.S.: The Stewart Platform manipulators. Mechanism and Machine Theory 35(1), 15–40 (2000)
2. Griffis, M., Duffy, J.: A forward displacement analysis of a class of Stewart platforms. J. Robot. Syst. 6(6), 703–720 (1989)
3. Innocenti, C., Parenti-Castelli, V.: Direct position analysis of the Stewart platform mechanism. Mech. Mach. Theory 26(6), 611–621 (1990)
4. Tsai, M.S., Shiau, T.N., Tsai, Y.J.: Direct kinematic analysis of a 3-PRS parallel mechanism. Mech. Mach. Theory (38), 71–83 (2003)
5. Sreenivasan, S.V., Waldron, K.J., Nanua, P.: Closed form direct displacement analysis of a class of a 6-6 platform. Mechanisms and Machine Theory 29(6), 855–868 (1994)
6. Yang, J., Geng, Z.J.: Closed form forward kinematics solution of a class of hexapod robots. IEEE Trans. Robot. Automat. 14, 503–508 (1998)
7. Ji, P., Wu, H.: A closed form forward kinematics solution for 6-6P Stewart platform. IEEE Trans. Robotics and Automation 17(4), 522–526 (2001)

Forward Kinematics for a Parallel Platform Robot

X.G. Huang

School of Mechanical and Electrical Engineering,
North China University of Technology, Beijing, China
huangxiguang@gmail.com

Abstract. This paper proposes a method to calculate the forward kinematics of parallel link manipulators. We calculate the forward kinematics numerically. The method is simple and can make the computer program easy to write.

Keywords: Robot, Forward Kinematics, Non-singular solutions.

1 Introduction

Parallel platform is a spatial mechanism that consists of a stationary base platform connected to a moving platform by six ball-joint-ended extensible legs. In the past decades, parallel platforms have received a great deal of attention from many researchers due to their inherent advantages over the conventional serial mechanism, such as, simpler structure, higher stiffness, better accuracy, heavier loading ability.

As parallel platforms are applied in wider fields, related studies such as kinematics issues including direct kinematics problem and inverse kinematics problem have been hotly preceded for decades. The direct kinematics problem, which is to determine the positions and orientations of the moving platform given the lengths of the six legs, is more difficult. This is in contrast to serial chain manipulators where the opposite is true. The motion planning and control of a parallel platform robot calls for the solution of the direct kinematics problem, which is a basic and challenging problem as well.

A significant number of researchers have been involved in this problem. Lazard, Ronga and Mourrain have proven that the general 6-6 hexapod FKP has 40 complex solutions using Greobner bases [1], Chern classes of vector bundles [2] and explicit elimination techniques [3]. Wen-Liang [4] and Zhang-Song [5] gave the closed-form solutions for the SP with planar base and platforms. For the general case, Lee [6] derived a set of 6th-dgree polynomial equations which lead to a 40th-degree univariate equation.

In this paper, we propose a method to calculate the forward kinematics of parallel link manipulators. We calculate the forward kinematics numerically. The method is simple and can make the computer program easy to write.

2 Forward Kinematics

2.1 Kinematic Model

Fig. 1 shows the geometric model of a 4SPS-2CCS parallel platform robot. The absolute local frame system $\boldsymbol{O}_1$-$\boldsymbol{X}_1\boldsymbol{Y}_1\boldsymbol{Z}_1$ is fixed to the point $\boldsymbol{A}_1$ on the base platform

L. Qi (Ed.): ISIA 2010, CCIS 86, pp. 529–532, 2011.

and the relative moving frame system $\boldsymbol{O}_2$-$\boldsymbol{X}_2\boldsymbol{Y}_2\boldsymbol{Z}_2$ fixed to the point $\boldsymbol{B}_1$ on the moving platform. The direct kinematics problem is to find the position and orientation of the moving platform supposing that the pose of the base platform is known and values for the six constraints connecting to the base and the platform are given.

Let $\boldsymbol{A}_i$ denote the coordinates of point A_i in the absolute frame $\boldsymbol{O}_1$-$\boldsymbol{X}_1\boldsymbol{Y}_1\boldsymbol{Z}_1$ and $\boldsymbol{B}_i$ denote the coordinates of point B_i in the moving frame $\boldsymbol{O}_2$-$\boldsymbol{X}_2\boldsymbol{Y}_2\boldsymbol{Z}_2$. The lengths of $\boldsymbol{A}_i\boldsymbol{B}_i$ are denoted as L_i. After performing a rotation $\boldsymbol{R}$ and a translation $\boldsymbol{P}$ to the moving platform, the image of points $\boldsymbol{B}_i$ is $\boldsymbol{RB}_i$+$\boldsymbol{P}$.

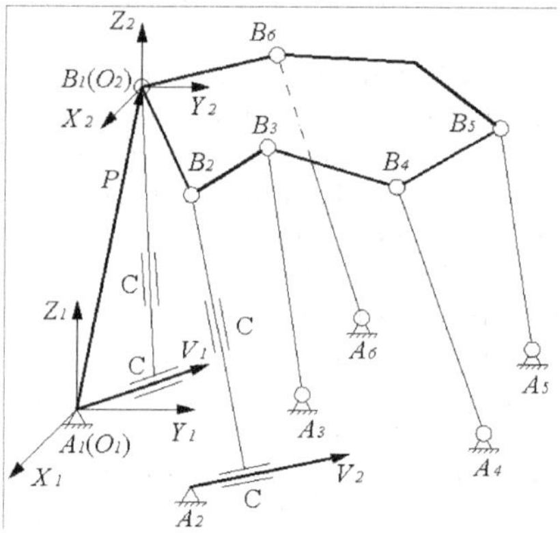

Fig. 1. The geometric model of a 4SPS-2CCS parallel platform robot

For four SPS legs, with given leg lengths, the constraint equations corresponding to the conditions of constant length of each leg are as follows:

$$L_i^{\ 2} = (\boldsymbol{P} + \boldsymbol{RB}_i - \boldsymbol{A}_i)^T (\boldsymbol{P} + \boldsymbol{RB}_i - \boldsymbol{A}_i) \quad i=3, 4, 5, 6 \tag{1}$$

For two CCS legs, with given leg lengths, the constraint equations corresponding to the conditions of constant length of each leg are as follows:

$$L_j^{\ 2} = [\boldsymbol{P}+\boldsymbol{RB}_j-(\boldsymbol{A}_j+x_j\boldsymbol{V}_j)]^T[\boldsymbol{P}+\boldsymbol{RB}_j-(\boldsymbol{A}_j+x_j\boldsymbol{V}_j)] \quad j=1, 2 \tag{2}$$

Since two axes of cylindrical joints are perpendicular to each other, the following condition exists:

$$0 = [\,\boldsymbol{P} + \boldsymbol{RB}_j - (\boldsymbol{A}_j + x_j\boldsymbol{V}_j)] \cdot \boldsymbol{V}_j \quad j=1, 2 \tag{3}$$

In the representation, $\boldsymbol{V}_i$ denotes the unit vectors of the axes of cylindrical joints on the base and x_i denotes the perpendicular distance from $\boldsymbol{A}_i$ to the axes of cylindrical joints.

For (6), (7), (8), we can obtain

$$(\boldsymbol{g}*\boldsymbol{q}'+\boldsymbol{q}*\boldsymbol{B}_i*\boldsymbol{q}'-\boldsymbol{A}_i)*(\boldsymbol{g}*\boldsymbol{q}'+\boldsymbol{q}*\boldsymbol{B}_i*\boldsymbol{q}'-\boldsymbol{A}_i)'=L_i^{\ 2} \quad i=3, 4, 5, 6 \tag{4}$$

$$(\boldsymbol{g}*\boldsymbol{q}'+\boldsymbol{q}*\boldsymbol{B}_j*\boldsymbol{q}'-\boldsymbol{A}_j-x_j\boldsymbol{V}_j)*(\boldsymbol{g}*\boldsymbol{q}'+\boldsymbol{q}*\boldsymbol{B}_j*\boldsymbol{q}'-\boldsymbol{A}_j-x_j\boldsymbol{V}_j)'=L_j^{\ 2} \tag{5}$$

$$[\boldsymbol{g} * \boldsymbol{q}' + \boldsymbol{q} * \boldsymbol{B}_j * \boldsymbol{q}' - (\boldsymbol{A}_j + x_j \boldsymbol{V}_j)] * \boldsymbol{V}_j' = 0 \text{, } j\text{=1,2} \tag{6}$$

(1)-(6) are consider as the forward kinematics equations of the 4SPS-2CCS parallel platform. There are 10 unknown variables g_0, g_1, g_2, g_3, q_0, q_1, q_2, q_3, x_1, x_2 in these 10 equations, the L_i are given as the input variables and the remainders are the structural parameters.

2.2 Equations Solving

Forward kinematics problems of parallel platform lead naturally to the system of nonlinear algebraic and/or transcendental equations which are very difficult to solve. The solution approaches for the Forward kinematics problem equations can be broadly divided into two classes: closed-form solutions and numerical methods. Closed-form solutions are based on analytical expressions and often require massive algebraic manipulations. The procedure of closed-form solutions is very complicated. In the last decade, homotopy continuation method developed into a convenient, reliable tool for solving Forward kinematics problems. Closed-form solutions tend to be faster and to have acceptable accuracy when the number of roots is moderate, but homotopy continuation method tends to be faster and more accurate when the number of roots is larger. Some publicly available software for polynomial continuation is available [7], [8]. In this section, we solve the Forward kinematics problem equations by using homotopy continuation method.

(1)-(3) are constructed by using Euler method. Substituting u=Tan(θ/2), cosθ= $(1-u^2)/(1+u^2)$, cosθ=$2u/(1+u^2)$ into (6)~(8), we get the nonlinear algebraic system. On the ordinary personal computer of Intel Pentium III 2.93 GHz and RAM 256 M, we solve the nonlinear algebraic system by using PHCpack [8] which is a kind of Homotopy continuation method software exploited by Illinois university. We find that the polynomials become increasingly so huge that the memory needed overflow increasingly the memory of the computer.

However, we get all non-singular solutions by using the method presented quickly. Substituting known parameters into (1)-(6), we obtain all non-singular solutions. The solving procedure presented in this paper has been translated into a Mathematica-5.2 computer program.

By this program and PHCpack, the 320 non-singular solutions including 12 real solutions of (14) in terms of values of g_0, g_1, g_2, g_3, q_0, q_1, q_2, q_3, x_1, x_2 are obtained by tracking 1024 solution paths. Because (g, q) and (-g,-q) denote the same rotation and position, the forward kinematics of the 4SPS-2CCS parallel platform have 160 solutions

3 Conclusions

We give the upper bound for the number of solutions of the forward kinematics problem of a parallel platform which has at most 160 non-singular solutions. We also obtain all 160 non-singular solutions of the forward kinematics problem in a numerical example by using homotopy continuation method.

References

1. Ronga, F., Vust, T.: Stewart platforms without computer. In: Proc. Int. Conf. on Real, Analytic and Algebraic Geometry, pp. 197–212 (1992)
2. Lazard, D.: On the representation of rigid-body motions and its application to generalized platform manipulators. Computational Kinematics 16(1), 175–181 (1993)
3. Mourrain, B.: The 40 generic positions of a parallel robot. In: Proc. ISSAC 1993, pp. 173–182 (1993)
4. Wen, F.A., Liang, C.G.: Displacement analysis of the 6-6 stewart platform mechanisms. Mech. Mach. Theory 29(4), 547–557 (1994)
5. Zhang, C., Song, S.M.: Forward position analysis of nearly general Stewart Platforms. ASME J. Mech. Des. 116, 54–60 (1994)
6. Lee, T.Y., Shim, J.K.: Forward kinematics of the general 6-6 Stewart platform using algebraic elimination. Mech. Mach. Theory 36(5), 1073–1085 (2001)
7. Morgan, A.P., Sommese, A.J., Watson, L.T.: Finding all isolated solutions to polynomial systems using HOMPACK. ACM Trans. Math. Software 15, 93–122 (1989)
8. Verschelde, J.: Algorithm: PHCpack: A general purpose solver for polynomial systems by homotopy continuation. ACM Transactions on Mathematical Software 25(2), 251–276 (1999)

Stochastic Bifurcation and Control of the Nonlinear Axial Compression System

Hong Yao[1], Tao Deng[1], Bi-Yue Li[2], and Guang-Jun Zhang[1]

[1] Department of Math. and Phys., Science College, Air Force Engineering University, 710051 Xi'an, ShaanXi, China
yyyyaaohhong@sohu.com
[2] Institute of Software Tech. Northeast University, 110004, Shenyang, Liaoning

Abstract. Stochastic response and control of nonlinear axial compression systems are discussed in the paper. Based on equivalent method of nonlinear differential equation, the paper gives a deep theoretical analysis to random response of nonlinear axial compression systems and builds the nonlinear model of the system, as well as achieves approximate analysis solution of the system's response; and the unstable behaviors—surge and rotating-stall in the system are researched by numerical simulating. It theoretically provides possibility to achieve more reliable stability control of nonlinear axial compression systems.

Keywords: axial compression systems, stochastic response, stability control, forecast.

1 Introduction

According to the theory of nonlinear oscillations, there are many special non-linear phenomena, which are namely intrinsically, appearing in nonlinear systems, such as chattering, self-exited vibration, sub harmonic oscillation, higher mode of vibration, chaos and so on. The linear models in open literature have limited applications, and can't be applied to some particular condition Though many methods [1,2] have been excogitated to forecast stochastic response of nonlinear systems, none can successfully give an approach to the practice satisfactorily, especially in the nonlinear mode of high orders. Few researches of qualitative analysis have been done in nonlinear systems under stochastic disturbance, especially going with intrinsically non-linear phenomena. Therefore, researching stochastic response in nonlinear systems is still an important research direction in the future. Both stall flatter and rotating stall are complicated phenomena whose influence on the air engine can't be ignored under the effect of many factors. Compressor blades would collapse because of the buffeting caused by the sharp conversing flow when the engine runs into the state of stall flatter or rotating stall, and aircraft engine misses, which can seriously threaten the safety state, would happen. According to the analysis above, it is essential to avoid axial compressors run into that to achieve stability control. A uniform nonlinear model describing stall performance of axial compression systems in the state of stall flatter and rotating stall was introduced

L. Qi (Ed.): ISIA 2010, CCIS 86, pp. 533–538, 2011.

by Moore and Greitzer [3] in 1986. The theory mentioned in their report offers an effective approach to the research of the stability of axial compression systems nowadays.

Some scholars have been trying to analyze aerodynamics characteristics using nonlinear dynamics [4] internally and in aboard during recent years. The bunch that most researchers focus on is analyzing the MG mode by using the theory of bifurcation analysis to get the control laws of nonlinear systems. The research about stochastic response is rare, and the reports about its theory haven't yet been found so far. It is significant to study stochastic response and forecast the potential coming forth phenomenon in this system under stochastic disturbance. It is also helpful to ameliorate the design of aerodynamic controllers to achieve stability control.

2 Dynamic Model of Nonlinear Axial Compression Systems [3]

As nonlinear differential equations, Moore-Greitzer' model is used to describe aerodynamic characteristics of nonlinear axial compression systems. After parameter transformation, it can be expressed as follows.

$$\begin{aligned} \dot{\Psi} &= \frac{1}{\beta^2}(\Phi - \Phi_T(\Psi)) \\ \dot{\Phi} &= -\Psi + \psi_c(\Phi) \end{aligned} \tag{1}$$

Where ,

$$\Phi_T(\Psi) = \sqrt{\gamma\Psi} - 1, \quad \gamma > 0 \tag{2}$$

$$\psi_c(\Phi) = c_0 + c_1\Phi + c_2\Phi^3 \tag{3}$$

And, Ψ stands for lift coefficient, $\Phi = (\bar{\phi}/W) - 1$, $\bar{\phi}$ denotes the local average coefficient of flow, W is a numeric constant. And γ is the quantity of engine throttle. β is parameter. Assume that, $c_0 + c_1 + c_2 > 0$ $0 < c_0 < -10c_3$ and $c_3 < 0$. We define $x_1 = \Psi$, $x_2 = \Phi$, Eq.(1) can be written:

$$\begin{aligned} \dot{x}_1 &= \frac{1}{\beta^2}(x_2 - \sqrt{\gamma x_1} + 1) \\ \dot{x}_2 &= -x_1 + c_0 + c_1 x_2 + c_3 x_2^3 \end{aligned} \tag{4}$$

Hence, the axial compression system of aero-engine can be expressed to be nonlinear differential equations with parameters.

3 Equivalent Method of Nonlinear Differential Equations [6]

Equivalent method of nonlinear differential equations is the popularization of equivalent linear method. The central idea is that the given nonlinear system can be replaced by another nonlinear system with a precise stable solution if the difference between them is minimal in the meaning of average or statistic. We suppose the nonlinear damping system be given as follows.

$$\ddot{Y} + b(Y,\dot{Y})\,\mathrm{sgn}(\dot{Y}) + Y = \xi(t) \tag{5}$$

Where, $\xi(t)$ stands for Gansion white noise whose intensity is 2D. Where

$$b(Y,\dot{Y}) = \sum_{i,j}^{N} b_{ij} \left|Y^i \dot{Y}^j\right| \tag{6}$$

Now, the approximately eq. (5) is replaced with a nonlinear damping system depending on energy.

$$\ddot{Y} + f(H)\dot{Y} + Y = \xi(t) \tag{7}$$

And choose,

$$f(H) = \sum_{i,j=1}^{N} C_{ij}(2H)^{(i+j-1)/2} = \sum_{i,j=1}^{N} C_{ij} A^{(i+j-1)} \tag{8}$$

Where,

$$2H = Y^2 + \dot{Y}^2 = A^2 \tag{9}$$

The difference between equivalent numerical value and initial value is calculated.

$$e = f(H)\dot{Y} - b(Y,\dot{Y})\,\mathrm{sgn}(\dot{Y}) \tag{10}$$

In order to achieve the best substitution, we make e yield the term.

$$E[e^2] = E[\{f(H)\dot{Y} - b(Y,\dot{Y})\,\mathrm{sgn}(\dot{Y})\}^2] \to \min \tag{11}$$

Then N^2 equations with certain parameters C_{ij} are decided. After parameter substitution,

$$Y = A\cos\phi\,, \quad \dot{Y} = A\sin\phi \tag{12}$$

We get the explicit expression of C_{ij}. By using eq.(13), we rewrite eq.(11).

$$E[\{f(A^2/2)A\sin\phi - b(A\cos\phi,\, A\sin\phi)\cdot \mathrm{sgn}(\sin\phi)\}A^{i+1}\sin\phi] = 0 \quad i,j = 1,2,\cdots,N \tag{13}$$

There is separation of probability density existing between A and ϕ, and uniform distribution of ϕ lying at the range $[0,2\pi)$. Choose

$$\begin{aligned} C_{ij} &= b_{ij}\int_0^{\pi/2}\cos^i\phi\sin^{j+1}\phi d\phi \Big/ \int_0^{\pi/2}\sin^2\phi d\phi \qquad i,j = 1,2,\cdots,N \\ &= \frac{2}{\pi} b_{ij}\Gamma\{(j+2)/2\}\Gamma(i+1)/2\}/\Gamma\{(i+j+3)/2\} \end{aligned} \tag{14}$$

so that eq. (13) comes into existence. The precise stable solution of eq. (7) is approximately equal to the solution of eq. (5) after obtaining the arithmetic expression of $f(H)$, $f(u)$ and C_{ij} can be known respectively from the arithmetic expression of eq. (8) and eq.(14). The constant number C is got due to the term of normalization.

$$\rho_s(a,\varphi) = C\frac{1}{2\pi}\exp(-\frac{1}{D}\int_0^{a^2/2} f(u)du) \tag{15}$$

4 Stochastic Response of Nonlinear Axial Compression Systems

Considering stochastic axial compression system by Gansion white noise,

$$\begin{aligned} \dot{x}_1 &= \frac{1}{\beta^2}(x_2 - \sqrt{\gamma x_1} + 1) \\ \dot{x}_2 &= -x_1 + c_0 + c_1 x_2 + c_3 x_2^3 + \xi(t) \end{aligned} \tag{16}$$

After contrasted with the following normative expression (16),

$$\ddot{Y} + b(Y,\dot{Y})\operatorname{sgn}(\dot{Y}) + Y = \xi(t) \tag{17}$$

That is,

$$\ddot{x}_1 - c_3\beta^4\dot{x}_1^3 - 3c_3\beta^2\dot{x}_1^2 + 3c_3\gamma\dot{x}_1 x_1 - (c_1 + 3c_3)\dot{x}_1 + f(x_1) = \xi(t) \tag{18}$$

The system can be replaced approximately with a nonlinear damping system depending on energy according to the equivalent method of nonlinear differential equations.

$$\ddot{Y} + f(H)\dot{Y} + Y = \xi(t) \tag{19}$$

According to $b(Y,\dot{Y}) = \sum_{i,j}^{N} b_{ij}\left|Y^i\dot{Y}^j\right|$, $f(H) = \sum_{i,j}^{N} C_{ij}(2H)^{(i+j-1)/2}$, where,

$2H = Y^2 + \dot{Y}^2 = A^2$, $C_{ij} = \frac{2}{\pi}b_{ij}\Gamma\{(j+2)/2\}\Gamma\{(i+1)/2\}/\Gamma\{(i+j+3)/2\}$,

Thus $b_{01} = -3c_3 - c_1$, $b_{02} = -3c_3\beta^2$, $b_{03} = -c_3\beta^4$, $b_{11} = 3c_3\gamma$, the others,

$b_{ij} = 0$; $C_{01} = b_{01}$, $C_{11} = \frac{2}{3\pi}b_{11}$, $C_{02} = \frac{4b_{02}}{3\pi}$, $C_{03} = \frac{3b_{03}}{2}$, the rest $C_{ij} = 0$.

Hence, $f(H) = C_{01} + \sqrt{2}(C_{02} + C_{11})H^{\frac{1}{2}} + 2C_{03}H$. The equivalent nonlinear system depending on energy of the system (18) of can be expressed as follows:

$$\ddot{Y} + (C_{01} + \sqrt{2}(C_{02} + C_{11})H^{\frac{1}{2}} + 2C_{03}H)\dot{Y} + Y = \xi(t) \tag{20}$$

The probability density of eq. (18)'s precise stable solution approximately equals to eq. (16)'s. It is

$$\rho_s(a) = C\frac{1}{2\pi}\exp[(-\frac{1}{D})(\frac{C_{01}}{2}a^2 + \frac{C_{02}+C_{11}}{3}a^3 + \frac{C_{03}}{4}a^4) \quad (21)$$

Where, the constant number C is decided due to the term of Normalization. And

$$\frac{1}{C} = (\frac{D}{2\pi C_{01}} + \frac{(3D)^{\frac{2}{3}}\Gamma(\frac{2}{3})}{6\pi(C_{02}+C_{11})^{2/3}} + \frac{\sqrt{\pi D}}{4\pi C_{03}^{1/2}})$$

Thus, its stable probability density approximately equals to

$$\rho_s(a) = C\frac{1}{2\pi}\exp[(-\frac{1}{D})(\frac{C_{01}}{2}a^2 + \frac{C_{02}+C_{11}}{3}a^3 + \frac{C_{03}}{4}a^4) \quad (22)$$

5 Numerical Simulating

Where $c_1 = 1.5$, $c_3 = -0.5$, the results of numerical simulating of the system are shown as figure 1, figure 2. Thus, the unsteady behaviors are predicted by a deep theoretical analysis to random response of nonlinear axial compression systems, and the stability control is advanced by the different matching parameters.

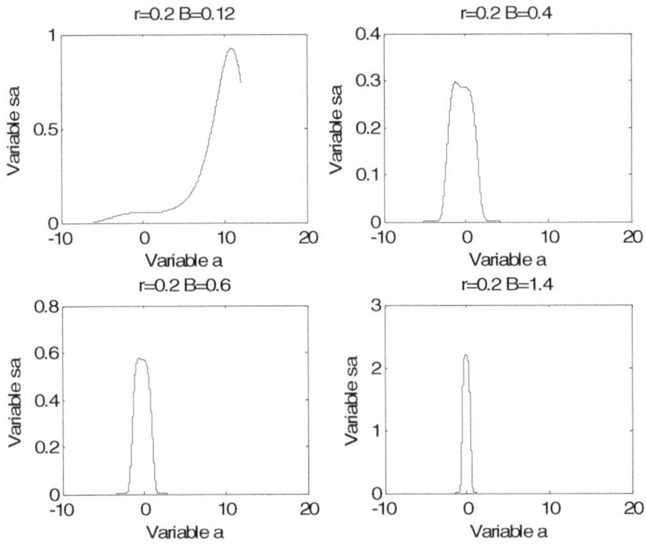

Fig. 1. $\gamma = 0.2$, $B = 0.12, 0.4, 0.6, 1.4$

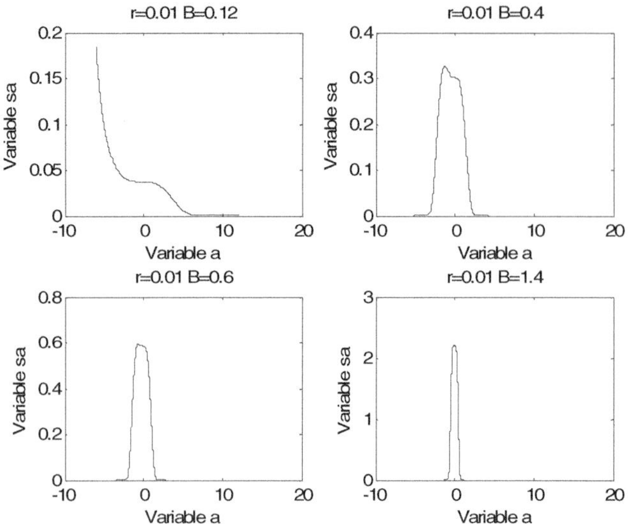

Fig. 2. $\gamma = 0.01$, $B = 0.12, 0.4, 0.6, 1.4$

6 Conclusions

After the above analytical process, we know the possibility of theoretically forecasting the response by using present linear methods directly in intrinsically nonlinear axial compression systems is unpractical because of the complexity and strong nonlinear characteristic of aerodynamic force.

The equivalent method of nonlinear differential equations is used to analyze the system to get the approximate analytic solution of stable probability density of the response in theory. Thus, it is significant to study the random vibration of nonlinear axial compression systems to forecast the response and possible phenomena under stochastic disturbance. It is also helpful to ameliorate the design of aerodynamic controller and achieve stability control. It gives an approach to forecast the response of the nonlinear system of high orders under stochastic disturbance.

References

1. Caughcy, T.K.: On the response of nonlinear oscillators to stochastic excitation. Prob. Eng. Mech., 2–4 (1986)
2. Zhu, W.Q., Yu, J.S.: The equivalent nonlinear system method. J. Sound Vib. 7, 385–395 (1989)
3. Moore, F.K., Greitzer, E.M.: A Theory of Post-Stall Transients in Axial Compression Systems, Part I-D evelopment of Equations, and Part II-Application. ASME J. of Engineering for Gas Turbines and Power 108 (1986)
4. Piccardi, C.: Bifurcaton Analysis via Harmonic Balance in Periodic Systems with Feedback Structure. Int. J. Control 62(6), 57–65 (1995)
5. Guckenheimer, J., Holmes, P.: Nonlinear oscillations dynamical systems, and bifurcation of vector fields. Springer, New York (1991)
6. Zhu, w.: Random Vibration. Science Press, Beijing (1998) (in Chinese)

Energy Efficient Medium-Voltage Circuits Based on Adiabatic CPL

Jianping Hu and Binbin Liu

Faculty of Information Science and Technology, Ningbo University
315211 Ningbo City, China
nbhjp@yahoo.com.cn

Abstract. Voltage scaling is the most effective solution in building energy efficient digital circuits. Reduction of the supply voltage (with a fixed threshold voltage) results in a quadratic reduction of dynamic energy at the expense of decreased performance. In this work, we investigate performances of CPAL (Complementary Pass-transistor Adiabatic Logic) circuits operating in near-threshold and super-threshold regions. The impacts on both energy efficiency and robustness are also discussed. Compressor 4:2 circuits using standard static CMOS, medium-voltage static CMOS, standard CPAL and medium-voltage CPAL are verified. The results show that the energy consumption of the medium-voltage CPAL circuits is greatly reduced with high robustness and reasonable speed.

Keywords: Medium-voltage circuits, Adiabatic CPL, Low power, Low voltage.

1 Introduction

As the density and size of the chips and systems continue to increase, the power consumption has been of great concern. CMOS circuits using the nominal source voltage can reach high operation frequencies with large energy consumptions. A direct solution for reducing energy consumption is to scale down supply voltage, since the dynamic energy is reduced quadratically [1]. Scaling supply voltage to sub-threshold region can reach minimum energy consumption but only suits for ultra low power design (f =10KHz to 5MHz) [2]. In order to attain more extensive application, scaling supply voltage to medium-voltage region is an attractive approach especially suiting for mid performances (f =5MHz to 100MHz) [2]. Low supply voltage near the threshold voltage is called near threshold techniques [3]. For many applications, the performance penalty of the near-threshold logic circuits is tolerable [3].

Adiabatic logic utilizes AC power-clock to recover effectively the charge delivered by the clock instead of being dissipated to the ground, resulting in lower dynamic power, thus the energy consumption is much smaller than that of the conventional CMOS [4]. However, the previously proposed adiabatic logic families focus mainly on nominal voltage circuits. Presently, an investigation for PAL-2N circuits operating on medium-voltage region has been reported in [5]. The results shown that PAL-2N circuits that operating on medium-voltage region can not only keep reasonable speed but also reduce greatly energy consumptions.

L. Qi (Ed.): ISIA 2010, CCIS 86, pp. 539–545, 2011.

This paper investigates the performances of CPAL (Complementary Pass-transistor Adiabatic Logic) circuits in near-threshold and super-threshold regions. The effects of lowering the voltage supply on the energy efficiency of CPAL circuits are investigated. The impacts on both energy efficiency and robustness are also discussed. 4:2 compressors using standard static CMOS, medium-voltage static CMOS, standard CPAL and medium-voltage CPAL are verified. The results show that the energy consumption of the medium-voltage CPAL circuits is greatly reduced with high robustness and reasonable speed.

2 CPAL Overview

The basic structure of the CPAL buffer is shown in Fig 1 (a) [4]. The detailed description on the CPAL circuits can be found in [4]. Fig. 1 (b) shows its simulation waveforms using 45nm BSIM4 model technology [6] at 1.0V supply voltage.

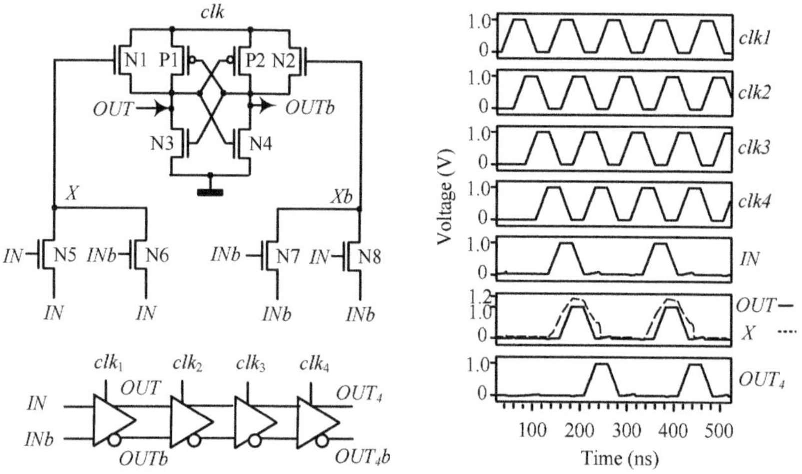

Fig. 1. CPAL buffer and its simulated waveforms

3 Voltage Scaling for CPAL Circuits

The total energy dissipation per cycle of the CPAL buffer operating in near-threshold and super-threshold regions can be expressed as

$$E_{total} = 8\frac{RC_L}{T}C_L V_{DD}^2 + a\frac{1}{2}C_{boot}(V_{DD} - V_{TN})^2 + \frac{1}{2}V_{DD}I_{Leak}T, \quad (1)$$

where R is turn-on resistance of the transmission gates (P1 and N1 or P2, and N2) that is approximately inversely proportional to the source voltage (V_{DD}), C_L is load capacitances of the CPAL buffer, V_{DD} is peak-peak voltage of power clocks, T is period of the power-clock, a is signal active ratio, C_{boot} is the capacitance of the bootstrapping node X or Xb, V_{TN} is threshold voltage of NMOS transistors, and I_{Leak} is average leakage current of the CPAL buffer, respectively.

In (1), the first term is adiabatic energy dissipation that is approximately proportional to source voltage for a given frequency, the second term is non-adiabatic energy dissipation that scales down quadratically with the supply voltage, and the third term is leakage energy dissipation caused by leakage current (I_{Leak}).

In current CMOS technologies, gate leakage current and PN-junction leakage of MOS transistors can be negligible relative to sub-threshold one because of the relatively low V_{TH}. The total leakage current of MOS transistors is mostly composed of the sub-threshold leakage, which is independent of the power source for near-threshold and super-threshold regions [1]. Therefore, the leakage dissipation of the CPAL circuits is approximately proportional to source voltage for a given frequency. From (1), the total energy consumption of the CPAL circuits would be cut down quadratically with the supply voltage for a given frequency.

In order to investigate the performances of the CPAL circuits in near-threshold and super-threshold regions, the CPAL buffer chain is simulated by varying the source voltage (V_{DD}) from 0.2V to 1.0V with 0.1V step using PTM (Predictive Technology Model) 45nm process. The energy dissipations of the CPAL buffer in different supply voltages are shown in Fig. 2 at 1MHz. As shown in Fig. 2, for a given frequency, the total energy consumption is quadratically reduced as supply voltage scales down, approximately.

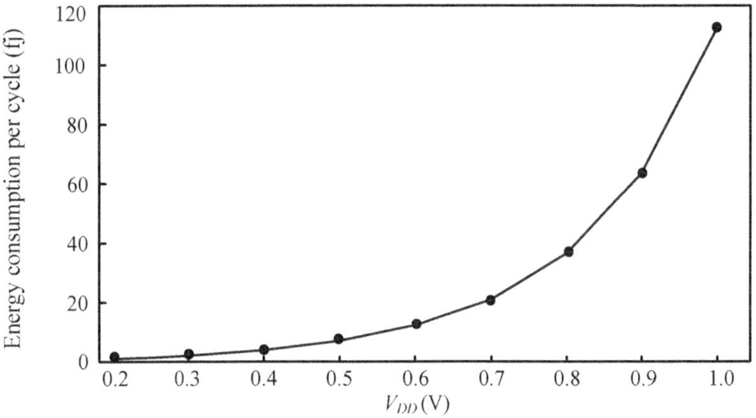

Fig. 2. Energy dissipations of the CPAL buffer in different voltages at 1MHz

Scaling the supply voltage would result in the performance loss, thus the robustness of the circuits must be considered. The fraction of V_{DD} at output is an important index of the robustness, as shown in Fig. 3. When the supply voltage (V_{DD}) exceeds 0.3V, the fraction of V_{DD} at output is 1, indicating that there is no wave distortion at the output. From the Fig 3, we can see that the wave distortion gets diminishing as the supply voltage increases.

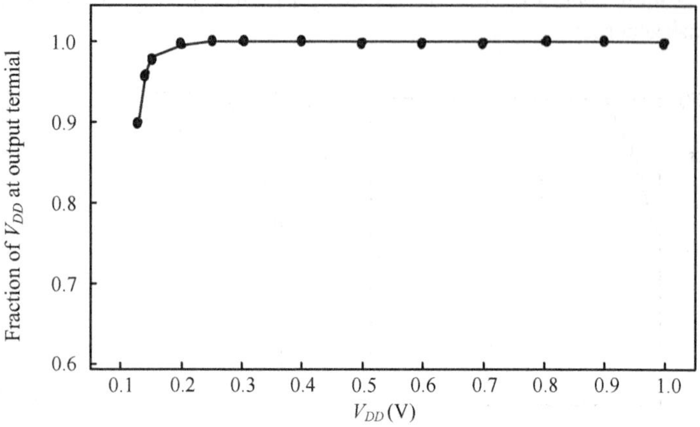

Fig. 3. Robustness of the CPAL buffer in different voltages at 1MHz

At each supply voltage, the maximum operating frequency is obtained, where the CPAL buffer has correct logic function. The maximum operating frequencies at different supply voltages are shown in Fig. 4. As shown from this result, the maximum operating frequency of the CPAL buffers operating in near-threshold and super-threshold regions reduces from 700MHz to 6MHz quadratic approximately, as the supply voltage decreases from 0.9V to 0.3V.

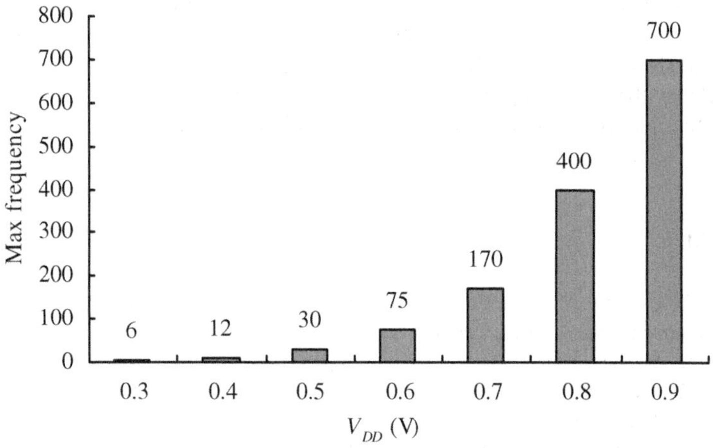

Fig. 4. Max operating frequency of the CPAL buffer in different voltages

Fig 5. shows energy consumption comparisons of the buffer/inverter based on standard static CMOS (normal source voltage with 1.0V), medium-voltage static CMOS (0.7V), standard CPAL (normal source voltage with 1.0V peak-peak power-clocks) and medium-voltage CPAL (medium source voltage with 0.7V peak-peak power-clocks). From the simulation result, the medium-voltage CPAL buffer obtains energy savings of 76.7%, 48.1% and 35.1% at 100MHz compared with standard static CMOS, medium-voltage static CMOS, and standard CPAL, respectively. We can see that CPAL and static CMOS buffers operating at medium source voltage with 0.7V are more efficient than the ones working at normal source voltage with 1.0V. The medium-voltage CPAL buffer has lowest energy consumptions at all the frequencies.

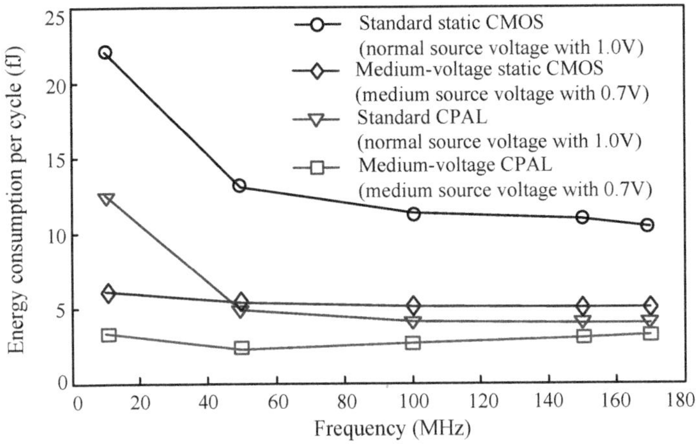

Fig. 5. Energy dissipation comparisons of static CMOS and CPAL buffers/inverters at 0.7V and 1V source voltages

4 4-2 Compressor Operating at Medium-Voltage Region

4-2 compressor is an important component in tree multipliers. The 4-2 compressor based on CPAL basic gates is shown in Fig. 6 [7].

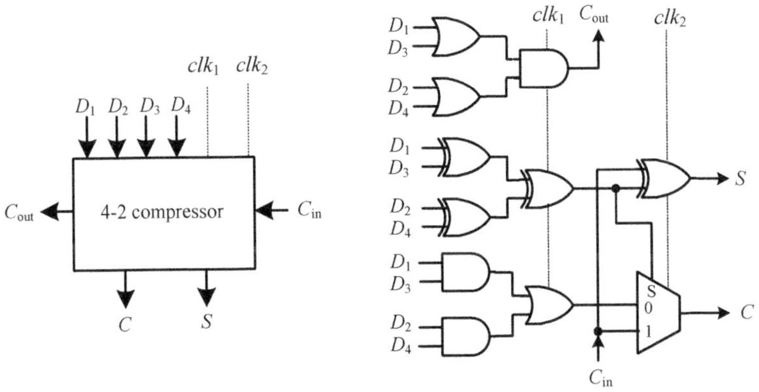

Fig. 6. 4-2 Compressor based on gate level

Fig 7 shows the simulation waves of the 4-2 compressor based on medium-voltage CPAL (medium source voltage with 0.7V peak-peak power-clocks). HSPICE simulation shows that the 4-2 CPAL compressor using medium source voltage has correct logic functions and ideal waveforms.

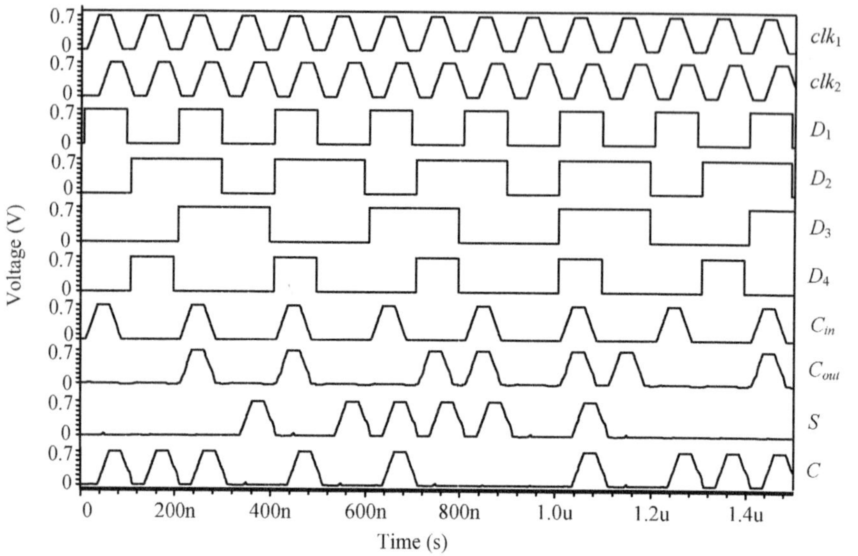

Fig. 7. Simulation waveforms of the 4-2compressor based on medium-voltage CPAL

Energy consumption comparisons of the 4-2 compressor based on standard static CMOS (1.0V), medium-voltage static CMOS (0.7V), standard CPAL (1.0V) and medium-voltage CPAL (0.7V) are shown in Fig. 8.

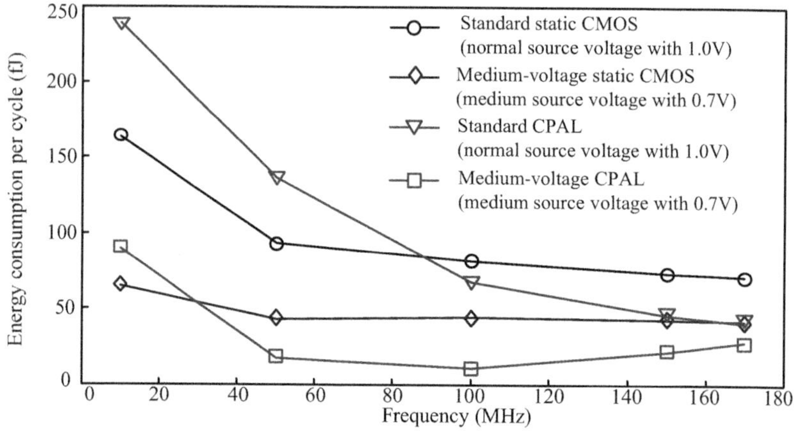

Fig. 8. Energy dissipation comparisons of static CMOS and CPAL 4-2 compressor at 0.7V and at 1V source voltage

From this simulation result, the CPAL compressor 4-2 working at V_{DD}=0.7V can operate well and saves considerable energy compared with the CPAL one at V_{DD}=1.0V and the CMOS one. The medium-voltage CPAL obtains energy savings of 85.5%, 76.6% and 84.8% at 100MHz compared with standard static CMOS, medium-voltage static CMOS, and standard CPAL, respectively.

5 Conclusions

The performances of CPAL circuits operating in near-threshold and super-threshold regions have been investigated. The impacts on both energy efficiency and robustness are discussed. The results show that the medium-voltage CPAL circuit has superiority in the energy efficient performance compared with other circuits. Scaling supply voltage of the CPAL circuits to medium-voltage region is an attractive approach especially suiting for mid performances.

Acknowledgments. Project is supported by National Natural Science Foundation of China (No. 61071049), Scientific Research Fund of Zhejiang Provincial Education Department (No. Z200908632), and Ningbo Natural Science Foundation (No. 2009A610066), and supported by the Scientific Research Foundation of Graduate School of Ningbo University in 2009 and the Excellent Dissertation Foundation of Graduate School of Ningbo University (No. PY2009006).

References

1. Wang, A., Calhoun, B.H., Chandrakasan, A.P.: Sub-threshold Design for Ultra Low-Power Systems, pp. 12–102. Springer, Heidelberg (2006)
2. Bol, D., Flandre, D., Legat, J.-D.: Technology Flavor Selection and Adaptive Techniques for Timing-Constrained 45nm Subthreshold Circuits. In: Proc. ACM/IEEE Int. Symp. Low-Power Electron, pp. 21–26 (2009)
3. Dreslinski, R., Wieckowski, M., Blaauw, D., Sylvester, D., Mudge, T.L.: Near-Threshold Computing: Reclaiming Moore's Law Through Energy Efficient Integrated Circuits. Proceedings of the IEEE 98, 253–266 (2010)
4. Hu, J.P., Xu, T.F., Li, H.: A Lower-Power Register File Based on Complementary Pass-Transistor Adiabatic Logic. IEICE Transactions on Information and Systems E88-D, 1479–1485 (2005)
5. Hu, J.P., Yu, X.Y.: Near-Threshold Adiabatic Flip-Flops Based on PAL-2N Circuits in Nanometer CMOS Processes. In: 2010 Pacific-Asia Conference on Circuits, Communications and System, PACCS 2010 (2010)
6. Zhao, W., Cao, Y.: New Generation of Predictive Technology Model for Sub-45nm Design Exploration. In: Proc. ISQED, pp. 585–590 (2006)
7. Liu, B.B., Hu, J.P.: Tree Multipliers with Modified Booth Algorithm based on Adiabatic. In: 12th International Symposium on Integrated Circuits (ISIC), pp. 302–305 (2009)

Design of Virtual Spinal Fixation Surgery System Architecture

Huazhu Song[1], Bin Zhao[1,2], and Bo Liu[1]

[1] School of Computer Science and Technology, Wuhan University of Technology, Wuhan, Hubei, China, 430070
mingjs@vip.163.com
[2] School of Computer Science and Engineering, Beihang University, Beijing, China, 100083

Abstract. The hierarchical-component architecture was given to efficiently implement 3D rendering and 3D interaction in virtual spinal fixation surgery environment. It consisted in data layer, user logic layer and components layer, and each layer had definite functional division. The data layer provided data support and uniform access interface for the system; the user logic layer adopted singleton pattern design; the third layer was comprised of the front-end component and haptic feedback component, where the component oriented method makes components mutually independently, improves reusability of components and contributes to the system function expansion. The hierarchical-component architecture design makes each layer or component be implemented independently and concurrently, and supports the system incremental development pattern. The finished system proves its rationality.

Keywords: spinal fixation, data visualization, three-dimensional interaction, haptic feedbac.

1 Introduction

Spine is the principle axis of human skeleton and it is composed by vertebras, it is the action control center, which supports human body and is the key branch to protect spinal cord. Now the spinal fixation surgery has become one of the required skills for a neurosurgeon. Along with the medical technology improved continuously, virtual reality technology shows good prospect in improving the accuracy and security of spinal fixation surgery and providing with better service to patients.

Many researchers or companies have studied and developed some surgery simulation software applied to clinical and teaching training, such as VolView of American Kitware Company, Amira of Germany Visage image Company, Mimics of Belgium Materialise Company and the VolVis[1] of visualization library of State University of New York, data integration processing platform for 3D medical imaging (3Dmed) developed by Automation Institute of Chinese Academy of Sciences, AccuRad pro 3D medical image advanced treatment system developed by Xian YingGu technology Co., LTD and so on.

At present, surgery simulation systems are mostly involved with orthopedic surgery [2], neurosurgery [3], virtual endoscope and ears-nose and maxillofacial surgery [4]

L. Qi (Ed.): ISIA 2010, CCIS 86, pp. 546–553, 2011.

and so on. Orthopedics focuses on joint surgery. 3D reconstruction in virtual reality working platform can be done with image data of CT, MRI and so on which are collected before surgery, and also 3D visualization can be implemented, some of which partly simulate rotation, move, simulate incisive bone, osteotomy and implanted function and so on. However, most of them draw slowly and aren't real-time; operators can not feel load-carrying condition during operation process and they are difficult to simulate the reality situation of spinal fixation well; in addition, simulation training software is very expensive and do not provide secondary development interface, and also they have high requirements for computer and other hardware.

Therefore, based on the current research fruits and repeatedly experiments, this paper aimed at proposing a good architecture on efficient right volume rendering method for spinal fixation surgery simulation and training to provide simple and convenient 3D interactive operation and implement well haptic perception. And the architecture should be easy to be expanded and provide more secondary development interface for the different operators. Section 2 gave the system architecture after introducing the system data source; data layer, logical layer and component layer were discussed from section 3 to section 5; and section 6 shown the conclusion.

2 System Architecture

2.1 System Data Sources

According to sources, data processed in system were divided into two kinds: external file and internal process unit. The former mainly came from CT tomoscan image stored in file format according to DICOM (Digital Imaging and Communications in Medicine III) standard 3.0. The latter were structured volume data, whose sources were a set of CT tomoscan images read into internal memory. Most CT images were converted into the symmetrical grid structure of volume data and the regular grid structure of volume data. However, with the different sampling interval and instruments, sometimes rectangular grid structure of volume data was used.

2.2 Hierarchical-Component Architecture of the System

Since the system was designed for the planning and training of spinal fixation surgery, the key point of system was how to accurately deal with the spine visualization and 3D interaction and better provide preoperative training and surgery planning to operators. In order to separate system data, user logic and system specific function components and in the meantime ensure the relative independence between visualization and 3D interaction, the system architecture adopted the hierarchical-component model to design.

As system architecture displayed in Fig. 1, longitudinally divided into data layer, user logic layer and component layer, and the component layer was divided into two independent components: front-end component and physical subsystem.

The arrow direction in Fig. 1 described the direction of message transmission among components, the message was directed from source to destination; the message destination was transparent relative to message source. It means the source action is not affected by the existence of the message destination. Parts connected by dotted arrow could transmit message indirectly, it is to say, while the two components both existed, they can transmit message to another indirectly, and if one of message source and message receiver did not exist, the system still ran correctly with the support of the existing sub-system.

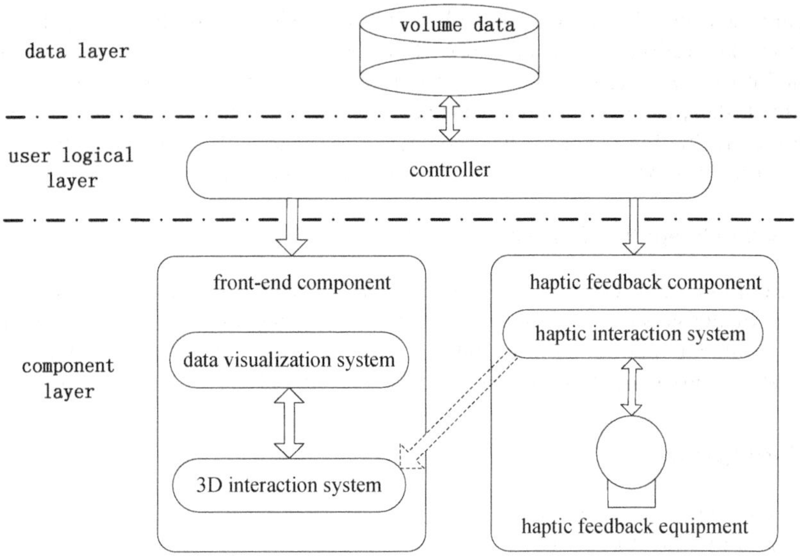

Fig. 1. Virtual spinal fixation surgery system architecture

3 Data Layer

The major duty of data layer was providing volume data accessing for other parts of system especially the component layer. Since other parts of system could not modify volume data, as constrains in design, data layer should be read-only to reduce data error in design and development to some extent. The data layer interface was defined as Fig. 2:

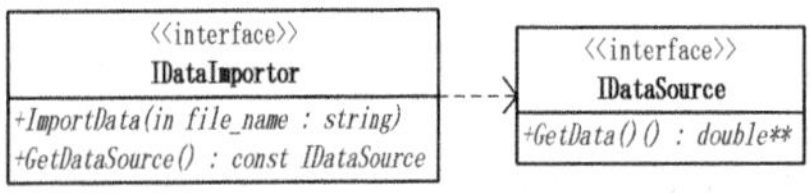

Fig. 2. Data layer interface

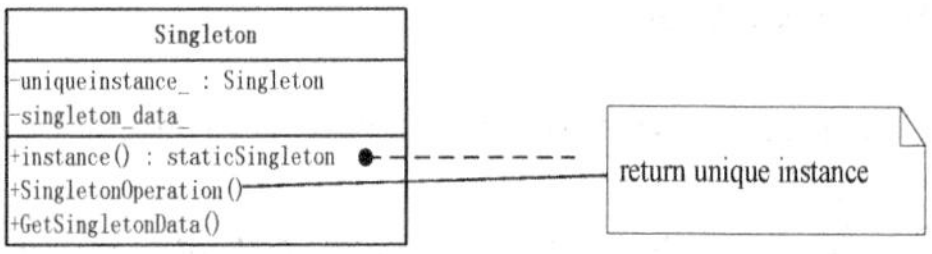

Fig. 3. Singleton pattern structur

In IDataImportor, Import(file_name) import data from specified path, GetDataSource() return object of implement IDataSource interface, force caller which using GetData() method access volume data in read-only.

4 User Logic Layer

From the system entire architecture diagram, we saw that user logic layer (for short: logic layer) mainly served as a bridge between data layer and component layer, and it had only one instance in system, so the logic layer design adopted singleton pattern. The purpose of using singleton pattern is as following:

(1) Logically, only one instance existed can ensure the correctness of logic.
(2) It can reduce the usage of internal memory through share, which is regarded as a uniform data mechanism in application.
(3) It can reduce performance loss caused by frequent constructing process in application.

Structure of singleton pattern displayed as Fig. 3, it obtained the unique instance in global by constructors of hide class and instance() method.

In addition, using singleton pattern also can ensure controlled access of system to the unique instance, since Singleton class encapsulates its unique instance, it can strictly control the customers how and when to access it. Meanwhile, singleton pattern application can reduce namespace and avoid the namespace polluted by global variable which be used to store the unique instance. Singleton class can has sub class, and it is very easy to use this extension class to allocate an instance, so user can do allocation while the instance of necessary class is running.

Logic layer implemented the uncoupling between component layer and data layer through defining a series of interfaces. Moreover, the logic layer also maintained the status of the whole system, such as obtaining and setting the position of 3D pointer (GetCursorPosition()), obtaining the volume data value of specific point (x,y,z) in space (GetVoxelValue(x, y, z)). System controller interface were displayed in Fig. 4:

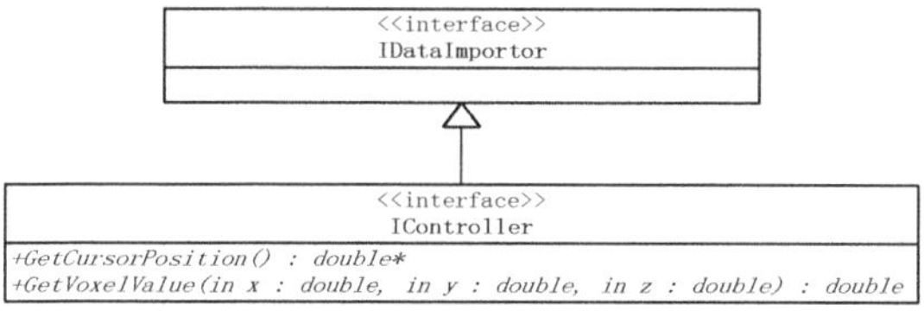

Fig. 4. System controller interface

Since the volume data in system mainly adopted 2 bytes CT tomoscan image which stored according to DICOM standard, we could do appropriate simplification for data layer and component layer while implementing.

5 Component Layer

Different from traditional object oriented design, this component granularity was larger, each component encapsulated specific action and logic independently and it had specific responsibility. 3D interactive research refers to two kinds of researches, one is interaction among virtual objects in visualization scene, also called virtual-virtual interaction; another is interaction with objects in virtual scene through haptic feedback equipment, also called virtual-real interaction. Virtual-virtual interaction and 3D reconstruction both occur in visual window, so the research and implementation methods independent of peripheral equipment; but the virtual-real interaction needs the support of 3D input and output equipments such as haptic feedback equipment. So we divided the interactive system into "front-end component" and "haptic feedback component" based on whether existed the peripheral equipment. The developing method based on component is benefit for the system function expansion in future.

5.1 Front-End Component Design

Front-end component provided system with spine data visualization and platform of research and developing for 3D interactive component, and the system graphic user interface which was a direct interface between user and system. Therefore, we must ensure the coupling between them to separate user performance from logic during front-end component design.

(1) Graphic user interface design
The system was to provide a 3D visualization and training system for spinal fixation surgery. In order to accord with operation habits of operators and reduce the system learning time, the system referenced common professional visual software while designing graphic user interface, such as 3D med, Mimics and so on. Through investigation, we found that the common layout of general medical visual software includes: menu bar, toolbar, status bar, control panel and render region and so on. Hereinto, menu bar and toolbar provide quick access to various kinds of functions and commands, and status bar displays the information of current system or operation.

Users mainly focus on control panel and render region in data visualization and 3D interaction. As the view of algorithm parameters adjustment, the control panel displays current function or operation property, so the operators can adjust and modify the corresponding property through control panel. Visualization and 3D interaction were all in render region, and the classify layout of render region included 1×3 (displayed in Fig. 5) and 2×2 (displayed in Fig. 6). Rendering window was showing the final result and interactive region, and it was divided into two kinds: volume rendering window and 2D slice rendering window. The 2D slice mainly displayed CT slice image in Axial View, Sagittal View and Coronal View.

These two different window layouts both were adopted in system design, and the operator can dynamically switchover window layout during running. The 1×3 layout highlighted volume rendering window shown in Fig. 4, for this design mainly is used to observe the 3D reconstruction result and visualization showing of training policy. The 2×2 layout partition view was divided as "+" shape shown in Fig. 5, and it was mainly used to observe different view of CT slice image.

(2) Data visualization
Data visualization focused on visualization for CT tomoscan data, including 3D bone tissue reconstruction displayed in Fig.7 and slice data rendering displayed in Fig.8.

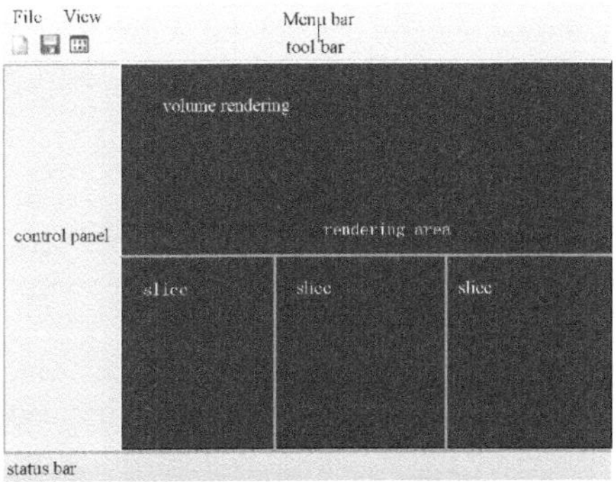

Fig. 5. 1 X3 graphic user interface layout

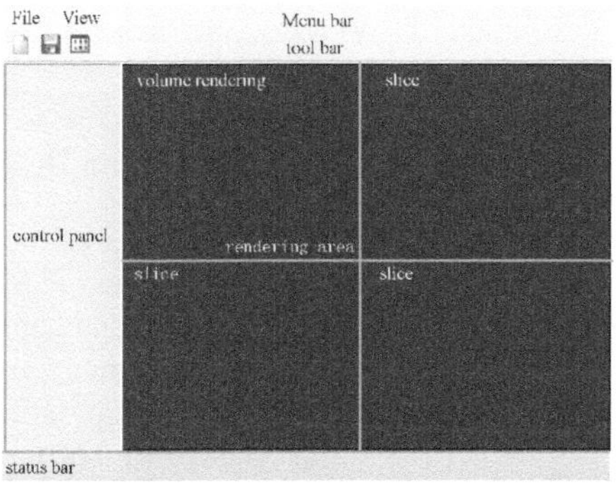

Fig. 6. 2 X 2 graphic user interface layout

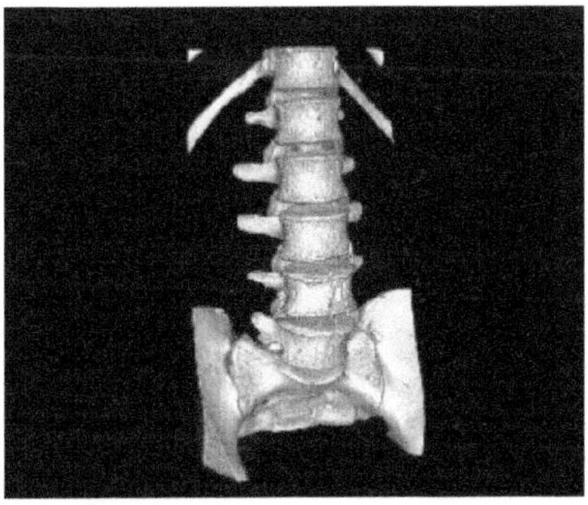

Fig. 7. 3D bone tissue reconfiguration (skeleton without surrounding soft tissue)

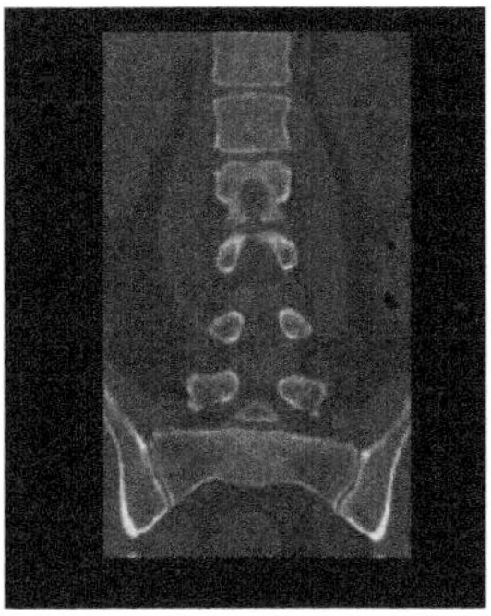

Fig. 8. Coronal surface view of heavy slice data rendering (skeleton with surrounding soft tissue)

Although visualization result were different, they were both final rendering results get from a series calculation for CT data, so the 3D reconfiguration and slice rendering had the same interface in abstract layer. Fig. 9 described data visualization abstract interface.

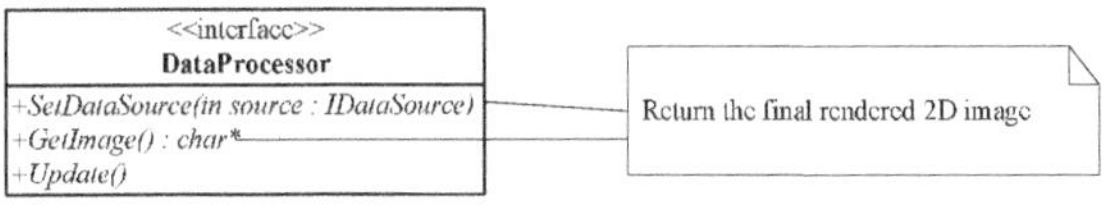

Fig. 9. Data visualization interface

(3) 3D interaction

The 3D interaction of front-end component was mostly virtual-virtual interaction: using image transmission equipments such as mouse and keyboard to implement interaction with other objects in 3D scene based on 3D Widgets, for example, control and display spine and slice which have been reconfigured. Meanwhile, the system implemented graphic user interface adjusting parameters of 3D Widgets by expanding traditional 3D Widgets, and the operator could dynamically control the Widgets while it is running.

5.2 Haptic Feedback Subsystem Design

The haptic feedback subsystem was the virtual-real interactive part in 3D interaction system. It mainly included haptic feedback equipment communication, image virtual pointer synchronous displaying and haptic feedback rendering.

The haptic feedback equipment in system was PHANToM Desktop, which connects with computer through parallel port EPP pattern and whose coder in needle can induct range of 6 degree of freedom and can output 3 degree of freedom power. OpenHaptics was used to design and develop the haptic feedback system.

The constitution of OpenHapitcs developer's kit was displayed in Fig.10. In haptic feedback system, the pencil-type probe of PHANToM equipment match to virtual probe in 3D scene, user interact with 3D bone tissue in scene by probe of equipment, this process was implemented as displayed in Fig.11.

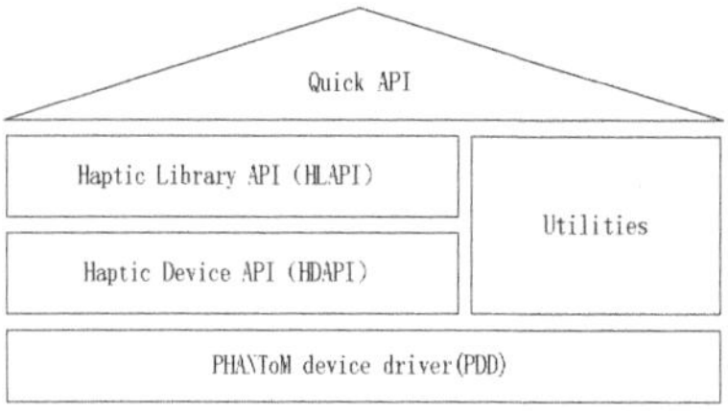

Fig. 10. OpenHaptics Toolkit architecture

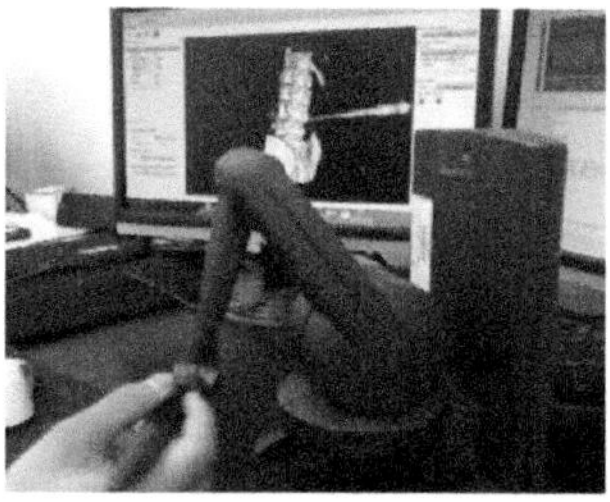

Fig. 11. Implemented software

6 Conclusion

This paper is for spinal fixation surgery; in this paper we give the architecture design of a virtual spinal fixation surgery system, which was hierarchical-component architecture. It divided system into data layer, logic layer and user layer components layer which included front-end component and haptic feedback component according to function partition. The architecture makes the system reduce degree of coupling among system parts, improve reusability of components and expanded easily.

Acknowledgements. We would like to be very grateful to the professor Xukun Shen and the National Key Lab for Virtual Reality Technology of Beihang University, for their careful guidance, help and supports.

References

1. Dai, Y.-j., Cao, J., He, F.: Present status and application of orthopedics virtual surgery system. Journal of Clinical Rehabilitative Tissue Engineering Research 12(30), 5957–5960 (2008)

2. Pham, A.M., Rafii, A.A., Metzger, M.C., Jamali, A., Bradley Strong, E.: Computer modeling and intraoperative navigation in maxillofacial surgery. Otolaryngology–Head and Neck Surgery 137, 624–631 (2007)
3. Beyer, J., Hadwiger, M., Wolfsberger, S., Buhler, K.: High-Quality Multimodal Volume Rendering for Preoperative Planning of Neurosurgical Interventions. IEEE Transactions on Visualization and Computer Graphics (13) (2007)
4. Caloss, R., Atikins, K., Stella, J.P.: Three-dimensional imaging for virtual assessment and treatment simulation in orthognathic surgery. Oral Maxillofac. Surg. Clin. North Am. 19(3), 287–309 (2007)

Study on Stability of Vehicle Mass Analysis System

Wei Shi[1], Shusheng Xiong[1,*], Chaoshan Zhang[2], Yaohua Jiang[1], Wei Li[1], Xijiang Wu[1], Xiaoshuai Ren[1], Wenhua He[1], Kailai Xu[3], and Ji Zhou[3]

[1] The Department of Energy Engineering, Zhejiang University, Hangzhou 310027, China
[2] Zhejiang Technology Institute of Economy, Hangzhou 310000, China
[3] JuYan Technology Co., Ltd., Hangzhou 310000, China
xiongss@zju.edu.cn

Abstract. A transient emission measurement system with VMAS is set up. The advantages, components and theory of VMAS are introduced. Twenty tests were made when the experimental vehicle was running in the best emission condition. In this paper, the system's stability is discussed based on test results. The VMAS' stability is great, and the data detected with VMAS can reflect the real situation of vehicle emissions. Then the importances are discussed about adoption the VMAS testing method for emission inspection of in-use petrol vehicles in China.

Keywords: VMAS; in-use vehicles; emission measurement.

1 Introduction

The lessons and experiences tell us that the most effective measure to control exhaust pollution is the implementation of stringent I / M (Inspection / Maintenance) (testing / maintenance) system. In the implementation process, a transient emission testing technology for in-use vehicle exhaust pollutants under several driving mode condition is named VMAS (Vehicle Mass Analysis System) which is developed in 1998 by Sensors, Inc. The advantages of system with VMAS are as follows. First of all, it is low-cost transient mass emissions measurement equipment. Secondly, in all of the detection method, it has the highest environmental benefits. Thirdly, the result tested by this system has a great correlation with that tested by IM240 which is considered as the most accurate testing method system. Currently, Beijing, Hangzhou, Nanjing and other cities have begun using the measurement system with VMAS. In actual operation, the system's stability is concerned by the car owner. A standard test line of VMAS was constructed in lab. The components and theory of VMAS are introduced, and the system's stability is discussed based on test results in this paper. Then the possibility and importance are discussed about adoption the VMAS testing method for emission inspection of in-use petrol vehicles in China.

2 VMAS Layout and System Theory

The measurement system mainly consists of a chassis dynamometer, an exhaust analyzer and an exhaust flow analyzer. Test vehicle is running on the chassis

* Corresponding author.

L. Qi (Ed.): ISIA 2010, CCIS 86, pp. 554–559, 2011.

dynamometer which is controlled by central controller to simulate in-use petrol vehicle running under several driving modes in the city road. At the same time, CO, HC and NOx, the mainly pollution gases in emission, were dynamically tested and recorded. Fig.1 is the system layout.

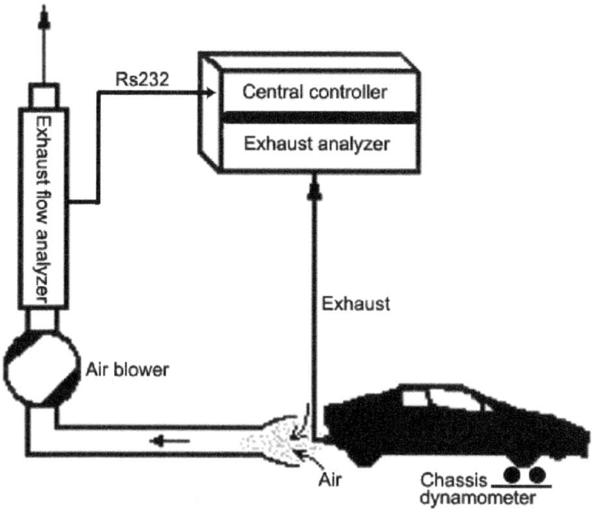

Fig. 1. Structure layout of the system

The devices apart from chassis dynamometer, exhaust analyzer and exhaust flow analyzer in the system are used for processing the data of pollution gases, recording environment variables and data communication.

Currently, the idle mode is used widely because of its simplicity and low cost, but it can not reflect real emission effectively. Different with the idle mode, the value of the pollution gas given in volume concentration, the unit of the pollution gases is g/km in VMAS. In the test, the system measures and records the quality of emissions every second and the result is send to the central controller. In the emission data processing, the quality of each pollutant is counted in accordance with the formula (1).

$$M=C\times\rho\times Q \tag{1}$$

Where M is the quality of each pollution gas (g/km), C measured by the exhaust analyzer is the volume concentration, ρ is the density (kg/m3), Q is the exhaust flow (m3/s). Among them, the Q value is gained by the calculation:

$$Q=Q_2\times\mu \tag{2}$$

$$\mu=(C_{O_0}-C_{O_2})/(C_{O_0}-C_{O_1}) \tag{3}$$

Here, Q_2 is the flow diluted with air, C_{O_0} is the oxygen concentration in air, C_{O_2} is the oxygen concentration in the exhaust diluted with air, and C_{O_1} is the oxygen concentration in the exhaust. C_{O_1} is detected by exhaust analyzer, value is usually between 0-5%; C_{O_2} and Q_2 are tested by exhaust flow analyzer.

3 Experiment

The experimental vehicle is a Santana Vista produced by Shanghai Volkswagen in 2007 in this paper. The car is running well and suitable for this experiment.

Note that the exhaust gas flowed through tow different pipes and was analyzed twice by system devices at different times. Data obtained from different devices needs to be adjusted. In this paper, the delay times of exhaust analyzer and exhaust flow analyzer are determined. And they can be found in Table 1.

Table 1. Devices' delay times

Device	Chassis dynamometer	Exhaust analyzer	Exhaust flow analyzer
Delay time	Datum(0s)	8s	12s

Atmospheric parameters affect the test results enormously. In order to ensure the comparability of measurement results, the result should be corrected by dilution factor and humidity correction factor. Formula for calculating these coefficients can be found in GB18285-2005. In this paper, all the result was corrected and the test parameters are shown in Table 2.

Table 2. Test parameters

Parameter	Lab temperature	Relative humidity	Air pressure
Value	18°C	40.2%	101.9kPa

To verify the stability of the system, twenty tests were got in the experiment. Each test was carried out one by one. In accordance with the test standard, each test continued 195 seconds. And the driving cycle is shown in Fig.2.

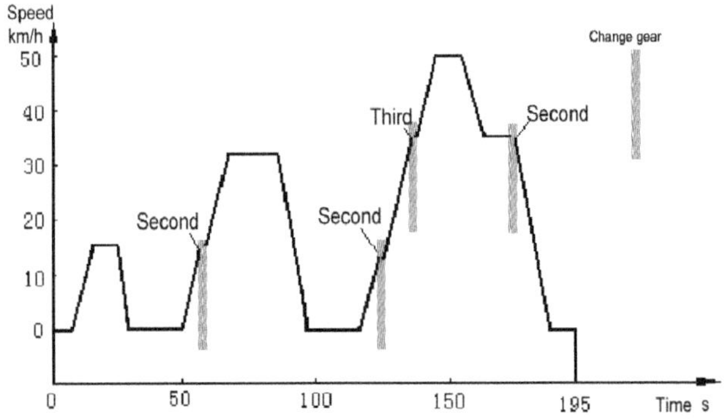

Fig. 2. The driving cycle of VMAS

The black line is idle speed in Fig.2. According to the specific requirements of national standards, the speed error must less than ±2km/h.

The average speed is 19km/h and the driving distance is 1.013km calculated in the ideal case. Additionally, the driver of the experimental vehicle was asked to throttle the car up and down smoothly. In order to eliminate the impact of environmental parameters, the experimental vehicle had run more than 15 minutes. Then the vehicle was running in the best emission condition.

4 Result and Discussion

Different with other measurement system, exhaust flow is necessary in VMAS system. Despite the value of exhaust flow is calculated indirectly and test parameters affect the result enormously, the trend of exhaust flow could still demonstrate the data reflected the real situation of vehicle emissions. The data of exhaust flow with speed in twenty tests is shown in Fig.3.

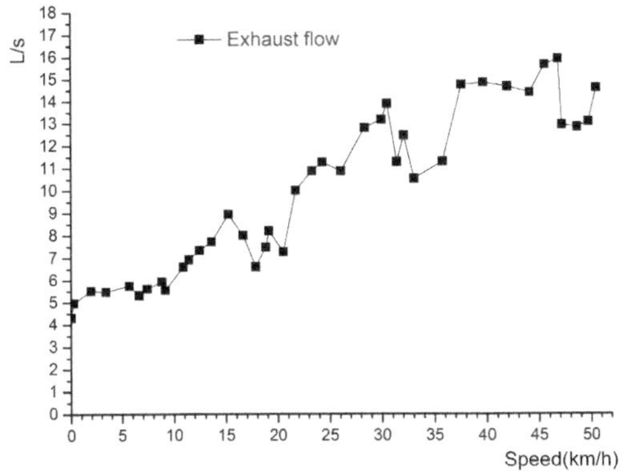

Fig. 3. The values of exhaust flow with speed in twenty tests

The exhaust gases leave piper in the form of pulses. The value of exhaust flow is changed after decompression and temperature reduction. So, it is different between measured and calculated in the ideal state. The black square in the Fig.3 is the average of flow with the same speed in twenty tests.

In a driving cycle, the experimental vehicle is an 8-second under 15km/h, 13-second under 35km/h and 12-second under 50km/h. They are the best conditions to verify the stability of VMAS. And Fig.4, Fig.5 and Fig.6 are the data of CO and sums of HC and NOx which the values of CO were marked with black squares, and sums were red dots.

The experimental vehicle is furnished with the three-way catalytic converter. After four tests, about twelve minute, the experimental vehicle was running in the best emission condition. In twenty tests, the test data of CO is from 0.003 to 0.025. And the data of sums of HC and NOx is from 0.0004 to 0.0134.

The maximum and minimum values of CO were 0.012 and 0.0025 in Fig.4, and data of sums of HC and NOx were 0.0003 and 0.0005. The relative errors, 0.0075 and

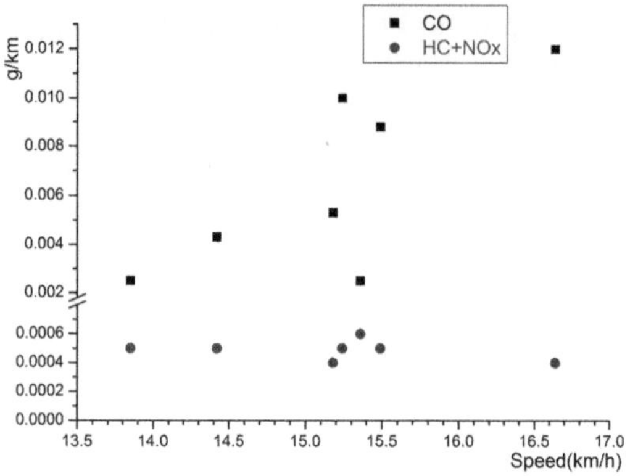

Fig. 4. The data of CO and sums of HC and NOx under 15km/h

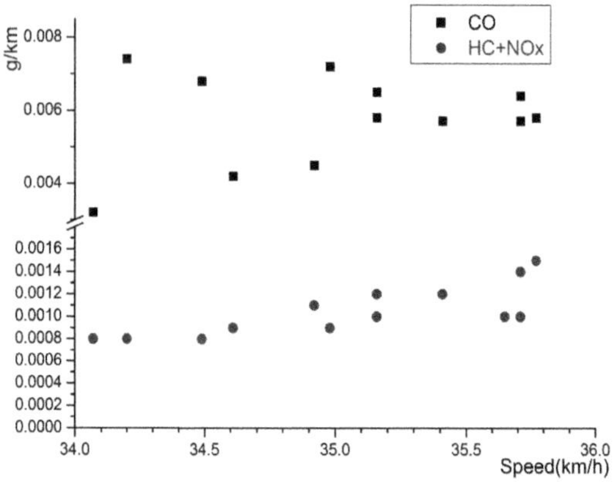

Fig. 5. The data of CO and sums of HC and NOx under 35km/h

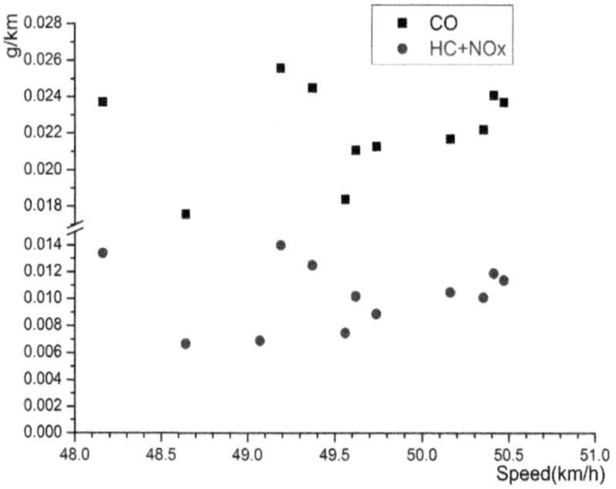

Fig. 6. The data of CO and sums of HC and NOx under 50km/h

0.0002, were big judging from the data. But compared with the emission limits, the relative errors were very small. The CO limit value is 0.0615g/km in DB33/660-2008, and that of sum of HC and NOx is 0.023g/km.

The distribution of values in Fig.5 and Fig.6 is similar with that in Fig.4. According to these three figures, it is clear to see the VMAS' stability is great. At the same time, the experimental results show that the data is larger with increase of speed. It is the direct response to the engine power. It is also prove that the data detected with VMAS can reflect the real situation of vehicle emissions.

5 Conclusion

In this paper, the experimental vehicle had mainly experienced there states: 15km/h, 35km/h and 50km/h. The VMAS' stability is great whether in low-speed condition or high-speed state. The data detected with VMAS can reflect the real situation of vehicle emissions.

The values were corrected by the test parameters, but the big relative errors were not removed. Exhaust gas treatment has a tremendous impact on the results. So, it is necessary to know whether the in-use vehicle furnished with the three-way catalytic converter or not.

There is a lot of work to do about adoption the VMAS testing method. To eliminate the impact of environmental parameters, owners whose in-use cars have to be tested every year should keep vehicles running more than fifteen minutes. Then the vehicle will be running in the best emission condition. And the result report will be true and reliable.

References

1. Min, Y.-j.: Research and engineering application of test technology for exhaust pollutants from vehicles
2. Xiong, S.-s., Xu, C.-s.: Study on S. I. LPG Engine by Rapid Compression-Expansion Machine to Improve Engine Performance. Transactions of CSICE 20(1) (2002)
3. Xiong, S.-s., Jin, H.: Combustion Characteristics and Power Recovery of JV481Q Engine Running on LPG. Chinese Internal Combustion Engine Engineering 22(1) (2001)
4. Vmas Service Manual Sensors, Inc. (2002)
5. Limits and measurement methods for exhaust pollutants from vehicles equipped ignition engine under two-speed idle conditions and simple driving mode conditions, GB18285-2005
6. Limits for exhaust pollutants from in-use vehicle equipped ignition engine in short transient loaded mode, DB33/660-2008

Nonlinear PID-Predictive Control for Multivariable Nonlinear System

Yan Zhang[1], Yanbo Li[1], Liping Yang[2], and Peng Yang[1]

[1] School of Control Science and Engineering, Hebei University of Technology, Tianjin 300130, China

[2] Department of Personnel, Hebei University of Technology, Tianjin 300130, China

yzhangzz@yahoo.com.cn

Abstract. A nonlinear PID controller is proposed based on neural network, which can overcome the difficulty of tuning the parameters of conventional PID controller. In the control process of nonlinear multivariable system, a decoupling controller is constructed, which takes advantage of multi-nonlinear PID controllers in parallel. Under the idea of predictive control, the multi-step predictive cost energy is adopted to train the weights of the decoupling controller. Simulation examples are given to show effectiveness of the proposed decoupling control.

Keywords: nonlinear PID control; predictive control; decoupling control, MIMO system.

1 Introduction

Today in process industries, an increasing number of process variables are being measured and stored. In particular, variables in most industrial processes exhibit strong temporal and spatial correlations. Techniques to capture these correlations from available data in the form of mathematical models would be highly valuable. Process control of a multi-input multi-output (MIMO) system is difficult, especially when the system exhibits high non-linearity [1]-[2]. Over the past several years, neural networks (NN) have been widely applied to identification and control for nonlinear systems[3]-[5]. Jin et al.[6] developed two neural network decoupling methods for multivariable systems by using the principle of feed forward compensation[6]. Shu et al.[7] investigated a new type PID neural network, and utilized this network to control the nonlinear multivariable system.

Motivated by this, a novel NN based PID type predictive control approach is proposed for MIMO nonlinear nonaffine discrete systems in this paper. The remainder of this paper is organized as follows. Section 2 presents the problem formulation and preliminaries. A Radial Basis Function (RBF) neural network is utilized to model and predict the controlled plant. A nonlinear PID controller design is described in Section 3. Section 4 shows the simulation results of the proposed method. Conclusions are given in Section 5.

L. Qi (Ed.): ISIA 2010, CCIS 86, pp. 560–566, 2011.

2 Multivariable Nonlinear System Representation

2.1 Multivariable Nonlinear System Description

Suppose the controlled nonlinear MIMO process with n inputs and n outputs can be described by a discrete-time equation,

$$Y(k)=f\ [Y(k-1),\cdots,Y(k-n_y-1),U(k-1),\cdots,U(k-n_u-1)] \tag{1}$$

where $Y(k)=[y_1(k),\cdots,y_n(k)]^T$, $U(k)=[u_1(k),\cdots,u_n(k)]^T$, $f(\cdot)$ is a smooth nonlinear function, and n_y and n_u are the order of outputs and inputs. In general, the mathematical model of a plant is unknown and an identifier based on Neural Network Identifier (NNI) is used:

$$\hat{Y}(k)=f_{NN}[Y(k-1),\cdots,Y(k-n_y-1),U(k-1),\cdots,U(k-n_u-1),W(k)] \tag{2}$$

where $f_{NN}(\cdot)$ is the nonlinear mapping function provided by the network, $W(k)$ is the weight vector of the whole network, and $\hat{Y}(k)$ is the output of NNI. Here, recursive predictive scheme is applied to achieve the multi-step prediction values.

$$\hat{Y}(k+1|k)=f_{NN}[Y(k),\cdots,Y(k-n_y),U(k),\cdots,U(k-n_u),W(k)] \tag{3}$$

$$\vdots$$

$$\begin{aligned}\hat{Y}(k+N|k)=f_{NN}[&Y(k+N-1),\cdots,Y(k+N-n_y-1),\\ &U(k+N-1),\cdots,U(k+N-n_u-1),W(k)]\end{aligned} \tag{4}$$

2.2 RBF Neural Network

RBF neural network is a kind of three-layer feed-forward network with single hidden layer. It conquered the problems which back-program(BP) network has, such as local optimum, slow convergence velocity and inefficiency. And the network structure figure is shown in Fig.1.

Assume that the outputs of the nerve cells in hidden layer is:

$$h_i(\mathbf{X})=R_i(\|\mathbf{X}-\mathbf{C}_i\|) \qquad i=1,2,\cdots,H \tag{5}$$

Network output is:

$$y_i=f(\mathbf{X})=\sum_{i=1}^{H}w_ih_i(\mathbf{X}) \tag{6}$$

Where $\mathbf{X}$ is the input vector, $\mathbf{C}_i$ is the central vector, w_i is the weight, $R_i(\cdot)$ is the Gauss function, namely:

$$R_i(\mathbf{X})=\exp\left(-\frac{\|\mathbf{X}-\mathbf{C}_i\|^2}{2\delta_i^2}\right) \tag{7}$$

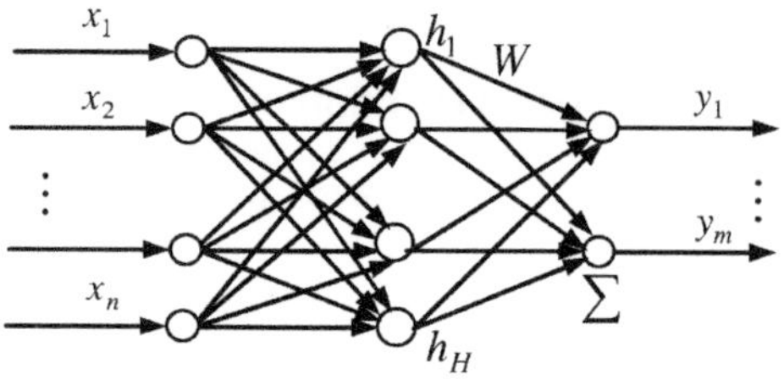

Fig. 1. RBF neural network

Where $R_i(\cdot)$ is very sensitive in part. The output can be obtained highly only in the central of the Gauss function which is near the $\mathbf{C}_i$, and the output is nearly zero if it is far from the centre. If the centre can be chosen suitably, just a few nerve cells can obtain great performance.

3 PID Type Predictive Control

3.1 Nonlinear PID Controller

If we set the input vector of RBF to $X(k)=[e_i(k),\Delta e_i(k),\Delta^2 e_i(k)]^T$, with $e_i(k)= r_i(k)- y_i(k)$, $r_i(k)$ is the i th reference trajectory. The output of RNN is the corresponding control signal $u_i(k)$. This can be expressed as $u_i(k)=g_{NN}[e_i(k),$ $\Delta e_i(k),\Delta^2 e_i(k),V_i(k)]$, where $V_i(k)$ is the weight vector of the whole network. This network whose structure is same as aforementioned can be regarded as a nonlinear PID controller. In the multivariable plant control, n nonlinear PID controllers are adopted in parallel. Assume that the input for the i th sub-nerwork is:

$$X_i(k)=[e_i(k),\Delta e_i(k),\Delta^2 e_i(k)]^T \tag{8}$$

Where $e_i(k+1)= r_i(k+1)- y_i(k+1)$. $r_i(k+1)$ is the i th desired trajectory. $y_i(k+1)$ is the i th real output. The outpur of the network is the i th control signal $u_i(k)$. The following objective function is optimized for the i th controller:

$$J_i=\frac{1}{2}[r_i(k)-y_i(k)]^2+\frac{1}{2}\sum_{j=1}^{n,j\neq i}[r_j(k)-y_j(k)]^2+\frac{\lambda_i}{2}[\Delta u_i(k)]^2 \tag{9}$$

Where $\lambda_i>0$ is the weighed control factor, $i=1,\cdots n$, $\Delta u_i(k)=u_i(k)- u_i(k-1)$. Let $R(k)=[r_1(k),\cdots,r_n(k)]^T$, Eq. (9) can be rewritten as:

$$J_i=\frac{1}{2}[R(k+1)-Y(k+1)]^T[R(k+1)-Y(k+1)]+\frac{\lambda_i}{2}u_i^T(k)u_i(k) \tag{10}$$

3.2 Multi-predictive Control for MIMO System

The control objective is to synthesize an output feedback control law, such that the output $y_i(k)$ tracks a smooth bounded desired trajectory $r_i(k)$ with an acceptable accuracy, where $i=1,\cdots n$. Here, multi-step prediction cost function is utilized:

$$J_i = \frac{1}{2}\sum_{j=1}^{N}[Y_r(k+j)-Y(k+j)]^T[Y_r(k+j)-Y(k+j)] + \frac{\lambda_i}{2}\sum_{j=1}^{N}[\Delta u_i(k+j-1)]^2 \tag{11}$$

Where $Y_r(k+j)=[y_{r1}(k+j),\cdots,y_{rn}(k+j)]^T$, $j=1,\cdots N$, is the given soft sequence matrix of $R(k)$. N is the predictive horizon.

Here, the output prediction is obtained by the iterative method as aforementioned. Assume that both the weights matrix of NNI and the weights matrix of he controller are kept invariable during the iterative proedure. The detailed steps of the calculation can be described as follows:

Step 1: Let step=1.

Step 2: Assuem that the input of the NNI is $Y(k),\cdots Y(k-n_y),U(k),\cdots,\ U(k-n_u)$. Calculate the one-step-ahead prediciton output $\hat{Y}(k+1)$ and obtain the error vector: $E(k+1)=[e_1(k+1),\cdots,e_n(k+1)]^T$. And apply the error vector in the decoupling controller to generate the control signal $U(k)=[u_1(k),\cdots,u_n(k)]^T$.

Step 3: Let step=step+1. Update the input variables of the NNI: $U(k-m)=U(k-m+1),\cdots,\ U(k)=U(k+1)$, $Y(k-n)=Y(k-n+1),\cdots$, $Y(k)=Y(k+1)$, and calculate the prediction output $\hat{Y}(k+2)$. Do the same as Step 2 to obtain the correspoding error vector, and apply the error vector in the decoupling controller to generate the control signal $U(k+1)$ for the next step.

Step 4: Repeat the above Step 3 and obtain the prediction output $\hat{Y}(k+1),\cdots,\hat{Y}(k+N)$.

Step 5: The weights of each sub-network are learned by minimizing the cost function of Eq.(11):

$$\Delta V_i(k)=V_i(k+1)-V_i(k)=-\eta_i\frac{\partial J_i}{\partial V_i(k)} \tag{12}$$

Where

$$\frac{\partial J_i}{\partial V_i(k)}=\sum_{j=1}^{N}[\frac{\partial[Y_r(k+j)-Y(k+j)]^T}{\partial V_i(k)}\cdot\frac{\partial J_i}{\partial[Y_r(k+j)-Y(k+j)]}+\frac{\partial u_i(k+j)}{\partial V_i(k)}\frac{\partial J_i}{\partial u_i(k+j)}]$$

As

$$\frac{\partial[Y_r(k+j)-Y(k+j)]^T}{\partial V_i(k)}=\frac{\partial[Y_r(k+j)-Y(k+j)]^T}{\partial u_i(k+j)}\frac{\partial u_i(k+j)}{\partial V_i(k+j)}$$

Thus

$$\frac{\partial J_i}{\partial V_i(k)} = \sum_{j=1}^{N} \left\{ -\frac{\partial Y^T(k+j))}{\partial u_i(k+j)} [Y_r(k+j) - Y(k+j)] + \lambda_i \cdot \Delta u_i(k+j-1) \right\} \frac{\partial u_i(k+j)}{\partial V_i(k)}$$

Combine this Eqution with Eq. (12), we can obtain

$$\Delta V_i(k) = -\eta_i \cdot \sum_{j=1}^{N} \left\{ -\frac{\partial Y^T(k+j))}{\partial u_i(k+j)} [Y_r(k+j) - Y(k+j)] + \lambda_i \cdot \Delta u_i(k+j-1) \right\} \frac{\partial u_i(k+j)}{\partial V_i(k)} \quad (13)$$

4 Simulation Results

Example 1: We consider the following system with 2-inputs and 2-outputs:

$$y_1(k) = 0.4 y_1(k-1) + \frac{u_1(k-1)}{1 + u_1^2(k-1)} + 0.2 u_1^3(k-1) + 0.5 u_2(k-1)$$

$$y_2(k) = .0.2 y_2(k-1) + \frac{u_2(k-1)}{1 + u_2^2(k-1)} + 0.4 u_2^3(k-1) + 0.2 u_1(k-1)$$

In the hidden layer of NNI there is 9 units. The structures of NNC are all selected as 3-8-1. For the system, Select the predictive step $N = 5$ and the weighted control factor $\lambda = [3.5, 5.2]^T$, respectively. It is shown that the proposed control system renders successful control results as shown in Fig. 2.

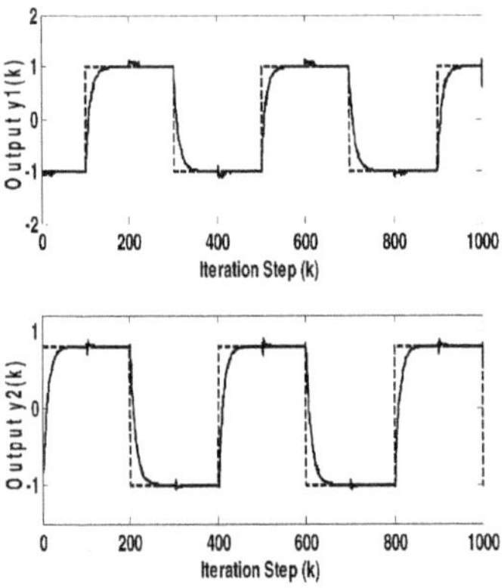

Fig. 2. Multi-variable system output curve for example 1

Example 2: Mini-type boiler nonlinear control system

In the experimentation, the control object is a mini-type boiler system. This boiler is made by stainless steel, including two layers: the heating layer and cooling layer. There is a 4.5KW heating cord in the heating layer and the input voltage of the heating cord is adjusted by the phase-shifting voltage-adjusting equipment of three-phase SCR. There are cooling water cycling in the cooling layer and the flux of water passed the three-phase magnetic pump can be adjusted by the transducer. The objective control values of the system include the temperature of boiler heating layer and the temperature of cooling layer, which is separately controlled by the output voltage of the phase-shifting voltage-adjusting equipment of three-phase SCR and the flex of cycle waters passed in the cooling layer. Because the phase-shifting voltage-adjusting equipment is a strong nonlinear equipment with a output as the figure “s”, the objective system can be considered as a strong-coupling nonlinear system with double inputs and double outputs.

Firstly, we sample 500 groups of data in open loop from the real system, and use them to train the RBF neural network off line and assume that the number of hidden layer of the identification network is 12. Then we put the neural network into the close-loop multi-step predictive control. We choose the predictive step $N = 5$, \ the weighting factor $\lambda = [20, 30]^T$, the sample time of the system is 5s, separately, and adjust to 65°C and 45°C after the system stabilized. The real-time response curves of the system showed in Fig.3 for the case of the temperature of the heating layer and the cooling layer in the initial system are set at 60°C and 40°C, respectively. In the Fig.3 the broken line is the system output of the identification network, the real line is the actual output of the system.

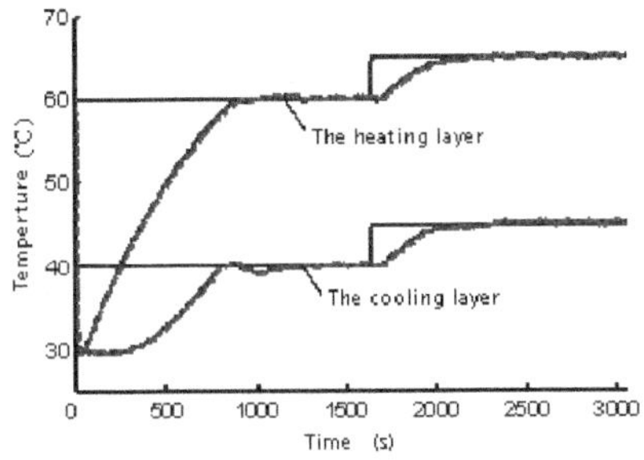

Fig. 3. The response curve of the boiler temperature

5 Conclusions

A nonlinear PID-predictive control scheme has been proposed in this paper. A RBF neural network was used to identification the controlled nonlinear MIMO system. Under the theory of recursive prediction, multi-step-ahead predictive value of the system can be calculated. On the base of conventional PID controller, a nnolinear NN-PID controller was constructed. This controller can tune the parameters of PID controller in real time. Simulations have shown that the algorithm is effective and practical for nonlinear MIMO process control.

References

1. Petlenkov, E.: NN-ANARX structure based dynamic output feedback linearization for control of nonlinear MIMO systems. In: Mediterranean Conference on Control and Automation, Athens, Greece, pp. 222–229 (2007)

2. AstrÖm, K.J., Johansson, K.H., Wang, Q.G.: Design of decoupled PID controllers for MIMO systems. In: Proceedings of the American Control Conf., Arlington, VA, pp. 25–27 (2001)
3. Zhang, Y., Chen, Z.Q., Yang, P., Yuan, Z.Z.: Multivariable nonlinear proportional-integral-derivative decoupling control based on recurrent neural networks. Chin. J. Chem. Eng. 12(5), 677–681 (2004)
4. Fu, Y., Chai, T.Y.: Neural-Network-Based Nonlinear Adaptive Dynamical Decoupling Control. IEEE Trans. on Neural Networks, China 17, 1–5 (2007)
5. Juang, C.F., Chen, J.S.: A recurrent fuzzy-network-based inverse modeling method for a temperature system control. IEEE Trans. on Systems, Man, and Cybernetics—part C 37, 410–417 (2007)
6. Jin, Q.B., Zeng, D.N., Wang, Y.H.: New decoupling method based on neural network for multivariable system. Journal of Northeastern University (Natural Science) 20, 250–253 (1999)
7. Shu, H.L.: Analysis of PID neural network multivariable control systems. Acta Autom. Sinica 25(1), 105–111 (1999)
8. Keel, L.H., Rego, J.I., Bhattacharyya, S.P.: A new approach to digital PID controller design. IEEE Trans. on Automat. Contr. 48, 687–692 (2003)

Predictive Control of Nonlinear System Based on MPSO-RBF Neural Network

Yan Zhang, Li Zhang, Guolin Xing, and Peng Yang

School of Control Science and Engineering, Hebei University of Technology, Tianjin 300130, China
yzhangzz@yahoo.com.cn

Abstract. A new predictive control scheme for nonlienar system is propoesed in this paper. In order to generate a set of optimization variables which have the same number of chaotic variables first, and at the same time to enlarge the scope of chaotic motion to the range of optimization variables, a new mixed particle swarm optimization (MPSO) algorithm is constructed. Then, this method is used to train the parameters of RBF neural network (NN). This NN can identify nonliear system with an acceptable accuracy, which can be seen from the simulation example. Furthermore, a direct multi-step predictive control scheme based on the MPSO-RBF neural network is proposed for nonlinear system. Simulation results manifest that the proposed method is effective and efficient.

Keywords: mixed particle swarm optimization algorithm, Radial Basis Function neural network, direct multi-step predictive control.

1 Introduction

Particle swarm optimization (PSO) which is a kind of evolutionary computation based on swarm intelligence technique is proposed by American scholars Eberhar and Kennedy in 1995[1]. PSO is a group of particles through the cooperation and competition among groups to complete the search for intelligent optimization. PSO shares many similarities with evolutionary computation such as genetic algorithm (GA). However, PSO has no evolution operators such as crossover or mutation, this is different from GA. So PSO algorithm is easier to be implemented. In recent years, the PSO has been applied in various problems of power systems. The active research focuses on the improvement of the algorithm in areas such as parameter selection and hybrid algorithms combined with other algorithms. Since the PSO is proposed, it attracted wide attention from scholars.

Many scholars are committed to improving the performance of PSO algorithm and proposed a variety of improved algorithm since PSO has emerged. A way to speed up factor set to improve time-varying parameters of the PSO strategy to improve the particle in the early search global search capability[2]. A median particle swarm optimization algorithm using the populations in each of the individual experience of the value of the particle swarm algorithm to change its vector of individual extreme value method, also changed the search rules of PSO[3]. It makes particles easily fall

L. Qi (Ed.): ISIA 2010, CCIS 86, pp. 567–573, 2011.

into local optimum, while the stability of convergence of the algorithm has also been improved to some extent. An idea of the chaos optimization to particle swarm optimization algorithm is introduced in [4].

In the RBF network training, initial value of network weights, the center of Gaussian function vector and the base width is difficult to determine. If these parameters are improperly selected, it will cause a decline in accuracy of approximation, even divergence of RBF network. So in this paper a new MPSO is proposed to optimize the RBF neural network parameters, which could obtain higher accuracy, better stability. Then RBF neural networks based on MPSO are used as direct multi-step prediction models and a closed-loop predictive control algorithem for nonlinear systems is proposed. Simulations have shown the algorithm to be effective and practical for nonlinear process control.

2 Particle Swarm Optimization Algorithm

2.1 A Basic Particle Swarm Optimization Algorithm

Assume that there are m particles in D -dimensional search space. X and V denote the particle's position and its velocity in the search space. The term velocity represents the change in the position of each particle. Thus, the position of the particle in D -dimensional space is represented as $x_i = (x_{i1}, x_{i2}, \cdots x_{iD})$ .The velocity of the particle in D -dimensional space is represented as $v_i = (v_{i1}, v_{i2}, \cdots v_{iD})$. The best previous position explored by the particle is recorded and denoted as $p_i = (p_{i1}, p_{i2}, \cdots p_{iD})$. Another value that is tracked and stored by PSO is the best value obtained so far by any particle in the population. This best value is a global best and is denoted by $p_g = (p_{g1}, p_{g2}, \cdots p_{gD})$. Each particle changes its position based on its current velocity and its distance. The modification can be represented by the concept of velocity and can be calculated as shown in the following formulas:

$$v_{id}(k+1) = v_{id}(k) + c_1 r_1 (p_{id} - x_{id}(k)) + c_2 r_2 (p_{gd} - x_{id}(k)) \tag{1}$$

$$x_{id}(k+1) = x_{id}(k) + v_{id}(k+1) \tag{2}$$

where k denotes iterations number of particles. $i = 1, 2, \cdots m$; $d = 1, 2, \cdots D$; c_1 and c_2 are the acceleration coefficients which are used to determine how much the particle's personal best and the global best influence its movement; r_1 , r_2 is random number from 0 to 1. Note that the terms $c_1 r_1 (p_{id} - x_{id}(k))$ and $c_2 r_2 (p_{gd} - x_{id}(k))$ in Eq. (1) are called the cognition and social terms respectively. The cognition term takes into account only the particle's own experience, whereas the social term signifies the interaction between the particles.

2.2 Formulas Mixed Particle Swarm Optimization

In each iteration, a fitness function is evaluated for all the particles in the swarm. The velocity of each particle is updated by keeping track of two best positions. One is the best position a particle has traversed so far. It is called "pBest". The other is the global best and is accordingly called "gBest". Based on the basic PSO, inertia weight introduced. Hence, a particle's velocity and position are updated as follows:

$$v_{id}(k+1) = wv_{id}(k) + c_1 r_1 (p_{id} - x_{id}(k)) + c_2 r_2 (p_{gd} - x_{id}(k)) \tag{3}$$

Suitable selection of inertia weighting factor w provides a balance between global and local explorations, thus requiring less iteration on average to find a sufficiently optimal solution. As originally developed, w often decreases linearly from about 0.9 to 0.4 during a run. In general, the inertia weighting factor w is set according to the following equation. Linear decreasing inertia weight formula is as follows:

$$w = w_{max} - \frac{w_{max} - w_{min}}{T_{max}} T \tag{4}$$

where w is called the "inertia weight" that controls the impact of the previous velocity of the particle on its current one; T_{max} is the maximum number of iterations (generations), and T is the current number of iterations.

In MPSO, the acceleration coefficients is set to time-varying parameters. So in this method, c_1 and c_2 will change over time. Initially $c_1 > c_2$, particles will be able to tend to the optimal population. In the final search stage $c_1 < c_2$,which will contribute to particle converges to the global optimal solution. formula such as equation (5) and (6) shows:

$$c_1 = (c_{1f} - c_{1i}) \frac{iter}{MAXITR} + c_{1i} \tag{5}$$

$$c_2 = (c_{2f} - c_{2i}) \frac{iter}{MAXITER} + c_{2i} \tag{6}$$

Where c_{1i} , c_{2i} , c_{1f} and c_{2f} are constants; *MAXITER* is the maximum number of iterations (generations), and *iter* is the current number of iterations. Value will then be introduced to the velocity of PSO update the equation to replace the equation in the extreme of individual particles p_{id} . The algorithm of median particle is $p_v = (p_{v1}, p_{v2}, \cdots, p_{vD})$,while p_v contains the information of all the particles experience. Median particle swarm optimization's velocity is updated as follows[5]:

$$v_{id}(k+1) = wv_{id}(k) + c_1 r_1 (p_{vd} - x_{id}(k)) + c_2 r_2 (p_{gd} - x_{id}(k)) \tag{7}$$

To prevent some of the particles in the iteration to a Stagnation phenomenon ,so the algorithm will use the unique ergodicity of chaotic variables. a chaotic iteration sequence will produce after the particle swarm current search to the global optimum value . Best particle position will replace the current position of particle randomly, which avoid the algorithm prematurity due to particle stagnation. A typical chaotic system is Logistic equations[6]:

$$y_{k+1} = \mu y_k (1 - y_k)(k = 0,1,2,\cdots,0 \le \mu \le 4) \tag{8}$$

Where μ is the controlling parameters; k is the number of iterations.

Hybrid particle swarm optimization algorithm as follows:

Step 1: PSO initialization

Step 2: Calculate fitness value, update the individual particles and global extremum

Step 3: Calculate inertia weight w and learning factors c_1 , c_2 , the average extreme p_v .

Step 4: According to the formula (2), (7) update the velocity and position.

Step 5: Chaotic optimization of the he optimal location of particle swarm p_g , then get the best possible solution fitness vector $\boldsymbol{x}_g^{*k}$.

Step 6: select a particles from the current swarm randomly ,at the same time $\boldsymbol{x}_g^{*k}$ replaced with the location of selected particle position vector.

Step 7 :Go to Step 2 until the algorithm reaches the maximum number of iterations, or get enough satisfactory solution.

3 MPSO Training in RBF Neural Network

The RBF neural network is generally composed of three layers: input layer, hidden layer and output layer[7]. Its main feature is the input layer to hidden layer is nonlinear mapping, while hidden layer to output layer is linear mapping.

In this paper, we use the proposed MPSO to optimize RBF neural network's weights, basis function centers and the width. MPSO uses the form of real number coding, a particle corresponds to a feasible solution RBFNN. Particle code includes weight, center of basis function vector, base width, particle velocity and the fitness function value. The key of the problem lies in mapping between the establishment of MPSO particle dimension and parameters of neural network.

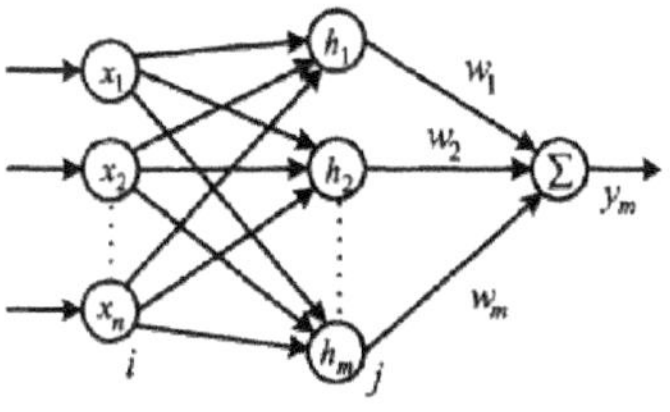

Fig. 1. The strcuture of neural network

By the RBF network that shows as Fig. 1, all the parameters of RBF network encode individual of real digital string, each particle's position vector is $X = (w_1, \cdots w_m, c_1, \cdots c_m, b_1, \cdots b_m)$. In optimization process, when we strike the fitness function value of groups' particles. First, position vector information of individual is decoded which expresses the parameters forms of the RBF neural network. Different individuals express different parameters of neural networks. Then, follow the formula of RBF neural network. Get output value of neural network by sample input, And then calculate the fitness function value of particles. Finally, the fitness function values of particles are compared to obtain individual extreme and global extreme.

Mean square error of Neural network is as MPSO fitness function, by powerful search performance of MPSO minimizes the mean square error of the network. Fitness function is as follows:

$$J = \frac{1}{2N}\sum_{k=1}^{N}\left(y(k) - y_m(k)\right)^2 \tag{9}$$

N is the training set sample, when the No. k sample is inputed, $y(k)$ and $y_m(k)$ respectively express the neural network 's desired output and actual output.

4 Direct Multi-step Predictive Controller Based on MPSO-RBF Neural Network

4.1 Algorithm Design

Suppose that the discrete system of a NARMAX model can be describe as:

$$y(k) = f[y(k-1),\cdots y(k-n),u(k-1),\cdots u(k-m)] \tag{10}$$

Where $f(.)$ is a continuous complex nonlinear function. n and m are the orders of output $y(k)$ and input $u(k)$, respectively. This model is usually unknown, neural network can be used for model identification as follows:

$$\hat{y}(k) = f_{NN}[y(k-1),\cdots,y(k-n),u(k-1),\cdots u(k-m),w] \tag{11}$$

Where f_{NN} is MPSO-RBF neural network. W is the weight matrix which is composed of all the weight values of neural network. $\hat{y}(k)$ is the output of NN to approximat the system output $y(k)$.

Here, a direct predictive algorithm is proposed[8]:

$$\hat{y}(k+j|k) = f_j[y(k-1),\cdots y(k-n),u(k-1),\cdots u(k-m)] \tag{12}$$

Where $f_j(.)$ is the No. j neural network model to predict the No. j step forward output of the controlled system at time k. MPSO-RBF neural network is used as prediction models. The objective function is to choose the weight parameters of NNs:

$$J = \frac{1}{2}\sum_{j=1}^{N_1}[y_r(k+j) - \hat{y}(k+j)]^2 + \frac{\lambda}{2}\sum_{i=2}^{N_2}[\Delta u(k+j-1)]^2 \tag{13}$$

Where $y_r(k+j)$ is soften sequence of setting values. N_1 is the maximum prediction step, N_2 is the control of time-domain. $\lambda > 0$ is control weighting factor.

$$\Delta u(k+j-1)=u(k+j-1)-u(k+j-2) \tag{14}$$

Setting sequence is:

$$\begin{cases} y_r(k)=y(k) \\ y_r(k+j)=\alpha y_r(k+j-1)+(1-\alpha)y_r(k) \end{cases}$$

Where α is the softness factor, and $0\leq\alpha<1$. By minimizing the objective function J, we can get control volume of algorithm $u(k)$, as follows:

$$u(k)=u(k-1)+\frac{1}{\lambda}\sum_{j=1}^{N_1}[y_r(k+j)-\hat{y}(k+j)]^2\frac{\partial\hat{y}(k+j)}{\delta u(k-1)}$$

RBF neural networks are used as prediction models in this paper, so

$$\frac{\partial\hat{y}(k+j)}{\delta u(k-1)}=\sum_{j=1}^{m}w_jh_j\frac{c_{1j}-x(1)}{b_j^2}$$

Where $x(1)=u(k-1)$. Then the increment of the parameters are given by minimizing the objective function as follows:

$$w_j(k)=w_j(k-1)+\eta\left[y(k)-y_m(k)\right]h_j+\alpha_t\left[w_j(k-1)-w_j(k-2)\right]$$

$$b_j(k)=b_j(k-1)+\eta\Delta b_j+\alpha_t\left[b_j(k-1)-b_j(k-2)\right],\quad \Delta b_j=\left[y(k)-y_m(k)\right]w_jh_j\frac{\left\|x-c_j\right\|^2}{b_j^2}$$

$$c_{ji}(k)=c_{ji}(k-1)+\eta\Delta c_{ji}+\alpha_t\left[c_{ji}(k-1)-c_{ji}(k-2)\right]$$

$$\Delta c_{ji}=\left[y(k)-y_m(k)\right]w_j\frac{x_j-c_{ji}}{b_j^2}$$

Where η is learning rate, α_t is momentum factor.

4.2 Simulation

Consider the following nonlinear system:

$$y(k)=\frac{y(k-1)y(k-2)}{1-y^2(k-1)+y^2(k-2)}+u(k-1)+1.5u(k-2)$$

The structure of MPSO-RBF neural network is 4-9-1. Sampling time is $t_s=0.5$, and N =1300, The learning rate is set $\eta=0.55$, the softness factor is $\alpha_t=0.05$, iterations of MPSO are 100, forecast steps $N_1=2$. Get forecasting model of system f_1, f_2. Corresponding MPSO training curves are respectively in Fig. 2. It shows that the output $y(k)$ is very close to that of the system desired signal with acceptable approximation errors. The proposed direct predictive control method shows good tracking performance for the nonlinear system.

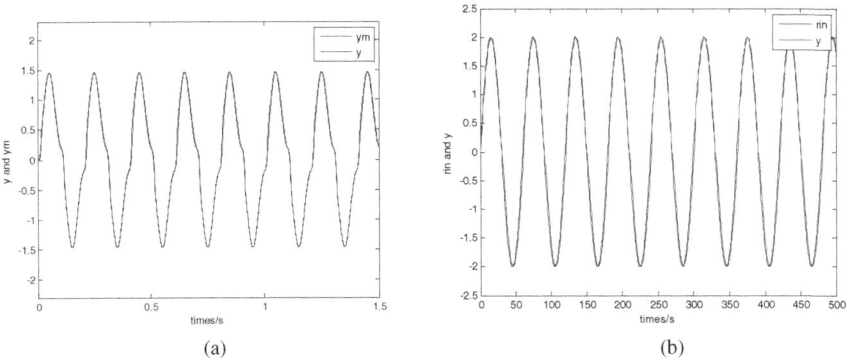

Fig. 2. (a) System identification curve of MPSO-RBF (b) System response curve

5 Conclusion

For neural network training problem, a new mixed particle swarm optimization algorithm which is based on many improved algorithms is proposed. Use particle swarm optimization algorithm to optimize neural network, it can solve problem of the neural network's practical application,and the problem includes the slow convergence, the learning does not have global search capabilities, easily falling into local minimum. In this paper, we use MPSO to optimize the parameters of RBF neural network. Get RBF neural network which has strong generalization ability and performance stability. we see MPSO-RBF neural network as the forecast model of directed multi-step predictive control algorithm, Nonlinear systems to be used direct multi-step predictive control algorithm. Simulation has shown that the algorithm is effective and practical in controlling nonlinear dynamic systems.

References

1. Kennedy, J., Eberhart, R.C.: Particle swarm optimization. In: Proc. IEEE International Conference on Neural Networks, Perth, Australia, Piscataway, pp. 1942–1948 (1995)
2. Asanga, R., Halgamuge, S.K., Watson, H.C.: Self-Organizing Hirrarchical Particle Swaem Optimizer with Time-Varying Acceleration Coefficients. IEEE Transaction on Evolutionary Computation 3(8), 240–255 (2000)
3. Li, B., Jiang, W.: Chaos optimization method and its application. Control Theory and Applications 18(4), 613–615 (1997)
4. Sheng, Z., Yin, Q.: Equipment condition monitoring and fault diagnosis technology and its application. Chemical Industry Press, Beijing (2003)
5. Zhang, S., Li, K., Zhang, S., et al.: A design program based on RBF nonlinear system's inverse control. System Simulation Technology 18(9), 2688–2690 (2006)
6. Zhang, R., Wang, S.: Multi-step predictive control based on neural network's nonlinear systems. Control and Decision 20(3), 332–336 (2005)

Image Segmentation Method Based Upon Otsu ACO Algorithm

Kanglin Gao[1], Mei Dong[2], Liqin Zhu[1], and Mingjun Gao[1]

[1] School of Computer Information Engineering, Shandong University of Finance, No. 40 Shungeng Road, 250014, Jinan, Shandong, China
gkl_sd@126.com

[2] School of Information Science and Engineering, Jinan University 250001, Jinan, Shandong, China
mei_dong_2004@126.com

Abstract. Image segmentation is a very important research content in the fields of computer vision and pattern recognition. As the basis for image understanding and image analysis it is always receives high attention. At present in those commonly used segmentation methods of contrast ratio, margin and grayscale detection the threshold processing is one of the most effective. The threshold methods can also be divided into Otsu, minimum error thresholding, optimum histogram entropy and minimum cross entropy, etc. By combining the advantages of ACO (Ant Colony Optimization) the present paper has designed an ACO segmentation algorithm of solving extra-class variance maximum value to determine the optimum threshold value. The algorithm can quickly and steadily find the optimum segmentation threshold in a non-linear way so as to effectively segment the target and its background, and receive a best result in image segmenting.

Keywords: ACO; Image Segmentation; Otsu; Edge Extracting.

1 Introduction

Image segmentation is the key and firstly important step of automatic target segmentation. It aims to segment a digital image into mutually disjoint (non-overlapping) regions. That is to say in accordance with the similarity norms of some target features or characteristic set it groups the image pixels into clusters and divides the image into meaningful regions so as to much reduce the data at the advanced processing stages of subsequent target recognizing and tracking, and maintain the characteristic information of the target structure at the same time.

Image segmentation method can be divided into three categories. First, by using some features of already known regions or the features that can be obtained in processing it seeks all kinds of image patterns and studies various pixel groups such as analyses of gray threshold, region growing and texture structure, etc; Second, it focuses on maintaining edge properties and tracking the edge so as to form a profile such as all sorts of edge detection operators; Third, it segments an image by using a scenary's priori knowledge and statistical properties. The segmentation method based

L. Qi (Ed.): ISIA 2010, CCIS 86, pp. 574–580, 2011.

on knowledge integrates segmentation and explanation. It firstly makes an initial segmentation of an image and extracts its regional properties. Then it uses the regional knowledge to derive the regional explanation and merge the two in accordance with the explanation. Usually image segmentation methods include threshold value, edge detection and regional tracking among which threshold value is commonly used one. At present there are many threshold segmentation methods such as minimum error threshold, Otsu and optimum histogram entropy, etc.

2 Otsu Threshold Segmentation Method

The method put forward by Otsu is derived on the basis of judging and analyzing the least square method serving as a convenient threshold selecting approach. The basic idea of the algorithm is: divide an image into two groups by taking a certain grayscale of the image histogram as threshold value and calculate the variances of the two groups. When the variance between the two divided groups is maximum we can use the grayscale as threshold value to segment the image.

Suppose the image grayness range is (0,1,…., L-1) . The pixel of grayscale I is n,then the total pixels of the image is as follows: $N = n_0 + n_1 + ... + n_{L-1}$

And then the normalized histogram is: $P_i = \frac{n_i}{N}, \sum_{i=0}^{L-1} P_i = 1$

Select threshold value t, and divide it into two categories according to its grayscale: C_0: {0,1,…,t},C_1: {t+1,t+2,…,L-1}. Therefore the emerging probability of C_0 and C_1, and the mean value layer are respectively given in the below formulas:

$$\omega_0 = P_r(C_0) = \sum_{i=0}^{t} p_i = \omega(t) \qquad \omega_1 = P_r(C_1) = \sum_{i=t+1}^{L-1} p_i = 1 - \omega(t)$$

$$\mu_0 = \sum_{i=0}^{t} i p_i / \omega_0 = \mu(t) / \omega(t) \qquad \mu_1 = \sum_{i=t+1}^{L-1} i p_i / \omega_1 = \frac{\mu_r - \mu(t)}{1 - \omega(t)}$$

The class variance of C_0 and C_1 can be obtained from the below formula:

$$\sigma_0^2 = \sum_{i=0}^{t} (i - \mu_0)^2 p_i / \omega_0 \qquad \sigma_1^2 = \sum_{i=t+1}^{L-1} (i - \mu_1)^2 p_i / \omega_1$$

Define intra-class variance as : $\sigma_\omega^2 = \omega_0 \sigma_0^2 + \omega_1 \sigma_1^2$

And extra-class as: $\sigma_B^2 = \omega_0 (\mu_0 - \mu_r)^2 + \omega_1 (\mu_1 - \mu_r)^2 = \omega_0 \omega_1 (\mu_1 - \mu_0)^2$

When the inter-variance of the two segmented regions is maximum it is regarded as their optimum separate status,thus determining optimum threshold value T: $T = \max[\sigma_B^2(t)]$.It is thus clear that T is the function of segmentation threshold value t.

Suppose the grayscale value of an image as m and pixel number with grayscale value i as ni. Seek T by changing K value between l-m so as to have: $T = \max[\sigma_B^2(t)]$. Then take T as threshold value to segment the image so as to get an optimum segmentation effect. Obviously, in order to the maximum value T, it is required to make variance calculation of all grayscale values between l-m and finally get the maximum variance after comparison. Such calculation work is huge and therefore it is necessary to find an effective and rapid way of seeking solution.

3 The Fundamental Principles of ACO and Its Application in Function Optimization

3.1 The Fundamental Principles of ACO

The Ant System was put forward by Dorigo in his 1992 doctoral dissertation. In 1997 he improved the AS [5] and proposed the ACS (Ant Colony System). It was soon applied to TSP (Traveling Salesman Problem) [3]. In 1999 Dorigo teemed all solution-seeking combination optimization algorithms developed from AS or ACS as ACO (Ant Colony Optimization). Domestic researchers have also made deep study of ACO and achieved many academic research results[1,2,4].

ACO is a heuristic algorithm developed by imitating ants finding food and seeking optimal path. Animal behaviorists observe all most totally blind ants communicate with each other through pheromone. When ants look for food and continue to move forward they will release certain pheromone on the path they have walked through. The more pheromone the former ants release the more stimulus the latter will have to choose the path the former ants have walked through, that is the path that contains more pheromone and that has high probability to be chosen by the other ants. When the positive feedback behavior emerges repeatedly the shortest path between nest and food is determined because the ants that have chosen the shortest path can always go back and forth in a shortest period between nest and food (because the path is shortest). Relatively the pheromone on the path accumulates more quickly than that on other longer paths. Finally all the ants will choose the shortest path so as to reach the most effective.

Many researches have proved that AS algorithm has strong ability of seeking better solutions not only because it uses positive feedback principle and accelerates evolutionary process to certain degree but also because it is in nature a parallel algorithm. Its properties are as follows:(1) Relatively Strong Robust: With a minor alteration of ACO it can be applied to solving other problems.(2) Distributed Calculation: ACO is an evolutionary algorithm based on populations and is in nature paralleled so as to be easily realized.(3) Easy Combination with Other Methods.

3.2 ACO's Mathematical Model and Function Optimum Application

ACO includes two basic stages of adaption and collaboration. In the stage of adaption all candidate solutions steadily adapt themselves according to the accumulated information. When a path has more ants to walk through it has more pheromone. That means the path will be more easily chosen as the candidate solution. In the stage of collaboration all candidate solutions are expected to bring about better ones through exchanging information among them. In order to spell out the ACO's mathematical model let's take solving of the continued domain function optimization problem as an example:

$b_i(t)$ stands for the number of ants on the element i in the moment t, $\tau_{ij}^k(t)$ refers to the amount of information on the path (i,j) in the moment t, n represents the scale and m is the total number of ants, then $m = \sum_{i=1}^{n} b_i(t)$. The moving ant k decides its direction according to the amount of information on different paths. In search process the ant calculates probability of state transition according to the information and inspiring information on different paths. Let $p_{ij}^k(t)$ stand for the probability of the ant no. k transiting from element i to element j, then:

$$p_{ij}^k(t) = \begin{cases} \dfrac{\tau_{ij}^{\alpha}(t)\eta_{ij}^{\beta}(t)}{\sum_{s \in allowed} \tau_{ij}^{\alpha}(t)\eta_{ij}^{\beta}(t)} & j \in allowed \\ 0 & otherwise \end{cases}$$

During initial period the amount of information on different paths is equal. Suppose $\tau_{ij}(0) = c$ (c is constant), m is the total number of ants, n is the total number of paths, and the moving ant no. k decides its transferring direction according to the amount of information on different paths.α,β stand for respectively the accumulated information of the ant no. k in moving and the different role played by heuristic factor in ant's choosing paths. $\eta_{ij}(t)$ is heuristic function indicating the expectation degree of ant transferring from element i to element j. Its expression is: $\eta_{ij}(t) = 1/d_{ij}$,and $d_{ij}(t)$ represents the distance between neighbouring elements.

In order to prevent "early maturity" when every ant or all ants finish one cycle of path choice the pheromone on different paths need to be renewed in below formula:

$$\tau_{ij}(t+n) = \rho\tau_{ij}(t) + \Delta\tau_{ij} \qquad \Delta\tau_{ij} = \sum_{k=1}^{m} \Delta\tau_{ij}^k$$

Therein, parameter ρ stands for volatilization degree of pheromone and (1-ρ) indicates residual factor of pheromone (value ranging [0,1]). $\Delta\tau_{ij}^k$ stand for the information increment left by the ant no. k on the path (ij) after the present cycle and $\Delta\tau_{ij}$ indicate the total information increment left by all ants on the path ij after the present cycle.

Description and Designing of Continued Domain Function Optimization Algorithm:

Suppose optimization function as maxZ=f(X), m number of ants is randomly distributed in definitional domain. Each ant takes corresponding movements according its own probability of state transition and renew the amount of information after one cycle. Probability of state transition is defined as:

$$p_i(t) = \frac{\max(\tau(t-1) - \tau_i(t-1))}{\max(\tau(t-1))}$$

Let Act (i) be the movement chosen by the no. i ant, then:

$$Act(i) = \begin{cases} \text{local searching} & \text{if } p_i(t) < \rho_0 (0 < \rho_0 < 1) \\ \text{entire searching} & \text{else} \end{cases}$$

Renewing rules for amount of information:

$\tau_i(t+1) = (1-\rho) \times \tau_i(t) + f(x(t+1))$ (pheromone volatilization ratio $0<\rho<1$)

ACO's simulation experiment of solving continued domain function optimization:

$$f(x, y) = x + 10 * \sin(5 * x) + 7 * \cos(4 * x)$$

Experimental Parameters: number of ants=100, entire transition probability=0.7, pheromone volatilization ratio 0.2, proportional constant=1.After ten iterations in x=7.857138 we have obtained the entire optimum value y=24.855313.

4 ACO and Its Experimental Result in Otsu Image Segmentation

4.1 Main Steps of ACO in Otsu Image Segmentation

The Otsu solution-seeking process is to search an optimum solution in solution domain so as to make its variance maximum. ACO algorithm can quickly find the optimum solution K* and maximum variance in a non-linear way. The steps are as follows: The original image is of grayscale 256 and of size 256*256.

(1) Initialization Parameters: entire transition probability is 0.7, pheromone volatilization ratio 0.2, proportional constant 1, etc. In order to apply ACO we use real number of solution domain to encode. Because the grayscale consists of 0~255 values the actual rang of each ant [xmin, xmax] is [0, 255]. The present design selects M ants (adjustable).

(2) Initialization Population. Produce initial ants with a scale of M and calculate initialization solution domain so as to obtain candidate values with M different solutions. If a selected initial value is too much inclined to one side that will cause problems of slow convergence of optimum solution and longer period of calculation. Therefore the algorithm uses an inhibition method to randomly produce an initialization population between grayscale q and p. Experiment shows the method has a quicker convergence in usual situations.

(3) Iteration Process (times for an ant's movements)
While not termination conditions
For i=1 to n do /to n components cycling
For k=1 to m do /to m ants cycling
Seek the no. k probability of state transition.
End for k
For j=1 to m do /to m ants cycling

According to the given parameters of the ant's probability of state transition make local and entire searching so as to decide the ant's next movement. Therein the extra-class variance proper value function is used to find corresponding m candidate solutions.

End for j

Renew the amount of information of each candidate value in various candidate component groups.

$$\tau_i(t+1) = (1-\rho)\times\tau_i(t) + f(x(t+1)) \qquad \text{(pheromone volatilization ratio } 0<\rho<1\text{)}$$

Select each component with higher amount of information as new candidate value.

End termination conditions

(4) In the end select the component with maximum amount of information or that reaches termination conditions as the final solution.

4.2 The Result of Simulation Experiment

The usage of ACO in Otsu image segmentation is in fact to use its rapid optimization algorithm to solve the maximum variance and the corresponding threshold K*. A very good result is received after an experiment with Otsu ACO segmentation on Lena and camerama images.

On basis of the segmentation we have also made an experiment with the target image edge extraction and a comparison with the result of Canny algorithm. The result is good, too. See picture 3. The experimental parameters: number of ants=100, entire transition probability=0.7, pheromone volatilization ratio=0.2, proportional constant=1. Iteration is 5 times.

In Fig 1 and Fig 2, (a) is original images and (b) algorithmic segmenting image. (c) is Canny edge extracting image and (d) ACO algorithmic edge extracting image.

(a) (b) (c) (d)

Fig. 1. ACO Otsu Image Segmenting Result 1 (Threshold=103, 256*256)

(a) (b) (c) (d)

Fig. 2. ACO Otsu Image Segmenting Result 2 (Threshold=91,128*128)

5 The Analysis of the Experimental Result Shows

(1) About the ant evaluation function. The method each ant sets maximum class variance with serves as the ant evaluation function. That is the bigger variance G(x) get the more possibly it draws near to the optimum solution. According to the formulary definition solve for each ant's proper value and then its selective probability and cumulative probability so as to determine the next iteration population's individuals. Therein those with bigger adaptive value are more possible to be selected. Otherwise they are less possible. Thus doing one generation after another, each generation gains different adaptive value. The newer generation of ants has higher adaptive value and so its solution approaches maximum value.

(2) Through trial calculation of the actual image when multiplication arrives to 1~5 generation it can all get optimum threshold k* and maximum variance. In order to speed up the calculation the algorithm has been improved in the following two aspects. One is to fully use the obtained histogram data and there is no need to solve for function value to be reserved each time in operation of evaluation function. Reserve the calculated w_0,w_1,u_0,u_1 that correspond to K and just take out the calculated data for operating the next generation so as to speed up the whole algorithmic operation process. The other is to apply a restraint mechanism in producing initial and new ants so as to ensure the new ants are within the designated limits. Therefore by avoiding unnecessary operation and enhancing operation speed and efficiency, plus with its parallel calculation feature ACO algorithm operates more quickly.

6 Conclusion

When the ACO the paper discusses is applied to an Otsu threshold segmenting image it can quickly and steadily solve Otsu variance and grayscale threshold value in a non-linear way so as to effectively enhance image threshold segmenting speed and performance. Since ACO has a rapid parallel computing ability it can realize a parallel computation of image segmentation. Experiment proves the algorithm is of effectiveness and has a better result of segmentation.

References

1. Duan, H., Wang, D., Yu, X.: Research Status and Prospect Of ACO. China Engineering Sciences 9(2), 98–102 (2007)
2. Wu, Q., Zhang, J., Xu, X.: ACO with Variation Features. Journal of Computer Research and Development 36(10), 1240–1245 (1999)
3. Dorigo, M., Gambardella, L.M.: Ant colonies for the traveling salesman problem. Biosystem 43, 73–81 (1997)
4. Duan, H.: Theorem and Application of ACO. Science Press, Beijing (December 2005)
5. Dorigo, M.: Optimization, Learning and Natural Algorithm, Ph.D., Thesis, DEI, Politecnico di Milano, Italy (1992)

Sampling Matrix Perturbation Analysis of Subspace Pursuit for Compressive Sensing

Qun Wang and Zhiwen Liu

Department of Electronic Engineering, Beijing Institute of Technology
Beijing 100081, China
wq99102@yahoo.com.cn

Abstract. In this paper, the Subspace Pursuit (SP) recovery of signals with sensing matrix perturbations is analyzed. Previous studies have only considered the robustness of Basis pursuit and greedy algorithms to recover the signal in the presence of additive noise with measurement and/or signal. Since it is impractical to exactly implement the sampling matrix A in a physical sensor, precision errors must be considered. Recently, work has been done to analyze the methods with noise in the sampling matrix, which generates a multiplicative noise term. This new perturbed framework (both additive and multiplicative noise) extends the prior work of Basis pursuit and greedy algorithms on stable signal recovery from incomplete and inaccurate measurements. Our works show that, under reasonable conditions, the stability of the SP solution of the completely perturbed scenario was limited by the total noise in the observation.

Keywords: multiplicative noise; Subspace Pursuit algorithm; Sampling matrix perturbation component.

1 Introduction

Compressive sensing (CS) is a novel compression technique widely used in science and engineering involving large scale data [1]. The basic principle is that sparse or compressible signals can be reconstructed from a small number of non-adaptive random projections, which carry sufficient information to approximate the signal well.

The initial studies consider a sampling scenario $y = \boldsymbol{A}x$ where $\boldsymbol{A} \in \mathbb{R}^{m\times N} (m \ll N)$ is a sampling/sensing matrix. Recently, either signal or measurement perturbations are added, i.e. $y = \boldsymbol{A}x + e$, e is an additive stochastic /deterministic error or noise term. However, when applying compressive sensing technique to real-world problems, we must consider the influence when one decodes the measurements y with a slightly different sensing matrix $\boldsymbol{\Phi}$ instead of the original one $\boldsymbol{A}$.

The sampling matrix $\boldsymbol{A}$ actually represents a system which the signal passes through, or some other physical phenomenon. Whatever the setting may be, it is often the case that the true nature of this system is not known exactly. For instance, it is impractical to implement $\boldsymbol{A}$ exactly due to precision errors or human manipulation in screening for genetic disorders. When this happens, the system behavior is (perhaps unknowingly) approximated, or assumed to be represented, by a different matrix $\boldsymbol{\Phi}$.

L. Qi (Ed.): ISIA 2010, CCIS 86, pp. 581–588, 2011.

$$\boldsymbol{\Phi} = \boldsymbol{A} + \boldsymbol{E} \tag{1}$$

It is obvious that the encoding and decoding matrix $\boldsymbol{A}$ and $\boldsymbol{\Phi}$ can not be too different, but until recently there has been no analysis of the effect this difference has on reconstruction error. In particular, the perturbation of the sensing matrix creates a multiplicative noise of the form $\boldsymbol{Ex}$ in the system. This type of noise is fundamentally different than simple additive noise. For example, to overcome a poor signal-to-noise ratio (SNR) due to additive noise, one would typically increase the strength of the signal. However, if the noise is multiplicative this will not improve the situation, and in fact will actually cause the worse error. Thus the impact on reconstruction from the error in the sensing matrices needs to be analyzed. The Subspace Pursuit algorithm has low computational complexity and high reconstruction accuracy. In this paper, we will analyze the SP recovery of signals with sensing matrix perturbations.

The remainder of the paper is organized as follows. Section II introduces relevant concepts and terminology for describing the proposed CS reconstruction technique. Section III contains the sampling matrix perturbation analysis of SP algorithm. Concluding remarks are stated in Section IV.

2 Preliminaries

2.1 Compressive Sensing and the Restricted Isometry Property

First of all, we precisely formulate the problem for CS. Let $\mathrm{supp}(x)$ denote the index set of the nonzero coordinates of an arbitrary vector $x = (x_1, x_2, \cdots, x_N) \in \mathbb{R}^N$, and let $\|x\|_0 = |\mathrm{supp}(x)|$ denote the support size of x, or equivalently, its ℓ_0-norm. Assume K-sparse signal $x \in \mathbb{R}^N$ be a real value signal with K or fewer non-zero components, i.e. $\|x\|_0 \le K \ll N$. And then apply a sampling matrix $\boldsymbol{A} \in \mathbb{R}^{m \times N}$ to the signal x and acquire an observation $b = \boldsymbol{A}x$ via m linear measurements. Often, we must consider the influence of additive noise so that the measurements become

$$y = b + e = \boldsymbol{A}x + e \tag{2}$$

where e is a stochastic/deterministic error or noise term with bounded energy $\|e\|_2 \le \varepsilon$.

The problem of low-complexity recovery of the unknown signal x from measurement y must be concerned with. A natural formulation of the recovery problem is within a ℓ_0-norm minimization framework, which seeks a solution to the problem

$$\min \ \|x\|_0 \ \text{ subject to } \|\boldsymbol{A}x - y\|_2 \le \varepsilon \tag{3}$$

Unfortunately, the above minimization problem is NP-hard, and hence cannot be used for practical applications [1].

One way to avoid using the above computationally intractable formulation (2) is to consider a ℓ_1-minimization problem, which can be solved using convex optimization techniques and is thus computationally feasible [1]. The method simply seeks a solution to recover the signal x

$$\min \|x\|_1 \text{ subject to } \|\boldsymbol{A}x - y\|_2 \le \varepsilon \tag{4}$$

where $\|x\|_1 = \sum_{i=1}^{N} |x_i|$ denotes the ℓ_1-norm of vector x. In words, we look for a signal reconstruction that is consistent with the samples but has minimal ℓ_1-norm. The intuition behind this approach is that minimizing the ℓ_1-norm promotes sparseness to approximate recovery of compressible signals.

Candès and Tao show in [1] that if the signal x is sparse and the sensing matrix $\boldsymbol{A}$ satisfies a certain quantitative property, then equation (3) can reconstruct the original signal x exactly. The accuracy of the ℓ_1-minimization method is described in terms of the so called Restricted Isometry Property (RIP), formally defined below.

Definition 1 (Truncation). Let $\boldsymbol{\Phi} \in \mathbb{R}^{m \times N}$ and $I \subset \{1, 2, \cdots, N\}$. The matrix $\boldsymbol{\Phi}_I$ consists of the columns of $\boldsymbol{\Phi}$ with indices $i \in I$.

Definition 2 (RIP). A sampling matrix $\boldsymbol{\Phi} \in \mathbb{R}^{m \times N}$ is said to satisfy the RIP with parameters (K, δ) for $K \le m$, $0 \le \delta \le 1$, if for all index sets $I \subset \{1, 2, \cdots, N\}$ such that $|I| \le K$ and for all $q \in \mathbb{R}^{|I|}$, one has

$$(1-\delta)\|q\|_2^2 \le \|\boldsymbol{\Phi}_I q\|_2^2 \le (1+\delta)\|q\|_2^2$$

We define the RIP constant (RIC) δ_K, as the infimum of all parameters δ for which the RIP holds, i.e,

$$\delta_K := \inf\left\{\delta : (1-\delta)\|q\|_2^2 \le \|\boldsymbol{\Phi}_I q\|_2^2 \le (1+\delta)\|q\|_2^2, \forall |I| \le K, \forall q \in \mathbb{R}^{|I|}\right\}$$

It is well known that many random (e.g., Gaussian, Bernoulli) matrices satisfy the RIP when $m = O(K \log(N))$. It has been shown in [2] that if $\boldsymbol{A}$ satisfies the RIP with parameter $(2K, \sqrt{2}-1)$, then equation (3) recovers a signal $\tilde{x}$ that satisfies

$$\|x - \tilde{x}\|_2 \le C_0 \|x - x_K\|_1 / \sqrt{K} + C_1 \varepsilon$$

where x_K denotes the vector obtained from x by maintaining the K entries with largest magnitude and setting all other entries in the vector to zero.

2.2 The Subspace Pursuit Algorithm

Although the ℓ_1-minimization techniques play an important role in designing computationally tractable CS decoders and its recovery guarantees are strong, their complexity is still highly impractical for many applications. For this reason, much work in compressed sensing has been done to find faster methods. Recently, a family of iterative greedy algorithms received significant attention due to their low complexity and simple geometric interpretation. These are the Orthogonal Matching Pursuit (OMP) [3] and Subspace Pursuit (SP) algorithm [4]. The basic idea behind these methods is to find the support of the unknown signal sequentially.

The analysis in [4] shows that in the noiseless setting, the SP algorithm can exactly reconstruct arbitrary sparse signals provided that the sensing matrix satisfies the RIP with a small constant parameter. In the noisy setting and in the case that the signal is not exactly sparse, it can be shown that the mean-squared error of the reconstruction is upper-bounded by constant multiples of the measurement and signal perturbation energies. On the basis of those, we summarize and extend by the following theorem 1.

Theorem 1. (Stability under signal and measurement perturbations) Let $x \in \mathbb{R}^N$ be an arbitrary signal, and let $y = Ax + e$. Suppose that the sampling matrix $A \in \mathbb{R}^{m \times N}$ satisfies the RIP with parameter $\delta_{3K} < 0.083$, then

$$\|x-\hat{x}\|_2 \le (C''_K + 1)\|x - x_K\|_2 + C''_K \frac{\|x - x_K\|_1}{\sqrt{K}} + C'_K \|e\|_2$$

Where $C'_K = \dfrac{1+\delta_{3K}+\delta_{3K}^2}{\delta_{3K}(1-\delta_{3K})}$ and $C''_K = C'_K\sqrt{1+\delta_K}$.

Proof of Theorem 1
To prove the theorem, the measurement vector must be considered

$$y = Ax + e = Ax_K + A(x - x_K) + e$$

Theorem 9 and Lemma 4 in [4] show that

$$\|\hat{x} - x_K\|_2 \le C'_K \left(\|A(x-x_K)\|_2 + \|e\|_2\right)$$

$$\|A(x-x_K)\|_2 \le \sqrt{1+\delta_K}\left(\|x-x_K\|_2 + \frac{\|x-x_K\|_1}{\sqrt{K}}\right)$$

Therefore

$$\|\hat{x} - x\|_2 \le \|\hat{x} - x_K\|_2 + \|x - x_K\|_2 \le C'_K \left(\|A(x-x_K)\|_2 + \|e\|_2\right) + \|x - x_K\|_2$$

$$\le C'_K \left(\sqrt{1+\delta_K}\left(\|x-x_K\|_2 + \frac{\|x-x_K\|_1}{\sqrt{K}}\right) + \|e\|_2\right) + \|x-x_K\|_2$$

$$= \left(C'_K\sqrt{1+\delta_K} + 1\right)\|x-x_K\|_2 + C'_K\sqrt{1+\delta_K}\frac{\|x-x_K\|_1}{\sqrt{K}} + C'_K\|e\|_2$$

3 Sampling Matrix Perturbation Analysis of Subspace Pursuit

Before introducing the next theorem, let us introduce the submatrix consisting of an arbitrary collection of K columns. We use the superscript (K) to represent the extremal values of ℓ_2-norms over all K columns submatrices of $\boldsymbol{A}$.

$$\varepsilon_A^{(K)} := \frac{\|\boldsymbol{A} - \boldsymbol{\Phi}\|_2^{(K)}}{\|\boldsymbol{A}\|_2^{(K)}} = \frac{\|\boldsymbol{E}\|_2^{(K)}}{\|\boldsymbol{A}\|_2^{(K)}} \quad and \quad \kappa_A^{(K)} := \frac{\sqrt{1+\delta_K}}{\sqrt{1-\delta_K}} \tag{5}$$

The first quantity is the relative perturbation of K-column submatrices of $\boldsymbol{A}$ with respect to the ℓ_2-norm, and the second bounds ratio of the extremal singular values of all K-column submatrices of $\boldsymbol{A}$ (see [5] for more details). We also need a measure to characterize how "close" a signal x is to a sparse signal x_K, and therefore define

$$r_K := \frac{\|x - x_K\|_2}{\|x_K\|_2} \quad and \quad s_K := \frac{\|x - x_K\|_1}{\sqrt{K}\,\|x_K\|_2} \tag{6}$$

3.1 Sampling Matrix Perturbation Analysis in ℓ1-Minimization

Herman *et al* [5] first showed that a noisy sampling matrix can be successfully used to recover a signal using ℓ_1-minimization. They extended the previous results in ℓ_1-minimization by generalizing the additive noise term ε to a total noise term $\varepsilon_{A,K,b}$, which includes a multiplicative noise term in addition to the usual additive noise. The following Theorem 2 shows that the reconstruction error using ℓ_1-minimization is limited by this noise level. With regard to noise in $\boldsymbol{A}$, we see that the stability of the solution is a linear function of relative perturbations ε_A $\varepsilon_A^{(K)}$.

Theorem 2 (Stability from completely perturbed observation in ℓ_1-minimization [5]) Let x be an arbitrary signal with measurements $b = \boldsymbol{A}x$, corrupted with additive noise to form $y = \boldsymbol{A}x + e$. Assume the RIC for sampling matrix $\boldsymbol{A}$ satisfies

$$\delta_{2K} < \frac{\sqrt{2}}{\left(1 + \varepsilon_A^{(2K)}\right)^2} - 1$$

The general signal x satisfies

$$r_K + s_K < \frac{1}{\kappa_A^{(K)}} \tag{7}$$

Set the total noise parameter

$$\varepsilon_{A,K,b} := \left(\frac{\varepsilon_A^{(K)} \kappa_A^{(K)} + \varepsilon_A \alpha_A r_K}{1 - \kappa_A^{(K)} \left(r_K + s_K\right)} + \varepsilon_b \right) \|b\|_2 \tag{8}$$

Where the relative perturbations

$$\varepsilon_A = \frac{\|A - \Phi\|_2}{\|A\|_2} \quad \varepsilon_b = \frac{\|e\|_2}{\|b\|_2} \quad and \quad \alpha_A = \frac{\|A\|_2}{\sqrt{1 - \delta_K}}$$

Then the solution $\tilde{x}$ to the ℓ_1-minimization problem (3) with ε set to $\varepsilon_{A,K,b}$, and using the decoding matrix $\boldsymbol{\Phi}$ (instead of $\boldsymbol{A}$) obeys

$$\|x - \tilde{x}\|_2 \le C_0 \frac{\|x - x_K\|_1}{\sqrt{K}} + C_1 \varepsilon_{A,K,b}$$

for some well-behaved constants C_0, C_1.

3.2 Sampling Matrix Perturbation Analysis in Subspace Pursuit

Now let us extend the perturbation analysis to greedy algorithms, especial Subspace Pursuit (SP) algorithm. We will see that a result analogous to that of ℓ_1-minimization can be obtained in this case as well. Condition (7) is still used to characterize that the signal can be well approximated by a sparse signal. Theorem 3 below shows that under this assumption, the reconstruction error in Subspace Pursuit is again limited by the tail of the signal and total noise level in the observation.

Theorem 3 (Stability from completely perturbed observation in Subspace Pursuit) Let x be an arbitrary signal with measurements $b = \boldsymbol{A}x$, corrupted with additive noise to form $y = \boldsymbol{A}x + e$. Assume the sampling matrix $\boldsymbol{A}$ satisfies RIC

$$\delta_{3K} < \frac{1.083}{\left(1+\varepsilon_A^{(3K)}\right)^2} - 1 \tag{9}$$

Let $\hat{x}$ be the reconstruction from SP using the decoding matrix $\boldsymbol{\Phi}$ (instead of $\boldsymbol{A}$) on measurements y. Then if condition (7) is satisfied, the estimation satisfies

$$\|x-\hat{x}\|_2 \le (C_K''+1)\|x-x_K\|_2 + C_K'' \frac{\|x-x_K\|_1}{\sqrt{K}} + C_K' \varepsilon_{A,K,b} \|b\|_2$$

where the total noise parameter $\varepsilon_{A,K,b}$ is defined in eq. (8).

Before we prove our main contribution, Theorem 3, we first introduce a result from [5] which demonstrates that matrices which are "close" to each other will possess similar RIC.

Lemma 1 (RIP for $\boldsymbol{\Phi}$ [5]). For any natural number $K = 1, 2, \cdots$, assume the RIC δ_K associated with $\boldsymbol{A}$, and the relative perturbation $\varepsilon_A^{(K)}$ associated with (possibly unknown) matrix $\boldsymbol{E}$ in (4). Then the RIC $\hat{\delta}_K$ for matrix $\boldsymbol{\Phi}$ satisfies

$$\hat{\delta}_K \le (1+\delta_K)\left(1+\varepsilon_A^{(K)}\right)^2 - 1$$

Proof of Theorem 3

Applying Lemma 1 to condition (9), which implies that the RIC $\hat{\delta}_K$ for matrix $\boldsymbol{\Phi}$ satisfies $\hat{\delta}_K < 0.083$. Therefore, we can apply Theorem 1 to the following measurements $y = \boldsymbol{\Phi}x + \tilde{e}$ with $\tilde{e} = (A - \Phi)x + e$. It means the recovered result $\hat{x}$ satisfies

$$\|x-\hat{x}\|_2 \le (C_K''+1)\|x-x_K\|_2 + C_K'' \|x-x_K\|_1 / \sqrt{K} + C_K' \left(\|(A-\Phi)x\|_2 + \|e\|_2\right) \tag{10}$$

According to Lemma 4 in [4], the RIP implies that for an arbitrary signal x

$$\|\boldsymbol{A}x\|_2 \le \sqrt{1+\delta_K}\left(\|x\|_2 + \|x\|_1/\sqrt{K}\right)$$

As shown in Lemma 2 of [5], the RIP also implies that for an arbitrary signal x

$$\|Ax\|_2 \geq \sqrt{1-\delta_K}\left(\|x_K\|_2 - \kappa_A^{(K)}\left(\|x-x_K\|_2 + \|x-x_K\|_1/\sqrt{K}\right)\right)$$

Similar to Lemma 3 of [5], we then have that

$$\|(A-\Phi)x\|_2 \leq \frac{\left(\|(A-\Phi)\|_2^{(K)}\|x_K\|_2 + \|(A-\Phi)\|_2\|x-x_K\|_2\right)}{\sqrt{1-\delta_K}\left(\|x_K\|_2 - \kappa_A^{(K)}\left(\|x-x_K\|_2 + \frac{\|x-x_K\|_1}{\sqrt{K}}\right)\right)}\|b\|_2$$

$$= \frac{\left(\|(A-\Phi)\|_2^{(K)} + \|(A-\Phi)\|_2 r_K\right)}{\sqrt{1-\delta_K}\left(1-\kappa_A^{(K)}(r_K+s_K)\right)}\|b\|_2 \leq \frac{\varepsilon_A^{(K)}\kappa_A^{(K)} + \varepsilon_A\alpha_A r_K}{1-\kappa_A^{(K)}(r_K+s_K)}\|b\|_2$$

Condition (8) then guarantees that the total perturbation obeys

$$\|(A-\Phi)x\|_2 + \|e\|_2 \leq \left(\frac{\varepsilon_A^{(K)}\kappa_A^{(K)} + \varepsilon_A\alpha_A r_K}{1-\kappa_A^{(K)}(r_K+s_K)} + \varepsilon_b\right)\|b\|_2 = \varepsilon_{A,K,b}\|b\|_2$$

Combined with Eq. (10), this completes the claim.

If we apply Theorem 3 to the sparse case, we will immediately obtain the following corollary.

Corollary 1 (sparse case). Assume that A is a sampling matrix with RIC satisfying Eq. (9). Let x be a K-sparse signal with noisy measurements $y = b + e = Ax + e$. Let $\hat{x}$ be the reconstruction from SP using the decoding matrix Φ (instead of A) on measurements y. Then the estimation satisfies

$$\|x-\hat{x}\|_2 \leq C'_K\left(\varepsilon_A^{(K)}\kappa_A^{(K)}\|b\|_2 + \|e\|_2\right)$$

Furthermore, one should make sure that the requirements imposed by Theorems 2 and 3 are reasonable and make sense. For instance, Theorem 2 essentially requires that $\varepsilon_A^{(2K)} < \sqrt[4]{2}-1$, which addresses the question "how dissimilar can A and Φ be?" In other words, the answer is that the ℓ_2-norm of 2K-column submatrices of Φ cannot deviate by more than about 19% of ℓ_2-norm of 2K-column submatrices of A. And the corresponding condition (9) in Theorem 3 requires that $\varepsilon_A^{(3K)} < \sqrt{1.083}-1$, which translates to an approximate 4% dissimilarity between A and Φ. The condition (7) essentially requires that the signal can be well approximated by a K-sparse signal [5]. This is, of course, a common assumption in compressed sensing.

4 Conclusions

In conclusion, real-world applications often utilize different matrices (perhaps unknowingly) to encode and decode a signal. The perturbation of the sensing matrix creates multiplicative noise in the system. In this paper, we found the conditions under which SP could stably recover the original data. Our results show the effect of using different matrices to recover a signal in compressed sensing: the stability of the recovered signal is a linear function of the sensing matrix relative perturbation $\varepsilon_A^{(3K)}$, which must either be calculated or estimated in real-world applications. This work confirms that this is both the case for ℓ_1-minimization and greedy algorithm. Our analysis can extend easily to arbitrary greedy methods.

References

1. Candès, E., Tao, T.: Decoding by linear programming. IEEE Trans. Inf. Theory 51(12), 4203–4215 (2005)
2. Candès, E.: The restricted isometry property and its implications for compressed sensing. C. R. Math. Acad. Sci. Paris, Serie I 346, 589–592 (2008)
3. Tropp, J., Gilbert, A.: Signal recovery from random measurements via orthogonal matching pursuit. IEEE Trans. Inf. Theory 53(12), 4655–4666 (2007)
4. Dai, W., Milenkovic, O.: Subspace pursuit for compressive sensing signal reconstruction. IEEE Trans. Info. Theory 55(5), 2230–2249 (2009)
5. Herman, M., Strohmer, T.: General Deviants: An analysis of perturbations in compressed sensing. IEEE Journal of Selected Topics in Sig. Proc.: Special Issue on Compressive Sensing 4(2) (April 2010)

The Design of Rural Consumer Services Cooperatives Management System Based on E-Commerce

Miao Wang and Zhan Bin Che

Software College, Zhongyuan University of Technology, Zhengzhou, China

Abstract. According the actual needs of rural consumer services cooperatives information construction, in this paper, online order processing flow is designed, the overall system solutions is presented, detailed design is introduced, which store all order information, can help the cooperatives manage sales by order ,achieve vertical management and transparency of work.

Keywords: Management System, Online Ordering, Order Processing, System Architecture, E-Commerce.

1 Introduction

Now E-commerce is booming at home and abroad. With technical problems solving and secure network transaction, the application system based on e-commerce mushroomed. Many enterprises are gradually promoting the transformation from traditional business to e-commerce business. In rural areas it is difficult to transport goods and materials, and the transit cost is high. An intermediate platform based on E-Commerce, is proposed to help material information dissemination and material flow, which achieves information sharing. Farmers can understand the current price level of materials needed and material transfer process at the first time. They are allowed to directly contact with the manufacturer, avoiding the additional transport costs arising because of the several material transport. This system is the information superhighway between the material manufacturers and farmers. With the help of the intermediate platform, consumer services cooperatives joint farmers, merchants and manufacturers together, help manufacturers sell goods at ex-factory price. Farmers buy high-quality products at low price, grassroots vendors reduce purchasing costs, eliminate fake and shoddy products and increase economic benefits, and manufacturers establish the sustainable market. The system will serve the cooperative finance department, the cooperative management information center, and workstations at all levels (including city-level workstation, county-level workstation, and township-level workstation) and collaboration manufacturers. Headquarters management information center manages and maintains it, is supervised by supervision departments at all levels. All township stations directly trade with headquarters, city-level and county-level workstation understand the lower station's business transactions, and the headquarters can know all level business conditions.

L. Qi (Ed.): ISIA 2010, CCIS 86, pp. 589–594, 2011.

2 Order Processing Flow

Order processing flow achieves the communication between all levels workstation and headquarters, financial supervision, real-time status feedback of goods and capital, report statistics. As the core of business activities, order is the main business processes in the rural consumer services cooperative management system[1]. Order processing flow is shown in Fig.1 [2, 3].

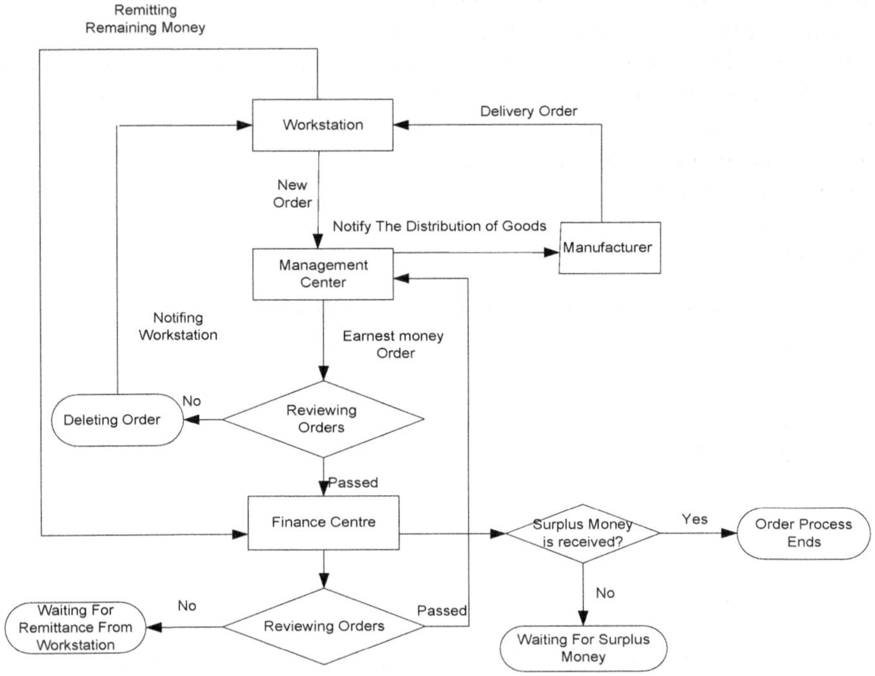

Fig. 1. Order Processing Chart

As can be seen from the chart, the system mainly has the following four categories of users:

1. Workstations at All Levels
Order generation is the basic functions at all levels of workstation. According to the needs of users, workstations application organize goods demand into online orders and submit them to the management information centers, earnest money is deposit into the designated bank account. When the manufacturers complete home delivery, the remaining money will be deposited into the designated bank account, submit orders to the finance department.
2. Management Information Center
According order processing flow, management information center follows several steps: First, it reviews orders from workstation at all levels, after confirmation, the orders is marked as verified , is submitted to the finance department to check out

earnest money; Second, through various review of the order , it notifies manufacturers to send goods.

3. Finance Department

Finance department deals with order according the following steps. First, it accepts orders which submitted by management information center after preliminary examination, checks out whether the earnest money is deposit into bank account, then modifies the status of orders Second, after goods receipt, workstation mark orders as received, finance department examine whether the remaining money is deposit into bank account, then modifies the status of orders; Lastly it transfer sale money through banks to the manufacturers account.

4. Manufacturers

After receiving delivery order, manufacturers sending the goods to designated locations in accordance with specifications, and then marked the order as shipped; when receiving the payment, they marked the order as complete.

3 The System Solutions

Based on relational database, rural consumer services cooperatives management system consists of five parts: system database, using data storage; workstation application, which help to generate orders online enquire goods flow information; manufacturer's application, which help manufacturers deal with order, send goods; based information management system, which help management information center to maintain basic data information about workstation user, goods, manufacturers and order; financial application, which is used to manage trading accounts information. The whole system topology is shown in Fig. 2[4].

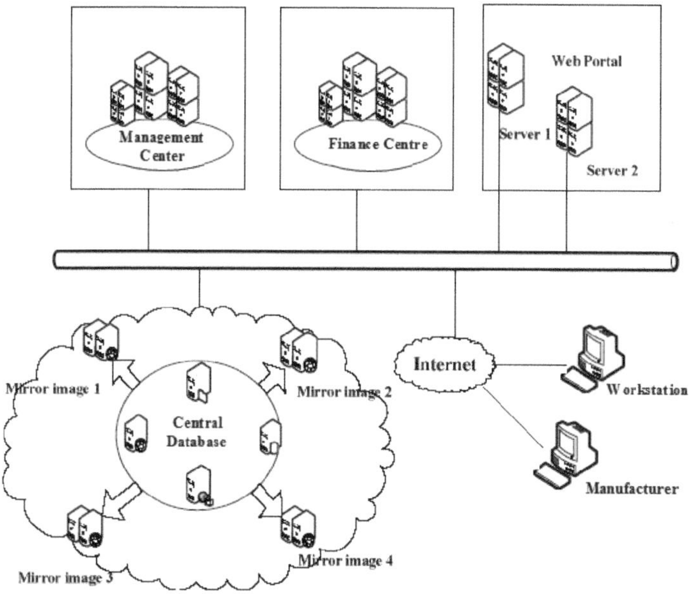

Fig. 2. The overall system solutions

Rural consumer services cooperatives management system uses two modes; one is browse/server mode, to develop front web program, the other is client/sever model, to develop background applications including financial management module and basic data management module

4 Database Design

Database design is the core and foundation for establishing database and application systems. Understanding the need analysis of the whole system and the functions to be achieved, analyzing the overall data concept, we draw the E- R diagram, which is shown in Fig. 3.

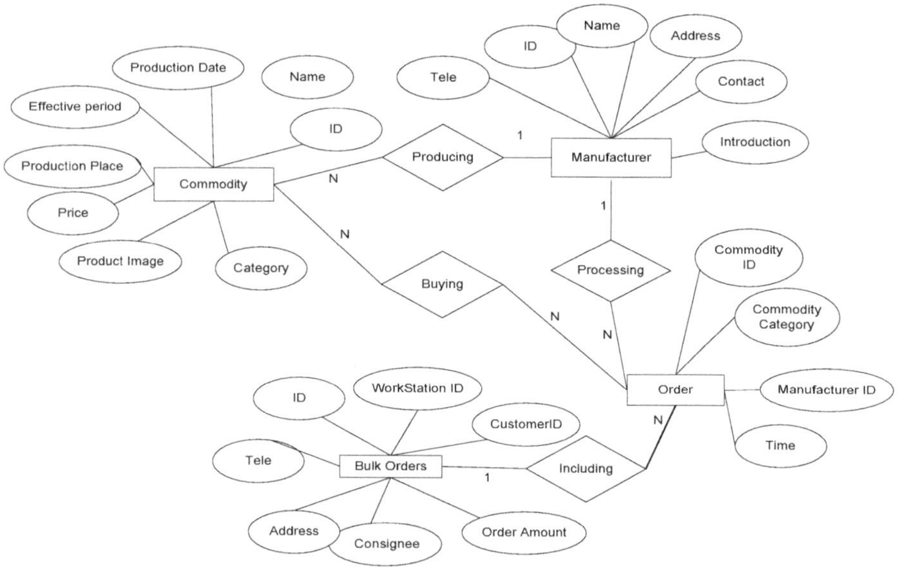

Fig. 3. Entity Relationship Diagram

5 System Functional Structure

The whole system is composed of four subsystems, and they are independent at development stage, complementary at operational stage.

5.1 Workstation Application

Workstation application is at the beginning of the order process. With its help, the workstation user browses all commodity from all manufacturers, carry out shopping operation to new orders. In the shopping process, the user can view the shopping cart to find the goods added and can remove them from the cart. They can view the status of their orders, generate their past transactions statistics reports. Fig. 4 shows its main functions.

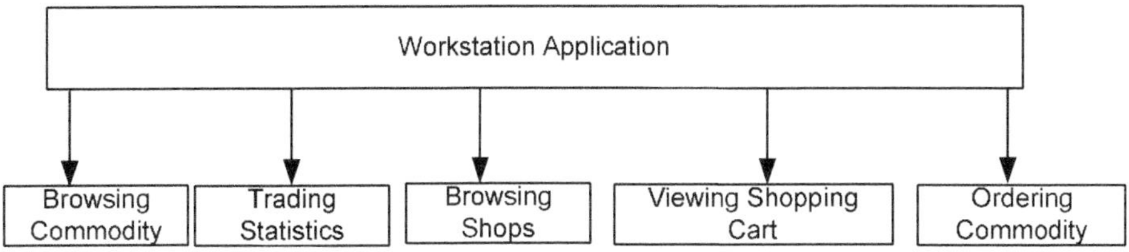

Fig. 4. The function modules of workstation application

5.2 Manufacturer Application

Manufacturer application is in the middle of one end of the order process. Manufacturer can view own their goods, verify their information, and browse their order and consider whether send the goods to the workstation according the order status. In addition, manufacturer can view their trade history and statistics reports. Fig.5 shows its main functions.

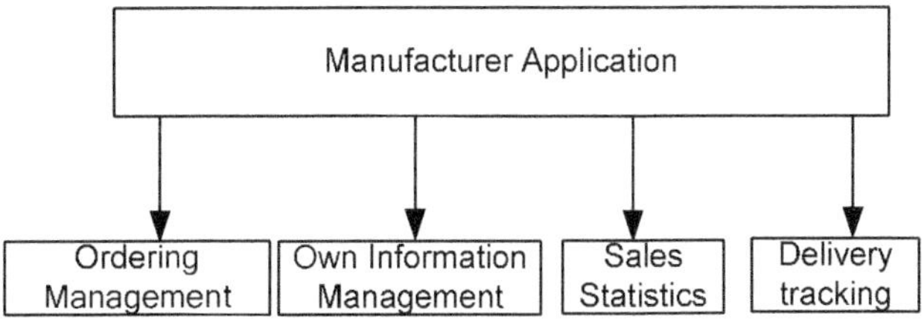

Fig. 5. The function modules of Manufacturer Application

5.3 Basic Information Management Subsystem

The subsystem uses client/server architecture. Firstly, it connects to the database, manages and maintains the information of all level workstation and their users, manufacturer, commodities and their category. Secondly it deals with orders and cancels order. Lastly it generates reports for workstation, manufacturer and a particular commodity. Fig. 6 shows its main functions.

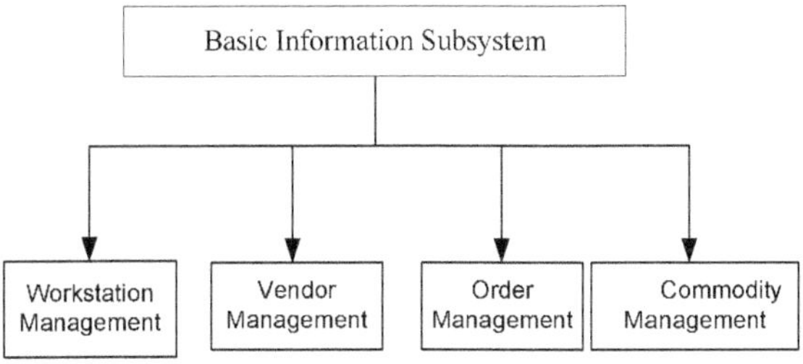

Fig. 6. The function modules of Basic information subsystem

5.4 Financial Application

Financial application uses client/server architecture. Its main functions includes dealing with orders and canceling order, setting the commission percentage, sharing profit among workstation, manufacturer, summarizing periodic(monthly, quarterly, half year, year) financial situation and printing all kind of transaction reports and commission reports. Fig. 7 shows its main functions.

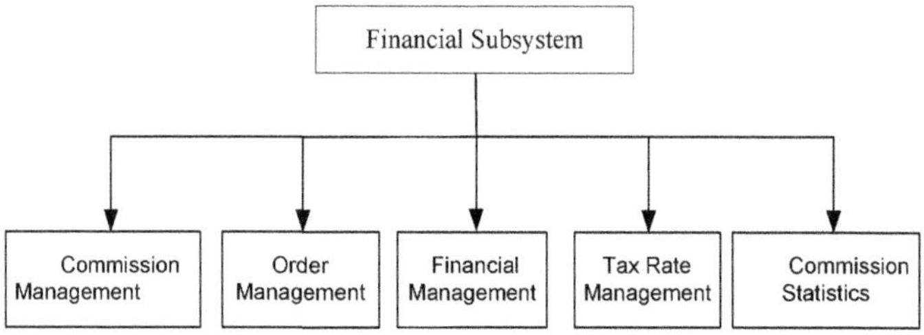

Fig. 7. The function modules of Financial application

6 Summary

Online ordering process is designed for the rural consumer services cooperatives management system, which manages purchase by order, store all order information to the headquarters server, guarantees information consistency and security. The system cleares the division of labor, responsibility and authority at all levels to strengthen monitoring and supervision.

References

1. Chu, J.H., Song, B.H.: Workflow Coordinated Process Modeling. Application Research of Computers 24(4), 120–124 (2007)
2. Xu, Z.C.: Research on Real-Time and Intelligent System of E-Commerce Order Processing. Dalian University of Technology (2006)
3. Gao, L., Wang, R.X., Jiang, X.P., Zhou, Z.F.: Study on the Order Process in Supply Chain Management. Application Research of Computers (8), 185–186 (2005)
4. Wang, H.L., Zheng, Q.S.: The Design of Rural Integrated Information platform. In: Proceedings of the International Conference on Intelligent Computation Technology and Automation, ICICTA (2009)

The Design and Implementation of County-Level Land and Resource Management System Based on Workflow

Miao Wang and Zhan Bin Che

Zhongyuan University of Technology, Zhengzhou, China

Abstract. According the business process of county-04land and resource management system, in this paper, system design ideas is presented, two business process models is designed, system functional structure and detailed design is introduced, which enhances the efficiency and the image of government departments.

Keywords: Land and Resource Management, Business Process Model, Workflow, E-Government.

1 Introduction

With the development of information technology and the universal application of Internet technology, e-government is becoming an important factor to upgrade department integrated administrative capacity. Speeding up the e-government construction is the only way to achieve modernization of management of land resources. E-government is the important part of county-level land and resources information system. County-level land and resources administration is the grassroots organizations, is the front window of government for the public. County-level e-government success is related to the overall work of land and resources, so measures must be taken to protect E-government healthy and orderly development in the grassroots land and resources administration

2 Business Process Analysis of County-Level Land and Resources Administration

According to the section business needs, combined with the practical handling, the basic processes of each business are designed [1].

2.1 The Business Process about State-Owned Land

The state-owned land business involves state-owned land use right allocating and changing, residential land registration, state-owned land mortgage registration and writing off.

The Process described as follows : (1) A staff input and submit the relevant information about a land management business, then it is transferred automatically to

L. Qi (Ed.): ISIA 2010, CCIS 86, pp. 595–599, 2011.

the secretary accepted. In order to effectively supervise the whole business process, the computer records the submit time. (2) After the system login, the secretary accepted instructs the business, the business is automatically transferred to the relevant sections.(3) The sections will check out business information, after confirmation the business process is completed, and the related notes are stored, the sections staff will transfer the files to business office.

2.2 The Business Process about Rural Homestead and Collective Land

County-level land and resources administration has branches in villages and towns, these branches are responsible for rural homestead allocating and changing, Collective construction land approval, Animal breeding land audit. The business processes is described as follows:

(1) Staff member of township land office enter the correct data according to business registration required, then send them to the relevant sections, where the leader puts forward comments. After the initial opinion, the business is transferred automatically to the deputy secretary, which will view and instruct it. (2) Finally, this business is sent to township land office, where it will be written off.

2.3 Petition Business Processes

The petition process is described as follows: (1) A staff in the letters and visits section registers all letters and visits in detail, and then section leader puts forward comments. After the initial opinion, the petition is transferred automatically to the deputy secretary, which will view and instruct it.(2) Because of the specificity of people visiting, the petition will finally be transferred automatically to the secretary, which sign it and send it to the letters and visits section.

3 System Design Ideas

Compared with general office automation system, this land and resource management system brings a new office experience, its features include:

3.1 System Integration and Collaborating Together

To monitor business processes and approval, automated office, to transfer and exchange information automatically, to track approval process, automated office system (OA) and management information system (MIS) should be integrated. All users get the information and complete the necessary operating through a unified platform and interface without switch between applications; the staff can collaborate to realize the cross-sectoral, multi-staff, multi-task work under unified planning and scheduling [2].

3.2 Implementing Strict Security Control

The office automation environment is a flexible, open environment, so system must provide a variety of security mechanisms to ensure data confidentiality and integrity, which includes authentication, role assignment, user action monitoring, information security level settings. According to different business, different tasks responsibility and different sections, workflow process is adjusted to ensure business data security. Some business data is restricted to read only can't be changed. The business is transferred according to specification [3].

3.3 Using Browse/Server Structure, Running on the Internet

The system is deployed cheaply without complex facilities, office platform can be constructed in the LAN or WAN. User only needs a browser to achieve remote office and uninterrupted service.

4 Two Business Process Models

There are two business process models, one is the free process model which is mainly used for work e-mail module to realize flow processing of internal e-mail, so that

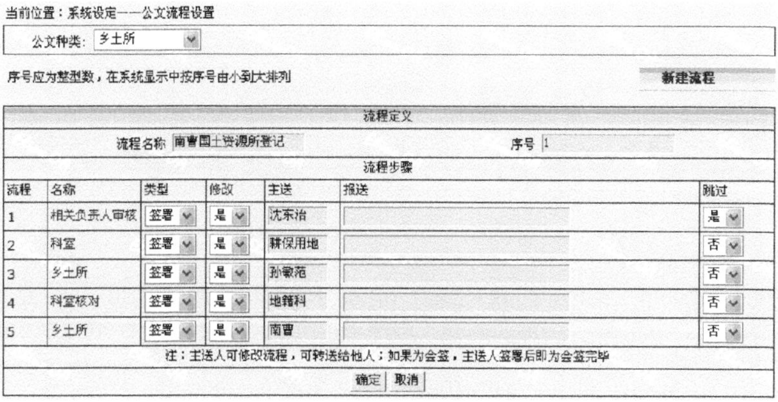

Fig. 1. Office Fixed Process

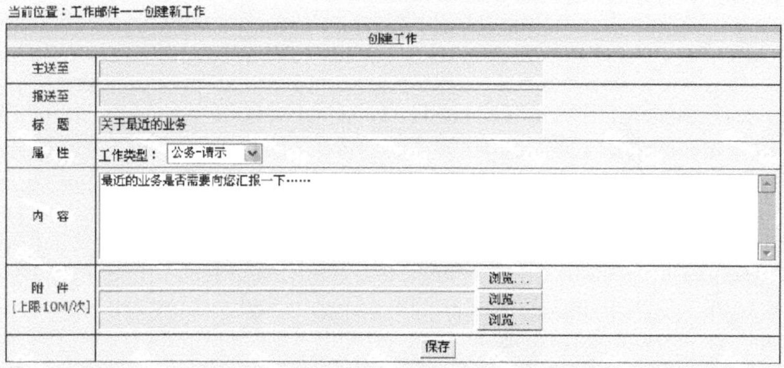

Fig. 2. Office Internal Free Process

related personals can view the office and business processes easily to meet the diversified needs in the daily work. The other is the fixed process model, which can achieves standardized process, schedule control, efficiency and transparency, constrict and guide staff to complete their jobs more standardized.

Fig. 1 shows some township land office fixed registration process, and fig.2 shows how to create a free internal work e-mail.

5 System Functional Structure

The whole system is composed of two subsystems: office system and business system. Fig. 3 shows its main functions [4].

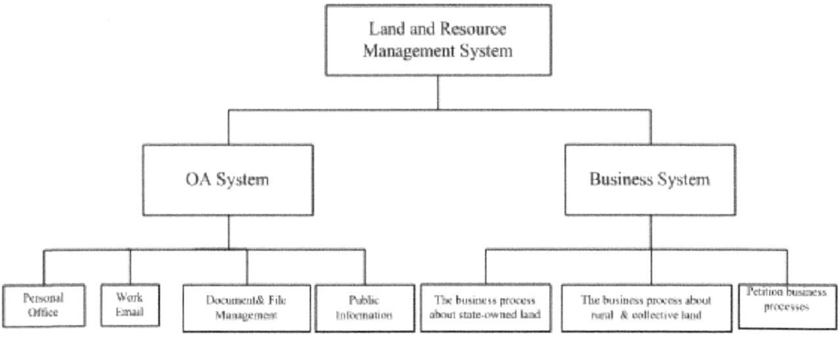

Fig. 3. System Functional Structure

Office system includes personal office, work e-mail, official document management, archives management. Personal office module is responsible for managing and scheduling personal work and affairs. Work e-mail module is often used to handle all internal office affairs, can facilitate the exchange of work between employees. Official documents module can send and receive documents. Archives management module provides logging and archiving file. Public information module manages notification bulletin, announcement, policies and regulations for staff search. System Configuration module can initialize basic data such as sector settings, permissions settings.

Business system can help to achieve external social public interest functions, includes state-owned land, rural homestead, collective land management, petition and monitoring management. State-owned land module is mainly responsible for state-owned land business reception, registration, transfer, supervise and inquiries, which is open to the business office; rural homestead, collective land module is responsible for rural homestead allocating and changing, collective construction land approval, animal breeding land audit, which is open to the township land office. Petition module can deal with the reception and registration of letters and visitors, which is open to the letters and visits section, providing visitors with legal and policy advisory services. Monitoring module can help the staff punished violation cases according to law, assist the judiciary to impose administrative penalties.

6 Summary

County-level land resources management system integrates office system and business system together, is based on the internet, achieve e-government, improve the government efficiency, enhance government's image, will promote social and economic development.

References

1. Jiao, C.Z., Wang, G., Liu, G.Q.: Design and research of workflow engine base on UML. Microcomputer Information 22(24), 127–129 (2006)
2. Tian, L., Lu, G.D.: Design and implementation of OA system for science and technology bureau. Computer Engineering and Design 26(04), 1056–1058 (2005)
3. Xu, F., Wu, X.C.: The research of security problem in the E-government of land and resources. Science of Surveying and Mapping 32(01), 132–134 (2007)
4. Shu, F.Y., Gou, M.X.: Design and Realization of Multilayered Land and Resources e-government System Based on Service Flow. Land and Resources Informatigation 06, 21–24 (2005)

Sliding Mode Control as Applied to Drilling Rotary System

Fubin Shi, Nurzat Rasol, and Lin Li

Key Laboratory of Drilling Rigs Controlling Technique,
Xi'an Shiyou University, Xi'an, China
{Kjcsfb,kjclilin}@xsyu.edu.cn, altiy@126.com

Abstract. This paper, using torsional multi-DOF model of drilling rotary system, presents the double surface sliding mode PID controller with exponent reaching law in order to handle the system failure caused by stick-slip oscillation. The sliding mode controller is applied two discontinuity surfaces as sliding surface, one of them is used to suppress the stick-slip oscillation at the drill bit, and the other surface makes the bit speed follow the rotary speed and achieves bit moving with a constant speed. The developed controller is a double surface sliding mode PID controller that can further improve dynamic and static characteristics of the drilling rotary system. The comparative analysis of simulation results show the control way not only has good robustness for the uncertainties of rock formation and drilling string, but also enhances the stability and reliability of the system.

Keywords: sliding mode control; stick-slip oscillation; PID control; drilling rotary system.

1 Introduction

Oilwell drilling is a mechanical system with complex dynamical phenomena, the operation characteristics of drilling rig, the uncertainties of rock formation, the changes of drillingstring result in non-desired oscillations that cause equipments failure, reduce penetration rate and increase drilling operation costs. Stick-slip oscillation appearing at the bottom-hole assembly (BHA) is particularly harmful for the bit, there are two major forms: a) the bit unable to rotate; b) the top-rotary system to move with a constant rotary speed, whereas the bit rotary speed varies between zero and up to six times the rotary speed at the surface. This phenomenon will cause great harmful oscillation around the drill strings and damage the drill bit and drill strings.

The drillstring with stick-slip oscillation has interested many researchers who developed models to adequately describe the phenomenon and proposed various solutions. Several lumped-parameter torsional models are of one degree [4], two degree [2]-[3] and multi-degree of freedom [5]-[6]. Such models consider the effect of friction appeared between the drillstring components and between the drillstring and formation

L. Qi (Ed.): ISIA 2010, CCIS 86, pp. 600–608, 2011.

[7]. For the control way, there are also several ways. For example, [8] and [9] proposed nonlinear friction compensator and reduced order proportional integral compensator, [10] and [11] used H_∞ technique and optimal control technique, respectively. Also, [12] and [13] propose the sliding mode PID controller with single surface based on two-degree model for rotary drilling system and the sliding mode control with double surface of a multi-DOF oilwell drillingstring.

The paper presents the double surface sliding mode PID controller with exponent law based on sliding mode control of multi-DOF oilwell drillingstring, which can effectively suppress the stick-slip oscillation. Numerical simulations have carried out to verify the idea.

2 Torisional Model of Drilling Rotary System

The rotary drilling rig is an essential part of oil drilling which provides enough torque and rotary speed for the bit and the drilling devices. The basic components of a rotary drilling rig are the derrick and hoist, swivel, kelly, turntable, drill pipes, bit, and pump.

Two-degree-of-freedom model consists of two damped inertias mechanically coupled by an elastic intertialess shaft (drillstring), only focus on the velocities of drill bit and turntable. Fig.1 depicts multi-DOF torsional model of the drillstring. It consists of four kinds of elements: the top-rotary system (J_r), the drill pipes (J_p), the drill collars (J_l), the bit (J_b). The inertias are connected one to each other by linear springs with tensional stiffness (K_t, K_{tl}, K_{tp}) and tensional damping (C_t, C_{tl}, C_{tb}). A viscous damping torque is considered at the top-drive system (T_{ar}) and at the bit (T_{ab}). A dry friction torque (T_{fb}) is considered at the bit.

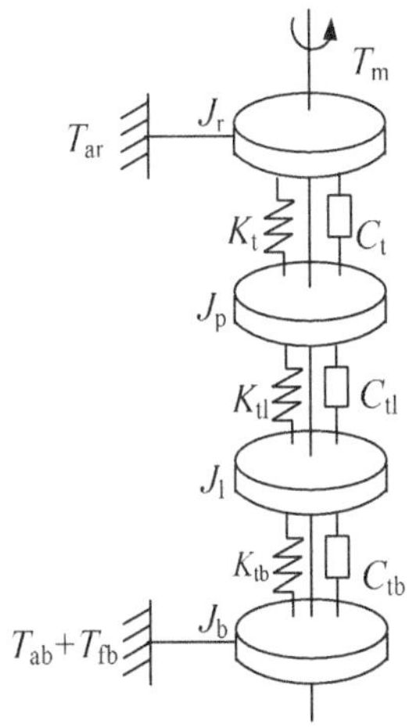

Fig. 1. Drilling rotary system model

Then, the equations of motion are given in [13] as:

$$\begin{aligned}
\ddot{\varphi}_r &= -\frac{C_t}{J_r}(\dot{\varphi}_r-\dot{\varphi}_p)-\frac{K_t}{J_r}(\varphi_r-\varphi_p)+\frac{T_m-T_{ar}(\dot{\varphi}_r)}{J_r}\\
\ddot{\varphi}_p &= \frac{C_t}{J_p}(\dot{\varphi}_r-\dot{\varphi}_p)+\frac{K_t}{J_p}(\varphi_r-\varphi_p)-\frac{C_{tl}}{J_p}(\dot{\varphi}_p-\dot{\varphi}_l)-\frac{K_{tl}}{J_p}(\varphi_p-\varphi_l)\\
\ddot{\varphi}_l &= \frac{C_{tl}}{J_l}(\dot{\varphi}_p-\dot{\varphi}_l)+\frac{K_{tl}}{J_l}(\varphi_p-\varphi_l)-\frac{C_{tb}}{J_l}(\dot{\varphi}_l-\dot{\varphi}_b)-\frac{K_{tb}}{J_l}(\varphi_l-\varphi_b)\\
\ddot{\varphi}_b &= \frac{C_{tb}}{J_b}(\dot{\varphi}_l-\dot{\varphi}_b)+\frac{K_{tb}}{J_b}(\varphi_l-\varphi_b)+\frac{T_b(\dot{\varphi}_b)}{J_b}
\end{aligned} \tag{1}$$

where φ_i, $\dot{\varphi}_i (i\in\{r,p,l,b\})$ are the angular displacements and angular velocities of drillstring elements, respectively. T_m is the torque coming from the electrical motor at the surface. The actuator dynamics is not considered, and $T_m=u$, u is the control input. $T_{ar}=c_r\dot{\varphi}_r$, c_r is the viscous damping coefficient.

x is the system state vector defined as:

$$\begin{aligned}
x &= (\dot{\varphi}_r,\ \varphi_r-\varphi_p,\ \dot{\varphi}_p,\ \varphi_p-\varphi_l,\ \dot{\varphi}_l,\ \varphi_l-\varphi_b,\ \dot{\varphi}_b)^T\\
&= (x_1,\ x_2,\ x_3,\ x_4,\ x_5,\ x_6,\ x_7)^T
\end{aligned} \tag{2}$$

The torque on the bit is:

$$T_b(x_7)=T_{ab}(x_7)+T_{fb}(x_7) \tag{3}$$

where T_{ab} is the influence of the mud drilling on the bit behaviour, $T_{ab}=C_b x_7$; T_{fb} is the friction modeling the bit-rock contact, it is defined as [14]:

$$T_{fb}(x_7)=\begin{cases} T_{eb}(x_7) & |x_7|<D_r\ \ |T_{eb}|\le T_{sb}\\ T_{sb}\,\mathrm{sgn}(T_{eb}(x_7)) & |x_7|<D_r\ \ |T_{eb}|>T_{sb}\\ W_{ob}R_b\mu_b(x_7)\,\mathrm{sgn}(x_7) & |x_7|>D_r \end{cases} \tag{4}$$

where $D_r>0$, T_{eb} is the reaction torque, T_{sb} is the static friction torque, $T_{sb}=W_{ob}R_b\mu_{sb}$; $R_b>0$ is the bit radius, $W_{ob}>0$ is the WOB. $\mu_b(x_7)$ is the bit dry friction coefficient considered as:

$$\mu_b(x_7)=\mu_{cb}+(\mu_{sb}-\mu_{cb})e^{-\frac{\gamma_b}{v_f}|x_7|} \tag{5}$$

where μ_{sb}, $\mu_{cb}\in(0, 1)$ are the static and Coulomb friction coefficients associated with J_b; $0<r_b<1$ and $V_f>0$. T_{eb} is:

$$T_{eb}=C_{tb}(x_5-x_7)+K_{tb}x_6-T_{ab}(x_7) \tag{6}$$

Using (2), system (1) can be written as:

$$\begin{aligned}
\dot{x}_1 &= \frac{1}{J_r}[-(C_t + C_r)x_1 - K_t x_2 + C_t x_3 + u] \\
\dot{x}_2 &= x_1 - x_3 \\
\dot{x}_3 &= \frac{1}{J_p}[C_t x_1 + K_t x_2 - (C_t + C_{tl})x_3 - K_{tl}x_4 + C_{tl}x_5] \\
\dot{x}_4 &= x_3 - x_5 \\
\dot{x}_5 &= \frac{1}{J_l}[C_{tl}x_3 + K_{tl}x_4 - (C_{tl} + C_{tb})x_5 - K_{tb}x_6 + C_{tb}x_7] \\
\dot{x}_6 &= x_5 - x_7 \\
\dot{x}_7 &= \frac{1}{J_b}[C_{tb}x_5 + K_{tb}x_6 - (C_{tb} - C_b)x_7 - T_{fb}(x_7)]
\end{aligned} \tag{7}$$

Equation (7) can simplify as:

$$\dot{x}(t) = Ax(t) + Bu + T_f(x(t)) \tag{8}$$

where A, B are constant matrices depending on system parameters and T_f represents the torque on the bit.

From (7), the following input state vector is obtained:

$$u_{in} = C_t(x_1 - x_3) + K_t x_2 + C_t x_1 \tag{9}$$

3 Design of Sliding Mode Control

In this section, we designed the double surface sliding mode PID controller with exponent law based on multi-DOF model.

A. *Double surface sliding mode controller*

The double surface sliding mode controller has two discontinuity sliding surfaces, one of them is used to control the stick-slip oscillation at the drill bit, and the other surface makes top-rotary system velocity tend to the reference value velocity and the bit velocity follow the top-rotary velocity ($x_1 \to x_3$, $x_3 \to \Omega_{ref}$). Therefore, the system with strong robustness can suppress the stick-sliding oscillation of drill bit and ensure the stability of system.

The sliding surface is defined as:

$$s = (x_1 - \Omega_{ref}) + \lambda[\int_0 x_1(\tau) - \Omega_{ref} d\tau + \int_0 x_1(\tau) - x_7(\tau) d\tau] \tag{10}$$

when s=0, $x_7 \to x_1$, $x_1 \to \Omega_{ref}$.

The reaching law is chosen as:

$$\dot{s} = -\eta \operatorname{sgn}(s) \tag{11}$$

The state vector can be chosen according to the input state linearization control law

$$u = u_{\mathrm{in}} - u_{\mathrm{s}} = C_{\mathrm{t}}(x_1 - x_3) + K_{\mathrm{t}}x_2 + C_{\mathrm{r}}x_1 \\ - \varepsilon[\lambda(x_1 - \Omega) + \lambda(x_1 - x_7) + \eta\,\mathrm{sgn}(s)] \tag{12}$$

where u_{s} is sliding mode controller.

B. Double surface sliding mode PID *controller with exponent reaching law*

Double surface sliding mode PID controller with exponent law combines double surface sliding mode control with PID control. The sliding surface is still two discontinuity surfaces for suppressing stick-slip oscillation and controlling top-rotary system, reaching law is exponent reaching law which can be faster the reaching speed and make the system has better robustness [15]. PID control can provide the faster dynamics and decrease steady state error. So the controller can increase the accuracy and robustness of system and achieve the control goal.

The sliding surface is given in (10), the reaching law is chosen as

$$\dot{s} = -k|s|^{\alpha} \cdot \mathrm{sgn}(s) - ks \tag{13}$$

Now, the input vector is chosen as:

$$u = \frac{1}{J_{\mathrm{r}}} u_{\mathrm{in}} - u_{\mathrm{s}} - u_{\mathrm{PID}} \tag{14}$$

where sliding mode control u_{s} is:

$$u_{\mathrm{s}} = \varepsilon[\lambda(x_1 - \Omega_{\mathrm{ref}}) + \lambda(x_1 - x_7) + k|s|^{\alpha} \cdot \mathrm{sgn}(s) + ks] \tag{15}$$

PID control is:

$$u_{\mathrm{PID}} = K_{\mathrm{p}}(1 + \frac{1}{T_{\mathrm{i}}s} + T_{\mathrm{d}}s) \tag{16}$$

4 Simulation Results and Discussion

The typical parameters, used for the simulation, are taken from [13]. Fig.2 shows the step response of the bit and the top-rotary angular velocity with double surface sliding

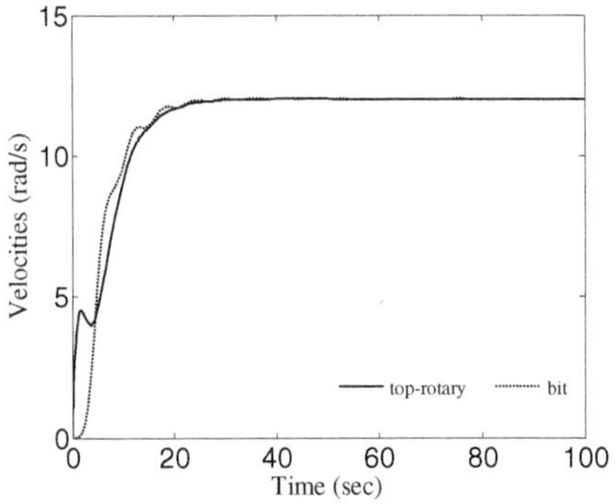

Fig. 2. The step response of the bit and top –rotary system angular velocity

mode PID controller with exponent reaching law. In simulation sliding mode controller parameters λ=0.15, k=0.2, ε=2; PID controller parameters K_p=0.1, $K_i = K_p/T_i$=0.002, $K_d = K_p T_d$ =0.51.

As seen in Fig.2, rise time $t_r \approx$15s, settling time $t_s \approx$18s, overshoot $\sigma\%$=0, steady-state error e_{ss}=0 for top-rotary system and drill bit. There are no more stick-slip oscillations, drill bit has good tracking ability.

5 Robust Analysis

Sliding mode control with strong robustness is a variable structure control system. There is verified the robustness of double surface sliding mode controller and double surface sliding mode PID controller with exponent reaching law.

The Fig.3(a) shows the step response of bit angular velocity with double surface sliding mode controller. The broken line is the system step response with changed tensional stiffness. As seen the Fig.3(a), the parameter changes bring some oscillations to system but it can be suppressed by sliding mode control.

The Fig.3(b) shows the step response of bit angular velocity with double surface sliding mode PID controller with exponent reaching law under same circumstance. As

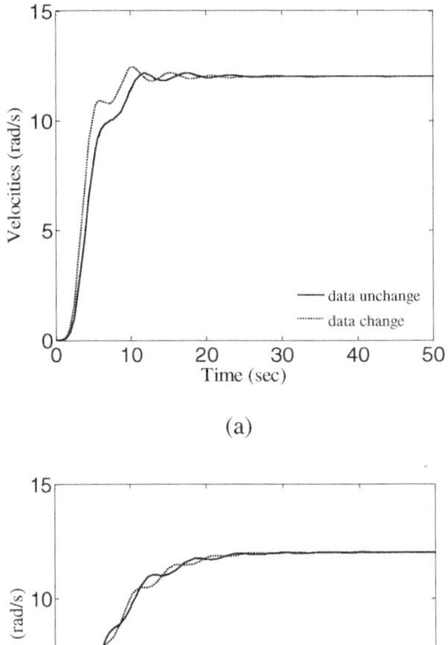

(a)

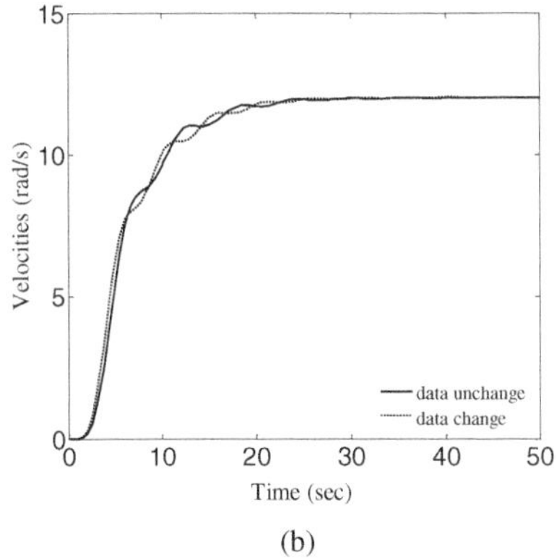

(b)

Fig. 3. The influence of parametric changes

seen the Fig.3(b), there are no more stick-slip oscillations, because exponent reaching law can be faster the reaching speed, PID control provides good dynamic characteristics, it has less settling time. All in all, double surface sliding mode PID controller with exponent reaching law shows more adjustable ability and robustness compared with double surface sliding mode controller.

When drill bit moves from soft formation to hard, friction torque is increased with complex rock formation, it increases the probability of stick-slip oscillation. However, the sliding mode controller has strong robust to rock formation uncertainties. As seen the Fig.4, the influence of rock formation changes to rotary drilling system with double surface sliding mode controller and double surface sliding mode PID controller with exponent reaching law, respectively. Seeing Fig.4(a), the double surface sliding mode controller can suppress stick-slip oscillation produced by the rock formation changes, but steady-state error is existed. While seeing Fig.4(b), the system with double surface sliding mode PID controller with exponent reaching law can effectively suppress stick-slip oscillation, the steady-state error can be reduced, thus it has more the accuracy and robustness.

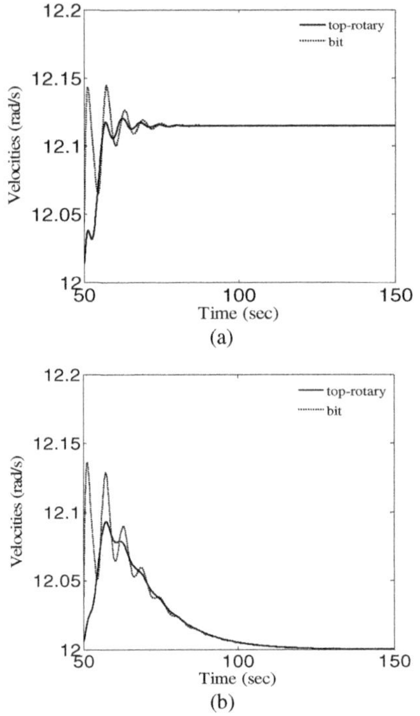

Fig. 4. The influence of rock formation

6 Conclusion

In this paper, we have proposed a double surface sliding mode PID controller with exponent reaching law for handling the stick-slip oscillation and increasing drill bit tracking ability, from comparing the simulation results, we can get as follows:

a) Drilling rotary system uses nonlinear multi-DOF model that has practical means.
b) Double surface sliding mode PID controller combines the advantage of two controllers, so it has good dynamic and static response, and enhances the system stability and reliability.
c) Double surface sliding mode controller with exponent reaching law can shorten the reaching time of the sliding surface, be in sliding motion as soon as, improve the robustness of the system.

Acknowledgment

This work was supported in part by a grant from the CNPC Science and Innovate Foundation Project: No.2009D-5006-03-07 and Key Project of Science and Technology Department of ShaanXi.

References

[1] Yigit, A.S., Christoforou, A.P.: Coupled Torsional and Bending Vibrations of Actively Controlled Drillstring. Journal of Sound and Vibration 234(1), 67–83 (2000)
[2] Navarro-López, E.M., Suárez-Cortez, R.: Vibraciones Mecánicas en una Sarta de Perforación: Problemas de Control. Revista Iberoamericana de Automática e Informática Industrial 2(1), 43–54 (2005)
[3] Brett, J.F.: The Genesis of Torsional Drillstring Vibrations. SPE Drilling Engineering, 168–174 (September 1992)
[4] Lin, Y.Q., Wang, Y.H.: Stick-slip Vibration of DrillStrings. Journal of Engineering for Industry 113, 38–43 (1991)
[5] Navarro-López, E.M., Cortés, D.: Avoiding Harmful Oscillations in a Drillstring through Dynamical Analysis. Accepted in Journal of Sound and Vibration, 152–171 (2006)
[6] Navarro-López, E.M., Suárez, R.: Practical Approach to Modelling and Controlling Stick-slip oscillations in oilwell drillstrings. In: Proceedings of the 2004 IEEE International Conference on Coutrol Applications, Taipei, Taiwan, September 2-4, pp. 1454–1460 (2004)
[7] Serrarens, A.F.A.: H_∞ control as applied to torsional drillstring dynamics. Msc. Thesis, Eindhoven University of Technology (2002)
[8] Abdulgalil, F., Siguerdidjane, H.: Nonlinear Friction Compensation Design for Suppressing Stick Slip Oscillations in Oil Well Drillstrings. In: 7th IFAC DYCOPS, Massachusets, USA (July 2004)
[9] Al-Harthi, M., Yaz, E.E.: Reduced Order Proportional Integral Compensator for Disturbance Suppression in Oil Well Drill-Strings. In: Proceedings of the 2002 IEEE International Conference on Control Applications, Glasgow, Scotland, U.K., September 18-20 (2002)
[10] Serrarens, A.F.A., van de Molengraft, M.J.G., Kok, J.J., van den Steen, L.: H_∞ Control for Suppressing Stick-slip in Oil Well Drillstrings. IEEE Transactions on Automatic Control 18(2), 19–30 (1998)
[11] Smit, A.T.: Using of Optimal Control Techniques to Dampen Torsional Drillstrings Vibrations, PhD thesis, University of Twente (March 1999)

[12] Abdulgalil, F., Siguerdidjane, H.: PID Based on Sliding Mode Control for Rotary Drilling System. Serbia & Montenegro, Belgrade, November 22-24 (2005)
[13] Navarro-López, E.M., Cortés, D.: Sliding-mode control of a multi-DOF oilwell drillstring with stick-slip oscillations. In: Proceedings of the 2007 American Control Conference, New York City, USA, July 11-13 (2007)
[14] Karnopp, D.: Computer Simulation of Stick-slip Friction in Mechanical Dynamic Systems. ASME Journal of Dynamics Systems, Measurement, and Control 107(1), 100–103 (1985)
[15] Zhang, Q.-z., He, Y.-y., Li, L.: Sliding Mode Control of Rotary Drilling System With Stick Slip Oscillation. In: The 2nd International Workshop on Intelligent Systems and Applications (ISA 2010), Wuhan, China, pp. 30–33 (May 2010)

Tower Crane Effective Life Assessment Based on Tower Crane Fleet Monitoring System*

Zeguang Han, Min Hu, Xinfang Song, Ruiqing Hao, and Xijian Zheng

School of Transportation & Mechanical Engineering, Shenyang Jianzhu University, Shenyang, Liaoning, 110168, China
{hanzeguang1,huxinshu2005,haoruiqingI}@163.com, songxinfang25@126.com, xijianzheng@sina.com

Abstract. How to obtain and make full use of reliable working information of tower crane fleet effectively is still a problem. Here, a monitoring system was built to help acquire tower crane working parameters and assess effective life. Firstly, based on Visual Basic and SQL Server technology, parameters were collected through wireless transmission, and then classified, saved and processed to get applicable load spectrum. Secondly, effective life was assessed and displayed directly through calling daemons. By using nominal stress method, effective life of dangerous points was calculated and real-time effective life assessment was realized, which could be taken as a reliable reference to take measures. Finally, an effective life calculation example of tower crane hoisting mechanism was provided. Results show that the system is accurate and reliable. The main advantage of the new method is its simplicity of implementation, with no need of extensive knowledge of load spectrum and nominal stress.

Keywords: Tower Crane; Monitoring System; Effective Life; VB; SQL.

1 Introduction

Tower crane is commonly designed according to load spectrum grade from standards, but suffering quite different load and use frequency in practice, which brings many unpredictable factors to running safety and much inconvenience to safety testing. Effective life is an important index to safety evaluation and load spectrum is the important basis of effective life calculation. Hence, actual load spectrum of on-site tower cranes means a lot to assessing effective life and monitoring safety status. Nowadays, most of foreign crane products match with safety monitoring device, which, however, is not applicable to domestic products directly. Manual recording mode exists widely at home, characterized with incorrect data, hard to save and low efficiency. A more advanced intelligent product can monitor working condition and work as "black box" [1-4], which could provide overload records for analysis after accident, but could hardly connect working parameters with online effective life assessment and help for further data application.

* This work is partially supported by National Key Science and Technology R&D Program in the Eleventh Five-Year Plan Grant #2008BAJ09B02-2 to Zeguang Han and Xijian Zheng.

L. Qi (Ed.): ISIA 2010, CCIS 86, pp. 609–616, 2011.

In this work, tower crane working parameters were acquired, classified, saved and processed to get reliable applicable load spectrum. Such functions as real-time display, searching and statistics were also available. With obtained load spectrum, effective life was calculated and displayed. In brief, tower crane fleet real-time effective life assessment was realized, which could be taken as a reliable reference for safety operation and management and for damage reason analysis.

The paper is organized as follows: Section 2 deals with the obtainment of tower crane load spectrum based on monitoring system. In Section 3 the effective life assessment process is shown. The proposed method is used to solve tower crane hoisting mechanism effective assessment problem in Section 4 as an example. Section 5 discusses the results. Finally, Section 6 summarizes the conclusions of the paper.

2 Obtainment of Tower Crane Load Spectrum

2.1 Structure of Tower Crane Fleet Online Monitoring System

Online monitoring system was designed in this paper to obtain real reliable load spectrum. Working parameters of on-site tower cranes were collected through Zigbee wireless transmission. With commercial software Visual Basic [5, 6] and SQL Server [7], the system was developed to classify, save and maintain acquired data automatically, which could greatly reduce labor intensity and ensure the reality, integrity and reliability of data.

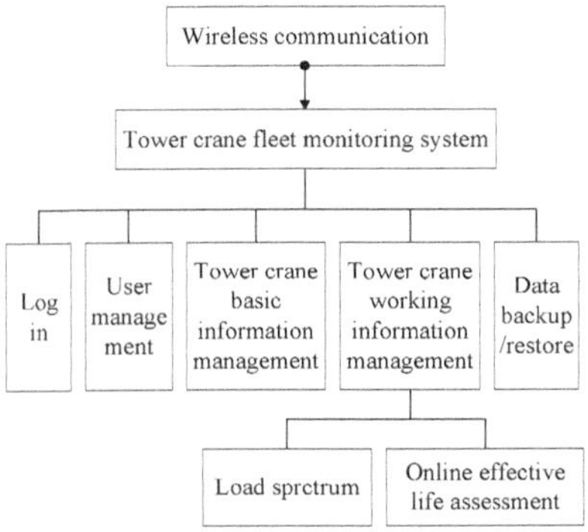

Fig. 1. Function diagram of monitoring system

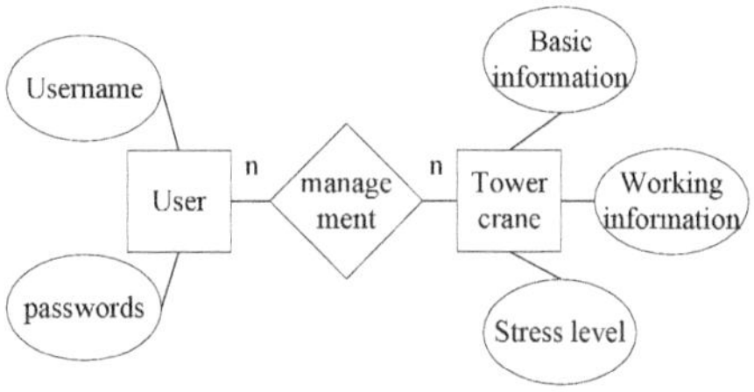

Fig. 2. E-R chart of monitoring system

The system is consisted of log in, user management, tower crane basic information management, working information management and data backup/restore modules, as shown in Fig.1.

The tower crane fleet monitoring system could not only record working parameters but also control safe operation based on effective life assessment with actual load spectrum and anti-collision calculation. So the crane database was established, including the basic information database base, working status database state, stress level database stress and user information database purview, the E-R chart was shown in Fig.2, part databases were shown in Fig.3.

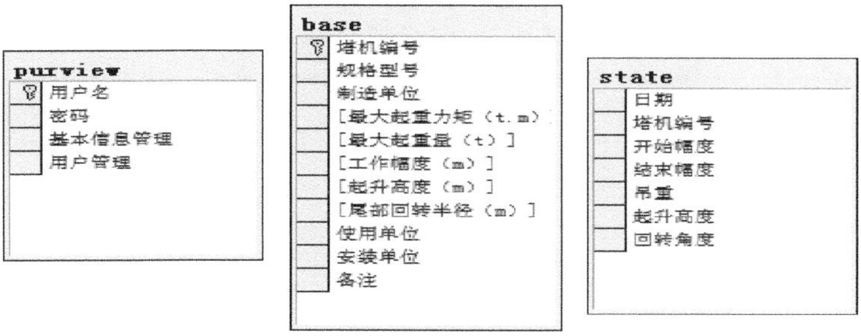

Fig. 3. Structure of database/datasheet

Tower crane working information management module, the most important one in the system, connected to datasheet state through widget ADO, to unpack, classify and save packet files of tower crane fleet working status. Working frequency and actual working load spectrum were also available through the statistics of running number of tower crane in every working condition.

2.2 Protocol Format of Communication Packets and Unpacking

The monitoring system could process acquired data dynamically. The packet sending to center control room is defined as a character mode of 52 bytes, as shown in Fig.4. The protocol format is shown in table 1.

```
文件(F) 编辑(E) 格式(O) 查看(V) 帮助(H)
0002-01B153837006A18581B0600F00020003C0000001020011601C6
0003-02B153837006A18581B0600F00020003C0000001020011601C6
0002-01B153837006A18901A0600F00020003C0000001020011601C6
0002-01B153837000000D00C0600F00020003C0000001020011601C6
0003-02B153837000000CC0B0600F00020003C0000001020011601C6
0002-01B1538370461E70170600F00020003C0000001020011601C6
0002-01B153837000000D0070600F00020003C0000001020011601C6
0002-01B15383700A00F70170600F00020003C0000001020011601C6
0002-01B153837000000DC050600F00020003C0000001020011601C6
0002-01B15383700AA23B41406 00F00020003C0000001020011601C6
0002-01B1538370000003818 0600F00020003C0000001020011601C6
0002 01B15383700AA235014 0600F00020003C0000001020011601C6
0002-01B153837000000B80B0600F00020003C0000001020011601C6
0002-01B15383700AA23EC130600F00020003C0000001020011601C6
0002-01B153837000000DC050600F00020003C0000001020011601C6
0002-01B1538370088139411 0600F00020003C0000001020011601C6
0002-01B153837000000DC050600F00020003C0000001020011601C6
```

Fig. 4. Tower crane packet file

Data files are all saved into 'txt file' and then unpacked and saved into the datasheet state. The working status data is determined according to the flag bits in every line of the document.

2.3 Searching of Working Status and Statistics of Running Time

Two searching keywords are available, namely date and tower crane number. User input one of them and then relevant results will be displayed in a window. A statistics

Table 1. Communication packet protocol format

Name	Type	Length
Start bit	character	1
Monitoring system address	character	2
Packet type	character	1
Time (hr, min, sec)	decimal integer	6
backup	character	2
Load, amplitude, torque ratio, height, angle and load ratio	Hexadecimal integer	Each 4
Warning code	character	2
backup	character	2
Time (year, month, day)	decimal integer	6
Flag code	character	4
Check code	character	2
End bit	character	1

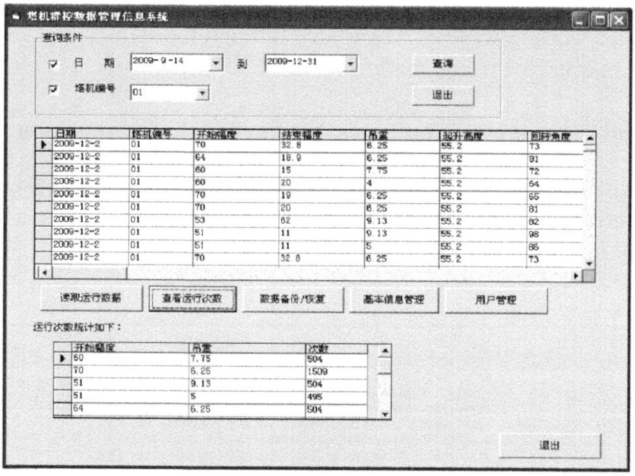

Fig. 5. Tower crane working status window

program was specially prepared to check working frequency. The running number in each working condition could be shown in Fig.5.

2.4 Working Load Spectrum of Tower Crane

Load status reflects the loading degree of hoisting mechanism and stress changes of dangerous position. Working under variable stress, tower crane has no regular show in the short term but an unstable variable stress characteristic for stress changes in the long term like a year. Therefore the actual stress changes spectrum can be gained by statistics. The module for obtaining actual load spectrum is provided. By Visual Basic programming, the classification, storage and statistics can be realized to obtain such load spectrum as shown in Fig.6.

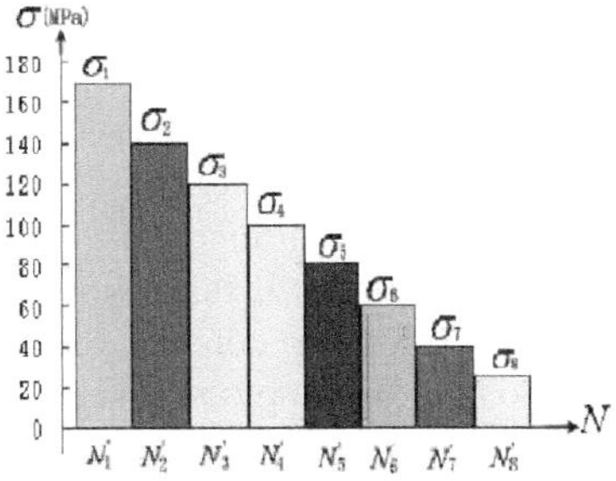

Fig. 6. Example of load spectrum

3 Online Effective Life Assessment of Tower Crane

3.1 Effective Assessment Method

Tower crane works under variable stress and fatigue damage comes. The common fatigue life estimation methods are nominal stress, local strain and simplified local stress-strain method. Here we choose the first one, calculating steps are shown in Fig.7. The fatigue calculation under unstable variable stress is based on fatigue damage cumulative hypothesis.

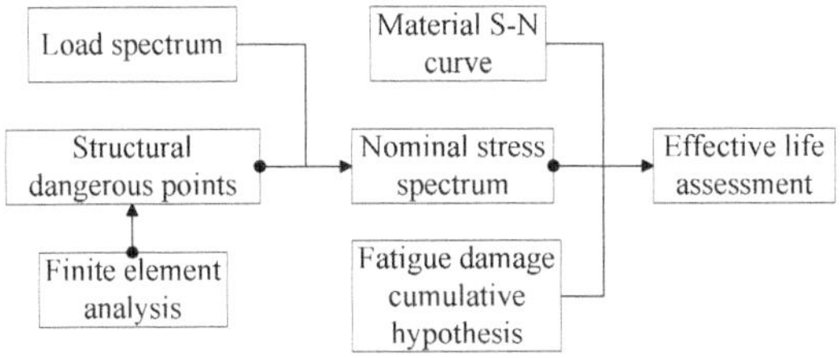

Fig. 7. Steps of effective life calculation

3.2 Dangerous Points of Tower Crane Hoisting Mechanism

Tower crane fatigue damage is mainly in key structures as tower body, balance and hoisting boom. Those parts can be described through practical detecting and finite element analysis.

3.3 Equivalent Work Cycle Index of Dangerous Points

Tower crane has different combinations of load status and use level on one working level. Generally speaking, life grows when load declines. According to the test curve of material limited life, the relation between fatigue limit and stress cycle index can be described as following:

$$\sigma_i^{m_1} N_i = \sigma_{rk}^{m_1} N_0 \quad (1)$$

Take the maximum stress of dangerous point in different working conditions σ_{max} as the fatigue limited stress σ_{eq}, the equivalent work cycle index N_i corresponding to stress σ_i can be obtained as following:

$$N_i = (\sigma_{eq}/\sigma_i)\, N_0. \tag{2}$$

Where N_0 is the material cycle base number, setting as 10^7, m is material constants, usually a value between 4.0 and 4.5, here m=4.0.

3.4 Effective Life Calculation

The damage of a dangerous point in stress level σi, suffering cycles n_i, is $D_i = n_i/N_i$. According to Miner criteria, in k stress level σi, after cycles n_i, the total damage can be described as:

$$D = \text{sum}\,(n_i/N_i)\ (i = 1,2,\ldots,k)\,. \tag{3}$$

The reciprocal of D in Eq.3 is the total life S_{totali} of point i. Set the spent life as S_i, then the effective life S_{effi} can be shown as following:

$$S_{effi} = S_{totali} - S_i. \tag{4}$$

If there are n dangerous points, the actual effective life of tower crane is

$$S_{eff} = \min\{\, S_{eff1}, S_{eff2}, \ldots\ldots, S_{effn}\}. \tag{5}$$

3.5 Effective Life Calculation

A daemon was prepared according to above principle, combined with finite element analysis, displaying the effective life on the screen. Part codes are as following:

```
...
If Option1.Value = True Then
      Adodc1.RecordSource = "select SUM (stress index /
equivalent index of dangerous point 1) as d FROM
[VIEW2]"
      Adodc1.Refresh
      D = Adodc1.Recordset.Fields(0)
      S = 1 / D
      S = Format(S,  "0.0")
      Text3.Text = S
      If Text1.Text = "" Then
        MsgBox "Input spent years: "
      Else
        a = Val(Text1.Text)
        b = S - a
        Text2.Text = b
      End If
...
```

4 Example of Tower Crane Online Effective Life Assessment

We take the hoisting boom of tower crane as example. After loading to tower crane model and solving, we get 17 key points as shown in Fig.8. Through data analysis, points 1, 3, 5, 11, and 12 were selected as dangerous points.

Take 8 working conditions as example, as shown in table 2, we calculate effective life of tower crane. By monitoring tower crane working status, working amplitude and lifting weights and choosing corresponding network to identify stress according to the working status, stress σ_i of all dangerous points at any time, any working condition can be obtained.

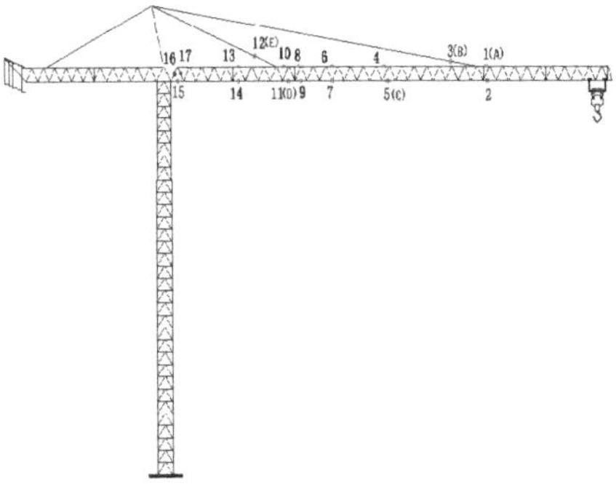

Fig. 8. Key and dangerous points

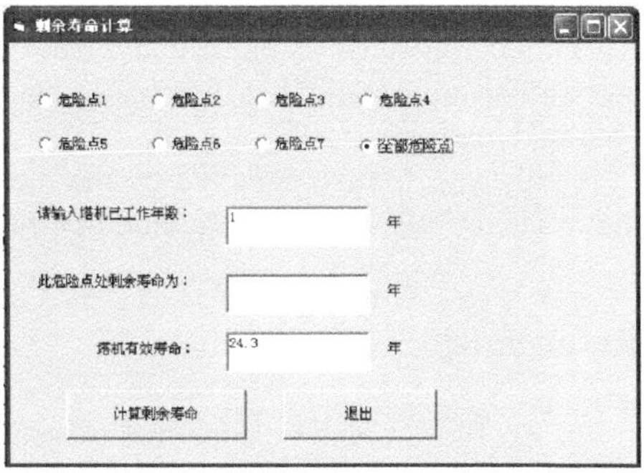

Fig. 9. Result of effective life assessment

5 Results

Here, we take point one as example, the equivalent work cycle index and stress are shown in table 2. According to the stress spectrum in Fig.6, in working condition one (70m, 6.25t), σ_1 = 180MPa, σ_2 = 150MPa and in table 2 σ_i = 171.14MPa, we choose σ_i = 180MPa because $\sigma_1<\sigma_i<\sigma_2$ and increase work cycle index with one. The same, we choose σ_i = 120MPa in condition two (60m, 4t) and σ_i= 80MPa in condition three (53m, 9.13t). For conditions 5 (52m, 9.13t), 7(51m, 5t) and 8 (45m, 5t), set σ_i = 40MPa and increase work cycle index with three. Once there is σ_i, including corresponding stress levels, statistics will be taken.

Table 2. Stress of dangerous point one at 8 working conditions (MPa)

Working Amplitude (m)	70.00	60.00	60.00	53.00	52.00	51.00	51.00	45.00
Lifting Weights (t)	6.25	7.75	4.00	9.13	9.13	9.13	5.00	5.00
Equivalent Work Cycle Index ($\times 10^7$)	0.559	1.24	2.30	4.77	11.6	36.8	186	1221
Stress (MPa)	171.14	109.39	76.89	48.86	38.66	31.52	31.58	30.49

According to above working conditions, the working status records in datasheet state and stress index in datasheet stress, the effective life of tower crane with the spent life as one year is 24.3 years, as shown in Fig.9.

6 Conclusions

In this paper, we showed a new method to tackle actual load spectrum obtainment problem. We defined the packet protocol format at first, then received working parameters through wireless communication and unpacked, classified and saved automatically with a daemon. The work uses the nominal stress approach to assess effective life of a tower crane hoisting mechanism.

One of the features of the method is that there is few manual operation in acquiring working parameters, as the daemon can deal with receiving, classification, unpacking and saving automatically. It means that labor intensity is greatly reduced and the obtained load spectrum is reliable.

Another advantage depicted by the method is its simplicity of implementation and that it is possible to use the method in other tower crane fleet safety assessment.

References

1. Li, S.: The Current Situation and Development of the China's Tower Crane Industry. J. Construction Mechanization, 9–11 (2000)
2. Sun, Z.: Applied Technology of Tower Cranes. China Building Materials Industry Press, Beijing (2003)
3. Yu, F., Wang, J., Shen, X., Zang, Y.: Development of Data Recorder for Tower Crane. J. Chinese Journal of Scientific Instrument., 275–276 (2006)
4. Yuan, L., Zhang, G., Yan, Q., Gao, H.: Design and Implementation of a Multifunction Black-box for Tower Crane. J. Construction Machinery, 73–78 (2007)
5. Sun, Y., Zhang, L.: Self-learning Tutorial for Visual Basic 6.0 Database Development. Posts & Telecom Press, Beijing (2002)
6. Jiang, B., Ouyang, L., Yang, C.: Visual Basic 6.0 Programming. Electronic Industry Press, Beijing (2001)
7. Wang, Z., Li, J.: Practical Guide for SQL Server 2005. Tsinghua University Press, Beijing (2006)

A Novel Watermark Algorithm for Document Protection Based on XML

Zaihui Cao[1] and Dongxian Yu[2]

[1] Department of Art and Design, Zhengzhou Instiute of Aeronautical Industry Management 450015 Zhengzhou, China
[2] Department of Computer Science and Technology, Zhengzhou Huaxin College 451150 Xin Zheng, P.R. China
czhhn@126.com, ydxtx1226@163.com

Abstract. Since networks progressed remarkably, XML documents created as digital contents or used for the data exchange between companies are becoming general. Resolving copyright problem and realization of communication privacy become more important in the scene using XML documents. This paper presented a novel watermark algorithm for document protection based on XML. The display effectiveness and the length of document are not changed by the algorithm, but watermarking can't easily be detected and attacked. Experimental results show that the proposed scheme is robust against various forms of attacks.

Keywords: Digital Watermarking; XML Document; Error Control; Information Capacity; Extensible Markup Language (XML).

1 Introduction

With the rapid development of Internet technologies, the amount of information sent and received electronically is increasing greatly. As the technology of transmitting information on network in secure, the importance of information security came to be recognized widely. Information hiding is a field of information security, and it includes methods creating covert channel where specification of transceiver is difficult, methods hiding the existence of information itself, and methods for digital watermarking. These technologies have lately attracted considerable attention as solution to copyright problems and the protecting method for communication privacy. With the XML technology development and application, the XML file is becoming the Internet data exchange standards, it security has also been a growing concern. This article has analysis text digital watermarking technology and XML document format, and proposed digital watermarking model and algorithm based on XML source document. This is a good try for the Internet Web security protection[1].

Text-based information hiding methods are the technics of information hiding using text as cover data and stego data. Most of methods for hiding data into text process texts as image essantially. XML are used widely for data exchange, and expected as a language of Web pages and digital contents. To develop the methods of

L. Qi (Ed.): ISIA 2010, CCIS 86, pp. 617–622, 2011.

information hiding using XML makes realize the way to establish secret communication channel using XML documents, and the ability to trace the source of unauthorized copies. We frist introduce the proposal of the methods of information hiding using XML then we propose general model of information hiding on XML[2]. The concrete example of the technics will be presented at the next section.

2 XML-Based Language for Digital Watermarking

XML is derived from Standard Generalized Markup Language (SGML). It is much more flexible than HTML, which is also a derivation of SGML. In HTML, only a set of predefined tags can be used, while in XML we can define new tags to customize our needs. Actually, XML is meta-language. For a specified application, we first define XML elements and document structure, namely DTD (Document Type Definitions) and XML-Schema. According to these definitions, we, further, design SAX (Simple API for XML) or DOM (Document Object Model). Then we can store and transmit information by legal XML documents consistent with these definitions[3]. In other words, information is exchanged through such XML documents and interpreted by the SAX or DOM parser interface for the corresponding application.

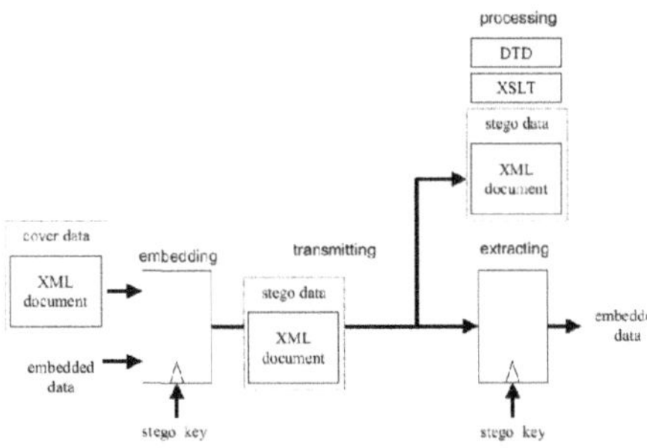

Fig. 1. Information hiding in XML data exchange

The first problem we must resolve is the Error control. Because the digital watermark embedding and extraction process can be thought of as data communication signals are sent and received, in the process of digital text communication the watermark tampering, deformation and other operations similar to the communication channel interference and noise caused by the error. As the similarity between the two, in this paper, using the communication of error control principle, we propose error control coding techniques which is applied to the watermarking process. Embedding, the original watermark with BCH error control code to bind, and then embedded in XML documents; extraction, through the codes on the integrity of the extracted watermark detection and error correction, in order to achieve watermark text containing watermarks tampering security control and recovery[4].

The second is the watermark encryption and decryption. Analyzing several common attack methods for watermark, the attack against watermarking algorithm and explain can be targeted to remove the watermarks and counterfeit watermark. One way to improve digital watermarking technology is the introduction of digital encryption systems. In this paper, we use the RSA public key algorithm, internationally recognized, the most practical value, to encrypt the watermark information, signature and its inverse operation[5].

According to XML encoding rules, when using quotation marks, double quotes and single quotes make the same effect, and therefore that can be used as the watermark embedding space. While encoding single quotation marks "0", double quote, said "1."then present the corresponding insertion algorithm (see Fig.2) and detection algorithm (see Fig.3).

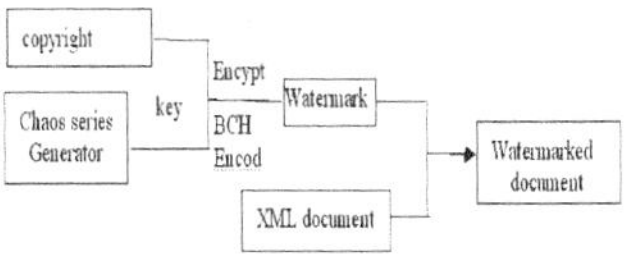

Fig. 2. The watermarking embedding model

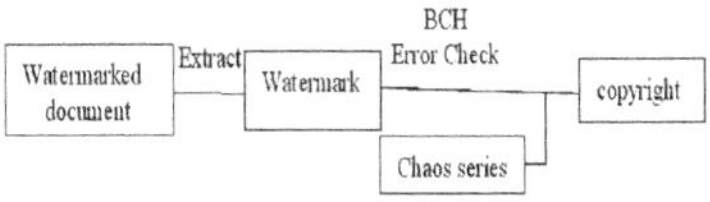

Fig. 3. The watermarking extract model

3 Algorithm for Watermarking XML-Document

Existing XML digital watermarking methods have been insufficient, the algorithm makes the following requirements:

① not change the XML document display;
② not change the length of the original XML document;
③ easily overlooked;
④ higher information capacity.

The first step is the process of watermark generation. The detailed process is as follows: 1) copyright signal convert. We first transform a meaningful watermark W (Plain Watermark) into a bit flow I of certain length. 2) encrypt the watermark. In our scheme, we use the Logistic chaotic system to encrypt the copyright series. The chaos model possesses some property of one-way cryptographic Hash function. So, by using the chaos model, the security of the watermarking system is strengthened. Acting as a randomized series generator, the Logistic chaos equation is:$X_{n+1} = \mu X_n(1-X_n)$ $\{\mu \in [1,4]\ n=0,1,2,\ldots\}$, We set $\mu=4$ in this computation. We select a threshold Q change the chaos series to the binary sequences P(i), The copyright is encrypted as $E(i) = W(i) \oplus P(i), i=1,2,\ldots, n$. $\oplus$= XOR. The Logistic's chaotic system is iterated until all the elements in the set B= {B1,B2,...,BN×N} is encrypted. Then every element in the encrypted we set E ={E1,E2,...,EN×N}[6].

3.1 Insertion Algorithm

We embed the watermark into the XML document by altering the case of letters in XML tags (calledUpper-Lower Coding, or ULC). Obviously, this will not increase the file size of XML documents, which makes our scheme superior to the other approach. Embedded point selection function Location (H), based on the characteristics of XML document, can be used to find information on the location of embedded watermark in the XML document. Location (H) function can be defined according to its own need, using the same function in the embedding and extraction process[7].

1) Initialization. Judging the string contained by every quotation marks in the H of the XML document, if the string itself contains quotation marks, then because XML does not allow the string contains the string used to surround the same quotation marks, to avoid syntax error, must unchanged. If it does not include quotation marks, you can change all the double quotes into single quotes;

2) Using Location (H) to identify all the embedding positions where watermark can be embedded in the H, namely K series k_1, k_2,... kn, n is the maximum number of embedded locations;

3) Coding the watermark M into M', i.e., the original watermarks, by MD5 algorithm for encryption and binding with error control codes, form the watermark redaction. M'is a k-bit binary bit stream m_1, m_2, ..., m_n. If n> k, then error out;

4) Set i=1;

5) If the m_i=1, then h[k_i]=double[h[k_i]], double is a function which change the single quotes into double quotes in thelocation K;

6) i=i+1; if i> n, then quit, otherwise turn to step 5).

stego example:

<event month="MONTH" date='DATE'>EVENT</event> ... 0

<event date='DATE' month="MONTH">EVENT</event> ... 1

3.2 Detection Algorithm

To check whether a watermarked XML document has been tampered or not, we first generate a watermark for it as described above. Afterward, the XOR operation is made between the generated watermark and the watermarked XML document as in the embedding process. As a result, we obtain another XML document and we can claim that if the watermarked XML document is integral, then the obtained document should be identical with the original one while a tampered XML document will cause the XML parser to report an error because the result of XOR is an illegal XML document[8]. Thus we can judge the integrity of XML documents effectively. The watermark detection algorithm was given as follows:

1) Using Location (H) to identify all the embedding positions where information can be hidded in the H, namely K series k_1, k_2,...k_n, n is the maximum number of embedded locations;

2) Set i=1;

3) If H [k_i] is a double quote, then m_i=1,Otherwise m_i=0;
4) i=i+1; if i $\leq$ n, go to step 3);

checking the extraction of the watermark information by error control code, if error, extracting after recovery. Otherwise, separating the watermark ciphertext from error control code and decrypting, then get a watermark plaintext.

4 Experiments and Analysis

Based on the above algorithm, We perform experiments on a computer running Windows XP with 2.4 GHz CPU and 256MB RAM and use the Java programming language to achieve all related functions. We summarized below:

Disguised. XML files can be arbitrarily downloaded the source document can be viewed. However, there are so many places using quotation marks, the differences between single quotes may be thought to be hand-written, and that generally do not attract attention. At the same time, the file size does not change after embed, web pages without distortion,therefore, better concealment[9].

The watermark capacity. General formatting text watermarking algorithm, there is hardly robust balance between watermarking capacity, concealment and contradictions. With the amount of watermark information increase, the watermark text's quality and robustness are declining. However, the watermarking algorithm for this study, Because the invisibility of the watermark is based on XML language encoding rules, nothing to do with the capacity and robustness, Thus the increase in capacity does not affect the watermark hidden properties, only reduce Watermarking robustness. After several rounds of embedding experiments, and getting statistical data.

Robustness. In order to verify the robustness of the XML-based digital watermark , we have made attack experiment on the web page embedding watermarks, as shown in Fig.4.

Since the embed technology in this article belongs to text Watermarking, it is weaker than multimedia information hidden in terms of robustness. But binding watermark information coding and error control codes together, using error codes and error correction can increase the robustness of the watermark. From experimental results in Figure 3, the watermarking algorithm designed in this paper is robust against format attack, not subject to format changes; a little fragile for substitution and cropping attacks. However, the proportion of Web sabotage reached 20% to 25%, the extraction of the watermark can still complete. Albeit the embedded watermark code element were bound with error-correcting codes, it is linear correlation between

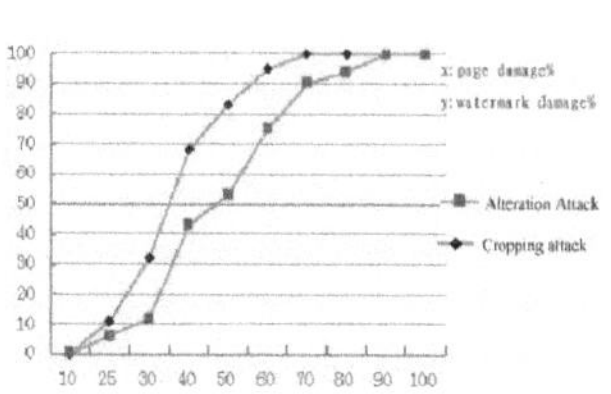

Fig. 4. Attacking experiments

them. The watermark is dependent on the order, when the sequence of the embed code element changes, the destruction for the watermark will be the largest and recoverability is very poor.

5 Conclusions

In this paper, a digital watermarking scheme based on XML language is put forward. And a digital watermarking algorithm based on XML document is proposed, taking advantage of the XML document representing redundanc--single and double quotes are interchangeable. In the absence of change in the length of the original XML source document and display effect, a higher volume of information can be hidden. The experimental data statistics indicated the hidden amount of information is about 1.4% of the source document information, with good security, and not readily detectable. The program involves only the network application protocols, and network device-independent, easy to implement and widely used. If it combines with hidden information for multimedia linked with the XML document, the outcome will be more prominent, and the applications may also be broader.

References

1. Swanson, M.D., Kobayashi, M.K., Tewfika, H.: Multimedia data-embedding and watermarking technologies. Proc. of the IEEE 86(6), 1064–1087 (1998)
2. Johnson, N.F., Jajodia, S.: Exploring steganography seeing the unseen. IEEE Computer 31(2), 26–34 (1998)
3. Moulin, P., Sullivan, J.A.: Information theoretic analysis of information hidding. IEEE Transaction on Information Theory 49(3), 563–593 (2003)
4. Chapman, M., Davida, D.: Hiding the Hidden: A Software System for Concealing Ciphertext as Innocuous Text. In: Luby, M., Rolim, J.D.P., Serna, M. (eds.) FC 1997. LNCS, vol. 1318, pp. 335–345. Springer, Heidelberg (1997)
5. Shibuya, R., Kaji, Y., Kasami, T.: Digital watermarking for PostScript and PDF documents. In: Symposium on Cryptography and Information Security, SCIS 1998, 9.2.E (January 1998) (in Japanese)
6. Spammimic - hide a message in spam, http://www.spammimic.com/index.shtml
7. Kuribayashi, M., Tanaka, H.: Video Watermarking of Which Embedded Information Depends on the Distance between Two Signal Positions. IEICE Trans. on Fundamentals of Electronics, Communications and Computer Sciences E86-A(12), 3267–3275 (2003)
8. Dick, K.: XML: A Manager's Guide, 2nd edn. Pearson Education, Inc., London (2003)
9. Damiani, E., di Vimercati, S.D.C., Samarati, P.: Towards securing XML Web services. In: Proceedings of the 2002 ACM Workshop on XML Security, pp. 90–96 (November 2002)

Model of Supply Chain Incentive Penalty Contract Based on the Linear Quality Demand

Jun Hu

College of Computer Science and Information Engineer
Zhejiang Gongshang University, Hangzhou, 310018
Junny_hu@163.com

Abstract. As quality management being of great importance on many companies increasingly, quality coordination has become the new one of supply chain coordination. When demand is in the linear correlation with quality in supply chain member, the traditional wholesale price has not coordinated the supply chain. In this paper, the incentive penalty contract is put forward under the linear quality demand and the coordination can be achieved through it. Also the decision variables such as price, quantity and quality are obtained.

Keywords: Linear Demand; Incentive penalty Contract; Supply Chain Quality.

1 Introduction

With rapid development of information technology and e-business, enterprises are facing unprecedented challenges in their management and development. Facing all the challenges, such as diversification of customer demand, shorter product life cycles, changes in market demand uncertainty, increasing competition among enterprises, companies have to change the operation mode, which single enterprise has, and they coordinate their relationship through supply chain contract.

This article will discuss quality, price and quantity coordination in supply chain when demand is uncertain and is in linear correlation with price, quantity and quality. As a single wholesale price contract is no longer able to achieve supply chain coordination, this paper applies the incentive penalty contract to coordinate supply chain. When the supply chain is coordinated, the members not only make decisions such as price and quantity, but also achieve a mutual level of quality wanted. The article will be organized as follows: Model description: it is to describe the decision-making process, variables and background in the model; Wholesale price model: it proves that the coordination can not be achieved under the wholesale price contract. Incentive penalty contract model: by the introduction of incentive penalty contract, supply chain coordination is achieved and decision variables such as quality, price and quantity among the supply chain members are obtained. Numerical simulation: the incentive penalty contract is proved effective in practice through a numerical simulation.

L. Qi (Ed.): ISIA 2010, CCIS 86, pp. 623–630, 2011.

2 Model Description

The basic model is described as follows: In a two-echelon, there are two participants in the supply chain: the manufacturer and the retailer. The manufacturer controls product quality and determines the quality level, θ, with the quality control cost, $c(\theta)$. The manufacturer sells the products to the retailer by the wholesale price, w . The retailer can order to the manufacturer, Q according to the demand, D and inspects the manufacturer's product. If the effort of the retailer's inspection is e, resulting about the inspection costs, $c(e)$, to the retailer. After that, the retailer benefits from selling products to the final consumers. Among them, the manufacturer and the retailer are risk neutral and completely irrational, the two sides information is symmetric, and all parties make decision according to the principle of expected profit maximization. The manufacturer is a leader, the retailer is a follower and they involve in the Stackelberg game.

In this section, there are the following assumptions: the retailer's order quantity, Q, is equal to the market demand, D, and the demand have a linear correlation with price, quantity and quality. The demand is not only influenced by the retail's price, p, quality inspection, e, but also by the manufacturer's quality control effort, θ.

$$D = a - p + \gamma e + \lambda\theta \tag{1}$$

The retailer's quality inspection costs: $c(e) = \dfrac{\eta e^2}{2}$ (2)

Among which, η is the retailer's ability coefficient of quality inspection cost. The larger the coefficient is, it shows the same effort to result in the higher cost.

The manufacturer's quality control costs: $c(\theta) = \dfrac{\zeta\theta^2}{2}$ (3)

Among which, ζ is the manufacturer's ability coefficient of quality control cost. The larger the coefficient is, it shows the same effort to result in the higher cost.

3 Wholesale Price Model

3.1 Model Description

The wholesale price contract is the simplest contract form and its model description is similar to the basic model described above. The manufacturer sells to the retailer by the wholesale price, w, and determines its own quality control effort, θ. The retailer sells to the consumer by the retail price, p, and determines its own inspection effort, e. There is not any other contract between both sides. As we all know, the wholesale price contract is not able to achieve supply chain coordination in the basic supply chain model. In the circumstances described in this article, the wholesale price can do the supply chain coordination?

From the basic model of supply chain, we can see the manufacturer's expected profit function is:

$$\Pi_M = Max_{(p,e)}(w-c)\cdot D - \frac{\zeta\theta^2}{2} \tag{4}$$

The retailer's expected profit function is:

$$\Pi_R = Max_{(p,e)}(p-w)\cdot D - \frac{\eta e^2}{2} \tag{5}$$

3.2 Model Solution

Inference: in wholesale price contract, the coordination of supply chain can not be achieved under the decentralized decision-making.

Proof

(1) Decentralized decision-making

In the decentralized control conditions, the retailer and the manufacturer make their decision on the principles of their expected return maximization.

Do the partial derivatives of p, e in the formula (5), we can obtain:

$$p = \frac{(a+\lambda\theta)\eta + (\eta-\gamma^2)w}{(2\eta-\gamma^2)} \tag{6}$$

$$e = \frac{\gamma}{(2\eta-\gamma^2)}(a+\lambda\theta-w) \tag{7}$$

Suppose $K = \dfrac{\eta}{(2\eta-\gamma^2)}$, Substitute the formula (6 , 7) into the expected profit function, Π_R, of the retailer and the market demand, D. We can get:

$$\Pi_R(w,\theta) = \frac{K}{2}\cdot(a+\lambda\theta-w)^2 \tag{8}$$

$$D(w,\theta) = a - p + \gamma e + \lambda\theta = K(a+\lambda\theta-w) \tag{9}$$

Substitute the formula (19) into the formula (4) of the manufacturer's expected profit function, we can get:

$$\Pi_M = Max_{(w,\theta)}(w-c)\cdot D - \frac{\zeta\theta^2}{2} = K(w-c)\cdot(a+\lambda\theta-w) - \frac{\zeta\theta^2}{2} \tag{10}$$

Do the derivative of w , θ in the formula (10), we can obtain w^* , θ^* p^* , e^* :

Substitute w^* , θ^* p^* , e^* into the expected profit function of the manufacturer, Π_M , we can get:

$$\Pi_M^* = \frac{\zeta K(a-c)^2}{2(2\zeta - \lambda^2 K)} \tag{11}$$

Substitute the above formula into the expected profit function of the retailer, Π_R , we can get:

$$\Pi_R^* = \frac{K}{2}\left[\frac{\zeta(a-c)}{(2\zeta - \lambda^2 K)}\right]^2 \tag{12}$$

In the decentralized control mode, the total profit of both retailer and manufacturer, Π_T^* , is:

$$\Pi_T^* = \Pi_M^* + \Pi_R^* = \frac{\zeta K(a-c)^2}{2(2\zeta - \lambda^2 K)} + \frac{K}{2}\left[\frac{\zeta(a-c)}{(2\zeta - \lambda^2 K)}\right]^2 \tag{13}$$

(2) Centralized decision-making

Assume under the centralized control condition, the total income of supply chain, Π_C^* , is:

$$\Pi_C = \underset{\theta,p,e}{Max}[a - p + \gamma e + \lambda\theta]\cdot(p-c) - \frac{\zeta\theta^2}{2} - \frac{\eta e^2}{2} \tag{14}$$

Do the derivative of p , e and θ to Π_C , we can get:

$$\Pi_C^* = \frac{\zeta K(a-c)^2}{2(\zeta - \lambda^2 K)} = B \tag{15}$$

To compare the formula (13) and (15) , and suppose the formula (13) is less than the formula (15), we can get:

$$\frac{\zeta K(a-c)^2}{2(2\zeta - \lambda^2 K)}\cdot\frac{3\zeta - \lambda^2 K}{(2\zeta - \lambda^2 K)} \leq \frac{\zeta K(a-c)^2}{2}\cdot\frac{1}{(\zeta - \lambda^2 K)} \tag{16}$$

That is: $4\zeta \geq 3\zeta$.

Because $\zeta \neq 0$ and $\zeta > 0$, that is (22) < (28) .It shows that the supposition is true.

It shows that the supply chain coordination can not be achieved in the decentralized decision-making mode as in the centralized decision-making mode with the mechanism of the wholesale price.

Suppose β is the ratio of the member's revenue, $\beta = \frac{\Pi_T^*}{\Pi_C^*}$ and $1 > \beta > 0$. We can get:

$$\lim_{\zeta \to \infty} \beta = \lim_{\zeta \to \infty} \frac{\Pi_T^*}{\Pi_C^*} = \lim_{\zeta \to \infty} \frac{(3\zeta - \lambda^2 K)(\zeta - \lambda^2 K)}{(2\zeta - \lambda^2 K)^2} = \frac{3}{4} \tag{17}$$

That is: $0 < \lim_{\zeta \to \infty} \beta \le \frac{3}{4}$.

4 Incentive-Penalty Contract Model

4.1 Model Description

Based on the basic model described in the paper, the manufacturer applies the incentives and punitive measures to the retailer's ordering behavior. When the retailer's order quantity, $Q > T$, the manufacturer offers the incentive, $\tau(D-T)$, to the retailer when the retailer's order quantity exceeded T; when the retailer's order quantity $Q < T$, the manufacturer gives the penalty over the uncompleted order quantity.

Among them, the variable definition is:

τ --penalty incentive factor, $0 < \tau < 1$;

T --Incentive and penalty standard.

In the incentive-penalty contract, the retailer's expected profit function is:

$$\Pi_R = Max_{(p,e)}(p-w)\cdot D + \tau(D-T) - \frac{\eta e^2}{2} \tag{18}$$

Upon maximizing the expected profit of the retail, we can get from the partial derivatives of p, e:

$$\frac{\partial \Pi_R}{\partial p} = 0, \quad \frac{\partial \Pi_R}{\partial e} = 0 \tag{19}$$

That is:

$$p = \frac{(a + \lambda\theta)\eta + (\eta - \gamma^2)w}{(2\eta - \gamma^2)} \tag{20}$$

$$e = \frac{\gamma}{(2\eta - \gamma^2)}(a + \lambda\theta - w) \tag{21}$$

To substitute p, e into Π_{RP}, we can get:

$$\Pi_{RP}^{*} = \frac{K}{2} \cdot (a + \lambda\theta_P - w_P)^2 + \tau_P[K(a + \lambda\theta_P - w_P) - T] \tag{22}$$

Among them, the manufacturer provides the contract, τ_S, θ_S, w_S, to the retailer.

To suppose the retailer's reservation profit is $\underset{-R}{U}$, that is:

$$\Pi_{RP} \geq \underset{-R}{U} \tag{23}$$

4.2 Solving the Model

If the contract can achieve coordination in supply chain, there are the same as the centralized decision-making for p, θ and e. That is:

$$\theta_P^* = \theta_C^* = \frac{K\lambda(a-c)}{(\zeta - \lambda^2 K)} \tag{24}$$

$$p_P^* = p_C^* = c + \frac{K\zeta(a-c)}{(\zeta - \lambda^2 K)} \tag{25}$$

$$e_P^* = e_C^* = \frac{K\gamma\zeta(a-c)}{\eta(\zeta - \lambda^2 K)} \tag{26}$$

Among them, p_P^*, θ_P^* and e_P^* is p, θ and e in the incentive-penalty contract.

From the formula, we can get:

$$D_P^* = \frac{\zeta K(a-c)}{(\zeta - \lambda^2 K)} \tag{27}$$

D_P^* is the demand When supply chain is coordinated.

Assuming that under the incentive-penalty contract, supply chain coordination can be achieved, there is Π_C^* equal to Π_{TP}^* (total profit of the members under the decentralized decision-making in supply chain). When $\Pi_{RP}^* = \underset{-R}{U}$ and $w_P^* = c$, the hypothesis of supply chain coordination is true. At this time, the retailer's expected revenue and incentive-penalty coefficient is:

$$\frac{K}{2} \cdot (a + \lambda\theta_P - w_P)^2 + \tau_P[K(a + \lambda\theta_P - w_P) - T] = \underset{-R}{U} \tag{28}$$

$$B + \frac{\zeta(\theta_C^*)^2}{2} + \tau_P[K(a + \lambda\theta_C^* - c) - T] = \underset{-R}{U} \tag{29}$$

$$\tau_P = \frac{B + \frac{\zeta(\theta_C^*)^2}{2} - \underset{-R}{U}}{K(a + \lambda\theta_C^* - c) - T} \tag{30}$$

That is: $\Pi^*_{RP} = \underset{-R}{U}$, $\Pi^*_{MP} = B - \underset{-R}{U}$ and $\Pi^*_{TP} = \Pi^*_C = B$.

Under the incentive-penalty contract model, total profit of supply chain under centralized control is equal to that under the centralized. When the retailer gets the reservation revenue, the other revenue belongs to the manufacturer.

The conclusion is: when $w^*_P = c$ and $\tau_P = \frac{B + \frac{\zeta(\theta^*_c)^2}{2} - \underset{-R}{U}}{K(a + \lambda\theta^*_c - c) - T}$, the incentive-penalty contract can achieve the coordination of supply chain.

5 Numerical Simulation

Assume that there are two members in a supply chain: the manufacturer and the retailer. The environment is described in the section. The decision-making process in the incentive penalty contract is listed below. First of all, the background variables are assigned; then, the decision variables and revenue functions are solved under all kinds of contract models; Finally, the factors affecting the supply chain revenue are analyzed (Table 1).

Table 1. Variables in the Incentive penalty Contract under Supply Chain Coordination

Incentive penalty contract	Decision varibles	Manufacturer	w^*	6.00
			θ^*	3.29
			T	25.00
			τ	41.43
			Π^*_M	108.35
		Retailer	p^*	25.76
			e^*	7.41
			Q^*	19.76
			Π^*_R	30.00
		Supply chain	D^*	19.76
			Π^*_T	138.35

The enterprises in supply chain often face demand uncertainty. When demand is in the linear correlation with quality of the members. The only wholesale price contract has not achieve supply chain coordination. Through the incentive penalty contract, the overall coordination of supply chain is not only achieved, but also the levels of the members' quality are determined. That means that two goals of coordination and quality selection are met in supply chain.

Acknowledgment

We thanks support of Ministry of Education, Humanities and Social Science Foundation Project (09YJC630200), 973 Pre-research Project (2009CB326 -204).

References

[1] Ju, H., Liu, Y.-l.: Research of supply chain coordination. Logistics Engineering and Management 11(30), 56–57 (2008)
[2] Vickery, S.K., Calantone, R., Dröge, C.: Supply Chain Flexibility: An Empirical Study. Journal of Supply Chain Management 35(3), 16–24 (1999)
[3] Chen, S.-c., Zhong, W.-f.: Research of demand change in supply chain. Agriculture Economy and Science 20(4), 47–48 (2009)
[4] Pan, H.-j., Zaang, J., Qi, J.: Review of supply chain coordination under uncertainty market demand research. Finance and Economy 22, 143–144 (2007)
[5] Marvel, H., Peck, J.: Demand Uncertainty and Returns Policies. International Economic Review 36(3), 691–714 (1995)
[6] Emmons, H., Gilbert, S.: Returns Policies in Pricing and Inventory Decisions for Catalogue Goods. Management Science 44(2), 276–283 (1998)

Study on the Characteristic of Electrical Impedance Spectroscopy of Soybean Seeds and the Detection of Seed Viability

Qiong Zhang[2], Dazhou Zhu[1], Ruifeng Hou[1], Dayu Pan[1], Xiaodong Wang[1], Zhihui Sun[2], and Cheng Wang[1,*]

[1] National Research Center of Intelligent Equipment for Agriculture, Beijing 100097, P.R. China
[2] School of Mechanical Engineering, University of Science and Technology in Beijng, Beijing 100083, P. R. China
wangc@nercita.org.cn

Abstract. Fast, non-destructive, and low cost detection of seed viability is important for seed storage and agricultural production. The present study analyzed the electrical impedance characteristics of soybean seed and investigated the feasibility of detecting seed viability by electrical impedance spectroscopy (EIS). Seeds were processed by two method- boiling, high temperature and humidity processing. Two different impedance analyzer were separately adopted, and the frequency ranges were 100Hz-700KHz and 60KHz-2MHz, respectively. The results of experiments demonstrated that EIS could reflect the change of seed internal structure and thus detect seed viability. The impedances of non-viable seeds were less than that of viable seeds, and the curve radiuses of viable seeds were larger than that of non-viable seeds. Preliminary results of this study indicated that EIS could reflect the degree of seed viability. It's promising to realize the fast and non-destructive detection of single soybean seed viability based on EIS.

Keywords: Electrical impedance spectroscopy, Seed viability, Bio-impedance, Soybean.

1 Introduction

Seed viability indicates the potential ability of germination and seedling growth in agriculture and forestry. Current technologies for detecting seed viability mainly include tetrazole coloring reduction test, X-ray photographic technique, leak of electrolytes. Tetrazole coloring reduction test, a mature approach for prediction of seed viability, distinguishes dead or normal seed according to the dyeing graph of embryo structure. The detection error of tetrazole coloring is less than 3%~5% compared to stand germination test. However, TTC reduction test would destruct seed and also need skilled operator [1]. X-ray photographic techniquelies on different

* Corresponding author.

L. Qi (Ed.): ISIA 2010, CCIS 86, pp. 631–636, 2011.

X-ray absorptions when X-ray passed though different selectively-permeable cell membrane between dead and normal seed, and therefore achieves quick and accurate test on seed viability. However X-ray photographic technique is only effective for seed which have clear difference between embryo and endosperm, such as some tree seeds [2]. As a simple, rapid detection method, leak of electrolytes is successfully applied to crop seeds such as corn and soybean [3]. However, seeds are generally soaked in water, resulting in a hypoxic environment that is injurious to certain species. Moreover, a semi-permeable layer in the seed coat of most species restricts leakage and thus confounds the relation ship of leakage with seed quality [4].

As a fast and non-destructive technology, electrical impedance spectroscopy (EIS) analyzes the impedance of objects for a certain frequency band. EIS has been applied in different fields of plant science, including the estimation of plant vigor [5], the resistance to coldness for Scots pines [6], the fresh degree of fruits and so on[7]. On the aspect of seed viability, T.Repo examined snap bean viability by EIS [8].Hydrated seeds exhibited two impedance arcs in the complex plane at the frequency range from 60Hz to 8MHz, and impedance spectra of viable and non-viable seeds differed. It was found that viability of larger volume snap bean could be detected by EIS. However, the volume of seeds will influence the measurement of impedance and the detection of small volume seeds still needs further investigation. The objective of this study was to explore the feasibility of detection soybean seeds viability which have smaller volume by EIS.

2 Materials and Methods

In order to analyze the impedance characteristic of seeds with different seed viability, normal seeds should be artificially shriveled into dead seeds. Two methods including the high temperature and humidity and the boiling method were applied to process the soybean seeds. For boiling method, seeds were put into boiled water for 7min and then dried for 48h till their water capacity is stable. For high temperature and humidity method, seeds were placed in a 40°C, 95% RH temperature and humidity chamber for 5d. The seeds though High temperature and humidity processing is equivalent to aging seeds under natural conditions [9]. Germination of each seed was determined according to the Chinese standard (GB/T 3543.4-1995). The detail treatment of two experiments and the measurement of EIS were as follows.

2.1 Boiling Processing and the Measurement of EIS

Nine intact plump seeds were selected, and five of them that suffered from boiling processing were labeled as dead seeds. All of nine seeds should be soaked in distilled water for 1h. Impedances of dead and normal seeds were more different after soaked 1h because different integrity of their cell membrane absorbed dissimilar water content. After soaking the surface flowing water was dried by a hair drier.

The impedance spectra (IS) were measured by an impedance analyzer (PV50A, Beijing Band ERA, China), two Ag/AgCl electrodes, electrode holders and electrode gel (Signalgel, Parker Laboratories, Fairfield, NJ, USA). Electrode holders (see Fig.1) were composed of horizontal stand and screw stem. Electrodes were fixed in horizontal stand by screw stem, and horizontal stand was freely moved to the desired position on an anti-vibration desk. The frequency was measured between 100Hz and 700KHz.The frequency range was divided logarithmically into 60 frequency point. The input voltage level of the sinusoidal signal was 1V (rms). One EIS sweep took approximately 40s.

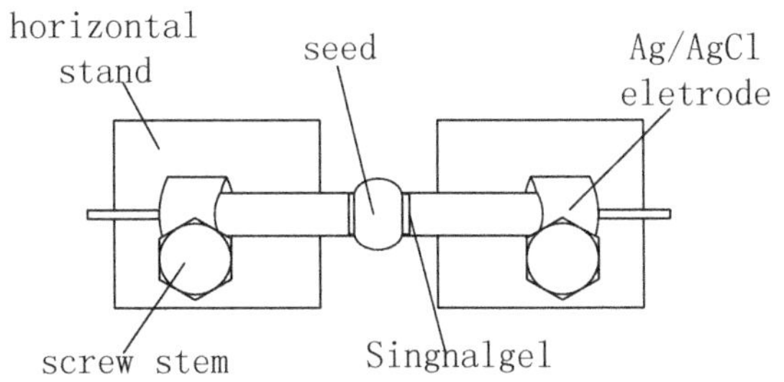

Fig. 1. Electrode holders diagram of EIS

2.2 High Temperature and Humidity Processing

Eighty intact plump soybean seeds (Zhonghuang 13) were selected, and forty of them were treated to be non-viable seeds. Each seed was weighed to get initial weight m_1. Non-viable seeds were processed in high temperature and humidity conditions. Moisture contents of single seeds were adjusted to 45% before EIS sweep.

The objective of adjusting to the same moisture contents was to eliminate moisture content influence on impedance. The method of increase the water content of individual seeds to 45% employed the following steps:

1) Storage moisture content $\omega_1\%$ was determined by grinding seeds and then drying at 103±2°C for 8 hours (GB/T 3543.6-1995).

2) Adjusted seeds were individually placed in 10ml glass vials. The amount of additional water in ml was calculated as follows:

$$\omega_a = \frac{m_1(\omega_1\% - \omega_2\%)}{1-\omega_2\%} - \Delta \tag{1}$$

Where ω_a was additional water, m_1 was initial weight, $\omega_2\%$ was target moisture content. Δ represented the influence of moisture from aging, for viable seeds $\Delta=0$ and non-viable seeds $\Delta = |m_3 - m_1|$. m_3 was the weight after aging.

3) The vials were sealed with Parafilm and then kept at 25°C for 24 hours in an incubator.

4) The final moisture was calculated as follows:

$$\omega_3\% = \frac{m_2 - m_1(1-\omega_1\%)}{m_2} \tag{2}$$

Where $\omega_3\%$ was the final moisture and m_2 was the final weight.

The measurement of EIS sweep was similar to the boiling experiment, but a more precise impedance analyzer (6500B, WAYNE KERR, UK) was used. In order to collect more impedance information about seeds, the frequency was measured between 100Hz and 1MHz.The frequency range was divided into 52 frequency point. The input voltage level of the sinusoidal signal was 100mv (rms). One EIS sweep took approximately 60s.

3 Results and Discussion

3.1 The EIS Characteristic of Seeds from Boiling Processing

Fig.2 was Nyqsuit diagram of seeds in boiling processing. Abscissa axis represented real part of impedance called R, and ordinate axis represented imaginary part called X. R was equal to resistance and X was equal to capacitance.

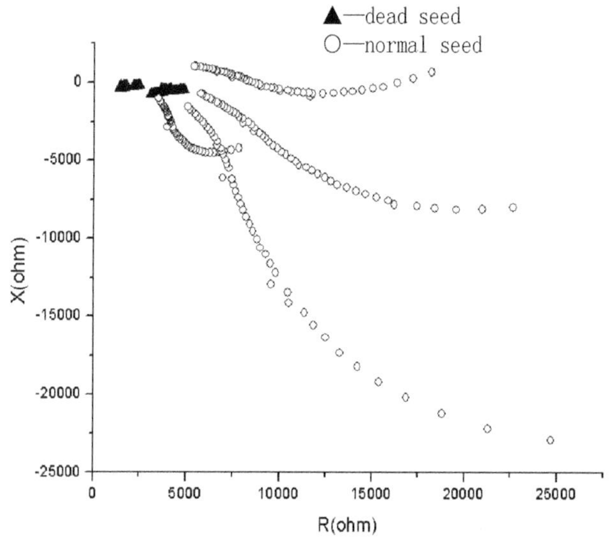

Fig. 2. Nyqsuit diagram of seeds in boiling processing

From Fig.2, it could be seen that the Nyquist curves between normal and dead seeds had obvious difference. The curve radiuses of normal seeds were larger than that of dead seeds. The impedances of normal seeds were between 1KΩ to 15MΩ, while the impedances of dead seeds were below 1KΩ. The reason may resulted from that structures and cell membranes of seeds were spoiled in boiling processing thus water inclined to pass though seeds and impedances subsequently decreased.

3.2 The EIS Characteristic of Seeds from High Temperature and Humidity Processing

The germination percentage of non-viable samples after processing was 72% which was far from aging objective 0%, which may be resulted from that the aging condition was not enough. The humidity of temperature and humidity chamber was often less than 95%RH. Successor experiment could extend aging time for complete loss of viability. The germination percentage of viable samples was 85%.

The non-germinated seeds after aging processing were labeled as non-viable seeds, and the germinated seeds without any processing were labeled as viable seeds. Their moisture contents were between 41% and 44% and had little differences. Fig 3 showed the impedance of some viable and non-viable seeds in high temperature and humidity processing.

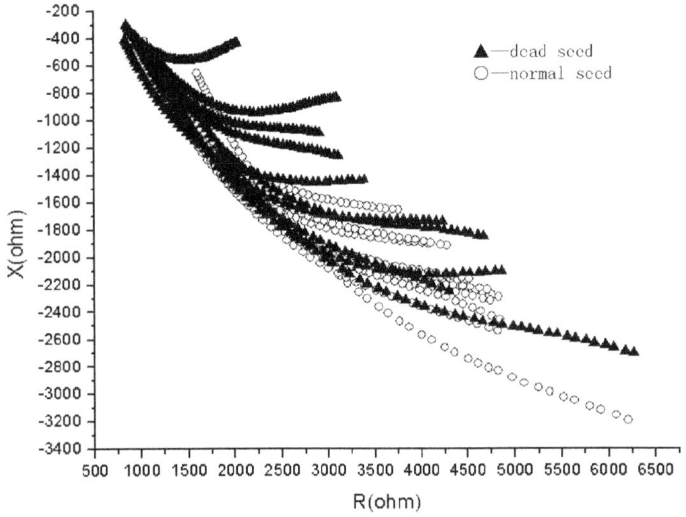

Fig. 3. Nyqsuit diagram of seeds in High temperature and humidity processing

From Fig 3, it could be seen that viable and non-viable seeds did not have clear separation in Nyquist curves of impedance, but the impedance tendency of difference between non-viable and viable seeds was still similar. With the increase of frequency, the impedance of seeds decreased. The impedances of non-viable seeds were between 1KΩ and 5KΩ while the viable seeds were between 1KΩ and 7KΩ. An apparent boundary of non- viable and viable seeds didn't exit in impedance, which might be caused by diversity of volume seeds and germination operational errors. However, the impedance tendency of difference between non-viable and viable seeds was still clear. Most non-viable seeds were distributed on the upper half of Fig 3, indicting that the impedances were smaller, while most viable seeds were distributed on the lower half which indicted that the impedances were greater. The tendency agreed with boiling experiment.

The EIS results of boiling experiment were more obvious than these of high temperature and humidity processing. It might be that boiling processing completely spoiled structures and cell membranes of seeds.

4 Conclusion

EIS could reflect seed internal structure and showed the difference of seed viability. The impedances of non-viable seeds were less than viable seeds and the curve radiuses of viable seeds were larger than non-viable seeds. EIS results of boiling experiment were more conspicuous than that of high temperature and humidity processing. It's promising to realize the fast and non-destructive detection of single soybean seed viability based on EIS, The present preliminary study would be beneficial to evaluate seed quality and planting value scientifically, properly, and rapidly.

Acknowledgment

This work was supported by the 948 Ministry of Agriculture project (2006-G63(5), 2010-S20) and Beijing key technology program (D101105046310002).

References

1. Ying, Y., Shen, Y., Li, L.: Study on the Testing of Seed Viability of Three Broad-leaved Trees with TTC Method. Seed 24(1), 32–35 (2005)
2. Li, J., Zhong, S.: Research on Seed Viability Determination of Pinus Armandii by X-ray Photographic Technique. Yunan Forestry Science and Technology 85, 21–27 (1998)
3. Shi, H., Ke, Y.: Optimized Conditions of the Electrical Conductivity Method for Determination of Seed Vigor in Maize. Seed 27(5), 7–10 (2008)
4. Taylor, A.G.: Seed storage, Germination and Quality. In: Wein, H.C. (ed.) The Physiology of Vegetable Crops. CAB International, Wallingford, pp. 1–36 (1997)
5. MacDoug Gall, R.C., Thompson, R.G., Piene, H.S.: Electrical Capacitance and Resistance Measurements as Related to Total Foliar Biomass of Balsam Fir Trees. Can. J. For. Res. 17, 1071–1074 (1987)
6. Repo, T., Zhang, G.: The Relation Between Growth Cessation and Frost Hardening in Scots Pines of Different Origins. Trees 14, 456–464 (2000)
7. Nelson, S.O., Forbus Jr., W.R., Lawrence, K.C.: Trans. ASA.E 38, 579 (1995)
8. Repo, T., Paine, D.H.: Electrical Impedance Spectroscopy in Relation to Seed Viability and Moisture Content in Snap Bean. Seed Science Research (12), 17–29 (2002)
9. Fan, L.: Effects of Three Artificial Aging Methods on Vigor Hysiological and Biochemical Characteristics of Soybean Seeds, pp. 12–22. Shanxi University, China (2007)

An EST-Based Automatic Route Shortening in Dynamic Source Routing Protocol

Li Xia[1], Shilin Jiang[1], Zhenglong Song[1], and Guangyan Sun[2]

[1] College of Information Science and Engineering, Northeastern University, Shenyang, China
[2] Shenyang Institute of Engineering, Shenyang, China
xiali@ise.neu.edu.cn, {jongsten,a3996358}@163.com,
sungy0921@sina.com

Abstract. The area of ad hoc network has been receiving increasing attention among researchers in recent years. The dynamic routing protocol is its key technology. The Dynamic Source Routing (DSR) protocol is a widely discussed routing protocol for ad hoc network. To enhance the performance of automatic route shortening in DSR protocol, the paper proposes an improved scheme which named Automatic Route Shortening based on Expected Sending Times (EST) of links in DSR, abbreviated EST-ARS. The simulation results indicate that EST-ARS outperform DSR protocol at packet delivery ratio and end-to-end delay.

Keywords: Ad Hoc network; Dynamic Source Routing; Automatic Route Shortening; Expected Sending Times.

1 Introduction

Ad Hoc network is a kind of wireless network, which does not depend on any fixed infrastructure. In Ad Hoc networks, the mobile nodes are not only hosts, but also routers. It is completely self-organization, and it is a new network that constituted by wireless nodes [1]. In the areas military, rescue, and distance education, the research and application of Ad Hoc network has great strategic significance.

Nodes' mobility in ad hoc network determines that traditional Internet-based routing protocol can not be applied to the network without any modification, so it is necessary to study specific routing protocol for ad hoc network. Significant research in this area has been focused on the design of efficient routing protocols. In recent years, researchers proposed many routing protocols for ad hoc network. Among them, Dynamic Source Routing (DSR) protocol [2] and Ad hoc On-demand Distant Vector (AODV) protocol [3] are two widely studied reactive protocols.

In this paper we focus on DSR protocol. To promote the performance of DSR protocol, researchers have presented a variety of optimization strategies for different parts in DSR protocol. Automatic Route Shortening is one of routing real-time optimization strategies. Automatic Route Shortening mechanism in DSR protocol, which is based on hops, is designed to obtain the shortest path. This mechanism can guarantee that the generated path has the least number of hops, but does not considerate its quality. Therefore, the path obtained through this mechanism may not

L. Qi (Ed.): ISIA 2010, CCIS 86, pp. 637–644, 2011.

be a stable and efficient route in many instances [4]. To solve these problems, we present an improved automatic route shortening mechanism which named based on Expected Send Times Automatic Route Shortening (EST-ARS).

This paper is organized as follows. Hops-based automatic route shortening scheme in DSR is described in section 2, and our EST-based automatic route shortening scheme is described in section 3. Simulation and results analysis is presented in section 4. Finally, conclusions with future research works are presented in section 5.

2 Hops-Based Automatic Route Shortening

The changes of network topology are inevitable in mobile ad hoc networks, which could lead to one or even more intermediate nodes in the route become no longer necessary [5]. If a node overhears a packet carrying a source route, then it examines the unexpended portion of that source. If this node is named in the later unexpended portion of the route, then it starts automatic route shortening process to increase transmission efficiency. The automatic route shortening in DSR protocol is based on the principle of minimum hops, which is the new route with less hops instead of the active route with more hops. The Fig.1 below illustrates an example about automatic route shortening based on hops.

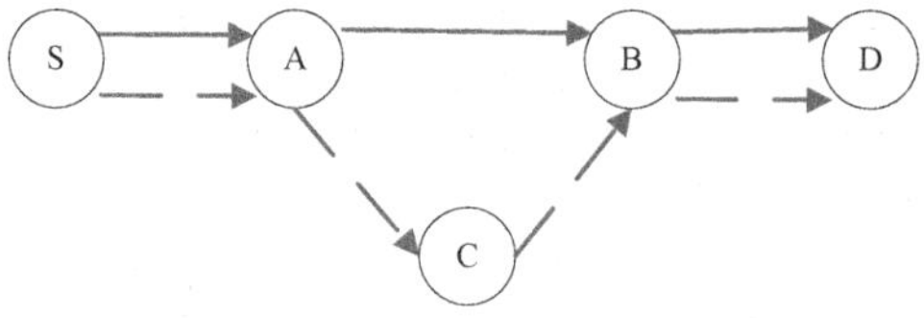

Fig. 1. Automatic Route Shortening Based on Hops

As shown in Fig.1, node B has overheard a data packet being transmitted from A to C, for later forwarding to B and to D. Then node B finds that it is not the intended next-hop destination for the packet and is named in the later unexpended portion of the packet's source route. It infers that the intermediate nodes between A and B are unnecessary. In this case, node B should return a gratuitous Route Reply to the original node S. The route returned in gratuitous Route Reply message sent from B to S gives the new shorter route as the sequence of hops from S to A to B to D. The original node S will update its Route Cache after receiving the Route Reply.

The path can be quickly switched to the shorter route by using hop-based automatic route shortening mechanism. But this is risky, especially when strong network mobility and large node density. Under high mobility, the new route obtained by this mechanism is likely to fail, which lead to a new route discovery process. This will greatly affect the performance of the network. When the node density of network is large, data packet in transmission will frequently trigger the automatic route shortening process. Finally, the large number of gratuitous Route Reply packets will be full of the entire network. Therefore, we propose automatic route shortening based on EST.

3 EST-Based Automatic Route Shortening

In this paper, we define the so-called Expected Sending Times(EST) to evaluate the quality of links, which is used to perform automatic route shortening process. And we have expanded the scope of automatic route shortening in DSR. As mentioned in related work, the route shortening in our proposal is not limited to the active route as automatic route shortening in DSR does.

3.1 Expected Sending Times Metric Design

The EST or Expected Sending Times of a link is the expected number of data transmissions required to send a packet over that link. The EST of a route is the sum of the EST for each link in the route. To calculate the EST, we use HELLO messages which are periodical broadcasted. Fig.2 illustrates propagation path of HELLO packets among the adjacent nodes.

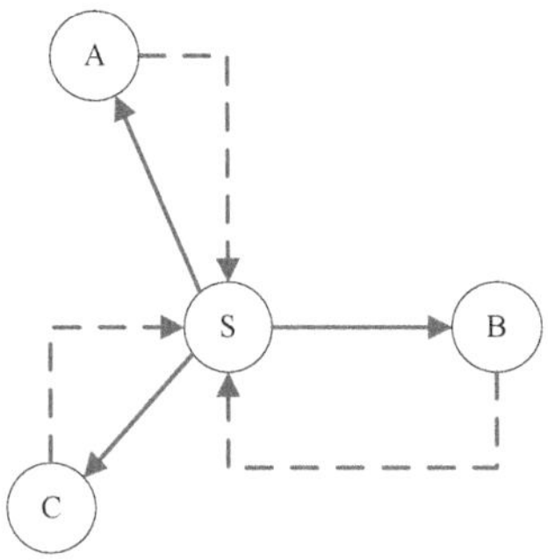

Fig. 2. Path of Hello Packets Transmission

In the process of HELLO messages dissemination, it is assumed that node S sends a HELLO message to its neighbors during each period r. For example, node A makes a record after each receives a HELLO message sent by node S. Thus node A will take count of the number of received HELLO messages from node S in the past period w. So the success rate of transmitting data packets from node S to node A can be calculated as:

$$r(t) = \frac{count(t-w,t)}{w/r} \tag{1}$$

Where count (t-w, t) is the number of HELLO packets received by node A during the past period w. And w/r is the number of HELLO messages that should have been received. In the case of the link S→A, this method allows A to measure d_s (the success rate of transmitting data packets from node S to node A), and S to measure d_r (the success rate of transmitting data packets from node A to node S). Calculation of a link's EST requires both d_s and d_r. The expected probability that a transmission is successfully received and acknowledged is $d_s \times d_r$. If a packet is not successfully acknowledged, the sender will retransmit this packet. As an attempt to transmit a packet can be considered a Bernoulli trial [6], the Expected Sending Times is:

$$EST = \frac{1}{d_s \times d_r} \tag{2}$$

A low value of EST means that it consumes less time and energy to send a packet successfully, and vice versa. Obviously, the smaller EST of a link is, the better the link quality. It is the same to a route.

3.2 EST-Based Automatic Route Shortening on Source Route

In EST-ARS, we consider the EST parameter of a new route as a measure of the route's quality to reduce the adverse impact of hop-based automatic route shortening. Only when the ne route's EST is smaller than original route's can it be considered valid in EST-based automatic route shortening mechanism. This can ensure that the new route is better if not better than the original should be same as the original. Fig.3 shows the process of EST-based automatic route shortening on source route.

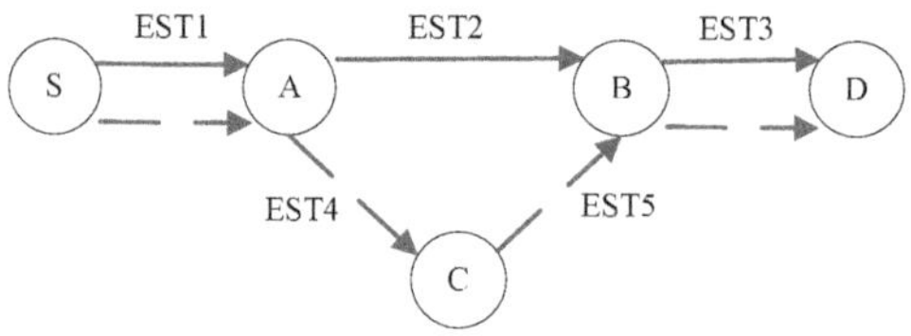

Fig. 3. EST-Based Automatic Route Shorting on Source Route

In Fig.3, node B has overheard a data packet being transmitted from A to C, for later forwarding to B and to D. Then node B finds that it is not the intended next-hop destination for the packet and is named in the later unexpended portion of the packet's source route. It infers that the intermediate nodes between A and B are unnecessary. In this case, node B should return a Request of Routing Comparison to the original node S. The route returned in Request of Routing Comparison message sent from B to S gives the new shorter route as the sequence of hops from S to A to B to D. After receiving the Request of Routing Comparison message the source node S will launch the comparison process about ESTs of the original route (from S to A to C to B to D) and the candidate route (from S to A to B to D). Then the destination node D will send the result of comparison to node S. If the result is that the new route's EST is smaller than original route's EST, the source node S will replace the original route with the candidate route to update its Route Cache. Otherwise it does not do this.

3.3 EST-Based Automatic Route Shortening Outside Source Route

Mobile ad hoc networks are envisioned to have dynamic, random, multi-hop topologies. So their network topologies change frequently [7]. Automatic route shortening limited to source route is not enough, so this paper proposes automatic route shortening outside source route. In this method, after a node outside source route overheard a data packet, if it finds that there is a shorter route to the destination node of the packet in its own Route Cache, it should return a Request of Routing Comparison which gives the full path to the original node. Then the source node will launch the comparison process about ESTs of the original route and the candidate route. And the next process is the same as described in the previous section. Fig.4 shows the case of automatic route shortening outside source route.

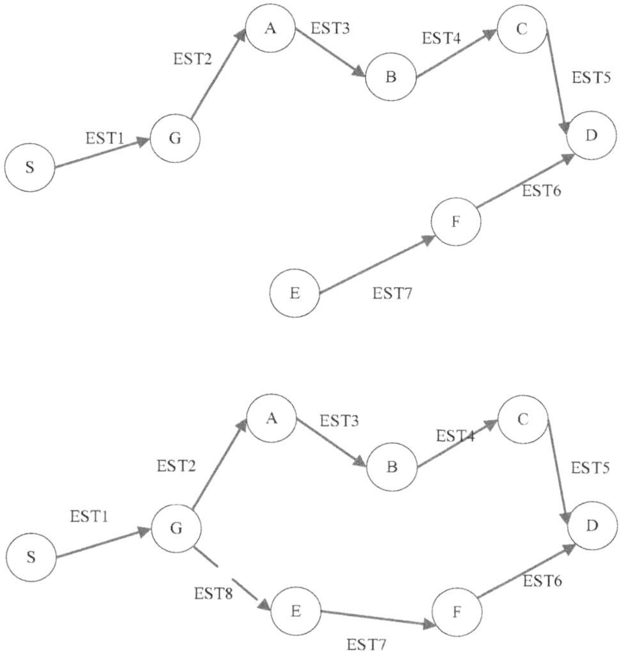

Fig. 4. Automatic Route Shortening outside Source Route

As shown in Fig.4, the sequence of source route is from S to G to A to B to C to D. There is a route to node D in node E's Route Cache. As nodes' mobility, node E reached the scope of node G's wireless transmission. And node E overheard a data packet which is sent to node A from node G. Then it finds that there is a shorter route to the destination node D of the packet in its own Route Cache compared to the source route. Therefore, node E will send a Request of Routing Comparison to the source node S. The Request of Routing Comparison message carries the candidate route which is from S to G to E to F to D. The source node S will launch the comparison process about ESTs of the original route (from S to G to A to B to C to D) and the candidate route (from S to G to E to F to D) after receiving the Request of Routing Comparison message. If the result of comparison returned by node D is that the candidate route's EST is smaller than original route's, the source node S shall replace the original route with the candidate route to update its Route Cache. Otherwise it does not do anything and the follow-up data packets would be still transmitted along the original route.

4 Simulation Results

Due to the difficulties associated to real tests, the benefits of EST-ARS have been verified by the simulation. The main objectives of simulation are collecting the average packet delivery ratios and end-to-end delays respectively using DSR protocol and EST-ARS under various conditions. Simulation parameters are given in Table 1.

Table 1. Simulation common parameter

Simulator	NS 2
Scenario Dimension	1000m*1000m
Mobile Nodes	50 nodes
Transmission Range	250m
Simulation Time	1000 seconds

The node movements are based on the Random Waypoint Model. Thirty of the 50 mobile nodes send traffic to each other. The traffic will be Constant Bit Rate (CBR) with a transmission rate of 10 packets per second, and the packet length is 512 bytes.

In Fig.5 and Fig.6, end-to-end delay and packet delivery ratio are respectively shown in several scenarios with different mobile speed when the pause time of nodes is fixed at 50 seconds. Each point in figures represents the mean value of 20 runs with equivalent simulation conditions.

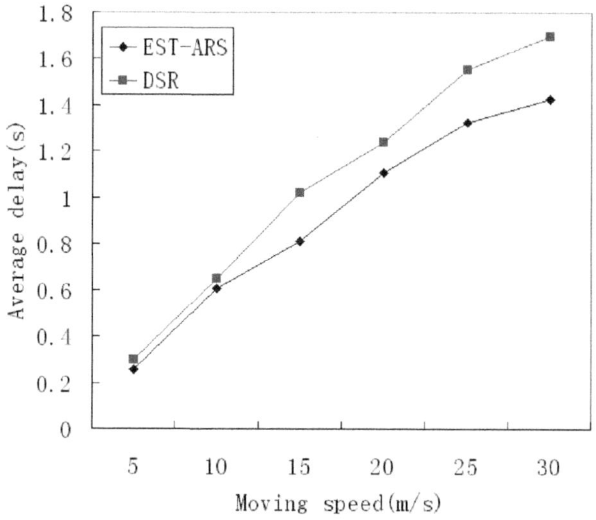

Fig. 5. End-to-End Delay

As we can see in Fig.5 and Fig.6, the end-to-end delay and the packet delivery ratio of EST-ARS are slightly improved compared with typical DSR protocol when the

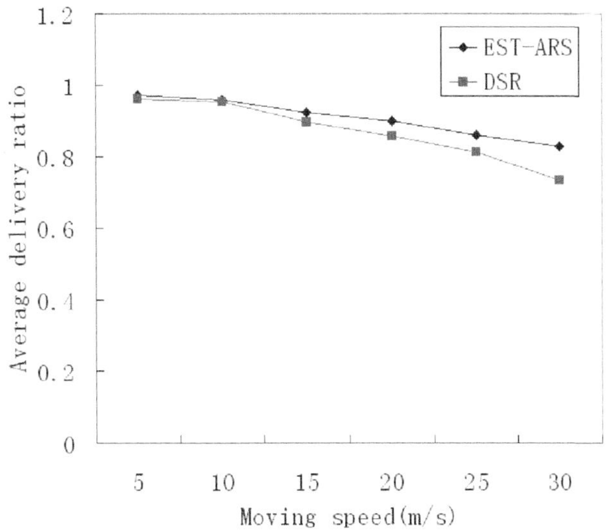

Fig. 6. Packet Delivery Ratio

mobility rate of nodes is relatively low. And the benefits (relative to DSR) of using EST-ARS are enhanced when the mobility rate of nodes increases. This is because EST-ARS can effectively block the bad automatic routing shortening in the case of higher rate. This benefit is obtained by increasing the routing overhead, and it is worthwhile in some cases.

5 Conclusion

This paper has presented an EST-based Automatic Route Shortening(EST-ARS) mechanism. Particularly, the route shortening in our proposal is not limited to the current route as automatic route shortening in traditional DSR does. Apart from theoretical investigation, EST-ARS is implemented in NS-2 in several scenarios with different mobile speed. And the simulation results show that the proposed scheme improves the end-to-end delay and packet delivery ratio.

References

1. Perkins: Ad Hoc Networking. Addison Wesley, Reading (2000)
2. Johnson, D., Hu, Y., Maliz, D.: The dynamic source routing protocol for ad hoc networks (DSR) [J/OL] (2004), `http://www.ietf.org/internet-drafts/draft-ietf-manet-dsr-10.txt`
3. Perkins, C., Belding-Royer, E., Das, S.: Ad hoc on-demand Distance Vector (AODV) routing. RFC3561 (2003)

4. Johnson, D., Hu, Y., Maliz, D.: The Dynamic Source Routing Protocol (DSR) for Mobile Ad Hoc Networks for IPV4 (2007), `http://www.rfc-editor.org/rfc/rfc4728.txt`
5. Ahmad, S., Awan, I., Waqqas, A., Ahmad, B.: Performance Analysis of DSR & Extended DSR Protocols. In: Asia International Conference on Modeling & Simulation, pp. 191–196 (May 2008)
6. Decouto, D.S.J., Aguayo, D., Bicket, J., Morris, R.: A High-Throughput Path Metric for Multi-Hop Wireless Routing. In: MobiCom 2003, September 14-19 (2003)
7. Seet, B.-C., Lee, B.-S., Lan, C.-t.: Route discovery optimization for dynamic source routing in mobile ad hoc networks. 9th IEEE Electronics letters IEEE 2000 (November 2000)

Prediction of Bacterial Toxins by Feature Representation of Position Specific Scoring Matrix and IB1 Classifier Fusion

Chaohong Song

College of Science, Huazhong Agricultural University, Wuhan 430070, China
chh_song@mail.hzau.edu.cn

Abstract. Successful prediction of bacterial toxins directly from primary sequence is much benefited to further basic knowledge of cell biology or for medical research and application. In this paper, we proposed a new method to predict bacterial toxins by using the feature representation of position specific scoring matrix and IB1 classifier fusion. The jackknife cross-validation is applied to test predictive capability of the proposed method. The predictive results showed that the total prediction accuracy is 96.62% for bacterial toxins and non toxins, which is higher than previous methods. Furthermore, we also discriminated endotoxin and exotoxin by the proposed method, and obtained satisfactory result with a total prediction accuracy 95.33%.

Keywords: prediction; bacterial toxin; position specific scoring matrix; IB1 classifier.

1 Introduction

As we all know, bacterial toxins are a major cause of some diseases during infection [1, 2], but if we could successfully predict bacterial toxins directly from primary sequence, we could make them benefit to further basic knowledge of cell biology or for medical research and application. For example, cholera toxin and the related labile-toxin of E. coli, as well as B. pertussis toxin, have been used as biologic tools to understand the mechanism of adenylate cyclase activation [3-5] and strong mucosal adjuvants have been used in experimental models [6]. So, it's desirable to get the knowledge of bacterial toxins for indepth understanding their genomic regulation. However, it is costly and time-consuming to assay whether a protein sequence is a bacterial toxin, or whether a bacterial toxin is an exotoxin or endotoxin by biology experiments. Thus it is of great practical significance to develop computational approaches for identifying bacterial toxins.

The present study have developed some new methods for predicting bacterial toxins, for example Saha and Raghava [7] used support vector machines (SVM) connected with amino acids and dipeptides composition to predict the bacterial toxins on a dataset which contains 150 bacterial toxins, in their research, they achieved an accuracy of 96.07% and 92.50%, respectively. Moreover they discriminated entotoxins and

L. Qi (Ed.): ISIA 2010, CCIS 86, pp. 645–649, 2011.

exotoxins using the same method and achieved an accuracy of 95.71% and 92.86%, respectively. On the same dataset, Yang and Li [8] also predicted bacterial toxins, using Increment of diversity and support vector machines; they achieved higher MCC for endotoxins and exotoxins than that of [7]. Encouraged by their research, in this paper, a new method was proposed to predict bacterial toxins by the feature representation of position specific scoring matrix and IB1 classifier fusion. We hoped that our method could play a complementary role to other existing methods for predicting bacterial toxins.

2 Method and Material

2.1 Dataset

Two data sets that we used in this paper were collected form Swiss-Prot database [9] and the dataset used by [7], all these could be download form http://www.imtech.res.in/raghava/btxpred/supplementary.html.

The first dataset is used for the classification of bacterial toxins and non-toxin. We used the cd-hit soft [10] to remove sequences that have more than 90% sequence identity and deleted the sequences which the length is less than or equal to 100, than we got the 141 bacterial toxins and 303 non- toxins.

The second dataset is used for the classification of exotoxins and endotoxins, we also used the cd-hit soft to remove sequences that have more than 90% sequence identity, and finally the dataset contains 73 exotoxins and 77 endotoxins.

2.2 Position Specific Scoring Matrix

Position Specific Scoring Matrix (PSSM) is a commonly used representation of motifs in biological sequences [11]. In this paper we used PSI-BLAST [12] to generate PSSM profile as a training feature, in each iteration of PSI-BLAST, a PSSM is generated from a multiple alignment of the high scoring hits by calculating position specific scores for each position in alignments. The PSSM generated in each step is used to perform next iterative search, thereby increasing the sensitivity of the search in each step. After three iterations, a PSSM with the highest score is generated. If the length of the target sequence is N, Then the matrix contains 20 times N elements, and each element represents the frequency of occurrence of each of the 20 amino acids at a particular position in the alignment. Subsequently, using the following sigmoid function, the final PSSM was normalized, and each matrix element was scaled to a range 0-1.

$$f(x) = \frac{1}{1+e^{-x}} \tag{1}$$

Then we summed up all the rows in the PSSM corresponding to the same amino acid in the sequence, followed by division of each element by the length of the sequence.

2.3 IB1 Classifier

IB1 classifier is one of Instance-based learning classifiers [13]. These classifiers use some functions that map instances to certain categories. They generally consist of the following steps:

(1) First use the following equation to normalize each numeric attribute's values.

$$nomr(x_a\ a) = \frac{x_a - \min a}{\max a - \min a} \tag{2}$$

Where $\min a$ and $\max a$ are the lowest and the highest values of some attribute a, which describes an instance x_a, whenever new information appears on the input, re-normalization it again.

(2) For two protein sequences $x = (x_1, x_2, \ldots x_n)$ and $y = (y_1, y_2, \ldots y_n)$, calculate their similarity as the following:

$$sim(x\ y) = \sqrt{\sum_{i=1}^{n} f(x_i, y_i)} \tag{3}$$

Where $f(x_i, y_i) = (x_i - y_i)^2$ if $x_i \neq y_i$, else $f(x_i, y_i) = 1$.

(3) Compute a weighted-similarity of its k-most similar instances' target values [13]:

$$Tval(K,t,k) = \sum_{i=1}^{k} \frac{sim(K_i) \times K_{it}}{\sum_{j=1}^{k} sim(K_i)} \tag{4}$$

Where K_i is one of the k-most similar stored instances, K_{it} is instance K_i's value for target attribute t, and $sim(K_i)$ is K_i's pre-computed similarity with the current test instance x.

If $Tval(K,t_j,k) = \max_i Tval(K,t_i,k)$, we deemed the instance x belongs to class t_j.

Generally, the larger values of k could reduce the effect of noise in the process of classification, but also make boundaries between classes less distinct. In practice, $k = 1$ often provides very good or acceptable results. So in this paper we selected $k = 1$, i.e. IB1classifier.

2.4 Evaluation of the Performance

Here we selected four parameters to evaluate the correct prediction rate and reliability of our method, the sensitivity (Sn), specificity (Sp), Matthew's correlation coefficient (MCC) and the overall prediction accuracy (Ac) were defined by:

$$Sn = TP/(TP + FN) \tag{5}$$

$$Sp = TP/(TP + FP) \tag{6}$$

$$MCC = (TP \times TN - FP \times FN) / \sqrt{(TP + FP)(TP + FN)(FN + TN)(TN + FP)} \quad (7)$$

$$Ac = (TP + TN) / M \quad (8)$$

Here TP denotes the numbers of the correctly recognized positives, FN denotes the numbers of the positives recognized as negatives, FP denotes the numbers of the negatives recognized as positives, and TN denotes the numbers of correctly recognized negatives. M is the total number of protein sequences.

3 Results and Discussion

From Table 1, we could see the performance of various methods developed for discriminating the bacterial toxins from non-toxins. Our method using position specific scoring matrix and IB1 classifier fusion was able to predict toxins with the total accuracy 96.62%, which were higher than that of the previous results.

Table 1. The performance of various methods in prediction of bacterial toxins

	Sensitivity	Specificity	MCC	Accuracy
Our method	93.83%	97.16%	0.9237	96.62%
Amino Acids[a]	92.14%	100%	0.9293	96.07%
Dipeptides[a]	86.43%	98.57%	0.8612	92.50%

a comes from [7]

The performance of various methods developed for predicting whether a bacterial toxin is an exotoxin or an endotoxin had been shown in Table 2. From Table 2 we could see that the total accuracy was 95.33%, although it was some little lower than that of amino acid composition, it maybe our method is poor for little data. However, the performance was also satisfactory, which showed that our method could play a complementary role to other existing methods for predicting bacterial toxins.

Table 2. The performance of various methods in discrimination of exotoxins and endotoxins

	Sensitivity	Specificity	MCC	Accuracy
Our method	98.57%	94.52%	0.9066	95.33%
Increment of diversity[b]	92.91%	99.24%	0.9428	
Amino Acids[a]	100%	91.43%	0.9293	95.71%
Dipeptides[a]	94.29%	91.43%	0.8612	92.86%

a comes from [7], b comes from [8]

Because we were trained and tested on a non-redundant dataset of 150 bacterial toxins that only included 77 exotoxins and 73 endotoxins, so in this paper, we did not discuss further segmentation of exotoxins. We hope that we will make further research with the data set increasing.

References

[1] Böhnel, H., Gessler, F.: Botulinum toxins – cause of botulism and systemic diseases? Vet. Res. Commun. 29, 313–345 (2005)
[2] Blackall, D.P., Marques, M.B.: Hemolytic uremic syndrome revisited: Shiga toxin, factor H, and fibrin generation. Am. J. Clin. Pathol. 121, 81–88 (2004)
[3] Harnett, M.M.: Analysis of G-proteins regulating signal transduction pathways. Methods Mol. Biol. 27, 199–211 (1994)
[4] Bokoch, G.M., Katada, T., Northup, J.K., Hewlett, E.L., Gilman, A.G.: Identification of the predominant substrate for ADP-ribosylation by islet activating protein. J. Biol. Chem. 258, 2072–2075 (1983)
[5] Neer, E.J.: Heterotrimeric G proteins: organizers of transmembrane signals. Cell 80, 249–257 (1995)
[6] Bagley, K.C., Abdelwahab, S.F., Tuskan, R.G., Fouts, T.R., Lewis, G.K.: Cholera toxin and heat-labile enterotoxin activate human monocyte-derived dendritic cells and dominantly inhibit cytokine production through a cyclic AMP-dependent pathway. Infect. Immun. 70, 5533–5539 (2002)
[7] Saha, S., Raghava, G.P.: BTXpred: Prediction of bacterial toxins. Silico. Biol. 7, 405–412 (2007)
[8] Yang, L., Li, Q.-z., Zuo, Y.-c., Li, T.: Prediction of Animal Toxins Using Amino Acid Composition and Support Vector Machine. Journal of Inner Mongolia University 40, 443–448 (2009)
[9] Boeckmann, B., Bairoch, A., Apweiler, R., Blatter, M.-C., Estreicher, A., Gasteiger, E., Martin, M.J., Michoud, K., O'Donovan, C., Phan, I., Pilbout, S., Schneider, M.: The Swiss-Prot protein knowledgebase and its supplement TrEMBL in 2003. Nucleic Acids Res. 31, 365–370 (2003)
[10] Li, W.z., Godzik, A.: Cd-hit: a fast program for clustering and comparing large sets of protein or nucleotide sequences. Bioinformatics 22, 1658–1659 (2006)
[11] Ben-Gal, I., Shani, A., Gohr, A., Grau, J., Arviv, S., Shmilovici, A., Posch, S., Grosse, I.: Identification of transcription factor binding sites with variable-order Bayesian networks. Bioinformatics 21, 2657–2666 (2005)
[12] Schaffer, A.A., Aravind, L., Madden, T.L., Shavirin, S., Spouge, J.L., Wolf, Y.I., Koonin, E.V., Altschul, S.F.: Improving the accuracy of PSI-BLAST protein database searches with composition-based statistics and other refinements. Nucleic Acids Research 29, 2994–3005 (2001)
[13] Aha, D.W.: A Study of Instance-based Algorithms for Supervised Learning Tasks: Mathematical, Empirical, and Psychological Evaluations. PhD dissertation, Dept. of Information and Computer Science, Univ. of California, Irvine (1990)

An Efficient Memetic Algorithm for Job Scheduling in Computing Grid

Luo Zhong, ZhiXiang Long, Jun Zhang, and HuaZhu Song

School of Computer Science and Technology, Wuhan University of Technology, Luoshi Road 122, Wuhan 430070, P.R. China
Longzhixiang2004@sina.com

Abstract. Grid job scheduling is an NP complete problem, concerning the large-scale resource and job scheduling, and the adoptive and efficient job scheduling algorithm is required. Genetic algorithms show good capability to solve the problem of the small-scale, but with the increase in the number of jobs and resources, genetic algorithm is hard to convergence or slow convergence. This paper proposed a Memetic Algorithm which designed crossover operators and mutation operator with hill-climbing algorithm and Tabu search algorithm for processing grid job scheduling. Hill Climbing scheduling usually can enhance processor utilization, and Tabu search algorithm have shorter completion times for job scheduling in computing grid. And then the algorithms' search ability and convergence speed were compared. The simulation results shown that the proposed algorithm can effectively solve the grid job scheduling problem.

Keyword: Computing Grid; Job scheduling; Memetic Algorithm; Hill-Climbing algorithm; Tabu search algorithm.

1 Introduction

To meet the increasing demand for computing capacity, grid computing and grid technologies appear today. Computing grid job scheduling problem is hard to be calculated and it has been proved that to find the optimal scheduling problem is a NP hard problem in the complex system. Among Numerous articles on the grid scheduling, the researchers have proposed a lot of algorithms to improve the grid scheduling problem. Reference [1] proposed structured Cellular MAs for independent scheduling under the ETC model. Reference [2] investigated the results of 11 kinds of scheduling algorithm, where Min-min algorithm calculated the minimum execution time of all scheduling jobs not dispatched. Reference [3] proposed a modified genetic algorithm which provided a preferable performance, but it made a not fast convergent speed. Many of these algorithms are hard to convergence or slow convergence to apply to the job scheduling of computing grid.

Memetic algorithms can be used to solve the above problem. This paper gave computing grid job scheduling problem based on Memetic algorithms, including distribution of client, application of client operations to the grid resources.

L. Qi (Ed.): ISIA 2010, CCIS 86, pp. 650–656, 2011.

In section 2 we briefly introduced the computational grid job scheduling problem; section 3 described the Memetic algorithm, and describes its operator; the experimental studies and results were in section 4, and section 5 summarized our work and discussed the direction of future work.

2 Job Scheduling of Computing Grid

2.1 Problem Description

In order to describe the job scheduling problem in computing grid, we need the estimation of the calculation load of each job, computing capacity of each resource, and the priority of each resource. This is the ETC matrix model - Expected Time to compute matrix model. So we had a hypothesis, we know computing capacity of each resource, prediction and estimation of the calculation load of each job, and the priority of each resource. With the computing capacity of each resource and job workload, a matrix model ETC[*t*][*m*] is built, that is the expected time that based on the resources m to complete the job t. The problem is described as follows:

a) A number of independent jobs to be scheduled, and job must be in implementation of only one resource.
b) A number of heterogeneous machines candidates to participate in the planning.
c) The workload of each job (in millions of instructions).
d) The computing capacity of each machine (in mips).
e) Ready time $ready_m$ indicates when machine m finish the previously assigned jobs.
f) The expected time to compute (ETC) (*number_job X number_machines*) matrix in which ETC[*i*][*j*] is the expected execution time of job *i* in machine *j*.

2.2 Fitness

In general, there are two criteria to evaluate the fitness. The first criterion is to minimize the makespan (the latest job completion time), the second criterion is to minimize the grid system of flowtime (completion of all jobs in the final time), that is:

$$\text{Minimize makespan: } \min_{Si \in Sched} \{\max_{j \in Jobs} F_j\} \quad (1)$$

$$\text{Minimize flowtime: } \min_{Si \in Sched} \{\Sigma_{j \in Jobs} F_j\} \quad (2)$$

In which, F_j is the final completion time of job *j*, *Sched* refers to all possible dispatch, *Jobs* refers to the work collection which must be dispatched. The makespan subject to any specific resources and any particular order of execution, when in order to minimize the resource's flowtime, the job should be in accordance with the calculation of their expected time to rise the foreword execution. We can know that makespan and flowtime are relatively objectives, trying to minimize one of them and the policy may not be suitable for another target.

The makespan uses completion time to express, making vector completion is the machine quantity size, and comp[*m*] is that machine m will eventually implement the jobs previously assigned time, that is, the machine m has been assigned the job of execution time. The comp[m] can be calculated as follows:

$$comp[m] = ready\_times[m] + \sum_{\{j \in Jobs \mid sched[j]=m\}} ETC[j][m] \tag{3}$$

The makespan objective can be expressed in the following form:

$$Makespan=max\{comp[i] \mid i \in Machines\} \tag{4}$$

The smaller the makespan value means that the scheduler of the resource allocation job carries out the higher efficiency. On the other hand, the flowtime's value of minimizing means to reduce the grid system of the average response time. More importantly, we used an intelligent load to balance the performance of the maximum grid system.

Fitness function is defined as follows:

$$Fitness=\lambda \cdot makespan+(1-\lambda) \cdot mean\_flowtime \tag{5}$$

Here is λ=0.75, which means the parameter makespan is very important.

3 Memetic Algorithm for Job Scheduling of Computing Grid

3.1 Memetic Algorithm

Memetic algorithm is a subset of evolutionary algorithms. Compared with other evolutionary algorithms, MAs has some special characteristics, and MAs synthesized the evolution search and partial search of advantage, from this point, MAs are hybrid evolutionary algorithm.

The performance of an MA not only depends on its operator and local search methods for the design and implementation, but also it associated with chromosomes. Memetic Algorithm description is shown in Fig.1.

In the case of the algorithm proposed in this work, mutation and recombination operators are applied to individuals independently of each other, and in different orders. After each recombination (or mutation), a local search step is applied to the newly obtained solution, which is then evaluated. If this new solution is better than the current one, it replaces the latter in the population.

3.2 MA Operator

Given the MA template showed in Figure 1, the different parameters and operators are proposed that will use it for solving the problem of job scheduling in grids. In order to solve the problem efficiently, we have to particularize the template with operators incorporating some special knowledge of the problem at hand. The objective is to design an algorithm for optimizing the QoS and productivity of grid systems.

Memetic Algorithm for the chromosome is shown in Fig.2 which using double chromosome, and chromosome left layer is the job level. Where J_i on behalf of the *ith* scheduled job, and the right level is the grid nodes. The direction of the line expresses

```
step1: Initialize the mesh of n individuals P(t=0)
step2: Initialize permutations recombination_order and mutation_order
step3: For each i∈P, LocalSearch(i)
step4: Evaluate(P)
step5: while(not stopping condition) do
        for j= 1..recombinations do
          SelectToRecombine S ⊆NP[recombination_order.current] ;
          i'= Recombine(S);
          LocalSearch(i');
         Evaluate(i');
         Replace P[recombination_order.current];
           recombination_order.next();
         end for
        for j= 1..mutations do
           i= P[mutation_order.current()];
           i'= Mutate(i);
           LocalSearch(i'); Evaluate(i');
           Replace P[rec_order.current] by i';
           recombination_order.next();
         end for
         Update recombination_order and mutation_order;
     end while
```

Fig. 1. MA algorithm template

J_1	P_1
J_2	P_2
J_3	P_3
J_4	P_4
…	…
J_{i-1}	P_{i-1}
J_i	P_i
J_{i+1}	P_{i+1}
…	…
J_n	P_n

Fig. 2. Chromosome code

the scheduling order, and the direction of the line expresses the scheduling relations of P_j and J_i, so the two form a gene.

Population initialization: The first individual was produced by the Longest Job to Fastest Resource-Shortest Job to Fastest Resource (LJFR-SJFR) method, and the remaining individuals were obtained through a large number of disturbances. Because it is simultaneously minimizing the makespan and the flowtime, we chose the LJFR-SJFR method, in which LJFR minimizes makespan and SJFR minimizes flowtime.

Selection operator: We use the roulette the way to carry on the choice to the chromosome, the probability of an individual is selected for the

$$p(x_i) = fit(x_i) / \sum_{i=1}^{N} fit(x_i) \tag{6}$$

Crossover operator: Using two cross, f two parent individuals a_1 and a_2 which were selected by the Selection operator, randomly generated two cross-points c_1 and c_2, In the cross-generated child individual, the grid nodes are before c_1 and after c_2, whose job queue are inherited by the a_1, and between the c_1 and c_2, grid nodes' job queue are inherited by the a_2. As shown in Fig. 3.

Mutation operator: We use the mutation operator is carried out by balancing the load of the machine. In the experiment, we found that some machines in the scheduling may exceed the load, while the other machine does not exceed the load. In accordance with machines' completion time, the time is ascending order, and first 1/4 of the machine is not considered beyond the load. Machine's load factor is represented by the *load_factor(m)*.

$$Load\_factor(m)=comp[m] \,/\, makespan(load\_factor(m) \in (0,1]) \tag{7}$$

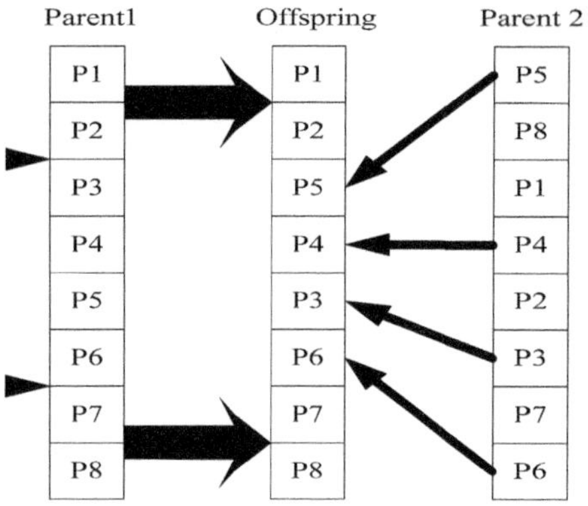

Fig. 3. Crossover

```
Hill Climbing (HC):
Require: a population collection P, local, local_v
Ensure: An improved population P
  step1:  t←0,  Initialization individual bestPop
  step2:  while (t!= Max){
             local←false;
             rand_select(local_v);
             Evaluate(local_v);
             while(local){
                 Best_Evalue(vm);
                 if(vm>v) v←vm;
                 else local←true; }
          t←t+1;
          if(v>bestPop) bestPop←v;
          return bestPop;
```

Fig. 4. Hill Climbing algorithm

Balance the load in two steps: First of all, the *mth* machines beyond the load state were randomly selected, then we determined the two jobs j and j', so j was assigned to the machine m'; the ETC value under condition that j was assigned to the machine m is less than or equal to the ETC value under condition that j' was assigned to the machine m'. Then j and j' are exchanged. The second step, if load balancing is not feasible, we can move to balance.

Updated population: Each crossover and mutation generating new solution is added to the population. There are two guidelines to be followed, one of which is that the new generated solution is only added to the population, the second is that the added new solution must be better than population's in current solution sufficiency.

Local search algorithm: Proper selection of local search is appropriate or not largely depending on the performance of the MA. In Figure 1, we can see every individual must be improved through local search. Here we use two kinds of modern heuristic algorithms, and compare them.

The Hill Climbing algorithm as shown in Fig.4 will terminate when not find better k-domain solution, and this k-domain solution from the current solution is randomly selected. Thus, if we use the k value is small, the hill climbing is quickly over, but the algorithm can not guarantee a good degree of convergence. On the contrary, if we set a very large k value, then the search algorithm will detect too many solutions, so that this will require a longer time to converge. So we set k value is through a large number of experiments.

Tabu Search (TC) is a meta-heuristic to solve optimization problems raised by Fred Glover in 1986. It is a local neighborhood search for an extension and simulates the human thinking process of a method, and it uses the taboo table to Storage the local optimal solution which has been searched and to mark, so that search process in the future avoids circuitous search for these solutions, and escapes from local optima; at the same time, some of the forbidden fine state are pardoned by flouting norms, because there is a certain probability of acceptance of inferior solutions, in order to ensure an effective search for different ways to explore, and ultimately realize global optimization, so the solution quality is high.

```
Tabu Search:
Require: a population collection P ,maxiter, d
Ensure: An improved population P
step1: tabuList←  , iter←0, C*=0, P*= ;
step2: Evaluate C(P); iter←1, C*=C, P*=P;
step3: while(iter≤maxiter) do
        Generate neighbor allocations using the move operators;
        Select the best move;
        Ignore the tabu status by aspiration criterion if such move generates a best solution;
        Made the best move by adding the selected bid into P;
        Based on the conflict graph, remove any conflicting bids in P;
        Insert the best move in the tabuList;
        Evaluate C(P);
        if C>C* then {P*=P; C*=C; iter_best=iter;}
        if iter - iter_best≤d then Diversification step;
        iter=iter+1;
    return the best population found;
```

Fig. 5. Tabu Search Algorithm

4 Experimental Study

Some experiments were done for testifying the validation of our proposed algorithm. We use the grid simulator to test our two different Memetic Algorithm with different local search algorithms, Hill-Climbing (HC) and Tabu search. And then the comparation experiment with Braun et al.'s GA was done.

The parameter shows the result as table 1:

Table 1. The parameters in the experiment

parameter	value
Population size	50
Chromosome size	30
Probability of Crossover	0.75
Probability of mutation	0.1
Population initialization	LJFR-SJFR
Selection method	Tournament
Recombination operator	Two-point recombination
Mutation operator	Rebalance
Local search method	HC / Tabu Search
Max generation	10000
Termination condition	Maximumal generation running

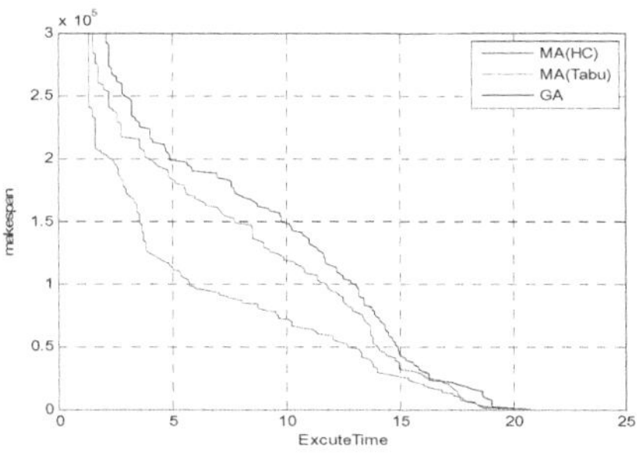

Fig. 6. Job-Time curve

Fig.6. depicted the time span of three algorithms with the job of curve size. We can confer in this figure that Compared to MA(HC) and GA, MA(Tabu) in performance has significantly improved.

The results of makespan parameter comparison: We gave the computational results for the makespan objective in Table2, where the first column indicates the best makespan obtained by Braun et al.'s GA, the second one the best makespan by our

MA(HC) implementation, the third one gives the difference (in%) between the best makespan reported by the Braun et al. GA and MA(HC), the next column the best makespan by our MA(Tabu) implementation, and the last one gives the difference (in%) between the best makespan reported by the Braun et al. GA and MA(Tabu).

Table 2. The comparison of makespan parameter

Braun et al. GA	MA(HC)	△(%)	MA(tabu)	△(%)
8050844.5	7966547.42	1.05	7759658.51	3.61
156249.2	156102.64	0.94	155457.45	0.51
258756.77	252213.37	2.53	251458.63	2.82
5272.25	5223.66	0.92	5204.46	1.29
3104762.5	3106755.87	-0.64	3106021.76	-0.41
75816.13	75847.49	-0.41	75831.28	-0.20
107500.72	110354.68	-2.65	109846.73	-2.18
2614.39	2622.31	-0.3	2618.65	-0.16
4566206	4508754.15	1.26	4458647.7	2.36
98519.4	98323.54	0.2	98302.43	0.22
130616.53	130147.98	0.36	130095.62	0.40
3583.44	3542.74	1.14	3526.49	1.59

We could see from the table 2, MA (Tabu) and MA (HC) in performance are better than the Braun et al. GA. This observation is interesting if the Grid characteristics were known in advance, MA (Tabu) and MA (HC) seems to be more appropriate for consistent and semi-consistent Grid scenarios.

5 Conclusion

This paper analyzed the characteristics of computing grid resources, and proposed a Memetic Alogrithm for the job scheduling of computing grid. In our MA, which using two local search algorithms, Hill-Climbing and Tabu search algorithm, and compare with Braun et al.'s GA on the experimental results. Our experimental study showed that the MA (Tabu) used in computing grid job scheduling is feasible, optional is another characters. It can provide a very short time high-quality strategy. Our future work is to extend the experimental study, consider other operators and local search algorithm in the job scheduling of computing grid.

References

1. Xhafa, F., Alba, E., Dorronsoro, B., Duran, B.: Efficient batch job scheduling in grids using cellular memetic algorithms. Journal of Mathematical Modelling and Algorithms 7(2), 217–236 (2008)
2. Braun, T.D., Siegelh, H.J., Beck, N.: A comparison of eleven static heuristics for mapping a class of independent tasks onto heterogeneous distributed computing systems. Journal of Parallel and Distributed Computing 61(6), 810–837 (2001)
3. Li, W., Yuan, C.: Research on Grid Scheduling based on Modified Genetic Algorithm. In: Pervasive Computing and Applications, ICPCA 2008 (2008)

Syndromes Classification of the Active Stage of Ankylosing Spondylitis in Traditional Chinese Medicine by Cluster Analysis of Symptoms and Signs Data

Kuiwu Yao[1], Liangdeng Zhang[1,2,*], Jie Wang[1], and Ji Zhang[2]

[1] Department of Cardiology, Guang'anmen Hospital, China Academy of Chinese Medical Sciences, 100053 Beijing, China

[2] School of Acupuncture and Moxibustion, Beijing University of Chinese Medicine, 100029 Beijing, China

yaokuiwu@yahoo.com.cn,
zldeng20102011@sohu.com

Abstract. Cluster analysis is a popular method for statistical classification for data mining. It is introduced to traditional Chinese medicine (TCM) for quantifying and normalizing the clinical practice objectively. The present study reported that TCM syndromes classification and diagnosis of 163 cases of ankylosing spondylitis (AS) active stage through cluster analysis were feasible. And 32 symptoms and signs of AS active stage were clustered and discriminated clearly. The results showed four syndromes and their corresponding therapy methods were compatible to guide TCM clinical practice, which integrated valuable experience and modern methodology preferably. Cluster analysis for AS information excavation in TCM is worthily manipulable as well as the use of which in other TCM fields.

Keywords: clustering analysis, ankylosing spondylitis active stage, syndromes classification and diagnosis, traditional Chinese medicine.

1 Introduction

Cluster analysis is an exploratory technique that can be used to reveal unknown heterogeneity, focuses on the inherent differences between cases rather than variables [1, 2]. In this method, groups of individuals are defined in terms of aggregate patterns empirically in complex data involving many variables [3]. Cluster analysis creates patterns that are mutually exclusive, as each subject in the present study can belong to only one cluster [4]. Stability and robustness are important issues in cluster analysis with its data and cluster dependent [5]. Cluster analysis has been a popular method for statistical classification, including biology, medicine, biotechnology, etc. [6]. Traditional Chinese medicine (TCM) induces cluster analysis to quantify the classification and diagnosis information objectively that contributes to uncover its black box theory.

Ankylosing spondylitis (AS) is the second most common chronic inflammatory joint disease after rheumatoid arthritis [7]. AS is often progressive, primarily injures

[*] Corresponding author.

L. Qi (Ed.): ISIA 2010, CCIS 86, pp. 657–663, 2011.

the sacroiliac joints of the pelvis and spine, affecting quality of life including role and physical functioning, psychological well-being, and social interactions [8, 9]. Patients' condition is serious in AS active stage. The use of TCM in the treatment of AS is effective substantially. However, it's difficult to assess the clinical results in TCM reported between different clinicians for lacking normalization, especially syndrome differetiation. Syndromes classification of AS active stage is an important requirement for proper management of the disease in TCM. The present study investigated the syndromes classification of AS active stage through cluster analysis based on 163 perspective patients' 32 variances of symptoms and signs.

2 Materials and Methods

2.1 General Materials

A total of 163 cases of AS active stage confirmed by activity parameters which collected from Beijing University of TCM affiliated Guoyitang Hospital, from June 2005 to December 2008. There were 117 (71.8%) males and 46 (28.2%) females, with average age 47.03±12.33 years and average course of disease 6.66±5.99 years. The positive and negative patients of HLA-B27 antigen are 120 (73.6%) cases, 43 (26.4%), respectively. And 81 (49.7%) cases had a family medical history of AS. The symptoms of 117 (71.8%) cases could be relieved quickly by non-steroidal anti-inflammatory drug. There were 136 (83.4%) and 72 (44.2%) cases had bilateral sacroiliitis equaling or exceeding III class examined by X-ray, bilateral sacroiliitis equaling or exceeding II class examined by computed tomography, respectively. The inflammation of sacroiliac articulation was confirmed in 114 (69.9%) cases by pathological examination. The activity directions of lumbar vertebrae including anterior curvature, lateral curvature and back stretch, were restricted in 76 (46.6%) cases. The average of erythrocyte sedimentation rate (ESR) and C-reactive protein (CRP) was 61.14±32.07 mm/1h, 43.28±25.25 mg/L, respectively.

2.2 Diagnostic Criteria

AS and its active stage were diagnosed in reference to the diagnostic criteria formulated by ARA, 1984 [10], National AS Seminar in Shantou and National Ministry of Health, 2002 [11].

2.3 Inclusion Criteria and Exclusion Criteria

The inclusion criteria contained patients who conformed to the above-mentioned criteria of AS active stage, agreed to be investigated with informed consent, and with age ranging from 16 to 75 years. The exclusion criteria contained patients who complicated with severe heart, lung, liver, kidney, blood and endocrine system diseases, mental disorder, and women who were in the stage of pregnancy or lactation.

2.4 Quality Control and Calculation of Sample Size

The unified diagnostic criteria, questionnaire of clinical epidemiology, and clinicians training were carried out and assessed in all the process by following to evidence-based medicine principle. And the assessment of symptoms concordance rate in consistency checking was greater than 0.75. The questionnaire of clinical epidemiology contained 32 symptoms and signs associated with TCM syndromes of AS active stage. Sample size was at least 5 times more than all the variance numbers, which calculated by common multiplicity rule. And 163 recruited patients in this research met the requirement.

2.5 Data Mining

Double people and computers were assigned to build the database with 32 TCM symptoms and signs of 163 patients in EpiData 3.0 software. Meanwhile, all the data was preprocessed. Then, the database was imported in SAS 6.12 software for statistical analysis, including frequency statistics, cluster analysis in Enterprise Miner module. The number of clustering identified as four classifications, and the clustered results of data were analyzed combined with TCM and data mining methodology.

3 Results

3.1 Frequency Statistic of the Symptoms and Signs of AS Active Stage

The most frequent symptoms and sings of AS active stage included lumbosacral area or lumbar and back pain seriously, serious pain in activity, irritabily, thick and big tongue with indentation, lassitude loin and legs, pain at night, palpitation and morbid forgetfulness, restriction of joints activity, palpable fever of joints, serious pain after exertion, yellowish urine, thirst with desire to drink, weak and thready pulse (Table 1).

Table 1. Frequency statistic of the symptoms and sings of AS active stage (%)

Variable	Symptom/sign	Frequency
X1	Lumbosacral area or lumbar and back pain seriouly	159 (96.4%)
X2	Serious pain in activity	119 (72.1%)
X5	Irritabily	112 (67.9%)
X7	Thick and big tongue with indentation	106 (65.0%)
X8	Lassitude loin and legs	104 (63.0%)
X10	Pain at night	103 (62.4%)
X14	Palpitation and morbid forgetfulness	101 (61.2%)
X15	Restriction of joints activity	97 (59.5%)
X16	Palpable fever of joints	93 (56.4%)
X20	Serious pain after exertion	91 (55.8%)
X23	Yellowish urine	89 (53.9%)
X28	Thirst with desire to drink	86 (52.1%)
X31	Weak and thready pulse	83 (50.3)

3.2 Cluster Result of Symptoms and Signs of AS Active Stage

The discrimination principle of Syndromes of AS active stage in TCM involved symptoms and signs, identify variables integrated with position and qualitation diagnosis in TCM. Then, the cluster method within 2 to 4 types of categories in SAS 6.12 software was used to analyze the 32 variances in the light with the principle. The results showed that the dispersibility of four categories clustered among the three classifications was best to TCM information of four diagnostic methods for clinical practice. Therefore, the name and classification of syndromes of AS active stage were defined, including syndrome of dampness and heat blocking collaterals, syndrome of stagnation of pathogen converted into heat, syndrome of deficiency of both the liver and kidney, and syndrome of stagnation of turbid phlegm and blood stasis (table 2). The weights of primary symptoms and signs of each syndrome contributed to make the classification and diagnosis clearly (table 3).

Table 2. Abstract of four categories of the syndromes diagnosis for AS active stage in TCM

Cluster	Name of syndrome	Cluster variables	Proportion explained	Second eigenvalue
1	Dampness and heat blocking collaterals	12	0.3269	1.5749
2	Stagnation of pathogen converted into heat	14	0.2574	1.5799
3	Deficiency of both the liver and kidney	12	0.2803	1.4453
4	Stagnation of turbid phlegm and blood stasis	10	0.2904	1.4186

3.3 Analysis on the Primary Points of Diagnosis of AS Active Stage

The primary points of clustered results to syndromes diagnosis with corresponding homologous weights of symptoms and signs conformed to TCM clinical practice quantitatively (Table 3). On the basis of that, classification and diagnosis of syndromes were discriminated easily and clearly. The syndrome of dampness and heat blocking collaterals is primarily consisted of lumbosacral area or lumbar and back pain, palpable fever of joints, heavy and fatigue, greasy or yellowish fur, swell and red of joints of lower extremity, constipation, red tongue, slippery and frequent pulse. The syndrome of stagnation of pathogen converted into heat is primarily consisted of lumbosacral area or lumbar and back pain seriously, pain aggravated in cloudy and rainy day, irritability, red tongue, depression and insomnia, yellowish urine, yellowish fur, and stringy pulse. The syndrome of deficiency of both the liver and kidney is primarily consisted of lumbosacral area pain seriously, lassitude loin and legs, few fur, restriction of forward bending and backward extending, heel pain or other tendoperiostosis pain, dizziness and tinnitus, erythroic tongue, weak and thready pulse. The syndrome of stagnation of turbid phlegm and blood stasis is primarily consisted of heavy and prinking pain of lumbar and back, dark or petechia tongue, excessive and yellowish phlegm, serious pain at night, fullness in stomach, thick fur, and unsmooth or moisten pulse.

Table 3. The primary points of syndromes diagnosis of AS active stage based on cluster analysis

Variable	Syndrome diagnosis	Weight
	Syndrome of dampness and heat blocking collaterals	
X16	Palpable fever of joints	0.4018
X3	Red tongue	0.1574
X4	Greasy fur	0.3187
X6	Swell and red of joints of lower extremity	0.4741
X10	Pain at night	0.3724
X2	Serious pain in activity	0.3405
X28	Thirst with desire to drink	0.2694
X9	Constipation	0.3110
X11	Slippery and frequent pulse	0.2027
	Syndrome of stagnation of pathogen converted into heat	
X12	Pain aggravated in cloudy and rainy day	0.1485
X13	Aversion to wind and cold	0.3459
X23	Yellowish urine	0.4650
X17	Back stiffness	0.6742
X18	Yellowish fur	0.5472
X19	Stringy pulse	0.6557
X21	Depression and insomnia	0.5109
X1	Lumbosacral area or lumbar and back pain seriouly	0.1135
X5	Irritability	0.1068
	Syndrome of deficiency of both the liver and kidney	
X31	Weak and thready pulse	0.4631
X22	Heel pain or other tendoperiostosis pain	0.0091
X15	Restriction of joints activity	0.4257
X8	Lassitude loin and legs	0.5143
X14	Palpitation and morbid forgetfulness	0.4287
X24	Erythroic tongue	0.3986
X20	Serious pain after exertion	0.1605
X25	Short breath and fatigue	0.4218
	Syndrome of stagnation of turbid phlegm and blood stasis	
X26	Dark or petechia tongue	0.4135
X27	Discomfortableness relief after activity	0.4682
X7	Thick and big tongue with indentation	0.2344
X29	Moisten and slow pulse	0.0833
X30	Excessive and yellowish phlegm	0.1071
X32	Heavy and prinking pain of lumbar and back	0.4172

3 Conclusion

Cluster analysis is one of the most typical statistical methods in data mining that is used to classify objects $X_i = (x_{i1}, x_{i2}, \cdots, x_{ip}) \in A(i = 1,2,\cdots,n)(x_{ij}(j = 1,2,\cdots,p)$ presents various characters of X_i) from the data space A into relative groups, objects to discover a system of n organizing observations from A into several groups, where members of the groups share properties in common and groups have no intersection [12]. TCM is fuzzy in syndromes differentiation and classification. Cluster analysis is feasible to quantize TCM information including syndromes, diagnosis, therapies, and other qualitative theory and clinical practice, such as valuable experience of famous veteran teran doctors of TCM. Western conventional therapeutic regimes have traditionally been insufficient to control symptoms and signs of AS, and have failed to halt disease progression [13]. TCM plays an important role for AS treatment in China, especially improving quality of life. Meanwhile, modern methodologies are induced to confirm the effectivity and safety treatment in TCM that normalization and internationalization of TCM will come true in future.

The present study through cluster analysis indicated that the syndromes diagnosis of AS active stage could be classify as dampness and heat blocking collaterals, stagnation of pathogen converted into heat, deficiency of both the liver and kidney, and stagnation of turbid phlegm and blood stasis. Of that, the syndromes of dampness and heat blocking collaterals, and stagnation of pathogen converted into heat, were most common in AS active stage as compared with clinical practice. Therefore, therapeutic methods of eliminating dampness and cooling to stop pain, cooling and expelling stagnation to eliminate pathogen, were so important. Besides, replenishing liver and kidney, strengthening healthy qi to eliminate pathogen, expelling phlegm and dispersing blood stasis, were also common treatment to realize personalized medicine. In a word, the clustering categorized results were congruent to guide TCM clinical practice, which combined valuable experience and modern methodology and promoted the progress of integrative medicine preferably.

AS active stage treated in TCM resulted with significant effect. And the induced methodology of cluster analysis upgraded to the therapeutic effect assessment in TCM, especially clustering for syndromes diagnosis objectively and normatively. Cluster analysis could be used widely for TCM information data mining, and further study is necessary to elevate the evidence-based class, including multicentre and large sample of randomized control trial on TCM and integrative medicine.

Acknowledgments. This material is the result of work supported with resources and the use of facilities at Beijing University of Chinese Medicine, Beijing, China. The study is supported by grant from "Research Fund of Capital Medical Development" of China (#SF-2007-I-04) to Pro. Jie Wang. And the authors are grateful to Dr. Qingyong He for his suggestion on data analysis.

References

1. Luke, D.A.: Getting the Big Picture in Community Science: Methods That Capture Context. Am. J. Community Psychol. 35, 185–200 (2005)
2. Shaw, S.Y., Shah, L., Jolly, A.M., Wylie, J.L.: Identifying Heterogeneity Among Injection Drug Users: a Cluster Analysis Approach. Am. J. Public Health 98, 1430–1437 (2008)
3. Wirfalt, A.K.E., Jeffery, R.W.: Using Cluster Analysis to Examine Dietary Patterns: Nutrient Intakes, Gender, and Weight Status Differ Across Food Pattern Clusters. J. Am. Diet. Assoc. 97, 272–279 (1997)
4. Hearty, A.P., Gibney, M.J.: Comparison of Cluster and Principal Component Analysis Techniques to Derive Dietary Patterns in Irish Adults. Br. J. Nutr. 101, 598–608 (2009)
5. Hennig, C.: Dissolution Point and Isolation Robustness: Robustness Criteria for General Cluster Analysis Methods. J. Multivar. Anal. 99, 1154–1176 (2008)

6. Cheong, M.Y., Lee, H.: Determining the Number of Clusters in Cluster Aanalysis. J. Korean Stat. Soc. 37, 135–143 (2008)
7. Younes, M., Jalled, A., Aydi, Z., Zrour, S., Korbaa, W., Salah, Z.B., Letaief, M., Bejia, I., Touzi, M., Bergaoui, N.: Socioeconomic Impact of Ankylosing Spondylitis in Tunisia. Joint Bone Spine 77, 41–46 (2010)
8. Haywood, K.L., Garratt, A.M., Jordan, K., Dziedzic, K., Dawes, P.T.: Disease-specific, Patient-assessed Measures of Health Outcome in Ankylosing Spondylitis: Reliability, Validity and Responsiveness. Rheumatology (Oxford) 41, 1295–1302 (2002)
9. Ward, M.M.: Quality of Life in Patients with Ankylosing Spondylitis. Rheum. Dis. Clin. North Am. 24, 815–827 (1998)
10. Ye, R.G., Lu, Z.Y.: Internal Medicine (in Chinese). People's Health Public House, Beijing (2004)
11. Zheng, Y.Y.: Clinical Research Guidelines for New Drugs of Traditional Chinese Medicine (in Chinese). China Medical Science and Technology Press, Beijing (2002)
12. Chen, J., Li, Y., Li, G., Li, Y.: Period Selection of Traffic Impact Analysis Based on Cluster Analysis. J. Transpn. Sys. Eng. & IT 9, 63–67 (2009)
13. Brandt, J., Marzo-Ortega, H., Emery, P.: Ankylosing Spondylitis: New Treatment Modalities. Best Pract. Res. Clin. Rheumatol. 20, 559–570 (2006)

Research on Detection of Instant Messaging Software

Hao Zhang[1], Guangli Xu[1], Jianmin Li[2], and Lili Wang[1]

[1] College of Light Industry, Hebei Polytechnic University, Tangshan, Hebei, China
mox012@163.com
[2] Institute of Electronics, Chinese Academy of Sciences, Beijing, China
282540005@qq.com

Abstract. Instant messaging software (IMS) provides a platform for the communication of information. It convenient for people to communicate, at the meantime enterprises and institutions, companies, families, local area network has security implications. Such as access to user's personal information and the company's trade secrets; dissemination of pornography, reactionary remarks; provide attack corridors for Trojans and worms. Therefore, study the detection and blocking techniques of instant messaging software to protect the security of local area network has great application value.

Through analyzing and summing up the network architecture and the communication protocol of IMS, a universal detection technology of IMS has been presented, which is detect communication protocols of IMS. When detecting the instant messaging software, dynamic link library of WinPcap systems in Windows systems should be used.

Keywords: instant messaging software; network security; detection; WinPcap.

1 Introduction

With the rapid development of Internet, instant messaging software such as QQ, MSN, Fetion and others have become more and more used by net users, it is not only as a chat tool, but also to be a device to provide voice, video and data transmission services. However, the harm caused by instant messaging software can not be ignored: firstly, it can through the firewall, which will cause the firewall lost its protective effect; secondly, LAN users to use data of this kind of software, video chat and other services, which take up a lot of bandwidth and affect the normal speed of the network; last, but not least, it provided a platform for obscene, reactionary remarks. It is very important to research on detection of instant messaging software according to network infrastructure and communication features of instant messaging software, which will be a meaningful thing to local area network security and information security.

2 The Instant Messaging Environment

TCP/IP is standard protocol of the Internet. With the rapid development and popularization of Internet, TCP/IP has become the most widely used protocol all over

L. Qi (Ed.): ISIA 2010, CCIS 86, pp. 664–669, 2011.

the world in the Internet. TCP/IP protocol is a heterogeneous networked communication protocol, it also applies to a local area network to achieve different kinds of Internet communication between computers. Architecture of TCP/IP is shown as figure 1. The layer 1 to 4 are application layer, transport layer, network layer, physical and data link layer.

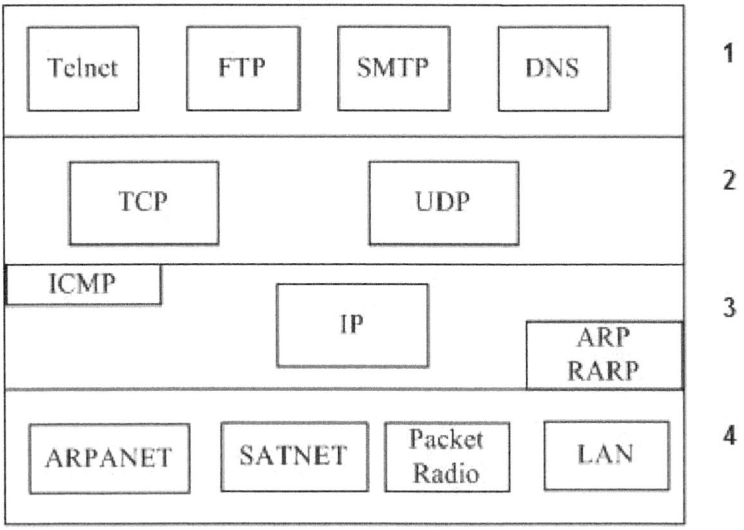

Fig. 1. Architecture of TCP/IP

This paper's Hardware environment is shown as Figure 2: 3 computers, an Ethernet hub station, a gateway. the LAN consist of PC A and B and a hub, computer C is the external network PC.

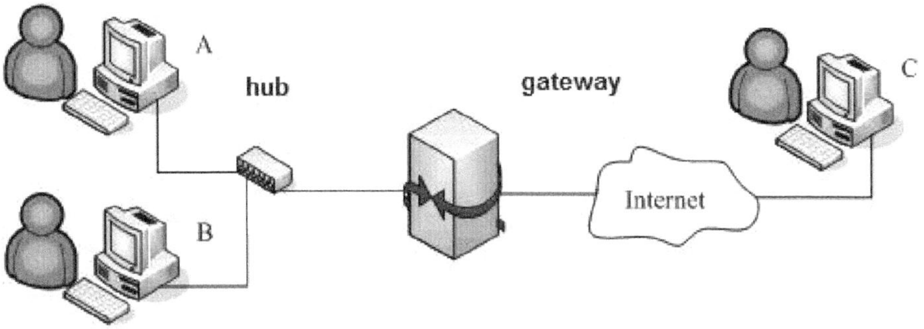

Fig. 2. The protocol analysis hardware environment

Software environment, including instant messaging software QQ2008, MSN, Kaspersky Internet Security 7.0 and WinPcap 4.0.1, and network protocol analysis tool Ethereal (Ethereal 0.99.0). All clients are using Windows XP SP3 operating system. The client A uses Ethereal, the adapter of capture packets in Ethereal is set to promiscuous mode, and three clients, close other applications in the network. A can capture and record all transfers within the LAN packets. To rule out the other unrelated packets, only concerning to capture and record instant messaging software packets by setting Ethereal filter the rules. QQ, MSN and Kaspersky Internet Security 7.0 beta are installed to the client B and C.

3 Windows Platform Packet Capture Analysis

WinPcap system is the better data packet capture and filtering procedures under the Windows platform, WinPcap network system will not affect the normal speed of the Internet, detection function of the system is based on this technology.

WinPcap (windows packet capture) is a free public network access system under Windows platform. It is for the Win32 platform network packet capture and analysis of open source libraries. The package includes a kernel mode filter for network packet capture and filtering functions, also known as NPF (Netgroup Packet Filter) packet driver; a low-level dynamic link library (packet.dll), which provides developers a low-level programming interface; a high level does not depend on the library (wpcap.dll), which provides developers to develop a higher level interface, the dynamic link library is independent of the system. Figure 3 is WinPcap's structure and location of the operating system.

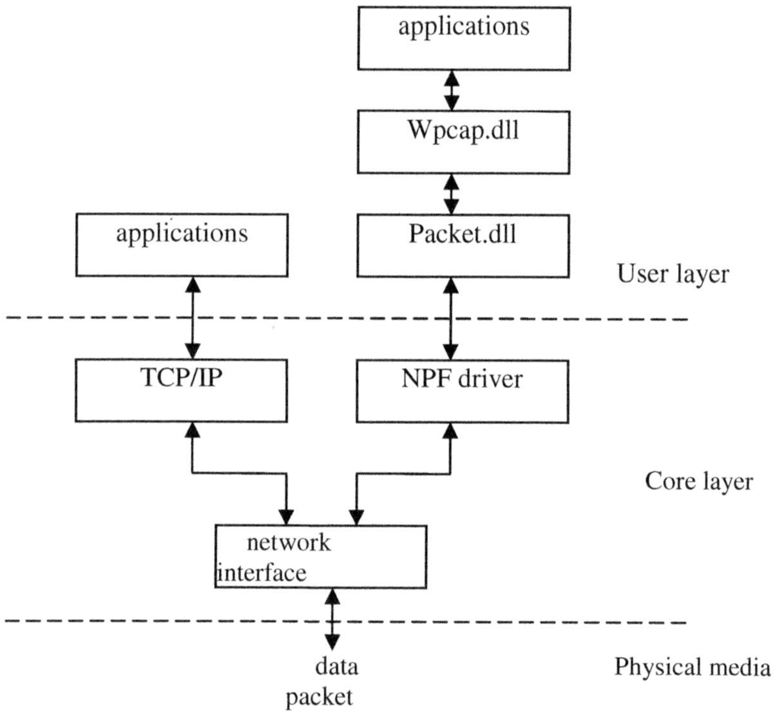

Fig. 3. Architecture of WinPcap

WinPcap provides the following functions: to capture the raw data packet, whether it sends or receives data packets or change data packets between the host in the net; before the packet is sent to the application, it should be filtered specified packet in accordance with the users' some special rules; the original packet is sent to the network; to collect and process statistical network traffic flow information. WinPcap is only in the addition of a bypass data link layer processing on the sending and receiving data and then filtering, buffering packets and other related processing. It did not affect network performance. WinPcap packet provides provide a platform to capture data packets, the interface of WinPcap should be used, it can capture and filter out wanted packets. Some related functions of WinPcap packet capture are as follows:

Get a list of network adapters connected to the function: pcap__findalldevs ();
Open the selected network adapter interface function: pcap__open__live ();
Compile BPF filter rules function: pcap__compile ();
set filter rules function: pcap__setfilter ();
LAN type of search functions: pcap__datalink ();
Network packet callback function: pcap__loop ().

4 Detection Module Design

The main task of detection module is to test instant messaging software, combining method of the intrusion detection technology and testing instant messaging software communication protocol, which can meet the functional requirements of detection module. The test procedures on detection module, IP packet unpack and universal detection technology in the detection module of the application are introduced as follows:

1) Instant messaging software testing process.

A network capture program packet, including the user and core part. Kernel is responsible for the capture data from the network, can also filter data; the task of user part is to performs packet formatting, protocol analysis, data conversion and processing and so on. Users part also can filter some packets. Network packet capture program programming processes is shown as Figure 4:

(1) select the network adapter.
Find all connected network adapter, and select the network interface to capture data. IP addresses can be used to represent the equipment, or use string to represent this device.
(2) capture program initialization.
Mainly for setting the selected network interface, such as the length of the packet capture is set to 65536, time out is 1 second and so on. To open the network adapter in promiscuous mode, capture all packets of other hosts. Because of the need analysis of all data packets, so you can not set the kernel filtering rules.
(3) capture the data.
Through callback function or directly through the packet capture function to get the captured packets.

(4) analysis of the data packet.
Using protocol analysis techniques to capture data packets according to the head of the packet layer and to find application layer data, using misuse detection to detect application layer data packet.
(5) test the rule base line data packet and send to other processing modules.
The data storage module stored rule base line test data, statistical information module on the packet IP address and port number and other information for statistical analysis, some of the information displayed to the detection program interface.

Where (4) and (5) can be achieved by the multi-threading, it can increase the efficiency of detection and analysis system, and reduce the probability of packet loss at the same time.

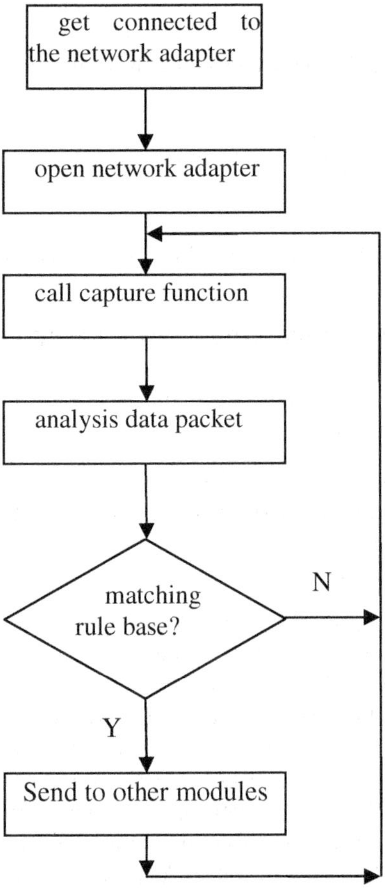

Fig. 4. The flow chart about detection

2) In order to detect more kinds of instant messaging software, the system stored the rule base in the configuration file. This instant messaging software testing method that is updating configuration file can increase scalability and availability in the system.

Adding a few simple rules in the configuration file can achieve WebQQ detection. Figure 5 is a detection system on the LAN instant messaging software testing process part of the results.

5 Conclusion

According to the detection method of IMS, a detection architecture of IMS is presented and developed at the Windows platform. The rule base store in the configuration file, users can add their own IMS, detection the rules, users can also add other applications, features, so that the system has better scalability and practicality.

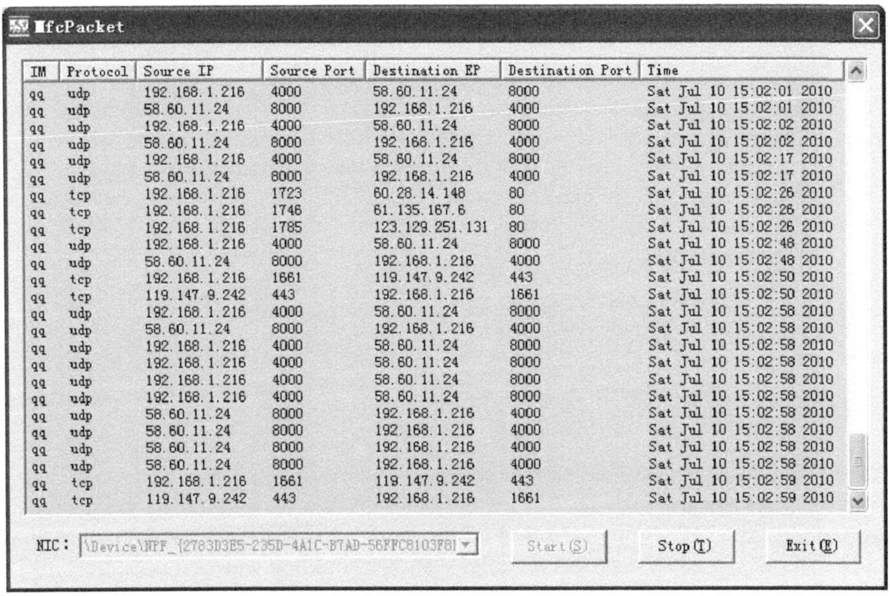

IM	Protocol	Source IP	Source Port	Destination EP	Destination Port	Time
qq	udp	192.168.1.216	4000	58.60.11.24	8000	Sat Jul 10 15:02:01 2010
qq	udp	58.60.11.24	8000	192.168.1.216	4000	Sat Jul 10 15:02:01 2010
qq	udp	192.168.1.216	4000	58.60.11.24	8000	Sat Jul 10 15:02:02 2010
qq	udp	58.60.11.24	8000	192.168.1.216	4000	Sat Jul 10 15:02:02 2010
qq	udp	192.168.1.216	4000	58.60.11.24	8000	Sat Jul 10 15:02:17 2010
qq	udp	58.60.11.24	8000	192.168.1.216	4000	Sat Jul 10 15:02:17 2010
qq	tcp	192.168.1.216	1723	60.28.14.148	80	Sat Jul 10 15:02:26 2010
qq	tcp	192.168.1.216	1746	61.135.167.6	80	Sat Jul 10 15:02:26 2010
qq	tcp	192.168.1.216	1785	123.129.251.131	80	Sat Jul 10 15:02:26 2010
qq	udp	192.168.1.216	4000	58.60.11.24	8000	Sat Jul 10 15:02:48 2010
qq	udp	58.60.11.24	8000	192.168.1.216	4000	Sat Jul 10 15:02:48 2010
qq	tcp	192.168.1.216	1661	119.147.9.242	443	Sat Jul 10 15:02:50 2010
qq	tcp	119.147.9.242	443	192.168.1.216	1661	Sat Jul 10 15:02:50 2010
qq	udp	192.168.1.216	4000	58.60.11.24	8000	Sat Jul 10 15:02:58 2010
qq	udp	58.60.11.24	8000	192.168.1.216	4000	Sat Jul 10 15:02:58 2010
qq	udp	192.168.1.216	4000	58.60.11.24	8000	Sat Jul 10 15:02:58 2010
qq	udp	192.168.1.216	4000	58.60.11.24	8000	Sat Jul 10 15:02:58 2010
qq	udp	192.168.1.216	4000	58.60.11.24	8000	Sat Jul 10 15:02:58 2010
qq	udp	192.168.1.216	4000	58.60.11.24	8000	Sat Jul 10 15:02:58 2010
qq	udp	58.60.11.24	8000	192.168.1.216	4000	Sat Jul 10 15:02:58 2010
qq	udp	58.60.11.24	8000	192.168.1.216	4000	Sat Jul 10 15:02:58 2010
qq	udp	58.60.11.24	8000	192.168.1.216	4000	Sat Jul 10 15:02:58 2010
qq	udp	58.60.11.24	8000	192.168.1.216	4000	Sat Jul 10 15:02:58 2010
qq	tcp	192.168.1.216	1661	119.147.9.242	443	Sat Jul 10 15:02:59 2010
qq	tcp	119.147.9.242	443	192.168.1.216	1661	Sat Jul 10 15:02:59 2010

Fig. 5. Running interface of detecting system

References

1. Joe, L.: Best practices for instant messaging in business. Network Security (2005)
2. Harlan, C.: Instant messaging investigations on a live Windows XP system. Digital Investigation (2004)
3. Stevens, R.W.: Special edition using TCP/IP, 2nd edn. Que Publishing, American (2003)
4. Yu, W., Chellappan, S., Wang, X., et al.: Peer-to-Peer system-based active worm attacks: modeling, analysis and defense. Computer Communications (2008)

A Research of P2P IPTV Node Measurement Based on Packets Analysis

Jian-biao Zhang, Qi Zhang, Han Zhang, and Li Lin

College of Computer Science and Technology,
Beijing University of Technology, 100124 Beijing, China

Abstract. The P2P IPTV has developed extremely fast, but it also has brought inevitable problems: underpass for illegal video spread and etc. Supervision and management of P2P IPTV is called by the governments. However most of P2P IPTV applications are closed source and the p2p network has the features of decentralization and anonymity, making analysis and measurement inconceivably difficult. This paper declared an approach of how to identified packets which contains node list though the protocol is closed source by using packet signature analysis. The results of measuring some typical P2P IPTV applications approved feasibility of the approach. It could be a reference for any research of P2P IPTV node measurement.

Keywords: P2P IPTV; Node Measurement; Packet Signature; Matching Ratio.

1 Introduction

P2P IPTV had its soaring development. Recent years, several of popular applications gave its advent to us: PPLive, PPStream, and QQLive. As a result of the features of decentralization and anonymity of P2P network [1, 8], illegal videos could make undiscovered spread. Moreover, lots of TV shows could be watched unauthorized [2]. This would compromise the benefits of valid companies. A solution of P2P IPTV supervision and management was called by governments. Different from C/S model, however, nodes, or called peers, provide content fragments instead of server. Consequently, no single node takes the responsibility of illegal video spread, increasing difficulty of supervision [3]. Therefore, obtaining the sockets, combined with IP address and port, is the first step of P2P IPTV management: to know whom was spreading.

2 Introduction of Correlation Techniques

Ordinarily, there are two of main aspects of P2P measurement research: open source and the opposite, closed source, which P2P IPTV just stands on this side. The measurement of open source applications, like BitTorrent and eMule, is primarily building up a client that obeys the principle and communication sequence of the protocol and then either running it in real environment or the simulated [5]. Measurement of closed source goes differently. A mainstream approach is same as

L. Qi (Ed.): ISIA 2010, CCIS 86, pp. 670–677, 2011.

network measurement, which is firstly summarizing the regularity of its communication, then recognizing the signature of its packet, finally finding the packets that containing nodes [6].

Ref. [3] and Ref. [10] explained the flow signature. Ref. [4] showed an approach of recognizing and controlling P2P steam by three of ways, which are TCP/UDP checking, (IP Port) checking, and signature checking. These papers contributed a basic inspiration to the measurement research which was elaborated in this article.

This paper analyzed the topology and theory of P2P IPTV and declared an approach of measurement based on packet collection and analysis. It mainly discussed two of signatures of packets: video stream and nodes container. The former signature was concludes by experiment statistic; the latter was come out by a concept of matching ratio calculation. After logical analysis, practical node measurement would be demonstrated by running a P2P IPTV application.

The organization of this paper is as follow, part 3 gave the topology and theory of P2P IPTV, part 4 elaborated the approach, part 5 was experiment by using 'PPStream', the last part was conclusion.

3 Topology and Theory of P2P IPTV

The topology of P2P IPTV network was illustrated in Fig. 1. The backbone networks were divided into several sections according to their geography location. Each section was made up by districts. For each section, as well as districts, there are some servers. Severs of section was called Region server, abbreviated as RS. Servers of district was called District server, marked as DS.

The primary functionality of RS was payloads, which was of DS, supervision and averaging. Query was supported. RS handled the cross-section-query of DS. Job of DS was as an agency of its local nodes, to handle the login and logout messages of user's client, or to store and update the index values of hot videos [13]. Furthermore, DS organized the list of local nodes and made the real-time update of it.

All nodes which were located in a same district sent resource and nodes update request to their DS. Video fragments were shared between nodes.

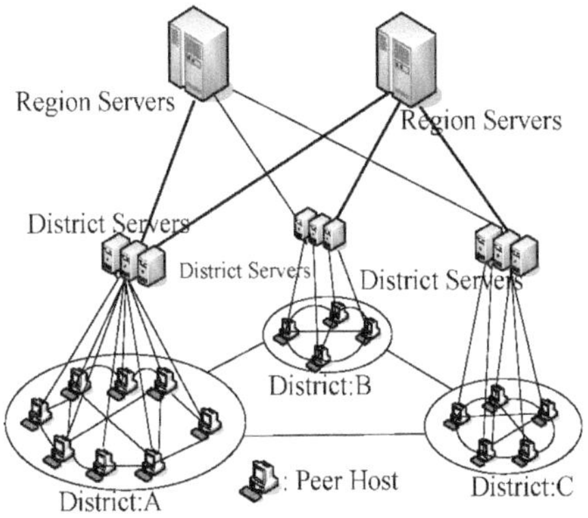

Fig. 1. Topology of P2P IPTV Network

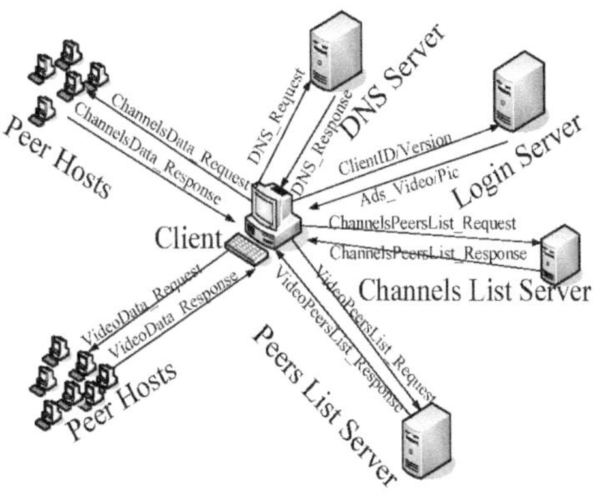

Fig. 2. Theory of P2P IPTV

According to Fig. 2, the theory of a P2P IPTV could be explained as following steps.

1) The Client sent DNS request for resolving the server's name.

2) Client sent the login message to server. Server inserted socket of the client into its node list and marked the client as active. After this procedure, snapshots and lists of hot videos were replied. TCP was used in this step.

3) Client required the latest programs. Server replied in two ways. One is to send the latest program list directly. Another is to give an active node list, the 'ChannelsPeersList_Response' message, back to the client. Client parsed the list, and then pushing further requests to nodes which were listed. Comparing with step 2, UDP was used. Fig. 2 showed the latter manner of reply.

4) Client sent an index of video file to the server. Server check the validity of index, then replying with active nodes which contained the fragments of requested video. Replying procedure was called 'VideoPeersList_Response'. UDP was used in this step.

5) Client sent video fragments download request to known nodes. Nodes replied.

6) Client arranged fragments which were sent from remote nodes, composing them into video stream, buffering, and then playing it.

7) Client replied other nodes' fragments download request, sending its owned fragments to them.

4 Node Measurement

4.1 Node Measurement Tactic

According to the analysis in part 3, the method of node measurement by utilizing packet analysis was shown in Fig. 3.

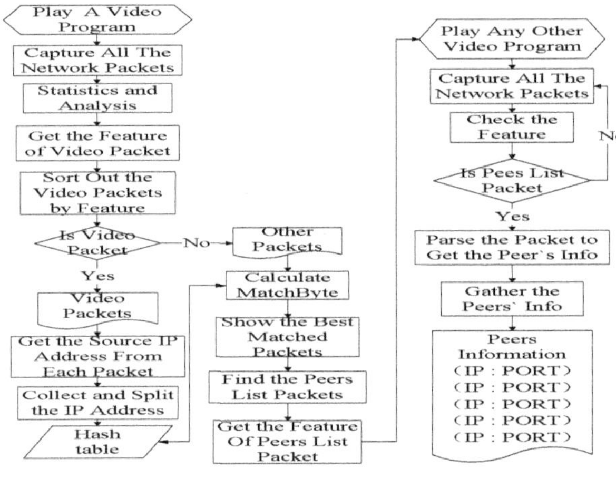

Fig. 3. Procedure of Node Measurement

First, open a P2P IPTV application and play a video. Start a packet sniffer to capture all packets which the P2P IPTV communicated. Concluded from experiment and analysis, packets of video fragments were of major proportion. How to recognize the signature of video packets would be found at section 4.2.

Applying the video packet signature, two types of packets could be classified: the video packets, and the unknown. Nodes list were concealed themselves in the latter. From theory of P2P IPTV, in part 3, worked out a fact that the node, which was sending video fragments, must be listed in the 'VideoPeersList_Response', so, by using matching ratio calculation, packets of nodes list would be uncovered. Matching ratio calculation was introduced in section 4.3.

Once packets of node list being identified, obtaining the signature of these packets was a piece of cake. Having the signature of node lists, the node lists of any video program would be easily discovered. The remaining work is just to get the information of nodes by parsing the packets of node list.

4.2 Approach of Video Packet Signature Analysis

According to the theory of P2P IPTV, video file was divided into several fragments [11]. Before playing, client had to store some fragments to fulfill the minimum requirement of playing, simultaneously uploading its owned fragments to other requesting client [7, 12, 15]. Throughout the entire procedure of playing, transfer of packets of video fragments wouldn't stop. This was result in that video packets were of major proportion of all transfered data. Looking deep into the statistic of transferred data, signature of video packets was to be uncovered. Take 'PPStream' as an example.

Most of P2P IPTV application had its signature [3, 5]. The signature of PPStream is UDP [11]=0x43. Formula P[i] =V means the value of the byte of i+1 of protocol P is V. The 14th byte of PPStream UDP packets meant the functionality of the packet, like control or video transfer. The statistic of the UDP [13] of PPStream is illustrated in Fig. 4.

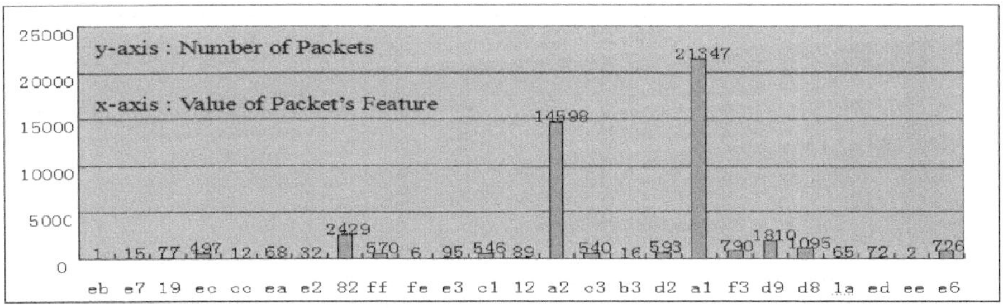

Fig. 4. Classification Charts of UDP

The x-axis meant values of UDP[13], where y meant the counts of packets. From this chart, obviously, UDP[13]=0xa1 and UDP[14]=0xa2 together shared more than half of the roundtable. These packets were video fragments. One of these was request, mostly a short packet. The other was the response, which seemed much bigger.

Applied the approach mentioned above to some popular protocols, a result list were to show in Table 1. In Table 1, '0x**' meant the value of UDP[x] was variant.

Table 1. P2P IPTV Video Packet's Feature Value Statistics

Name	Transfer Protocol	Feature Values of the Video Packet ([x] means UDP [x])				Function
PPStream	UDP	[10]:0x43	[11]:0x00	[12]:0x00	[13]:0xa1	request
PPStream	UDP	[10]:0x43	[11]:0x00	[12]:0x00	[13]:0xa2	response
UUSee	UDP	[8]:0x14	[9]:0x07	[10]:0x0b	[11]:0x04	response
PPLive	UDP	[9]:0x**	[10]:0x**	[11]:0x**	[12]:0x56	response
QQlive	UDP	[8]:0xfe	[9]:0x19	[10]:0x04	[11]:0x00	response
PPfilm	UDP	[9]:0x**	[10]:0xcc	[11]:0x**	[12]:0x52	response

4.3 Matching Ratio Calculation

The paper had discussed that by knowing the node which transfer video data to the client, it could calculate the matching ratio of the 'unknown' one of the two types of packets. Goal of matching ratio calculation was to discover the packet that contained node list.

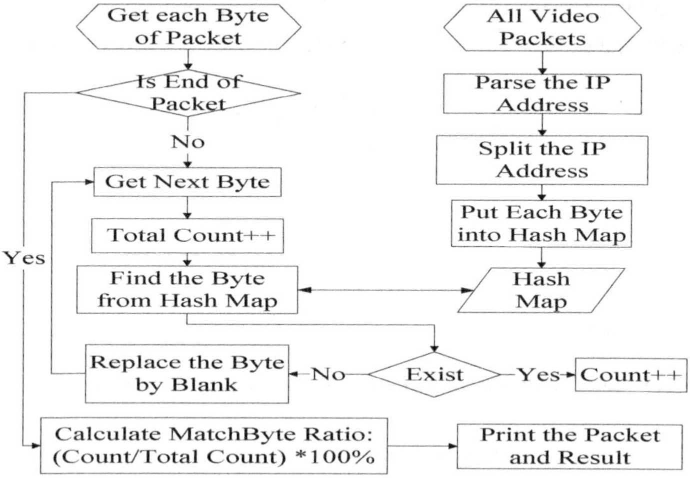

Fig. 5. Procedure of Calculate Matching Ratio

There is a fact that, nodes list always has nodes which have already offline, since user's computer, or client, could be powered off at any time. The nodes list update frequency of DS has latency. From this point of view, if used a four-byte-set of IP address for matching ratio calculation, it could lose remarkable amounts of nodes list packets which contains a considerable number of offline nodes. However, the purpose of this article was to introduce an approach of finding node lists, no matter active or offline. Therefore, IP address were byte separated. For example, '123.115.109.236', when calculating the matching ration, was divided as 123, 115, 109 and 236. All IP was byte separated and stored into a map. The procedure of matching ratio calculation was illustrated in Fig. 5.

5 Experiment

To guarantee the experiment gains an exact result, a popular P2P IPTV, the 'PPStream' or 'PPS' for short, was chosen. Applying the approach and tactic in part 4, a node list packet were found and shown at Fig. 6.

As shown in Fig. 6, leading 34 bytes were respective header of Ethernet packet. UDP[10]-UDP[13] were PPStream protocol header. From UDP[15] lasted 20 bytes were GUID of the requesting video. UDP[40]=0x26 meant there were 38 nodes in the lists. Each node was made up with a set of a four-byte IP address, two-byte port, and a four-byte nonsense data (rectangle=IP, underline = nonsense data). When playing program, client approximately received 3 to 8 packets, containing 200 to 300 nodes.

Measurement experiments were taken under the circumstance of more than playing 100 different videos by PPStream. See Fig. 7. X-axis meant the time taken of measurement in seconds. Y meant nodes counts. Averagely, 130 nodes were found in 0.3s.

```
0000  00 1a 6b 4c 0e be 08 10  17 30 9c d1 08 00 45 00
0010  02 13 39 58 00 00 73 11  38 e6 3c 1c d5 b7 c0 a8
0020  01 20 45 7c 0a 9c 01 ff  8b 66 f7 01 43 42 17 ff
0030  00 14 6c b5 47 05 3d d5  0a 28 71 6c ea b7 a6 43
0040  28 80 0f 09 72 52 01 84  01 00 00 00 00 28 26 0a
0050  7b 71 c3 e1 7c 06 cd 0a  0b 0a 3d 95 56 99 87 0f
0060  bc 0a 0b 0a 3a f5 61 91  e9 28 c3 0a 0b 0a dd db
0070  cb bf 14 0f b4 0a 0b 0a  7d 22 27 d0 0d 29 b7 07
0080  0b 0a 7b 73 cd 5e fe 0f  b2 0a 0b 0a 3d 30 04 4a
0090  9b 30 b1 0a 0b 0a dd cd  b1 5e 7d 08 9e 0a 0b 0a
00a0  76 58 c4 5b e4 26 af 0a  0b 0a 77 31 6a b0 47 31
00b0  a2 0a 0b 0a 7a 8c 01 cf  7e 04 a1 0a 0b 0a 7a 8c
00c0  04 87 7b 21 ac 09 0b 0a  da 1b a5 cc a8 25 99 0a
00d0  0b 0a dd 08 91 1e bb 2d  a4 0a 0b 0a 7a 9f f6 8a
00e0  aa 0a 91 0a 0b 0a dd d1  c0 56 3b 2f 96 0a 0b 0a
00f0  3c db 92 f1 99 2a b3 0a  0b 0a 7d d3 5d a2 07 09
0100  b2 0a 0b 0a da 08 f9 ef  fc 21 93 0a 0b 0a dd d0
0110  84 50 40 12 ac 09 0b 0a  7a 9d 26 03 06 2e 97 0a
0120  0b 0a 7a 9d 89 73 96 46  90 0a 0b 0a 3c 0e a1 31
0130  5e 2b 8f 08 0b 0a dd ce  a0 d5 fe ec 88 0a 0b 0a
0140  dd ce a5 40 34 06 8f 0a  0b 0a 7a 9c 10 0e 71 3c
0150  a4 0a 0b 0a 7a 9c 12 7f  15 08 97 0a 0b 0a 3d 89
0160  f2 c0 a3 05 8a 0a 0b 0a  71 e5 9a 07 1f 0e 95 0a
0170  0b 0a 77 73 2d d3 2c 20  88 0a 0b 0a 74 03 22 2b
0180  8a 0e 87 0a 0b 0a 77 76  b6 69 03 0a 86 09 0b 0a
0190  77 73 ca 20 1a 31 91 0a  0b 0a 7c 5e 20 b1 3b 07
01a0  8a 0a 0b 0a 77 74 b4 3a  f5 2c 8f 0a 0b 0a 3c 12
01b0  56 9e 3f 08 94 0a 0b 0a  71 e4 f8 f4 d6 1f 81 0a
01c0  0b 0a 7b bc 49 de c8 47  9a 0a 0b 00 00 03 00 00
01d0  32 1c 1c 00 00 24 00 00  00 00 46 14 00 00 00 00
01e0  00 00 00 00 00 00 00 00  00 00 00 00 00 00 00 00
01f0  ae 47 2b 42 0e 6a db 40  63 26 bc 3f 93 18 e4 3f
0200  00 00 00 00 00 00 00 00  18 00 00 00 5d 08 00 00
0210  53 03 00 00 b4 03 00 00  56 01 00 00 00 00 00 00
0220  00
```

Fig. 6. Peers List Packet of PPStream

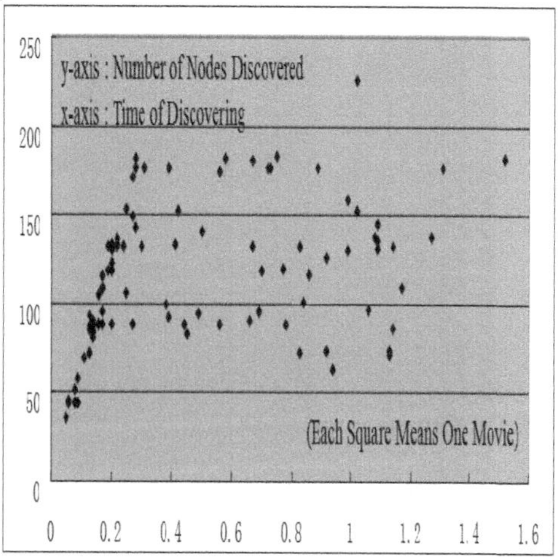

Fig. 7. The Result of Node Measurement on One Host

For advance experiment, multiple hosts were established in a LAN. It considered three of variant, the host IP address, the video file, and the measurement time. H stood for host; V stood for video, T stood for the measurement time. Symbol 'IX' meant between two experiments, the variants 'X' were identical, where 'DX' meant different. Like DH meant different hosts; IV meant the same video.

Fig. 8 to Fig. 10 respectively showed the overlap ratio of node measurement result of IHIVDT, DHIVDT and different WAN IP hosts with the same video.

Using single host to measure, an average of 130 nodes were found in 0.3s. When measuring an identical video in different time, 25% of nodes were overlapped. Four video were taken 20 times of experiment, finding averagely 1100 nodes which overlapped ones were excluded, per host a day.

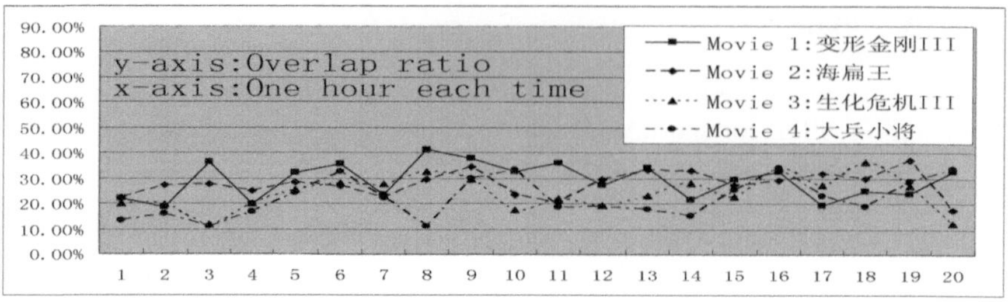

Fig. 8. Overlap Ratio of Node Measurement on One Host in Different Time

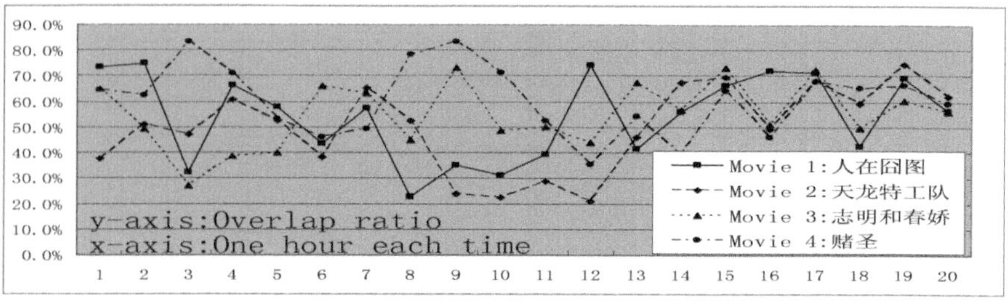

Fig. 9. Overlap Ratio of Node Measurement on Two Hosts in LAN

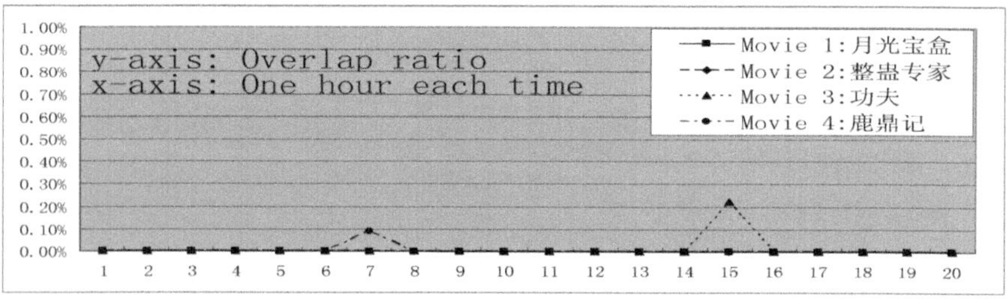

Fig. 10. Overlap Ratio of Node Measurement on Two Host in WAN

When the experiment used DHIVDT, establishing two hosts in a LAN, overlapped ratio was 60%. Finally, two different hosts were under WAN IP. It was an approximately 0.0% overlapped ratio.

It could draw a conclusion that overlapped ratio were influenced by WAN IP address. More nodes, low overlapped ratio, more WAN hosts to establish. If better performance was required, multiple hosts witch different WAN IP addresses according to the topology of P2P IPTV Network should be established.

The same experiment had been made with UUSee. Results showed identically.

6 Conclusion

Recent years, solutions of P2P IPTV supervision and management are called by governments due to the anonymity and the vast-and-fast development of P2P network. This paper began with the introduction of theory of P2P IPTV applications and then declared an approach of node measurement of closed source applications based on packet analysis. Discussions and experiment were elaborated in this paper. Results approved the feasibility of it. This paper could be as a reference of measurement of other closes source applications.

References

1. Zou, D.-y., Song, M.-n.: A District Management Strategy For P2P Chord Model. J. Journal of Beijing University of Posts and Telecommunication 31(3), 54–55 (2008)
2. Peng, Z.-l., He, B.: A Discussion of Developing Trend And Contend Supervision of P2PTV. J. Modern Television Technology, 34–36 (August 2008)
3. Zhou, L.-j., Li, Z.-t.: Proposition And Certification Of P2P Media Streaming Traffic Feature. J. Journal of Chinese Computer Systems 30(8), 1478–1479 (2009)
4. Yu, J.-h., Liao, X., Sun, L.: A Research And Implement Of P2P Flow Identification. J. Guangdong Communication Technology, 2–5 (April 2007)
5. Kuibak, Y.: The Emule Protocol Specification. In: DANSS Lab School of Computer Science and Engineering. The Hebrew University of Jerusalem, Jerusalem (January 17, 2005)
6. Xu, Z.-l., Mo, S.-h.: Identification Of P2P Streaming Traffic Using Application Signatures. J. Application Research of Computes 26(6), 2214–2216 (2009)
7. Yang, J., Li, R.-z.: P2P Streaming Media Broadcast System Caching Algorithm Based on Time-Interval. J. Computer Engineering and Design 31(1), 90–92 (2010)
8. Zhang, Z.-s., Luo, D.-l., Yang, Z.-d.: New tendency of P2P Stream Media. Journal of Hebei North University (Natural Science Edition) 23(1) (Feburary 2007)
9. Zhang, W.-w., Fan, X.-l.: Node Selection Algorithm for P2P Streaming Media Based on Fuzzy Theory. J. Computer Engineering 35(23), 88–89 (2009)
10. Gong, J., Sun, Z.-x., Chen, E.-y.: A Kind of P2P Streaming Media Identification Method Based on Traffic Behavior Analysis. J. Computer Technology and Development 19(9), 129–131 (2009)
11. Wang, S.-q., Jiang, X.-z., Tian, F., Li, N.: ON Cache at Clients in a Peer-to-peer Based Streaming Media Video-On-Demand System. J. Computer Applications and Software 26(9), 219–221 (2009)
12. Hu, P., Nie, P.-p., Lu, J.-d.: Typical P2P Streaming Media Model and Its Key Techniques. J. Computer Engineering 35(3), 60–62 (2009)
13. Peng, Z., Lu, G.-z., Liang, J., Yang, Z.-k.: Survey on Peer-to-Peer Video-on-demand Streaming Protocols. J. Computer Science 35(12), 9–14 (2008)
14. Fang, W., Wu, M.-h., Ying, J., Zhang, Y.: Research On Peer-to-peer Architecture and Algorithm for Streaming Live Media. J. Computer Applications and Software 22(5), 35–37 (2005)
15. Jiang, T., Zhong, Y.-p.: Design of Buffer Management Scheme for Multicast P2p Streaming Media Based on Peercast. J. Computer Applications and Software 26(6), 213–215 (2009)

Research of Community Discovery Algorithm Guided by Multimodal Function Optimization

Ma Rui-xin* and Wang Xiao

Software school of Dalian University of Technology, Dalian 116022
teacher_mrx@126.com

Abstract. This paper introduces the concept of community seed, comes up with a novel algorithm which based on the multimodal function optimization idea. Generally, the relationship between\\nodes in the same community is much closer than nodes in different communities. We use different sizes of network structures Zachary and Dolphins to test our algorithm, the experimental results show that this method is able to finish dividing the network in low time complexity, high efficiency without any priori information.

Keywords: Community discovery, priori information, community seed, multimodal function optimization.

1 Introduction

With the development of web service technology, users become more and more interest in participating, sharing and interacting to gain personalized information.. Communities are the reflection of network's modularization and heterogeneity. It is of great value to look for and discover communities in large networks by deeply researching in the framework of networks. E.g. communities in social network are used to reveal the groups in which users have the same interests and habits. Social structures' discovery in WWW is good for increasing the efficiency and accuracy of search engineer, achieving the goal of filtering information, tracing hot topics and analyzing information.

This paper introduces the concept of community seed which leads other nodes to locate around it. A node's position decides its importance in the entire network.

The remainder of this paper is organized as follows: Section 2 introduces some existed discovery algorithms; Section 3 introduces the concept of community seed; Section 4 discuss the validity of the proposed approach. Section 5 compared different algorithms' time complexity. At the end of this paper, we provide a conclusion.

2 Common Algorithms in Community Division

The research of community discovery in large networks springs from the study of sociology scholars Girvan and Newman. Some famous community mining algorithms

---

* Ma Rui-xin, born in 1975, teacher of DLUT, specialized in Data mining, E-commerce and Personalized Recommendation System.

L. Qi (Ed.): ISIA 2010, CCIS 86, pp. 678–683, 2011.

include G-N, Kernighan and so on. G-N uses the idea of splitting. Kernighan is a kind of greedy algorithm [1]. Because it firstly divides the network into two parts under the premise of knowing the scale of the network, it is not practical in reality and implementation. Radichii and his fellows came up with a split algorithm on the basis of G-N. However this algorithm'll not work well unless the numbers of lashing triangles are much enough for calculating the coefficient of edges in this network.

Analyzing those discovery algorithms, we know that it is essential to design a question unrelated algorithm with low time complexity, high accuracy and without prior information to solve the community discovery problems in isomerous networks.

3 Community Discovery Algorithm Based on MFO

We assume that in this algorithm, each node belongs to only one community. The core levels of the nodes decide whether their positions are good or not.

Definition 1: If a node is the first member of a community, it'll lead the following nodes to locate around it. The number one node becomes the seed of this community.

The steps to choose community seeds are as show below.

Firstly, all nodes are sorted in decreasing order of degrees, which constitute a list, *Clist*. The set of community seeds *S* is initially set to empty. Then, the nodes are checked in turn from the beginning to the end of the list. If a node does not have connection with all the seeds in *S*, the node becomes a new seed and be added to S. Because the node with more degrees is checked first, the community seed in each set must be the best node in this community. These community seeds respectively guide the rest nodes that are in the same community to locate multiple optima [2].

The association between node *I* and community *SN* is calculated as formula (1).

$$R^i_{SC_k} = \frac{\sum_{j \in SC_k} [E_{i,j} \times f_j]}{\sum_{k \in \{SC\}} [E_{i,l} \times f_l]} \quad (1)$$

$E_{i,j}$ is a bipolar threshold function [4]. When node *I* has adjacent with node *J*, $E_{i,j}$ is 1, otherwise is 0. f_i is the degree of node I. In this paper, position of a node is decided by its degree. For a node *I* in *Clist* but outside of every existed community *S* (we call it free node), the adjacent value between *I* and *S* is only decided by the nodes to which *I* connected.

For a network with *n* nodes, the calculation of this algorithm's time complexity is calculated as show below.

Step one: Compute each node's degree and put them into decrease order to form *Clist*, the time complexity is $O(n^2)$.

Step two: Compute the adjacent value between free nodes and existed communities. For the number *n* node, it's just need to check the association between *n* and the former *n-1* nodes in *Clist*, so the complexity is n.

Step three: Decide which community does node *I* belongs to, time complexity is n. Therefore, the time complexity of this algorithm is $O(n^2)$.

4 Analysis and Test of Experimental Results

We used two classical social networks, Zachary club and dolphins network to prove that our algorithm is operable and efficient.

4.1 Zachary Club

Zachary Club is a common experimental network in social network's analysis [3]. This network includes 34 nodes and 78 edges. As a real social network, Zachary club is often used to test the efficiency of community discovery algorithms.

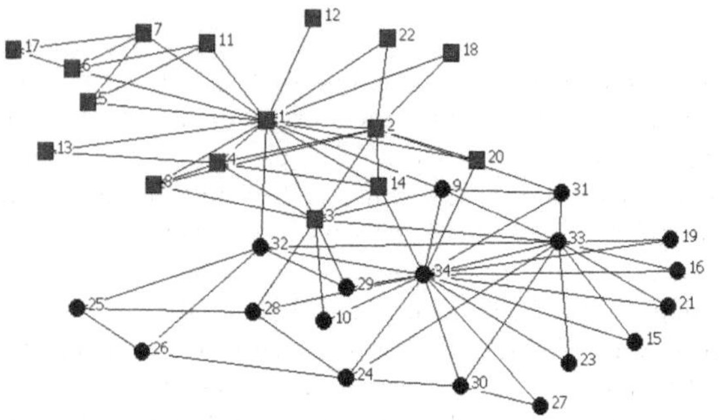

Fig. 1. Node 1, node 33 and node 34 are the most central nodes (the possible community seeds) in this network. This figure shows the division result of G-N algorithm.

The procedure of dividing Zachary Club is as follows. First, we take node ㉞ as the first seed of community *A*. Second, as for node ①, there isn't an edge $E_{34,1}$, so node ① becomes a seed of a new community *B*. Third, according to *Clist*, we check nodes in turn from the beginning to the end to calculate the association between free node *I* and any existed community $S(S\in\{SC\})$. The framework after division is as figure 2 shows.

Table 1. Clist of Karate Club

Rank	Node	Degree	Rank	Node	Degree	Rank	Node	Degree
1	㉞	17	**13**	⑭	4	**25**	⑬	2
2	①	16	**14**	⑮	4	**26**	⑯	2
3	㉝	12	**15**	㉘	4	**27**	⑰	2
4	③	10	**16**	㉚	4	**28**	⑱	2
5	②	9	**17**	㉛	4	**29**	⑲	2
6	④	6	**18**	⑤	3	**30**	㉑	2
7	㉜	6	**19**	⑪	3	**31**	㉒	2
8	⑨	5	**20**	⑳	3	**32**	㉓	2
9	㉔	5	**21**	㉕	3	**33**	㉗	2
10	⑥	4	**22**	㉖	3	**34**	⑫	1
11	⑦	4	**23**	㉙	3			
12	⑧	4	**24**	⑩	2			

Let's take node ③ and ⑨as an example to calculate R_A^3, R_B^3, R_A^9, R_B^9.

$R_A^3 = 0.4$; $R_B^3 = 1 - 0.4 = 0.6$; $R_A^9 = 35/51$; $R_B^9 = 1 - R_A^9 = 16/51$

$R_A^3 \langle R_B^3 \quad R_A^9 \rangle R_B^9$,so node ⑨ belongs to community *A*, node ③belongs to community *B*.

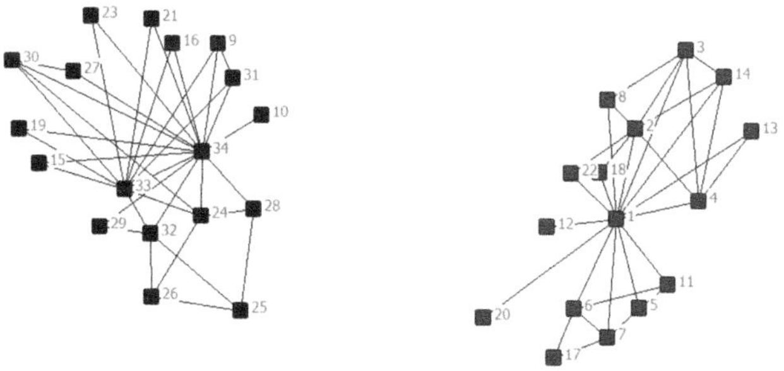

Fig. 2. Based on MFO, we instructed all nodes located around the community seeds, calculated the coefficient between free nodes and the existed communities

Using Karate club as the test data, our algorithm is able to mine and divide network accurately without any prior information. The accuracy can run up to 100%.

4.2 Dolphin Network

Dolphin network is also a usual network in researching social network [4,5]. Lusseau and his fellows conducted systematic surveys in Doubtful Sound, Fiordland, New Zealand. The survey route has remained constant over the 7-year period and covered the entire home range of the Doubtful Sound population. The entire network includes 62 nodes and 159 edges. In figure 3, each node represents a dolphin, if two dolphins have regular contact, adds an edge to link them.

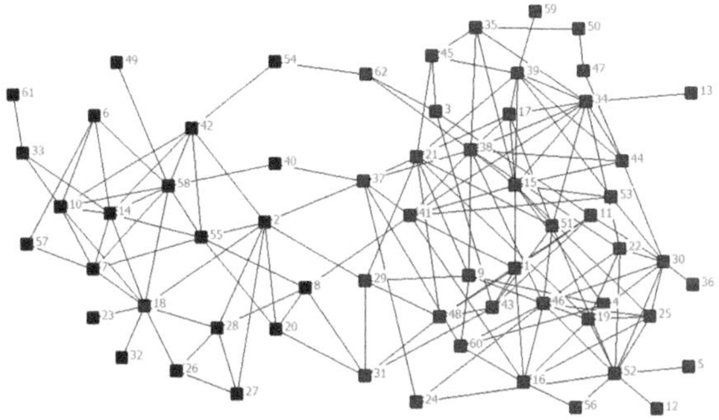

Fig. 3. It's hard to find the seeds in dolphin networks by eyes, G-N algorithm divides dolphins into two parts

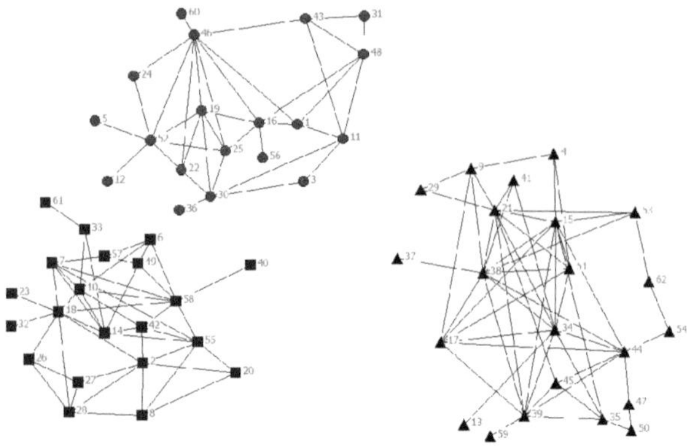

Fig. 4. This algorithm divides the entire network into three parts

5 Algorithm Complexity Contrast

Table 2 inspects different documents as well as compares the time complexity of different algorithms.

The Community discovery algorithm based on multimodal function optimization is able to finish community structures' division without any artificial interfere. Compared with Radichii's algorithm, it not only be adapted to netlike networks but also can be applied to treelike networks. Therefore, this algorithm has great development space.

Table 2. Comparison of Different Algorithm's Time Complexity

name	reference	complexity
N-G	[3]	$O(m^2n)$
G-N	[2]	$O(n^2m)$
BB	[7]	$O(n^3)$
Radicchi	[5]	$O(m^4/n^2)$
Algorithm based on MFO	This paper	$O(n^2)$

6 Conclusions

Community discovery includes plenty of information about network patterns. General community discovery algorithms need users to provide prior information, which cannot effectively reveal the real community structures inside. This algorithm is capable of discovering community structure accurately, efficiently and effectively. Moreover, this algorithm is easy to work with other algorithms by using different weighting functions to expand the sphere of this algorithm's application.

References

1. Li, X.: Adaptively choosing neighbourhood bests using species in a particle swarm optimizer for multimodal function optimization. In: Deb, K., et al. (eds.) GECCO 2004. LNCS, vol. 3102, pp. 105–116. Springer, Heidelberg (2004)
2. Gan, W.-y., He, N., Li, D.-y.: Community Discovery Method in Networks Based on Topological Potential. J. Journal of Software 20, 2241–2254 (2009)
3. Zachary, W.W.: An information flow model for conflict and fission in small groups. N. Journal of Anthropological Research 33, 452–473 (1977)
4. Lusseau, D., Schneider, K., Boisseau, O.J.: The bottlenose dolphin community of doubtful sound features a large proportion of long-lasting associations—Can geographic isolation explain this unique trait? J. Behavioral Ecology and Sociobiology 54, 396–405 (2003)
5. Lusseau, D., Newman, M.E.J.: Identifying the role that animals play in their social networks. J. Proc. of the Royal Society B: Biological Sciences 271, 477–481 (2004)

Energy Efficiency Evaluation for Iron and Steel High Energy Consumption Enterprise

Gongfa Li*, Jianyi Kong, Guozhang Jiang, Hua Zhang, Zhigang Jiang, Gang Zhao, and Liangxi Xie

College of Machinery and Automation, Wuhan University of Science and Technology, 430081 Wuhan, China
ligongfa@yahoo.com.cn

Abstract. In order to optimize the allocation of energy resources, reduce enterprise's energy consumption and achieve the goals of energy conservation and emission reduction, energy efficiency evaluation indicator structure and evaluation method which are fit for iron and steel high energy consumption enterprise are proposed, an energy efficiency evaluation system is designed for iron and steel high energy consumption enterprise to simulate the dynamic behaviors of energy use and consumption during the production process. The system has the function of data acquisition, statistical analysis and prediction. During the coke oven production process, this system is applied to the process from coal material to coke cake, and its feasibility is shown.

Keywords: high energy consumption; energy efficiency evaluation; evaluation index; evaluation system; iron and steel enterprise.

1 Introduction

Highly energy consumption enterprise is the principal end user of energy consumption. Modeling enterprise energy consumption process and simulating the dynamic behavior of energy using are of great importance. It not only can realize the qualitative analysis and quantitative evaluation of enterprise energy consumption, but also is an effective means of enterprise energy efficiency evaluation.The key of simulation and analysis is the construction of model. Up to now, there have been input-output model, statistic regression analysis model, system dynamics model and system identification model for enterprise energy consumption analysis at home and abroad. To some extent, these methods, which mainly utilized the modern mathematical achievements, reflect the enterprise production scale and energy consumption status. Especially, input-output model, as a good tool for cost accounting and analysis, can quantitatively reflects the consumption components of product, but it is on the foundation of historical data analysis to account the energy consumption, when the industrial structure and technology are changed, it hasn't the predictive

* Corresponding author.

L. Qi (Ed.): ISIA 2010, CCIS 86, pp. 684–690, 2011.

ability, and meanwhile, input-output model can't directly reflect the production process [1]. System dynamics model occupies preferable predication performance in the system of society, economy and ecology, but it is unsuitable for the quantitative analysis of enterprise energy consumption process with diverse components [2]. Moreover, statistic regression analysis model and system identification model are usually used to build up the mathematical model of energy consumption equipment or unit energy consumption activity, but don't adapt to the global and complex process of enterprise energy consumption [3]. Anyhow, the above methods have their own advantages, but they can't establish the global and direct model of enterprise energy consumption process, which integrated with mathematical and visual character and reflecting the collaborative work between energy consumption activities.

Enterprise energy consumption process describes energy transfer in energy transmission pipe net and energy conversion through energy consumption equipment. It has the characteristics of obvious process, uncertainty, concurrence, and asynchrony etc. Most of the energy consumption processes are continuous; meanwhile, there are also some discrete events in it, such as the equipment start/stop. Aiming at the shortage of the means to evaluate enterprise's efficiency of high energy consumption of our country at present, from the angle of reducing emission and energy-conservation, energy efficiency evaluation index system of high energy consumption trade is set up at many levels of many angles, the energy efficiency index of quantization to examine can be offered, an objective basis achievement to examine in saving energy and reducing the cost for enterprises is offered. While the index system is studied, the enterprise energy consumption evaluation method on the basis of the model is studied, the integrated model of enterprise's energy consumption system through the study on production procedure is set up, the scientific and rational assessment means of the energy consumption in enterprise's production process are got using modern analysis and optimization technology of simulation and assessment, prediction analysis is offered for such enterprise's energy consumption as direct energy consumption, indirect energy consumption and complete energy consumption of the enterprise unit product, decision support is offered for enterprise's energy resources rational distribution, the balance of energy between energy supply and energy demand, high efficiency of energy utilization, data support is supplied for the fact that enterprises carry on real energy-saving and cost-reducing facilities and improvement, acquisition, disposition of energy-saving and cost-reducing decision .

2 Energy Efficiency Evaluation Index System for Iron and Steel Enterprise

According to the characteristics of energy consumption, the energy-conservation focal points of the iron and steel enterprise are optimization of production procedure and renewal of the equipment of mainly product consuming energy, comprehensive utilization of mainly product consuming energy, remaining energy recovery and energy pollutant. So, enterprise's efficiency comprehensive evaluation is considered from not only the economic benefits but also such different fields as the energy consumption activity management, production technology energy-conservation and environmental protection. Energy efficiency evaluation indexes are fellow the following principles of iron and steel high energy consumption enterprise.

Integrality principle: They reflect the true state of the energy consumption system of every enterprise as in an all-round way as possible.

Concise principle: The index concept is clear, the data are apt to examine, the effectiveness is strong.

Importance principle: It should be the important index in all fields, through reducing the total amount of indexes, enable investigating the economy of measuring is feasible.

Level principle: It should be able to reflect the inherent structure of enterprise's energy consumption system, so it must have certain level nature.

Comparability principle: In order to compare, the evaluation indexes are required to have certain comparability on the time and space.

According to above-mentioned principles, through analyzing the basic data materials of enterprises, statistical data of energy utilization of enterprises and production data that are easy to obtain and examine, energy efficiency evaluation index type of iron and steel enterprise is 4 parts, namely economic indexes, technical management indexes, energy management indexes and environmental efficiency indexes . Its index system is shown as Fig. 1.

(1) Economic indexes: They include value of industrial output and value-added .

(2) Technology management indexes: They include energy efficiency of production and exchange unit, energy efficiency of transportation and distribution unit, energy efficiency of product and energy efficiency of equipment consuming energy.

(3) Energy management indexes: They include accomplishment ratio of energy supply plans, guarantee degree of stock volume to production at then end of term, proportion in good condition of the measuring apparatus, economy ratio of remaining heat recovery, energy-conserving potentiality, skill and investment of energy-conserving.

(4) Environmental efficiency indexes: They include air pollutant displacement, water pollutant displacement, solid pollutant displacement and air quality.

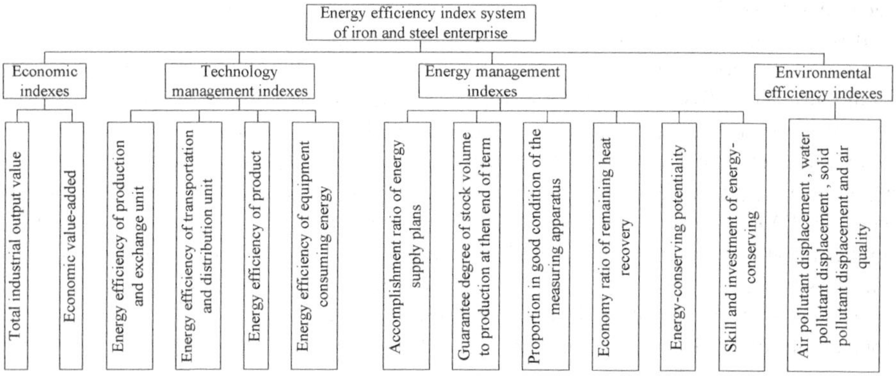

Fig. 1. Energy efficiency index system of iron and steel enterprise

Every index meaning is shown as following:

Total industrial output value reflects the total scale of industrial production and total level in certain time.

Economic value-added reflects the input, output and economic benefits situation of the industrial enterprise. It's the basis to improve the production and management of industrial enterprise.

Energy efficiency of production and exchange unit, namely enterprises use the energy and change the energy production equipment energy consumption level, closely related to enterprise's efficiency level.

Energy efficiency of transportation and distribution unit is an important link of every terminal department, for most enterprises, the energy losses of transportation and distribution does not forms the main part of enterprises total energy losses, but has reflected the promotion level of energy-conserving potentiality.

Energy efficiency of product is a energy utilization ratio of the products, reflects the ratio of theory effective energy consumption to real net energy consumption producing the unit product.

Energy efficiency of equipment consuming energy reflects the proportion of total effective energy consumption of equipment to total amount of energy supplying.

Accomplishment ratio of energy supply plans is calculated according to the material object amount of energy mainly, analyzed the reason not hitting the target .

Guarantee degree of stock volume to production at then end of term is calculated according to some main energy material object amount and equipment generally, reflects the impact of energy stores change on economic result .

Proportion in good condition of the measuring apparatus can weigh the accuracy and feasibility of enterprise's energy data .

Economy ratio of remaining heat recovery reflects enterprise retrieve actually lost heat to economize the energy.

Energy-conserving potentiality is a goal that the energy-conservation demands to reach to reflect the disparity that actual efficiency compares with advanced level.

Skill and investment of energy-conserving, in order to reduce energy consumption, economize the energy, enterprises need to carry on technological transformation or reconstruct to some equipment, so the measure investment of the energy technology should carry on, this index reflects the economic benefits that the energy-conserving skill arranges .

3 Energy Efficiency Evaluation System for Iron and Steel Enterprise

Energy efficiency evaluation system for iron and steel enterprise is designed as following demand.

(1) Data warehouse of energy information, namely data set facing theme, integration, relatively stable and reflecting history changes is set up, is used for supporting enterprise's energy administrative decision.

(2) Evaluation index system of energy efficiency in iron and steel enterprise is established from such a lot of angles as energy resources, environment, enterprise's benefit.

(3) Synthetically analysis of energy efficiency index is realized, using the system engineering method, variation tendency of energy consumption in enterprises and every department is predicted, structure of energy consumption of all products in every department is optimized, efficiency of energy consumption is improved, economic benefits and environmental benefit are comprehensive evaluated, support for making policy is offered.

The systematic structure frame is shown as Fig. 2, the function of every part is described as following.

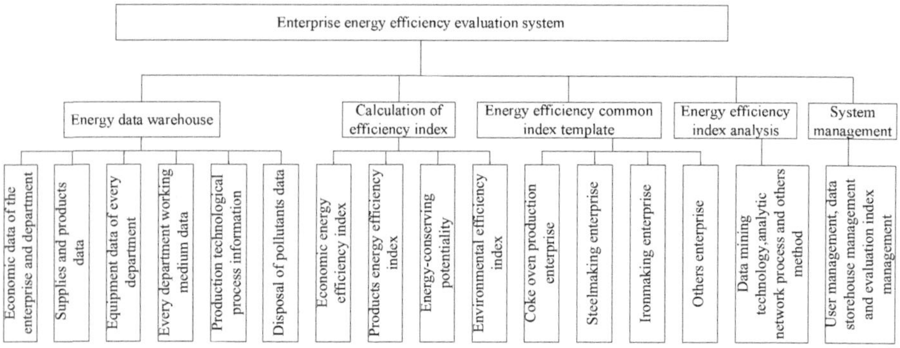

Fig. 2. Systematic Structure Frame of Iron and Steel Enterprise Energy Efficiency Evaluation

Energy data warehouse is used for storing historical data and present data that the enterprise energy uses, it is divided into economic data of the enterprise and department, the supplies and products data, the equipment data of every department, every department working medium data, production technological process information, disposal of pollutants data.

Calculation of efficiency index, namely centering on economic energy efficiency index, products energy efficiency index, energy-conserving potentiality, environmental efficiency index, according to firsthand information and data of warehouse, adopting country normal computing technology, quantization appraises the individual event index of enterprise's efficiency.

Energy efficiency common index template. According to the trade characteristics of iron and steel high energy consumption enterprise, the efficiency index template of the trade is made, an efficiency index in common use to the trade is offered.

Energy efficiency index analysis. Using data mining technology, through excavating to the energy data warehouse, build the model is built to predict the variation tendency of the energy uses while one period, support is offered in energy plan decision.

System management includes user management, management of data storehouse and management of evaluation index.

This system has the following characteristics.

(1) The basic operation of the figure user's interface is supported, for example users can fully understand, freedom handled, the systematic models of energy consumption is established in a flexible way ;

(2) Offering the communication frame that users defined every model element property.

(3) Supporting the dynamic simulation of the energy consumption course, offering the corresponding simulation pictures.

(4) Assessing the difference of indexes According to different energy efficiency evaluation index, energy efficiency evaluation result is produced by simulation.

(5) Supporting graphics compared analysis of simulation result of different models.

4 Application for Coke Oven Production

Study and application of comprehensive assessment of energy efficiency in coke oven production process is carried on in order to verify its feasibility. The life cycle of this system is shown as Fig. 3.This system collects the primal production data at first, evaluation index system of energy efficiency is established, through analyzing, the integrated model of the energy consumption system is set up, then the simulation of systematic energy consumption is carried on. Evaluation of simulation result of the system according to energy efficiency evaluation index system, after making the efficiency evaluation value, making comparison with expected index, in order to obtain ideal value, optimization is got through the aid decision system, optimization result is obtained as feedback, energy consumption system integrated model is adjusted until expectation index that efficiency assesses is got[4].

The systematic operation result shows that the integrated model of system can describe the energy consumption system from stage construction, much angles, give full play to the visual advantage combining with analysis; The simulation method of enterprise energy consumption system based on model can offer a series of dynamic performance analysis datum of energy consumption system, as important basis of systematically analysis and optimization. The parameter disposition of energy consumption system is evaluated by energy efficiency evaluation method based on model, decision support is offered for high efficiency operation of energy consumption system. Application has verified the systematic feasibility of comprehensive evaluation of enterprise's efficiency of high energy consumption.

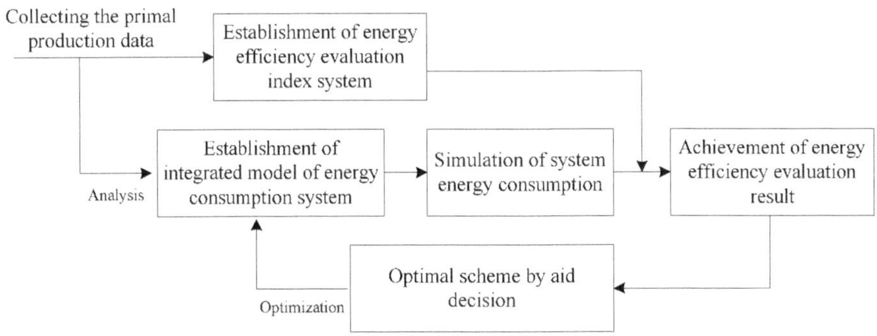

Fig. 3. Life Cycle of Cole Oven Production Energy Efficiency Evaluation Application System

5 Conclusion

The industrial enterprise is a subject of energy consumption, energy-conserving potentiality is enormous, is necessary and urgent for enterprise's energy efficiency to carry on comprehensive evaluation. Using the analytical method of the system engineering, efficiency comprehensive evaluation index system and evaluation method has been put forward, enterprise's efficiency comprehensive evaluation system have been designed, quantitative analysis and optimization means for high-efficiently energy use in high energy consumption enterprise are provided through launching enterprise's efficiency comprehensive evaluation of high energy consumption omni-directionally. It has remedied the deficiency of systematic enterprise energy development tactics, realized digitization and systemization of enterprise's energy efficiency evaluation, the obvious economic benefits to enterprises is brought. This method has the characteristics of science and succinct, can offer basis to the enterprise energy decision policy, thus improve the utilization efficiency of the enterprise energy.

Acknowledgement

This research reported in the paper is supported by National Natural Science Foundation of China (70971102).This support is greatly acknowledged.

References

1. Cai, J., Du, T.: Input-Output Model & Analysis of Energy Consumption Crude Steel and Environmental Loads on Steel Enterprise. Chinese Journal of Gold Journal 3, 306–312 (2001)
2. Jia, H., Ding, R.: System dynamics-Analysis of Feedback Dynamic Complexity. Higher Education Press, Beijing (2002)
3. Sohlberg, B.: Grey box modelling for model predictive control of a heating process. Journal of Process Control 13, 225–238 (2003)
4. Li, G., Kong, J., Jiang, G.: Research and Application on Compound Intelligent Control System for Coke Oven Heating. Chinese Journal of Iron and Steel 43, 89–92 (2008)

Research on Dynamic Connectivity of Urban Road Network

Bing Su[1,2], Yanmei Shen[1], and Changfei Ge[1]

[1] School of Economics and Management, Xi'an Technological University,
Xi'an 710032, China
[2] The State Key Lab for Manufacturing Systems Engineering,
Xi'an 710049, China

Abstract. Connectivity of road network is an index to evaluate the rationality of urban road network planning. In the past, it was defined as the ratio of the number of edges to the number of nodes from static perspective. In practice, the network is unreliable; some roads may be blocked at certain times. So dynamic connectivity of road network is put forward from two points of view in this paper: (1) based on the scanty two forms—blockage and non-blockage of each edge, dynamic connectivity of binomial distribution obeyed by each blocked edge is presented; (2) based on the number of edges, dynamic connectivity of random distribution obeyed by the number of unblocked edges is introduced.

Keywords: road network; binomial distribution; random distribution; dynamic connectivity.

1 Introduction

Urban road network which is comprised of all the roads within city is vector of social economic activities and transportation. Connectivity of road network is a significant index to evaluate the rationality of the network planning [1, 2]. In the past, it was defined as $J = 2m/n$ with m the number of edges and n the number of nodes [3, 4, 5, 6]. Due to the certainty of edges and nodes, this paper called it as static connectivity. But in practice, the network is unreliable; some roads may be blocked at certain times (e.g. blocked by unexpected events such as snowfall or traffic accidents) [7, 8] such that the number of edges in the network may be changing. So a new index — dynamic connectivity of road network is put forward from two points of view in this paper: (1) based on the scanty two forms— blockage and non-blockage of each edge, dynamic connectivity of binomial distribution obeyed by each blocked edge is presented; (2) based on the number of edges, dynamic connectivity of random distribution obeyed by the number of unblocked edges is introduced. Definitions and properties of the index are given. Finally, an application of dynamic connectivity for a local road network in Xi'an is presented.

L. Qi (Ed.): ISIA 2010, CCIS 86, pp. 691–696, 2011.

2 Dynamic Connectivity of Binomial Distribution Obeyed by Each Blocked Edge

Due to the scanty two forms—blockage and non-blockage of each edge, each blocked edge obeys binomial distribution [9], so dynamic connectivity of binomial distribution obeyed by each blocked edge is presented.

2.1 Definition of Dynamic Connectivity of Binomial Distribution Obeyed by Each Blocked Edge

Let $G(V, E)$ denote an urban road network with $|V| = n$ nodes and $|E| = m$ edges, k denote the number of blocked edges, and a denote the number of reduced nodes because of blockage at certain time point. In order to discuss the problem, we make the assumption $a = 0$, so $k \in \{0, 1, 2, \ldots, m-n+1\}$.

Let J^k denote dynamic connectivity under k -strip of blocked edges, J_{RDC} denote dynamic connectivity of binomial distribution obeyed by each blocked edge, and p ($0 < P < 1$) denote probability of blockage occurrence of each edge.

Definition 1. $J_{BDC} = E(J^k) = J^0 P\{x=0\} + J^1 P\{x=1\} + \ldots + J^{m-n+1} P\{x = m-n+1\}$

$$= \sum_{k=0}^{m-n+1} \frac{2(m-k)}{n} P\{X = k\} = \frac{2m - 2E(k)}{n} \tag{1}$$

In the concrete solving, we use formulas in table I to solve J^k and $P\{X = k\}$.

Table 1. Dynamic connectivity and probability under k -strip of blocked edges

	$k=0$	$k=1$	...	$k = m-n+1$
$P\{X=k\}$	$C_m^0 p^0 (1-p)^m$	$C_m^1 p^1 (1-p)^{m-1}$	...	$C_m^{m-n+1} p^{m-n+1} (1-p)^{n-1}$
J^k	$2(m-0)/n$	$2(m-1)/n$	...	$2(m-m+n-1)/n$

Equation (1) can reflect mature degree of urban road network under k -strip of blocked edges at certain time point. The higher the value of J_{RDC} is, the more the number of edges averagely connected with each node are, and the better the network is.

2.2 Properties of Dynamic Connectivity of Binomial Distribution Obeyed by Each Blocked Edge

Property 1. In any urban road network, the value of J_{RDC} is lower than that of J.

Proof. If there isn't any blocked edge in urban road network, connectivity of road network is static connectivity, and $J = \frac{2m}{n}$;

If there are some blocked edges in urban road network, connectivity of road network is dynamic connectivity, and $J_{BDC} = E[\frac{2(m-k)}{n}] = \frac{2m - 2E(k)}{n}$;

Since $0 < P < 1$, and $E(k) = (m-n+1)p > 0$;

Hence, $J_{BDC} = \frac{2m - 2E(k)}{n} < \frac{2m}{n} = J$.

Property 2. Let $G_1(V_1, E_1)$ and $G_2(V_2, E_2)$ denote two networks with the same sections and nodes, p_1 and p_2 respectively denote probability of blockage occurrence of each edge in G_1 and G_2. If $1 > p_1 \geq p_2 > 0$, then $J^1_{BDC} \leq J^2_{BDC}$.

Proof. According to equation (1), $J^1_{BDC} = \frac{2m - 2E(k)}{n} = \frac{2m - 2(m-n+1)p_1}{n}$

and $J^1_{BDC} = \frac{2m - 2E(k)}{n} = \frac{2m - 2(m-n+1)p_2}{n}$;

Since $1 > p_1 \geq p_2 > 0$, and $2(m-n+1)p_1 \geq 2(m-n+1)p_2$;

Hence, $J^1_{BDC} = \frac{2m - 2(m-n+1)p_1}{n} \leq \frac{2m - 2(m-n+1)p_2}{n} = J^2_{BDC}$.

3 Dynamic Connectivity of Random Distribution Obeyed by the Number of Unblocked Edges

At certain times, the number of unblocked edges in the network may be changing and obey random distribution [10], so dynamic connectivity of random distribution obeyed by non-blocked sections is introduced.

3.1 Definition of Dynamic Connectivity of Random Distribution Obeyed by the Number of Unblocked Edges

Let l denote the total number of unblocked edges, $f(l)$ denote probability density of l, and J_{RDC} denote dynamic connectivity of random distribution obeyed by the number of unblocked edges. We make the assumption $a = 0$, so $n-1 \leq l \leq m$.

Definition 2. $J_{RDC} = E(\frac{2l}{n}) = \frac{2E(l)}{n} = \frac{2\int_{n-1}^{m} l \times f(l) d_l}{n}$ (2)

Equation (2) analyzes dynamic connectivity from the perspective of the change of the number of unblocked edges in urban road network at certain times. The higher the value of J_{RDC} is, the more the number of edges averagely connected with each node are, and the better the network is.

Concrete type of random distribution obeyed by l is influenced by some Factors (such as geographical position or traffic operation status). We can know concrete distribution law by observing massive data that is relative of those factors.

If l obeys normal distribution with μ, δ^2, then $f(l) = \frac{1}{2\pi\delta} e^{-\frac{(l-\mu)^2}{2\delta^2}}$ [10],

$$\text{so } J_{RDC} = \frac{2\int_{n-1}^{m} l \times f(l) d_l}{n} = \frac{2\int_{n-1}^{m} l \times \frac{1}{2\pi\delta} e^{-\frac{(l-\mu)^2}{2\delta^2}} d_l}{n} = \frac{2\mu}{n}. \quad (3)$$

3.2 Properties of Dynamic Connectivity of Random Distribution Obeyed by the Number of Unblocked Edges

Property 3. In any urban road network, the value of J_{RDC} is lower than that of J.

Property 4. Let $G_1(V_1, E_1)$ and $G_2(V_2, E_2)$ denote two networks with the same edges and nodes. Supposed l_1 in G_1 and l_2 in G_2 respectively obeys normal distribution with μ_1, δ_1^2 and μ_2, δ_2^2. If $\mu_1 \geq \mu_2$, then $J^1_{RDC} \geq J^2_{RDC}$.

Proof. According to equation (3), $J^1_{RDC} = \frac{2\mu_1}{n}$, and $J^2_{RDC} = \frac{2\mu_2}{n}$;

Since $\mu_1 \geq \mu_2$; Hence, $J^1_{RDC} = \frac{2\mu_1}{n} \geq \frac{2\mu_2}{n} = J^2_{RDC}$.

4 An Application of Dynamic Connectivity

Figure1 describes the same local road network in Xi'an in 2005 and 2008. Through figure I , we could find that this local road network in 2008 has been increased some new edges. In order to explain definitions and properties of dynamic connectivity, we calculate and compare the value of dynamic connectivity in the two years.

In order to do a better explanation, we make the following assumptions. 1) The probability of blockage occurrence of each edge both in 2005 and 2008 is $p = 0.5$; 2) Unblocked edges obey normal distribution with $\mu = 10$ in 2005 and $\mu = 15$ in 2008.

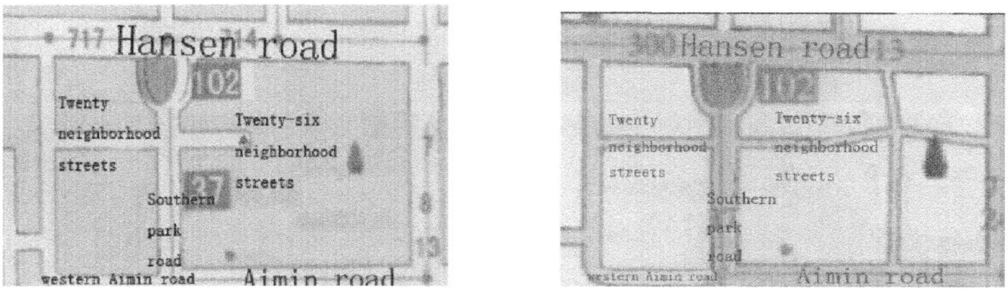

(a) In 2005 (b) In 2008

Fig. 1. The same local road network in Xi'an in 2005 and 2008

Table 2 gives the value of J_{BDC} , J_{RDC} and J in G of 2005 and 2008.Through analyzing table 2, we obtain the following conclusions. 1) Compared to G of 2005, edges have been increased by 5 and nodes have been increased by 3 in G of 2008. 2) The value of J is higher than the value of J_{BDC} or J_{RDC} . In other words, the value of static connectivity is higher than the value of dynamic connectivity in the same G . 3) The value of J_{BDC} or J_{RDC} in 2008 is higher than that in 2005. In other words, dynamic connectivity is better in G which has increased new edges.

Table 2. Tthe value of J_{BDC} , J_{RDC} and J I n 2005 and 2008

	m	n	J	J_{BDC}	J_{RDC}
In G of 2005	12	9	2.67	2.22	2.22
In G of 2008	17	12	2.83	2.33	2.50

5 Conclusions

Connectivity of road network is a significant index to evaluate the rationality of urban road network planning. This paper gives definitions and properties of dynamic connectivity of road network from two points of view. Through the conclusions of analysis, edge blockage has great influence on connectivity of road network. Decision-maker could calculate the value of dynamic connectivity under different situations to make rational road network planning. This paper still has deficiency. How to seek changed law of nodes in road network is the problem in future.

Acknowledgements. The research was supported by Education Department Fund from Shanxi Provence under Grants 09JK495, and Principal Fund from Xi'an Technological University under Grants XGYXJJ0539.

References

1. Wang, W., Xu, R.Q., Yang, T.: Theory and Application of Urban Transportation Planning, pp. 211–215. Dongnan University Press, Nanjing (1998)
2. Yang, T., Xing, Y., Peng, A.X.: Technological Research on Quality Evaluation of Urban Traffic Network. Journal of Urban Road and Flood-Control (3), 9–16 (1994)
3. Zeng, S., Yang, P.K.: A Study of Evaluation of Urban Road Network by Relative Index. Journal of Highway and Transport 13(3), 93–96 (2000)
4. Lu, J., Wang, W.: Planning Indexes System of Urban Road Network. Journal of Traffic and Transportation Engineering 4(4), 62–67 (2004)
5. Feng, S.M., Gao, H., Guo, C.X.: Evaluation of Structural Types of Urban Road Network. Journal of Harbn Institute of Technology 39(10), 1610–1613 (2007)
6. Qian, X.J.: Discussion on Evaluating Method of City Road Networks Structure. Journal of Traffic Technology and Economy 9(2), 88–93 (2007)
7. Stefan, P., Maria, G.S.: A New Algorithm for Reoptimizing Shortest Paths when the Arc Costs Change. Operations Research Letters (31), 149–160 (2003)
8. Liu, L.X., Yang, H.F.: Traffic Pathes' Selection under Emergency and Complex Conditions. Journal of Beijing Union University (Natural Sciences) 18(3), 67–70 (2004)
9. Yang, R., Zheng, W.R., Wang, B.Y.: Probability Theory and Mathematical Statistics, pp. 39–47. Tsinghua University Press, Beijing (2005)
10. Liu, J.K., Wang, J.S., Zhang, Y.H.: Applied Probability and Statistics, pp. 59–68. Science Press, Beijing (2004)

Modeling and Analyses of the N-link PenduBot

Yuan Shao-qiang and Li Xin-xin

Departmen of Automation Science and Electrical Engineering, Beihang University, Beijing, China
ysq1960@sina.com, lhxwoneng@163.com

Abstract. As the foundation of the multi-link PenduBot control, the mathematical model should be established first. Based on the method of kinematics and dynamics, the N-link PenduBot mathematical models are established in this paper, including the nonlinear model, the linear model and the generalized model which considers the motor characteristic. The natural characteristic of the system is analyzed. By using the condition number of the controllability matrix, the control difficulty for the higher order systems is compared. From the comparison result, the control law of the different links PenduBot is obtained. Increasing with the count of PenduBot links, the control difficulty is growing dramatically. Only a strong control measure can make it stable. So, further study on multi-link PenduBot can be carried on.

Keywords: N-link PenduBot; mathematical model; differential equation; control difficulty.

1 Introduction

PenduBot, which is also called Arm Driven Inverted Pendulum or Rotary Inverted Pendulum, is abbreviation of Pendulum and Robot. As a kind of valuable equipment in the automatic control theory research, the inverted pendulum is of simple structure, small size and low cost. It is a very complex and fast-responding system which is multivariable, strong coupling, typically nonlinear, high-order and natural unstable. In recent years, PenduBot attaches more attention of researcher around the world. In 2005, the stabilization control of the three-link PenduBot is realized [1]. Since then, the research on the control problem of the multi-link PenduBot started.

In most papers, researchers only studied on the specific links of the PenduBot. In this thesis, as the foundation of the research on the multi-link PenduBot, the multi-link PenduBot mathematical model is established, including the nonlinear model, the linear model and the generalized model and deeper analysis is carried on.

2 The N-link PenduBot Modeling

Two-link PenduBot is the most common PenduBot. As an example, the two links of the PenduBot are defined as Link 1 and Link 2 [2]. Similarly, the links of the N-link PenduBot are defined as Link 1, Link 2,$\cdots$, Link N.

L. Qi (Ed.): ISIA 2010, CCIS 86, pp. 697–703, 2011.

2.1 The Kinematics Equation of N-link PenduBot

It is assumed that all components of the PenduBot are rigid bodies. As is shown in Fig.1, a Cartesian coordinate system is established with the motor shaft as the origin[3][4]. The definition and values of the symbols is in Table 1.

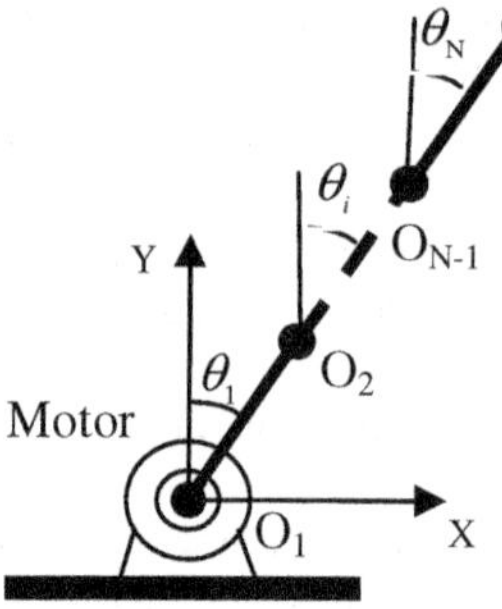

Fig. 1. Schematic of N-link PenduBot

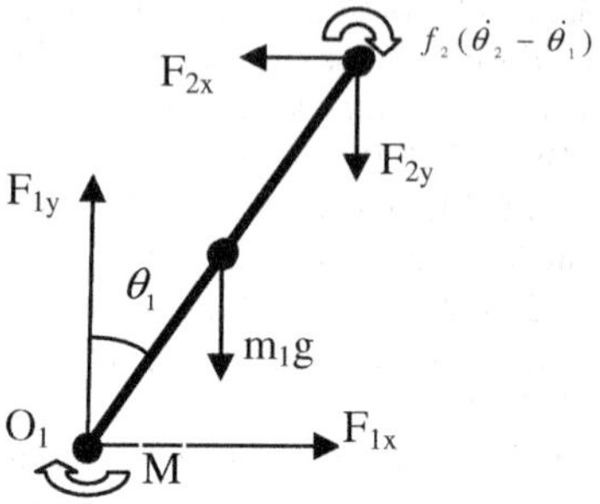

Fig. 2. Forces analysis diagram of Link 1

Table 1. Definition and value of symbols

Symbol	Meaning	Unit	Values
M	The motor drive torque	N*m	
m_i	Mass of Link i	Kg	0.1332 (i=1) 0.114 (i=2~N)
L_i	Length of Link i	m	0.20 (i=1) 0.28 (i=2~N)
O_i	Joint of Link i-1 and Link i		
G_i	Center-of-mass of Link i		
l_i	Distance of Oi to Gi	m	0.10 (i=1) 0.14 (i=2~N)
f_i	Friction torque coefficient of Link i around Oi	N/m*s	0.04 (i=1) 0.026 (i=2~N)
J_i	Inertia of Link i relative to center-of-mass	Kg*m^2	0.0024 (i=1) 0.005 (i=2~N)
θ_i	Angular of Link i relative to vertical position	rad	
$\dot{\theta}_i$	Angular velocity of Link i	rad/s	
F_{ix}	Force of Link i from Link i-1 at the X axis	N	
F_{iy}	Force of Link i from Link i-1 at the Y axis	N	

The N-link PenduBot kinematics equation is obtained as follows:

$$\begin{aligned}
&O_1=(O_{1x},O_{1y})=(0,0)\\
&O_i=(O_{ix},O_{iy})=(\sum_{i=1}^{N-1}L_i\sin\theta_i,\sum_{i=1}^{N-1}L_i\cos\theta_i)\\
&G_i=(G_{ix},G_{iy})=(O_{ix}+l_i\sin\theta_i,O_{iy}+l_i\cos\theta_i)\\
&i=1,2,\cdots,N
\end{aligned}\tag{2.1}$$

2.2 The Dynamic Equation of N-link PenduBot

The external force which the Link 1 is subject to, is shown in Fig.2. So, the dynamic equation in the X-axis direction and the Y-axis direction are:

$$\begin{aligned} F_{1x} - F_{2x} &= m_1 (G_{1x})'' \\ F_{1y} - F_{2y} - m_1 g &= m_1 (G_{1y})'' \end{aligned} \tag{2.2}$$

With O_1 as the origin, the torque equation of Link 1 is:

$$(J_1 + m_1 l_1^2)\ddot{\theta}_1 = M + m_1 g l_1 \sin\theta_1 + f_2(\dot{\theta}_2 - \dot{\theta}_1) - f_1\dot{\theta}_1 - F_{2x} L_1 \cos\theta_1 + F_{2y} L_1 \sin\theta_1 \tag{2.3}$$

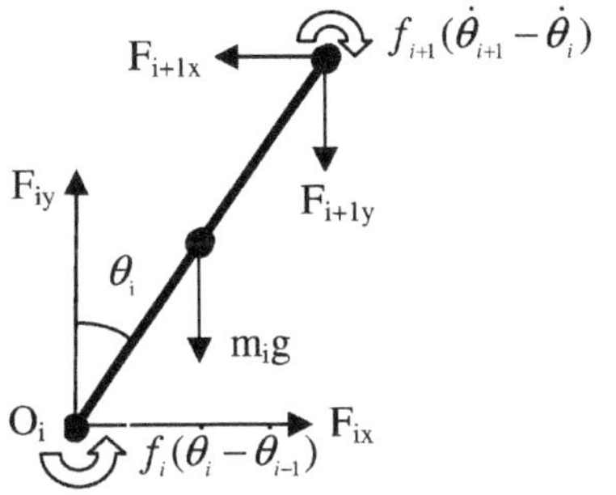

Fig. 3. Forces analysis diagram of Link i

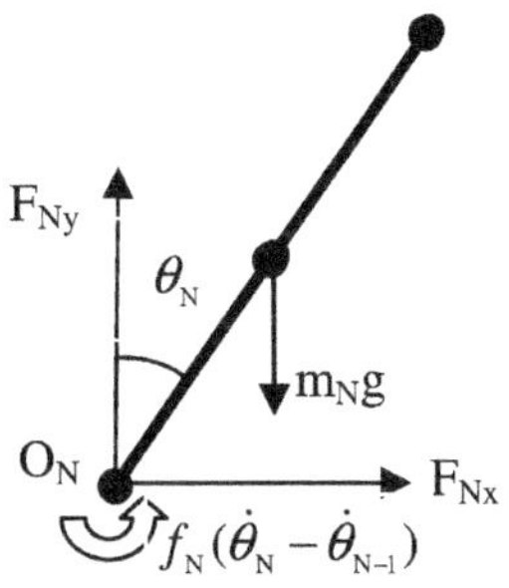

Fig. 4. Forces analysis diagram of Link N

Similarly, as is shown in Fig.3, the dynamic equation of Link I in the X-axis direction and the Y-axis direction are:

$$\begin{aligned} F_{ix} - F_{i+1x} &= m_i (G_{ix})'' \\ F_{iy} - F_{i+1y} - m_i g &= m_i (G_{iy})'' \end{aligned} \tag{2.4}$$

With O_i as the origin, the torque equation of Link i is:

$$\begin{aligned} (J_i + m_i l_i^2)\ddot{\theta}_i &= m_i g l_i \sin\theta_i - f_i(\dot{\theta}_i - \dot{\theta}_{i-1}) + f_{i+1}(\dot{\theta}_{i+1} - \dot{\theta}_i) \\ &- F_{i+1x} L_i \cos\theta_i + F_{i+1y} L_i \sin\theta_i - m_i (O_{ix})'' l_i \cos\theta_i + m_i (O_{iy})'' l_i \sin\theta_i \end{aligned} \tag{2.5}$$

As is shown in Fig. 4.The dynamic equation of Link N in the X-axis direction and the Y-axis direction are:

$$\begin{aligned} F_{Nx} &= m_N (G_{Nx})'' \\ F_{Ny} - m_N g &= m_N (G_{Ny})'' \end{aligned} \tag{2.6}$$

With O_N as the origin, the torque equation of Link N is :

$$(J_N + m_N l_N^2)\ddot{\theta}_2 = m_N g l_N \sin\theta_N - f_N(\dot{\theta}_N - \dot{\theta}_{N-1}) - m_N (O_{Nx})'' l_N \cos\theta_N + m_N (O_{Ny})'' l_N \sin\theta_N \tag{2.7}$$

2.3 Differential Equations of N-link PenduBot

Based on the above analysis, with z = $[\,\theta_1\ \theta_2\ \theta_3 \cdots \theta_{N-1}\ \theta_N\,]^{\mathrm{T}}$ as the variables, the nonlinear differential equations of N-link PenduBot is obtained as follows [3,4]:

$$H_1 \ddot{z} = H_2 \dot{z} + H_3 + H_0 M \tag{2.8}$$

where

$$H_0 = \begin{bmatrix} 1 \\ 0 \\ 0 \\ \vdots \\ 0 \\ 0 \end{bmatrix}, H_2 = \begin{bmatrix} -f_1 - f_2 & a_2 L_1 \dot{\theta}_2 \sin(\theta_2 - \theta_1) + f_2 & a_3 L_1 \dot{\theta}_3 \sin(\theta_3 - \theta_1) & \cdots & a_{N-1} L_1 \dot{\theta}_{N-1} \sin(\theta_{N-1} - \theta_1) & a_N L_1 \dot{\theta}_N \sin(\theta_N - \theta_1) \\ -a_2 L_1 \dot{\theta}_1 \sin(\theta_2 - \theta_1) + f_2 & -f_2 - f_3 & a_3 L_2 \dot{\theta}_3 \sin(\theta_3 - \theta_2) + f_3 & \cdots & a_{N-1} L_2 \dot{\theta}_{N-1} \sin(\theta_{N-1} - \theta_2) & a_N L_2 \dot{\theta}_N \sin(\theta_N - \theta_2) \\ -a_3 L_1 \dot{\theta}_1 \sin(\theta_3 - \theta_1) & -a_3 L_2 \dot{\theta}_2 \sin(\theta_3 - \theta_2) + f_3 & -f_3 - f_4 & \cdots & a_{N-1} L_2 \dot{\theta}_{N-1} \sin(\theta_{N-1} - \theta_2) & a_N L_3 \dot{\theta}_N \sin(\theta_N - \theta_3) \\ \vdots & \vdots & \vdots & \ddots & \vdots & \vdots \\ -a_{N-1} L_1 \dot{\theta}_1 \sin(\theta_{N-1} - \theta_1) & -a_{N-1} L_2 \dot{\theta}_2 \sin(\theta_{N-1} - \theta_2) & -a_{N-1} L_3 \dot{\theta}_3 \sin(\theta_{N-1} - \theta_3) & \cdots & -f_{N-1} - f_N & a_N L_{N-1} \dot{\theta}_N \sin(\theta_N - \theta_{N-1}) + f_N \\ -a_N L_1 \dot{\theta}_1 \sin(\theta_N - \theta_1) & -a_N L_2 \dot{\theta}_2 \sin(\theta_N - \theta_2) & -a_N L_3 \dot{\theta}_3 \sin(\theta_N - \theta_3) & \cdots & -a_N L_{N-1} \dot{\theta}_{N-1} \sin(\theta_N - \theta_{N-1}) + f_N & -f_N \end{bmatrix}$$

$$H_1 = \begin{bmatrix} b_1 & a_2 L_1 \cos(\theta_2 - \theta_1) & a_3 L_1 \cos(\theta_3 - \theta_1) & \cdots & a_{N-1} L_1 \cos(\theta_{N-1} - \theta_1) & a_N L_1 \cos(\theta_N - \theta_1) \\ a_2 L_1 \cos(\theta_2 - \theta_1) & b_2 & a_3 L_2 \cos(\theta_3 - \theta_2) & \cdots & a_{N-1} L_2 \cos(\theta_{N-1} - \theta_2) & a_N L_2 \cos(\theta_N - \theta_2) \\ a_3 L_1 \cos(\theta_3 - \theta_1) & a_3 L_2 \cos(\theta_3 - \theta_2) & b_3 & \cdots & a_{N-1} L_3 \cos(\theta_{N-1} - \theta_3) & a_N L_3 \cos(\theta_N - \theta_3) \\ \vdots & \vdots & \vdots & \ddots & \vdots & \vdots \\ a_{N-1} L_1 \cos(\theta_{N-1} - \theta_1) & a_{N-1} L_2 \cos(\theta_{N-1} - \theta_2) & a_{N-1} L_3 \cos(\theta_{N-1} - \theta_3) & \cdots & b_{N-1} & a_N L_{N-1} \cos(\theta_N - \theta_{N-1}) \\ a_N L_1 \cos(\theta_N - \theta_1) & a_N L_2 \cos(\theta_N - \theta_2) & a_N L_3 \cos(\theta_N - \theta_3) & \cdots & a_N L_{N-1} \cos(\theta_N - \theta_{N-1}) & b_N \end{bmatrix}, H_3 = \begin{bmatrix} a_1 g \sin\theta_1 \\ a_2 g \sin\theta_2 \\ a_3 g \sin\theta_3 \\ \vdots \\ a_{N-1} g \sin\theta_{N-1} \\ a_N g \sin\theta_N \end{bmatrix}$$

$$a_i = m_i l_i + \sum_{j=i+1}^{N} m_j L_i, \quad b_i = J_i + m_i l_i^2 + \sum_{j=i+1}^{N} m_j L_i^2, \quad i = 1, 2, \cdots, N$$

In the straight-arm state, the N-link PenduBot can be linearized in a small neighborhood of the up-right position where θ_i is zero degree. So, $\sin\theta_i$ and $\cos\theta_i$ can approximate to 0 and 1 respectively. With X = $[z \quad \dot{z}]$ as the state vector, the linear mathematical equation is simplified as follows:

$$\dot{X} = \begin{bmatrix} 0 & I_N \\ \bar{H}_1^{-1}\bar{H}_2 & \bar{H}_1^{-1}\bar{H}_3 \end{bmatrix} X + \begin{bmatrix} 0 \\ \bar{H}_1^{-1}\bar{H}_0 \end{bmatrix} M \tag{2.9}$$

where: I_N is an N-order unit matrix

$$\bar{H}_0 = \begin{bmatrix} 1 \\ 0 \\ 0 \\ \vdots \\ 0 \\ 0 \end{bmatrix}, \bar{H}_1 = \begin{bmatrix} b_1 & a_2 L_1 & a_3 L_1 & \cdots & a_{N-1} L_1 & a_N L_1 \\ a_2 L_1 & b_2 & a_3 L_2 & \cdots & a_{N-1} L_2 & a_N L_2 \\ a_3 L_1 & a_3 L_2 & b_3 & \cdots & a_{N-1} L_3 & a_N L_3 \\ \vdots & \vdots & \vdots & \ddots & \vdots & \vdots \\ a_{N-1} L_1 & a_{N-1} L_2 & a_{N-1} L_3 & \cdots & b_{N-1} & a_N L_{N-1} \\ a_N L_1 & a_N L_2 & a_N L_3 & \cdots & a_N L_{N-1} & b_N \end{bmatrix},$$

$$\bar{H}_2 = \begin{bmatrix} -(f_1 + f_2) & f_2 & 0 & \cdots & 0 & 0 \\ f_2 & -(f_2 + f_3) & f_3 & \cdots & 0 & 0 \\ 0 & f_3 & -(f_3 + f_4) & \cdots & 0 & 0 \\ \vdots & \vdots & \vdots & \ddots & \vdots & \vdots \\ 0 & 0 & 0 & \cdots & -f_{N-1} - f_N & f_N \\ 0 & 0 & 0 & \cdots & f_N & -f_N \end{bmatrix}, \bar{H}_3 = \begin{bmatrix} a_1 g & 0 & 0 & \cdots & 0 & 0 \\ 0 & a_2 g & 0 & \cdots & 0 & 0 \\ 0 & 0 & a_3 g & \cdots & 0 & 0 \\ \vdots & \vdots & \vdots & \ddots & \vdots & \vdots \\ 0 & 0 & 0 & \cdots & a_{N-1} g & 0 \\ 0 & 0 & 0 & \cdots & 0 & a_N g \end{bmatrix}$$

2.4 Generalized Model of N-link PenduBot

The generalized model is a model of the entire system, including the N-link mechanical system, the torque motor, the sensors, and the connecting mechanism. Instead of the torque acting on the Link 1(the driving arm), the system input is the input voltage of the torque motor. It is necessary to use the generalized model in the actual control of the N-link PenduBot. To establish the generalized model of the N-link PenduBot, it is critical to find the relation between the motor input voltage U_a and the motor output torque M.

It is assumed only to consider the role of CEMF (Counter-Electromotive Force) and ignore the impact of the inductance of the motor. The main symbols are defined in Table 2.

Table 2. Defination of symbols

Symbol	*Meaning*
U_a	Armature voltage
U_0	Dead voltage
T_{em}	Electromechanical torque
T_d	Motor stall torque
ω	Motor rotational angular velocity
J_0	Inertia of the motor rotator
J_C	Inertia of the motor connecting mechanism

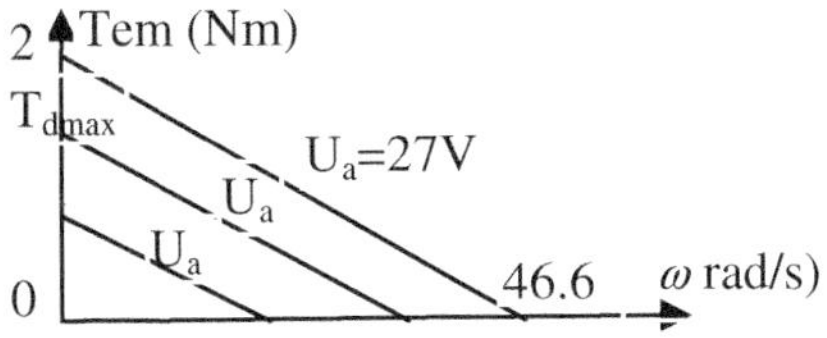

Fig. 5. The mechanical characteristic curve for DC torque motor

As is shown in Fig. 5, the mechanical characteristic curve for DC torque motor is a series of downward-sloping parallel lines. Each line can be expressed as:

$$T_{em} = T_d - K_\omega \omega = K_u U_a - K_\omega \omega \tag{2.10}$$

Subtracted from the accelerating torque of the motor rotor and the connector part, the motor output torque equation is obtained as follows:

$$M = T_{em} - (J_0 + J_c)\ddot{\theta}_1 = K_u U - K_\omega \dot{\theta}_1 - (J_0 - J_c)\ddot{\theta} \tag{2.11}$$

Substituted (2.11) into the linear model (2.9). With X = $[z \quad \dot{z}]$ as the state vector, the generalized model of the N-link PenduBot is derived as follows:

$$\dot{X} = \begin{bmatrix} 0 & I_N \\ \bar{H}_1^{'-1}\bar{H}_2 & \bar{H}_1^{'-1}\bar{H}_3 \end{bmatrix} X + \begin{bmatrix} 0 \\ \bar{H}_1^{'-1}\bar{H}_0 \end{bmatrix} U_a \tag{2. 12}$$

where: I_N is an N-order unit matrix

$$\bar{H}_0' = \begin{bmatrix} K_u \\ 0 \\ 0 \\ \vdots \\ 0 \\ 0 \end{bmatrix}, \bar{H}_2' = \begin{bmatrix} -(f_1+f_2)-K_\omega & f_2 & 0 & \cdots & 0 & 0 \\ f_2 & -(f_2+f_3) & f_3 & \cdots & 0 & 0 \\ 0 & f_3 & -(f_3+f_4) & \cdots & 0 & 0 \\ \vdots & \vdots & \vdots & \ddots & \vdots & \vdots \\ 0 & 0 & 0 & \cdots & -f_{N-1}-f_N & f_N \\ 0 & 0 & 0 & \cdots & f_N & -f_N \end{bmatrix}$$

$$\bar{H}_1' = \begin{bmatrix} b_1+J_0+J_c & a_2L_1 & a_3L_1 & \cdots & a_{N-1}L_1 & a_NL_1 \\ a_2L_1 & b_2 & a_3L_2 & \cdots & a_{N-1}L_2 & a_NL_2 \\ a_3L_1 & a_3L_2 & b_3 & \cdots & a_{N-1}L_3 & a_NL_3 \\ \vdots & \vdots & \vdots & \ddots & \vdots & \vdots \\ a_{N-1}L_1 & a_{N-1}L_2 & a_{N-1}L_3 & \cdots & b_{N-1} & a_NL_{N-1} \\ a_NL_1 & a_NL_2 & a_NL_3 & \cdots & a_NL_{N-1} & b_N \end{bmatrix}, \bar{H}_3' = \begin{bmatrix} a_1g & 0 & 0 & \cdots & 0 & 0 \\ 0 & a_2g & 0 & \cdots & 0 & 0 \\ 0 & 0 & a_3g & \cdots & 0 & 0 \\ \vdots & \vdots & \vdots & \ddots & \vdots & \vdots \\ 0 & 0 & 0 & \cdots & a_{N-1}g & 0 \\ 0 & 0 & 0 & \cdots & 0 & a_Ng \end{bmatrix},$$

Thus, the N-link PenduBot mathematical models, including the nonlinear model, the linear model and the generalized model, are established. An arbitrary links PenduBot model can be derived from them easily.

3 Analysis of the PenduBot Natural Characteristics

After the establishment of N-link PenduBot model, varieties of high-order pendulum model can be obtained. But what is the common and different feature among different links PenduBot? How to compare the control difficulty between them?

When analyzing the control system, the controllability matrix of the system is defined as $Q_c = \begin{bmatrix} B & AB & A^2B & \cdots & A^{n-1}B \end{bmatrix}$. Whether Q_c is full rank is a standard to judge the controllability of the system. To further discuss the degree of the system controllability, it is necessary to study on the characteristic roots of Q_c.

As a rule, the closer the characteristic root to the origin, that is, the smaller the norm of the characteristic root, the worse the controllability of the system. So, the norm of the characteristic root can be used as the indicator to judge the system controllability. Specifically, for a controllability matrix, as long as all its characteristic roots are calculated, choose one with the smallest norm as the measure of the degree of the system controllability.

Additionally, the controllability matrix of the multivariable-system can have multiple characteristic roots. The greater the gap between the norm of the characteristic roots, the more difficult the system controls. Among the characteristic roots of the controllability matrix, the biggest and the smallest characteristic root have the greatest impact on the controllability of the system. The difference of their norms represents the difference of the system's regulation speed. As is known to all, the singular values of a matrix can be used to describe the characteristic roots. Therefore, the condition number of the controllability matrix, that is, the ratio of the maximum singular value and the minimum singular value, is chosen to judge the degree of the system controllability [5] or the difficulty to control the system. The definition of condition number is:

$$Cond(Q_c) = \overline{\sigma}(Q_c) / \underline{\sigma}(Q_c)$$

Substitute corresponding numeric values of PenduBot parameters in Table 1. Thus, the controllability matrix is obtained. Calculate the characteristic roots and the condition number and fill into Table 3.

From the analysis of Table 3, some rules are obtained:

- Pole counts = differential equation counts = double of the PenduBot link counts.
- PenduBot link counts = naturally unstable link counts (The center of gravity is above the pivot) = positive real pole counts = negative real pole counts.
- Exist pairs of positive and negative real poles with a symmetrical trend. The number of the pairs is equal to the PenduBot link counts.
- The maximum positive real pole is increasing with the PenduBot link counts. The condition number of the controllability matrix is increasing dramatically with the PenduBot link counts.
- The naturally unstable link count is equal to the indirect-driven link counts.

Table 3. Natural Characteristics of Straight-Arm PenduBot

Links of Straight-arm PenduBot	*Pole counts*	*Indirect-driven link counts*	*Unstable link counts*	*Positive real pole counts*	*Positive real poles*	*Negative real poles*	*Cond*(Q_C)
2	4	1	2	2	5.8131	-12.353	4.883*
					4.0461	-4.8957	10^4
3	6	2	3	3	8.1679	-18.3721	4.4342*
					5.7206	-7.4058	10^7
					3.6824	-4.1479	
4	8	3	4	4	10.9085	-20.7250	5.0185*
					7.1669	-15.0555	10^{10}
					3.3682	-7.6565	
					3.9640	-3.6212	
5	10	4	5	5	13.7602	-22.7386	9.5233*
					10.5858	-13.5741	10^{13}
					7.7780	-9.2867	
					5.5015	-5.9676	
					3.1318	-3.2289	

4 Conclusion

Based on the N-link PenduBot model, an arbitrary links PenduBot model can be established more easily. Increasing with the count of PenduBot links, the control difficulty is growing dramatically. After the realization of the three-link PenduBot stabilization control, the four-link straight-arm PenduBot control problem will be the subject of challenge in future. Only a strong control measure can make it stable.

References

1. Yuan, S.Q., Kang, X.W., Miao, M.C., Li, X.S.: Design and Analysis of Three Links Rotational Inverted Pendulum. In: 2006 International Conference on Information & Control Technology, Shenzhen, pp. 362–365 (2006)
2. Yuan, S.Q., Wang, D., Li, X.S.: Research on Control Problem of PenduBot Based on PSO Algorithm. In: 2009 International Conference on Computational Intelligence and Natural Computing, pp. 346–349. IEEE Press, Wuhan (2009)
3. Li, H.X., Wang, J.Y.: Modeling of n-order inverted pendulum. J. Fuzzy Systems and Mathematics. 16, 251–257 (2002)
4. Sun, Y.E., Wang, Y.: Analyses of the Multiple Rotational Inverted Pendulum. In: The 5th World Congress on intelligent Control and Automation, pp. 814–818. IEEE Press, Hangzhou (2004)
5. Wang, D.J., Cong, S., Qin, Z.Q.: Survey of Research on Inverted Pendulum Control System. J. Control Engineering of China 10, 9–13 (2003)

Study on PID Neural Network Decoupling Control of Pneumatic Membrane Structure Inflation System

Qiu-shuang Liu[1,*], Xiao-li Xu[1,2], and Yong-feng Chen[3]

[1] School of Mechanical Engineering, Beijing Institute of Technology
100081 Beijing, China
qiuemail@sina.com
[2] Key Laboratory of Modern Measurement & Control Technology (Ministry of Education), Beijing Information Science & Technology University
100192 Beijing, China
xuxiaoli@bistu.edu.cn
[3] China TransInfo Technology Corp.
100191 Beijing, China
xd_boy163@hotmail.com

Abstract. In order to solve the strong coupling problem existing between the frequency converter-fan-pressure difference loop and the return air damper-CO_2 content loop in the pneumatic membrane structure inflation system, this paper studies PID neural network decoupling control algorithm based on neural network theory and PID theory, establishes double-variable dual-output PID neural network decoupling control system model. The application results show that the PID neural network decoupling control algorithm is effective on decoupling of two loops of pneumatic membrane structure inflation system, gets better control effect, and improves the system real-time control.

Keywords: PID; neural network; pneumatic membrane structure.

1 Introduction

Today, pneumatic membrane structure has been applied in all types of building structures: stadiums, recreational centers, exhibition centers. Large-span pneumatic membrane buildings take air-supported membrane structure as the main body, and adopt high-strength flexible membrane material. Its principle is as follows: fix the membrane material on the periphery of ground base structure, use the inflation system to raise the indoor air pressure to a certain level, use the pressure difference between inside and outside roof to resist external forces; because it is supported by using air pressure, no beam or column is needed, leading to greater clear building space.

In this paper, PID neural network decoupling controller is used to make a series of improvements on the conventional PID controller, make the controller use neuron's self-learning ability, comply with some optimal indicators, and automatically adjust the PID controller parameters, thus solving the strong coupling issue of the air pressure difference and CO_2 content in air in the inflation loop.

---

* Corresponding author.

L. Qi (Ed.): ISIA 2010, CCIS 86, pp. 704–710, 2011.

2 PID Neural Network Control Model

PID neural network is constituted through incorporating PID control law into the neural network, and is a multi-layer neural network composed of proportional (P), integral (I), differential (D) neurons[1], [2]. The number, connection code, connection weight value of neurons at each layer are determined according to PID control law basic principle and established experience, so as to ensure system stability and fast convergence. As shown in Fig. 1, PID neural network control model consists of input layer, hidden layer and output layer. Input layer has two neurons, each receiving rated quantity r and regulated quantity y; hidden layer has three neurons, and it input and output functions are proportional (P), integral (I), differential (D) functions respectively; the output layer has a neuron designed to export the controlled quantity required by the object [3], [4].

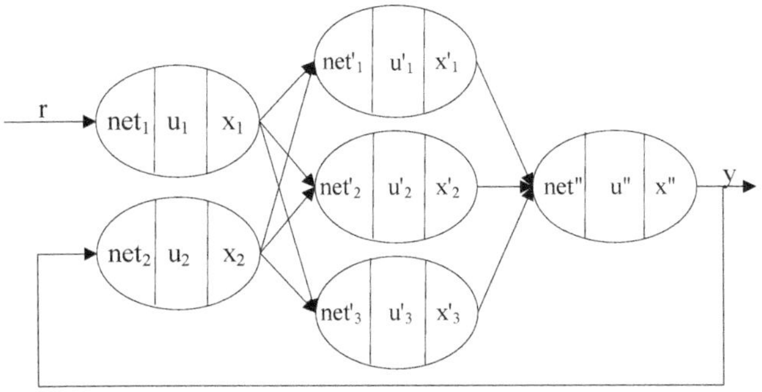

Fig. 1. Single-output neuron network basic structure form

1) Input layer

The input layer has two same neurons, the neurons input is as follows:

$$\begin{cases} net_1(k) = r(k) \\ net_2(k) = y(k) \end{cases}. \tag{1}$$

The state of neuron:

$$u_i(k) = net_i(k). \tag{2}$$

The output of neuron at input layer

$$u_i(k) = \begin{cases} 1, u_j(k) > 1 \\ u_i(k), -1 \le u_j(k) \le 1 \\ -1, u_j(k) < -1 \end{cases}. \tag{3}$$

In (1) ~ (3), r(k) is the given value for the system; y(k) is the system controlled quantity; i is the number of subnet input layer (i = 1,2); k is the sampling time .

2) Hidden layer

PID neural network hidden layer consists of three neurons: one proportional element, one integral element and one differential element respectively. Their total input value calculation formulas are the same, that is,

$$net'_j(k)=\sum_{i=1}^{2} w_{ij}x_i(k). \tag{4}$$

There are three kinds of state functions for neurons at the hidden layer, namely:
The state of proportional element:

$$u'_1(k)=\begin{cases}1, u'_1(k)>1\\ net'_1(k), -1\le u'_1(k)\le 1\,\mathrm{x+y=z}\\ -1, u'_1(k)<-1\end{cases}. \tag{5}$$

The state of integral element

$$u'_2(k)=\begin{cases}1, u'_2(k)>1\\ u'_2(k-1)+net'_2(k), -1\le u'_2(k)\le 1\,\mathrm{x+y=z}\\ -1, u'_2(k)<-1\end{cases}. \tag{6}$$

The state of differential element:

$$u'_3(k)=\begin{cases}1, u'_3(k)>1\\ u'_3(k)-net'_3(k-1), -1\le u'_3(k)\le 1\\ -1, u'3(k)<-1\end{cases}. \tag{7}$$

There are three kinds of neurons input and output functions at hidden layer:
The output of proportional element:

$$x'_1(k)=u'_1(k). \tag{8}$$

The output of integral element:

$$x'_2(k)=x'_2(k-1)+u'_2(k). \tag{9}$$

The output of differential element:

$$x'_3(k)=u'_3(k)-u'_3(k-1). \tag{10}$$

In (4) ~ (10), j is the number of neurons at hidden layer in the subnet (j = 1,2,3); w_{ij} is connection weight value from the input layer to hidden layer in the subnet; $x_i(k)$ is the output value of neurons at input layer in the subnet; the variable marked with "′ "represents the variable at hidden layer.

3) Output layer

The output layer of PID neural network has one neuron, and forms one output volume. The input of neuron at output layer is the weighted sum of output values of all neurons at hidden layer.

$$u''_h(k) = \sum_{j=1}^{3} w'_{jh} x'_j(k) . \tag{11}$$

The output of neurons at output layer:

$$x''_h(k) = u''_h(k) . \tag{12}$$

These output values are the control input of the object $y_h(k)$. In which, h= 1 is the number of neurons at the output layer, w'_{jh} is the connection weight value from the hidden layer to output layer. The variable marked with “"”represents the variable of output layer.

3 PID Neural Network Control of Membrane Structure Inflation System

3.1 Control of Membrane Structure Inflation System

VAV inflation system is a real-time intelligent control system which changes the air volume to adapt to the parameter changes such as indoor and outdoor air pressure, indoor temperature, air quality CO_2 content, etc. VAV inflation system adjusts the air volume according to real-time dynamics, thus reduces fan transmission power consumption, and receives a significant energy-saving effect.

The inflation system can be divided into two parts: fresh air supply system, and the end section. Fresh air supply system mainly consists of frequency converter, fan and return air damper. The end section is the air inlet and return air inlet of membrane structure. Two fans blow the return air mixed with fresh air in mixed air box into inside the buildings. The outlet of fan is installed with airflow check valve, and the end of return air inlet is installed with control electric valve. Indoor air is transmitted back to the mixed air box under the negative pressure, after hot and cold treatment, transmitted into indoors through fan pressurization, proceed with such circulation. In order to simplify the analyzed problem, we should establish the material and energy ties between the two parts through bridge-duct in the middle, consider the coupling effect of two loops: frequency converter - fan – air pressure difference loop, and the return air damper-CO_2 content loop, and use PID neural network to make decoupling control of the two loops [5].

3.2 Control Structure of Membrane Structure Inflation System

the model of double-variable PID neural network decoupling control system of pneumatic membrane structure is shown in Fig. 2. PID neural network decoupling control system adjusts network output based on the set value and the system’s actual output value. In the figure, the system setting values are $r1$ and $r2$, respectively indicating the air pressure difference setting value (P) and the CO_2 content setting value (W), $v1$ and $v2$ stand for the output values of PID neural network, used to control the frequency converter and return air damper, $y1$ and $y2$ stand for the system output values, respectively the system's actual pressure difference and CO_2 content [6].

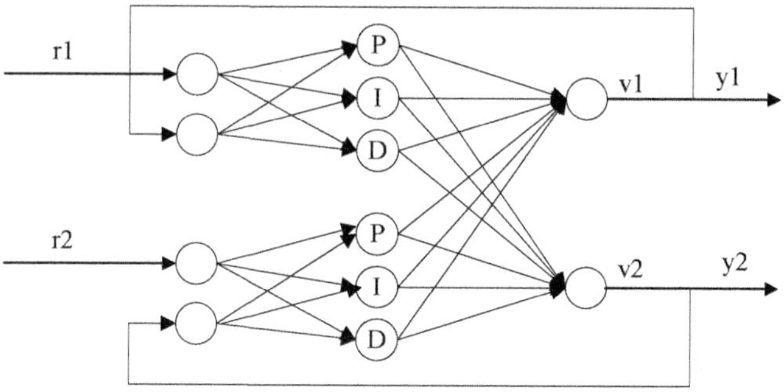

Fig. 2. Double-variable PID neural network decoupling control system of pneumatic membrane structure

4 Actual Operating Results and Analysis

Under the condition of stable operation of inflation system, keep other conditions unchanged, respectively change the air pressure difference and the CO_2 content setting value, and then measure the air pressure difference and CO_2 content in order to verify the decoupling control effect of the trained PID neural network decoupling control system.

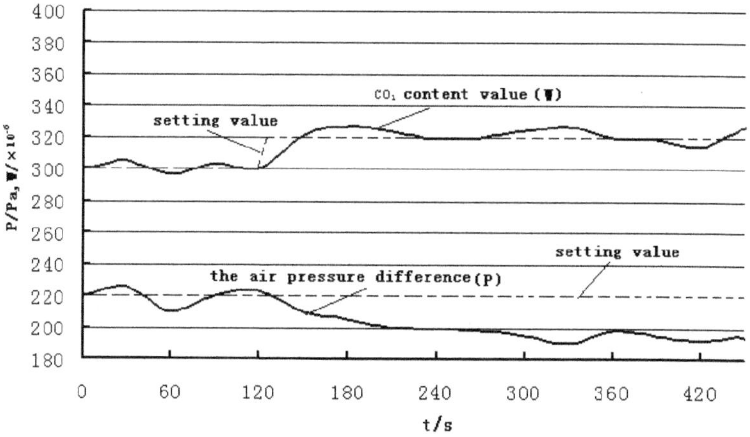

Fig. 3. The control result of traditional PID controller

The control result of traditional PID controller is shown in Fig. 3. At 120s, CO_2 content setting value is changed from 300×10^{-6} to 320×10^{-6}. As the CO_2 content setting value is large, the opening of return air damper gets smaller so as to reduce the return air volume, the air pressure difference can not maintain the former stable value, but get smaller and smaller as the return air damper gets increasingly small, and finally deviates from the air pressure difference setting value, down below 200Pa from the original 220Pa.

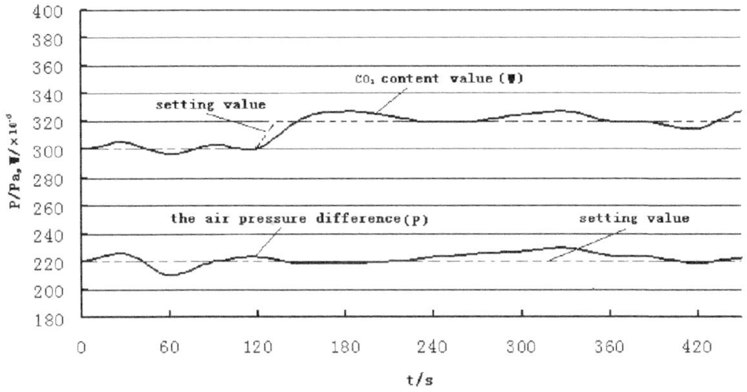

Fig. 4. The result of PID neural network decoupling control

The result of PID neural network decoupling control is shown in Fig. 4. At 120s, CO_2 content setting value is changed from 300×10^{-6} to 320×10^{-6}. As the CO_2 content setting value gets smaller, the opening of return air damper gets smaller in order to reduce the return air volume. PID neural network adjusts the frequency of frequency converter according to the opening of return air valve, so as to control the fan speed, and to keep the air pressure difference. The air pressure difference maintains at 220Pa, the change in the opening of return air damper has no impact on air pressure difference; after 120s, the CO_2 content also reaches the setting value 320×10^{-6}, and the whole system keeps stable. The above results show that the addition of PID neural network decoupling control can eliminate the impact of return air damper-CO_2 content loop on the frequency converter - fan – air pressure difference loop.

5 Conclusion

The controller designed by this paper incorporates PID control law into the neuron network, thus has the advantages of neuron network and PID control, overcomes the traditional control methods and the shortcomings of general neuron network, mainly represented as: (1) applicable to large-lag, nonlinear, time-varying system; (2) the structure is based on the requirements of PID control law, relatively simple and standard; and (3) the PID neural network decoupling control can effectively eliminate the impact of the opening of return air damper on air pressure difference, thus leading to stable operation of converter-fan-pressure difference loop and the return air damper- CO_2 content loop in the VAV inflation system; it has a good decoupling control effect, and offers effective method for decoupling control of variable-air-volume inflation system.

Acknowledgment

The author wishes to thank Key Project of Science and Technique Development Plan Supported by Beijing Municipal Commission of Education "KZ200910772001", Funding Project for Academic Human Resources Develoment in Institutions of Higher Learning Under the Jurisdiction of Beijing Municipalipality "PHR20090518" Open Project Supported by Beijing Key Laboratory on Measurement and Control of Mechanical and Electrical System "KF20091123206, KF2009112302 KF20101123204".

References

1. Zaheer-uddin, M., Tudoroiu, N.: Neuro-PID Tracking Control of a Discharge Air Temperature System. Energy Conversion & Management 45, 2405–2415 (2004)
2. Chen, J.H., Huang, T.H.: Applying Neural Networks to on-line update PID controllers for Nonlinear Process Control. Journal of Process Control 14, 211 (2004)
3. Guo, B.J., Yu, J.S.: A Single-neuron PID Adaptive Multicontroller Scheme Based on RBFNN. Transactions of the Institute of Measurement & Control 27, 243–259 (2005)
4. Dong, W.J., Liu, C.H., Song, H.: Application Contrast on Servo Electromotor Model between RBF and PIDNN. Control Engineering of China 15, 113–115, 118 (2008)
5. Wang, H.L., Huang, J., Zi, B.: Design for Temperature Controller Using PIDNN Based on DSP. Electric Transmission 36, 40–43 (2006)
6. Ding, X.G., Liu, G.J.: Study on Identification Parameters of Wastewater Treatment System Based on PIDNN. Computer Technology and Development 18(5), 200–202 (2008)

An Improved Reversible Watermarking Algorithm Based on Random Sequence

Jun Tang

Department of Information Engineering, Hunan Urban Construction College
411101 Xiangtan, China
xttangjun@163.com

Abstract. We proposed an improved digital watermarking algorithm based on random sequence into reversible watermarking algorithm. Therefore, this method will use the robust watermarking algorithm of the well-known random sequence as embedding approach. Sobel edge detection algorithm is employed to extract the pixel value of edges from the watermarked image. And the final watermarked image is produced by replace the original image corresponding to the pixel value of the edges for the purpose of embedding watermark. Because the robust watermarking algorithm can tolerate the image which is destroyed to protect the copyright, there is no watermarking examination problem although this approach causes some loss of watermark information according to the experiment result. Moreover, it can not only examine whether the image has embedded watermarks, but also restore the original image.

Keywords: digital watermarking, random sequence, reversible, sobel edge detection.

1 Introduction

Based on tamper resistance and user needs, the current research of watermarking algorithms can be divided into three areas. The first is focused on robust watermarking [1]. This kind of algorithms can better resist malicious attacks such as rotation, cutting, compression, blurring, and sharpening. After extraction and verification of watermarks, ownership of copyright can be immediately identified. The second type is called fragile watermarking [2], which is mainly characterized by the high fragility of watermarks. Any slight tampering of image pixels can result in a serious damage of embedded watermarks. Therefore, whether the embedded watermarks can be successfully extracted indicates the integrity of a watermarked image. The third is semi-fragile watermarking [3]. It can detect malicious tampering, locate tampered regions, and further extract watermarks from intact regions. Based on the extracted watermarks, which regions of the image have been tampered can be effectively identified.

In Chapter 2, some related works on image reversible algorithm. Chapter 3 presents a reversible watermarking algorithm based on random sequence. Finally, some conclusions and future works of this thesis will be presented in Chapter 4.

L. Qi (Ed.): ISIA 2010, CCIS 86, pp. 711–717, 2011.

2 Reversible Watermarking Algorithm

Digital watermark has been widely used to protect the ownership of digital media. However, the quality of watermarked image becomes worse when the watermark is embedded into the image. Therefore, many specialists and scholars advance reversible watermarking algorithm in an effort to alleviate the problem. With the help of the reversible watermarking algorithm, watermark can be embedded into a variety of multiple media such as digital audio, digital image and digital video. In decode stage, it not only extracts the watermark but also can restore the original image from the watermarked image.To ensure that the reproduction of your illustrations is of a reasonable quality, we advise against the use of shading. The contrast should be as pronounced as possible.

Fridrich et al. [4] applies lossless compression algorithm and least significant bit (LSB) replacement to develop several reversible schemes. Celik et al.'s scheme [5] provided a solution to the defects of Fridrich et al.'s scheme and ensured a higher embedding capacity and image quality.

Celik et al. also used this concept to propose a data hiding scheme and image authentication scheme [6]. Hu and Jeon [7] proposed a reversible visible watermarking scheme to satisfy a new application scenario where the visible watermark serves as a ownership identifier, but it can be completely removed to resume the original image data. Xuan et al. [8] embeds the watermark by modifying the high and middle frequency wavelet coefficients and applies similar concept to compress these coefficients. Leest et al. [9] also proposed a similar idea to embed watermark into digital images.

Tian [10] developed a reversible watermarking scheme that employs the integer transformation and difference expansion to generate values to embed the bits of watermark. Alattar [11] proposed another integer transformation scheme to improve Tian's scheme. This scheme improves the embedding capacity and does not require as much computation efforts as Tian's scheme and applies it to color image. Kuribayashi et al. [12] also proposed a similar reversible watermarking scheme in this area. Thodi and Rodriguez [13] proposed a histogram shifting algorithm to embed the location map. This scheme improves the distortion performance at low embedding capacities and mitigates the capacity control problem. Coltuc and Chassery [14] developed a simple integer transform scheme called reversible contrast mapping (RCM) that applies to pairs of pixels.

Vleeschouwer et al. [15] developed an improved version of circular reversible watermarking scheme by using the bijective transformations. This scheme shifts two bin positions at most and subsequently avoids the serious distortion. Yang et al. [16] also proposed histogram expansion scheme to increase the embedding capacities. This scheme embeds data by modifying those integer discrete cosine transform (DCT) coefficients with peak bin in each coefficient histogram. Ni et al. [17] proposed a reversible data hiding scheme that utilizes the zero or the minimum bins of the histogram of an image and slightly modifies the pixel values to embed data into the image.

It provides a higher quality of the watermarked image generated by this type. The image quality is the highest of all reversible watermarking algorithms literature, in addition the computational complexity is low and the execution time is short. However, in these schemes additional information is needed to extract the watermark and to restore the original image. The embedding capacities in this type are fewer but the robustness is the best. Despite of the low embedding capacities, its robustness is good enough to meet the requirements of the reversible watermarking algorithm.

3 An Improved Reversible Watermarking Algorithm

The algorithm we use here is based on the process using random sequence as the watermark which is embedded into the frequency domain of an image [18]. Usually, a key is used to produce random sequence as watermark. According to the embedding approach, the watermark is embedded by modifying frequency domain coefficient, and then the watermarked image is generated through the inverse transform.

The pixel value of the original image is transformed into frequency domain coefficient such as DCT or DWT. Then, the rule between the original image and watermark is tried to find out according to the hiding method that proposed by the expert. The frequency domain coefficient that are selected on the original image are modified to replace the original coefficient so that watermarked image can be derived from the inverse transform.

The watermark that is generated by the default testing key is calculated repeatedly. The frequency domain coefficient of detection image are selected through the rules set by the embedding approach, and selected frequency coefficient and watermark are calculated to obtain the individual similarity values, which form a similarity diagram. The similarity diagram is used for proving the existence and uniqueness of watermark. Different keys can get a different similarity values respectively. Among them, the similarity value of the correct key will be displayed on the top spot, and the rest of the similarity value will be near 0. If the highest similarity value SV exceeds the threshold TH, and then there is embedded watermark in the detection image. The interested reader is encouraged to get the details from.

According to the watermark algorithm mentioned above, this chapter combines it with edge detection to design a reversible watermarking algorithm. This section is divided into two parts to illustrate the process of watermark embedding and examining, and restoring original images.

The steps of watermark embedding process are as follows:

Step 1. Extract edge information from O and obtain a mask M_s through Sobel edge detection, and then extract the pixel value of O, which corresponds to the mask M_I, thereby forming a new diagram called E_o, in which contains only the edge pixel value of O. The rest pixel values are 0.

Step 2. Use the algorithm of hiding random sequence to get the watermarked image W, and use the same mask M_s to get the edge pixel image E_w from W.

Step 3. Subtract the corresponding pixel value of images E_o and E_w to get a different map D.

Step 4. The new watermarked image W n is produced by O(i, j)- E_o (i, j)+ E_w (i, j), where (i, j) represents the position of image pixel, so the final embedded watermarked image W_n is accomplished. Later, the restored information D and watermarked image W_n is utilized to reach the goal of restoring O.

The steps of how to examine and restore watermarked image are as follows:

Step 1. The receiver get W_n and D, and then use watermarked detection algorithm to do the watermark test of W_n and get a similarity diagram. If the highest similarity value SV in the diagram exceeds the threshold value we set, there is watermark in W_n. If not, there is no watermark in W_n.

Step 2. After verifying watermarked image, the validation of the image origin is proved, and then add the pixel value of D and W_n correspondingly, such that O is restored, namely that O(i, j)= W_n (i, j)+D(i, j), where (i, j) represents the position of image pixel.

4 Experimental Results

The experiment is conducted by the grey-level images of Lena, Baboon and Peppers, with the size of 256×256, as shown in Figure 1, Figure 3 and Figure 5. Aimed on the watermark with the size of 200×200 that is generated by the 200 different keys, this experiment utilizes the Robust Associative Watermarking Algorithm proposed by Shen and Hsu [19] to embed them into Lena, Baboon and Peppers respectively, thereby getting 200 embedded images, so the total number of the embedded images is 600. Experimental parameter cited from. First, the new image of embedded watermark is derived from these 600 images by using the approach in this chapter, and then is testified by watermark. The result from Table 1 shows that the watermark is verified correctly after being modified by our approach. There are two standards for the judgement: False-negative errors and False-positive errors. False-negative errors means there is embedded watermark in the image, but this cannot be successfully examined. False-positive errors means there's no embedded watermark in the image, but be judged mistakenly there is.

The random sequence that is generated by the 100th key is embedded into Lena, Baboon and Peppers, and the embedded image is obtained by the approach of the chapter, as shown in Figure 2, Figure 4 and Figure 6. It shows that, in the 100th position, the similarity values of the three images are the highest, other values are near 0. For D, it is compressed by arithmetic code, and 4.41 kb, 8.15 kb and 4.27 kb are obtained for Lena, Baboon, and Peppers, respectively.

Fig. 1. Grey-level images of Lena

Fig. 2. Watermarked Lena

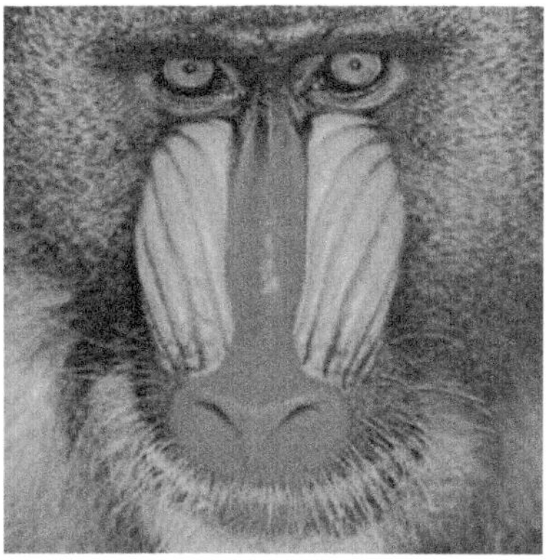

Fig. 3. Baboon

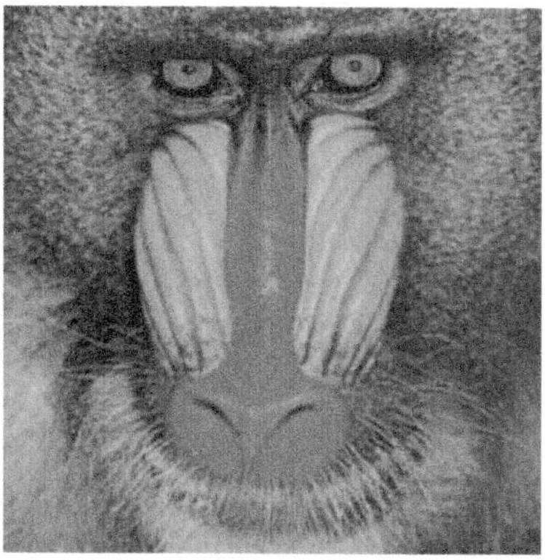

Fig. 4. Watermarked Baboon

Fig. 5. Peppers

Fig. 6. Watermarked Peppers

5 Conclusions

In this Paper, we have proposed a reversible watermarking algorithm and a image tamper detection and recovery based on vector quantization. The images are watermarked by a robust watermarking algorithm in order to prove their originality and authority, but the user needs the original images for many applications such as medical images, military images and images for the juridical which requires accurate images. Therefore, the concepts of random sequence as watermark and Sobel edge detection were utilized to propose a reversible watermarking algorithm. According to the experiment results demonstrate that our method only needs to pay less extra overheads for restoring original images. Furthermore, the proposed method can transform the robust watermarking algorithm into reversible watermarking algorithm.

References

1. Huang, H.C., Wang, F.H., Pan, J.S.: Efficient and robust watermarking algorithm with vector quantization. Electronics Letters 37(13), 826–828 (2001)
2. Celik, M.U., Sharma, G., Saber, E., Tekalp, A.M.: Hierarchical watermarking for secure image authentication with localization. IEEE Trans. Image Process. 11(6), 585–594 (2002)
3. Maeno, K., Sun, Q., Chang, S.F.: New semi-fragile images authentication watermarking algorithms using random bias and nonuniform quantization. IEEE Trans. Multimed. 8(1), 32–45 (2006)
4. Fridrich, J., Goljan, J., Du, R.: Invertible authentication. In: SPIE Proceedings of Security and Watermarking of Multimedia Content, San Jose, pp. 197–208 (2002)
5. Celik, M.U., Sharma, G., Tekalp, A.M., Saber, E.: Lossless generalized-lsb data embedding. IEEE Transactions on Image Processing 14(2), 253–266 (2005)
6. Celik, M.U., Sharma, G., Tekalp, A.M., Saber, E.: Reversible data hiding. In: Proceedings of the International Conference on Image Processing, NY, USA, pp. 157–160 (2002)
7. Hu, Y., Jeon, B.: Reversible visible watermarking and lossless recovery of original images. IEEE Transactions on Circuits and Systems for Video Technology 16(11), 1423–1429 (2006)
8. Xuan, G., Yang, C., Zhen, Y., Shi, Y.Q., Ni, Z.: Reversible data hiding based on wavelet spread spectrum. In: Proceedings of the IEEE 6th Workshop on Multimedia Signal Processing, Italy, pp. 211–214 (2004)
9. Leest, A., Veen, M., Bruekers, F.: Reversible image watermarking. In: Proceedings of the ICIP International Conference on Image Processing, Barcelona, Spain, vol. 3, pp. II-731-4 (2003)
10. Tian, J.: Reversible data embedding using a difference expansion. IEEE Transactions on Circuits Systems and Video Technology 13(8), 890–896 (2003)
11. Alattar, A.M.: Reversible watermark using the difference expansion of a generalized integer transform. IEEE Transactions on Image Processing 13(8), 1147–1156 (2004)
12. Kuribayashi, M., Morii, M., Tanaka, H.: Reversible watermark with large capacity based on the prediction error expansion. IEICE Trans. Fundamentals E91-A(7), 1780–1790 (2008)
13. Thodi, D.M., Rodriguez, J.J.: Expansion embedding algorithms for reversible watermarking. IEEE Trans. Image Process. 16(3), 723–730 (2007)

14. Coltuc, D., Chassery, J.: Very fast watermarking by reversible contrast mapping. IEEE Signal Processing Letters 14(4), 255–258 (2007)
15. Vleeschouwer, C.D., Delaigle, J.F., Macq, B.: Circular interpretation of bijective transformations in lossless watermarking for media asset management. IEEE Transactions on Multimedia 5(1), 97–105 (2003)
16. Yang, B., Schmucker, M., Niu, X., Busch, C., Sun, S.: Reversible image watermarking by histogram modification for integer dct coefficients. In: Proceedings of the IEEE 6th Workshop on Multimedia Signal Processing, Siena, Italy, pp. 143–146 (2004)
17. Ni, Z., Shi, Y.Q., Ansari, N., Su, W.: Reversible data hiding. IEEE Trans. Circuits Syst. Video Technol. 16(3), 354–362 (2006)
18. Chang, C.C., Fan, Y.H., Tai, W.L.: Four-scanning attack on hierarchical digital watermarking method for image tamper detection and recovery. Pattern Recognition 41, 654–661 (2008)
19. Shen, J.J., Hsu, P.W.: A robust associative watermarking algorithm based on similarity diagrams. Pattern Recognition 40(4), 1355–1367 (2007)

Research on Achieving of VME Bus by VIC068A

Li Ji-sheng[1,2] and Liu rong[1,3]

[1] College of Physics and Information Technology, Shaanxi Normal University, Xi'an, Shaanxi, 710062

[2] College of Electrical Engineering, Xi'an Jiaotong University, Xi'an, Shaanxi, 710049

[3] School of Electronics and Information, North western Poly technical University, Xi'an, Shaanxi, 710072

Abstract. Using Cypress company's VIC068A chip, the link interface circuit of VME bus and DSP are designed. The result indicates that this circuit may better achieve communication from DSP to CPU of Motorola 68K series and solve the problem of incompatibility between DSP and CPU of Motorola 68K series.

Keywords: VME bus; VIC068A; Bus Achieve; Circuit.

1 Introduction

VME bus is a kind of computer structure, and the term VME stands for Versa Module Eurocard, which is a standard co-defined by the group made up of its three manufacturers Motorola, Mostek and Signetics. During the late 1970s, people were excessively in pursuit of using the power of CPU in the design of multiple-processor computer system, causing all difficulties. As a result, it is clearly needed to establish a bus that has nothing to do with microprocessor. VME bus, in such a condition, is developed on the basis of VERSAbus defined by Motorola for 6800 microprocessor, which is standardized by IEEE std－P1014- in 1987 (Revision C.1). The system flow chart is shown in Figure 1.

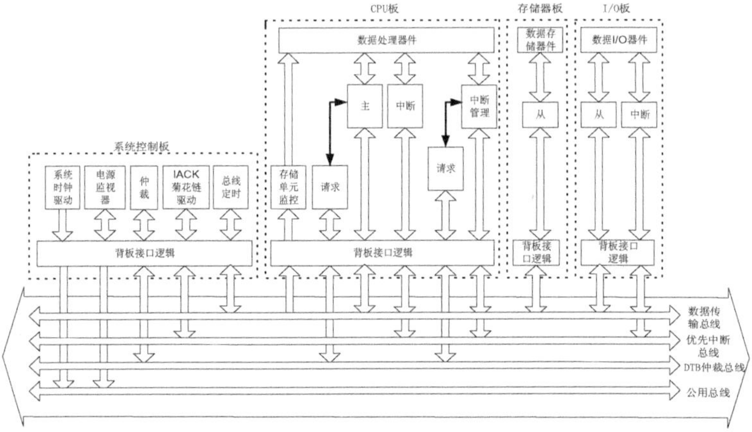

Fig. 1. The structure of VME bus

L. Qi (Ed.): ISIA 2010, CCIS 86, pp. 718–722, 2011.

For developing the application module based on VMEbus, we must consider the issues how to connect with the interface of VMEbus according to specific requirements. Generally speaking, users can choose the existing IC from professional company because of their comprehensive function. VIC068A of Cypress Company is an early introduced VME bus bridge controller, full-featured, with comprehensive main module/subordinate ordinate module functionality, which can also be used as system controller. VIC068A is compatible with VIC64 in function and pin, but it adds supports for VIC64 standard. Also, because the local bus of the interface chips in VIC068A is designed by Motorola for 68K series' processors, therefore, for other types of processors that are incompatible with the bus of 68K series' processors, it is required to complete the interface conversion of processors' bus with the used type of processors when using VIC068A [1,2,3].

2 Schematic Diagram and Pin of VIC068A

VIC068A owns full interface capability, that is, it can conduct the standard transfer and block transfer of A32、A24、D32、D16、D8 main module and subordinate module, including 5 kinds of bus releasing modes, which can lend support to write the self-test process, self-defining AM code, self-defining bus timing and other functions. Its block transfer can achieve the transfer whose length is larger than 256 bytes by virtue of appropriate external circuit. Meanwhile it owns the dual-channel characteristics. If the VIC068A is tested that SCON pin is in the state of low effect when power-on, it will be selected as the system controller to achieve the bus arbitration, bus timing, interrupt, interrupt management, IACK daisy chain driver, SYSCLK driver, and etc.

The reset methods of VIC068A are shown as follows: ①Internal reset. It is the most common reset, which is used to conduct reset for the selected registers and internal logic. ②System reset. It can conduct reset through VME backplane, that is, VIC068A can product the SYSRESET signal with the help of writing configuration registers. ③Global reset. It means to reset all configuration registers, which is often used as power-on reset.

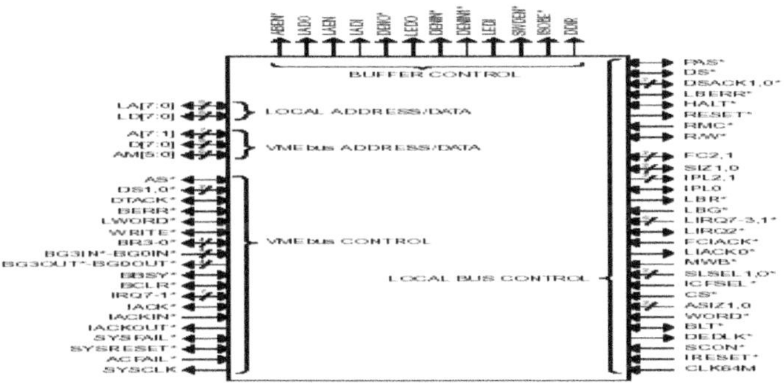

Fig. 2. VIC068A pin map

Figure 2 shows the VIC068A pin map and the following is the induction of signal pins which are related to VIC068A data transferring.

MWB, CS: MWB is the local functional module, for example DSP sends the request for VMEbus. CS means the chip selection that local bus writes VIC068A internal register. The designer can utilize different address decoding to select the MWB and CS in order to distinguish whether this operation is conducted on VME bus or VIC068A. LBR, LBG is the local bus request and enabling signal. PAS and DS, R/W: PAS is the local bus address strobe, DS is the local data strobe, and R / W means the local data direction.

DSACK1, 0: local data bit width recognition signal. Similar to DS of VMEbus, VIC068A can accept DSACK input when conducting subordinate module transfer or DMA transfer to confirm the end of this period. When carrying out the main module operation, this signal is output to confirm the end of VMEbus master transfer to the local bus. Besides, this signal can also be used together with WORD, SIZ1,0 signal to indicate the bit width of the data.

A set of buffer control signals. LADI, LAEN, LADO, ABEN are respectively input local address latch, enable, out put VMEbus address latch and enable. LEDI, DENIN, LEDO, DENO are respectively input VMEbus data latch, enable, out put VMEbus latch and enable. The other signals such as LBERR, FC2, FC1, BLT, DEDLK can be viewed from Cypress handbook.

3 VIC068A Achieves VME Bus

(1) The standard transfer operations of main module. When the local bus (local host processor) makes MWB low effective and if PAS is also effective, then it will send the request of single data or block transfer; after having received the two signals effectively, if VIC068A isn't the current VMEbus master, it will apply for VMEbus access. When satisfying: ① AS is invalid since the last cycle ② DTACK and BERR are invalid③ The reception degree of BGiIN is low ④ After the appropriate delay, the control right to start the data transfer is obtained.

(2) The standard transfer operations of subordinate module. VIC068A registers related to operations and configuration of subordinate module are shown as follows: SSOCR0(bit0-5)、SSOCR1、SS1CR0、SS1CR1、LBTR、AMSR. Signal SLSEL1 and SLSEL0 are output by off chip VMEbus address decoding circuit. When VIC068A detects an effective level in the state of low SLSELi, moreover if satisfying ①AS is effective(output by another main module). ② Dsi in current cycle is effective. ③DTACK or BERR is withdrawn, then VIC068A will check A32/A24/A16 and the transfer type. If SSiCR0 is configured and allows users to conduct subordinate module visit specified by AM codes, then the subordinate module visit will start and it will drive LBR = 0 immediately to apply for local bus. If the configuration register of VIC068A doesn't allow the specific visit (directed by the AM codes), the VMEbus request will be ignored and the LBR will not be produced.

(3) Block transfer operation. The main block transfer of VIC068A has two ways: MOVEM block transfer and local DMA block transfer. MOVEM means the local processor drives the data transfer and posses the local bus control right; while in

DMA way, VIC068A is the controller of local bus and it can use the data of local resources visited by DMA.
(4) Interrupt handling function of VIC068A. VIC068A provides a comprehensive capability of generating and managing VMEbus interruption and local interruption, besides, the capability of status or error interruption is also included.

Interrupter module is responsible for producing VMEbus interrupt signal IRQ [7..0], and we control the generation and revocation of interrupt by setting register VIRSR, while VIVBR1-7 sets the status/id code of each interrupt. The local processor needs relevant register to inform the interrupt panel of generating corresponding level of interrupt. When the interrupt handler module which is in charge of the current interrupt starts the interrupt recognized cycle, the interrupt module is also responsible for putting the status/id code of this interrupt on D[7..0].

4 Circuit Implementation

Figure 3 shows the connection diagram between floating-point SHARC DSP (such as ADSP2106x) and VIC068A from AD Company. When VIC068A is used to achieve the connection between the bus of other types of processors and VMEbus interface, it is generally needed to add the logic conversion of circuit implementation between the control line of VIC068A and the local processors [4], where we use CPLD device such as MAX7256S (PQFP208) from Altera Company as the interface circuit. Notice that the sequence order of SHARC's data bus byte uses little endian (Intel data format), while the sequence order of VMEbus data is big endian, same with Motorola68k, therefore, the data bus requires cross-linked.

Same with VME bus, VIC068A is asynchronous operation, that is, when the main module is in the process of transfer, it needs relevant subordinate ordinate module to send responsive signals to complete the data transfer. Address strobe, data strobe and responsive signals have no specific sequential relationship with clock. However, SHARC bus is synchronous working, that is, ADSP2106x directly drives the read-write signals, accomplishing transfer within specified period [5]. Therefore, the interface circuit needs to finish synchronous/ asynchronous conversion. In addition to that, it also requires to generate the external control logic in block transfer and to respond to VIC068A interrupt management function, converting its request into ADSP interrupt input as well.

The access addresses of ADSP for VIC068A internal registers and VME bus are different [6], so the HI address decoder is used to drive VIC068A's pins CS, MWB and PAS, and RD, WR are used to produce R / W and DS. When the VME bus accesses to local bus (VIC068A subordinate module read-write operation), ADSP2106x is connected to VIC068A using host interface. This experiment has applied asynchronous invocation DSP.

VIC068A is a comprehensive bus interface controller with package ways of 144,160-pin TQFP, 145-pin PGA and etc. Cypress also offers a VMEbus subordinate interface controller with a relatively low cost. If what the designer wants is only to accept the visit of VMEbus, then CY7C960 /CY7C961maybe a good choice. The VME electrical specification not only requires large signal line current but also needs the coordination of external logic circuit and driving circuit (such as bus transceiver.

External logic, address latch and counters needed by block transfer) in order to achieve the complete controller circuit. CY7C964 specifically provided by Cypress can finish this function coordinated with VIC068A/VIC64. It is a kind of bus interface local circuit with flexible configuration, including 8-bit transceiver, latch, and counter and so on, however, it also needs the help of 3 CY7C964 and corresponding development tools. Users can compare those two options to make their own choices.

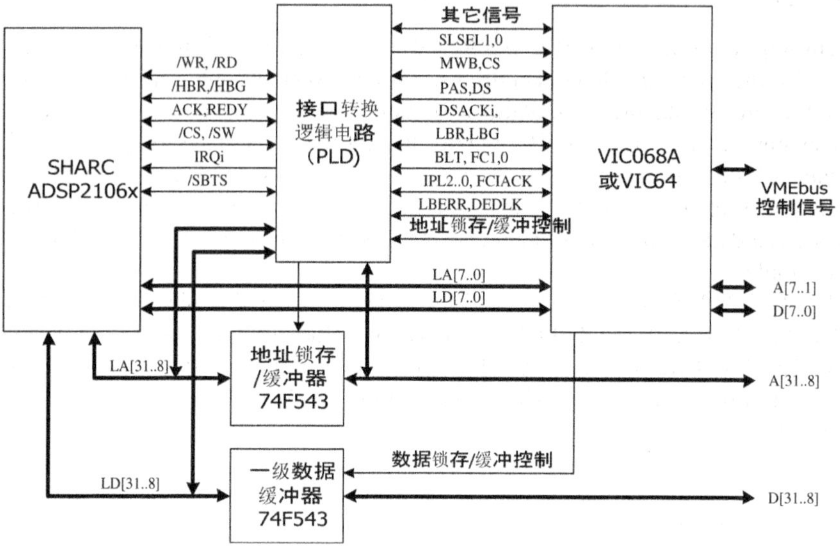

Fig. 3. The circuit schematic diagram of using VIC068/VIC64 to achieve bridging between SHARCDSP and VMEbus

References

1. Brydon, G.: Decoupling Opens Bottlenecks In VME64-based DSP systems. Electronic Design, vol. (3), pp. 147–149 (1997)
2. Wang, H., Gao, M., Han, Y.: The design and implementation of SHARC parallel processing system based on VME bus. Beijing University of Technology 20(4), 480–484 (2000)
3. Xiang, B.: Use FPGA to implement VME bus interface. Acoustics and Electronic Engineering (3), 40–44 (2000)
4. Chiechi, B.: MFLOPS Dictates Diligent Board Design. Electronic Design (3), 52–55 (1996)
5. Grehan, R.: DSPs get parallel. Computer Design (9), 83–98 (1997)
6. Ludemann, J., Ressing, D., Wurth, R.: A SHARC DSP Cluster as HERA-B DAQ Building Block. IEEE Transaction On Nuclear Science 44(3), 403–406 (1997)

Dynamic Router Real-Time Travel Time Prediction Based on a Road Network

Wenting Liu and Zhijian Wang

Hohai University, Nanjing, Jiangsu 210098 China
liuwt@hhu.edu.cn

Abstract. This paper is concerned with the task of dynamic router real-time travel time prediction for an arbitrary origin-destination pair on a map. The predicting travel time is based on the historical travel time and the current travel time. The historical travel time is calculated by speeds. The traffic pattern similar to the current traffic are searched among the historical patterns and closest matched patterns are used to extrapolate the present traffic condition. The method is combined the historical traffic patterns with real-time traffic data as a linear.A router is chosen from a few candidate routers based on the prediction technique. The resulting model is tested with realistic traffic data, and is found to perform well.

Keywords: Data Mining, Pattern Match, Traffic Rules.

1 Introduction

The travel time predication has always been an important part of intelligent transportation systems(ITS) research domain. An accurate time-time predication of travel time can be a crucial part of the driver information or the traffic management system. With the improvement of geographic positioning technologies and popularity of communication methods, there are huge traffic data accumulated by people while developing a lots application. So the pattern related to the stated of the transport network from traffic data to be drawn, to improve road traffic capacity. In these studies, there are a lot of researches on travel time prediction but most of them based on the segment of urban traffic network,not router, and based on historical data and static route mainly. Road networks are dynamic and stochastic systems, real-time information directly affect the accuracy of travel time prediction. It becomes the key issues to the resolved to monitoring the state of traffic network by taking advantage of the data.

Our contributions can be summarized as follows:

1) We propose a new travel time predication model based on historical and real-time information, for improving on the accuracy of predicting in a road network.
2) we propose the pattern of traffic rules, for improving on the effectiveness of predicting in a road network.

L. Qi (Ed.): ISIA 2010, CCIS 86, pp. 723–729, 2011.

The rest of the paper is organized as follows. Section 2 surveys the related work. Section 3 details the scheme. Section 4 shows the experiment results. We conclude this paper in Section 5.

2 Related Work

At present, there have been numerous methods for travel time predication, including those based on time series estimation methods, artificial intelligence method, hybrid of data fusion and regression methods and so on. The analyzed problem belongs to the field of travel time predication based on a pair of Original and Destination(OD). Studies based on similar input exist in the literature [1], but our problem has a few specific features that make it different from most other research in the area. The autoregressive MA(moving average) and artificial neural networks are often compared [2]. In many researches, Kalman Filtering model are often regard as an efficient method, which can support the past, current and future state [3,4]. There are some precise prediction methods using time series data of travel time between two points, on freeways [5]. Zhang et al. proposed methods based on linear regression model [7]. Nishiuma et al. proposed a route travel time prediction method using singular value precomposition [6]. Park et al. used neural networks [8] and Liu et al. proposed a method with clustering [9,10]. The features of freeways are that more accurate prediction is possible,because there is little change in traffic between two points.

However, the majority of the method focus on statical spatial network and historical travel time predication. Because the traffic network and the choices of routers are all complicated. The predication of travel time must be high accuracy and fast through abstracting the traffic pattern from huge historical data and revising the result according to the real-time information.

3 Applied Method

The dynamic router based real-time travel time predication technique is used for travel time predication in this section. There are three tasks to be completed: the candidate routers generation, travel time predication, dynamic router generation.

The task of the candidate routers generation is the basic task. If there are no well paths in our method, the dynamic routers travel time predication is in vain. But the router choice problem in the urban road network is more complex than in freeway. Existing many work for path computation on a given OD pair in the urban network has been focused on the shortest-path first, expressway first or signal less path first, and so on. In this paper, we don't research the existing router selection problem further. We select a few most frequency routers for a given OD pair from a large of historical data, because the drivers of the location based services application are the router exports in the urban network really.

The task of travel-time prediction for a given origin-destination (OD) pair on a map is one of the fundamental tasks in traffic modeling. Much effort has been devoted since the 90s when ITS appeared. In general, there are two views in traffic modeling, which we call the observer's view and driver's view [11]. This paper is based on the drivers view.

3.1 Problem Setting

This section summarizes our problem setting for travel time prediction.

Definition 1. (Link) *A link is a road segment between two neighboring intersections.* □

Definition 2. (Router) *A router is a sequence of links, where any two consecutive links share an intersection.* □

Link is the fundamental element of the routers for a given origin-destination (OD) pair on a map. For the routers we can predicate the travel time respectively.

Definition 3. (Travel Time Predication) *Some parameters decided for some candidate routers are used for the selection of traffic rules, the travel time can be estimated by combination of historical and real-time travel time.* □

For a given OD pair, we offer some candidate routers as Fig.1. In Fig 1, for a given OD pair, there are 4 candidate routers(r_1, r_2, r_3, r_4),the lines represent links and the circles are intersections. The main task is predicate the travel time for any router, then choose the fastest router.

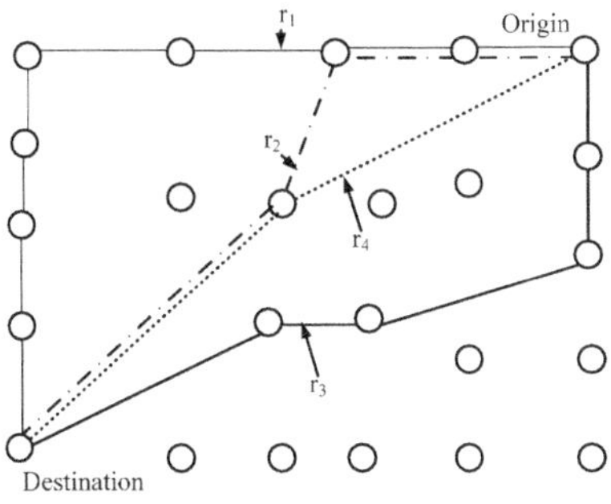

Fig. 1. Candidate Routers of a OD pair

3.2 Travel Time Predication

The basic idea of travel time predication can be estimate by using the linear combination of historical and real-time predication,as shown in (1),where Origin(O),Destination(D) and journey start time(t) are the input parameters of the predication formula,t' is the historical start time, T_c and T_h are the current and historical travel time predication results, α, β are the weighted combination variables for real-time and historical travel time predication.

$$T(O, D, t) = \alpha * T_c(O, D, t') + \beta * T_h(O, D, t) \ where \ \alpha + \beta = 1 \quad (1)$$

[Historical Travel Time Predication]. In order to reduce the computation time on classified huge historical database, we group temporal and spatial dimension to present some traffic rules and patterns [3]. So that only similar segments of the historical database are search. However, if the searching time-window is too large, the timely performance will be reduced. For example, weekend traffic patterns obviously are different from the workday's. Therefore, predicating the travel time of weekend can only search all historical traffic patterns of weekend in a year. Thus predication time can be reduced to 2/7.(2 days a weekend for one week).

The classification method in temporal dimension are grouped into year, season, month, workday, weekend, holiday, hour and half an hour and so on, the spatial dimension are classified zone, router, link and so on. For a city, road are classified different statues level. The average speed of collected records between 0-5km/h is defined as level 1 and 6-10km/h as level 2, and so on. If the speed is greater than 40, the level is defined as 9. According the category of temporal and spatial dimensions, the historical traffic rules and patterns are generated by data mining technology. The traffic rules, the mining results, are stored in historical traffic database.

For an example, the traffic rules are shown as (2). The rules means the conditions include date(May 20,2007), weekend, time(8:30), road id(A1),direction(from west to east), we can get the traffic status is level 5, i.e, the speed is 21-25km/h. At last we can calculate the travel time use the speed.

$$IF\ 20070520\ \&\ weekend\ \&\ 8:30\ \&\ A1\ \&\ D\ THEN\ Level5\ldots \quad (2)$$

[Current Travel Time Predication]. According to the traffic rule, we can only get the historical travel time. We must consider the real-time event. Table 1 lists all the parameters of predication. The real-time rules are decided by the traffic management. That is α and β values changeable in the formula (1).

Table 1. The Parameters of Predication

Parameters	Notes	Parameters	Notes
Time	the index of time,$(1 \cdots 48)$	Link	the index of link
Direction	the direction of float car	link Level	link traffic status
Length	the length of link	Default pattern	the default pattern if not existing historical pattern
Event	real-time event	DataType	the date type
Rule	real-time rules		

[Dynamic Router Travel Time Predication]. According to the results of historical and current predication, dynamic router travel time predication includes the following 3 processes:

1) The driver inputs the parameters(Origin,Destination, start-time) ⇒ Candidate Paths Generation
2) Match the stored traffic rules ⇒ historical travel time predication
   Real-time traffic information⇒ current travel time predication.
3) Using formula (1)⇒ the fastest candidate router⇒ the suggested router.

The processes of travel time predication are shown as Fig.2.

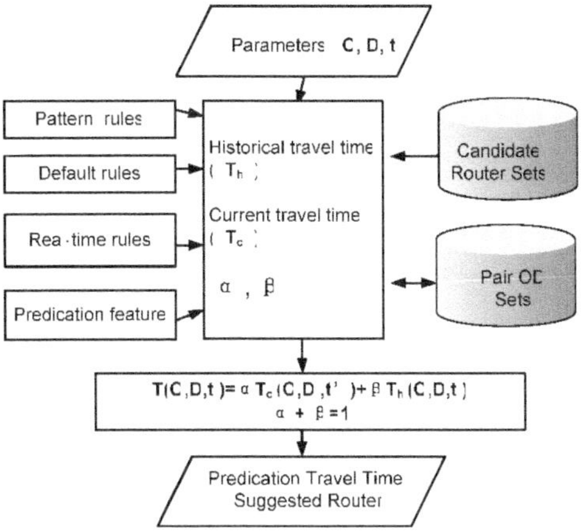

Fig. 2. The process of travel time predication

4 Experiments

4.1 Data and Methodology

We use the real map of the city Ningbo, Zhejiang province of China. And we obtained one month the GPS floating car data(2007/5/1-2007/5/31) that had aready been map-matched,i.e. it had a form of velocity and event time bound to a passage of a specific road segment in a given direction. We only have one month, so we use the data of first 18 days for mining the traffic pattern and the data of the last 6 days is for testing the travel time predication,except for the fist 7 day (the seven-day holidays in China, the traffic flows are greater than any other day,obviously). The two performance indices: relative mean errors(RME) and root mean squared errors(RMSE) are used to compare the predictors and listed as Eqs.(3)and(4)where n is the number of prediction, X_i and $\overline{X_i}$ present the travel time and prediction time,respectively.

$$RME = \frac{1}{n}\sum_{i=1}^{n}|\frac{X_i - \overline{X_i}}{X_i}| \quad (3)$$

$$RMSE = \sqrt{\frac{1}{n}\sum_{i=1}^{n}|\frac{X_i - \overline{X_i}}{X_i}|^2} \quad (4)$$

4.2 Results

In the initial experiments, the predication methods based on historical data and current time were applied. We compare our method to the two methods, and then show the integrated experiment results at last. We design two different kinds of traffic patterns,i.e, workday and weekend, for predicting the real-time. In the experiments, we random choose different OD pairs at twenty to 8, peek hours, on May 10th, 2007(Thursday), and then we calculated the error values of each method. The results show our dynamic predicator has the lowest RME and RMSE values in Table 2.

Table 2. RME and RMSE of Different Predication Methods on Workday

	Historical Predicator	Real-time Predicator	Dynamic Predicator
RME	20.50%	13.00%	10.80%
RMSE	24.82%	18.13%	15.92%

At the same way ,we random choose different OD pairs at 9, on May 20th, 2007(Sunday). We list the results in Table 3.

Table 3. RME and RMSE of Different Predication Methods on Weekend

	Historical Predicator	Real-time Predicator	Dynamic Predicator
RME	29.60%	21.00%	20.00%
RMSE	35.78%	32.21%	30.55%

According to the results, We found that the error values are seemed higher, because we didn't have enough raw data, but our dynamic predicator perform well.

5 Conclusion

In this paper, we first present a travel time prediction technique based on historical pattern and real-time traffic events and give a predication model. The model is combined the historical traffic patterns with real-time traffic data as a linear, overcomes the lack of the existing models which ignore the impact of real-time traffic information and only use the historical information, uses the data mining to find the historical rules which are decided the historical travel time. The resulting model is tested with a pair of Original and Destination(OD)on a road network, and is found to perform well.

In the future, we plan to further investigate the schema of multi-source historical data fusion, find the patterns of multi-source history data fusion and create the traffic rules based on multi-source historical data.

Acknowledgment

This research is supported by the Fundamental Research Funds for the Central Universities (Hohai University) 2010B06714 (Research on Dynamic Router Real-time Travel Time Prediction).

References

1. Idé, T., Kato, S.: Travel-Time Prediction using Gaussian Process Regression: A Trajectory-Based Approach. In: Proc. of the 9th SIAM international conference on Data Mining (SDM), pp. 1185–1196 (2009)
2. Mining Traffic Data Form Probe-car System for Travel Time Prediction. In: Proceedings of the ACM SIGKDD International Conference on Knowledge Discovery and Data Mining, pp. 906–911 (2004)
3. Chung, E.: Classification of Traffic Pattern. In: Proc. of the 11th World Congress on ITS, pp. 687–694 (2003)
4. Yang, J.S.: Travel Time Predication Using the GPS Test Vehicle and kalman Filtering Techniques. In: Proc. of the 2005 American Control Conference, pp. 2128–2133 (2005)
5. Ueno, H., Ohba, Y., Kuwahara, M.: The Comparison of Two Type Travel Time Prediction Methods Using Toll Collection System Data. In: The Paper of Technical Meeting on Intelligent Transport Systems, IEE Japan, ITS-02-18-20, pp. 7–11 (2002)
6. Nishiuma, N., Goto, Y., Kumazawa, H.: Prediction of Route Travel Time Using Singular Value Precomposition. In: Proc. of the 47th Japan Joint Automatics Control Conference, vol. (701) (2004)
7. Zhang, X., Rice, I.: Short Term Travel Time Prediction. Transport Res., Ser. C 11, 187–210 (2003)
8. Park, D., Rilett, L.: Multiple-period Freeway Link Travel Times Using Modular Neural Networks. Transport Res. Rec. 1617, 1–334 (1988)
9. Liu, W., Wang, Z., Feng, J.: Continuous Clustering of Moving Objects in Spatial Networks. In: Lovrek, I., Howlett, R.J., Jain, L.C. (eds.) KES 2008, Part II. LNCS (LNAI), vol. 5178, pp. 543–550. Springer, Heidelberg (2008)
10. Liu, W., Feng, J., Wang, Z.: Constrained Clustering Objects on a Spatial Network. In: WRI World Congress on Computer Science and Information Engineering (CSIE 2009), pp. 636–639 (2009)
11. Kriegel, H.-P., Renz, M., Schubert, M., Zuefle, A.: Statistical density prediction in traffic networks. In: Proc. SIAM Intl. Conf. Data Mining, pp. 692–703 (2008)

A Novel Duality and Multi-Level Security Model Based on Trusted State

WeiPeng Liu

Beijing Information Institute, Beijing,100094, China
weipengliu0209@163.com

Abstract. In order to develop high security level operating system that applies to trusted computing platform, at the same time to meet the requirements for confidentiality and integrity protection, a novel Duality and Multi-Level Security Model based on Trusted State(DMLSMTS) is proposed in this paper. It protects the confidentiality of information based on BLP model and the integrity based on Biba model. It introduces "trusted measurement function" and uses the mechanism of trusted measurement which is unique in trusted computing platform to estimate the trusted state of subject or object, and based on the evaluation result to call trusted agent to grant the access which validates the BLP model or Biba model. It gives the formal description of the model. Furthermore it proves that the system remains in a secure state after executing the security rules.

Keywords: Trusted Computing, Trusted State, Duality and Multi-Level Security Model.

1 Introduction

Security model plays a very important role in the development of security operating system. However, developing system security model is considered to be very complicated work, it will consume much financial and material resource, and so researchers choose to improve the existed models according to actual security requirements, and then prove the improved model is secure.

The current security operating systems, including DTOS [1], Xenix [2] and SE-Linux [3] and so on, mainly base on BLP model to control information flow of the system. BLP is the earliest security model proposed by D.E. Bell and L.J. LaPadula in 1973 [4][5], which is revised and perfected in 1976 [6]. BLP model is considered as the base to define Multi-Level Security (MLS) and also as the widely accepted basic security axiom, its influence in MLS support is known as the same place as Hilbert's axiom in Euclidean Geometry [7].

But from the time when the BLP model is proposed to now, there has great development on the research of computer security, and with the constant changing of security threat and environment, there are two main problems that must deal with when applying BLP model into current computing environment:

1) Lack of the description to the trusted state of subject or object in system. Usually in actual security system, subject's security attribute (level) inherits the user

L. Qi (Ed.): ISIA 2010, CCIS 86, pp. 730–738, 2011.

who starts the process, the process which is started by the user of high security level has high security level, but noticeably, it is not always trusted , here "secure" is not equal to "trusted", even if it is trusted, when accessing untrusted objects, its trusted state maybe be destroyed, so it is necessary to describe the trusted state of entity.

2) Lack of the protection of integrity to subject or object in system. BLP model is proposed to solve the confidentiality problem in military affairs system, it only supports one-way information flow and can prevent information flowing from high security object to low security object. But with the constant changing of security requirement and security threat, integrity has become another keystone that needs to be protected in security operating system, so it is necessary to add a novel element for BLP model which is used to protect the integrity of system.

The rest of this paper is organized as follows: In section 2, it shows the formalized description of the model; In section3 , it proves the security of the model; This paper is concluded with a outlook of future research direction in the section 4.

2 The Novel Model

In this section, it will describe novel design idea, important definitions, security rules and adjustment rules.

2.1 Novel Design Idea

In TCG specification [8] about trusted computing platform, it defines what is "trusted", it is that if an entity achieves the expected goal in expected manner, then the entity is trusted. TCG also gives the concept of trusted computing platform, it is that a platform which can be trusted by local user, remote user or entity. In practical engineering implementation, entity's "trusted" is usually obtained by trusted measurement, in other words, to estimate according to the integrity measurement value of entity through cryptography hash function, and then compare the measurement value with the expected value. If the measurement value is consistent with the expected measurement value, then the entity is "trusted", or it is "untrusted". So the trusted measurement is one of important functions of trusted computing platform.

Based on the measurement to entity's state in system provided by trusted computing platform, we import the entity's state in system as a variable to the novel model, and add a mapping function of entity's state. We can use this function to determine the current state of subject or object which mainly includes confidentiality trusted state, integrity trusted state, untrusted state and unchecked state.

In the novel model, it protects information confidentiality through BLP model and information integrity through Biba model, moreover, authorize or refuse current access through judging the current state of subject or object in the case that the current access disobeys BLP or Biba model. If both of the subject and object are in confidentiality trusted state, even when the current access disobeys the rules of BLP model, but obeys the rules of Biba model, this access also can be executed with the help of trusted agent. If both of the subject and object are in the trusted state of integrity, even when the current access disobeys the rules of Biba model, but obeys the rules of BLP model, then this access also can be executed with the help of trusted agent.

2.2 The Preliminary of Model

A formal security model is essential when reasoning about the security of a system. Without an unambiguous definition of what security means, it is impossible to say whether a system is secure. Security model usually compromises of a set of security definitions and rules which are used to describe the security policy precisely and unambiguously, based on the main design idea of the model narrated above, we will give formally illustrate our novel model.

Definition 1. Subject, object and operation: S is subject set, O is object set, and $A=\{r,w,a,e\}$ is a set of access mode from subject to object, r is "read" access, w is "write" access, a is "append-write" access, e is "execute access".

Definition 2. Trusted state set: $T=\{C\_trustedstate, I\_trustedstate, untrustedstate, unchecked\}$ is trusted state set of subject or object, $C\_trustedstate$ is confidentiality trusted state , $I\_trustedstate$ is integrity trusted state, $untrustedstate$ is not trusted state and $unchecked$ is not checked state. $GR=\{g,r\}$, g means "get" access right, r means "release" access right. Request set is $RQ=\{GR\times S\times O\times A\}$, $\forall rq\in RQ$ means subject request or release some access right of object.

Definition 3. Trusted measurement function of subject: $S\_SM:S\rightarrow State$, $\forall s\in S$, $S\_SM(s)=C\_trustedstate$ means that subject is in confidentiality trusted state; $S\_SM(s)=I\_trustedstate$ means that subject is in integrity trusted state; $S\_SM(s)=utrustedstate$ means subject is in untrusted state; $S\_IM(s)=unchecked$ means subject is in unchecked state.

Definition 4. Trusted measurement function of object: $O\_SM:O\rightarrow State$, $\forall o\in O$, $O\_SM(o)=C\_trustedstate$ means object is in confidentiality trusted state; $O\_SM(o)=I\_trustedstate$ means object is in integrity trusted state; $S\_SM(o)=utrustedstate$ means object is in untrusted state; $O\_IM(o)=unchecked$ means object is in unchecked state.

Definition 5. Confidentiality level set: L_C is confidentiality level set, $L_C=\{(c,k_c),c\in C_f,k_c\subseteq K_c\}$, C_f is positive integer confidentiality function set, $\forall c_m,c_n\in C_f$, $c_m>c_n$ means confidentiality c_m is higher than c_n. $K_c=\{k_1,k_2,\ldots\ldots,k_n\}$ is non-graded confidentiality category, $\forall k_1,k_2\in K_c,k_1\subseteq k_2$ means k_1 is included in k_2. Define "dominate" relation $\geq$ which satisfies partial order relation in L_C, supposing $l_1=(c_1,k_1)\in L_c$,

$l_2 = (c_2, k_2) \in L_c$, if $l_1 \geq l_2$ only when $c_1 \geq c_2$, $k_1 \supseteq k_2$. C_System_High is the highest system confidentiality level.

Definition 6. Integrity level set: L_I is integrity level set, $L_i = \{(i, k_i), i \in I_f, k_i \subseteq K_I\}$, I_f is positive integer integrity function set, $\forall i_m, i_n \in I_f, i_m > i_n$ means integrity i_m is higher than i_n. $K_c = \{k_1, k_2, ..., k_n\}$ is non-graded integrity category, $\forall k_1, k_2 \in K_i$, $k_1 \subseteq k_2$ means k_1 is included in k_2. Define "dominate" relation $\geq$ which satisfy partial order relation in L_I, $l_1 = (c_1, k_1) \in L_i$, $l_2 = (c_2, k_2) \in L_i$, if $l_1 \geq l_2$ only when $i_1 \geq i_2$, $k_1 \supseteq k_2$. I_System_High is the highest system integrity level.

Definition 7. System state: system state v, the element $v \in V = \{B \times M \times F \times I \times T \times H\}$ of set V is a system state:

Current Access Set B **:** to $b = (s \times o \times a) \in B$, $s \in S, o \in O, a \in A$ means subject s access object o in the manner of a;

Access Control Matrix M **:** M ={ M is matrix | $m_{ij} \in M$ is access right set from subject s_i to object o_j };

Confidentiality Level Function F **:** it consists of three subfunctions, they are $f = \{f_s, f_c, f_o\}$, f_s is the Max confidentiality level function of subject, $f_s(s) \in L_c$ means the Max confidentiality level of subject; f_c is current confidentiality level function of subject, $f_c(s) \in L_c$ means current confidentiality level of subject, $f_s(s) \geq f_c(s)$; f_o is confidentiality level function of object, $f_o(s) \in L_c$ is confidentiality level of object.

Integrity Level Function I **:** it consists of three subfunctions, they are $I = \{I_s, I_c, I_o\}$, I_s is the MAX integrity level function of subject, $I_s(s) \in L_i$ means the highest integrity level of subject; I_c is current integrity level function of subject, $I_c(s) \in L_i$ means current integrity level of subject, $I_s(s) \geq I_c(s)$; I_o is integrity level function of object, $I_o(s) \in L_i$ is integrity level of object.

Trusted State T **:** $T = \{C\_trustedstate, I\_trustedstate, untrustedstate, unchecked\}$.

Hierarchy of Object H **:** $H = \{h \mid h \in P(O)^o \cap attribute1 \cap attribute2\}$.

Attribute 1: $\forall o_i \in O, \forall o_j \in O(o_i \neq o_j \Rightarrow H(o_i) \cap H(o_j) = \phi)$;

Attribute 2: $!\exists\{o_1, o_2, o_3, ...o_w\} \subseteq O[\forall r(1 \leq r \leq w \Rightarrow o_{r+1} \in H(o_r))$
$\cap(o_{w+1} \equiv o_1)]$.

Definition 8. Confidentiality check function: $CCF: O \rightarrow \{\text{Yes,No}\}$, it is a mapping function from object to $\{\text{Yes, No}\}$ set, $Yes = 1, No = 0, Yes$ means that obey the confidentiality check rule, and No means that disobey the confidentiality check rule.

Definition 9. Integrity check function: $ICF: O \rightarrow \{\text{Yes,No}\}$, it is a mapping function from object to $\{\text{Yes, No}\}$ set, $Yes = 1$, $No = 0$, Yes means that obey the integrity check rule, and No means that disobey the integrity check rule..

Definition 10. Trusted agent set: $SA \subset S_t$, trusted agent is a trusted subject started up when access between subject and object disobeys the security rules of BLP model or Biba model, $\forall s_a \in SA$, it has $f_s(s_a) = C\_System\_High$, $I_s(s_a) = I\_System\_High$. The MAX confidentiality level of trusted agent is the highest system confidentiality level C_System_High, the MAX integrity level of security agent is the highest system integrity level I_System_High.

2.3 Security Axioms

We adopt the description method presented in literature [4][5], because this method is classical and easily understandable.

Rule 1: Trusted extended discretionary security property
A state $v = (b \times m \times f \times i \times t \times h)$ satisfies trusted extended discretionary security, if and only if when $(s_i, o_j, x) \in b \Rightarrow x \in M_{ij}$, $S\_IM(s) \neq untrusted$, $O\_IM(o) \neq untrusted$.

Rule 2: Trusted extended simple security property
A state $v = (b \times m \times f \times i \times t \times h)$, to subject set S, $s \in S$, satisfies trusted extended simple security, if and only if when $(s, o, x) \in b \Rightarrow$

a) $x = e$, and $S\_IM(s) \neq untrusted$, $O\_IM(o) \neq untrusted$;

b) $x = r$, and $f_s(s) \geq f_o(o)$, $S\_IM(s) \neq untrusted$, $O\_IM(o) \neq untrusted$;

c) $x = a$, and $I_s(s) \geq I_o(o)$, $S\_IM(s) \neq untrusted$, $O\_IM(o) \neq untrusted$;

d) $x = w$,and $f_s(s) \geq f_o(o)$, $I_s(s) \geq I_o(o)$, $S\_IM(s) \neq untrusted$, $O\_IM(o) \neq untrusted$.

Rule 3: Trusted extended read security property
A state $v = (b \times m \times f \times i \times t \times h)$, to untrusted subject S_{ut} , $s \in S_{ut}$, satisfies trusted extended read security, if and only if when $(s, o, r) \in b \Rightarrow$

a) $f_c(s) \geq f_o(o)$, $I_c(s) \leq I_o(o)$, and $S\_IM(s) \neq untrusted$, $O\_IM(o) \neq untrusted$;

b) $f_c(s) \geq f_o(o)$, $I_c(s) > I_o(o)$,and $S\_IM(s) = I\_trustedstate$, $O\_IM(o) = I\_trustedstate$, then it needs to satisfy Adjustment Rule 1;

c) $f_c(s) < f_o(o)$, $I_c(s) \leq I_o(o)$,and $S\_IM(s) = C\_trustedstate$, $O\_IM(o) = C\_trustedstate$, then it needs to satisfy Adjustment Rule 2.

Rule 4: Trusted extended append-write security property
A state $v = (b \times m \times f \times i \times t \times h)$, to untrusted subject S_{ut} , $s \in S_{ut}$, satisfy trusted extend read security, if and only if when $(s, o, a) \in b \Rightarrow$

a) $f_c(s) \leq f_o(o), I_c(s) \geq I_o(o)$, and $S\_IM(s) \neq untrusted$, $O\_IM(o) \neq untrusted$;

b) $f_c(s) > f_o(o)$, $I_c(s) \geq I_o(o)$, and $S\_IM(s) = C\_trustedstate$, $O\_IM(o) = C\_trustedstate$, then it needs to satisfy Adjustment Rule 3;

c) $f_c(s) \leq f_o(o)$, $I_c(s) < I_o(o)$,and $S\_IM(s) = I\_trustedstate$, $O\_IM(o) = I\_trustedstate$, then it needs to satisfy Adjustment Rule 4.

Rule 5: Trusted extended write security property
A state $v = (b \times m \times f \times i \times t \times h)$, to untrusted subject S_{ut} , $s \in S_{ut}$, satisfy trusted extend write security, if and only if when $(s, o, w) \in b \Rightarrow$

a) $f_c(s) = f_o(o)$, $I_c(s) = I_o(o)$;

b) $f_c(s) = f_o(o)$, $I_c(s) < I_o(o)$,and $S\_IM(s) = I\_trustedstate$, $O\_IM(o) = I\_trustedstate$, then it needs to satisfy Adjustment Rule 4;

c) $f_c(s) = f_o(o), I_c(s) > I_o(o)$, and $S\_IM(s) = I\_trustedstate$, $O\_IM(o) = I\_trustedstate$, then it needs to satisfy Adjustment Rule 1;

d) $f_c(s) > f_o(o)$, $I_c(s) = I_o(o)$,and $S\_IM(s) = C\_trustedstate$, $O\_IM(o) = C\_trustedstate$, then it needs to satisfy Adjustment Rule 3;

e) $f_c(s) < f_o(o)$, $I_c(s) = I_o(o)$,and $S\_IM(s) = C\_trustedstate$, $O\_IM(o) = C\_trustedstate$, then it needs to satisfy Adjustment Rule 2.

2.4 Adjustment Rules

Adjustment Rule 2: With rule 3's c), if $(s,o,r) \in M$, to the requirement: $rq = (g,s\ ,o,r)$, it uses below steps to deal with:

(1) System start up trusted agent s_a , $s_a \in S_t$, create function f^1 , to make $f_s^1 = f_s$, $f_c^1 = f_c$, $f_o^1 = f_o$; create function I^1 , $I_s^1 = I_s$, $I_c^1 = I_c$, $I_o^1 = I_o$; create $t^1 = t$; create H^1 , to make $H^1 = H$. If $(s_a,o,r) \in M$, then $M^1 = M$; or else authorize (s_a,o,r) , to make $M^1 = M \cup (s_a,o,r)$, after execute requirement (g,s_a,o,r) , $b^1 = b \cup (s_a,o,r)$, system goes into state $v^1 = (b^1, M^1, f^1, i^1, t^1, H^1)$;

(2) Trusted agent s_a create object $o`$ , create $f^2$ , to make $f_s^2 = f_s^1$ , $f_c^2 = f_c^1$ , $f_o^2 = f_c$ ; create $t^2$ , to make $t^2 = t^1$ ; create $H^2$ , to make $H^2 = H^1 \cup (o`)$, system goes into state $v^2 = (b^2, M^2, f^2, i^2, t^2, H^2)$, $i^2 = i^1 ; b^2 = b^1 ; M^2 = M^1$;

(3)Authorize $(s_a,o`,a)$ , to make $M^3 = M^2 \cup (s_a,o`,a)$, after execute the requirement $(g,s_a,o`,a)$ , $b^3 = b^2 \cup (s_a,o`,a)$, trusted agent s_a write the content which read from o into $o`$, system goes into state $v^3 = (b^3, M^3, f^3, i^3, t^3, H^3)$, $M^3 = M^2 ; f^3 = f^2 ; i^3 = i^2 ; t^3 = t^2 ; H^3 = H^2$;

(4) If $CCF(o`) = Yes$ , then go to next step; or else $s_a$ deletes $o`$, system goes into state $v^5 = (b^5, M^5, f^5, i^5, t^5, H^5)$, $b^5 = b^3 - \{(s,o`,a),(s_a,o`,r)\}$, $M^5 = M^3 - \{(s,o`,a),(s_a,o`,r)\}$, $f^5 = f^1 ; i^5 = i^1 ; t^5 = t^1 ; H^5 = H^3 - (o`)$, refuse (g,s,o,r) ;

(5) Authorize $(s,o`,r)$ , to make $M^4 = M^3 \cup (s,o`,r)$, after execute requirement $(g,s,o`,r)$ , $b^4 = b^3 \cup (s,o`,r)$, system goes into state $v^4 = (b^4, M^4, f^4, i^4, t^4, H^4)$, $f_s^4 = f_s$, $f_c^4 = f_c$, $f_o^4 = f_o^2$, $i^4 = i^2 ; t^4 = t^3$; $H^4 = H^3$;

(6) s_a detels $o`$ , system goes into state $v^5 = (b^5, M^5, f^5, i^5, t^5, H^5)$ , $b^5 = b^4 - \{(s_a,o`,a),(s,o`,r)\}$ , $M^5 = M^4 - \{(s_a,o`,a),(s,o`,r)\}$ , $f^5 = f^2$ , $i^5 = i^2$ , $t^5 = t^2$ , $H^5 = H^4 - (o`)$.

In the nature, Adjustment Rule 1,3,4 are similar with Adjustment Rule 1 in the basic adjustment idea, due to the confine of paper length, they are omitted here.

3 Security Analysis and Proof

Theorem 1. A state $v_i = (b \times M \times f \times i \times t \times H)$ is a security state if and only if it satisfies the trusted extended discretionary security property, trusted extended simple security property, trusted extended read security property, trusted extended append-write security property, and trusted extended write security property.

Theorem 2. If a state $v_i = (b \times M \times f \times i \times t \times H)$ is a security state which satisfies the theorem1, then with the adjustment rule 1, the system still goes to another security state.

Now, we prove that theorem2 is correct.Because adjustment rule1 is only to the confidentiality level, but integrity level is not changed in the state transition process, so it only needs to prove that it obeys the security of BLP model.Suppose with adjustment rule 2, system can not go to security state, v is not trusted state, then in the whole process that from v^1 to v^5 there at leas has a not security state .

1) If v^1 is a not trusted state, according to adjustment rule 2's (1), $(s_a, o, r) \in M$ or $(s_a, o, r) \in M^1$, then satisfy $ds-property$; because $s_a \in SA$, $f_s^1(s_a) = C\_System\_High$, then $f_s^1 \geq f_o$, so satisfy $ss-property$; so v is a trusted state, to $s \in S_{ut}$ is satisfy $*-property$, but to adjust rule 2's (1), there is not any operation to $s \in S_{ut}$, so it is satisfy $*-property$ that v^1 to $s \in S_{ut}$, all above, v^1 satisfy $ds-property$, $ss-property$ and $*-property$, so v^1 is a security state;

2) If v^2 is a not trusted state, then according to adjustment rule 2's (2), $b^2 = b^1$, $M^2 = M^1$, $f_c^2 = f_c^1 = f_c$, $f_o^2 = f_c$, v^1 is a security state, after system has created $o`$, v^2 still satisfy three pieces of security rules, so it is still a security state;

3) If v^3 is a not security state, according to adjustment rule 2's (3), $(s_a, o`, a) \in M^3$, obviously satisfy $ds-property$; $s_a \in SA$, satisfy $ss-property$; because v^2 is a security state, so it is satisfy $*-property$ to $s \in S_{ut}$, but to adjust rule 2's (3), there is not any operation to $s \in S_{ut}$, so it is satisfy $*-property$ that v^3 to $s \in S_{ut}$, all above, v^3 satisfy $ds-property$, $ss-property$ and $*-property$, so v^3 is a security state;

4) If v^4 is a not trusted state, then according to adjustment rule 2's (5), $(s_a, o`, a) \in M^4$, satisfy $ss-property$; $s \in S_{ut}$, and $f_c^4 = f_c$, $f_o^4 = f_o^2 = f_c$, satisfy $*-property$; v^3 is a security state, satisfy $ss-property$, so v^4 is still a security state;

5) If v^5 is a not trusted state, because v^4 is a security state, then according to adjust rule 2's (6), if $(s_i, o_j, x) \in b^5$, then $x \in M_{ij}$, v^5 still satisfy three security rules after deletes $o`$, so it is still a security state.

Above all, states from v^1 to v^5 are all security states, then the assumption is wrong, original proposition is right.

4 Conclusion and Future Work

This paper proposes a novel duality and multi-level security model based on trusted state suitable for trusted computing platform. In the future research, we'll mainly study how to label all the subjects and objects in system based on this model, and should make sure that the operation between any subject and object will be entirely covered by this access control policy in implementation. Moreover, the formal specification and verification the model is also an important and necessary step for next work.

References

1. Secure Computing Corporation. DTOS Generalized Security Policy Specification. DTOS CDRL A019, Secure Computing Corporation, Roseville, Minnesota (June 1997)
2. Gligor, V.D., Burch, E.L., Chandersekaran, C.S., Chapman, R.S., Dotterer, L.J., Hecht, M.S., Jiang, W.D., Luckenbaugh, G.L., Vasudevan, N.: On the Design and the Implementation of Secure Xenix Workstations. In: Proceedings of the 1986 IEEE Symposium on Security and Privacy, pp. 102–117 (April 1986)
3. Loscocco, P., Smalley, S.: Integrating Flexible Support for Security Policies into the Linux Operating System. Technical report, NSA and NAI labs (January 2001)
4. Bell, D.E., La Padula, L.J.: Secure Computer Systems: Mathematical Foundations. Hanscom AFB, Bedford, MA, Rep.ESD-TR-73-278, vol. 2. ESD/AFSC (1973)
5. Bell, D.E., La Padula, L.J.: Secure Computer Systems: Mathematical Foundations. Hanscom AFB, Bedford, MA, Rep.ESD-TR-73-278, vol. 2. ESD/AFSC (1973)
6. Bell, D.E., La Padula, L.J.: Secure Computer System: Unified Exposition and MULTICS Interpretation. MTR-2997 Rev. 1. The MITRE Corporation, Bedford, MA, USA (March 1976)
7. Lin, T.Y., Bell, D.E., Lapadula, L.J.: Axioms: A "New" Paradigm for an "Old" Model. Paper of the 1992 NSPW (September 1992); Proceedings on the 1992-1993 ACM SIGSAC New Security Paradigms Workshop, Little Compton, Rhode Island, USA, pp. 82–93 (August 1993)
8. TCG Specification Architecture Overview Specification Revision 1.2 (April 28, 2004)

Analysis of Single-phase APF Overtone and Idle Current Examination

Yang Li[1], Kai Wang[1], and Ning Xin[2]

[1] Department of Electrical and Automatic Engineering, Nanchang University Nanchang, China
yangliedu@163.com, wangkai19832002@yahoo.com.cn
[2] Department of Electrical Engineering, Liming Vocational University Quanzhou, China
nxdoudou@163.com

Abstract. This paper introduced three kind of single-phase APF overtones and idle current examination method first,then compare with MATLAB simulation based on these methods,receives a satisfactory conclusion.

Keywords: single-phase; APF; idle current.

1 Introduction

In recent years, along with the misalignment electric power electronic device is widely applied day by day, the electrical network has produced the massive overtones. The electricity iron load is the typical misalignment overtone source, the Zhejiang and Jiangxi electric railway's clear the electrical network has brought certain influence for Jiangxi, therefore, it is necessary to carry on the analysis to the Zhejiang and Jiangxi iron overtone, and proposed that related suppressed the strategy. This paper studies is suitable for electric railway's single-phase APF[1], carries on the analysis to its harmonic current examination method.

2 Instant Reactive Power Single-phase Overtone Idle Work Examination Method

The inspection based on the instant reactive power theory's three-phase circuit harmonic current examination method, discovered will always examine the first three-phase signal becomes the mutually perpendicular $\alpha\beta$, then further calculates again. Regarding the single-phase circuit, simplifies above method, only need construct interphase current and the actual electric current lags T/4 (here T is again power frequency cycle), the direct production supposition's two phase coordinates are signals. [2]

Supposes the network voltage spurt value is:

$$u_s(t) = \sqrt{2}U \sin \omega t \tag{1}$$

L. Qi (Ed.): ISIA 2010, CCIS 86, pp. 739–746, 2011.

The electrical network current spurt value is:

$$i_s(t) = \sum_{n=1}^{\infty} \sqrt{2} I_{2n-1} \sin[(2n-1)\omega t + \varphi_{2n-1}] \tag{2}$$

In $\alpha\beta$ coordinate system, designated $i_s(t)$ for α signal, lag T/4 it to construct β signals, obtains the imaginary $\alpha\beta$ signals, as follows:

$$\begin{bmatrix} i_\alpha \\ i_\beta \end{bmatrix} = \begin{bmatrix} \sum_{n=1}^{\infty} \sqrt{2} I_{2n-1} \sin[(2n-1)\omega t + \varphi_{2n-1}] \\ \sum_{n=1}^{\infty} \sqrt{2} I_{2n-1} \sin[(2n-1)\omega (t - T/4) + \varphi_{2n-1}] \end{bmatrix} \tag{3}$$

According to the reactive power theory, may calculate:

$$\begin{bmatrix} i_p \\ i_q \end{bmatrix} = \begin{bmatrix} \sin \omega t & -\cos \omega t \\ -\cos \omega t & -\sin \omega t \end{bmatrix} \begin{bmatrix} i_\alpha \\ i_\beta \end{bmatrix} = \begin{bmatrix} \bar{i}_p \\ \bar{i}_q \end{bmatrix} + \begin{bmatrix} \tilde{i}_p \\ \tilde{i}_q \end{bmatrix} = \begin{bmatrix} \sqrt{2} I_1 \cos \varphi_1 \\ \sqrt{2} I_1 \sin \varphi_1 \end{bmatrix} + \begin{bmatrix} \tilde{i}_p \\ \tilde{i}_q \end{bmatrix} \tag{4}$$

Above various in the formula, $\sin \omega t$ and $\cos \omega t$ are the voltage signal which obtains after the zero crossing synchronization and the phase-locked loop. Above the variable "-" the expression direct component, "~" the expression AC component, $\bar{i}_p$ and $\bar{i}_q$ express the electric current separately active and the idle work direct component. The inverse transformation may obtain the fundamental current $i_{\alpha f}$ and $i_{\beta f}$ [3] .

$$\begin{bmatrix} i_{\alpha f} \\ i_{\beta f} \end{bmatrix} = \begin{bmatrix} \sin \omega t & -\cos \omega t \\ -\cos \omega t & -\sin \omega t \end{bmatrix}^{-1} \begin{bmatrix} \bar{i}_p \\ \bar{i}_q \end{bmatrix} = \begin{bmatrix} \sqrt{2} I_1 \sin(\omega t + \varphi_1) \\ \sqrt{2} I_1 \sin(\omega t - T/4 + \varphi_1) \end{bmatrix} \tag{5}$$

Subtracts the fundamental wave component from the full current,then obtain the harmonic current component:

$$i_{sh} = i_\alpha - i_{\alpha f} \tag{6}$$

Based on the instant reactive power's single-phase overtone examination control diagram, as shown in Figure 1. LPF is the low pass filter, PLL is the network voltage synchronization phase-locked loop[4], the cosine signal has the electric circuit combined action to produce and the network voltage which in transformation matrix C needs with the phase sinusoidal signal and the corresponding cosine signal. When separates the q channel, after examining the overtone idle work adds together, then simultaneously examines the harmonic current and the idle current.

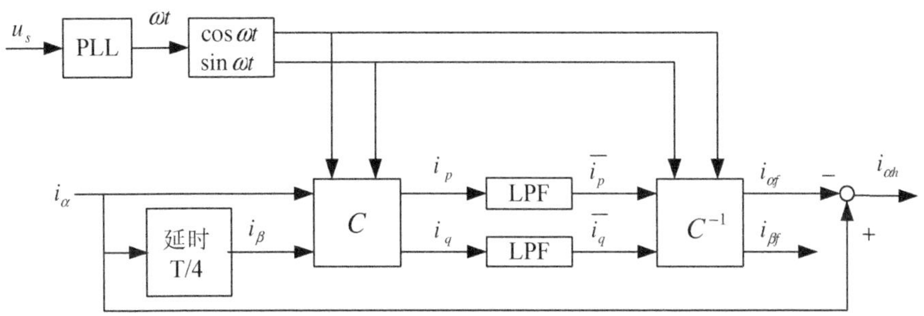

Fig. 1. Based on instant reactive power single-phase overtone examination control diagram

3 Wattful Current Separation Overtone and Idle Current Examination Law

In this method, uses with the network voltage frequency unit cosine, the sinusoidal signal multiples directly separately with the electrical network electric current, and after low pass filter, obtains in the electrical network electric current instantaneous fundamental wave wattful current and the instantaneous fundamental wave idle current, then obtains the instant harmonic current.

Supposes the network voltage same type (1), the electrical network electric current is:

$i_q = \sqrt{2}I_q \sin\varphi\cos\omega t$ is electrical network instant fundamental wave idle current

$i_h = \sum_{n=2}^{\infty}\sqrt{2}I_{2n-1}\sin[(2n-1)\omega t + \varphi_{2n-1}]$ is electrical network instant harmonic current

$$\begin{aligned}
&i_s(t) = \sqrt{2}I_1\sin(\omega t + \varphi) + \sum_{n=2}^{\infty}\sqrt{2}I_{2n-1}\sin[(2n-1)\omega t + \varphi_{2n-1}] = \\
&\sqrt{2}I_1\cos\varphi\sin\omega t + \sqrt{2}I_1\sin\varphi\cos\omega t + \sum_{n=2}^{\infty}\sqrt{2}I_{2n-1}\sin[(2n-1)\omega t + \varphi_{2n-1}] = \\
&\sqrt{2}I_p\sin\omega t + \sqrt{2}I_q\cos\omega t + \sum_{n=2}^{\infty}\sqrt{2}I_{2n-1}\sin[(2n-1)\omega t + \varphi_{2n-1}] = i_p + i_q + i_h
\end{aligned} \tag{7}$$

The type (7) nearby two simultaneously is multiplied by $2\sin\omega t$:

$$\begin{aligned}
&2i_s(t)\sin\omega t = 2\sqrt{2}I_p\sin^2\omega t + 2\sqrt{2}I_q\cos\omega t\sin\omega t + \\
&\sum_{n=2}^{\infty}2\sqrt{2}I_{2n-1}\sin[(2n-1)\omega t + \varphi_{2n-1}] = \sqrt{2}I_p - \sqrt{2}I_p\cos 2\omega t + \\
&\sqrt{2}I_q\sin 2\omega t + \sum_{n=2}^{\infty}\sqrt{2}I_{2n-1}\{\cos[2(n-1)\omega t + \varphi_{2n-1}] + \cos(2n\omega t + \varphi_{2n-1})\}
\end{aligned} \tag{8}$$

Likewise, the type (7) nearby two with rides $2\cos\omega t$:

$$2i_s(t)\cos\omega t = \sqrt{2}I_q + \sqrt{2}I_q\cos 2\omega t + \sqrt{2}I_p\sin 2\omega t + \\ + \sum_{n=2}^{\infty}\sqrt{2}I_{2n-1}\{\sin(2n\omega t - \varphi_{2n-1}) + \sin[2(n-1)\omega t + \varphi_{2n-1}]\} \tag{9}$$

Based on the wattful current separation's overtone examination control diagram like Figure 2, according to the above analysis, the electrical network electric current expands two times, multiplied by $\sin\omega t$, is lower than two time of electric current base frequency after the cut-off frequency low pass filter (LPF) may obtain $\sqrt{2}I_p$ [5], multiplied by $\sin\omega t$ obtains i_p ; Likewise may obtain i_q . This may act according to the user the choice, simultaneously carries on the compensation to the overtone and the idle current or only carries on the compensation to the overtone. If simultaneously carries on the compensation to the overtone and the idle current, may separate i_q channels, or adopts $i_p + i_h$.

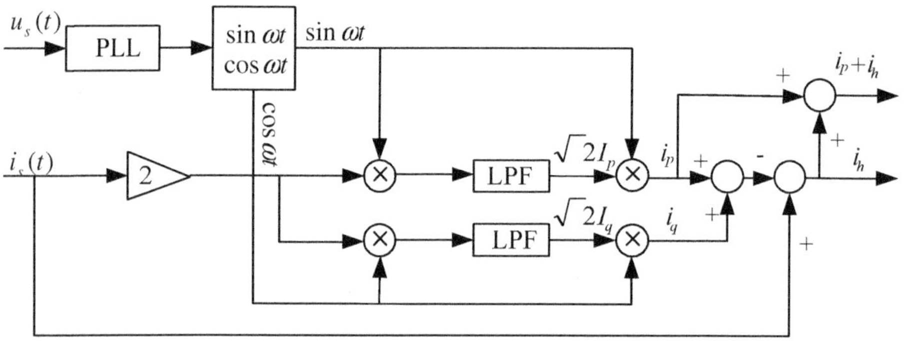

Fig. 2. Wattful current separation overtone examination control diagram

4 Specific Power Factor Overtone and Idle Current Examination Law

Supposes the network voltage expression same type (1), after supposing the compensation the electrical network electric current is:

$$i_s^{'}(t) = ku_s(t) = k\sqrt{2}U\sin\omega t \tag{10}$$

Before supposing the compensation, electrical network electric current Fourier expansion is:

$$i_s(t) = \sum_{n=1}^{\infty} i_n \sin(n\omega t + \varphi_n) = ku_s(t) + i_q(t) + i_h(t) \tag{11}$$

Above equation 1 does not have the active power, namely satisfies:

$$\frac{1}{T}\int_{t=0}^{T} u_s(t)[i_q(t)+i_h(t)]dt = 0 \tag{12}$$

(10) substitution type (11):

$$k = \frac{\frac{1}{T}\int_{t=0}^{T} u_s(t)i_s(t)dt}{\frac{1}{T}\int_{t=0}^{T} u_s^2(t)dt} = \frac{\overline{u_s(t)i_s(t)}}{\overline{u_s^2(t)}} \tag{13}$$

We can obtain the improvement by the above analysis based on the specific power factor harmonic current examination method, like Figure 3. This method directly examine electrical network's idle work and the harmonic component by the network voltage and the electric current.

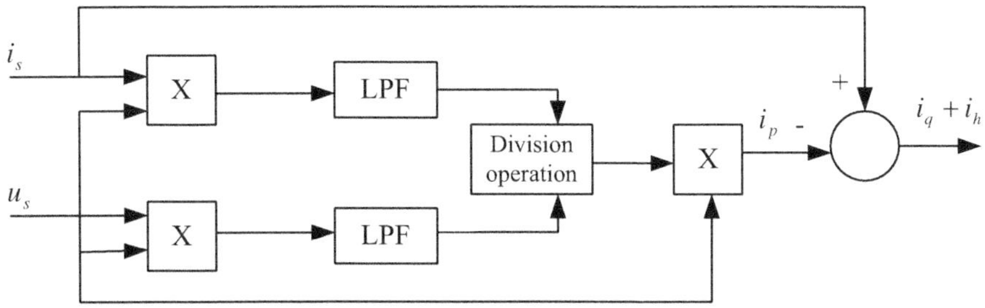

Fig. 3. Improvement based on specific power factor harmonic current examination diagram

5 MATLAB Simulation

Selection for the electricity iron single-phase APF overtone idle current examination method, applied in MATLAB7.0 simulink to analyze three kind of single-phase APF.

The supposition electricity iron tows arm's load voltage (the unit: V), electric current (unit: A) may use the type (14), (15) to simulate[6]:

$$u_s(t) = 27500\sin\omega t + 500\sin(3\omega t + 30°) \tag{14}$$

$$\begin{aligned} i_s(t) = {} & 490[\sin(\omega t + 30°) + 0.22\sin(3\omega t + 40°) + 0.10\sin(5\omega t + 18°) + \\ & 0.07\sin(7\omega t + 30°) + 0.05\sin(9\omega t + 60°) + 0.02(11\omega t + 46°) + \\ & 0.01\sin(13\omega t + 83°) + 0.01\sin 15\omega t + 0.01\sin(17\omega t + 138°) + \\ & 0.01\sin(19\omega t + 20°) + 0.01\sin(21\omega t + 138°) + 0.01\sin 23\omega t] \end{aligned} \tag{15}$$

We may obtain the simulation result by the type :

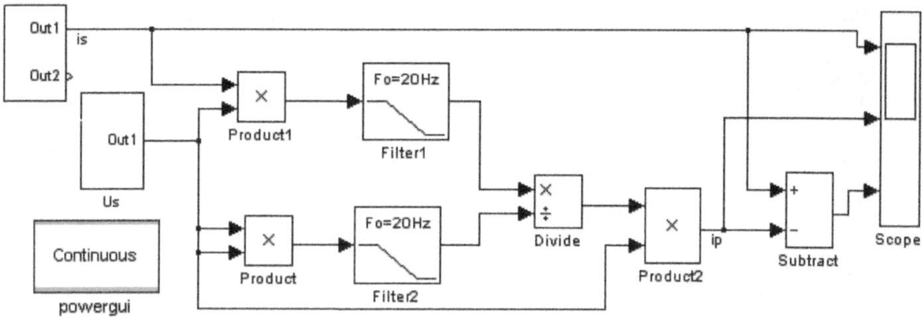

Fig. 4. Based on specific power factor overtone examination simulation model

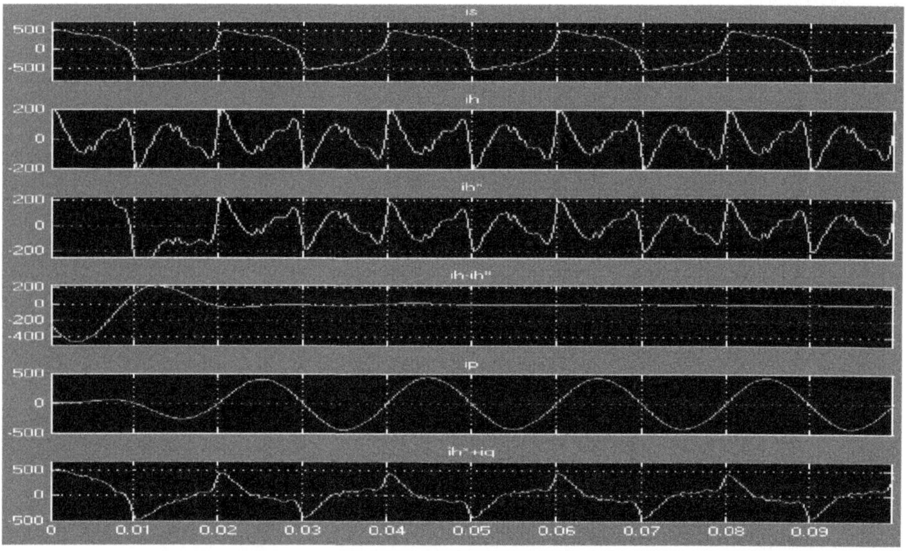

Fig. 5. Instant reactive power theory examination method simulation result

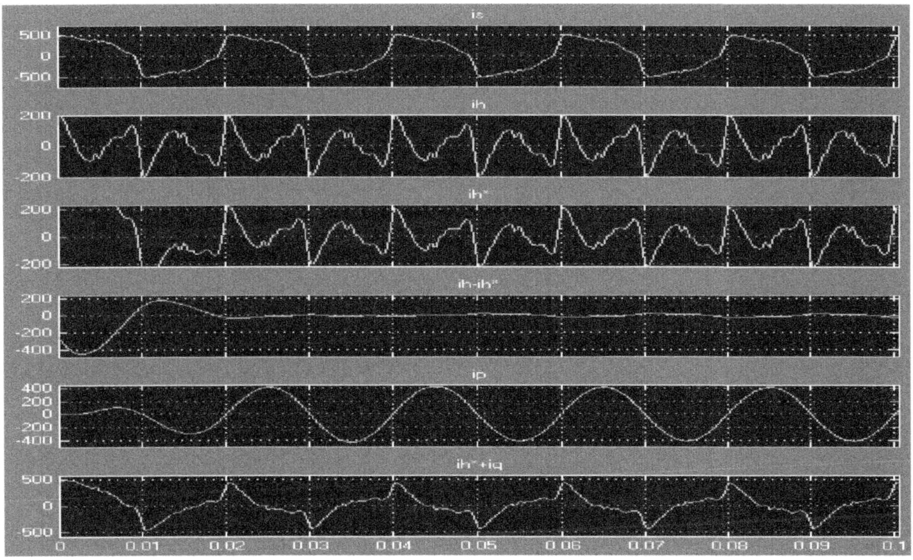

Fig. 6. Wattful current separation overtone examination simulation result

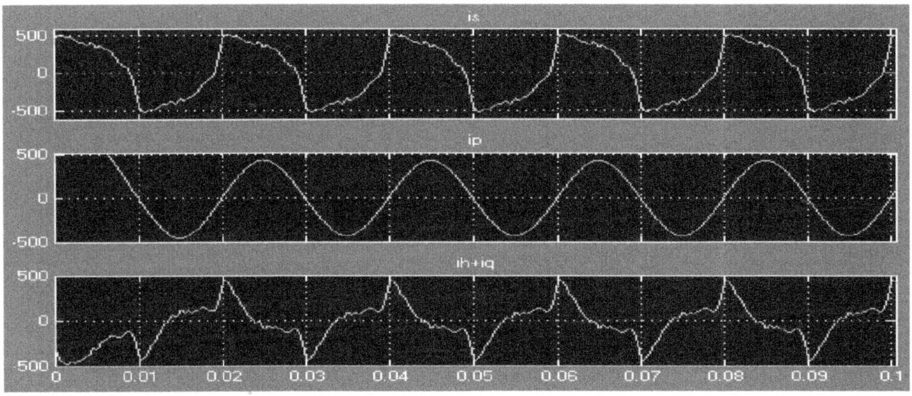

Fig. 7. Specific power factor overtone examination simulation result

Carries on the comparison to three kind of examination method's simulation result, gets the following conclusion:

(a) the idle current based on the instant reactive power theory examination law can only examine sum of the overtone and the idle current based on the specific power factor's overtone examination law;

(b) Three kind of examination's method structures is getting more and more simple, the examination precision to be getting more and more low, timeliness to be getting better and better.

6 Conclusions

From overall evaluation, take the electrification railroad hauling power supply system as compensating the object, uses the wattful current separation examination law effect is best. Therefore, this paper uses the overtone and idle current's examination based on the wattful current separation examination method.

References

[1] Xie, G.R. (ed.): Power system grounding technology. Water Conservancy and Electric Power Press, Beijing (1991)
[2] High-Voltage Power System, Tsinghua University compiled technical expertise. The impact of high-current technology. Science Press, Beijing (September 1978)
[3] Shen, L.S.: High-Speed Data Acquisition System Theory and Application. People's Posts & Telecom Press, Beijing (1995)
[4] Liu, H.P.: TMS320LF240X DSP C language development and application. Beijing University of Aeronautics and Astronautics Press, Beijing (2003)
[5] Liu, H.P.: TMS320LF240X DSP structure, principle and application. Beijing University of Aeronautics and Astronautics Press, Beijing (2002)
[6] Chen, Y.Y.: MATLAB signal processing Xiang Jie. People's Posts & Telecom Press, Beijing (2001)

Designer of Unified Power Flow Controller

Wen Jin Dai and Kai Wang

Department of Electrical and Automatic Engineering, Nanchang University Nanchang, China
{dwj480620,wangkai19832002}@yahoo.com.cn

Abstract. This paper introduced singular value decomposition(SVD) elementary theory, Proposed one kind of Unified Power Flow Controller(UPFC) based on SVD, reduces or eliminates active and tendency interaction between the idle current.With the MATLAB simulation of software, obtain the satisfactory results, thus proving the usefulness of this method.

Keywords: SVD; UPFC; idle current.

1 Introduction

SVD[1] is one kind of orthogonal matrix resolution law, also each kind of matrix divides in the solution method is the most reliable one kind. In the science and the engineering calculation, the singular value minute solves (SVD) is a powerful tool, widely applies in linear questions and dynamic system's identification, optimal approximation stable state and tentative data processing. In electrical power system, also study the human to take seriously specially. This article with the input and output data which obtains from the synchronous machine actual movement process embark, divides the cleavage theory using the singular value, according to the inspection data matrix's singular value, determined that the electrical machinery electric circuit model medium group plans leg's number, the use smallest singular value carries on the evaluation to the second-level voltage control effect, through increases system's smallest singular value to enhance system's stability.

2 SVD Rationale

A. Matrix singular value

Matrix $A\in C^{m\times n}$ singular value is with the matrix $A^H A$ and AA^H related concepts, before establishment singular value concept, discusses the matrix first $A^H A$ and AA^H related nature.

When $A\in C^{m\times n}$, $A^H A\in C^{m\times n}$ and $AA^H \in C^{m\times n}$ are the Hermit matrices, thus is the normal matrix.

The theorem 1.1 suppose $A\in C^{m\times n}$, then the matrix $A^H A \in C^{m\times n}$ and the matrix $AA^H \in C^{m\times n}$ have the following nature:

L. Qi (Ed.): ISIA 2010, CCIS 86, pp. 747–754, 2011.

Order (A) =Order $(A^H A)$ =Order $(AA)^H$,

$A^H A$ and AA^H non-vanishing characteristic value is equal;

$A^H A$ and AA^H are half Zhengding matrix, when order (A) = n , $A^H A$ are the Zhengding matrices, when order (A) = m , AA^H are the Zhengding matrices.

From this, may define the matrix the singular value.

B. Matrix singular value decomposition

The theorem 1.2 suppose $A \in C^{m \times n}$, order (A) =r, then the tenth Earthly Branch matrix $U \in C^{m \times n}, V \in C^{m \times n}$ to cause:

$$U\begin{bmatrix} \Delta & 0 \\ 0 & 0 \end{bmatrix}V^H, \quad \Delta = \begin{bmatrix} \delta_1 & & \\ & \ddots & \\ & & \delta_r \end{bmatrix}, \quad \sigma_1 \geq \sigma_2 \geq \cdots \geq \sigma_r \rangle 0$$ are A singular values. The unitary matrix V row vectors are AA^H standard orthogonal eigen vector/feature vector/proper vector, is called A right strange vectors; V first r rows are corresponding $A^H A$ r non-vanishing characteristic value eigen vector/feature vector/proper vector. The unitary matrix U row vectors are AA^H standard orthogonal eigen vector/feature vector/proper vector, is called A left strange vectors. [2]

C. Singular value resolution in control system's application

Inputs the m output regarding m the system, its transfer function matrix G(s), may make the following transformation:

$$\mathrm{G(s)} = Z(s) \bullet \Lambda(s) V^T(s) \tag{1}$$

$\Lambda(s)$ by $G(s)$ singular value constitution's opposite angle, may express is:

$$\Lambda(s) = diag[\sigma_1(s)\sigma_2(s), \cdots, \sigma_m(s)]$$

Respectively be about strange vector quantity, may express is:

$$Z(s) = [z_1(s)z_2(s)z_3(s)\ldots z_m(s)] \quad V(s) = [v_1(s)v_2(s)v_3(s)\ldots v_m(s)]$$

They respectively be the constitution unitary space standard orthogonal basis, such may be represented as:

$$G(s) = \sum_{i=1}^{m} \sigma_i(s) z_i(s) v_i^T(s) = \sum_{i=1}^{m} \sigma_i(s) w_i(s) \tag{2}$$

Also because of $\mathrm{G(s)} = Z(s) \bullet \Lambda(s) V^T(s)$, then:

$$Y(s) = Z(s) \bullet \Lambda(s) V^T(s) \bullet U(s) \tag{3}$$

The transfer function $G(s)$ [3]decomposes three parts, respectively be by the input space which $\{v_i(s)\}$ is composed, $\{\sigma_i(s)\}$ is composed the increase space by the singular value, and by the output space which $\{z_i(s)\}$ are composed. The output singular value matrix $Z(s)$ and input singular value moment $V(s)$ have instructed the system input - output direction strong and the weak relation. The singular value decomposition schematic drawing, as shown in Figure 1.

The system kth output regarding during the lth input value's gain is:

$$g_{kl}(s) = \frac{y_k(s)}{u_l(s)} = \sum_{i=1}^{m} \sigma_i \langle W_i(s), E_{kl} \rangle \tag{4}$$

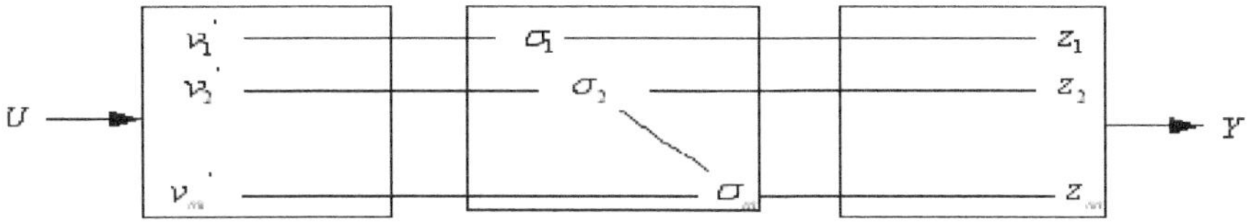

Fig. 1. Based on SVD transfer function decomposition

3 Design of UPFC Controller Based on SVD

A. UPFC model

The UPFC model, as shown in Figure 2. It is composed of two potential source converters back to back, between them through a direct-current capacitor connection. The series connected converter enters the alternating-current system through the series transformer string. The parallel converter unites into the system through the shunt transformer. Through the adjustment series side injecting voltage V_{se}, may control transmission system's compound tidal current quantity (Pr + jQr). The series connected converter pours into the compound power relies on the output voltage and the transmission electric current. The parallel converter provides the active power which through the direct-current connection alignment series converter needs. On the other hand, each converter may independent absorb and provide the active power to the system. The parallel converter's reactive power may use for to adjust the shunt transformer junction the voltage size.

Three-phase UPFC[4] system's single-phase chart, as shown in Figure 3. Series connected and the parallel converter separately v_{sev} and v_{shv} indicated by the potential source, underneath letter V express three-phase some (a, b or c). R and L expressed separately in the series transformer and the transmission line resistance and the inductive reactance value, Rsh and Lsh express shunt transformer's resistance and the reactance value separately.

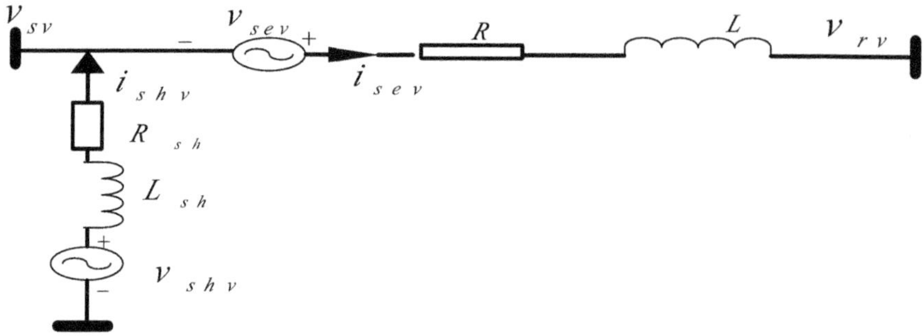

Fig. 2. Loaded with the unification tidal current controller's electrical power system

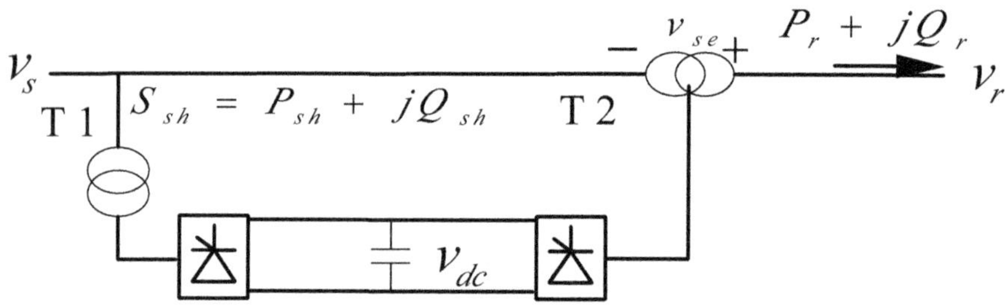

Fig. 3. Three-phase unification tidal current system's representative chart

B. Design of Singular value controller

Assigns the transfer function matrix may decompose the singular value form:

$$F = U\Sigma V^H \tag{5}$$

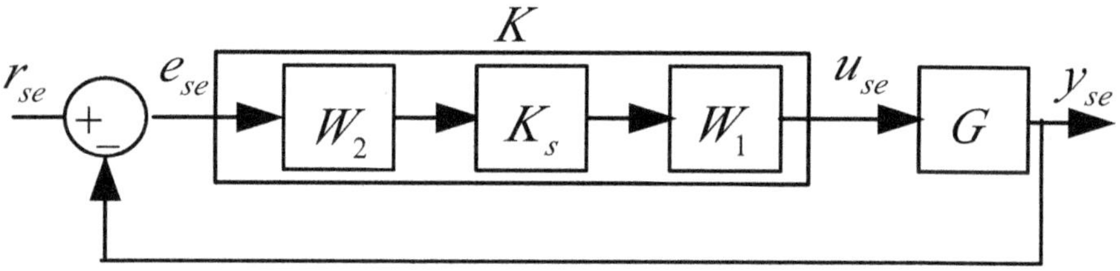

Fig. 4. The singular value decomposes controller's model

All controller's matrixing function relations are: $K = W_1 K_s W_2$ (6)

Matrix K from G^{-1} singular value decompositions, the compensating system $G_{new} = GK$ has become the opposite angle form like this completely. When matrix K carries on the dynamic appraisal when each kind of frequency, it is called the dynamic damping compensator.

$$G_0^{-1} = \begin{bmatrix} 0.048 & -0.240 \\ 0.240 & 0.048 \end{bmatrix} \tag{7}$$

In order to eliminate the static error, the opposite angle controller K_s is regarded as an integrator, the form is as follows:

$$K_s = \begin{bmatrix} \frac{k_i}{s} & 0 \\ 0 & \frac{k_i}{s} \end{bmatrix} \tag{8}$$

Constant k_i by Σ_0^{-1} decisions, and may change by obtains the ideal rising time. By above analyzes the knowledge, the singular value decomposition controller and the system model is opposite. Thus, the singular value decomposition controller may also regard as the conventional controller structure instead to set.

C. UPFC control system

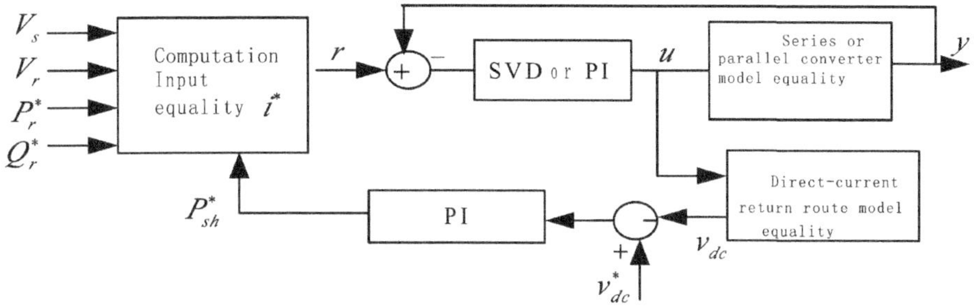

Fig. 5. Unification tidal current control system's block mold

The UPFC control system's control module, as shown in Figure 5. Controller's function is the control chart 3.1 receiving ends active and the reactive power. Using the receive terminal voltage and the line magnitude of current, active and the reactive power may write is:

$$P_r = \frac{3}{2}(v_{rd} i_{sed} + v_{rq} i_{seq}) \quad Q_r = \frac{3}{2}(-v_{rd} i_{seq} + v_{rq} i_{sed}) \tag{9}$$

When active and idle work performance number, the series connected converter's straight axle and hands over the axis the corresponding magnitude of current (i_{sed}^* , i_{seq}^*) to be possible to obtain by (9) , the form is as follows:

$$i_{sed}^* = \frac{2}{3}\frac{(P_r^* v_{rd} - Q_r^* v_{rq})}{\Delta} \qquad i_{seq}^* = \frac{2}{3}\frac{(P_E^* v_{rq} + Q_r^* v_{rd})}{\Delta} \tag{10}$$

The perturbation period's main purpose, to maintain the direct-current return route voltage is constant. In the direct-current return route reference voltage (v_{dc}^*) and the actual voltage (v_{dc}), will produce the error signal. The error signal through an independent proportion integral controller transmission, will obtain a command signal. Afterward, uses this command signal, infers the parallel converter the active power reference value (P^*_{sh}), then control parallel converter straight axle electric current peak-to-peak value.

4 Simulation Result

Figure 6 contrasted separately when the singular value decomposed the controller and the static decoupling controller, system's dynamic response situation. When 2s the active power grows from 1.278pu to 2.278pu, to theory system not any uncertainty. From the chart in 5.6 discovered when uses the SVD controller, the reactive power not big (is smaller than in the instantaneous change peak-to-peak value 0.1pu). But uses time the static decoupling controller, the reactive power instant's change peak-to-peak value is very actually big, indicated that has the very strong dynamic interaction in active and the idle work tidal current between. Obviously, proposed the singular value decomposition controller may suppress the dynamic interaction effectively.

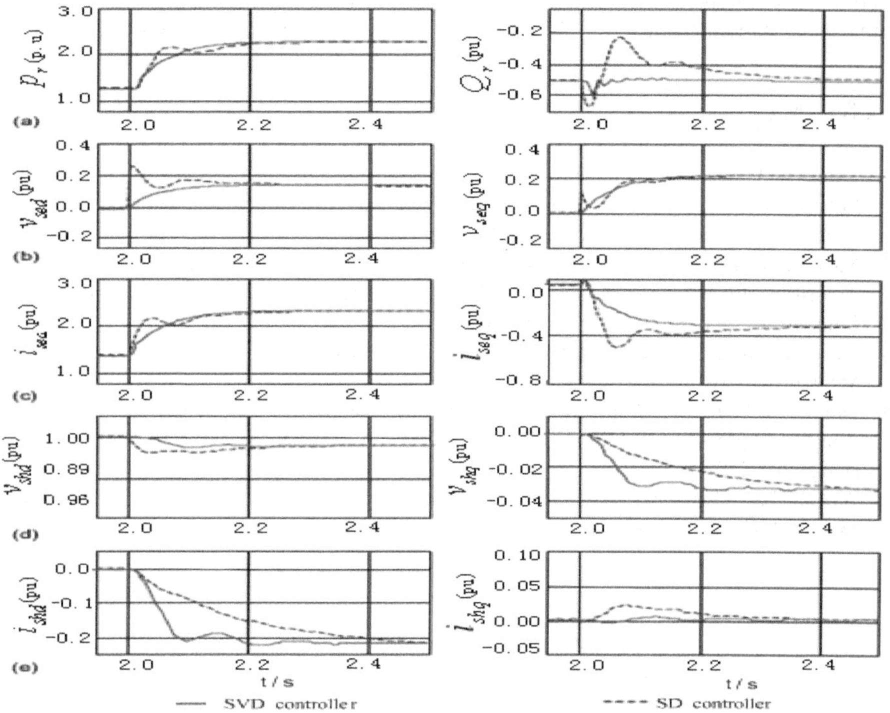

Fig. 6. Unification tidal current controller's definite response

Figure 7 indicated that when 2s the systematic active power length of stride increases 50%. At the same time, the reactive power reduced 50%, and inputs time 20% error transfer function 10% error's response. Figure 4.2 pair of singular value controller and the proportion integral controller's performance carries on the comparison, thus it may be known, when has the disturbance proportion integral controller's control and output variable withstanding sudden change, but the singular value controller's change actually very much relaxes.

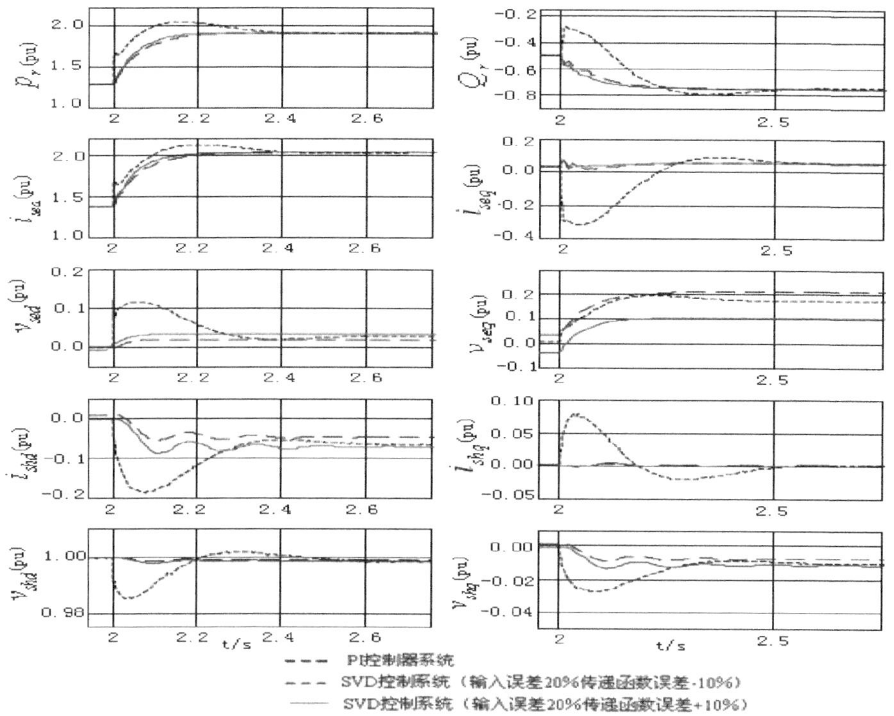

Fig. 7. Inputs 20% errors to output a 10% erroneous common time-base signal tidal current controller's response

5 Conclusions

In the article proposed designs UPFC with SVD, reduces active and interaction between the idle work tidal current. The result indicated that it may reduce active and interaction effectively between the idle work tidal current, may improve the unification tidal current controller's dynamic property.

References

[1] Xie, G.R.: Power system grounding technology. Water Conservancy and Electric Power Press, Beijing (1991)
[2] High-Voltage Power System, Tsinghua University compiled technical expertise. The impact of high-current technology. Science Press, Beijing (September 1978)
[3] Shen, L.S.: High-Speed Data Acquisition System Theory and Application. People's Posts & Telecom Press, Beijing (1995)
[4] Liu, H.P.: TMS320LF240X DSP C language development and application. Beijing University of Aeronautics and Astronautics Press, Beijing (2003)
[5] Liu, H.P.: TMS320LF240X DSP structure, principle and application. Beijing University of Aeronautics and Astronautics Press, Beijing (2002)
[6] Chen, Y.Y.: MATLAB signal processing Xiang Jie. People's Posts & Telecom Press, Beijing (2001)

Author Index